BIOCHEMISTRY RESEARCH TRENDS

CALCIUM PHOSPHATE

STRUCTURE, SYNTHESIS, PROPERTIES, AND APPLICATIONS

BIOCHEMISTRY RESEARCH TRENDS

Additional books in this series can be found on Nova's website
under the Series tab.

Additional E-books in this series can be found on Nova's website
under the E-book tab.

BIOCHEMISTRY RESEARCH TRENDS

CALCIUM PHOSPHATE

STRUCTURE, SYNTHESIS, PROPERTIES, AND APPLICATIONS

ROBERT B. HEIMANN
EDITOR

NOVA BIOMEDICAL

New York

For permission to use material from this book please contact us:
Telephone 631-231-7269; Fax 631-231-8175
Web Site: http://www.novapublishers.com

Library of Congress Cataloging-in-Publication Data

Calcium phosphate : structure, synthesis, properties, and applications / [edited by] Robert B. Heimann.
pages cm
Includes index.
ISBN 978-1-62257-299-1 (hardcover)
1. Calcium phosphate. I. Heimann, R. B. (Robert Bertram), 1938- editor of compilation.
TP245.C3C354 2012
546'.39324--dc23

2012021408

Published by Nova Science Publishers, Inc. † New York

CONTENTS

PREFACE

Calcium phosphates play a paramount role on Earth as one of the basic building blocks of life. The gravity-defying bony skeletons of all vertebrates owe their mechanical strengths and resilience to hydroxyapatite-collagen scaffolds as do the dentine and enamel materials of teeth. These apatite/organic composites provide strength, flexibility, a porous structure allowing exchange of essential nutrients, and a biologically compatible resorption and precipitation behavior under the right physico-chemical conditions.

However, considering the low abundance of phosphorus in the Earth's crust of less than 0.1 wt% it is an odd choice of Nature to build most of the critical pathways of photosynthesis and animal metabolism around this rare element. The photosynthetic pathway called the Calvin cycle and the intercellular energy transfer both rely on adenosine triphosphate (ATP) that releases energy when reduced to adenosine diphosphate (ADP). The unique energetic contribution of the phosphate groups is central to the function of ATP, arguably the most abundant biomolecule in nature. Furthermore, the DNA owes its double helix structure to phosphate ester bridges that link the two strands of the helix.

The calcium phosphate mineral most ubiquitously occurring in nature is apatite in its variations hydroxyapatite, fluoroapatite, chlorapatite etc. In the past there has been widespread agreement that these minerals act as the most important reservoirs supplying life on Earth with essential phosphorus. However, more recently another hidden source of phosphorus came into focus: feldspars. In feldspars, P is able to replace tetrahedrally coordinated Si by coupled substitution with Al to maintain charge balance, i.e. $2\ Si^{4+} \leftrightarrow Al^{3+} + P^{5+}$. Given the abundance of feldspars in the Earth's crust, and the easy accessibility for plants and soil biota of their P-containing weathering products, feldspars may indeed be a more significant source of P than apatite.

In light of the overwhelming importance of calcium phosphates to sustain life it comes to no surprise that the content of the majority of the chapters in this volume center on the biological aspects of calcium phosphates including bone replacement scaffolds, osseoconductive coatings of endoprosthetic implants, dental glass-ceramics, and bone and dental cements.

Chapter 1 describes the historical development of the scientific knowledge gained on calcium orthophosphates from the late 18th to the mid-20th centuries. This historical evidence clearly demonstrates that many scientific facts and modern experimental approaches to calcium phosphates have been known for a long time. In fact, a considerable quantity of relatively recent investigations is just a further development of results of earlier studies or a rediscovery of already forgotten facts.

Chapter 2 deals with the structure of carbonate-bearing hydroxyapatite (CHAP) that has been investigated extensively by diffraction methods and chemical spectroscopy. The important conclusions obtained from these studies are that type A-CHAP with carbonate occurring in the c-axis structural channels is present in far greater proportions than type B-CHAP in which carbonate substitutes for the phosphate group. This contradicts previous assumptions, and suggests that type A-CHAP is possibly the labile carbonate component mediating acid-base reactions in the body.

Chapter 3 summarizes and analyzes the slow acceptance, particularly from the medical and dental communities, of the notion that molecular H_2O resides at the few weight-percent level both in synthetic CHAP and in bone mineral. Hence this review documents the very convoluted pathway leading to the still incomplete acceptance of structurally incorporated water in apatite and to analyze why this recognition took so long. The existence of structural water in apatite initially was overlooked, undervalued, and subsequently avoided. The culmination of this story lies in the future, because there are always ramifications to structural-chemical variations in a mineral. Hence it is high time to start exploring the effects of crystallographically incorporated water on the properties of biological apatite. This chapter is an instructive example how tenaciously scientist adhere to preconceived notions and how reluctant they sometimes are to accept novel concepts and paradigmatic shifts of knowledge.

In Chapter 4 another contentious aspect of research into the structure of apatites is being discussed. Oxyhydroxyapatite and oxyapatite have been proposed as intermediates during dehydroxylation of hydroxyapatite prior to their thermal decomposition into tri- and tetracalcium phosphates. Hydroxyl ions in the c-axis channels of hydroxyapatite are most mobile and upon exposure to high temperatures, for example during thermal spraying, the first to leave the lattice to form oxyhydroxyapatite. There have been several attempts to obtain oxyapatite by heating, but its metastability and high dehydroxylation levels promote decomposition towards tri- and tetracalcium phosphates. However, upon cooling the hygroscopic nature readily leads to the return of hydroxyl ions into the lattice.

Phase-pure oxyapatite stable at room temperature has not been successfully obtained yet. The possible microstructures formed during dehydroxylation by different thermal processes are discussed with reference to either oxyhydroxyapatite solid solutions or mechanical mixtures of hydroxyapatite and oxyapatite.

Nano-sized particles and crystals of calcium phosphate described in Chapter 5 play an important role during formation of calcified tissues of vertebrates. Recent developments in biomineralization show that nano-dimensional crystals of biological apatite self-assemble into complex structural building blocks of bones and teeth. This process occurs under strict control by bioorganic matrixes. This chapter reports on current state-of-the-art and recent developments on the subject, starting from synthesis and characterization to biomedical and clinical applications.

In dentistry, glass-ceramics are mainly used as restorative biomaterials for the fabrication of inlays, crowns and bridges. Chapter 6 shows that engineered phase formation of fluoroapatite crystals with controlled optical properties is an important process in the development of dental materials. This controlled crystallization enables the translucency of the material to be adapted to that of the natural tooth structure. Therefore, fluoroapatite glass-ceramics are popular veneering materials in dentistry. This study examines the synthesis, microstructure, properties and main application areas of these materials.

Chapter 7 investigates the phase composition of hydroxyapatite coatings deposited onto Ti6Al4V substrates by atmospheric plasma spraying (APS). Laser-Raman, nuclear magnetic resonance (NMR), and electron energy loss (EEL) spectroscopy provide insight into the mechanisms of phase transformations the starting material undergoes when subjected to the hot plasma jet, and the pathways of reconstruction of the structure during exposure to simulated body fluid (SBF) for up to 12 weeks. Incubation in SBF advances the proportion of crystalline HAp since disordered structures such as oxyhydroxy/oxyapatite, thermal decomposition products such as tri- and tetracalcium phosphates, and amorphous calcium phosphate formed during plasma spraying, exhibit substantially higher solubility and will thus be removed from the phase assembly.

While in the medical industry deposition of calcium phosphates on implants still occurs predominately by plasma spraying, Chapter 8 provides information into deposition of calcium phosphate coatings on titanium alloy by electrochemical methods. This process occurs at low temperature and allows obtaining thin coatings with controlled thickness. Variation of current density is used to control the chemical composition of the coating. However, high current densities lead to mechanically weak coatings. To solve this problem, pulsed electro-deposition is proposed to obtain uniform coatings without the pinholes and craters usually observed with direct current electro-deposition. However, while now higher current densities can be employed, the chemical composition of the coating is that of calcium-deficient hydroxyapatite. Thus, hydrogen peroxide must be added to the electrolytic solution to obtain monophasic coatings of stoichiometric hydroxyapatite. The results of the study show that electro-deposited calcium phosphate coatings protect the titanium alloy against corrosion, and promote the precipitation of bone-like apatite thus confirming the bioactivity of this prosthetic material.

A further deposition technique of calcium phosphate coatings has been found in radio-frequency (RF) magnetron sputtering. Chapter 9 focuses on current achievements in the field of radio-frequency (RF) magnetron sputter deposition of biocompatible calcium phosphate--based coatings for medical implants with an emphasis on their composition, structure, mechanical and biological performances. The chapter shows that RF magnetron sputtering allows preparing dense, homogeneous, pore-free and highly adherent biocompatible coatings able to prevent leaching of cytotoxic ions from metal to the surrounding tissues, and capable of stimulating osteogenesis of new bone. The main advantages and limitations of this technique are presented as well as the most important challenges and critical issues highlighted.

While the previously summarized chapters deal with the structure and biological applications of hydroxyapatite, in several chapters consideration is also given to the roles tricalcium phosphate (TCP) and tetracalcium phosphate (TTCP) play as bone scaffolds, and bone and dental cements. Chapter 10 reviews synthetic scaffolds replacing the traditional autogenous bone grafts. The most commonly used synthetic scaffolds are calcium phosphate ceramics, mainly β-tricalcium phosphate (β-TCP). It is shown how shape, size and porosity of these biomaterials affect cellular activity en-route to bone regeneration. This chapter will also analyze the data of various studies characterizing the interaction between β-TCP and osteogenic cells, either *in vitro* or *in vivo*, the effect of β-TCP on tissue response to the implanted material, and the mechanism of bone healing with respect to gene expression, cell localization and β-TCP resorption.

Along these lines, Chapter 11 addresses calcium phosphate cements (CPC) with good biocompatibility and resorbability, excellent bioactivity, non-cytotoxicity, ability to develop osteoconductive pathways, and sufficient mechanical strength for a number of applications. The major advantages of the CPC include fast setting time, excellent moldability, outstanding biocompatibility, and easy manipulation. Therefore, the cements are commonly used for filling bone defects and in trauma surgeries as moldable paste-like bone substitute materials. At present, there are two types of CPC, characterized by the end-product of the reaction: apatite (AP) cements and dicalcium phosphate dihydrate (DCPD or brushite) cements. Ionic incorporations of Mg, Sr, Zn, Na, K, F and carbonates in calcium phosphates such as α- and β-TCP can alter a number of structural, physico-chemical and biological properties such as lattice parameters, crystallinity, solubility, and bone bonding capability. These advantages notwithstanding, there remains a need for improving the overall performance of CPC, especially in terms of injectability and mechanical properties aiming at specific clinical applications such as vertebroplasty and kyphoplasty (treatment of osteoporotic fractures of the spine).

Chapter 12 reports on the development and recent results related to dental cements based on tetracalcium phosphate (TTCP) that are widely used as either single- or multicomponent, mainly hydroxyapatite-forming cements or in conjunction with functional polymers to obtain organic-inorganic composites. The latter are formed in an acid-base reaction from alkaline TTCP with carboxylic acids or organic phosphate polymer moieties. The setting reaction of these cements occurs by deprotonation of the organic acid followed by the formation of Ca^{2+}-acid chelates and the precipitation of nano-crystalline hydroxyapatite. The high alkalinity of the setting reaction of TTCP cements provides a strong antimicrobial potency of interest to dental applications, e.g. as pulp capping agents or root canal fillers.

Calcium phosphates and related compounds can occur in phosphate kidney stones. In Chapter 13, the epidemiology, clinical significance, etiology, and compositional analysis of phosphate kidney stone disease are summarized. Comprehensive analysis of calcium phosphate stones, combining morphologic examination and compositional analysis, with identification of stone components, quantification of their respective proportions, and determination of the internal structure, contributes to improved etiologic diagnosis. An example of a regional Italian case study is provided and the results are compared to similar studies in Spain and France.

Calcium orthophosphates substituted with transition metal ions such as titanium and/or zirconium display remarkable properties including solid state ionic conductivity, low solubility at least one order of magnitude lower than that of hydroxyapatite, and very low thermal expansion. In Chapter 14 it is shown that suitable osteoconductivity render these ceramic materials possible novel biomaterials, for example as long-term stable coatings for endoprosthetic implants or osteosynthetic metallic supports, applied by plasma spraying technology. Owing to their ionic conductivities these ceramics could also be utilized as charge carriers for novel implant devices to speed up bone healing and bone ingrowth, respectively in the presence of a stationary electromagnetic field.

Milk and dairy products contain significant quantities of calcium and phosphate and thus are important providers of essential nutrients to human consumers. In Chapter 15 it will be shown that the quantity and quality of calcium and phosphate present in these products in the aqueous and micellar phases depend on the type of dairy products and especially on their physico-chemical conditions. In the aqueous phase, calcium is mainly associated with citrate and inorganic phosphate. This phase, containing less than 1 mM of calcium phosphate, is considered supersaturated. In the micellar phase, the calcium concentration is about 20 mM, associated with organic phosphate (phosphoseryl residues of casein molecules) and inorganic phosphate to form nanoclusters of calcium phosphate. Solubilisation of calcium phosphate from casein micelles is observed during milk acidification or after addition of calcium chelatants. On the other hand, calcium and phosphate co-precipitate when milk or whey liquids are heated and/or made alkaline during production of cheeses. Several dairy industries recover calcium phosphate contained in different dairy liquids such as cheese whey by heating and alkalinization. The chemical and physical properties of these milk-based calcium phosphates depend on the whey used and the parameters of manufacture. These mineral ingredients are later used for calcium enrichment of different food products.

Exposure of skin to solar radiation is one of the leading causes of skin cancer and malignant melanoma. Chapter 16 describes how inorganic sunscreens containing tricalcium phosphate may constitute a safe way to protect the skin from noxious effects of solar radiation. Tricalcium phosphate (ß-TCP) has potential as an active ingredient in sunscreens since it is biocompatible and non-toxic. Moreover when doped with appropriate metals it is able to absorb UVB and UVA radiations. Optical absorption spectra indicated that the incorporation of Fe^{3+} ions in the tricalcium phosphate matrix generated absorption in the ultraviolet region. Hence it was concluded that Fe^{3+}-doped tricalcium phosphate can be used as important active medium for sunscreens.

A completely different application of calcium phosphates has been found in remediation technologies to sequester toxic radioactive and non-radioactive heavy metal contaminants in the environment. Owing to the structural variability of apatite with the general formula $(A)_5[(XO_4)_3/(Y)]$ four possibilities for crystallo-chemical ion exchange exist: (i) variation of metal cations A, (ii) variation of anion type I, XO_4, (iii) variation of anion type II, Y, and (iv) complex combined exchange of cations and anions. In Chapter 17 different metal immobilization processes such as XLTITE (Crystal Tight), MRP (Mineral reservoir processes), PIMS (Phosphate-induced metal stabilization), and SUPERCALCINE will be described that use apatite mineral structures for immobilization. The crystallization of apatite and the apatite structure-type mineral ellestadite will be explained that occur in many different industrial residues subjected to combined thermal and hydraulic processes. In addition, apatite minerals in nature will be compared to synthetic apatites.

The quest for keeping safely away radioactive waste elements from the biosphere is high up on the agenda of research and development in environmental science. Chapter 18 describes the potential application of calcium phosphates to immobilize high- and intermediate-level radioactive waste. Apatites and sodium zirconium phosphate [NZP] structure-type ceramics have long been proposed as candidates for immobilizing a large variety of fission product and transuranium elements. Apatite is a complex structure with moderate radiation tolerance and several examples are found in nature with radioactive elements trapped within the crystal structure. However because of its greater flexibility in terms of solid solution substitutions, the sodium zirconium phosphate (NZP) structure has more recently been advocated for this purpose. The special attraction of the [NZP] structure is that it can accommodate almost all radionuclides present in high-level waste streams within its crystal structure. The substitution of alkaline earth ions for alkalis in [NZP] increases the leach resistance considerably, thus

$CaZr_4(PO_4)_6$ and $CaTi_4(PO_4)_6$ are preferred options. The chapter reviews the general background of the development of these materials as ceramic nuclear waste forms as well as their crystal chemistry, aqueous durability and radiation tolerances.

Tricalcium phosphate derived from bone ash is an important ingredient of bone china developed in Britain from the late 18th century onwards. In Chapter 19 the history and materials science of bone china has been reviewed. The ceramics is characterized by a high proportion of calcined animal bone that provides ß-TCP to the ceramic paste able to react during firing beyond 1,000°C with silica and alumina released from decomposing clay minerals, forming anorthite. Hence the ceramic body of bone china consists of about 40% ß-TCP, 20% calcic plagioclase (anorthite), and 40% vitreous phase. In contrast to European hard-paste porcelain of Meissen type, the rather low proportion of vitreous phase confers to this hybrid soft-paste porcelain ware several advantageous mechanical and thermal properties, making it popular tableware throughout the world.

Apart from the commanding role calcium phosphates play in biology, medicine, environmental science, and ceramic technology as described above, there is one aspect of overwhelming socio-economical importance that has not been covered in this volume: calcium phosphates are indispensible materials as constituents of agri- and silvicultural fertilizers. The dominant phosphate fertilizers are 'superphosphate' manufactured by reacting phosphate rock with sulfuric acid according to $Ca_{10}[(PO_4)_6/F_2] + 7H_2SO_4 + 5H_2O \rightarrow 3Ca(H_2PO_4)_2 \cdot H_2O + 7CaSO_4 \cdot 2H_2O + 2HF$ or 'triple superphosphate' manufactured by reacting phosphate rock with orthophosphoric acid according to $Ca_{10}[(PO_4)_6/F_2] + 14H_3PO_4 + 10H_2O \rightarrow 10Ca(H_2PO_4)_2 \cdot H_2O + 2HF$.

Phosphate rock (phosphorite) is the only economically viable source of phosphorus for producing fertilizers according to the simplified equations shown above. Five countries produce nearly 75% of global supplies: the United States, China, Morocco, Russia and Tunesia. Estimates suggest that 50% of the world's phosphate resources will be depleted during the next 60 to 70 years, owing to steeply increased population and increasing demand for meat and dairy products that require more fertilizer per unit weight of food produced. In contrast to crude oil for which alternative primary energy sources are potentially available there is no feasible replacement for phosphate. As a result the growing world population may face a dramatic nutritional crisis in the foreseeable future. To alert world governments to this threatening fact, a group of 30 internationally acclaimed chemists and materials scientists have recently issued a White Book under the title "A Sustainable Global Society" aimed at counseling societies and their leaders how to alleviate the pending crisis. One possible strategy involves recycling of phosphorus contained in municipal wastewater and sewage sludge by either chemical precipitation, i.e. addition of iron and/or aluminum chlorides or sulfates to wastewater, or biological phosphorus removal. The first process relies on precipitating of up to 90% of all influent P as complex non-stoichiometric iron or aluminum hydroxide phosphates. The second process is known as 'enhanced biological phosphorus removal' (EBPR) whereby a community of phosphorus-accumulating bacteria is established that can take up 20-30% of their dry weight as P. Increasing P levels in the resulting sludge lead to precipitation of struvite (ammonium magnesium phosphate, $(NH_4)Mg(PO_4) \cdot 6H_2O$) or hydroxyapatite from the sludge liquors by either selective ion exchange, precipitation in stirred reactors, or precipitation in fluidized bed or air-agitated reactors. Such phosphorus recovery plants operate already in Japan, are under construction in Canada, and operate as full-scale pilot plants in Italy and the United Kingdom. Electrochemical pilot plants that recycle phosphorus as struvite that can directly be used as fertilizer are working at present in Leoben, Austria and Knittlingen, Germany.

The present volume provides a modern, comprehensive and concise approach to the chemistry, crystallography, materials science, mineralogy, and biology of calcium phosphates and their application in reconstructive orthopedy as bone scaffolds, osseoconductive implant coatings and bone cements and in dentistry as glass-ceramics for inlays, crowns, bridges and veneering materials as well as dental cements. Sparingly soluble transition metal-substituted calcium orthophosphates are described that may have potential as novel biomaterials for long-term residence in the body. Further accounts on the prevalence, etiology and epidemiology of calcium phosphate-based urinary stones as well as the application of Fe^{3+}-doped ß-tricalcium phosphate in sunscreens are provided. The important role played by calcium phosphate in milk and cheese as carrier of nutritionally elements is highlighted. Moreover, owing to their compositional flexibility apatites and phosphates with [NZP] structure are scavengers of non-radioactive and radioactive toxic elements and can thus be used for remediation of contaminated soils and wastewaters, and for immobilization of high- and intermediate-level radioactive waste. Last but not least calcium phosphate from calcined animal bone has been used as replacement of feldspar in high-fired bone china since its invention in the late 18th century.

In: Calcium Phosphate
Editor: Robert B. Heimann

ISBN: 978-1-62257-299-1

Chapter 1

CALCIUM ORTHOPHOSPHATES AND MAN: A HISTORICAL PERSPECTIVE FROM THE 1770S TO 1940

Sergey V. Dorozhkin*
Moscow, Russia

ABSTRACT

The historical development of the scientific knowledge on calcium orthophosphates from the 1770s until 1940 will be described. Many forgotten and poorly known historical facts and approaches have been extracted from old publications and analyzed, systematized and reconsidered from the modern point of view. The chosen time scale starts with the earliest available studies of the 1770s - to the best of my findings, calcium orthophosphates were unknown before that date -, continues through the 19th century and finishes in 1940 since after then the number of publications on calcium orthophosphates rapidly increased and hence the subject becomes too unwieldy. Furthermore, since publications of the second half of the 20th century are easily accessible, a substantial amount of them has been already reviewed by other researchers. The reported historical findings clearly demonstrate that many scientific facts and experimental approaches have been known for a long time. In fact, a considerable quantity of relatively recent investigations on calcium orthophosphates is just a further development of the results of earlier studies or a rediscovery of already forgotten knowledge.

Keywords: apatite, calcium orthophosphate, calc phosphate, lime phosphate, history

1. INTRODUCTION

By virtue of their abundance in nature and presence in living organisms, calcium apatites [1] and other calcium orthophosphates (Table 1) appear to be a chemical compounds of

* Email: sedorozhkin@yandex.ru

special interest in many fields of science, including geology, chemistry, biology and medicine [2, 3]. As follows from the designation, calcium orthophosphates contain both calcium (Ca, atomic number 20) and phosphorus (P, atomic number 15) as major constituents. Concerning the history of both chemical elements, according to Wikipedia calcium (from Latin *calx*, genitive *calcis*, meaning "limestone", "lime" or "mortar") compounds were known as early as the 1st century CE, when the ancient Romans manufactured lime (calcium oxide) by burning limestone.

Table 1. Existing calcium orthophosphates and their solubilities [2, 3]

Ca/P ratio	Compound	Formula	Solubility @ 25°C, -log(K_s)	Solubility @ 25 °C, g/L	pH-stability range @ 25°C
0.5	Monocalcium phosphate monohydrate (MCPM)	$Ca(H_2PO_4)_2{\cdot}H_2O$	1.14	~ 18	0.0 – 2.0
0.5	Monocalcium phosphate anhydrous (MCPA or MCP)	$Ca(H_2PO_4)_2$	1.14	~ 17	[c]
1.0	Dicalcium phosphate dihydrate (DCPD), mineral brushite	$CaHPO_4{\cdot}2H_2O$	6.59	~ 0.088	2.0 – 6.0
1.0	Dicalcium phosphate anhydrous (DCPA), mineral monetite	$CaHPO_4$	6.90	~ 0.048	[c]
1.33	Octacalcium phosphate (OCP)	$Ca_8(HPO_4)_2(PO_4)_4{\cdot}5H_2O$	96.6	~ 0.0081	5.5 – 7.0
1.5	α-Tricalcium phosphate (α-TCP)	α-$Ca_3(PO_4)_2$	25.5	~ 0.0025	[a]
1.5	β-Tricalcium phosphate (β-TCP)	β-$Ca_3(PO_4)_2$	28.9	~ 0.0005	[a]
1.2 – 2.2	Amorphous calcium phosphates (ACP)	$Ca_xH_y(PO_4)_z{\cdot}nH_2O$, $n = 3 – 4.5$; 15 – 20% H_2O	[b]	[b]	~ 5 – 12 [d]
1.5– 1.67	Calcium-deficient hydroxyapatite (CDHA)[e]	$Ca_{10-x}(HPO_4)_x(PO_4)_{6-x}(OH)_{2-x}$ $(0<x<1)$	~ 85	~ 0.0094	6.5 – 9.5
1.67	Hydroxyapatite (HA, HAp)	$Ca_{10}(PO_4)_6(OH)_2$	116.8	~ 0.0003	9.5 – 12
1.67	Fluorapatite (FA or FAp)	$Ca_{10}(PO_4)_6F_2$	120.0	~ 0.0002	7 – 12
1.67	Oxyapatite (OA, OAp or OXA)[f]	$Ca_{10}(PO_4)_6O$	~ 69	~ 0.087	[a]
2.0	Tetracalcium phosphate (TTCP or TetrCP), mineral hilgenstockite	$Ca_4(PO_4)_2O$	38 – 44	~ 0.0007	[a]

[a] These compounds cannot be precipitated from aqueous solutions.

[b] Cannot be measured precisely. However, the following values were found: 25.7±0.1 (pH = 7.40), 29.9±0.1 (pH = 6.00), 32.7±0.1 (pH = 5.28). The comparative extent of dissolution in acidic buffer is: ACP >> α-TCP >> β-TCP > CDHA >> HA > FA.

[c] Stable at temperatures above 100°C.

[d] Always metastable.

[e] Occasionally called "precipitated HA (PHA)".

[f] Existence of OA remains questionable.

However, calcium sulfate (also known as Plaster of Paris or lime plaster) had been known much earlier. For example, three pre-pottery Neolithic (around 7200 BC) statues were discovered in a buried pit at 'Ain Ghazal in Jordan sculpted with lime plaster over a framework of reeds and twine. However, calcium in the elemental state was not isolated until 1808, when the famous British chemist and inventor Sir Humphry Davy (1778 – 1829) electrolyzed a mixture of lime and mercuric oxide [4, 5].To conclude this introduction, the earliest research paper that I have been able to find, containing the word "phosphorus" in the title was written by the English natural philosopher, chemist, physicist and inventor Robert Boyle (1627 – 1691) and published in 1693 [6], shortly after his death.

2. General Problems with Accessing Old Publications

Owing to the considerable problems to access the scientific literature published prior to the first half of the 19th century, a deep foray into the history of the subject still remains both fragmental and incomplete. During that time, the scientific concepts were rather different from the modern ones and chemical formulae of the substances had not been introduced yet. Furthermore, scientific journals were rare and just a few of them are currently available in electronic form.

Fortunately, many books published before the 20th century have now been scanned by Google as part of its project to make the world's books accessible online. This timely project by Google combined with the power of modern electronic databases of scientific publications allows reconstructing the major historical milestones on calcium orthophosphates, which was often impossible for earlier review writers.

For example, a paper by Driskell (1994) entitled: "Early history of calcium phosphate materials and coatings" [7] starts from the classical publication of 1920 by Albee (assisted by Morrison) [8]. In 1999, Shackelford published a paper: "Bioceramics – an historical perspective" [9], in which the same publication was mentioned as the earliest reference. The same is valid for the historical papers by Hulbert *et al.* [10, 11] and Shepperd [12]. Thus, it might create the false impression that calcium phosphates were unknown before 1920. Nevertheless, the precise sequence of the scientific discoveries happening in the 18th century still remains poorly restorable, while the historical time scale of even earlier scientific events remain almost completely irrecoverable.

This is mainly due to a lack of the quotation practice that existed in the scientific literature published in the 19th century and before. Besides, even today, only scientific books digitalized by Google are widely available for the entire 18th century and approximately the first quarter of the 19th century.

No journal papers devoted to apatites and/or any type of calcium phosphates published before the 1820s have been found. However, this does not mean that those papers do not exist; presumably, they either have not been scanned yet - this might particularly be valid for discontinued scientific journals - or the access to the scanned versions is restricted to a limited number of subscribers.

3. INVESTIGATIONS AND KNOWLEDGE ON CALCIUM ORTHOPHOSPHATES

3.1. In the 18th Century

According to the available literature, the history of calcium orthophosphates started in 1769. For example, in 1881 Roscoe and Schorlemmer [13] published the following statement: "*Gahn [14], in 1769, discovered the existence of calcium phosphate in bones, but it was not until this fact was published by Scheele [15] in 1771 that phosphorus was obtained from bone-ash, which has from that time invariably served for its preparation.*" (p. 458). Simultaneously, the presence of phosphoric compounds was discovered in blood serum, which follows from a publication of 1770 [16][1] stating that *"The ſerum conſiſts chemically of a coagulable matter, and water in which common ſal ammoniac and phoſphoric ammoniac, and generally common ſalt, and frequently ſelenites, and, fixed ammoniac, are dissolved; ..."* (p. 309). Furthermore, let me quote a publication of 1777 [17]: "*I have only been informed of this diſcovery, by the Gazette, Salutaire de Bouillon, October, 1775. It is there ſaid, that Mr. Henry Gahn, a phyſician at Stockholm, has communicated a proceſs for extracting from bones the ſaline matter in question; and that Mr. Scheele had aſcertained, that the earth of animals was compoſed of a calcareous ſubſtance united with the phoſphoric acid. This diſcovery, continues the author of the article of the Gazette, belongs to Mr. Gahn, and has been confirmed by later experiments."* (p. 383). Presumably, this quote might be considered one of the earliest mentioning of calcium phosphates in history.

Further, according to Shepperd [12], the German chemist Martin Heinrich Klaproth (1743 – 1817) and the French chemist Joseph-Louis Proust (1754 – 1826) also studied calcium orthophosphates. Unfortunately, Shepperd did not provide any references to the publications by those great chemists. Nevertheless, a search performed among the Google books has provided some results. For example, in a French book published in 1790 [18], one can find a subchapter entitled "*Apatite. Phosphate de Chaux*" which commences by the following sentence: *"C'est Mr. Klaproth, qui à découvert le premier cette combinaison de la terre calcaire avec l'acide phosphorique"* (p. 363). Further, *"Mr. Proust dans une Lettre à Mr. d'Arcet a donné l'Analyse de cette pierre, qui est une combinaison de la terre calcaire avec l'acide phosphorique*" (p. 366). One can see that, by that time, apatite was already recognized as a calcium phosphate, which is also confirmed by a German book published in 1789 [19], in which "Apatit" was called "phosphorsaure Kalkerde" (p. 2).

Furthermore, according to a book written by Lavoisier and translated into English [20], the production process of orthophosphoric acid by decomposition of calcined bones in sulfuric acid has been known since at least: *"The bones of adult animals being calcined to whiteneſs, are pounded, and paſſed through a fine ſilk ſiewe; pour upon the fine powder a quantity of dilute ſulphuric acid, leſs than is ſufficient for diſſolving the whole. This acid unites with the calcareous earth of the bones into a ſulphat of lime, and the phoſphoric acid remains free in the liquor"* (p. 205). Further, the production process of white phosphorus was described as follows: *"The liquid is decanted off, and the reſiduum waſhed with boiling water; this water which has been uſed to waſh out the adhering acid is joined with what was*

[1] Note the old-fashioned use of a long, medial or descending letter "ſ", which is a form of the minuscule letter "s" formerly used where "s" occurred in the middle or at the beginning of a word.

before decanted off, and the whole is gradually evaporated; the diſſolved ſulphat of lime criſtallizes in form of ſilky threads, which are removed, and by continuing the evaporation we procure the phoſphoric acid under the appearance of a white pellucid glaſs. When this is powdered, and mixed with one third its weight of charcoal, we procure very pure phoſphorus by ſublimation" (p. 206).

In the last decade of the 18th century, further investigations on calcium orthophosphates were performed by the two French chemists Antoine François, Comte de Fourcroy (1755 – 1809) and Nicolas Louis Vauquelin (1763 – 1829), who discovered the existence of acidic calcium orthophosphates, currently known as MCPM, MCPA, DCPD, DCPA (Table 1). For example, according to a book, published in 1816 [21], *"Super-PHOSPHAT OF LIME was diſcovered in 1795, by Fourcroy and Vauquelin. It had indeed been often formed before, but chemiſts had neglected to examine it"* (p. 458).

In addition, let me quote a book by Fourcroy, translated into English in 1804 [22]: *"A FEW years ago this ſalt, which I here denominate the acid phoſphate of lime, was unknown. Scheele very properly remarked that the ſaline earth of bones was diſſolved by an acid in human urine, but he did not obſerve that this union between phoſphoric acid and the bony phoſphate, makes a kind of permanent and particular ſalt different from the latter. It was in the year 1795, the third year of the Republic, that I diſcovered it, with Citizen Vauquelin, in a connected ſeries of experiments upon bony matters; wherein we proved that the calcareous phoſphate which conſtitutes the ſolid baſe, is only in part decompoſable by acids, and that the portion of phoſphoric acid which is ſeparated, retains in ſolution phoſphate of lime, which it then defends from all ſubſequent alteration by other acids"* (p. 347). Furthermore, in his book Fourcroy mentioned other contributors to early studies on calcium orthophosphates: *"Citizens Nicholas of Nancy, Pelletrier, Berniard, Bullion, in France; Weſtrumb, and ſeveral chemiſts in Germany; Bonvioſin, at Turin; Tenant, Pearſon, and ſome others in England, multiplied their inquiries and experiments"* (p. 337). Thus, it is evident that at the end of the 18th century calcium orthophosphates were extensively investigated by a number of researchers in several countries of Western Europe.

To conclude this section, one should recognize that the basic knowledge on calcium orthophosphates became available by the end of the 18th century.

3.2. In the 19th Century

During the 19th century, both scientific books as well as a small number of scientific journals became available. However, for the first half of the 19th century, journals are almost unavailable. Nevertheless, according to the available literature, a real explosion of the scientific knowledge on calcium orthophos-phates occurred in the 19th century. For example, Fourcroy [22] stated that presence of water in acidic calcium orthophosphates was known by 1804: *"The acid phoſphate of lime contains a remarkable quantity of water: the cryſtallization immediately ſoftens and becomes liquefied by the fire"* (p. 349). Unfortunately, based on this quotation, it is impossible to differentiate which type of the acidic calcium orthophosphates, MCPM or DCPD, Fourcroy had in mind. Knowledge on acidic calcium orthophosphates was further developed in 1805 [23]: *"Acid Phosphate of Lime is formed either by taking away a part of the base by sulphuric, nitric, or muriatic acid, or by superadding phosphoric acid to the last described phosphate"* (p. 195) and in 1806 [24]: *"In this way were distinguished among the salts two combinations, one neutral, and one with an excess of acid; and these were supposed to be determinate, as in sulphate and super-sulphate of potassa, or the phosphate and super-phosphate of lime*" (p. 38). More precise information on acidic calcium orthophosphates became available by 1819 [25]: "*XI. Biphosphate of Lime. – This salt may be formed by digesting phosphate of lime in phosphoric acid, dissolved in hot water*" (p. 327). "*XII. Quadriphosphate of Lime. – Glassy Phosphoric Acid of the Apothecaries. – This salt may be formed by digesting, for some time, finely powdered phosphate of lime in a quantity of sulphuric acid, sufficient to saturate all the lime of the phosphate, and afterwards diluting the mixture with a sufficient quantity of water, and filtering. Sulphate of lime remains on the filter, and a liquid quadriphosphate passes through*" (pp. 327-328). Presumably, aqueous solutions of DCPD and MCPM, respectively, were prepared; therefore, we can claim that MCPM and DCPD were distinguished already by 1819. Furthermore: "*XIII. Subphosphate of Lime. – This salt occurs native under the names of apatite or asparagus stone*" (p. 328). Thus, the differences between HA (subphosphate of lime) and TCP (phosphate of lime) were established by 1819.

The preparation process of the pure calcium orthophosphate currently known as CDHA was developed by 1813 [26]: "*a. Phosphat of lime, proper. As this salt constitutes the basis of bones, it is not necessary to prepare it artificially. It may be obtained in a state of purity by the following process: Calcine the bones to whiteness, reduce them to powder, and wash them repeatedly with water, to separate several soluble salts which are present. Dissolve the whole in muriatic acid, and precipitate by means of ammonia. The precipitate, when well washed and dried, is pure phosphat of lime.*" (section PHOSPHAT). One can see that the basics of this process are still used to produce chemically pure calcium orthophosphates from natural resources (phosphate ore and bones). In 1819, the major properties were described as follows [25]: "*4. Phosphate of lime is a white insoluble powder, destitute of taste, and unaltered by exposure to air. It is soluble in hydrochloric (muriatic) and nitric acids, and may be precipitated from solution in them by means of ammonia. When exposed to a very violent heat, it undergoes a kind of fusion, and is converted into a white semi-transparent porcelain.*" (p. 327). However, this knowledge of the thermal properties of phosphate of lime was based on Fourcroy's investigations [22]: "*The phoſphate of lime is extremely difficult to fuse; neverthele∫s, by ∫trong fire, ∫uch as that of a gla∫s-house, it ∫oftens and acquires a ∫emi-tran∫parence like the grain of porcelain.*" (p. 341). Thus, calcium phosphate ceramics has been known, since, at least, 1804.

In the 4th edition of the Encyclopædia Britannica (1810), applications of calcium orthophosphates were described as follows [27]: "*The pho∫phate of lime is of great importance in chemi∫try, for the purpo∫e of extracting pho∫phoric acid, to be decompo∫ed to obtain pho∫phorus. It is also employed for making cupels, for poli∫hing metals and precious ∫tones, and for removing ∫pots of grea∫y from linen, paper, and ∫ilk. It is u∫ed in medicine as a remedy for rickets, to correct the ∫uppo∫ed effects of acids in ∫oftening the bone*" (p. 585). Although it is slightly beyond the subject, one should mention, that calcium phosphite ($Ca_3(PO_3)_2$) is also mentioned in the Encyclopædia Britannica, as "pho∫phite of lime" (p. 586). A chemical term "tribasic phosphate of lime", which fully corresponds to α-TCP and β-TCP, has been known since, at least, 1832 [28, p. 90].

Concerning publications in scientific journals, in the first half of the 19th century, research papers on calcium apatites and other calcium orthophosphates were published by the Swedish chemist Jöns Jacob Berzelius (1779 – 1848) [29, 30], as well as by M. Baruel [31] and J.D. Smith [32]. For example, in a book by Bache, published in 1819 [25], the following statement on Berzelius' contribution was published: "*5. According to an analysis by Berzelius, calculated in numbers, in which the equivalent number for lime is assumed, phosphate of lime is composed of*

Phosphoric acid 34·3—

Lime 29·0—one atom;

so that it appears that the composition of this salt cannot be reconciled with the atomic theory" (p. 327). However, as early as 1804, Fourcroy [22] wrote: *"100 parts of phoſphate of lime contain, according to the analyſis of Citizens Fourcroy and Vauquelin,*

Phoſphoric acid 41

Lime 59" (p. 346),

while he described the chemical composition of acid phosphate of lime as follows: "*An accurate analyſis of this ſalt affords the following proportions of component parts,*

Lime 45

Phoſphoric acid 54" (p. 351).

Somewhat later, the German chemist Eilhard Mitscherlich (1794 – 1863), who was a student and friend of J.J. Berzelius and who is perhaps best remembered today for his law of isomorphism (1819), also worked in this area. For example, the earliest chemical formulae of calcium phosphates were published in 1844 on p. 69[a] of one of Mitscherlich's books [33] as $2.Ca^2$Ᵽ and $5.Ca^2$Ᵽ. Besides, in 1849, it was written that "*M. J. L. Lassaigne, at the meeting of the Academy of Sciences of Paris, of the 15th January, presented a memoir upon this subject, showing by experiments that the phosphate and carbonate of lime are introduced into plants in solution in water containing carbonic acid, which had before been shown as to the phosphate by M. Dumas, and has long been known as to the carbonate"* [34]. Thus, a higher solubility of both calcium orthophosphates and calcium carbonate in weak acids was already known in 1849.

The second half of the 19th century started with the book "Chemistry" by Wilson, published in 1850 [35]. The following parts from this book are worth quoting. "*797. Phosphates of Lime. – There are many compounds of lime and phosphoric acid, owing to the peculiarity of that acid in relation to the number of equivalents of base it combines with at once. The most interesting phosphate of lime is that which occurs in bones, and is distinguished as the bone-earth phosphate, $3CaO,PO_5$"* (p. 219). Thus, various types of calcium orthophosphates were already known by 1850. However, the preparation technique sounds unusual to a modern reader: *"The phosphorus combines in part with the oxygen of the lime, CaO, to form phosphoric acid, and this with undecomposed lime, to form phosphate of lime, CaO,PO_5. At the same time another portion of the phosphorus combines with the calcium of the lime, forming phosphuret of calcium, CaP"* (p. 164). In the volume

"Chemistry" by Brande and Taylor, published in 1863 [36], one can find the following statements: "*COMMON PHOSPHATE OF LIME; TRIBASIC PHOSPHATE OF LIME; BONE PHOSPHATE; (3(CaO),PO_5). – This salt occurs abundantly in bone-ash, and is found as a mineral product*" (p. 331). Furthermore, "*Native phosphate of lime (bone phosphate) occurs in apatite, moroxite, phosphorite, and asparagus stone; its primitive form is a six-sided prism: it also occurs in some volcanic products*" (p. 332). Thus, the similarity between the inorganic phase of bones and calcium orthophosphate rocks of natural origin (apatite and phosphorites) was known by 1863. *"When a solution of bone-earth in hydrochloric or nitric acid is boiled to expel all carbonic acid, and decomposed by caustic ammonia, the bone-phosphate separates in the form of a bulky precipitate, which, when perfectly dried, is white and amorphous"* (p. 331). This statement is really astonishing because it might be considered the first reference to ACP [37], 32 years before discovering x-rays in 1895 by the German scientist Wilhelm Conrad Röntgen (1845 – 1923)! The following quotation: *"The substance known under the name of coprolites, and which appear to be the excrements of fossil reptiles, also abound in phosphate of lime"* (p. 332) means that by 1863 researchers were aware of this fact.

The further historical description is based on journal publications only. Campbell Morfit [38], Robert Warington - there were two chemists by this name, presumably father (1807 – 1867) and son (1838 – 1907)) who performed the earliest well documented systematic studies of outstanding quality [39-42] -, and Carl R. Fresenius [43] also worked with calcium orthophosphates. Furthermore, let me quote a paragraph from Ref. [40]: "*Mitscherlich tells us, that when chloride of calcium is added to ordinary disodic phosphate, the latter being maintained in excess, the precipitate formed is tricalcic phosphate, while the solution becomes acid from the production of monosodic phosphate. Berzelius, on the contrary, states, that the precipitate formed under these conditions is not tricalcic phosphate, but the octocalcic triphosphate, which lie has elsewhere described. All experimenters agree, that when the operation is reversed, and disodic phosphate is poured into an excess of chloride of calcium, the precipitate is neither tricalcic nor octocalcic, but dicalcic phosphate."* (pp. 296-297). Thus, TCP, DCP and OCP have been known since at least 1866, maybe somewhat earlier. Furthermore, these researchers were already aware of the fact that the type of a precipitated calcium orthophosphate depends on a sequence of mixing the reagents.

Among the available publications written by the two Robert Waringtons [39-42], Ref. [40] by Robert Warington Jr. deserves both special attention and extensive quotations. For example, to prove that OCP indeed was already known in 1866, let us quote from Ref. [40]: "*Octocalcic phosphate can only be produced by the simultaneous formation of monosodic phosphate:*

$$8CaCl_2 + 5Na_4H_2P_2O_8 = Ca_8H_2P_6O_{24} + 16NaCl + 2Na_2H_4P_2O_8$$" (p. 300).

Notice the completely balanced chemical equation, fully identical to a modern one. It is hard to believe that it was already published in 1866! It is interesting to note that only 22 years appeared to be sufficient to perform the transition from the earliest primitive chemical formulae without oxygen, such as 2.Ca^2Ᵽ and 5.Ca^2Ᵽ [33] to the modern chemical equations. More to the point, the chemical formulae of hydrated forms of calcium orthophosphates were known by 1866: *"8·73 grs. of the vacuum-dried salt, lost on ignition 12·30 grains, or 26·35 per cent.; the formula $Ca_2H_2P_2O_8{\cdot}4H_2O$, demands 26·16 per cent. of water"* (p. 299).

Needless to explain that "$Ca_2H_2P_2O_8{\cdot}4H_2O$" represents two molecules of DCPD (see Table 1). Furthermore, *"It is interesting to observe that while disodic phosphate is of an alkaline nature, dicalcic phosphate possesses faint acid properties"* (p. 300). More to the point, the form and shape of DCPD crystals were described as well: *"The crystalline form of the dicalcic tetrahydrated phosphate has been examined by Professor Church. He describes the crystals as thin rhomboïdal plates, of which the diagonally opposite acute angles are sometimes truncated, hexagonal forms being thus produced. This truncation seems to be occasionally hemihedral, and then may proceed up to the diagonal between obtuse angles; from this change triangular forms arise. Other modifications are also met with"* (pp. 300-301). Another interesting conclusion might be found here: *"We may then safely affirm that whenever dicalcic phosphate, octocalcic triphosphate, or any phosphate of intermediate composition, is precipitated from solution by ammonia, the salt obtained will be the octocalcic triphosphate; a tricalcic phosphate cannot be obtained in this manner. The following is probably a type of the reaction:—*$4Ca_2H_2P_2O_8 + 6NH_3 = Ca_8H_2P_6O_{24} + 6NH_4.P_2O_8$*"* (pp. 301-302).

This appears to be the earliest mentioning of the fact that TCP cannot be precipitated from an aqueous solution; we know today that ACP or CDHA are precipitated instead. In addition, the following quotation from the same publication: *"It is quite possible that precipitated tricalcic phosphate may possess somewhat different solubilities, when prepared by different methods; this difference can, however, scarcely be great"* (p. 304) has two important aspects: (i) by 1866 this fact was still not quite clear, and (ii) it indirectly points to variability in the Ca/P ratio for the precipitated ACP or CDHA that is the major reason of different solubilities.

The latest available publication by Warington [42] was devoted to the hydrolysis of freshly precipitated TCP *i.e.*, either ACP or CDHA to stoichiometric HA under continuous (up to 50 hours) boiling in distilled water. From the results of numerous chemical analyses the author concluded that during boiling the aqueous suspension of TCP was slowly transformed to a suspension of $3Ca_3P_2O_8.CaOH_2O$ *i.e.*, HA (see Table 1) and soluble acidic calcium orthophosphates. The following conclusion was drawn: *"Since it appears that all phosphates of calcium less basic than apatite are unstable under the continued action of pure water, it seems probable that a more exact examination of natural phosphates would show that many phosphates now regarded as tricalcic are in fact of a more basic nature"* (p. 989). Thus, the apatitic nature of the majority of natural calcium orthophosphates has been predicted in 1873. The next available journal article on the hydrolysis of TCP was published only in 1929 [44].

The first accessible journal paper on detection and preservation techniques of various deposits, including calcium orthophosphates, was published in 1852 [45]. The presence of calcium orthophosphates in teakwood was established in 1862 [46]. Various analytical topics on calcium orthophosphate chemistry have been studied since at least 1863 [47] and remained to be a subject of active research approximately until the 1910s [48-57].

A popular superphosphate fertilizer, representing *"a mixture of the last-mentioned compound and sulphate of lime"* [58] has been known since at least 1803 [59]: *"The phofphoric acid feparated from that portion of lime, immediately combines with the reft of the phofsphate of lime, and forms fuper phofphate of lime, which is not farther decompofable by fulphuric acid"* (pp. 394-395). More than 20 research papers on superphosphate were published by the beginning of the 20th century, which indicate the importance of this chemical for human beings. To give appropriate tribute to those early investigators and restore the historical perspective, it is worthwhile quoting them all [60-82]. It is obvious that almost all these studies were performed and published by single researchers (see individual comments to Refs. [61, 68]) and also that the majority of the investigations were devoted to analytical chemistry. More to the point, let me quote from the 1906 publication by Wells [83]: "The apparent constancy of the proportion of carbonate and phosphate of calcium in bones made an impression on Hoppe-Seyler in 1862, and we find him speculating on the possibility of the components of the two salts being joined together to form one giant molecule: 3 (Ca_3 (Po_4)$_2$)-$CaCo_3$, which he imagined might be united in some such way" (p. 522) – see Figure 1. Further, Wells mentioned: "This formula is interesting chiefly from the historical standpoint, but it serves to emphasize the tendency of these salts to appear in nearly constant proportions in the animal body, a fact possibly of some importance" (p. 523). Obviously, the atomic arrangement shown in Figure 1 represents the earliest structural drawing of a single molecule of a calcium orthophosphate now known as carbonate-apatite. The attentive reader will notice two different types of calcium (now known as Ca(1) and Ca(2)) in this structure.

Although the vitrification properties of some forms of calcium orthophosphates during heating have been known since at least 1804 [22], the first journal paper on this topic accessible today was published in 1877 [84]. The modern chemical formula of perfectly transparent crystals of natural FA, $Ca_5(PO_4)_3F$ has been known since at least 1873 [85], while the major crystallographic faces of a natural calcium apatite were described in 1883 [86]. The chemical formulae of DCP described as 'Mono-Hydrogen Calcium Orthophosphate, $HCaPO_4$', and MCP described as 'Tetra-Hydrogen Calcium Phosphate, $H_4Ca(PO_4)_2$' have been known since at least 1879 [58, pp. 205-206]. Interestingly, in a publication of 1871 the chemical formulae of calcium orthophosphates were written in different ways, such as 3CaO PO_5 for apatite and CaO 2HO PO_5 for 'some acid phosphate of lime' [62]. Neutral phosphates of lime were further investigated in 1872 [87].

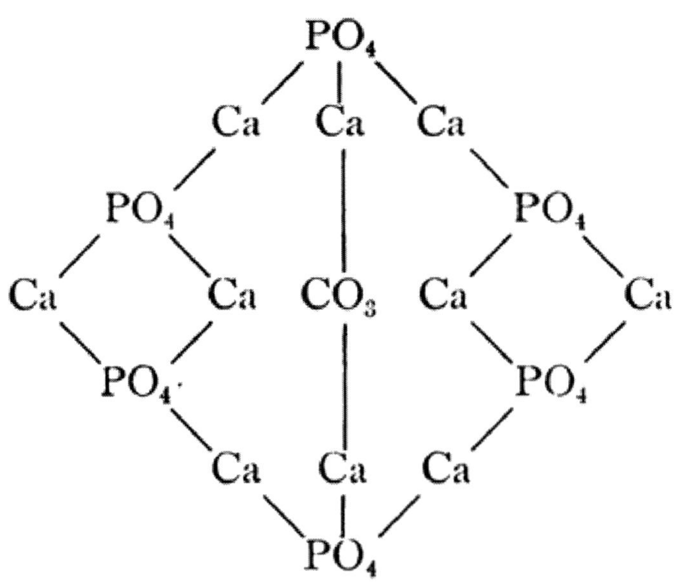

Figure 1. The first available structure of a bone mineral, currently known as carbonate-apatite. Reprinted from Ref. [83].

Besides, in the 19th century, calcium apatites were considered combined compounds, which results from this quote: *"Calcium phosphate, combined with calcium chloride or calcium fluoride, occurs in the well-known minerals, apatite and osteolite"* [58, p. 188]. One might guess that in the 19th century the atomic arrangement of a single molecule of carbonate-apatite (Figure 1) could have inspired researchers to compose similar drawings for single molecules of FA, HA and/or chlorapatite; however, I have not succeeded in finding anything on this matter. Chemical equations, describing various interactions between calcium phosphates and other chemicals have been known since at least 1863. For example, the afore-quoted production processes of both orthophosphoric acid and white phosphorus mentioned in Lavoisier's book [20] were written in 1863 using chemical equations [36] as follows: *"When bone-phosphate is digested in dilute sulphuric acid, it is resolved into sulphate of lime and (if a sufficiency of sulphuric acid be used) phosphoric acid:* $3(CaO),PO_5 + 3SO_3 = 3[CaO,SO_3] + PO_5$*"* (p. 331). Furthermore, various types of phosphates such as metaphosphate and acid phosphate, and the differences in their solubility were known by 1881 [13]: "*In order to prepare phosphorus, the bone-ash is first mixed with so much dilute sulphuric acid as to form the acid phosphate :–*

$$Ca_3(PO_4)_2 + 2H_2SO_4 = CaH_4(PO_4)_2 + 2CaSO_4.$$

The solution of this soluble acid phosphate is next poured off from the precipitated gypsum, and evaporated to dryness, after which, the solid residue being heated to redness, water is given off and calcium metaphosphate formed :–

$$CaH_4(PO_4)_2 = Ca(PO_3)_2 + 2H_2O.$$

This salt is then carefully mixed with charcoal, and heated to bright redness in earthenware retorts shown in Figure 147, when the following change takes place :–

$$Ca(PO_3)_2 + 10C = P_4 + Ca_3(PO_4)_2 + 10CO$$" (p. 460).

Quantitative analysis of a calcium orthophosphate was performed in 1884 [88], followed by remarks by C. Glaser in 1885 [89]. In the 1880s, the occurrence of calcium apatite [90] and TTCP [91-93] in metallurgical slags was discovered. An attempt of direct estimation of phosphoric acid as the tribasic phosphate of lime was performed in 1888 [94]. The chemical interaction of TCP with carbon dioxide and iron hydroxide was investigated in 1891 [95].

To conclude this section, one should notice that the major discoveries of the chemistry of calcium orthophosphates were made in the 19th century.

3.3. From 1900 Until 1940

As the modern scientific databases comprise scientific papers published since 1900, the history of calcium orthophosphates in the 20th century is based on journal publications only. This approach allows establishing the real chronological order.

The 20th century started with systematic studies performed by F. K. Cameron and co-workers [82, 96-102] and H. Bassett [103-106]. For example, in 1908 Bassett composed the first version of the solubility diagram $CaO–P_2O_5–H_2O$ [105]. According to the literature [13, 58, 107], by that time researchers already operated with individual calcium orthophosphates. Binary salts of calcium orthophosphates with orthophosphates of other chemical elements have been known since at least 1911 [108]. The well-known chemical route to prepare CDHA by mixing aqueous solutions of calcium nitrate and K_2HPO_4 (now $(NH_4)_2HPO_4$ is used instead) was investigated conductometrically in 1915 [109] while a neutralization reaction of H_3PO_4 by $Ca(OH)_2$ was investigated in 1923 [110]. The first paper on *in vitro* mineralization using aqueous solutions containing ions of calcium and orthophosphate was published in 1926 [111], followed in 1933 by another publication [112]. The latter studies might be considered predecessors of current mineralization investigations using simulated body fluid (SBF) and other artificial solutions.

The crystal structure of FA was established in 1931 [113], followed by a major study on the structural characteristics of apatite-like calcium orthophosphates of various origin in 1933 [114]. Such terms as hydroxyfluorapatite, $Ca_{10}(F,OH)_2(PO_4)_6$, OA and carbonate-apatite were already known. Besides, a currently unknown substance, 'a hydrate of tricalcium phosphate, $Ca_9(H_2O)_2(PO_4)_6$' with a molecular weight of 966.4 and apatite-like diffraction pattern (obviously, it was CDHA with $x = 1$, see Table 1) was mentioned as well [114]. Furthermore, the solubility data of several calcium orthophosphates were updated in 1931 [115, 116]. Additionally, between 1931 and 1933 Bredig *et al.*, [117], Trömel [118] and Schneiderhöhn (a reference was not found) distinguished α- and β-modifications of TCP (see Table 1) and created the initial versions of the high-temperature diagram for the binary system $CaO–P_2O_5$. Silicon- (or silica-) containing calcium orthophosphates, very popular at the turn of this millennium, in fact appear to be known since at least 1937 [119, 120], simultaneously with defluorination studies of natural phosphate rocks [120-122], while the earliest research paper on the application of calcium orthophosphates in chromatography was published in 1942 [123].

In 1940, the available level of knowledge on basic calcium orthophosphates (TCP, TTCP and apatites) was summarized into an extensive review [124] that might be considered the first comprehensive review on the subject.

Interestingly, the authors of this review have written about HA that its 'existence as a unique stoichiometric compound (is) doubtful' [124, p. 259, Table], which is a clear indication of the great difficulties to synthesize HA but not CDHA. At this time, the existence of OA was uncertain [124]. Besides, in 1940, a fundamental study on the equilibrium in the system $CaO–P_2O_5–H_2O$ was published [125].

To finalize this chemical section, mentioning should be made of the fact that by 1928 it was clearly known that TCP could not be prepared by wet-precipitation. The following quote attest to this uncertainty *"We have been unable to find any evidence of the existence of a molecular species with the formula $Ca_3(PO_4)_2$. Precipitates of this 'substance' rarely have the correct empirical composition and cannot be purified by recrystallization. The evidence is such as to lead us to suspect that there may be no such chemical entity as $Ca_3(PO_4)_2$. No one has succeeded in preparing it by precipitation (12, 35-48), bearing out the theoretical objections to such a reaction on the grounds that fifth order reactions do not occur (49). We do not take the position that there can be no compound with this formula. It may be found possible to synthesize it by other methods, but so far no one has succeeded in preparing it by precipitation."* [126, p. 128]. Similar conclusions were drawn by other researchers [127]. Nevertheless, this knowledge was not common then, as in 1935 a report was published, stating that *"Tricalcium phosphate monohydrate was prepared by the slow addition of calcium chloride solution to a constantly agitated alkaline solution of disodium phosphate, maintained at 65° to 70°C"* [128].

This controversy has been later explained in Ref. [124] as originating from a matter of definition: *"The terms "tricalcium phosphate" and "hydroxyapatite" are very widely and very loosely used. For example, some authors use the former for any precipitate more basic than dicalcium phosphate, although such precipitates have been frequently shown to have an apatite lattice or to be mixtures of dicalcium phosphate and an apatite. Others confine the use of the term to those precipitates with P_2O_5:CaO ratios approaching that of $Ca_3(PO_4)_2$"* (p. 265).

More recent (after 1940) publications on the chemistry of calcium orthophosphates are not considered, since they are already well known.

4. Early Publications on Calcium Orthophosphates of Geological Origin

As mentioned above, apatite was recognized as a calcium phosphate already by 1789 [18, 19]. As early as 1794, Kirwan [129] in his book on mineralogy wrote a section "*Phoſphorite.* Apatite of Werner" (pp. 129-130) stating *that "In Spain, it is found in extenſive ſtrata forming intire mountains; ... Sometimes alſo cryſtallized in Germany, and of greeniſh colour. In Saxony, it is found in low hexahædral priſms with truncated edges and angles, the planes of the ſides longitudinally ſeamed, or in hexangular tables heaped one on the other"*. Thus, calcium orthophosphates of geological origin has been known since the end of the 18th century. At the very beginning of the 19th century (in 1805), Nisbet [23] wrote: *"Phosphated Lime, apatite or phosphorite, has been found in an amorphous state in Hungary, and forming entire mountains in Spain. It is also found in truncated hexahedral, longitudinally striated prisms, laminated in their transverse fracture, and generally with tin and fluor. Klaproth found it to contain acid 45, lime 55"* (p. 195).

In 1809, Murray [130] wrote on natural phosphorite: *"Its colour is yellowish or greyish white, without lustre or transparency; rough to the touch; and having a specific gravity of 2.8. This substance was analysed by Pellerier, who found it to consist of lime 59, phosphoric acid 34, fluoric acid 2.5, silex 2, iron 1, with a trace of carbonic and muriatic acids"* (p. 477). Thus, a complex quantitative composition of natural phosphorites was established by 1809. Eleven years later, in 1820, Robert Jameson published in his fundamental book "A system of mineralogy" [131] a section entitled "Genus II. APATITE" (pp. 574 – 587) in which major information on physical, chemical, and crystallographic properties of natural apatites and phosphorites as well as the geographic locations of major deposits were summarized. Shortly afterwards Delvalle Lowry (married Varley) published in 1822 a catechism book "Conversations on mineralogy" [132] from which a self-explanatory 'discussion' is worth quoting [132]:

> "The family of phosphates of lime contains but two species, apatite and phosphorite.
> Frances. Are they very different from each other?

> Mrs. L. In appearance they are, but not in composition; apatite is generally crystallized, and strongly translucent, or transparent: but phosphorite is always amorphous and opake" (p. 187).

Thus, the key similarities and differences between two major types of calcium orthophosphates of geological origin were already established by 1822.

Concerning journal papers on the subject, the earliest publication on calcium orthophosphates of geological origin I have managed to find was published in 1873 [85], followed by another one dedicated to deposits of natural calcium orthophosphates that was published in 1889 [133]. Presumably, the latter publication might be considered as the first review paper on calcium orthophosphates, in which the author described various types of natural phosphate deposits, their formation, as well as possible utilization, mainly as fertilizers.

Next available scientific publications on natural calcium orthophosphates were published in 1895 [134], 1901 [135, 136], 1902 [137-139] and 1904 [140]. An important study on the presence of orthophosphates of various chemical elements, including calcium in soils was published in 1906 [141]. However, this subject is beyond the scope of this review.

5. Studies on the Presence of Calcium Orthophosphates in Living Organisms

5.1. Studies Performed in the 18th Century

As mentioned above, the presence of calcium orthophosphates in bones has been known since the 1770s [13, 16]. Although the presence of phosphorus in urine was established already in 1669 by Brand, the knowledge of dissolved calcium in urine is available since at least the 1780s [142]): *"We have ſeen that urine contains many peculiar ſalts; theſe ſalts are combinations of the phoſphoric acid with volatile alkali, ſoda, and lime, and the baſe of the calculus of the bladder"* (p. 353). According to another publication [143], the discovery of calcium orthophosphates in urine is related to the great chemists Scheele and Lavoisier: "*The immortal, and ever to be deplored, Lavoiſier ſuppoſed theſe ſubſtances to conſiſt of acidulous phoſphate of lime and animal matter, many of them being partially fuſible; but ſtill it was the unrivalled Scheele who diſcovered that the urine of healthy perſons contains ſuperphoſphate, or acidulous phoſphate, of lime; and who alſo indicated the experiment which verified his opinion, that phoſphate of lime is the baſis of bone"* (pp. 39-40).

A special chapter entitled "Concerning the bones of animals" (pp. 400-406) was published in 1788 as a part of de Fourcroy's book [142]. Several passages from this chapter are worth quoting: *"Theſe hard ſubſtances cannot be conſidered as paſſive in the animal economy, they are true ſecretory organs, which ſeparate from the blood and other humours a peculiar ſaline ſubſtance, of which they are the reſervoir, or place of depoſition"* (p. 400). This is the correct description of the role of bones in general metabolism. *"The bones of men and of quadrupeds, which alone have been hitherto examined by chemiſts, are not earthy matters, as was formerly ſuppoſed: they contain a certain quantity of gelatinous matter, diſperſed in ſmall cavities, formed by the interval between the ſolid parts which compoſy their texture; and the ſolid parts themſelves, though they ſeem to reſemble earthy ſubſtances by their inſolubility and firmneſs, have been for ſeveral years paſt acknowledged to be a true neutral ſalt, conſiſting of phoſphoric acid and lime"* (p. 401). Thus, the bioorganic/inorganic composite nature of bones has been known since at least 1788.

Concerning the level of knowledge on the presence of calcium orthophosphates in teeth and other calcified tissues of animals, available by the end of the 18th century, let me quote three self-explanatory sentences from a publication of 1797 [144]: *"More recently, Berniard (Journal de Physique, Octobre, 1781,) obtained phosphoric acid not only from fossil bones, from those of the whale and sea-horse, but from the tooth of the manati and the grinder of the elephant. It has also been extracted from ivory. In short, they who have experimentally attended to this subject have agreed, that animal bones are chiefly composed of phosphoric acid and calcareous earth, or are the phosphate of lime: and that their teeth consist in the main, of the like materials"* (p. 62).

5.2. Studies Performed in the 19th Century

Further investigations on calcium orthophosphates in living organisms were performed in the 19th century. For example, the structure, composition, properties and formation mechanisms of bones were described in detail in the London medical dictionary, published in 1809 [145]. The following statements from that dictionary are worth quoting: *"Bones consist of fibers crossing each other in different directions"* (p. 263), *"The substance of bones is now known to be a calcareous phosphat, which is deposited by exhalent arteries; and this substance is constantly renovated, while an equal portion is carried off by absorption"* (p. 263), *"Bones, we have said, owe their solidity to phosphorated lime. The bones of an adult contain the greatest proportion of earth; those of a fœtus, of gelatin. Fish bones, from Mr. Accum's experiments contain one-sixth part more of phosphoric acid than the bones of quadrupeds. Carbonated alkalis decompose the phosphat of lime in bones. They contain also a small proportion of selenite, carbonate of lime, and, according to Proust, of mineral alkali"* (p. 264). Thus, the presence of carbonates and other trace elements in the mineral phase of bones, and their self-renovating properties have been known since at least 1809. The following self-explanatory quotation from a book published in 1812 [146] is worth mentioning: *"It is remarkable, that though phosphate of lime is always found in the urine of adults, this salt is not evacuated by infants. The rapid formation of the bones, in the first periods of life, requires that there should be no waste of any of the phosphoric salts; and Nature, ever provident, has provided accordingly"* (p. 225).

Furthermore, according to de Fourcroy's book, published in 1804 [22], the presence of calcium orthophosphates in milk and blood was already known: "*Citizens Berthollet and Fourcroy confirmed the diſcovery of Scheele reſpecting the preſence of phoſphate of lime in human urine. They found it alſo in milk, in blood, in the muſcles, in jelly, andc."* (p. 537). Besides, as seen from a popular book by Sir Humphry Davy, published in 1814 [147], the general principles of bone and teeth formation (biomineralization) were already known: *"During the earliest period of animal life they consist almost entirely of a gelatinous membrane of the form of the bones, but of a loose spongy texture, the cells or cavities of which are destined to be filled with phosphat of lime; it is the gradual acquisition of this salt which gives to the bones their subsequent hardness and durability. Infants first receive it from their mother's milk, and afterwards derive it from all animal and from most vegetable food, especially farinaceous substances, such as wheat flour, which contain it in sensible quantities. A portion of the phosphat after the bones of the infant have been sufficiently expanded and solidified, is deposited in the teeth, which consist at first of only a gelatinous membrane or case, fitted for the reception of this salt; and which, after acquiring hardness within the gum, gradually protrude from it"* (p. 302). What is more, such disease as rickets (rachitis) was already known to be caused by deficiency of calcium phosphates: *"Pray, Mrs. B. does not the disease of the bones of children, called the rickets, proceed from a deficiency of phosphat of lime?"* and the response: *"The other cause of this disorder is an imperfect digestion and assimilation of the food, attended with an excess of acid, which counteracts the formation of phosphat of lime"* (p. 303). Thus, according to the available literature [148], one can conclude that the correct basic knowledge on the structure, chemical composition and properties of bones was obtained already by 1820.

The earliest journal publication I have been able to find, devoted to the importance of calcium orthophosphates to bone health was published in 1842 [149], followed by a study published in 1895 [150], while those on calcium phosphates of bones were published by Aeby in 1872 [151] and 1873 [152, 153], followed by papers written by other researchers [154, 155].

In spite of the aforementioned achievements, basic research on bones continued. For example, compositional differences between bones belonged to young and old individuals were extensively investigated in the middle of the 19th century. For example, let me quote a summary on this topic, published in 1928 [156]: *"In 1855 Frémy ... concluded that in humans there is, with increasing age, a relative increase of $CaCO_3$ with respect to calcium phosphate. Von Recklinghausen ... in 1858 concluded from his own analyses and from the work of Bibra (1844), of Stark (1845), of Heintz (1849), and of Frémy, that there was no observable difference in composition between the bones of young and those of old individuals. In 1860 Milne Edwards ... reported that his own analyses agreed with those of Bibra and of Frémy in showing that the proportion of carbonate in the bones of young animals was smaller than in adult bone. Zalesky...) in 1866 reviewed the literature and concluded that the bones of children of various ages have the same composition as those of adults, not only as a whole, but also as regards the proportion of individual constituents. He pointed out that the methods of determining CO_2 employed by previous investigators were defective. In 1872 Wildt ... analyzed the ashed bones of rabbits of different ages and found an increase in the proportion of CO_2 with increasing age. Weiske ... in 1889 analyzed the ashed bones of birds and found an increase in CaO and CO_2 with advancing age, while the P_2O_5 remained constant. Graffenberger ... in 1891 analyzed the ashed bones of rabbits and also found an increase of the ratio calcium carbonate : calcium phosphate with increasing age"* (pp. 155-156). This quotation clearly shows that the chemical composition of the inorganic phases of bones was a hot topic of research in the 19th century.

In 1861, the presence of calcium orthophosphates in urine was investigated in further details [157, 158], while the presence of calcium orthophosphates in the auditory system of mammals has been known since at least 1879 [159]. The earliest journal paper I have been able to find on the behavior of an undisclosed calcium orthophosphate in organisms of carnivores was published in 1883 [160].

Undesirable *in vivo* formed calcifications of biologically formed calcium orthophosphates have been known since at least 1809 [161]: *"IV. The calculi found in the urethra consist of ammoniaco-magnesian phosphate, and phosphate of lime, with a small portion of uric acid; ...*" and "*Mr. Brande, in the next section, has given the result of analysis of the calculi found in the horse, ox, sheep, rhinoceros, dog, hog, and rabbit. These were found mostly to consist of phosphate of lime and carbonate of lime in different proportions*" (p. 477). Furthermore, from a publication of 1814 [162]: "*2d. Fusible Calculi. – These are composed chiefly of phosphate of lime, and the triple phosphate of Magnesia and Ammonia*" and "*4th. Bone-earth Calculi. – These are composed chiefly of phosphate of lime*" (p. 94).

However, the earliest research paper I have been able to find on undesirable calcifications was published in 1856 [163], followed by a book [164] written by the German doctor, anthropologist, pathologist, prehistorian, biologist and politician Rudolf Ludwig Karl Virchow (1821 – 1902), in which he introduced a term 'metastatic calcification'. To the best of my knowledge, later journal publications on undesirable *in vivo* formed calcifications containing calcium orthophosphates were in print only in 1882 [165] and 1883 [166].

5.3. Studies Performed from 1900-1940

Research papers on the importance of calcium orthophosphates to bone health were published in 1906 [167] and 1910 [168]. Papers on the importance of calcium orthophosphates for lactation were published in 1914 [169], 1919 [170] and 1922 [171], and the fact that the optimum Ca:P ratio in a healthy diet is the same as that found in human bones or in mother's milk has been known by 1930 [172].

Research papers on undesirable *in vivo* formed calcifications were continued to be published in 1901 [173], 1906 [83] and 1913 [174]. An important conclusion was drawn in 1906 [83]: "*The composition of the inorganic salts in calcified areas in the body seems to be practically the same, if not identical, whether the salts are laid down under normal conditions (ossification) or under pathological conditions*" (p. 520). Thus, the similarity between the calcium orthophosphates of bones and those found in metastatic calcification has been known since at least 1906. Subsequent studies on a behavior of an undisclosed calcium orthophosphate in organisms were in print in 1912 [175] and 1921 [176]. A paper on the importance of calcium and phosphorus for growing children was published in 1913 [177], while that on the necessity of normal calcium and phosphorus metabolism was published in 1930 [178]. Investigations of the amounts of dissolved calcium orthophosphates in spinal fluid [179] and serum [180-182] were performed in the 1920s, and those in saliva [183] and in serum of pregnant women [184, 185] and diseased people [185, 186] were performed in the 1930s. To conclude this topic, we should also mention comprehensive reviews published in 1935 [187] and 1937 [188] summarizing the knowledge on calcium orthophosphates in living organisms available at this time.

6. Early Attempts to Treat Various Diseases by Administering Calcium Orthophosphates

Attempts to treat various diseases by administering calcium orthophosphates have been known since at least 1804. Initially, they were applied to treat rickets [22]: "*In medicine it has begun to be prefcribed in the Rachitis, to diminifh the effects of acids which foften the bones, andc."* (p. 346). Further details became available by 1819 [25]: *"4. Phosphate of lime has been employed as a remedy in rickets, a disease in which there appears to be a deficiency of this salt deposited in the bones. Its usefulness in this disease may be considered rather doubtful, unless the digestive powers of the system should be improved at the same time"* (p. 327).

However, the earliest journal articles I have been able to find were in print as early as 1851 [189, 190]. The author performed a thorough study on the biomedical influence of phosphate of lime that was *"taken with the breakfast, dinner, and supper, in order to have it mixed as well as possible with the food taken"* on treatment of various diseases. According to the author [189]: *"... the phosphate of lime proved most beneficial in children who suffered from scrofula, diarrhoea, ulcerations and excoriations of the skin and the bowels, general waste of the cellular tissue, loss of power, andc. In these cases, the mere use of six to ten grains per diem was often quite sufficient to effect the cure, and I have met with such striking and satisfactory results, as to leave no doubt that the want of the phosphate of lime was the real cause of the symptoms alluded to"* (p. 434, left column). Furthermore, the author states: *"As to other affections, in which I have tried the phosphate of lime, I have to mention rickets, caries, inflammations, and consecutive abundant suppurations of the cellular tissue, and also fractures of the bones. In all these cases, the administration proved most beneficial, and I would strongly advocate its further experimental use. With respect to fractures of the bones, I have to state in particular, that the consolidation of the callus took place in a much shorter period than is generally the case; however, too large doses of phosphate of lime must be shunned in these cases, as I have observed an abundant callus, causing a deformity of the bones, produced by the daily administration of twenty-four grains of the phosphate for a fortnight"* (p. 434, right column). In a second paper, the same author mentioned an interesting hypothesis that *"phosphate of lime contained in a certain quantity of blood*" (thus, the presence of calcium orthophosphates in blood was already known in 1851) might be partly *"originated from the wear and tear of the bones and the muscular tissue"*. Let me quote [190]: *"Supposing now the wear and tear of bones to be abnormally increased, an abnormal quantity of phosphates will consequently re-enter into the composition of the blood, and by analysing the latter we shall find an increased quantity of phosphates in it. The general waste of bones is therefore leading to an increase of the phosphates in the blood, and nothing really would be more erroneous, than to suppose the general quantity of phosphates in the system to be increased, whenever we find their quantity increased in the blood"* (p. 668, right column). The readers are strongly advised to scan both papers to rediscover the stunning medical ideas and approaches published 160 years ago!

The next medical study available to me involving phosphate of lime was published in 1874 [191]. An interesting idea to cure cancer *"by embedding a piece of bone material or calcium phosphate in a cancer and exposing it to a pencil of X rays ..."* to create secondary radiation in calcium orthophosphates (either natural or implanted ones) was announced in 1913 [192]. Strangely enough, a search in the scientific databases on further development of this idea has resulted in no follow-up publications as researchers introduced radioactive isotopes of other chemical elements instead [193-196]. Nevertheless, it might be worthwhile reinvestigating this approach.

The value of phosphorus and calcium in treatment of asthma, hay fever and allied diseases was investigated in 1928 [197]. The earliest paper I have been able to find on addition of calcium orthophosphates as food supplement in forage of animals was published only in 1953 [198].

7. Artificial Grafts Throughout History

The artificial generation of tissues, organs or even more complex living organisms was throughout the history of mankind a matter of myth and dream. Unfortunately, due to the practice of cremation in many societies, little is known about the prehistoric materials used to replace bones lost to accident or disease. Nevertheless, according to the available literature, introduction of non-biological materials into the human body was noted far back in prehistory. For example, the remains of a human found near Kennewick, Washington, USA, referred to as the 'Kennewick Man', were dated up to 9000 years old. This individual, described by archeologists as a tall, healthy, active person, wandered through the region now known as southern Washington with a spear point embedded in his hip. It had apparently healed in and did not significantly impede his activity. This unintended implant illustrates the body's capacity to deal with implanted foreign materials. The spear point has little resemblance to modern biomaterials, but just the same it was a tolerated foreign material implant [199].

Later, man's attempts to repair the human body with the use of implant materials were recorded in the early medical writings of the Hindu, Egyptian and Greek civilizations. The earliest successful implants were in the skeletal system. For example, applying modern pathology methods to a ~ 3,000-year-old mummy, researchers at Ludwig-Maximilian University, Munich, showed that ancient Egyptian physicians designed wooden prostheses to help their ailing patients, in this case a 50 – 60 years old woman whose toe was amputated [200]. Besides, the famous painting by Fra Angelico (ca. 1395 – 1455) 'The Healing of Justinian by Saint Cosmas and Saint Damian' (Figure 2), a visualization of the legend of the twins saints Cosmas and Damian (died ca. 287 AD) depicting the transplantation of a homograft limb onto an injured soldier, is one early instance of the vision of regenerative medicine.

Historically, the selection of the materials was based on their availability and the ingenuity of the individual making and applying the prosthetics [201]. Archaeological finds exhibited in museums showed that materials used to replace missing human bones and teeth included animal or human (from corpses) bones and teeth, shells, corals, ivory (elephant tusk), wood, as well as some metals (gold or silver). For instance, the Father of Western medicine, Hippocrates (ca. 460-ca. 370 BC) apparently used gold wire and linen thread for ligatures in the repair of bone fractures.

Aulus Cornelius Celsus (ca. 25 BC – ca. 50 CE) recommended filling of large cavities with lint, lead and other substances before attempting extraction to prevent the tooth from breaking under the pressure of the instrument. This may have been the beginning of filling materials for carious teeth. The Etruscans learned to substitute missing teeth with bridges made from artificial teeth carved from the bones of oxen, while in ancient Phoenicia loose teeth were bound together with gold wires for tying artificial ones to neighboring teeth. Popp states that ancient Egyptians also made artificial ears, noses and eyes [202].

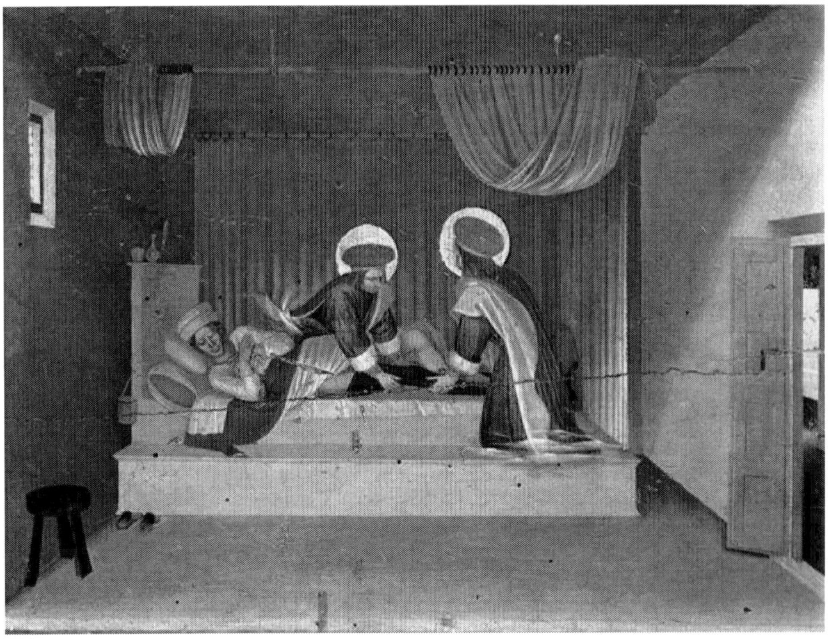

Figure 2. Fra Angelico (ca. 1395 – 1455) "The Healing of Justinian by Saint Cosmas and Saint Damian" (approx. 1439). Museo di San Marco, Florence, Italy.

The Chinese recorded the first use of dental amalgam to repair decayed teeth in the year 659 CE, while in the Americas the pre-Columbian civilizations used gold sheets to cover cranial cavities following trepanation [203]. Besides, while excavating Maya burial sites in Honduras in 1931, archaeologists found a fragment of a mandible of Mayan origin, dating from about 600 CE. This mandible, considered to be that of a woman in her twenties, had three tooth-shaped pieces of shell placed into the sockets of three missing lower incisor teeth. In 1970, a Brazilian dental academic Prof. Amadeo Bobbio studied the mandible specimen and took a series of radiographs. He noted compact bone formation around two of the implants, which led him to conclude that the implants were placed during life time [204]. This may be the first recorded use of dental implants. More to the point, an iron dental implant in a corpse dated ~ 200 CE was found in Europe. This implant, too, was described as properly integrated in bone [205].

In the Middle Ages, one of the first scientific descriptions of congenital and acquired defects of the maxilla and their treatments was given by Ambroise Paré (1510 – 1590) in his *Dix livres de la chirurgie, avec le magasin des instrumens necessaires à icelle*, published in Paris in 1564. He specifically described defects of the palate with bone destruction caused by arquebus shots, stab wounds or syphilitic gumma, describing also the accompanying speech deficiency and giving general principles of treatment. He used a flat, vaulted, metallic plate of gold or silver with a sponge attached to it. The sponge was introduced into the defect, where it expanded with readily absorbed nasal and oral secretions, thus holding the obturator base in position. Paré is also credited with having prepared artificial teeth from bones and ivory. In the 17th century, a piece of a dog's skull was successfully transplanted into the damaged skull of a Dutch duke.

In the 18th century, a common method to replace teeth was the homologous transplantation of teeth in humans. John Hunter (1728 – 1793) investigated in his pioneering work the effect of transplantation not only at a clinical level (he claimed, that homologous transplanted teeth lasted for years in the host) but also performed animal experimental work on the fate of transplants, thereby setting the basis for a scientific approach on transplantation medicine [206]. Besides, various restorative materials might be used for feeling defects, capping exposed pulps and dental cavities. Currently they include zinc orthophosphate, zinc polyacrylate (polycarboxylate), various amalgams, glass ionomer cement of complicated formulations, calcium hydroxide *etc.*; however, this is another story. Within the scope of this review, calcium orthophosphate-based formulations will be considered only. According to the available literature, Dr. Junius E. Cravens (1844 – 1920) from USA proffered creative concepts in pulp capping in the 1870s. He held the opinion that dentin-like material would be the best to keep the pulp vital. Therefore, Cravens used a calcium orthophosphate powder mixed with lactic acid to lower the viscosity. The result was a soluble calcium lactic orthophosphate which was applied to the exposed pulp tissue [207]. This pulp-capping agent was brought to the market by the S.S. White Company with the trade name "Lacto-Phosphate of Lime" (Figure 3).

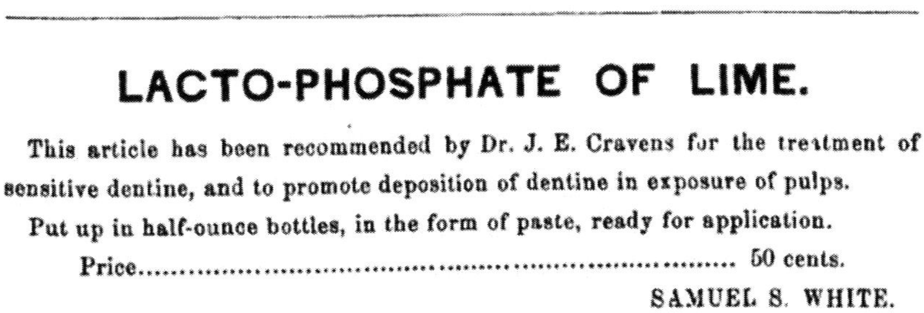

LACTO-PHOSPHATE OF LIME.

This article has been recommended by Dr. J. E. Cravens for the treatment of sensitive dentine, and to promote deposition of dentine in exposure of pulps.

Put up in half-ounce bottles, in the form of paste, ready for application.

Price.. 50 cents.

SAMUEL S. WHITE.

Figure 3. An advertisement of the S.S. White Company for "Lacto-Phosphate of Lime" 1873. Reprinted from Dent. Cosmos 1873, 15, 683.

Table 2. Chemical composition of two bone-ash samples taken from a publication of 1871 [41]

	Commercial bone-ash	Pure ox bone-ash
Moisture and volatile matter	6.70	1.86
Siliceous matter	9.69	0.51
Oxide of iron	0.58	0.17
Lime	43.37	52.46
Magnesia	1.14	1.02
Phosphoric acid	33.68	39.55
Carbonic acid, alkalies, and other substances undetermined	4.84	4.43
Total	100.00	100.00

Sugar-containing formulations were known as well [208]. This study by Cravens might be considered the first mentioning of artificial calcium orthophosphate-based biocomposites and hybrid biomaterials (reviewed in Ref. [209]). Thus, one might claim that the history of calcium orthophosphates as dental grafts started in 1870's.

One must stress that in the past many implantations failed due to various infections that tended to be exacerbated in the presence of implants, since they provided a region inaccessible to the body's immunologically competent cells. Thus, the use of implantable biomaterials did not become practical until the advent of an aseptic surgical technique developed by a British surgeon Sir Joseph Lister (1827 – 1912) in the 1860s.

Furthermore, there was a lack of knowledge about a toxicity of the selected materials. The exact chemical composition of normal calcified tissues of mammals has been known since at least 1870s, which might be concluded from Table 2 [41] and the following quotations: *"The bones and teeth of animals contain large amounts of calcium phosphate, together with some carbonate and fluoride"* [58, p. 188] and *"Calcium phosphate is also the chief inorganic constituent of bones, forming about 80 per cent of burnt bones; the other constituents being magnesium phosphate, calcium carbonate, and calcium fluoride"* [58, p. 205]. These data were updated in 1894 [210], and the apatitic structure of calcium orthophosphates of bones and teeth has been established by x-ray diffraction in 1926 [211] and confirmed in 1932 [212] and 1933 [213]. Since that time, bone mineral has been frequently identified as HA. In this context, the application of calcium orthophosphates as artificial grafts appears to be logical due to their similarity with the mineral phases of bones and teeth.

To conclude this topic, one should stress that the performance of living tissues is the result of millions years of evolution. In contrast to this, the performance of acceptable artificial substitutions designed by man to repair damaged tissues is known for only a few decades. This explains the great differences between natural and artificial biomaterials. To get a historical perspective on the development of artificial grafts prepared from other materials, the interested reader is referred to the specialized literature [7, 9, 10-12, 214-217].

8. Calcium Orthophosphates as Bone Graft Substitutes: A Historical Perspective

Historically, plaster of Paris (calcium sulfate) was the first widely tested artificial bioceramics. For example, according to Wikipedia, literature dating back to 975 CE notes that calcium sulfate was useful for setting broken bones. However, these were only *ex vivo* applications. According to the available literature, by the end of the 19th century surgeons already used plaster of Paris as a bone-filling substitute [218]. Nevertheless, it was the German surgeon Themistocles Gluck (1853 – 1942) who in 1890 performed the first well-documented ivory (virtually, pure biological apatite) knee replacement embedded in a calcium sulfate-based cement. This operation was followed by a total wrist replacement in another patient three weeks later [219]. Later in 1890, Gluck presented a further case of a total knee replacement to the Berlin Medical Society whereby at only 35 days after operation the patient was free of pain with active knee flexion and extension. All joint arthroplasties performed by Gluck were remarkably successful in the short term; however, all ultimately failed because of chronic infections [220, 221]. After the abovementioned case of lacto-phosphate of lime (Figure 3), this seems to be the second well-documented grafting application of calcium orthophosphates.

However, in the aforementioned cases, either the biomedical applications of biologically produced calcium orthophosphates (Gluck) or dental applications, not requiring any surgery (Cravens) have been described. According to the electronic databases and previous reviews on the subject [7, 9, 10-12, 214-217], the first attempt to implant a laboratory-produced calcium orthophosphate (TCP) as an artificial material to repair surgically created defects in rabbit bones was performed in 1920 [8] by the U.S. surgeon Fred Houdlette Albee (1876 – 1945), who invented bone grafting [222] and some other advances in orthopedic surgery. He injected either 0.5 or 1 cm^3 of 5 % slurries of TCP in distilled water (which was then sterilized for three successive days in an Arnold sterilizer at 60° C) into surgically created radial bone gaps of rabbits, leaving the periosteum intact [8]. Radiographic analysis revealed that the TCP-injected defect demonstrated more rapid bone growth and union than the control. The average length of time for bony union with TCP was 31 days, compared to 41 days for the controls. No appreciable bone growth was stimulated by injecting TCP beneath the periosteum in non-defective radii or into subcutaneous tissues. Although this seems to be the first scientific study on use of an artificially prepared calcium orthophosphate for *in vivo* repairing of bone defects, it remains unclear whether the TCP used was a precipitated or a ceramic material and whether it was in a powder or a granular form. Unfortunately, Albee published nothing further on this topic. In 1927, Hey Groves (1872 – 1944) described pure ivory hip hemi-arthroplasty for fracture [223]. In 1931, Murray also reported an acceleration of healing following implantation of calcium salts composed of 85% TCP and 15% $CaCO_3$ in canine long bone defects [224, 225].

In the1930s, the classic osteoinductive phenomenon was well defined by Huggins [226], who demonstrated that autoimplantation of transitional epithelium of the urinary bladder to abdominal wall muscle in dogs provoked ectopic bone formation. Somewhat later, Levander demonstrated that crude alcoholic extracts of bones induced new bone formation when injected into muscle tissue [227, 228].

Simultaneously, in the 1930s Haldeman and Moore [229], Stewart [230], Key [231] and Shands [232] discovered the fact that only certain types of calcium orthophosphates mentioned in Table 1 really influence the bone healing process. For example, Haldeman and Moore implanted various calcium orthophosphates such as MCP and DCP (it remains unclear whether they were in hydrated or anhydrous forms), TCP as well as calcium glycerophosphate as dry powdered salts into 0.5 to 1 cm defects in the radii of seventeen rabbits, while the opposite side served as control. Radiographic analysis demonstrated that in no case the presence of MCP, DCP or calcium glycerophosphate had a favorable influence on delayed healing compared to the control, whereas the presence of TCP at the fracture site appeared to favor the union [229]. Furthermore, Key [231] suggested that *"if a defect in bone could be filled by a non-irritating, slowly soluble mass, which was porous and which contained calcium phosphate and carbonate in a form in which they could be resorbed, it would be reasonable to expect osteoblasts to invade the mass, utilize the calcium, and build new bone which would replace the mass of calcium and cause the bone to be restored to its original form. The ideal material would appear to be rather dense cancellous bone from which a large percentage of the organic material had been removed."* (p. 176). However, Key found that *"Neither calcium phosphate and carbonate in the proportions in which they occur in bone, nor bone powder, made by removing the organic matter from bone, appear to stimulate osteogenesis of bone when implanted in a bone defect"* (p. 184). Stewart [230] concluded that *"1. Lime salts and boiled bone when placed into a bone defect with either traumatized muscle or fascia do not serve as a source of available calcium resulting in supersaturation of connective tissue and regeneration of missing bone. 2. Live bone chips placed in bone defects regenerate the missing bone"* (p.871). Shands [232] also reported conflicting effects of several calcium salts (calcium glycerophosphate, a mixture of TCP (3 parts) and $CaCO_3$ (1 part), bone ash and calcium gluconate) on bone repair. For example, in defects in the ulna of dogs, the investigated calcium salts appeared to stimulate bone formation, while in operations upon the spine, calcium glycerophosphate did not stimulate bone formation but rather appeared to exert an inhibiting influence. In 1948, Schram and Fosdick confirmed the fact that only certain types of calcium orthophosphates influence the bone healing process [233]. Similar conclusions were drawn by Ray and Ward [234] in 1951.

Due to the reasons mentioned in the abstract, more recent historical events are being reviewed only very briefly. For example, in 1950 the history of calcium orthophosphate cements was initiated by David Kingery [235]. The author of this important publication investigated mixtures of both oxides and hydroxides of various metals with aqueous solutions of orthophosphoric acid and discovered a number of cold-setting formulations. For example, he found that CaO, sintered at 1100 °C, did not set in H_3PO_4, while mixtures in liquid containing 9.6% CaO were found to set after ~ 12 h in the presence of H_3PO_4 [235]. The latter mixture might be considered the first prototype of self-setting calcium orthophosphate cements (reviewed in Refs. [236, 237]). A further publication on this subject appeared in 1975 [238]. However, the real history of the self-setting calcium orthophosphate cements started in 1982 [239, 240]. [See also Chapter 11 and 12 of the present book. The Editor].

The modern history of ACP [37] started in 1955 [241]. More than twenty years afterwards the first dental applications of a calcium orthophosphate (erroneously described as TCP) in surgically created periodontal defects [242] and the use of dense HA cylinders for immediate tooth root replacement [243] were reported. According to the available databases, the first paper with the term 'bioceramics' in its abstract was published in 1971 [244], while those with 'bioceramics' in the title were published in 1972 [245, 246]. However, application of the ceramic materials as prostheses had been known before [247-250]. On April 26, 1988, the first international symposium on bioceramics was held in Kyoto, Japan.

The history of calcium orthophosphate-based biocomposites and hybrid biomaterials (reviewed in Ref. [209]) arguably began in 1981, while that of calcium orthophosphate coatings (reviewed in Refs. [251, 252]) began in 1980 [253]. An extensive commercialization of the dental and surgical applications of calcium orthophosphate (mainly, HA) bioceramics occurred in the 1980s, largely due to the pioneering efforts by Jarcho [254-257] in the USA, de Groot [258-260] in Europe, and Aoki [261-264] in Japan. Shortly afterwards HA has become a bioceramic of reference in the field of calcium orthophosphates for biomedical applications. Preparation and biomedical applications of apatites derived from sea corals (coralline HA) [265-267] and bovine bone [268] have been reported at the same time [269], while the history of biphasic, triphasic and multiphasic calcium orthophosphates (reviewed in Ref. [270]) started in 1985 [271, 272]. Last but not least, the history of nanodimensional and nanocrystalline calcium orthophosphates (reviewed in Refs. [273, 274]) began in 1994 [275-279]. [See also Chapter 5 of the present book. The Editor].

To obtain further details on the recent history of calcium orthophosphates, bioceramics and biomaterials in general, the interested reader is referred to other reviews on the subject [7, 9, 10-12, 214-217].

Conclusion

The reported historical findings clearly demonstrate that the subject of calcium orthophosphates has been investigated for almost two and a half centuries. Furthermore, a substantial amount of both scientific facts and experimental approaches have been known for very many decades. In fact, a considerable quantity of relatively recent investigations on calcium orthophosphates is either just a further development of earlier studies or a rediscovery of the already forgotten knowledge. Nevertheless, first commercially produced bone grafts and other types of biomaterials consisting of calcium orthophosphates appeared just a few decades ago [254-264]. Furthermore, the biomedical properties of even the most advanced formulations are still unsatisfactory while frequently incompatible with those of natural bones and teeth. However, the advantages and obvious benefits of calcium orthophosphates remain both an inspiration and a gleam of hope for many researchers and clinicians. It is to be hoped that the historical overview reported above may encourage natural scientists and clinicians to rethink and reinvestigate the already forgotten and poorly known facts and approaches while further developing calcium orthophosphates and calcium orthophosphate-based formulations.

References

[1] As a mineral species, apatite was first recognized by the German Father of Geognosy, Abraham Gottlob Werner (1750 – 1817) in 1786 and named after the ancient Greek word απατάω (apatao) – "to mislead" or "to deceive", because it had previously been mistaken for other minerals, such as beryl, tourmaline, chrysolite, amethyst, fluorite, *etc.* Today apatite is the name for a group of minerals with the same crystallographic structure but does not indicate one chemical composition. That is why the term "calcium apatite" is used in this review.

[2] Dorozhkin, S.V. Calcium orthophosphates. *J. Mater. Sci.* 2007, 42, 1061-1095.

[3] Dorozhkin, S.V. Calcium orthophosphates in nature, biology and medicine. *Materials* 2009, 2, 399-498.

[4] Davy, H. The Bakerian lecture: On some new phenomena of chemical changes produced by electricity, particularly the decomposition of the fixed alkalies, and the exhibition of the new substances which constitute their bases; and on the general nature of alkaline bodies. *Phil. Trans. R. Soc. Lond.* 1808, 98, 1-44.

[5] Davy, H. Electro-chemical researches on the decomposition of the earths; with observations on the metals obtained from the alkaline earths, and on the amalgam procured from ammonia. *Phil. Trans. R. Soc. Lond.* 1808, 98, 333-370.

[6] A paper of the honourable Robert Boyl's, depoſited with the secretaries of the Royal Society, Octob. 14. 1680. and opened ſince his death; being an account of his making

the phosphorus, etc. *Phil. Trans.* 1693, 17, 583-584. (please note the old-fashioned replacement of letter "s" by "f" in some words).
[7] Driskell, T.D. Early history of calcium phosphate materials and coatings. In: Characterization and performance of calcium phosphate coatings for implants. ASTM STP 1196. Horowitz, E. Parr, J.E. (Eds.), American Society for Testing and Materials,
[8] Albee, F.H., assisted by Morrison, H.F. Studies in bone growth. Triple calcium phosphate as a stimulus to osteogenesis. *Ann. Surg.* 1920, 71, 32-39.
[9] Shackelford J.F. Bioceramics – an historical perspective. *Mater. Sci. Forum* 1999, 293, 1-4.
[10] Hulbert, S.F., Hench, L.L., Forbers, D., Bowman, L.S. History of bioceramics. *Ceram. Int.* 1982, 8, 131-140.
[11] Hulbert, S.F., Hench, L.L., Forbers, D., Bowman, L.S. History of bioceramics. In: Ceramics in Surgery. Vincenzini, P. (Ed.), Elsevier, Amsterdam, The Netherlands. 1983, pp. 3-29.
[12] Shepperd, J. The early biological history of calcium phosphates. In: Fifteen years of clinical experience with hydroxyapatite coatings in joint arthroplasty. Epinette, J.A., Manley, M.T. (Eds.), Springer, France, 2004, pp. 3-8.
[13] Roscoe, H.E., Schorlemmer, C. A treatise on chemistry. Volume I: The non-metallic elements. Macmillan and Co., London, 1881, 751 pp.
[14] Johan Gottlieb Gahn (1745 – 1818) was a Swedish chemist and metallurgist, who discovered manganese in 1774.
[15] Carl Wilhelm Scheele (1742 – 1786) was a German-Swedish pharmaceutical chemist, who discovered oxygen (although Joseph Priestley published his findings first), as well as identified molybdenum, tungsten, barium and chlorine before Humphry Davy did.
[16] The monthly review; or, literary journal: by several hands. Vol. XLII. Printed for R. Griffiths: and sold by T. Becket and P.A. de Hondt, in the Strand. London, 1770, 584 pp.
[17] A dictionary of chemistry. Containing the theory and practice of that science: its application to natural philosophy, natural history, medicine, and animal economy: with full explanations of the qualities and modes of action of chemical remedies: and the fundamental principles of the arts, trades, and manufactures, dependent on chemistry. Translated from the French. With notes, additions, and plates. The second edition. To which is added, as an appendix, a treatise on the various kinds of permanently elastic fluids, or gases. Vol. III. Printed for T. Cadell, and P. Elmsly, in the Strand. London, 1777, 666 pp.
[18] de Born, Catalogue methodique et raisonné de la collection des fossils de mlle. Éléonore de Raab. Tome premier. A Vienne. Aux Dépens de J.V. Degen, par I. Alberti, 1790, 504 pp.
[19] Crell, L. Chemische Annalen für die Freunde der Naturlehre, Ärztengelahrtheit, Haushaltungskunst und Manufakturen. Helmstedt und Leipzig, in der J.O. Müllerschen Buchhandlung, 1789, 568 pp.
[20] Lavoisier, Elements of chemistry, in a new systematic order, containing all the modern discoveries. Translated from the French, by Robert Kerr. Printed for William Creech, and sold in London by G.G. and J.J. Robinsons. Edinburgh, 1790, 511 pp.
[21] Encyclopaedia Perthensis; or Universal dictionary of the arts, Sciences, literature, andc. intended to supersede the use of other books of reference. Second edition, in twenty-

three volumes. Vol. XVII. Printed by John Brown, Anchor Close, for the proprietors, and sold by all the booksellers in the United Kingdom. Edinburgh, 1816, 720 pp.

[22] Fourcroy, A.F. A general system of chemical knowledge; and its application to the phenomena of nature and art. In eleven volumes. Translated from the original French by William Nicholson. Vol. III. London, 1804, 472 pp.

[23] Nisbet, W. A general dictionary of chemistry, containing the leading principles of the science, in regard to facts, experiments, and nomenclature. For the use of students. Printed for S. Highley (successor to the late Mr. John Murray). London, No. 24, Fleet-street, 1805, 415 pp.

[24] Murray, J. A system of chemistry. In four volumes. Vol. I. Printed for Longman, Hurst, Rees and Orme, London; and William Creech, and A. Constable and Co. Edinburgh. Edinburgh, 1806, 592 pp.

[25] Bache, F. A system of chemistry for the use of students of medicine. Printed and published for the author. William Fry, Printer. Philadelphia, 1819, 624 pp.

[26] Good, J.M., Olinthus G., Newton B. Pantologia: A new cyclopædia, comprehending a complete series of Essays, Treatises, and Systems, alphabetically arranged; with a general dictionary of arts, sciences, and words: the whole presenting a distinct survey of Human Genius, Learning, and Industry. Vol. IX. P – PYX. London, 1813.

[27] Encyclopædia Britannica; or, a dictionary of arts, sciences, and miscellaneous literature; enlarged and improved. The fourth edition. Vol. V. Printed by Andrew Bell, the Proprietor, for Archibald Constable and Company, Edinburgh; and for Vernor, Hood, and Sharpe, London. Edinburgh, 1810, 797 pp.

[28] Muhlenberg, W.F. Address in Hygiene. In: Transactions of the medical society of the state of Pennsylvania, at its thirty-third annual session, held at Titusville, May 10, 11 and 12, 1832. Vol. XIV. Times printing house. Philadelphia, 1832, pp. 81-102.

[29] Berzelius, J. Untersuchungen über die Zusammensetzung der Phosphorsäure, der phosphorigen Säure und ihrer Salze. *Ann. Physik* 1816, 53, 393-446.

[30] Berzelius, J. Ueber basische phosphorsaure Kalkerde. *Justus Liebigs Annalen der Chemie* 1845, 53, 286-288.

[31] Baruel, M. Analysis of a double phosphate of lead and lime. *J. Franklin Inst.* 1838, 25, 343.

[32] Smith, J.D. Ueber die Zusammensetzung verschiedener Arten von südamerikanischem Guano, nebst der Beschreibung einer neuen Methode, Ammoniak zu bestimmen, so wie Kalk und Magnesia, wenn sie an Phosphorsäure gebunden sind, zu trennen. *J. Prakt. Chem.* 1845, 35, 277-291.

[33] Mitscherlich, E. Lehrbuch der Chemie. Erster Band. *Die Metalloïde. Vierte Auflage.* Berlin, 1844, 609 pp.

[34] F. On the method by which the phosphate and carbonate of lime is introduced into the organs of plants. *J. Franklin Inst.* 1849, 48, 156.

[35] Wilson, G. Chemistry. William and Robert Chambers, Edinburgh, 1850, 316 pp.

[36] Brande, W.T., Taylor, A.S. Chemistry. Blanchard and Lea, Philadelphia, USA, 1863, 696 pp.

[37] Dorozhkin, S.V. Amorphous calcium (ortho)phosphates. *Acta Biomater.* 2010, 6, 4457-4475.

[38] Morfit, C. On Colombian guano; and certain peculiarities in the chemical behavior of "bone phosphate of lime". *J. Franklin Inst.* 1855, 30, 325-329.

[39] Warington, R. CXIX. On a curious change in the composition of bones taken from the guano. *Mem. Proc. Chem. Soc.* 1843, 2, 223-226.

[40] Warington, R., Jr. XXVII. – Researches on the phosphates of calcium, and upon the solubility of tricalcic phosphate. *J. Chem. Soc.* 1866, 19, 296-318.

[41] Warington, R. X. – On the solubility of the phosphates of bone-ash in carbonic water. *J. Chem. Soc.* 1871, 24, 80-83.

[42] Warington, R. XL. – On the decomposition of tricalcic phosphate by water. *J. Chem. Soc.* 1873, 26, 983-989.

[43] Fresenius, R. Ueber die Bestimmung der Phosphorsäure im Phosphorit nebst Mittheilung der Analysen des Phosphorits und Staffelits aus dem Lahnthal. *Zeitschrift für Analytische Chemie* 1867, 6, 403-409.

[44] Lorah, J.R., Tartar, H.V., Wood, L. A basic phosphate of calcium and of strontium and the adsorption of calcium hydroxide by basic calcium phosphate and by tricalcium phosphate. *J. Am. Chem. Soc.* 1929, 51, 1097-1106.

[45] Hassall, A. On the detection and preservation of crystalline deposits of uric acid, urate of ammonia, phosphate of lime, triple phosphate, oxalate of lime, and other salts. *Lancet* 1852, 59, 466-467.

[46] Abel, F.A. XVIII. – On the occurrence of considerable deposits of crystallized phosphate of lime in teak-wood. *J. Chem. Soc.* 1862, 15, 91-93.

[47] Stammer, C. Bestimmung kohlensauren Kalkes neben phosphorsaurem Kalk. *Zeitschrift für Analytische Chemie* 1863, 2, 96-97.

[48] Roussin, Z. Pru¨ fung des Wismuthsubnitrats auf eine Verfälschung mit Kalkphosphat. *Zeitschrift für Analytische Chemie* 1868, 7, 511.

[49] Janovsky, J.V. Ueber die verschiedenen Methoden der Phosphorsäure-Bestimmung neben Eisenoxyd, Thonerde, Kalk und Magnesia. *Zeitschrift für Analytische Chemie* 1872, 11, 153-167.

[50] Pellet, H. Die Zusammensetzung des Niederschlags, welcher durch Ammoniak aus sauren Lösungen von Phosphorsäure, Baryt, Kalk und Magnesia gefällt wird. *Zeitschrift für Analytische Chemie* 1882, 21, 261.

[51] Stokvis, B.J., Salkowski, E., Smith, W.G. Ueber die Löslichkeitsverhältnisse des phosphorsauren Kalks im Harn. *Zeitschrift für Analytische Chemie* 1884, 23, 273-274 (in reality, this publication represents a set of 3 short independent studies performed by 3 individual authors but combined under a general title).

[52] Ott, A. Die Löslichkeitsverhältnisse des phosphorsauren Kalks im Harn. *Zeitschrift für Analytische Chemie* 1886, 25, 279-280.

[53] Kennepohl, G. Zur Bestimmung von Eisenoxyd und Thonerde neben Kalk und Phosphorsäure. *Zeitschrift für Analytische Chemie* 1889, 28, 343.

[54] Immendorff, H., Reitmair, O. Zur Bestimmung des Kalks in Gegenwart von Phosphorsäure, Eisen, Thonerde und Mangan. *Zeitschrift für Analytische Chemie* 1892, 31, 313-316 (in reality, this publication represents a set of 2 short independent studies performed by 2 individual authors but combined under a general title).

[55] Fingerling, G., Grombach, A. Eine neue Modifikation der Bestimmung der zitratlöslichen Phosphorsäure in den Futterkalken nach Petermann. *Zeitschrift für Analytische Chemie* 1907, 46, 756-760.

[56] Schulze, B. Untersuchung des phosphorsauren Futterkalkes. *Zeitschrift für Analytische Chemie* 1911, 50, 126-127.

[57] Hinden, F. Anreicherungsmethode zur Bestimmung der Phosphorsäure in phosphorsäurearmen Kalksteinen. *Zeitschrift für Analytische Chemie* 1915, 54, 214-216.
[58] Roscoe, H.E., Schorlemmer, C. A treatise on chemistry. Volume II: Metals. Part 1. Macmillan and Co., London, UK, 1879, 504 pp.
[59] Duncan, A. Jr. The Edinburgh new dispensatory: containing, I. The Elements of Pharmaceutical Chemistry. II. The Materia Medica; or, the Natural, Pharmaceutical and Medical Hiſtory of the different Subſtances employed in Medicine. III. The Pharmaceutical Preparations and Compositions; including complete and accurate translations of the Octavo Edition of the London Pharmacopoeia, published in 1803. Illuſtrated and Explained in the Language and according to the Principles of Modern Chemistry. Printed for Bell and Bradfute; G. and J. Robinson, London; and Gilbert and Hodges, Dublin. Edinburgh, 1803, 710 pp.
[60] Fresenius, R. Zur Analyse der Superphosphate. *Zeitschrift für Analytische Chemie* 1868, 7, 304-309.
[61] Chesshire, J.A., Hughes, J., Sutton, F., Sibson, A. Ueber die Bestimmung des Betrags an "reducirten" Phosphaten in Superphosphaten. *Zeitschrift für Analytische Chemie* 1870, 9, 524-527 (in reality, this publication represents a set of 4 short independent studies performed by 4 individual authors but combined under a general title).
[62] Williams, C.P. On the solubility of some forms of phosphate of lime. *J. Franklin Inst.* 1871, 92, 419-423.
[63] Rümpler, A. Ueber eisen- und thonerdehaltige Superphosphate und deren analytische Untersuchung. *Zeitschrift für Analytische Chemie* 1873, 12, 151-163.
[64] Albert, H., Siegfried, L. Beiträge zur Werthbestimmung der Superphosphate. *Zeitschrift für Analytische Chemie* 1877, 16, 182-188.
[65] Albert, H., Siegfried, L. Beiträge zur Werthbestimmung der Superphosphate. *Zeitschrift für Analytische Chemie* 1879, 18, 220-224.
[66] Pavec, A. Zur maassanalytischen Bestimmung der Phosphorsäure im Superphosphat und Spodium mittelst Uranlösung. *Zeitschrift für Analytische Chemie* 1879, 18, 360-361.
[67] Mohr, C. Ein maassanalytisches Bestimmungsverfahren der in Rohphosphaten und Superphosphaten enthaltenen Phosphorsäure mit Uran bei Gegenwart von Eisenoxyd. *Zeitschrift für Analytische Chemie* 1880, 19, 150-153.
[68] Erlenmeyer, E., Wattenberg, H., Wein, E., Rösch, L., Lehmann, J., Johnson, S.W., Jenkins, E.H. Zur Analyse der Superphosphate. *Zeitschrift für Analytische Chemie* 1880, 19, 243-246 (in reality, this publication represents a set of 7 short independent studies performed by 7 individual authors but combined under a general title).
[69] Meyer, C.F. Weitere Mittheilungen über das Zurückgehen der eisen- und thonerdehaltigen Superphosphate – Berichtigung. *Zeitschrift für Analytische Chemie* 1880, 19, 309-311.
[70] Drewsen, S. Zur Bestimmung der löslichen Phosphorsäure in Superphosphaten. *Zeitschrift für Analytische Chemie* 1881, 20, 54-57.
[71] Mollenda, A. Eine neue Methode zur maassanalytischen Bestimmung der Phosphorsäure in den Superphosphaten. *Zeitschrift für Analytische Chemie* 1883, 22, 155-159.

[72] Phillips, W.B. Rate of reversion in superphosphates prepared from red navassa rock. *J. Am. Chem. Soc.* 1884, 6, 224-228.
[73] Wagner, P. Eine neue Methode zur Feststellung des Handelswerthes der Superphosphate. *Zeitschrift für Analytische Chemie* 1886, 25, 272-278.
[74] Emmerling, A. Eine Methode zur Bestimmung der wasserlöslichen Phosphorsäure in Superphosphaten auf maassanalvtischem Wege. *Zeitschrift für Analytische Chemie* 1887, 26, 244-247.
[75] Stoklasa, J. Bestimmung des Wassers in den Superphosphaten. I. *Zeitschrift für Analytische Chemie* 1890, 29, 390-397.
[76] Crispo, D. Belgische Methode zur Bestimmung der in Wasser löslichen Phosphorsa¨ ure in den Superphosphaten. *Zeitschrift für Analytische Chemie* 1891, 30, 301-303.
[77] Güssefeld, O. Eine Schüttelmaschine für die Analyse von Superphosphaten. *Zeitschrift für Analytische Chemie* 1892, 31, 556.
[78] Keller, A. Eine Schüttelmaschine für Superphosphaten. *Zeitschrift für Analytische Chemie* 1893, 32, 590-591.
[79] Kalmann, W., Meissels, K. Eine Methode zur maassanalytischen Schätzung der wasserlöslichen Phosphorsa ¨ ure in Superphosphaten. *Zeitschrift für Analytische Chemie* 1894, 33, 764-766.
[80] Glaser, C. Zur maassanalytischen Bestimmung der wasserlöslichen Phosphorsäure in Superphosphaten. *Zeitschrift für Analytische Chemie* 1895, 34, 768-769.
[81] Seib, O. Bestimmung der zitratlöslichen Phosphorsäure in Superphosphaten. *Zeitschrift für Analytische Chemie* 1905, 44, 397-398.
[82] Cameron, F.K., Bell, J.M. The phosphates of calcium, III; Superphosphate. *J. Am. Chem. Soc.* 1906, 28, 1222-1229.
[83] Wells, H.G. Pathological calcification. J. Med. Res. 1906, 14, 491-525.
[84] C. Vitreous phosphate of lime. *J. Franklin Inst.* 1877, 104, 315.
[85] Church, A.H. New analyses of certain mineral arseniates and phosphates. 1. Apatite; 2. Arseniosiderite; 3. Childrenite; 4. Ehlite; 5. Tyrolite; 6. Wavellite. *J. Chem. Soc.* 1873, 26, 101-111.
[86] L., W.J. On a fine specimen of apatite from Tyrol, lately in the possession of Mr. Samuel Henson. *Nature* 1883, 27, 608-609.
[87] Reichardt, E. Ueber neutralen phosphorsauren Kalk, Darstellung und Löslichkeit desselben. *Zeitschrift für Analytische Chemie* 1872, 11, 275-277.
[88] Mohr, C. Ueber die quantitative Bestimmung der zurückgegangenen Phosphorsäure und der Phosphorsäure im Dicalciumphosphat. *Zeitschrift für Analytische Chemie* 1884, 23, 487-491.
[89] Glaser, C. Bemerkungen zu der Abhandlung des Herrn Carl Mohr über die quantitative Bestimmung der zurückgegangenen Phosphorsäure und der Phosphorsäure im Dicalciumphosphat. *Zeitschrift für Analytische Chemie* 1885, 24, 180.
[90] Hutchings, W.M. Occurrence of apatite in slag. *Nature* 1887, 36, 460.
[91] Hilgenstock, G. Eine neue Verbindung von P_2O_5 und CaO. *Stahl und Eisen* 1883, 3, 498.
[92] Hilgenstock, G. Das vierbasische Kalkphosphat und die Basicitätsstufe des Silicats in der Thomas-Schlacke. *Stahl und Eisen* 1887, 7, 557-560.

[93] Scheibler C. Ueber die Herstellung reicher Kalkphosphate in Verbindung mit einer Verbesserung des Thomasprocesses. *Ber. Dtsch. Chem. Ges.* 1886, 19, 1883-1893.
[94] Tzschucke, H. Versuch einer directen Bestimmung der Phosphorsäure als dreibasisch phosphorsauren Kalk. *Angewandte Chemie* 1888, 1, 383-385.
[95] Georgievics, G.V. Über das Verhalten des Tricalciumphosphats gegen Kohlensäure und Eisenhydroxyd. *Monatshefte für Chemie* 1891, 12, 566-581.
[96] Cameron, F.K., Hurst, L.A. The action of water and saline solutions upon certain slightly soluble phosphates. *J. Am. Chem. Soc.* 1904, 26, 885-913.
[97] Cameron, F.K., Seidell, A. The action of water upon the phosphates of calcium. *J. Am. Chem. Soc.* 1904, 26, 1454-1463.
[98] Cameron, F.K., Seidell, A. The phosphates of calcium. I. *J. Am. Chem. Soc.* 1905, 27, 1503-1512.
[99] Cameron, F.K., Bell, J.M. The phosphates of calcium. II. *J. Am. Chem. Soc.* 1905, 27, 1512-1514.
[100] Cameron, F.K., Bell, J.M. The phosphates of calcium. IV. *J. Am. Chem. Soc.* 1910, 32, 869-873.
[101] Cameron, F.K., McCaughey, W.J. Apatite and spodiosite. *J. Phys. Chem.* 1911, 15, 463-470.
[102] Mebane, W.M., Dobbins, J.T., Cameron, F.K. The solubility of the phosphates of calcium in aqueous solutions of sulfur dioxide. *J. Phys. Chem.* 1929, 33, 961-969.
[103] Bassett, H., Jr. Beiträge zum Studium der Calciumphosphate. I. Die Hydrate der Calcium-Hydroorthophosphate. *Z. Anorg. Chem.* 1907, 53, 34-48.
[104] Bassett, H., Jr. Beiträge zum Studium der Calciumphosphate. II. Die Einwirkung von Ammoniakgas auf Calcium-Hydroorthophosphate. *Z. Anorg. Chem.* 1907, 53, 49-62.
[105] Bassett, H., Jr. Beiträge zum Studium der Calciumphosphate. III. Das System CaO – P_2O_5 – H_2O. Z. *Anorg. Chem.* 1908, 59, 1-55.
[106] Bassett, H., Jr. The phosphates of calcium. Part IV. The basic phosphates. *J. Chem. Soc.* 1917, 111, 620-642.
[107] Norton, T.H., Newman, H.E. LV. - On a soluble compound of hydrastine with monocalcium phosphate. *J. Am. Chem. Soc.* 1897, 19, 838-840.
[108] Rolfe, B.H. Autunite (hydrated uranium-calcium phosphate). *Lancet* 1911, 177, 766.
[109] Withers, W.A., Field, A.L. A conductivity study of the reaction between calcium nitrate and dipotassium phosphate in dilute solution. *J. Am. Chem. Soc.* 1915, 37, 1091-1105.
[110] Wendt, G.L. An electrometric study of the neutralization of phosphoric acid by calcium hydroxide. *J. Am. Chem. Soc.* 1923, 45, 881-887.
[111] Shipley, P.G., Kramer, B., Howland, J. Studies upon calcification *in vitro*. *Biochem. J.* 1926, 20, 379-387.
[112] Benjamin, H.R. The forms of the calcium and inorganic phosphorus in human and animal sera II. The nature and significance of the filtrable, adsorbable calcium-phosphorus complex. *J. Biol. Chem.* 1933, 100, 57-78.
[113] Náray-Szabó, S. The structure of apatite $(CaF)Ca_4(PO_4)_3$. *Zeitschrift für Kristallographie* 1931, 75, 387-398.
[114] Hendricks, S.B., Hill, W.L., Jacob, K.D., Jefferson, M.E. Structural characteristics of apatite-like substances and composition of phosphate rock and bone as determined from microscopical and X-ray diffraction. *Ind. Eng. Chem.* 1931, 23, 1413-1418.

[115] Clark, N.A. The system P_2O_5–CaO–H_2O and the recrystallization of monocalcium phosphate. *J. Phys. Chem.* 1931, 35, 1232-1238.
[116] Lugg, J.W.H. A study of aqueous salt solutions in equilibrium with solid secondary calcium phosphate at 40° C. *Trans. Faraday Soc.* 1931, 27, 297-309.
[117] Bredig, M.A., Franck, H.H., Fülnder, H. Beiträge zur Kenntnis der Kalk-Phosphorsäure-Verbindungen. *Z. Elektrochem.* 1932, 38, 158-164.
[118] Trömel, G. Beiträge zur Kenntnis des Systems Kalziumoxyd-Phosphorpentoxyd. Mitt. Kaiser-Wilhelm-Inst. Eisenforsch. Düsseldorf 1932, 14, 25-34.
[119] Nagelschmidt, G. A new calcium silicophosphate. *J. Chem. Soc.* 1937, 865-867.
[120] Hill, W.L., Hendricks, S.B., Jefferson, M.E., Reynolds, D.S. Phosphate fertilizers by calcination process. Composition of defluorinated phosphate. *Ind. Eng. Chem.* 1937, 29, 1299-1304.
[121] Marshall, H.L., Reynolds, D.S., Jacob, K.D., Tremearne, T.H. Phosphate fertilizers by calcination process. Reversion of defluorinated phosphate at temperatures below 1400 °C. *Ind. Eng. Chem.* 1937, 29, 1294-1298.
[122] Curtis, H.A., Copson, R.L., Brown, E.H., Pole, G.R. Fertilizer from rock phosphate conversion by fusion and treatment with water vapor. *Ind. Eng. Chem.* 1937, 29, 766–771.
[123] Moore, L.A. Activation of dicalcium phosphate for the chromatographic determination of carotene. *Ind. Eng. Chem.* 1942, 14, 707-708.
[124] Elsenberger, S., Lehrman, A., Turner, W.D. The basic calcium phosphates and related systems. Some theoretical and practical aspects. *Chem. Rev.* 1940, 26, 257-296.
[125] Elmore, K.L., Farr, T.D. Equilibrium in the system calcium oxide–phosphorus pentoxide–water. *Ind. Eng. Chem.* 1940, 32, 580-586.
[126] Shear, M.J., Kramer, B. Composition of bone. III. Physicochemical mechanism. *J. Biol. Chem.* 1928, 79, 125-145.
[127] Trömel, G., Möller, H. Die Bildung schwer löslicher Calciumphosphate aus wäßriger Lösung und die Beziehungen dieser Phosphate zur Apatitgruppe. *Zeitschrift für anorganische und allgemeine Chemie* 1932, 206, 227-240.
[128] Larson, H.W.E. Preparation and properties of mono-, di-, and tricalcium phosphates. *Ind. Eng. Chem. Anal.* Ed. 1935, 7, 401-406.
[129] Kirwan, R. Elements of mineralogy. Second edition with considerable improvements and additions. Vol. I. Earth and stones. Printed by J. Nichols, for P. Elmsly, in the strand. London, 1794, 510 pp.
[130] Murray, J. A system of chemistry. In four volumes. Second edition. Vol. III. Edinburgh. Printed for Longman, Hurst, Rees and Orme, and John Murray London; and for William Creech, Bell and Bradfute, and W. Laing, Edinburg. Edinburgh, 1809, 712 pp.
[131] Jameson, R. A system of mineralogy, in which minerals are arranged according to the natural history method. Third edition, enlarged and improved. Vol. II. Printed for Archibald Constable and Co. Edinburgh; and Hurst, Robinson and Co. Cheapside, London. Edinburgh, 1820, 632 pp.
[132] Lowry, D. Conversations on mineralogy. First American from the last London edition. Published and sold by Uriah Hunt, No. 239, Market Street. Philadelphia, Sept. 1822, 332 pp.
[133] Shaler, N.S. Nature and origin of phosphate of lime. *Science* 1889, 13, 144-148.

[134] Graber, H. XV. Diopsid und Apatit von Zöptau. *Tschermaks Mineralogische und Petrographische Mitteilungen* 1895, 14, 265-270.
[135] Church, A.H. Apatite in Ceylon. *Nature* 1901, 63, 464.
[136] Klein, C. Brushite from the island of Mona (between Haiti and Puerto Rico). *Sitzber. K. Preuss. Aka.* 1901, 720-725.
[137] Wolff, J.E. Palache, C. Apatite from Minot, Maine. *Proc. Am. Acad. Arts Sci.* 1902, 37, 517-528.
[138] Wolff, J.E. Palache, C. On apatite from Minot, Maine. *Zeitschrift für Krystallographie und Mineralogie* 1902, 36, 438-448.
[139] de Mercey, N. Sur des gites de phosphate de chaux de la Craie à Bélemnites, formés avant le soulèvement du Bray. *Comptes Rendus Hebdomadaires des Seances de l'Academie des Sciences*. 1902, 135, 1137-1138.
[140] Zimanyi, K. On the green apatite from Malmberget in Sweden. *Zeitschrift für Krystallographie und Mineralogie* 1904, 39, 505-519.
[141] Fraps, G.S. Availability of phosphoric acid of the soil. *J. Am. Chem. Soc.* 1906, 28, 823-834.
[142] Elements of natural history, and of chemistry: being the second edition of the elementary lectures on those sciences, first published in 1782, and now greatly enlarged and improved, by the Author, M. de Fourcroy, doctor of the faculty of medicine at Paris, of the Royal Academy of Sciences, andc. andc. andc. Translated into English. With occasional notes, and an historical preface, by the translator. Vol. IV. Printed for G.G.J. and J. Robinson, Pater-Noster-Row. London, 1788, 465 pp.
[143] Tilloch, A. The philosophical magazine: comprehending the various branches of science, the liberal and fine arts, agriculture, manufactures, and commerce. Vol. II. London, 1798, 442 pp.
[144] The New-York magazine, or literary repository. New series – Vol. II. Printed and ſold by T. and J. Swords, No. 99 Pearl ſtreet. New Your, 1797, 672 pp.
[145] Parr, B. The London medical dictionary; including under distinct heads every branch of medicine, viz. anatomy, physiology, and pathology, the practice of physic and surgery, therapeutics, and materia medica; with whatever relates to medicine in natural philosophy, chemistry and natural history. Vol. I. London, 1809, 786 pp.
[146] Parkes, S. The chemical catechism, with notes, illustrations, and experiments. The fifth edition. Printed for the author; and sold by Lackington, Allen, and Co. London, 1812, 562 pp.
[147] Davy, H. Conversations on chemistry, in which the elements of that science are familiarly explained and illustrated by experiments and plates. To which are added, some late discoveries on the subject of the fixed alkalies. Didnep's Press, 1814, 383 pp.
[148] Wilson, J. Lectures on the structure and physiology of the parts composing the skeleton, and on the diseases of the bones and joints of the human body, preceded by some observations on the influence of the brain and nerves. Printed for Burgess and Hill, London, 1820, 410 pp.
[149] Nasse, H. Ueber die Bestandtheile der Knochen in einigen Krankheiten. *J. Prakt. Chem.* 1842, 27, 274-280.
[150] Levy, M. Chemische Untersuchungen über osteomalacische Knochen. *Zeitschrift für physiologische Chemie* 1895, 19, 239-270.

[151] Aeby, C. Ueber die Constitution des phosphorsauren Kalkes der Knochen. *J. Prakt. Chem.* 1872, 5, 308-311.

[152] Aeby, C. Ueber die nähern Bestandtheile des Knochenphosphates. *J. Prakt. Chem.* 1873, 6, 169-171.

[153] Aeby, C. Ueber die Metamorphose der Knochen. *J. Prakt. Chem.* 1873, 7, 37-44.

[154] Wibel, F. Die Constitution des Knochenphosphates, insbesondere die Existenz und Bildung einer basischen Verbindung ($Ca_3P_2O_8$+xCaO). *J. Prakt. Chem.* 1874, 9, 113-132.

[155] Weiske, H. Beitrag zur Knochenanalyse. *Zeitschrift für Physiologische Chemie* 1883, 7, 474-478.

[156] Kramer, B., Shear, M.J. Composition of bone. IV. Primary calcification. *J. Biol. Chem.* 1928, 79, 147-160.

[157] Jones, H.B. II. – On the occurrence of deposits of crystallised phosphate of lime in human urine. *J. Chem. Soc.* 1862, 15, 8-15.

[158] Hassal, Phosphorsaurer Kalk im Urinsediment. *Zeitschrift für Analytische Chemie* 1862, 1, 125.

[159] Moss, Ueber das Vorkommen und die Bedeutung phosphorsaurer Kalkconcremente im Stamme des Gehörnerven. *Archiv für Psychiatrie und Nervenkrankheiten* 1879, 9, 122-128.

[160] Tereg, Arnold, Das Verhalten der Calciumphosphate im Organismus der Fleischfresser. Pflüger, Archiv für die Gesammte Physiologie des Menschen und der Thiere. *Pflügers Archiv Eur. J. Physiology* 1883, 32, 122-170.

[161] The monthly magazine; or, British register. Vol. 27, Part I. Printed for Richard Phillips, No. 6, Bridge-Street. London, 1809, 736 pp.

[162] Pemberton, C.R. A practical treatise on various diseases of the abdominal viscera. Third edition, revised and corrected. Printed by W. Bulmer and Co. Cleveland-row, St. James's; for G. and W. Nicol, booksellers to his Majesty, Pall-Mall. London, 1814, 201 pp.

[163] Under the care of Mr. Canton. Stone in the bladder of a boy for upwards of three years; lithotomy; extraction of a large oval calculus of phosphate of lime; recovery. *Lancet* 1856, 68, 650-651.

[164] Virchow, R. Untersuchungen über die Entwicklung des Schädelgrundes im gesunden und krankhaften Zustände und über den Einfluss derselben auf Schädelform, Gesichtsbildung und Gehirnbau. Druck und Verlag von Georg Reimer, Berlin, 1857, 128 pp.

[165] Jones, S. St. Thomas's hospital.: Lateral lithotomy; removal of large stone composed of oxalate of lime covered with phosphates; recovery. *Lancet* 1882, 119, 781-782.

[166] Hill, B. University college hospital.: Two cases of calculus of oxalate and phosphate of lime removed by lithotrity at a single sitting; singularly long duration of symptoms, with remarks upon an improved clover's evacuating bottle. *Lancet* 1883, 121, 59-60.

[167] Hotz, G. Phosphorsäure- und Kalkstoffwechsel bei Osteomalacie unter dem Einfluss der Phosphortherapie. *Zeitschrift für Experimentelle Pathologie und Therapie* 1906, 3, 605-632.

[168] Gaßmann, T., Chemische Untersuchungen von gesunden und rachitischen Knochen. *Zeitschrift für physiologische Chemie* 1910, 70, 161-170.

[169] Zuckmayer, F. U˙ ber die Frauenmilch der ersten Laktationszeit und den Einfluss einer Kalk- und Phosphorsa¨ urezulage auf ihre Zusammensetzung. *Pflüger's Archiv für die Gesammte Physiologie des Menschen und der Tiere* 1914, 158, 209-218.
[170] Meigs, E.B., Blatherwick, N.R., Cary, C.A. and with the collaboration of Woodward, T.E. Further contributions to the physiology of phosphorus and calcium metabolism of dairy cows. *J. Biol. Chem.* 1919, 40, 469-500.
[171] de Wesselow, O.L.V. The calcium and inorganic phosphorus content of the maternal blood during pregnancy and lactation. *Lancet* 1922, 200, 227-228.
[172] Schneider, J.Z. The calcium to phosphorus ratio as related to mineral metabolism. *Int. J. Orthodontia Oral Surg. Radiography* 1930, 16, 277-285.
[173] Fisher, T. A singular case of extensive deposit of phosphate of lime in the lungs. *Lancet* 1901, 157, 244-245.
[174] Moody, D.W.K. Note on a case of a large concretion composed mainly of calcium phosphate found inside the fistulous channel in a case of fistula in ano. *Lancet* 1913, 181, 684.
[175] Zuckmayer, F. Beitrag zur Aufnahme und Verwertung von Kalk und Phosphorsäure durch den Darm. *Pflüger's Archiv für die Gesammte Physiologie des Menschen und der Tiere* 1912, 148, 225-256.
[176] Blühdorn, K. Untersuchungen des Kalk- und Phorphorsäurestoffwechsels bei Verabfolgung großer Gaben von Kalk und Natriumphosphat. *Zeitschrift für Kinderheilkunde* 1921, 29, 43-55.
[177] Herbst, O. Calcium und Phosphor beim Wachstum am Ende der Kindheit. *Zeitschrift für Kinderheilkunde* 1913, 7, 161-192.
[178] Lowy, R. The necessity of normal calcium phosphorus metabolism. *Int. J. Orthodontia, Oral Surgery and Radiography* 1930, 16, 947-950.
[179] Barrio, N.G. Comparative studies in the chemistry of blood and cerebrospinal fluid. II. Calcium, magnesium, and phosphorus. *J. Labor. Clin. Med.* 1923, 9, 54-56.
[180] Wang, C.C., Felsher, A.R. The effect of hemolysis on the calcium and inorganic phosphorus content of serum and plasma. *J. Labor. Clin. Med.* 1925, 10, 269-272.
[181] Nitschke, A. Darstellung Einer den Calciumgehalt und Einer den Phosphatgehalt des Serum Senkenden Substanz - II. Mitteilung. Nachweis in Milz und Lymphknoten. *Klinische Wochenschrift* 1929, 8, 794.
[182] Lowenberg, C., Mattice, M.R. Note on the determination of inorganic phosphate of serum and spinal fluid on the supernatant fluid from calcium estimation. *J. Labor. Clin. Med.* 1930, 15, 598-600.
[183] Krasnow, F., Karshan, M., Krejci, L.E. The determination of calcium and phosphorus in saliva. *J. Labor. Clin. Med.* 1932, 17, 1148-1152.
[184] Timpe, O. Das Verhalten des Calcium- und Phosphorspiegels des Serums in der Schwangerschaft und bei Osteomalacie unter der Einwirkung von Vigantol. *Archiv für Gynakologie* 1931, 146, 240-247.
[185] Klauder, J.V., Brown, H. Study of the calcium-phosphorus ratio in the serum of syphilitic pregnant women. *Am. J. Obstetrics and Gynecology 1931, 22, 60-64.*
[186] Lierle, D.M., Sage, R.A. A study of the calcium, phosphorus and potassium of the serum and spinal fluid in asthma. *J. Allergy* 1932, 3, 325-331.
[187] Schmidt, C.L.A., Greenberg, D.M. Occurrence, transport and regulation of calcium, magnesium and phosphorus in animal organism. *Physiol. Rev.* 1935, 15, 297-434.

[188] Huggins, C. The composition of bone and the function of the bone cell. *Physiol. Rev.* 1937, 17, 119-143.
[189] Beneke, W. On the physiology and pathology of the phosphate and oxalate of lime, and their relation to the formation of cells. *Part I. Lancet* 1851, 57, 431-434.
[190] Beneke, W. On the physiology and pathology of the phosphate and oxalate of lime, and their relation to the formation of cells. *Part II. Lancet* 1851, 57, 668-670 and 699-700.
[191] Blanc, H. On the treatment of phthisis by the phosphate of lime and the juice of raw meat. *Lancet* 1874, 103, 831-832.
[192] Fessenden, R.A. Cancer and secondary rays from calcium phosphate. *Lancet* 1913, 182, 1804.
[193] Thomas, S., Assi, P., Marycel, B., Correa, M., Liberato, W., Brito, V. Yttrium 90-hydroxyapatite, a new radioisotope for chronic synovitis in hemophilia. *Haemophilia* 2008, 14, 77.
[194] Chinol, M., Vallabhajosula, S., Goldsmith, S.J., Klein, M.J., Deutsch, K.F., Chinen, L.K., Brodack, J.W., Deutsch, E.A., Watson, B.A., Tofe, A.J. Chemistry and biological behavior of samarium-153 and rhenium-186-labeled hydroxyapatite particles: potential radiopharmaceuticals for radiation synovectomy. *J. Nucl. Med.* 1993, 34, 1536-1542.
[195] Argüelles, M.G., Berlanga, I.S.L., Torres, E.A. Preparation of ^{153}Sm-particles for radiosynovectomy. *J. Radioanal. Nucl. Chem.* 1999, 240, 509-511.
[196] O'Duffy, E., Clunie, G., Lui, D., Edwards, J., Ell, P. Double blind glucocorticoid controlled trial of samarium-153 particulate hydroxyapatite radiation synovectomy for chronic knee synovitis. *Ann. Rheum. Dis.* 1999, 58, 554-558.
[197] Sterling, A. The value of phosphorus and calcium in asthma, hay fever, and allied diseases. *J. Labor. Clin. Med.* 1928, 13, 997-1005.
[198] de Franciscis, G. Bicalcium phosphate as adjunct in the feeding of domestic animals. [Il fosfato bicalcico quale integrante dell'alimentazione del bestiame.] *Bollettino della Società italiana di biologia sperimentale* 1953, 29, 1794-1796.
[199] Ratner, B.D. A history of biomaterials. In: Biomaterials science: an introduction to materials in medicine. Ratner, B.D., Hoffman, A.S., Schoen, F.J., Lemons, J.E. (Eds.), 2nd Ed. Academic Press, San Diego, CA, USA. 2004, pp. 10-19.
[200] Jacoby, M. Custom-made biomaterials. *Chem. Eng. News* 2001, 79, 30-35.
[201] Huebsch N., Mooney, D.J. Inspiration and application in the evolution of biomaterials. *Nature* 2009, 462, 426-432.
[202] Popp. H. Zur Geschichte der Prosthesen. *Med. Welt* 1939, 13, 961-964.
[203] Ring, M.E. Dentistry: an illustrated history. Harry N. Abrams, New York, USA, 1992, 320 pp.
[204] Bobbio, A. The first endosseous alloplastic implant in the history of man. *Bull. Hist. Dent.* 1972, 20, 1-6.
[205] Crubezy, E., Murail, P., Girard, L., Bernadou, J.P. False teeth of the Roman world. *Nature* 1998, 391, 29.
[206] Hunter, J. The natural history of the human teeth: explaining their structure, use, formation, growth, and diseases. Printed for J. Johnson, № 72. St. Paul's Church-yard, London, 1771, 191 pp.
[207] Cravens, J.E. Lacto-phosphate of lime; pathology and treatment of exposed dental pulps and sensitive dentine. *Dent. Cosmos* 1876, 18, 463-469; discussion 469-476.

[208] Pendleton, L.W. The lacto-phosphate of lime. *Transactions of the Maine Medical Association.* 1873, 4, 313-318.
[209] Dorozhkin, S.V. Biocomposites and hybrid biomaterials based on calcium orthophosphates. *Biomatter* 2011, 1, 3-56.
[210] Gabriel, S. Chemische Untersuchungen über die Mineralstoffe der Knochen und Zähne. *Zeitschrift für physiologische Chemie* 1894, 18, 257-303.
[211] de Jong, W.F. La substance minérale dans les os. *Recueil des Travaux Chimiques des Pays-Bas* 1926, 45, 445-448.
[212] Henschen, C., Straumann, R., Bucher, R. Ergebnisse röntgenspektrographischer Untersuchungen am Knochen - I. Mitteilung. Krystallitbau des anorganischen und des organischen Knochens. *Deutsche Zeitschrift für Chirurgie* 1932, 236, 485-514.
[213] Bredig, M.A. Zur Apatitstruktur der anorganischen Knochen- und Zahnsubstanz. *Hoppe-Seylers Zeitschrift für Physiologische Chemie* 1933, 216, 239-243.
[214] Weinberger, B.W. An introduction to the history of dentistry with medical and dental chronology and bibliographic data. Two volumes. The C. V. Mosby Company, St. Louis, USA, 1948, 992 pp.
[215] Baden, E. Prosthetic therapy of congenital and acquired clefts on the palate: an historical essay. *J. Hist. Med. Alld. Sci.* 1955, 10, 290-301.
[216] Ratner, B.D., Bryant, S.J. Biomaterials: where we have been and where we are going. *Annu. Rev. Biomed. Eng.* 2004, 6, 41-75.
[217] Dammaschke, T. The history of direct pulp capping. *J. Hist. Dentistry* 2008, 56, 9-23.
[218] Dreesmann, H. Ueber Knochenplombierung. *Beitr. Klin. Chir.* 1892, 9, 804-810.
[219] Gluck, T. Referat über die durch das moderne chirurgische. *Langenbecks Archiv für Klinische Chirurgie* 1891, 41, 187-239.
[220] Muster, D. Themistocles Gluck, Berlin 1890: a pioneer of multidisciplinary applied research into biomaterials for endoprostheses. *Bull. Hist. Dent.* 1990, 38, 3-6.
[221] Eynon-Lewis, N.J., Ferry, D., Pearse, M.F. Themistocles Gluck: an unrecognised genius. *BMJ* 1992, 305, 1534-1536.
[222] Albee, F.H. Bone-graft surgery. W.B. Saunders Company, Philadelphia and London, 1915, 417 pp.
[223] Hey Groves, E.W. Some contributions to the reconstructive surgery of the hip. *Br. J. Surg.* 1927, 14, 486-517.
[224] Murray, C.R. Delayed and non-union in fractures in the adult. *Ann. Surg.* 1931, 93, 961-967.
[225] Murray, C.R. The modern conception of bone formation and its relation to surgery. *J. Dent. Res.* 1931, 11, 837-845.
[226] Huggins, C. The formation of bone under the influence of epithelium of the urinary tract. *Arch. Surg.* 1931, 22, 377-408.
[227] Levander, G. On the formation of new bone in bone transplantation. *Acta Chir. Scand.* 1934, 74, 425-426.
[228] Levander, G. A study of bone regeneration. *Surg. Gynecol. Obstet.* 1938, 67, 705-714.
[229] Haldeman, K.O., Moore, J.M. Influence of a local excess of calcium and phosphorus on the healing of fractures. An experimental study. *Arch. Surg.* 1934, 29, 385-396.
[230] Stewart, W.J. Experimental bone regeneration using lime salts and autogenous grafts as sources of available calcium. *Surg. Gynec. Obstet.* 1934, 59, 867-871.

[231] Key, J.A. The effect of a local calcium depot on osteogenesis and healing of fractures. *J. Bone Joint Surg.* 1934, 16, 176-184.
[232] Shands, A.R., Jr. Studies in bone formation: the effect of the local presence of calcium salts on osteogenesis. *J. Bone Joint Surg.* 1937, 19, 1065-1076.
[233] Schram, W.R., Fosdick, L.S. Stimulation of healing in long bones by use of artificial material. *J. Oral Surg.* 1948, 6, 209-217.
[234] Ray, R.D., Ward, A.A., Jr. A preliminary report on studies of basic calcium phosphate in bone replacement. *Surg. Form.* 1951, 3, 429-434.
[235] Kingery, W.D. II, Cold-setting properties. *J. Am. Ceram. Soc.* 1950, 33, 242-246.
[236] Dorozhkin, S.V. Calcium orthophosphate cements for biomedical application. *J. Mater. Sci.* 2008, 43, 3028-3057.
[237] Dorozhkin, S.V. Calcium orthophosphate cements and concretes. *Materials* 2009, 2, 221-291.
[238] Driskell, T.D., Heller, A.L., Koenigs, J.F. Dental treatments. *US Patent* No. 3913229. October 21, 1975.
[239] LeGeros, R.Z., Chohayeb, A., Shulman, A. Apatitic calcium phosphates: possible dental restorative materials. *J. Dent. Res.* 1982, 61, Spec. Iss., 343.
[240] Brown, W.E., Chow, L.C. A new calcium phosphate setting cement. *J. Dent. Res.* 1983, 62, Spec. Iss., 672.
[241] Robinson, R.A., Watson, M.L. Crystal-collagen relationships in bone as observed in the electron microscope. III. Crystal and collagen morphology as a function of age. *Ann. NY Acad. Sci.* 1955, 60, 596-660.
[242] Nery, E.B., Lynch, K.L., Hirthe W.M., Mueller; K.H. Bioceramic implants in surgically produced infrabony defects. *J. Periodontol.* 1975, 46, 328-347.
[243] Denissen, H.W., de Groot, K. Immediate dental root implants from synthetic dense calcium hydroxylapatite. *J. Prosthet. Dent.* 1979, 42, 551-556.
[244] Blakeslee, K.C., Condrate, R.A., Sr. Vibrational spectra of hydrothermally prepared hydroxyapatites. *J. Am. Ceram. Soc.* 1971, 54, 559-563.
[245] Garrington, G.E., Lightbody, P.M. Bioceramics and dentistry. *J. Biomed. Mater. Res.* 1972, 6, 333-343.
[246] Cini, L., Sandrolini, S., Paltrinieri, M., Pizzoferrato, A., Trentani, C. Materiali bioceramici in funzione sostitutiva. Nota preventiva. (Bioceramic materials for replacement purposes. Preliminary note.). *La Chirurgia Degli Organi Di Movimento* 1972, 60, 423-430.
[247] Rivault, M.A. Evolution, conception et technologie des travaux de prothèse fixe , réalisés en céramo -métallique. (Evolution, conception and technology of fixed prosthesis made of ceramic and metal). *Revue Francaise d'Odonto-Stomatologie* 1966, 13, 1367-1402.
[248] Dumont, A., Appel, M., Favard, E. Prothèses plurales en céramique sur métal. Soudage et artifices de jonction. (Multiple prostheses made of ceramics on metal. Soldering and artifacts of the junction). *Annales Odonto-Stomatologiques* 1968, 25, 231-240.
[249] Hulbert, S.F., Young, F.A., Mathews, R.S., Klawitter, J.J., Talbert, C.D., Stelling, F.H. Potential of ceramic materials as permanently implantable skeletal prostheses. *J. Biomed. Mater. Res.* 1970, 4, 433-456.
[250] Hench, L.L., Splinter, R.J., Allen, W.C., Greenlee, T.K. Bonding mechanisms at the interface of ceramic prosthetic materials. *J. Biomed. Mater. Res.* 1971, 2, 117-141.

[251] Paital, S.R., Dahotre, N.B. Calcium phosphate coatings for bio-implant applications: materials, performance factors, and methodologies. *Mater. Sci. Eng. R* 2009, 66, 1-70.
[252] León, B., Jansen, J.A. (Eds.), Thin calcium phosphate coatings for medical implants. Springer, New York, USA, 2009, 326 pp.
[253] Ducheyne, P., Hench, L.L., Kagan II, Martens, A., Bursens, A., Mulier, J.C. Effect of hydroxyapatite impregnation on skeletal bonding of porous coated implants. *J. Biomed. Mater. Res.* 1980, 14, 225-237.
[254] Jarcho, M., Bolen, C.H., Thomas, M.B., Bobick, J., Kay, J.F. Doremus, R.H. Hydroxylapatite synthesis and characterization in dense polycrystalline form. *J. Mater. Sci.* 1976, 11, 2027-2035.
[255] Jarcho, M., O'Connor, J.R., Paris, D.A. Ceramic hydroxylapatite as a plaque growth and drug screening substrate. *J. Dental Res.* 1977, 56, 151-156.
[256] Jarcho, M., Salsbury, R.L., Thomas, M.B., Doremus, R.H. Synthesis and fabrication of β-tricalcium phosphate ceramics for potential prosthetic applications. *J. Mater. Sci.* 1979, 14, 142-150.
[257] Jarcho, M. Calcium phosphate ceramics as hard tissue prosthetics. *Clin. Orthop. Rel. Res.* 1981, 157, 259-278.
[258] Rejda, B.V., Peelen, J.G.J., de Groot, K. Tricalcium phosphate as a bone substitute. *J. Bioeng.* 1977, 1, 93-97.
[259] de Groot, K. Bioceramics consisting of calcium phosphate salts. *Biomaterials* 1980, 1, 47-50.
[260] de Groot, K. (Ed.), Bioceramics of calcium phosphate. CRC Press, Boca Raton, Fl, 1983, 146 pp.
[261] Aoki, H., Kato, K.M., Ogiso, M., Tabata, T. Studies on the application of apatite to dental materials. *J. Dent. Eng.* 1977, 18, 86-89.
[262] Kato, K., Aoki, H., Tabata, T., Ogiso, M. Biocompatibility of apatite ceramics in mandibles. Biomater. *Med. Dev. Artif. Organs* 1979, 7, 291-297.
[263] Akao, M., Aoki, H., Kato, K. Mechanical properties of sintered hydroxyapatite for prosthetic applications. *J. Mater. Sci.* 1981, 16, 809-812.
[264] Akao, M., Aoki, H., Kato, K., Sato, A. Dense polycrystalline β-tricalcium phosphate for prosthetic applications. *J. Mater. Sci.* 1982, 17, 343-346.
[265] Roy, D.M., Linnehan, S.K. Hydroxyapatite formed from coral skeletal carbonate by hydrothermal exchange. *Nature* 1974, 247, 220-222.
[266] Holmes, R.E. Bone regeneration within a coralline hydroxyapatite implant. *Plast. Reconstr. Surg.* 1979, 63, 626-633.
[267] Elsinger, E.C., Leal, L. Coralline hydroxyapatite bone graft substitutes. *J. Foot Ankle Surg.* 1996, 35,396-399.
[268] Shipman, P., Foster, G., Schoeninger, M. Burnt bones and teeth: an experimental study of color, morphology, crystal structure and shrinkage. *J. Archaeological Sci.* 1984, 11, 307-325.
[269] LeGeros, R.Z., LeGeros, J.P. Calcium phosphate bioceramics: past, present, future. *Key Eng. Mater.* 2003, 240-242, 3-10.
[270] Dorozhkin, S.V. Biphasic, triphasic and multiphasic calcium orthophosphates. *Acta Biomater.* 2012 (early view).
[271] Anuta, D.A., Richardson, D. Biphasic hydroxyapatite/beta-tricalcium phosphate granules bound in polymerized methyl methacrylate: bone substitute studies.

Transactions of the Annual Meeting of the Society for Biomaterials in conjunction with the Interna. 1985, 8, 62.
[272] Moore, D.C., Chapman, M.W., Manske, D.J. Evaluation of a new biphasic calcium phosphate ceramic for use in grafting long bone diaphyseal defects. *Transactions of the Annual Meeting of the Society for Biomaterials in conjunction with the Interna*. 1985, 8, 160.
[273] Dorozhkin, S.V. Nanosized and nanocrystalline calcium orthophosphates. *Acta Biomater*. 2010, 6, 715-734.
[274] Dorozhkin, S.V. Nanodimensional and nanocrystalline apatites and other calcium orthophosphates in biomedical engineering, biology and medicine. *Materials* 2009, 2, 1975-2045.
[275] Layrolle, P., Lebugle, A. Characterization and reactivity of nanosized calcium phosphate prepared in anhydrous ethanol. *Chem. Mater*. 1994, 6, 1996-2004.
[276] Cui, F.Z., Wen, H.B., Zhang, H.B., Ma, C.L., Li, H.D. Nanophase hydroxyapatite-like crystallites in natural ivory. *J. Mater. Sci. Lett*. 1994, 13, 1042-1044.
[277] Li, Y.B., de Wijn, J., Klein, C.P.A.T., de Meer, S.V., de Groot, K. Preparation and characterization of nanograde osteoapatite-like rod crystals. *J. Mater. Sci. Mater. Med*. 1994, 5, 252-255.
[278] Li, Y.B., de Groot, K., de Wijn, J., Klein, C.P.A.T., de Meer, S.V. Morphology and composition of nanograde calcium phosphate needle-like crystals formed by simple hydrothermal treatment. *J. Mater. Sci. Mater. Med.* 1994, 5, 326-331.
[279] Shirkhanzadeh, M. X-ray diffraction and Fourier transform infrared analysis of nanophase apatite coatings prepared by electrocrystallization. *Nanostruct. Mater*. 1994, 4, 677-684.

In: Calcium Phosphate
Editor: Robert B. Heimann

ISBN: 978-1-62257-299-1

Chapter 2

The Carbonate Ion in Hydroxyapatite and Biological Apatite

Michael E. Fleet[*]
Department of Earth Sciences, University of Western Ontario,
London, Ontario, Canada

Abstract

The structure of carbonate-bearing hydroxyapatite (CHAP) has been investigated extensively by diffraction methods and chemical spectroscopy. Although these studies established that the carbonate ion occurs both in the *c*-axis structural channel (type A) and as a substituent for the phosphate group (type B), the structural details and proportions of the A and B carbonate ions in the nano-scale crystals of apatite in bone and dental enamel remained elusive. More recently, relatively large single crystals of CHAP have been grown from carbonate-rich melts and investigated by X-ray structure and Fourier transform infrared spectroscopy (FTIR) methods. The important conclusions for biological apatite are that type A carbonate is present in far greater proportions than previously assumed and is possibly the labile carbonate component mediating acid-base reactions in the body. Previous studies are reviewed. Stereochemical details and substitution mechanisms for the new structures are discussed, and the weak out-of-plane bend (ν_2) FTIR band is proposed as an analytical tool for estimating the relative proportions of A and B carbonate and hydrogen carbonate in quenched, annealed and aged products.

1. Introduction

The calcium phosphate apatites [$Ca1_4Ca2_6(PO_4)_6X_2$; with X = F, OH, Cl, O, etc.] [1] have importance in both geochemistry and biology. Apatites sequester phosphorus, rare earth elements, actinides, and volatile elements in the Earth's crust and mantle. Also, fluorapatite is a practical host for the containment of high-level nuclear waste [2]. Within the biosphere,

[*] Email address: mfleet@uwo.ca

carbonate-bearing hydroxyapatite (presently referred to as carbonate apatite and abbreviated as CHAP) is by far the most important biomineral, accounting for up to about 65 wt% of cortical bone and 97 wt% of dental enamel [3,4]. Fluoride-bearing CHAP is the important anticaries component of dental enamel [5], and CHAP and carbonated fluorapatite (CFAP; also known as francolite) are the dominant minerals in phosphorites [6].

Calcium apatite minerals are dominantly solid solutions of hydroxyapatite [HAP; ideally $Ca_{10}(PO_4)_6(OH)_2$; $Z = 1$], fluorapatite [FAP; $Ca_{10}(PO_4)_6F_2$] and chlorapatite [CLAP; $Ca_{10}(PO_4)_6Cl_2$]. The natural phases all have the hexagonal space group $P6_3/m$, although pure, end-member HAP and CLAP crystallize in the monoclinic space group $P2_1/b$ [7,8]. The $P6_3/m$ structure of calcium apatites is well known. Apatite is an orthophosphate: isolated PO_4 tetrahedra centered at $z = (1/4, 3/4)$ are linked by Ca1 in ninefold (6+3) coordination and Ca2 in an irregular sevenfold (6+1) coordination. A prominent feature of the structure is the large c-axis channel that accommodates the X anion component (F,OH,Cl) and is defined by triclusters of Ca2 cations at $z = (1/4, 3/4)$ (Figure 1). In FAP, the F anion is located on the c-axis at $z = (1/4, 3/4)$ in the center of a tricluster of Ca cations. Hydroxyl ions in HAP and Cl ions in CLAP are displaced along the c-axis and have split atom positions with occupancy of 0.5: the hydroxyl oxygen is at $z = \pm(0.198, 0.302)$ and the much larger Cl anion is displaced further at $z = \pm(0.432, 0.068)$ [9]. Thus, the coordination of the X anion is equilateral triangular in FAP but near octahedral in CLAP. The structural role of carbonate in hydroxyapatite and fluorapatite has been investigated extensively by x-ray powder, x-ray single-crystal, and neutron powder diffraction methods [10-19], as well as by infrared, Raman and nuclear magnetic resonance spectroscopy [20-32] and theoretical simulations [33,34]. It has been established that the carbonate ion can be accommodated either in the c-axis structural channel or as a substituent for the phosphate group: the former carbonate is known as type A and the latter as type B.

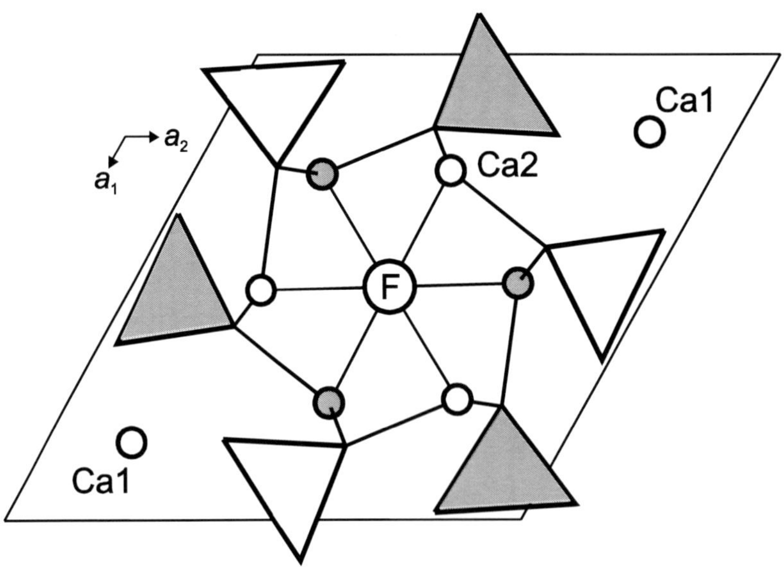

Figure 1. Structure of fluorapatite showing triclusters of Ca2 cations in c-axis channel wall: Ca2 and PO_4 groups (triangles) are located at $z = 1/4$ (shaded) and $z = 3/4$ (open); unit-cell outline is displaced by (-0.5, -0.5, 0.0).

Table 1. Synthesis experiments and amounts of Na, A and B carbonate (*pfu*[1])

Apatite	Expt.	*T* (°C)	*P* (GPa)	EPMA Na	X-ray structure A[2]	B	B/A	FTIR B/A
Na-free apatites								
CHAP	PC17	1400	3.0	-	-	0.17(2)	-	1.3
CHAP	PC16	1400	4.0	-	0.14(3)	-	-	0.3
CHAP	PC71	1400	2.0	-	0.75	0.11(2)	0.2	0.2
CHAP	PC18	1400	3.0	-	1.08(6)	0.49(2)	0.5	0.5
CHAP	PC55	1400	3.0	-	1.09(4)	0.56(2)	0.5	0.4
Na-bearing apatites								
CHAP	LM005	1200	0.5	0.87(3)	1.00(5)	0.77(3)	0.8	0.9
CHAP	LM006	1200	1.0	0.35(4)	0.52(3)	0.38(2)	0.7	0.8
CCLAP	LM169	1350	1.0	0.40(4)	0.37(3)	0.57(2)	1.5	1.5
CCLAP	LM171	1200	1.0	0.39(3)	0.46(4)	0.58(3)	1.3	1.6
CCLAP	LM173	1000	1.0	0.39(5)	0.42(3)	0.57(2)	1.4	1.4
CFAP	LM130	1250	1.0	0.09(3)	0.08(3)	0.23(1)	2.9	2.6
CFAP	LM136	1250	1.0	0.09(3)	0.16(3)	0.21(2)	1.3	1.2
CFAP	LM142	1150	1.0	0.11(5)	0.14(3)	0.23(1)	1.6	1.5

[1] *pfu* is per formula unit [$Ca_{10}(PO_4)_6X_2$].

[2] contents of channel carbonate for PC18 are A = 0.50(4) and A2 = 0.58(4), and for PC55 are A = 0.50(1) and A2 = 0.59(2).

In infrared spectra of natural and synthetic Na-free CHAP, type A carbonate is characterised by a doublet band at about 1545 and 1450 cm^{-1} and a singlet band at about 878 cm^{-1}, whereas type B carbonate has these bands at about 1455, 1410 and 871 cm^{-1}, respectively [3,21,22]. In these literature studies, end-member A-type CHAP was prepared by equilibrating HAP in an atmosphere of CO_2 at 900-1000°C, and B-type CHAP by precipitation in basic aqueous solution. More detailed structural analysis of carbonate apatite by diffraction methods has been frustrated by several factors, including: (1) the limited substitution of carbonate (especially of type B carbonate); (2) small (nano-scale) crystal size of biological apatite and apatite precipitated from aqueous solution and of francolite from phosphorites; (3) poor crystallinity; and (4) weak and overlapped electron density of carbonate atoms.

This chapter reviews recent single-crystal x-ray and FTIR studies on the accommodation of the carbonate ion in Na-free and Na-bearing CHAP, using crystals grown from carbonate-rich melts in Quickpress and end-loaded piston cylinder equipment [30, 35-41].

Other experimental details are given in Table 1 and references [30, 35-41]. The crystal products included Na-free type A and type AB CHAP, and Na-bearing type AB CHAP, carbonate fluorapatite (CFAP) and carbonate chlorapatite (CCLAP). A total of fifteen crystals were investigated by single crystal x-ray structure analysis, and crystal compositions corresponded approximately to the substitution formula $Ca_{10-(y+z)}Na_y vac_z[(PO_4)_{6-(y+2z)}(CO_3)_{y+2z}][X_{2-2x}(CO_3)_x]$, where *vac* is a large cation vacancy and $x \approx y$ up to 1.0 and $z \approx 0.0$ for

CHAP, $x \approx y \approx 4z \approx 0.4$ for CCLAP, and $x \approx y \approx 2z \approx 0.1$ for CFAP, for equivalent conditions of synthesis (Table 1).

2. Detection and Analysis of Carbonate by FTIR

The literature on the application of infrared spectroscopy to carbonate in apatite is voluminous, with Elliott [10], LeGeros et al. [22] and Bonel [21] usually cited as early pioneering studies. Fourier transform infrared (FTIR) spectra of carbonate apatites are dominated by a complex band at about 1000-1100 cm^{-1} for the asymmetric stretch vibration of the phosphate group. As noted above, the characteristic bands for the carbonate ion in apatite are a prominent doublet band at 1400-1600 cm^{-1} (asymmetric stretch vibration; ν_3) and a weak singlet band at 872-880 cm^{-1} (out-of-plane bend vibration; ν_2) (Figures 2, 3, 4). The ν_3 vibration is twofold degenerate for the free carbonate ion but is represented by a doublet in apatite spectra because the degeneracy is lifted for site symmetries lower than trigonal. There is also a weak band for the stretch vibration of hydroxyl structurally-bound in the apatite channel near 3572 cm^{-1}. Features consistent with the presence of the HPO_4^{2-} ion [24, 26] are absent in the present high-pressure synthesized CHAP spectra.

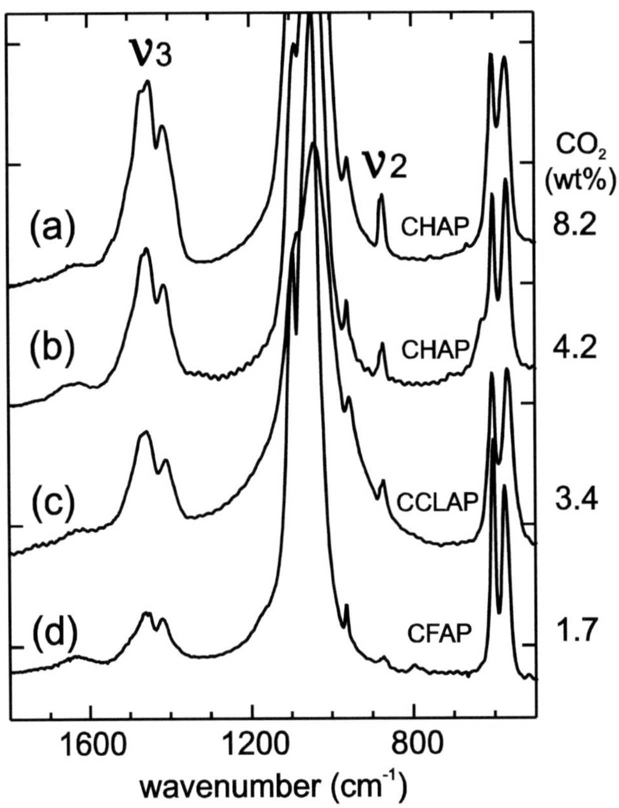

Figure 2. Fourier transform infrared (FTIR) spectra for Na-bearing type AB carbonate apatites synthesized at high pressure identifying bands due to asymmetric stretching (ν_3) and out-of-plane bending (ν_2) of carbonate ions: note that profile of ν_3 absorption band is independent of content of carbonate (reported as CO_2).

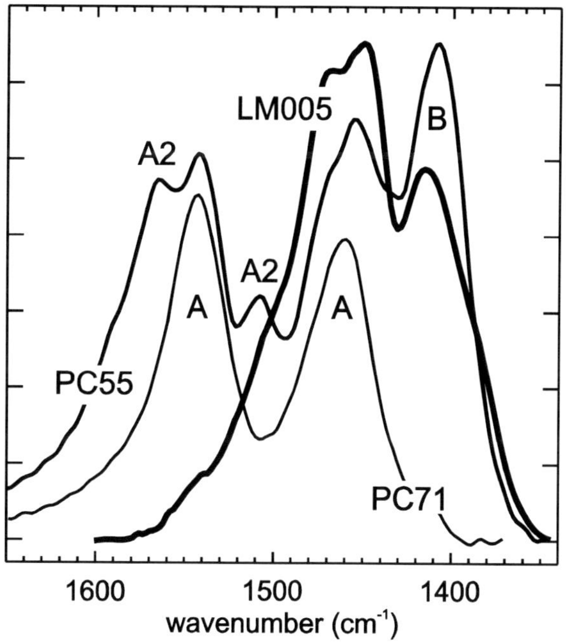

Figure 3. FTIR spectra in the asymmetric stretch (ν_3) region for type A CHAP (PC71), Na-free AB CHAP (PC55), and Na-bearing AB CHAP (LM005).

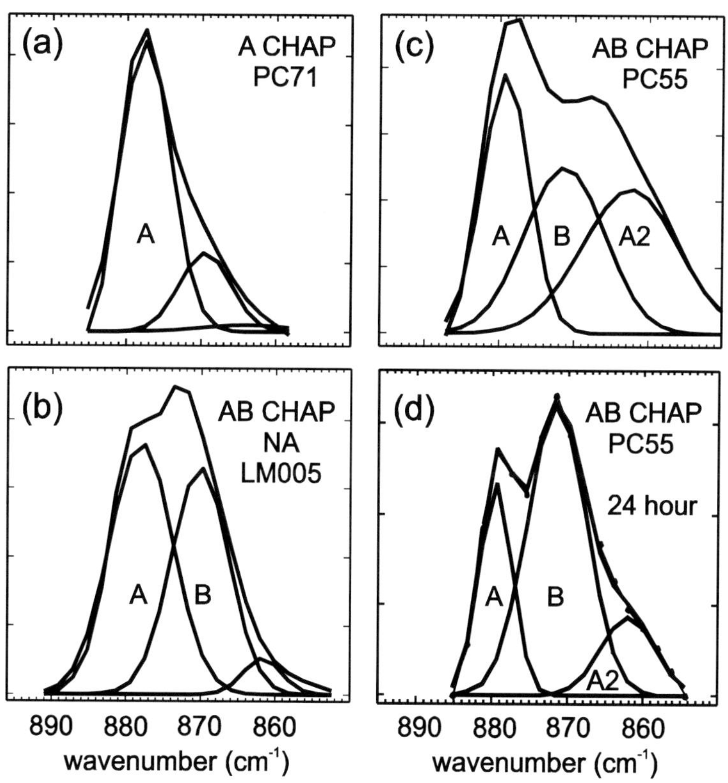

Figure 4. FTIR spectra in the out-of-plane bending (ν_2) region identifying fitted Gaussian components for A, B and A2 carbonate species, and (d) illustrating marked loss of channel carbonate (A, A2) on annealing PC55 in air for 24 hours (Table 2).

The ν_3 region of the infrared spectra of type AB CHAP is comprised of complexly overlapped contributions from A and B carbonate ion species. For Na-free type A CHAP synthesized at high pressure (e.g., PC71 in Table 1), the ν_3 region consists of a symmetrical doublet with limbs centred at 1544 and 1461 cm^{-1} [35] (Figure 3). The spectra of type AB CHAP from the same Na-free series of experiments (e.g., PC55 in Table 1) are complex but six individual bands are generally recognized by five definite peaks and a prominent shoulder [36, 30] (Figure 3). Initial assignments attributed the strong peak near 1409 cm^{-1} to one limb of the B-carbonate ion doublet enhanced by calcite contamination, but this interpretation was revised by Fleet [31], who attributed all of the intensity of the 1409 cm^{-1} band to B-carbonate. The 1540 cm^{-1} band is clearly type A but there is no clear basis for proportioning the absorption in the 1470-1455 cm^{-1} interval between A and B carbonate, even with knowledge of the actual (A, B and A2) site occupancies obtained from the X-ray structures (Table 1).

The two remaining doublet bands, at 1569 and 1507 cm^{-1}, are assigned to a second channel carbonate species labelled A2 and attributed to a stuffed and more weakly bound carbonate ion location in the apatite channel [30, 36].

In Na-bearing AB CHAP crystals synthesized at high pressure [38], the ν_3 doublet for the type A carbonate ion, which is expected from the x-ray structure results to be an important feature, is seemingly shifted to lower wavenumber into the region normally associated with type B carbonate (Figure 3). These spectra for the synthetic Na-bearing crystals are similar to those of CHAP precipitated from basic solution, dental enamel and cortical bone [3, 23, 24] and previously considered to represent the dominant presence of B carbonate.

Major change is also evident in the FTIR spectra of high-pressure synthesized Na-free CHAP annealed at high temperature in air. As reviewed in Fleet [31] aliquots of the high-pressure quenched product of PC55 (Table 1) were supported on platinum foil and heated in air at 1000°C for intervals of 15 and 30 minutes, and 12 and 24 hours [36]. The principal changes observed were a major loss in the intensity of the channel A2 bands and a transfer of absorption intensity from the 1409 cm^{-1} band of type B carbonate to the broad band at 1470-1450 cm^{-1}. Similar results were obtained for a second series of annealing experiments, on PC17 [30] (Table 1).

Overall, it is not possible to deconvolute the complex ν_3 spectra of carbonate apatite crystals synthesized at high pressure without further crystal-chemical insight, and in any case the ν_3 region is often of limited utility for study of biological apatites because it tends to be obscured by absorption bands of functional groups on proteins and glucosaminoglycans [24, 26, 27]. In contrast, the ν_2 band generally better reflects the true proportions of the principal carbonate species, and has been utilised extensively to investigate carbonate ion environments of apatite biomaterials [26-28].

In apatite biomaterials, the ν_2 region is normally deconvoluted to give a singlet band for type A carbonate at 878 cm^{-1} and bands for B carbonate at 871 cm^{-1} and 866 cm^{-1} [23,24], and the composite ν_2 band for laboratory synthesized low-pressure CHAP crystals has components for type A carbonate at 878-881 cm^{-1} and type B at 870-873 cm^{-1} [10,23]. The ν_2 spectra for high-pressure synthesized CHAP were interpreted in Fleet [31] by fitting Gaussian components to peaks and prominent shoulders (Figure 4). For Na-free type A CHAP (PC71) and Na-bearing type AB CHAP (LM005), the ν_2 spectra were fitted by two bands, one for type A carbonate at 878 cm^{-1} and one for B carbonate at 870 cm^{-1} (Figure 4a,b). The fitted ν_2 spectrum for PC71 confirms that this well-investigated type A CHAP does contain a small proportion of type B carbonate, which is not accounted for in the detailed X-ray structure investigations below [35,37]. The ν_2 spectrum for Na-free AB CHAP (PC55) was fitted to three bands, at about 879, 871 and 863 cm^{-1} assigned to A, B and A2 carbonate, respectively (Figure 4c). The assignment of the stuffed channel species A2 to the band at about 863 cm^{-1} in Na-free AB CHAP is based largely on the decrease in intensity of this band in the spectra of Na-bearing AB CHAP (Figures 3, 4), and confirmed by good agreement with the x-ray structures.

The resulting proportions of the A, B and A2 carbonate ion species from peak areas for PC55 are 31:34:35, and thus in very good agreement with the proportions of 30:34:36, respectively, obtained from the x-ray structure site occupancies. Moreover, the area ratios B/(A+A2) of the fitted ν_2 band components are in very good linear agreement with the corresponding quantities calculated from the x-ray structure site occupancies [31] (Figure 5).

Thus, the proportions of A and B carbonate ions in these high-pressure carbonate apatites are now firmly established by two independent methods, FTIR ν_2 band areas and x-ray structure analysis [35,36,38,30]. This agreement establishes FTIR ν_2 band areas as a viable independent method for estimating the proportions of channel (A) and B carbonate in CHAP precipitated from aqueous solutions, apatite biominerals and francolites, and it reinforces earlier studies on biological CHAP [23,24,26]. The ν_2 singlet band is generally better resolved and more sensitive to local stereochemical environment than the ν_3 doublet, being located at 878-880, 871-872, and 863-864 cm^{-1} for A, B and A2 carbonate, respectively. Note also the significant shifts in ν_2 band position between calcite (powder and single crystal; 871-876 cm^{-1} [42]), amorphous $CaCO_3$ (865 cm^{-1}), carbonate-phosphate glass (853 cm^{-1}) [43], and sodium hydrogen carbonate (837 cm^{-1}).

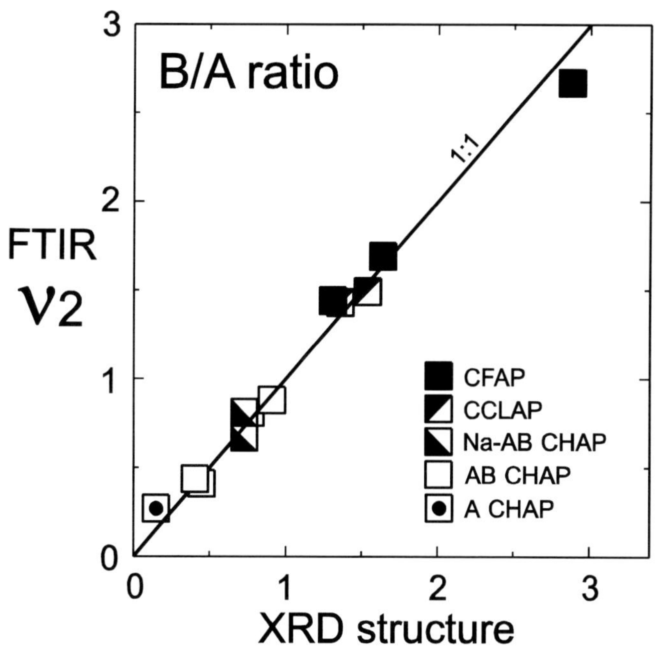

Figure 5. Comparison of the proportion of formula amounts of B and A carbonate (B/A) determined from FTIR ν_2 band areas and the x-ray structure method.

The infrared spectra are also useful in establishing that CO_3 and OH were the only channel anion (X) species present in significant amounts in the synthetic high-pressure CHAP crystals. The OH stretch band (3572 cm^{-1}) and OH libration band (near 631 cm^{-1}) show a systematic decrease in intensity with increase in carbonate content, with carbonate content estimated from both total ν_3 band area and x-ray site occupancies. This inverse correlation is well illustrated in Figure 2 of Fleet and Liu [38], which displays FTIR spectra for a series of Na-bearing CHAP compositions. The OH libration band is prominent for LM002 with 3.8 wt% CO_3^{2-} but absent for LM005 with 11.1 wt% CO_3^{2-} and all channel locations for the type A carbonate ion statistically filled (Table 1). In addition, for Na-free CHAP, Figure 3 of Fleet et al. [30] shows that the libration band is prominent for PC17 with no detectable type A carbonate content (Table 1) and reduced to a very weak shoulder for PC18 with (A+A2) > 1.0. Also, the OH libration band is absent in Na-bearing CFAP and CCLAP (Figure 2), revealing that the channel anions in both of these composition series were essentially only halide and carbonate.

3. Detection of Carbonate by Single-Crystal X-Ray Diffraction

The substitution of the carbonate ion into apatite is generally limited in extent, and, unlike atom-for-atom solid solution, introduces new atomic positions and lowers the local symmetry to $P1$. Therefore, for each A, B and A2 carbonate ion species there are six equivalent orientations for host structures with $P\bar{3}$ symmetry and twelve for $P6_3/m$ symmetry. Since only one orientation can be occupied in any given unit-cell, the maximum average site occupancies are 0.1667 and 0.0833, respectively. The carbonate ion substituents are replicated by the high symmetry of the host apatite structure and therefore contribute coherent Bragg scattering intensity to the single-crystal diffraction pattern. The diffraction pattern represents only the average structure, a composite of the host structure and the carbonate ions locally ordered in minimum energy locations and configurations. In addition, atoms of the host structure are locally displaced to accommodate the carbonate ions, and this displacement results in anomalous increase in anisotropic displacement parameters, especially for type AB CHAP containing significant amounts of both A and B carbonate ions [16, 30]. As a result of these problems, the x-ray structures of carbonate apatites present numerous challenges in interpretation, especially where carbonate contents are very low. In particular, electron densities appreciably less than that of a hydrogen atom and overlap of carbonate and host structure atoms limit the amount of structural information that can be extracted from X-ray structure analysis. Figure 6 shows the residual electron density map for a type A CHAP (PC71 in Table 1) computed at z = 0.5 and close to the channel axis, after refinement of the apatite host structure. The peak residual density locates the off-axis oxygen atom of the type A carbonate ion (labelled O5 elsewhere in this chapter); it is distributed over six equivalent positions by the hexagonal symmetry of the apatite host structure. The other two oxygen atoms of the type A carbonate ion are located similarly above and below O5 but close to the c-axis, confirming the early conclusion of Elliott [10] from polarized infrared spectra that the plane of the channel carbonate ion is parallel to the c-axis.

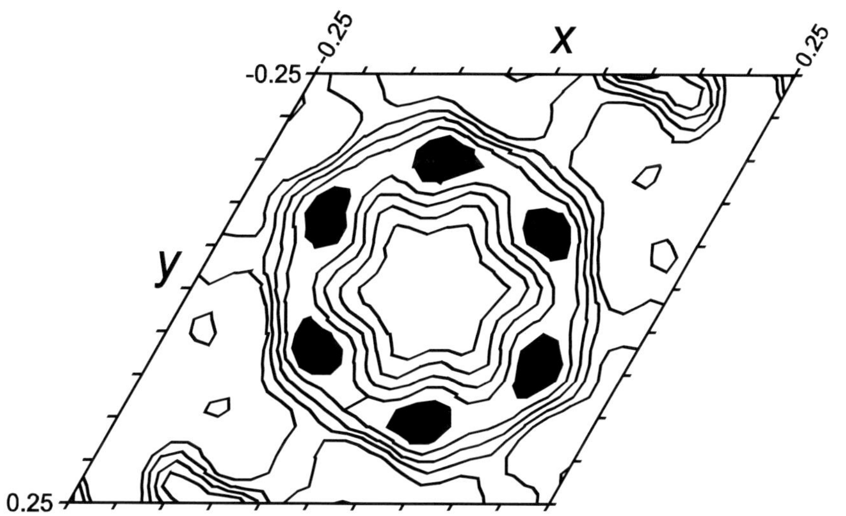

Figure 6. Residual electron density map calculated at height $z = 0.5$: density $>0.5\cdot10^{-3}$ e/nm^3 (black areas) locates the off-axis oxygen atom (O5) of the type A carbonate ion, which is positionally-disordered in the apatite channel.

Note that all crystals investigated were of high diffraction quality, and the diffraction patterns revealed no evidence of ordering of the carbonate ion defects in the form of superstructure reflections, diffuse scattering or anomalous reflection broadening.

Type A carbonate ion. Resolution of the channel (type A) carbonate ion near the *c*-axis of the x-ray structures is complicated by high multiplicity and overlap of the electron density for the X anions (OH, O, F, Cl), carbon atom, and two of the three carbonate oxygen atoms. For partial occupancy of the channel positions it is not possible to refine independently the occupancy, positional, and thermal parameters for these atoms. Fortunately, the channel is wholly occupied by carbonate in one sample of Na-bearing type AB CHAP (LM005 in Table 1) [38], and three-quarters occupied by carbonate in a Na-free type A CHAP (PC71) [30]. For partial occupancy of the apatite channel, carbonate ion site occupancies may be determined with some confidence by refining the occupancy of the off-axis oxygen O5. The electron density for O5 is well resolved in residual electron density maps, at $z = 0.5$ for type A CHAP (Figure 6) and at $z = 0.0$ and 0.5 for type AB carbonate apatites, even in Na-bearing CFAP structures that have less than 0.2 carbonate ions per unit cell [39].

Fleet and Liu [35] found that their synthetic Na-free type A CHAP (PC71 in Table 1) had a new space group ($P\bar{3}$), and that the carbonate ion was ordered along the apatite channel at $z \approx 0.5$, with the possible channel location at $z \approx 0.0$ vacant. A preferred structure for this $P\bar{3}$ type A CHAP was obtained by rigid body refinement of the channel carbonate ion, using a novel procedure for defining the ideal equilateral triangular geometry[37], and assuming the carbonate ion dimensions in calcite (e.g., O-O = 2.219 Å, C-O = 1.281 Å; [44]). It was established that two of the oxygen atoms (O6 and O7) lay close to the *c*-axis at about $z \approx 0.333$, 0.667, consistent with a substitution mechanism controlled by charge balance requirements of:

$$[{}^{A}CO_3 = 2(OH)] \quad (1)$$

The carbonate ion is rotated about 14° counter-clockwise relative to the ideal model structure [41] and canted 9° away from the *c*-axis: note that it is viewed in Figure 7a in one of the six equivalent orientations present in electron density maps. The carbonate ion is located in the apatite channel by six bonds to Ca2 atoms, giving Ca2-O bond distances consistent with an ideal distribution of valence units for the carbonate oxygen atoms (i.e., 4/3 to C and 1/3 to 2xCa2). The phosphate groups are displaced slightly from ideal HAP positions to dilate the apatite channel in the immediate vicinity of the carbonate ion (at $z \approx 0.5$), and contract it above and below (i.e., at $z \approx 0.0, 1.0$).

Suetsugu et al. [11] reported a different structure for type A CHAP crystals grown in a $CaCO_3$ flux under an argon gas pressure of 55 MPa. They reported that the space group was $P\bar{6}$ and the channel carbonate ion was oriented with only one oxygen atom close to the *c*-axis. However, they did not investigate the possibility that their crystal symmetry might be $P\bar{3}$ [41]. Very recently, type A CHAP crystals have been prepared by heating HAP in an atmosphere of CO_2 at 900°C and the structure reinvestigated by Rietveld powder x-ray diffraction in space group *Pb* [19]. The FTIR spectrum has ν_3 bands at 1539 and 1467 cm^{-1} and ν_2 at 879 cm^{-1}, in good agreement with PC71 [35].

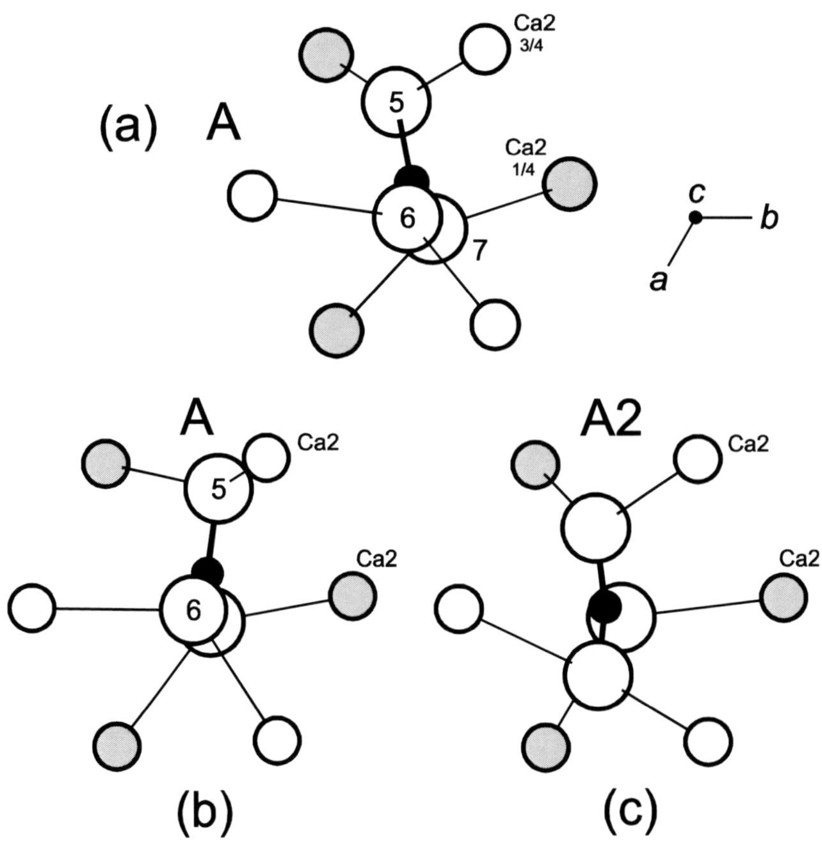

Figure 7. Orientation of channel carbonate ions in Na-free CHAP: (a) type A in A CHAP (PC71); (b) type A and (c) A2 in AB CHAP (PC55); apatite channel is defined by triclusters of Ca2 cations at heights $z = 1/4$ (shaded) and $z = 3/4$ (open).

The *Pb* structure has two independent channel carbonate ions, and both are oriented with two oxygen atoms close to the monoclinic *c*-axis, which is equivalent to the orientation of the type A carbonate ion in the $P\bar{3}$ structure of PC71 (Figure 6 of Fleet and Liu [41]). Note that discrepancies in detailed comparison between PC71 and the *Pb* structure of Tonegawa et al. [19] are attributable to the more limited resolution of the Rietveld powder refinement method.

The type A carbonate ion is located similarly to PC71 in the channels of the Na-free and Na-bearing AB CHAP (PC55 [36,37]; LM005 [38], respectively), but now rotated clockwise, opposite to the rotation for the type A CHAP (Figure 7b). The carbonate ion is rotated about 7° and canted 7° relative to the *c*-axis in the Na-free CHAP (PC55). In the Na-bearing AB CHAP (LM005) it is rotated 13° but remains parallel to the *c*-axis (tilt angle = 0.0°). These data were obtained for both structures by rigid body refinement of the carbonate ion, following the procedure of Fleet and Liu [37]. The orientations observed in our synthetic apatites are similar to those for channel carbonate ions in defect structures calculated using static lattice DFT [33] and energy minimization/molecular dynamics [34] methods, except that the axial oxygen atoms (O6 and O7) are more-or-less symmetrically positioned with respect to $z \approx 0.0$ and 0.5 in our studies. The lack of significant canting and the symmetrical positioning of the carbonate ion in the apatite channel of the Na-bearing AB CHAP reflects the absence of other channel constituents; the occupancy of the type A carbonate in this carbonate apatite being essentially 1.0. In comparison, carbonate ions were in competition with hydroxyl ions (and possible H_2O, as well) in the *c*-axis channels of the Na-free type A [35] and type AB CHAP [30,36] apatites. The channel carbonate of the Na-bearing AB CHAP (LM005) is displaced to one side of the apatite channel, with O6 and O7 lying almost on the *c*-axis (Figure 8) and O5 extending outwards close to the channel wall and very close (1.14 Å in the average structure) to an O3 atom of a substituted phosphate group. In comparison, in Na-free type A CHAP the carbonate ion (and carbon atom) is more centrally located, resulting in O5-O3 distances of 2.07 and 2.24 Å.

The $P\bar{3}$ structure of type A CHAP [35] provides a key piece of evidence for understanding the local environment of the channel carbonate ion in type AB apatites. In these complexly disordered structures, the apatite channel must be dilated in the vicinity of the bulky carbonate ion and contracted above and below it at $z \pm 0.5$. The anisotropic displacement parameters of O1, O2 and O3 for the average structures show that this is achieved largely by rotation of the PO_4 tetrahedron about the O1-P bond axis [30].

Type A2 carbonate ion. The A2 carbonate ion of Na-free AB CHAP from experiment PC55 [36] is rotated 2° counter-clockwise and canted 11° away from the *c*-axis relative to the ideal configuration [41] (Figure 7c). Then a slight downward shift to $z \approx 0.45$ for the carbon atom yields a fairly equitable distribution of Ca2-O bond distances. Refinement of the atom positions in the original x-ray structure study [36] was complicated because the A2 orientation resulted in twelve near-equivalent weak residual electron density maxima for both O8 and O9, with twelve centered at $z = 0.5$ and twelve at $z = 0.0$. Moreover, even though O8 and O9 were off-axis oxygen atoms, their electron density maxima in different equivalent orientations of the carbonate ion were in near coincidence, to the extent that a single atom position was refined in Fleet and Liu [36]. The two separate atom positions were resolved in Fleet [41] from a combination of refinement, residual electron density and assumption of ideal carbonate ion geometry.

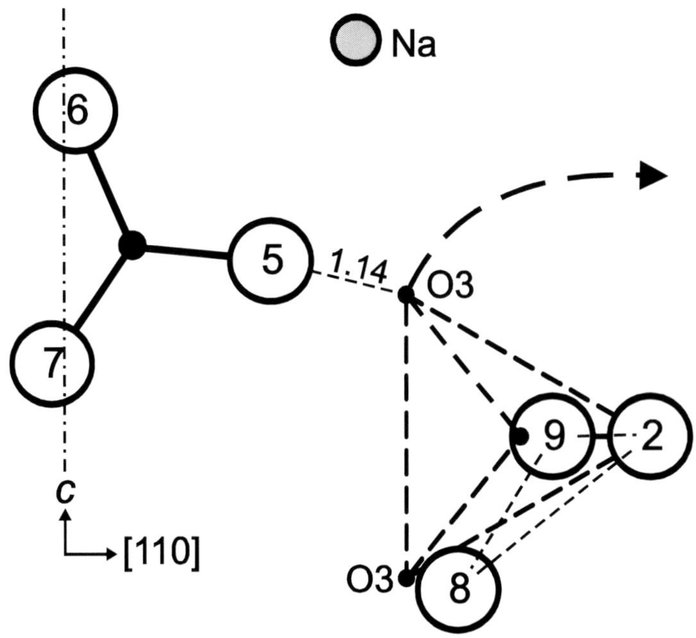

Figure 8. Fragment of the structure of Na-bearing CHAP (LM005) showing location of B carbonate ion close to the sloping face of a substituted phosphate tetrahedron: multiplicities due to $P6_3/m$ symmetry of the host apatite structure are not shown; short O5-O3 interatomic distance is in Å units.

Type B carbonate ion. Interpretation of the location of the B carbonate ion is more problematical, and has been controversial [14, 16, 36, 38-40]. Logically, the B carbonate ion is located in the vicinity of the substituted phosphate group and occupies as many of the phosphate oxygen sites as possible. Several studies have noted that the presence of B carbonate is indicated by significant reduction in the occupancy of the P position [13, 16, 45]. Indeed, for the Na-bearing apatites, the phosphorus occupancy decreases systematically with increase in total carbonate content through the series CFAP, CCLAP and CHAP. Also, our various studies have failed to detect any significant residual electron density in the vicinity of the phosphate group except for that associated with oxygen atoms of B carbonate lying close to sloping (i.e., inclined to *c*-axis) faces of the substituted phosphate tetrahedron [36,38]. In addition, the B carbon atom of CHAP precipitated from aqueous solution was located in the centre of the sloping faces in the neutron scattering study of Wilson et al. [14]. Further, indirect, evidence is that the coupling of A and B carbonate ions discussed below requires the omission of one of the two symmetry-related O3 atoms, as illustrated by the following carbonate-for-phosphate substitution mechanisms:

$$[0.5^{Ca}vac + {}^{B}CO_3 = 0.5Ca + PO_4] \quad (2)$$

and,

$$[Na + {}^{B}CO_3 = Ca + PO_4] \quad (3)$$

A preliminary location for the B carbonate ion in the present high pressure-temperature synthesized apatites was obtained from study of Na-free AB CHAP in the form of residual electron density near O3 [30,36]. This identified a probable B carbonate oxygen atom, which is presently labelled O8 (Figure 8). The other two oxygen atoms were assumed to be buried in the electron density of O1 and O2. Subsequent X-ray structure study of the Na-bearing CHAP (LM005) [38] resulted in a similar orientation for the B carbonate ion, but now O9 is resolved separately from O1 in the average structure. This structure confirmed that the B carbonate ion in CHAP is located near a sloping face of the substituted phosphate tetrahedron but inclined at an angle of 53° to the mirror plane; i.e., tilted 18° away from the sloping face. It is emphasized that there is some ambiguity in defining the precise positions of B carbonate oxygen atoms located very close to phosphate oxygen atoms, since electron densities extensively overlap in the average structures.

The location and orientation of the B carbonate ion in synthetic CFAP [39] is at variance with the conclusion of Leventouri et al. [16, 17] and Antonakos et al. [18], who proposed that the B carbonate was located on the (horizontal) mirror plane, with oxygen atoms at O1, O2 and a third location mid-way between the two symmetry related O3 atoms of the substituted phosphate group.

Their conclusions were based on Rietveld neutron powder diffraction study of francolite from Epirus, Greece and its heat-treated products. It is well known that oxygen positions are determined with poor precision in conventional Rietveld powder structure refinement of apatites [8, 46]. Our studies emphasized the unique advantage of single crystal diffraction methods in locating weak electron densities.

However, the relevant experimental observations for CFAP in Leventouri et al. [16], and Fleet and Liu [39] are not significantly discrepant, since both studies found that O1 and O2 are shifted on the mirror plane toward the centre of the phosphate tetrahedron. The neutron powder diffraction study of Leventouri et al. [16] simply detected the average position of O1 and O2 in the annealed francolite structure, whereas Fleet and Liu [39] resolved separate O1/O9 and O2/O8 positions for the phosphate and carbonate oxygen atoms in CFAP.

They found that substitution by type B carbonate results in observed atom shifts of 0.31 Å for O1 to O9 and 0.41 Å for O2 to O8. When these shifts are weighted by the proportion of type B carbonate estimated for the Epirus francolite (0.95 per formula unit), the corresponding apparent shortening of the P-O1 and P-O2 bond lengths in the average structure is calculated to be about 0.05 and 0.06 Å, which compares favourably with about 0.04 and 0.04 Å, respectively, in Leventouri et al. [16].

4. Coupling of Substituents

Fleet and Liu [38] observed that the sodium cation defect, and the A and B carbonate ion defects in Na-bearing CHAP must be locally coupled in the type AB apatite structure to facilitate charge compensation and minimize the effects of spatial accommodation. Sodium and vacancies at Ca sites are required for charge balancing the carbonate-for-phosphate substitution after the removal of the fourth oxygen atom (cf. above substitution mechanisms 2 and 3). Moreover, in the absence of a proximal B carbonate ion, the off-axis oxygen (O5) of the channel type A carbonate would be only about 1.1-1.2 Å from an O3 oxygen atom of a phosphate group (Figure 8). These factors strongly suggest that the A and B carbonate ions and Na are locally coupled in the structures of Na-bearing CHAP (and of Na-bearing CFAP and CCLAP also) to facilitate charge compensation and minimize the effects of spatial accommodation. The prohibitively short O5-O3 interaction is eliminated when the O3 oxygen atom in question is removed entirely from the structure and, logically, charge balance requirements favour placing the Na cation closest to this vacant O3 oxygen site. Note that an explanation for the near-linear 1:1 correlation between Na and type A carbonate (Table 1) is not intuitively obvious in the absence of a role for spatial accommodation, because Na is introduced to charge balance the introduction of B carbonate (substitution mechanism 3), but is not formally required to charge balance the type A carbonate in substitution mechanism 1 [i.e., $^{A}CO_3 = 2(OH, F, Cl)$].

FTIR spectra independently confirm that the A and B carbonate ions and Na are coupled locally as a defect cluster as depicted in Figure 8. It has been noted above that the ν_3 bands of type A carbonate in Na-bearing carbonate apatites are shifted to lower wavenumber and extensively overlap the spectrum of type B carbonate.

The resulting characteristic ν_3 band profile is similar for all Na-bearing carbonate apatite composition series (X = OH, F, Cl) and carbonate contents investigated [37, 39, 40] (Figure 2). This consistency shows that the chemical environments of the Na cation, and the A and B carbonate ions are similar in all of these apatites, i.e., these substituents are present as a defect cluster common to all the composition series investigated (Table 1), and randomly distributed within the host apatite structures.

It is possible that there exists a degree of intermediate range order, with several AB pairs organized into columnar *c*-axis domains, analogous to the columnar domains suggested for mixed X anion occupancy in the channel of Ca apatites [3, 4]. However, the x-ray diffraction patterns of these apatites do not reveal any evidence of extensive ordering and it seems unlikely that compositionally distinct domains analogous to those observed in Brazilian gem-grade apatite [47] exist in them.

We recognize that charge compensation through the introduction of vacancies at the large cation site (substitution mechanism 2) may involve a separate defect cluster in CFAP and CCLAP, because both of these apatite series have molar (B/Na) > 1, although this is not reflected by change in their ν_3 band profiles [39, 40].

5. Annealing and Aging Experiments

Two series of room-pressure experiments on high-pressure synthesized CHAP resulted in profound compositional change within the apatite channel, as revealed by FTIR spectroscopy. Type AB CHAP from experiment PC55 was annealed at high-temperature [36] and hydrogen carbonate-bearing type AB CHAP from several Na-bearing experiments was aged at room temperature and 110°C [48].

As noted above, small amounts of PC55 were heated in air at 1000°C for intervals of 15 and 30 minutes and 12 and 24 hours [36]. The ν_2 region of the FTIR spectra was overlooked in the earlier study and is interpreted here for the first time (Figure 4c, d; Table 2). Only minor change is evident in the ν_2 region after annealing for 15 and 30 minutes, but there is significant change after annealing for 12 and 24 hours in the form of narrower band widths and loss of channel (types A and A2) carbonate.

Table 2. Change in relative proportions of carbonate species with annealing at 1000°C and aging at room pressure

Annealing AB CHAP at 1000°C						
	A	B	A2			
PC55	31	34	35			
PC55, 15 minute	33	34	33			
PC55, 12 hour	22	58	20			
PC55, 24 hour	25	60	15			
Aging Na-AB CHAP containing HCO_3						
	A	B	A2	HCO_3	A+A2	HCO_3
LM003						
quenched	26	39	9	27		
aged	42	37	21	0		
relative change					+28	-27
LM013						
quenched	40	32	10	17		
aged	57	33	10	0		
relative change					+17	-17
LM092						
quenched	22	48	1	29		
aged	45	45	8	2		
relative change					+30	-27

The quenched high-pressure synthesized product (PC55) has an equitable distribution of A, B and A2 carbonate ions (Figure 4c), but the carbonate in PC55 after annealing for 24 hours is dominantly type B. Interestingly, the ν_3 region of the same FTIR spectra also indicated major loss of channel carbonate [36] but with significant differences in details. Figure 3 of Fleet and Liu [36] showed that significant change occurred in the ν_3 region after as little as 15 minutes. Also the inferred loss of type A carbonate after annealing for 24 hours was insignificant. However, Fleet and Liu [36] noted that interpretation of the ν_3 region was ambiguous, especially in regard to background fitting and overlap of A and B bands in the 1470-1455 cm^{-1} spectral interval. In comparison, the ν_2 profile of the 24 hour annealing experiment (Figure 4d) clearly resolves the three carbonate ion species previously assigned to A, B and A2 carbonate in the abundance sequence B >> A > A2.

These annealing experiments on PC55 demonstrate mobility of channel carbonate species (or of their decomposition products) in relatively large type AB CHAP crystals at 1000°C. High temperature promotes reorganization and "recrystallization" of the apatite and preferential loss of type A2 carbonate which is expected to be only loosely bound in the channel. A possible mechanism for this loss is the decomposition reaction:

$$CO_3^{2-} \rightarrow CO_2 + O^{2-} \quad (4)$$

Fleet and Liu [48] reported for the first time the presence of the hydrogen carbonate ion in Na-bearing type AB CHAP synthesized at 0.1-1 GPa, 800-1350°C and high $p(CO_2)$, and reasoned that it was loosely bound in the *c*-axis structural channel because it was lost on aging for up to 367 days at room-temperature in a dessicator. Although $NaHCO_3$ is well known to have a limited thermal stability at low pressure, decomposing in air at about 85°C with the simultaneous evolution of water and CO_2 [49], it has extensive stability at high pressure [50, 51]. FTIR spectra of the high-pressure type AB CHAP investigated (LM003; Figures 2 and 3 of Fleet and Liu [48]), indicated an appreciable content of hydrogen carbonate, but this was not quantified and interesting features of the ν_2 region were overlooked. The ν_2 spectrum for LM003 has four resolved bands, three for carbonate ions (A, B and A2) and a new band at 837 cm^{-1} representing vibration of the CO_3 cluster in the hydrogen carbonate ion (Figure 9a).

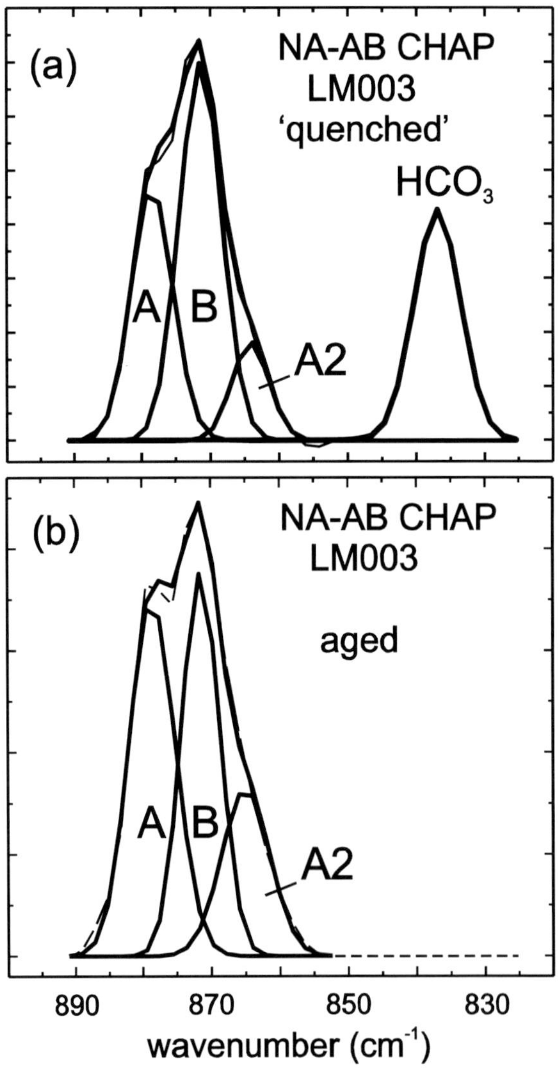

Figure 9. Out-of-plane bend (ν_2) region of FTIR spectra for Na-bearing AB CHAP (LM003) identifying band for hydrogen carbonate ion in the quenched sample (a) and loss of this band with concomitant increase in intensity of bands for other channel species (A and A2) in the sample aged at room temperature and 110°C (b) (see also Table 2).

Interestingly, the 837 cm^{-1} band disappeared completely on aging for 283 days at room-temperature and 55 days at 110°C (Figure 9b; Table 2). Note that all hydrogen carbonate AB CHAP samples presently labelled "quenched" were stored at room temperature for extensive periods (55-82 days) before FTIR study so that the actual proportions of hydrogen carbonate on quenching were no doubt greater than reported in Table 2. Also, aging conditions required to remove all hydrogen carbonate were a function of crystal size: large (100 μm) crystals required heating at 110°C to remove residual amounts of hydrogen carbonate but fine grained samples did not.

The limited results presently reported (Figure 9; Table 2) show that loss of hydrogen carbonate is marked by concomitant increase in the proportion of type A and A2 carbonate, with the proportion of type B carbonate remaining essentially unchanged. Thus, the carbonate ions formed on aging do not appear to be mobile in the apatite channel at room temperature, i.e. they do not leave the crystal system. Instead, the crystals progressively dehydrate through reactions simplified as:

$$HCO_3^- + (OH)^- \rightarrow H_2O + CO_3^{2-} \qquad (5)$$

Also, in an aqueous medium this apatite-channel system could act as a *p*H buffer, through the reaction:

$$HCO_3^- \rightarrow H^+ + CO_3^{2-} \qquad (6)$$

In summary, the present observations suggest that the hydrogen carbonate ion is incorporated in the *c*-axis channel of CHAP under the conditions of synthesis but is unstable at ambient pressure and temperature. Thus hydrogen carbonate CHAP is a potential *p*H buffer for restricted surficial geochemical environments. Perhaps more intriguingly, it seems probable that the channel constituents of biological apatites would influence acid-base reactions in bone cells and blood.

6. Apatite Biomineralization

The synthetic Na-bearing CHAP LM006 from Fleet and Liu [38] (with 3.5 wt % total CO_2) has a similar composition to the inorganic fraction of bovine bone and human and pig dental enamel (Table 3) and its FTIR spectrum (b in Figure 2) is similar in the ν_3 region to that of rodent bone mineral [23,52] and dental enamel [3,24,38]. All of these synthetic and natural materials are Na-bearing carbonate apatites, and their infrared spectra are characterized by ν3 absorption more-or-less confined to the 1400-1500 cm^{-1} spectral interval. Therefore, the structure of CHAP LM006 is probably a good approximation to the local structure of carbonated hydroxyapatite in bone and dental enamel. An important result of the Fleet and Liu [38] study is that the proportion of A and B carbonate in the synthetic Na-bearing CHAP samples (LM006 and LM005) was determined independently of infrared spectroscopy using the single-crystal x-ray structure method, and shown to contain approximately equal amounts of A and B carbonate.

Table 3. Compositions of synthetic Na-bearing carbonate apatite, inorganic fractions of bone, tooth enamel, and dentine [3], and ideal hydroxyapatite [$Ca_{10}(PO_4)_6(OH)_2$]

	Na-CHAP LM006 (wt %)	Bovine Bone (wt %)	Enamel (wt %)	Dentine (wt %)	Ideal HAP (wt %)
Ca	39.5	36.6	37.6	40.3	39.6
P	17.6	17.1	18.3	18.6	18.5
CO_2	4.2	4.8	3.0	4.8	-
Na	0.81	1.0	0.7	0.1	-
K	-	0.07	0.05	0.07	-
Mg	-	0.6	0.2	1.1	-
Sr	-	0.05	0.03	0.04	-
Cl	-	0.1	0.4	0.27	-
F	-	0.1	0.01	0.07	-

This conclusion was reinforced by detailed study of the ν2 spectral region of the FTIR spectra [31] which clearly resolves separate type A and B bands and significant amounts of type A carbonate, in good agreement with the x-ray structure results (Figures 4,5). In summary, we suggest that crystalline biological apatites appear to be Na-bearing type AB carbonate hydroxyapatites, the biological equivalent of LM006, with type A carbonate accounting for up to 50% of total carbonate.

Earlier studies [3, 24] concluded that biological apatites were dominantly type B carbonate hydroxyapatites, based on the appearance of the ν_3 region infrared bands. Recent studies [31, 38-40] suggest that the characteristic ν_3 region profile of biological apatites is the result of their Na content. The minor amount of Na in these apatites (Table 3) has a major influence on their infrared spectra because it is concentrated in the defect clusters (Figure 8). The ν_3 region is otherwise of limited utility for study of apatite extracted from bone because it tends to be obscured by absorption bands of functional groups on proteins and glucosaminoglycans [23]. However, weak and diffuse intensity beyond 1500 cm^{-1} could represent multiple channel sites for weakly bound carbonate ions, as suggested by Rey et al. [23] and now reinforced by the presence of a second channel carbonate species (A2) in Na-free AB CHAP [30,36]. Alternative explanations for the smearing and displacement of the ν_3 band intensity for the channel carbonate are multiple large-cation (Ca2) configurations around a single type A carbonate site, and destabilization of the A carbonate ion by close proximity of the O5 oxygen atom to the apatite channel wall (Figure 8).

The complex ν_2 band appears to better reflect the true proportions of the principal carbonate species in biological apatites, as demonstrated above for synthetic apatites using the results of the single-crystal x-ray structure method for reference (Figure 5). Rey and coworkers [23] reported B/A ratios from ν_2 region spectra ranging from 1.1 to 0.8 for pig enamel, and 1.4 to 1.2 for various bone samples (cow, human, chicken, rat, and rabbit) [23,26]. Based on their ν_2 region spectra, these biological apatites are all type AB carbonate hydroxyapatites. We recognize that, although the overall structure of crystalline biological apatites appears to have been reproduced, the present synthetic CHAP crystals differ markedly from bone mineral in respect to crystal size, reactivity of the surface layer, and absence of HPO_4^{2-} [23,26,53]. Also, although the apatite species of bone is generally understood to be CHAP, structurally-bound hydroxyl is commonly not detected by spectroscopic studies (including infrared, Raman, and nuclear magnetic resonance) on mineralized tissue extracted from bone [4]. However, these null results may be attributable to the procedures used for processing mineralized tissues from mammals [27, 52]. The present observations on change in the proportions of carbonate species with annealing and aging point to a possible role for the apatite channel in mediating acid-base reactions in the body [54]. The channel constituents of nano-scale crystals may account for a considerable proportion of the labile fraction of carbonate in biological apatites. The classic exchange study of Neuman and Mulryan [55] using CO_2 labelled with ^{14}C suggested that there were just two chemical forms of sequestered CO_2 in bone apatite; about 60% of the CO_2 (as carbonate) resided within the crystal lattice and 40%, presumed to be hydrogen carbonate, resided in the hydration shells of the crystals. However, the present study has shown that the A and B carbonate ions in synthetic CHAP not only have different crystal chemistries but they are functionally different in respect to reaction with fluid media. We have seen that the channel constituents could function as a solid *p*H buffer. When annealing and aging times are scaled to nano-scale dimensions this buffer could be as efficient as the postulated hydrogen carbonate-populated hydration shell. But these are questions for future study.

Conclusion

Single crystals of Na-bearing carbonate hydroxyapatite (CHAP), grown from carbonate-rich melts at high pressure and containing high contents of both type A and B carbonate ions, have similar chemical compositions and Fourier transform infrared (FTIR) spectra to the nano-scale apatite in bone and dental enamel. These natural CHAP materials are reinterpreted as Na-bearing AB carbonate apatites with channel (type A) carbonate accounting for up to about 50% of total carbonate. Deconvolution of the composite band for out-of-plane bend of the carbonate ion (ν_2) in FTIR spectra of synthetic CHAP yields proportions of A and B carbonate in very good quantitative agreement with site occupancies from single-crystal x-ray structure refinement, whereas the stronger ν_3 region absorption band is generally too complex to deconvolute. FTIR spectra of Na-free AB CHAP crystals have bands of a second and more weakly bound apatite channel carbonate species (A2). Room-pressure annealing experiments on Na-free CHAP and aging experiments on Na- and hydrogen carbonate-bearing CHAP reveal extensive mobility and reaction of channel carbonate species, with little or no loss of carbonate ions substituting for phosphate (type B), as estimated from change in ν_2 band spectra. Hence, channel carbonate species are labile under the conditions of the annealing and aging experiments and are probably labile in the natural systems as well. The Na cation, and A and B carbonate ion defects in synthetic CHAP are locally coupled to minimize the effects of charge compensation and spatial accommodation.

Acknowledgements

I thank the Natural Sciences and Engineering Research Council of Canada for financial support.

References

[1] Y. Pan and M. E. Fleet, "Phosphates: Geochemical, Geobiological and Material Importance", Reviews in Mineralogy and Geochemistry, 48, 2002 (eds: M. J. Kohn, J. Rakovan, J. M. Hughes) Mineralogical Society of America, Washington, pp13-49.

[2] R. Bros, J. Carpena, V. Sere and A. Beltritti, *Radiochim. Acta*, 74, 277 (1996).

[3] J. C. Elliott, "Phosphates: Geochemical, Geobiological and Material Importance", Reviews in Mineralogy and Geochemistry, 48, 2002 (eds: M. J. Kohn, J. Rakovan, J. M. Hughes), Mineralogical Society of America, Washington, pp427-453.

[4] B. Wopenka and J. D. Pasteris, *Mater. Sci. Engin.*, C25, 131 (2005).

[5] F. Brudevold, D. E. Gardner and F. A. Smith, *J. Dental Res.*, 35, 420 (1956).

[6] G. H. McClellan and J. R. Lehr, *Amer. Mineral.*, 54, 1374 (1969).

[7] J. M. Hughes and J. Rakovan, "Phosphates: Geochemical, Geobiological and Material Importance", Reviews in Mineralogy and Geochemistry, 48, 2002 (eds: M. J. Kohn, J. Rakovan, J. M. Hughes) Mineralogical Society of America, Washington, pp. 1-12.

[8] T. J. White, C. Ferraris, J. Kim and S. Madhavi, "Micro- and Meso-porous Mineral Phases", *Reviews in Mineralogy and Geochemistry*, 57, 2005 (eds: G. Ferraris, S. Merlino), Mineralogical Society of America and the Geochemical Society, Washington, pp. 307-373.

[9] J. M. Hughes, M. Cameron and K. D. Crowley, *Amer. Mineral.*, 74, 870 (1989).

[10] J. C. Elliott, *Structure and Chemistry of the Apatites and Other Calcium Orthophosphates*, Elsevier, Amsterdam (1994).

[11] Y. Suetsugu, Y. Takahashi, F. P. Okamura and J. Tanaka, *J. Solid State Chem.*, 155, 292 (2000).

[12] T. I. Ivanova, O. V. Frank-Kamenetskaya, A. B. Kol'tsov and V. L. Ugolkov, *J. Solid State Chem.*, 160, 340 (2001).

[13] R. M. Wilson, J. C. Elliott and S. E. P. Dowker, *Amer. Mineral.*, 84, 1406 (1999).

[14] R. M. Wilson, J. C. Elliott, S. E. P. Dowker and R. I. Smith, *Biomaterials*, 25, 2205 (2004).

[15] R. M. Wilson, S. E. P. Dowker and J. C. Elliott, *Biomaterials*, 27, 4682 (2006).

[16] Th. Leventouri, B. C. Chakoumakos, H. Y. Moghaddam and V. Perdikatsis, V., *J. Mater. Res.*, 15, 511 (2000).

[17] Th. Leventouri, B. C. Chakoumakos, N. Papanearchou and V. Perdikatsis, *J. Mater. Res.*, 16, 2600 (2001).

[18] A. Antonakos, E. Liarokapis and Th. Leventouri, *Biomaterials*, 28, 3043 (2007).

[19] T. Tonegawa, T. Ikoma, T. Yoshioka, N. Hanagata and J. Tanaka, *J. Mater. Sci.*, 45, 2419 (2010).

[20] J. C. Elliott, *Tooth Enamel* (eds. R. W. Fearhead and M.V. Stack) John Wright and Sons, Bristol, U.K. (1964), pp. 20-22.

[21] G. Bonel, *Annales de Chimie (Paris, France)*, 7, 65 (1972).

[22] R. Z. LeGeros, O. R. Trautz, E. Klein and J. P. LeGeros, *Experimentia*, 25, 5 (1969).

[23] C. Rey, B. Collins, T. Goehl, I. R. Dickson and M. J. Glimcher, *Calcif. Tissue Int.*, 45, 157 (1989).

[24] C. Rey, V. Renugopalakrishnan, M. Shimizu, B. Collins and M. J. Glimcher, *Calcif. Tissue Int.*, 49, 259 (1991).

[25] K. Beshah, C. Rey, M. J. Glimcher, M. Schimizu and R. G. Griffin, *J. Solid State Chem.*, 84, 71 (1990).

[26] H.-M. Kim, C. Rey and M. J. Glimcher, *Calcif. Tissue Int.*, 59, 58 (1996).

[27] G. Cho, Y. Wu and J. L. Ackerman, *Science*, 300, 1123 (2003).

[28] H. E. Mason, A. Kozlowski and B. L. Phillips, *Chem. Mater.*, 20, 294 (2007).

[29] H. E. Mason, F. M. McCubbin, A. Smirnov and B. L. Phillips, *Amer. Mineral.*, 94, 507 (2009).

[30] M. E. Fleet, X. Liu and P. L. King, *Amer. Mineral.*, 89, 1422 (2004).

[31] M. E. Fleet, *Biomaterials*, 30, 1473 (2009).

[32] R. C. Tacker, *Amer. Mineral.*, 93, 168, (2008).

[33] A. Peeters, E. A. P. De Maeyer, C. Van Alsenoy and R. M. H. Verbeeck, *J. Phys. Chem. B*, 101, 3995 (1997).

[34] S. Peroos, Z. Du and N. H. de Leeuw, *Biomaterials*, 27, 2150 (2006).

[35] M. E. Fleet and X. Liu, *J. Solid State Chem.*, 174, 412 (2003).

[36] M. E. Fleet and X. Liu, *J. Solid State Chem.*, 177, 3174 (2004).

[37] M. E. Fleet and X. Liu, *Biomaterials*, 26, 7548 (2005).

[38] M. E. Fleet and Xi Liu, *Biomaterials*, 28, 916 (2007).

[39] M. E. Fleet and Xi Liu, *Amer. Mineral.*, 93, 1460 (2008).

[40] M. E. Fleet and Xi Liu, *J. Solid State Chem.,* 181, 2494 (2008).

[41] M. E. Fleet, X. Liu and Xi Liu, *Amer. Mineral.*, 96, 1148 (2011).

[42] W. B. White, "The Infrared Spectra of Minerals", Monograph 4, 1974, (ed: V. C. Farmer) Mineralogical Society, London, pp227-284.

[43] M. E. Fleet, Xi Liu and X. Wang, *J. Non-Cryst. Solids*, 355, 2604 (2009).

[44] J. R. Smyth and D. L. Bish, *Crystal Structures and Cation Sites of the Rock-Forming Minerals*, Allen and Unwin, London (1988).

[45] H. Morgan, R. M. Wilson, J. C. Elliott, S. E. P. Dowker and P. Anderson, *Biomaterials*, 21, 617 (2000).

[46] P. H. J. Mercier, Y. Le Page, P. S. Whitfield and L. D. Mitchell, *J. Appl. Crystallogr.*, 39, 369 (2005).

[47] C. Ferraris, T. J. White, J. Plévert and R. Wegner, *Phys. Chem. Mineral.*, 32, 485 (2005).

[48] M. E. Fleet and Xi Liu, *Amer. Mineral.*, 92, 1764 (2007).

[49] L. Dei and G. G. T. Guarini, *J. Thermal Anal.*, 50, 773 (1997).

[50] H. Kagi, T. Nagai, K. Komatsu, T. Okada, C. Wada, J. S.Loveday and J. B. Parise, *J. Neutron Res.*, 13, 21 (2005).

[51] Xi Liu and M. E. Fleet, *J. Mineralogical Petrological Sci.*, 104, 25 (2009).

[52] S. Aparicio, S. B. Doty, N. P. Camacho, E. P. Paschalis, L. Spevak, R. Mendelsohn and A. L. Boskey, *Calcif. Tissue Int.*, 70, 422 (2002).

[53] A. L. Boskey, *Calcif. Tissue Int.*, 72, 533 (2003).

[54] D. A. Bushinsky, S. B. Smith, K. L. Gavrilov, L.F. Gavrilov, J. Li and R. Levi-Setti, *Amer J. Physiol.- Renal Physiol.*, 283, F1091 (2002).

[55] W. F. Neuman and B. J. Mulryan, *Calcif. Tissue Res.,* 1, 94 (1967).

In: Calcium Phosphate
Editor: Robert B. Heimann

ISBN: 978-1-62257-299-1

Chapter 3

STRUCTURALLY INCORPORATED WATER IN BONE APATITE: A CAUTIONARY TALE*

Jill Dill Pasteris#

Department of Earth and Planetary Sciences
and the Center for Materials Innovation
Washington University in St. Louis, St. Louis, Missouri, US

ABSTRACT

The calcium phosphate mineral hydroxylapatite [OHAp, $Ca_{10}(PO_4)_6(OH)_2$] exhibits remarkable properties from both a medical-biological and a mineralogical viewpoint. These physical and chemical properties were quantitatively evaluated over many decades of research. Even more time has been required to determine (or, at least, to infer) the physical-chemical mechanisms that control those properties, such as crystal shape, hardness, and solubility in synthetic materials, and the ability to change composition over time and to strongly adhere to collagen in the biological milieu. This chapter summarizes and analyzes the slow acceptance, particularly from the medical and dental communities, of the notion that molecular H_2O resides at the few weight-percent level both in synthetic carbonated OHAp and in bone mineral. The purpose of this review is to document the very convoluted pathway leading to the still-incomplete acceptance of structurally incorporated water in OHAp and to analyze why this recognition took so long. Some comparisons are made with related controversies over the chemical form in which carbonate occurs in bone and tooth and over the state of hydroxylation of biological apatites. This retrospective literature analysis, going back to the mid-1800s, provides an example of why one should not disregard reproducible but unexpected results simply because of the absence of a readily apparent model to account for them. The fact of structural water in apatite initially was overlooked and subsequently avoided. It currently has worked its way to the level of simply undervalued. The culmination of this story lies

* This chapter provides information on the development of knowledge specifically about biological apatite that has been referenced in Chapter 1 by Sergey Dorozhkin but could not be evaluated in much detail there. Hence these two chapters are designed to completely cover the complex, convoluted, controversial, comprehensive and labyrinthine ways in which acquisition of information and knowledge on one of the most important minerals on Earth has progressed. The Editor.

Email: pasteris@levee.wustl.edu

in the future, because there are always ramifications to structural-chemical variations in a mineral. It is time to begin to explore the effects of crystallographically incorporated water on the properties of bioapatite.

1. Introduction to Carbonated Biological and Synthetic Apatite

As reported in numerous mineralogy sources [1, 2], the mineral name 'apatite' is derived from the Greek for 'to deceive.' The textbook explanation for the nature of the deception is that samples of natural, geologic apatite can be mistaken for other minerals of similar appearance. Review of the chemical, mineralogic, and medical literature of the past one hundred or more years, however, suggests that the real deceptiveness of apatite comes from its ability to accept into its structure about half the elements in the periodic table [3,4], thereby causing substantial changes in its chemical and physical properties. The simplified mineral formula for the apatite group is $Ca_{10}(PO_4)_6(OH,F,Cl)_2$, representing its three most common natural species, i.e., hydroxylapatite (OHAp), fluorapatite (FAp), and chlorapatite (ClAp). The species most similar to that which occurs biologically in bone and tooth is hydroxylapatite. That said, there is a large and medically very important difference between biological apatite (bioapatite) and hydroxylapatite. Although many of these differences are now understood, in numerous cases after years of controversy, the model of bone apatite is still not complete.

Although in a sense it is only *details* that remain to be worked out, these very specific aspects of bone apatite could be of considerable importance to the understanding of, for instance, what changes occur in bone mineral during the ageing process and how such changes might be prevented or slowed, and what parameters of the mineral and the process of its fabrication might be tweaked for the purpose of producing better biomaterials.

The present chapter begins with an historical overview of the many steps to our current understanding of the presence of structurally incorporated water in bone apatite and in its synthetic carbonated apatite analogs. The conclusion of this documentation is that the notion of structural water in bioapatite still has not pervaded the major literature in the medical, dental, and related fields. A retrospective analysis follows, concerning how this important topic has remained as more of a backwater discussion than a raging controversy for well over one hundred years.

The associated research activities and minor debates summarized below illustrate the importance of models (as well as their limitations), demonstrate temporal changes in the notion of what constitutes interdisciplinary research, and suggest future pathways to enhanced understanding of bone and tooth tissue.

2. Setting the Stage: One Controversy after Another

Apatites follow the compositional template $X_4Y_6(PO_4)_6Z_2$, which in hydroxylapatite, OHAp, becomes $[Ca_{10}(PO_4)_6(OH)_2]$. Among calcium phosphate phases, apatites are recognized by their anion-filled channels oriented along the *c*-axis of the structure (see Figure 1). Those six of the ten calcium ions that define the boundary of the channels are labeled Ca(2), and the other four calcium ions within the structure are labeled as Ca(1).

One of the earliest revelations about *bio*apatite was that, in addition to phosphate, it also contains carbonate. Even within the past 60 years, however, there has been debate about whether the carbonate in the bulk bone material represents the coexistence of the calcium carbonate mineral calcite ($CaCO_3$) together with apatite or whether the carbonate is a constituent of the apatite phase [5-7]. It was ultimately established that carbonate is an actual substituent within the crystalline structure of the apatite [8], but that fact did not fully permeate the medical literature until many years later (see debate in [9,10]).

Subsequently, multiple research groups worked to show that bone apatite incorporates about 5-8 wt% carbonate $(CO_3)^{2-}$ and that the carbonate dominantly substitutes for phosphate $(PO_4)^{3-}$ tetrahedra in the crystal lattice, a substitution labeled as type B [11-13] [1]. From the viewpoint of the modeling of crystal behavior, one should note that this is actually an unlikely substitution. The phosphate and carbonate atomic clusters have greatly different sizes and shapes (3-D tetrahedral vs. 2-D triangular, respectively), as well as different ionic charges.

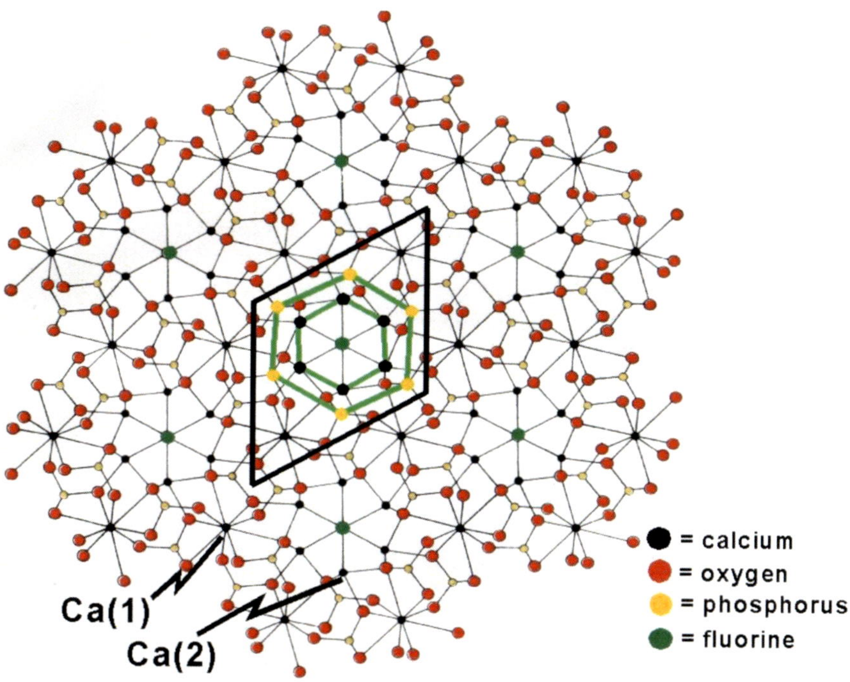

Figure 1. Atomic structure of the fluoride analog to hydroxylapatite, i.e., fluorapatite, $Ca_{10}(PO_4)_6F_2$, in which one unit cell is outlined with a black rhombus. The site for the channel anion (central green, filled circle), e.g., OH^- in OHAp, is outlined in a small hexagon. The black, filled circles at the apices of the inner hexagon represent the channel-defining Ca(2) cations. Ca(1) cations do not share any bonds with the channel anions. Figure from [40] (their figure 5), with some additions. (Reproduced with kind permission of the Mineralogical Society from a paper by Beevers and McIntyre (1946)).

[1] For more detailed information on carbonate substitution see Chapter 2. The Editor.

A much less common incorporation mechanism in bone mineral is $(CO_3)^{2-}$ substituting for OH^- in the channel sites of apatite, referred to as type-A substitution, which also necessitates additional charge balancing [11]. One therefore can understand some reluctance in the initial acceptance that large amounts of $(CO_3)^{2-}$ substitute into apatite, even though carbonate does not even appear in the OHAp formula. Intriguingly, there was an even longer period of uncertainty about the exact concentration of the hydroxyl ion (OH^-) in bone mineral, despite hydroxyl's explicit appearance in the formula of OHAp. Bone apatite's deficiency in hydroxyl compared to pure OHAp was first realized more than 35 years ago [14,15] and experimentally explored in synthetic calcium phosphate phases shortly later [16]. In spite of these data, a long-extended controversy later ensued concerning exactly how much/little hydroxyl ion (OH^-) exists in bone mineral and what controls its concentration [17-19]. The most recent analyses indicate that bone mineral contains only about 15-20% of the hydroxyl content in pure OHAp [20,21], i.e., about an 80-85% hydroxyl deficiency. To a large degree, the deficiency in hydroxyl can be explained by the need for charge balance as CO_3^{2-} ions substitute for PO_4^{3-}, as represented by the formula $Ca_{10-x}[(PO_4)_{6-x}(CO_3)_x](OH)_{2-x}$ [12,13,17,19,22,23]. As a result of these chemical-structural adaptations, bone mineral typically does not have a Ca:P ratio of 1.67. It also is hydroxyl-deficient, as recorded in a hydroxyl (OH^-) ion content of less than 2 per crystallographic stoichiometric unit cell. There is, however, yet another complexity in comparing bioapatite with pure OHAp. For over 120 years, chemical, geological, and medical researchers sometimes have argued, but more often simply dithered, about whether any of the molecular water repeatedly shown to be associated with both synthetic and bone-derived carbonated apatite is actually part of the latter's crystallographic lattice or just adsorbed on the mineral's surface. In summary, analyses and discussions that have spanned many decades of time, multiple scientific fields, and several generations of ever-improving analytical instrumentation somehow have never gained community-wide recognition of the fact that bone mineral contains as essential hydrogen-bearing components not only the hydroxyl ion (OH^-), but also molecular water (H_2O), in amounts as great as a few weight percent.

3. The Story that Would Not End: What about the Water?

'Modern' scientific investigation of bone mineral began about 150 years ago with the recognition that the composition of the solid phase in bone is dominated by calcium and phosphate [24] but that it also contains carbonate [25-27]. It was about 100 years later, however, before most researchers in the field could be said to agree that the solid, i.e., mineral, portion of bone was a *single crystalline* compound and that the compound could be broadly labeled as apatitic [28] (compare debate between [10] and [9]). This chapter is organized so that one may either read the complete decade-by-decade summary below or simply read the highlights of the research up through 1969 in Table 1 and then re-rejoin the summaries of modern post-1970 research concerning water in apatite.

Interest in the water content of whole bone and the perception that some portion of the water was contained in the bone crystals themselves began in the 1800s (e.g., [25-27]; also see the excellent summary, in German, by Gabriel [5]).

Table 1. Brief Overview of the History from 1842 to 1969 of the Analysis and Interpretation of Bone Mineral with Special Reference to its Water Content

Reference	Paper type*	Some specific contribution of data, interpretation, or explicit formula for bone mineral
Compiled by Hendricks and Hill 1942		Some published formulas for the inorganic component of bone. Berzelius (1845): $Ca_5H_2(PO_4)_4$; Gassman (1937): $Ca_{10}(PO_4)_6CO_3$; Gruner et al (1937): $(Ca,C)_{8\text{-}9}[(P,C)O_4]_6Ca(OH)_2$; Klement (1938): $Ca_{10}(PO_4)_6(OH)_2$
1840s: Books present new bone analyses and summarize older bone analyses carried out over several previous decades.		
von Bibra 1842 Book *Chemical Investigations on the Bones and Teeth of Humans and Vertebrates* (in German)	E/A	Remarkable work providing detailed chemistry of whole bones and of bone and tooth mineral of many animals and humans. Reported values for CaO, P_2O_5, $CaCO_3$, Fe, F, K, and other trace elements. Sketches of bone cross-sections as seen under microscope.
Gmelin 1849 Book *Handbook of Chemistry, Vol. III, Metals*	R	Analyses compiled on *bone ash*. Berzelius is cited as stating that this "salt" is 8CaO, $3PO_5$
1870s: Analysis and interpretation of bone has become more common, and the approaches and results are more sophisticated and well documented.		
Maly and Donath 1873 Contributions to the chemistry of bones (in German)	E/A	Investigation of solubility of bone mineral and other Ca phosphates in various solvents. Bone mineral has same solubility as synthetically precipitated Ca phosphates. Bone powder is extremely hygroscopic.
Wibel 1874 (in German) The constitution of bone phosphates, especially the existence and formation of a basic compound; ($Ca_3P_2O_8$ + xCaO)	R	Summary of the Ph.D. thesis and related work by Aeby. Strongly discredited Aeby's analytical approach and, thereby, his proposed formula for bone mineral as $[6Ca_3P_2O_8 + 2H_2O + 2CaO + CO_2] + 3H_2O$, which Wibel described as calcium phosphate + carbonic acid + basic water + water of crystallization (author's translation of p. 114).
Aeby 1875 (in German) On the chemistry of bones	R	Results of several approaches to investigating content of water in whole bone and bone mineral. Proposed more than 1 wt% H_2O as water of crystallization in bone mineral. Focused on bones "naturally calcined" by fossilization, which contain about 12 wt% H_2O (including water released below 100°C).
Watts 1875 Book *A Dictionary of Chemistry and of the Allied Branches of Other Sciences*	R	Report on the composition of geological francolite [carbonate fluorapatite] with 1.5 wt% H_2O. Summarized Aeby's analyses on fossil bone as showing that both CO_2 and water of crystallization can be expelled by heating to 450°C.
1890s: Analyses of both bone and other calcium phosphate phases.		
Gabriel 1894 (in German) Chemical investigations on the mineral components of bones and teeth	R, E/A	Summary and evaluation of bone mineral analyses of previous 50 years, plus Gabriel's own work. Gabriel chemically extracted organic components from bone before doing controlled heating experiments, which revealed about 2 wt.% crystallographically incorporated H_2O in bones and teeth.
1930s: Still much confusion about the nature of bone mineral. Mounting evidence for H_2O in bioapatite.		
Schleede et al 1932 (in German) On the understanding of calcium phosphates and apatites	E/A	Weight loss from 200 to 900°C of carefully prepared (at 100°C) OHAp was 2.5 wt%. Water was almost continuously released over this interval. Water believed to be strongly bound in the crystal. Channels large enough for both OH^- ions and H_2O molecules.

Table 1. (Continued)

Reference	Paper type*	Some specific contribution of data, interpretation, or explicit formula for bone mineral
Henricks et al 1932 The crystal structures of some natural and synthetic apatite-like substances	E/A	Report of "structural reasons for supposing ...that the phosphate compound of bone is hydrated" (p. 369). Bone mineral = $Ca_{10}(CO_3)(H_2O)(PO_4)_6$, p. 368.
Huggins 1937 The composition of bone and the function of the bone cell	R	Excellent summary of still-ongoing debate about WHAT bone mineral is. X-ray diffraction is making a huge contribution to this discussion. Huggins did not think bone mineral is a single compound of definable composition, but he believed water is present and held firmly in bone mineral.
McConnell 1938 The problem of the carbonate-apatites: A carbonate oxy-apatite (dahllite)	R	Report that geological, carbonated fluorapatites released approx. 1-2 wt% H_2O during heating at 110-300°C. McConnell believed that it is important to evaluate H_2O in the structural context of $Ca_{10}(PO_4)_6(OH,F)_2$ and that both OH^- and H_2O can be accommodated in the apatite channels.
1940s: Increasing research to link the chemistry (incl. H_2O) together with the actual structure of apatite.		
Eisenberger et al 1940 The basic calcium phosphates and related systems	R	Summary of the contributions of both chemists and mineralogists. Recognized the remarkable flexibility of apatite structure in accepting chemical substitutions.
Henricks et al 1942 The inorganic constitution of bone	E/A	Current (cf. 1932) support for a bone mineral formula of $[Ca_{8.50}Mg_{0.25}Na_{0.19}]\ [(PO_4)_{5.07}(CO_3)_{1.24}](H_2O)_{2.0}$. Provided structural diagrams showing how Ca^{2+} and $2F^-$ can be substituted by $2H_2O$.
1950s: Increasing contributions from dental researchers; hydration shell recognized to enclose apatite; more structural modeling of where H_2O resides.		
McConnell 1952 Problem of the carbonate apatites	E/A	Report that even when all anion sites in FAp are filled with F^-, channels still contain additional OH^- and H_2O. Proposed that H_2O substitutes for Ca(2) ions, and $(OH)_4$ substitutes for (PO_4).
Neuman et al 1952 Surface chemistry of bone	E/A	Report that "...considerable amounts of water are intimately associated with the crystals of hydroxylapatite and of bone" (p. 4239). Hydration shells surround apatite crystallites, being strongly held by their surfaces.
Burnett and Zenewitz 1958 Studies of the composition of teeth: VII The moisture content of calcified tooth tissues	E/A	"In spite of the many studies of the composition of human calcified tissues, their moisture content has neither been investigated extensively nor established conclusively" (p. 581). Measured moisture release in tooth materials up to 100 and 197°C. Not all such water could be restored during (even lengthy) rehydration.
Reference	Paper type*	Some specific contribution of data, interpretation, or explicit formula for bone mineral
Neuman and Neuman 1958 Book *The Chemical Dynamics of Bone Mineral*	R	Excellent summary and enlargement upon previous research on bone mineral, with insights into questions still to be answered and experiments to be done. "...composition of most apatite preparations does not agree with the composition predicted by the assumed structure...lose significant quantities of water in prolonged heating" (p. 41).
1960s: Additional and more sophisticated XRD analysis of enamel; NMR and IR analyses come into their own.		
Little and Casciani 1966 The nature of water in sound human enamel	E/A	Application of NMR to define 3 types of water in enamel, released 1) up to 100° C, 2) 100-900°C, 3) 900-1300°C. #2 and 3 types are both "caged" but "freely tumbling" water and, thus, "unbound."

Reference	Paper type*	Some specific contribution of data, interpretation, or explicit formula for bone mineral
Brophy and Nash 1968 Compositional, infrared, and X-ray analysis of fossil bone	E/A	Heating experiments at 1250 or 1400°C that released up to 6 wt% H_2O in fossil bone samples, i.e., carbonated (~3 wt% CO_2), fluoride-rich (~3 wt% F) apatite.
Simpson, 1968 Substitutions in apatite: Potassium-bearing apatite	E/A	Report that apatites synthesized at <100°C typically are Ca-deficient and water-rich. Can replace up to half of OH^- with H_2O. Recorded 4.5-6.5 wt% H_2O. H_2O incorporation accounts for increase in a-dimension of unit cell.
LeGeros et al 1969 Two types of carbonate substitution in the apatite structure	E/A	Meticulous XRD and IR analyses of synthetic carbonated OHAp. Can only substitute CO_3 for OH (in channels) in the absence of H_2O. "Entrapped water" in the channels could explain small increase in a-dimension of unit cell.
Paulik et al 1969 Investigation of the composition and crystal structure of bone salt by derivatography and IR spectrophotometry	E/A	Detailed TGA and IR analyses on crystallites from natural bone that had been defatted and deproteinated. Called bone mineral a "water- and carbonate ion-bearing so-called 'hydroxylapatite of non-stoichiometric composition'" (p. 421). IR monitoring of loss of much structural water by about 600°C.
Elliott 1969 Recent progress in the chemistry, crystal chemistry and structure of the apatites	R	Report of three conferences in 1967-68 focused on apatites. Recoreded need to understand apatites more generally in order to understand bioapatites better. Progress being made in growth of single crystals. Record of water in non-stoichiometric hydroxylapatites. Review of NMR data on different types of water in calcified tissues.

*Paper type: R = review paper; E/A = experimental and/or analytical paper.

Although the intimate relation between water and mineral already was recognized in the 1800s, the complex internal architecture of the whole bone and the extremely small sizes of its mineral and organic components clearly were not understood. This lack of detailed knowledge did not prevent several of the earliest researchers on bone from reporting, with obvious surprise, the extremely strong hygroscopic affinity of bone, especially of pulverized, clean bone material [27, 29].

4. A Short History of Water in Inorganic and Biological Apatites

4.1. Most of What We Needed to Know Was Revealed by the Mid-1890s

One of the major breakthroughs in the understanding of the complexity of bone mineral was made by German chemist Carl Aeby, whose several related publications in the early to mid-1870s also started a controversy concerning the appropriate means and samples by which to analyze the mineral component of bone [30]. Until Aeby's work, many of the published so-called bone analyses actually were made on bone ash, which is the residue after heating bone to at least several hundred degrees Celsius. Aeby [27] analyzed fresh bone, however, finely pulverizing it and applying a number of tests to establish its water content. He recognized not only the strongly hygroscopic nature of bone powder but also that the attraction of water increased as the mineral concentration in the bone increased. In addition, there was a rise in temperature as the bone powder took on water. Aeby viewed the latter response as evidence for a chemical interaction between the bone mineral and water. He also observed that the bone powder could not be fully desiccated at temperatures of even 130°C. Among his other insights, Aeby [27] also realized the advantage of being able to work with bone mineral that was free of organic matter. To his credit, he declined using bone ash, which he viewed as compromised by intense heating. Instead, he looked to bone material that was naturally devoid of organics, namely fossil material (on which he had written his doctoral dissertation). Aeby [27] acknowledged that one could not claim a quantitative understanding of bone mineral by analyzing fossil bone, which he referred to as metamorphosed. He did believe, however, that a better qualitative understanding of fresh bone mineral could be derived from such work.

One of his fossil bone samples released approximately 12 wt% water during heating to about 500°C. Aeby's [31] recognition that fresh bone samples had ~10 wt% of what he called "hygroscopic moisture" led him to understand that it was about 2 wt% rather than 12 wt% water that was of mineralogic interest. [It should be noted that these water values were specified to two places to the right of the decimal.] Aeby was fully convinced that actual 'crystal water', i.e., water of crystallization, was chemically bonded with the calcium phosphate mineral in bone. He [31] reported the composition of the mineral in bone to be [6 $Ca_3P_2O_8$ + 2 H_2O + 2 CaO + CO_2] + 3 H_2O (as formally expressed by [30]), thereby distinguishing two types of water.

Aeby was just one of several German chemists in the late 1800s who were analyzing bone material. Interestingly, it was a fellow German analyst of fresh and fossil bone material, F. Wibel, who most strongly took Aeby to task for the latter's conclusions about the chemical nature of bone material. In doing so, Wibel [30] both succinctly summarized Aeby's work and brought it to the attention of the English-speaking scientific community, where both men's work was summarized in such publications as The Journal of the Chemical Society [of London] in its section on Abstracts of Chemical Papers Published in Other Journals. Wibel [30] concentrated on critiquing two aspects of Aeby's analytical results: (1) the 10:3 molar ratio of CaO:P_2O_5 (i.e., 10:6 atomic Ca:P ratio) rather than the 3:1 ratio reported decades earlier by the well respected Swedish chemist, J.J. Berzelius, who published the relationship 3 CaO + P_2O_5 as the composition of bone ash (apparently equated with the composition of fresh bone mineral by many), and (2) the proposal of water and carbonate as explicit chemical components of bone mineral. Wibel questioned Aeby's strong reliance on heat-treatment (similar to modern-day thermogravimetric analysis) of *fossil* material, whose mineral component clearly could have changed in composition from that of fresh bone.

Dr. S. Gabriel, a later German chemist wrote an extensive publication 'Chemical investigations into the mineral component of bones and teeth' ([5], in German), which is regarded by some as the first 'modern' scientific article on that subject. The early analyses, which focused on CaO, MgO, P_2O_5, and CO_2, were done almost exclusively on ashed bone, as indicated above. Gabriel, just as Aeby twenty years earlier, realized the importance of isolating the bone mineral without heating the sample to vaporize the organic matter. Heating had always made it difficult to determine the carbonate concentration in the mineral, but, Gabriel further pronounced, it would make it impossible to evaluate the presence of "chemically bound water" (p. 274), which hardly anyone had experimentally undertaken to analyze.

Gabriel adopted the use of a 200°C mixture of glycerin and potassium hydroxide to strip the organic components from the bone, which then was further cleaned in boiling water. The residual bone showed no sign of darkening when heated, indicating its lack of residual hydrocarbons, but it was extremely hygroscopic. To remove the adsorbed water, Gabriel heated the material to 130°C and held it at temperature long enough to stabilize its weight. Additional weight loss was monitored as the sample was heated further to about 350°C. When the latter heating step was carried out in a small glass tube, the released volatiles could be condensed and shown to be water. Gabriel indicated that such water "must be regarded as chemically bound" (p. 276) to the apatite. Upon stepwise heating, the evolution of water alone was followed by that of carbon dioxide.

Gabriel was pleased that the notion of chemically bound water in fresh (unfossilized) bone had raised considerable interest, but he also was intrigued by Aeby's [27] report of water in elephant ivory. Using the same bone and tooth samples of his own choosing, Gabriel [5] compared the amount of chemically bound water between (1) the residue of his glycerin-treated bone and the typical ash that is a residue of bone heating and (2) between a bone sample and an enamel sample. In all four cases, the amount of bound water was found to be about 2.25 wt%. Subsequent analyses showed more variability in water content among bones from different animal species.

A recurring concern was that even the most trustworthy analyses of actual bone mineral typically summed to somewhat less than 99 wt%. The question was whether this deficit was a specific, but unrecognized component. The long-sought answer to the deficit problem was eventually found by Gabriel [5] when he determined through heating to very high temperature that the bone mineral also had a considerable hydroxyl (OH^-) component. He referred to the latter as 'constitutional water' to distinguish it from the 'crystal water' (water of crystallization) referred to above as chemically bound water. Gabriel believed his water analyses to be less accurate and reproducible than his determinations for other components in bone mineral, but estimated that the chemically bonded water amounted to about 1 mole of water of crystallization per mole of bone mineral. Thus, by 1894, it was, in principle, known that the mineral in bone (as well as in dentin and enamel) is a crystalline, hydrated, hydroxylated calcium phosphate phase.

Another important aspect of this time period is that information pertinent to bone mineral began to appear in the literature of other scientific fields. For instance, Gabriel [5] implied the usefulness of studying geological materials when he correctly concluded that natural phosphates, phosphorite rocks, and geological apatite were compositionally very similar to bone mineral. Watts [32] already had summarized earlier literature on the 1.5 wt% H_2O measured in a geological form of carbonated fluorapatite called francolite. Förster [33] followed with an in-depth chemical study of the calcium-phosphate system in order to better understand a process used in steelmaking to eliminate unwanted phosphorus from molten iron ore.

Researchers thus had defined remarkably well what bone mineral is – at least chemically – by the mid-1890s. Yet, over 100 years later, no review article on bone mineral (e.g., [13, 19, 34, 35]) listed molecular H_2O as an intrinsic component of bioapatite. As active, respected members of the community of bone and tooth researchers, the above authors were reflecting the apparent general rejection of the concept of water as an intrinsic component in bioapatite or the apparent widespread judgment that any such water was of little consequence.

4.2. Moving through the 1900s

The application of X-ray diffraction (XRD) to minerals in the early 1900s opened up the way for structural analysis of members of the apatite group. By the mid-1920s, the crystallographer de Jong [36] had used XRD on bone to irrefutably demonstrate that its mineral component was a type of apatite. Náray-Szabó [37] and Mehmel [38, 39] in the early 1930s had characterized fluorapatite crystallographically. Their work was followed 15 years later by Beevers and McIntyre's [40] small changes to fluorapatite's unit cell parameters and the exact positioning of phosphate in the structure, as reproduced in Figure 1 of this chapter. Meanwhile, the German chemists continued to play an active role in exploring the various phases in the CaO-P_2O_5-H_2O system, their solubilities in a variety of organic and inorganic acids, and the conditions under which they are stable, as well as in comparing them with the now-recognized mineral hydroxylapatite [41-43]. In their careful experiments, supported by XRD analysis of the solids, Schleede et al. [42] carried out thermogravimetric analyses (TGA) on hydroxylapatite synthesized at 100°C. During subsequent heating from 200 to 900°C, a total of approximately 2.5 wt% water was released almost continuously. Based on the very high release temperatures, the authors concluded that this water was strongly bound within the apatite. Schleede and colleagues also demonstrated that their carefully synthesized OHAp was thermally stable up to 1500°C. Inorganic chemist Klement [44] subsequently published his very meticulous analyses on bone mineral from a wide range of vertebrates, focusing on the components Ca, PO_4, and CO_3, especially their molar ratios. There was no discussion of water in the mineral, but repeated reference to hydroxylapatite's stoichiometric formula of $Ca_{10}(PO_4)_6(OH)_2$.

Apatite research was also being carried out on several fronts in the United States at this time. Soil and fertilizer chemists Hendricks et al. [45] were analyzing and comparing the properties of geological and synthetic apatite-related substances with a view toward understanding the most useful phosphate sources for agricultural fertilizer. They proposed that bone mineral was a member of a solid solution series of OHAp that has the formula $Ca_{10}CO_3(PO_4)_6 \cdot xH_2O$, in which x most likely is 1. Medical researcher C. Huggins [46] wrote a review article addressing the composition of bone, in which he accepted the hydrated formula proposed by Hendricks et al. [45], in part due to the obvious retention of water by apatite to temperatures of hundreds of degrees Celsius. Interestingly, though, Huggins was not sure that the solid component of bone consisted of only a single phase, and he therefore referred to the 'calcium phosphate complex' (p. 124).

In the 1930s, American mineralogist Duncan McConnell also began writing what would become an extensive list of articles on apatite, published in a wide range of scientific journals. His work, although shown later to be flawed in several significant ways, played an important role in changing the view of bone apatite and related carbonated apatites from that of merely *chemical* entities. In his papers, McConnell brought the most meticulous XRD analyses together with the highest quality chemical analyses (e.g., [1]). In doing so, McConnell forced several scientific communities to consider the structural as well as chemical implications of the crystallographic incorporation of carbonate into the apatite phase. Many of his samples at this time were geologic in nature, but his later work explicitly involved bone mineral and tooth enamel.

Even in his 1938 paper, McConnell already had identified some of the complications in the analysis for water in apatite that would plague researchers for decades to come. (1) Even excellent analysts who made TGA measurements on samples apparently of very similar material could derive water concentrations that differed by 100% or more from each other's. (2) It matters greatly which temperature intervals are used to distinguish between [as McConnell, 1938, defined them] a) hygroscopic water [released below 110°C], b) *OH-ions* released as H_2O [above 300°C], and c) *water* released that "is very firmly held by superficial forces but probably is not present as OH-groups in the crystal" (between 110 and 300°C, p. 301). (3) The choice of 300°C is quite arbitrary to distinguish the origin of different aliquots of released water. In any event, the amounts of water released at 110-300°C from his carbonated apatites ranged from 1.20 to 2.10 wt% [1], marking water as a clearly significant component.

Moreover, McConnell [1] sought to determine not only where in the structure the CO_3^{2-} ions resided (a contentious and complex issue in its own right), but also where the H_2O molecules resided. For the latter, he focused on the channel sites, but unfortunately did not clearly distinguishing between those TGA-derived H_2O contents that are actual water molecules in the structure and those that represent OH-ions. For the latter determination, McConnell should have selected a much higher heating temperature than 300°C for his lower bound.

In several of his papers, McConnell clearly stated that he did not believe apatite could be stable if there were large numbers of vacancies in the channels. This notion is both intriguing crystallographically (i.e., how *do* channels remain defined if they lack the contents to support them?) and very significant to the development of models for how water *could* be incorporated in apatite. For one of his carbonated apatite samples, McConnell [1] was concerned that "it does not contain sufficient water to fill the fluorine positions...[and even if all detected water] were assumed to enter the structure as hydroxyl ions there would still be a significant deficiency [less than complete fill of all the channel sites]..." (p. 301). Both the ideal formula $Ca_{10}(PO_4)_6(OH)_2$ and structure of OHAp had become parts of a model to which to adhere. From this time forward, others would try to assign sufficient anions and water molecules to completely fill the two channel sites per formula unit.

By the 1940s, Hendricks had investigated the chemical complexities of bone mineral further, arriving at a chemical formula that indicated more solid solution and a higher state of hydration than his previous formula, i.e.,

$[Ca_{8.50}Mg_{0.25}Na_{0.19}][(PO_4)_{5.07}(CO_3)_{1.24}](H_2O)_{2.0}$ [47].

Meanwhile, chemical engineers Eisenberger et al. [48] had further investigated the basic Ca-phosphates, as distinguished from acidic $(HPO_4)^{2-}$ and $(H_2PO_4)^{-}$ phases, with a view toward both theoretical and practical applications. They especially emphasized the crystallographic conclusion of Schleede et al. [42] that the structural channels in OHAp were of sufficient size to accommodate *both* their full complement of hydroxyl ions and water molecules. The stage appeared to be set for more detailed crystallographic investigation of the structural site of water molecules.

The 1950s brought more detailed structural models for the accommodation of water molecules in apatite as well as a more sophisticated view of the 'hygroscopic water' that had been recognized since the 1800s. Thus, McConnell [8] presented crystallographic diagrams of apatite (somewhat analogous to Figure 1 in this chapter) in which H_2O molecules were shown partially replacing Ca ions, especially those in the Ca(2) sites that define the channels in the apatite structure. The latter vacancies (absences of Ca^{2+}) were necessitated by a charge imbalance created when $(CO_3)^{2-}$ ions replaced $(PO_4)^{3-}$ ions in carbonated apatite. In addition, he proposed that some $(PO_4)^{3-}$ ions were substituted by $[(OH)_4]^{4-}$ ions. These specific sitings and O-H speciations have not withstood the test of time. More importantly, however, they were proposed in response to actual analyses that showed the presence of water in apatite.

Medical researchers W.F. Neuman and colleagues [49] meanwhile had done several meticulously controlled experiments on synthetic hydroxylapatite (unfortunately, not further described or characterized) and on powdered, de-fatted bone. Among other analyses, they did high-speed centifugation studies on synthetic OHAp in an aqueous slurry and on similar OHAp in which the aqueous phase had been isotopically spiked. The spiking allowed them to determine that the water held by the crystals does not contain the solutes of the bulk solution. Despite similar grain sizes, the particles of fresh bone became hydrated to only 40% of the level acquired by the synthetic OHAp. Based on their data, Neuman et al. [49] proposed that bone crystallites are encased in a hydration shell, much like that on natural, geological clay particles (typically less than 1 micrometer in diameter). These ideas about the surface chemistry of bone mineral were further developed and formalized in Neuman and Neuman's (1958) fine book *The Chemical Dynamics of Bone Mineral* [50].

By the end of the decade, however, dental researches Burnett and Zenewitz [51] still felt compelled to report: "In spite of the many studies of the composition of human calcified tissues, their moisture content has neither been investigated extensively nor established conclusively" (p. 581). Fortunately, the challenge of investigating water in bioapatite prompted the dental community to focus its research on enamel, which is a moderately close analog to bone. Enamel's larger and better crystalline mineral phase, however, made it more amenable than bone crystallites to chemical and structural analysis [51].

The 1960s brought a huge increase in our knowledge about the nature of bone and enamel apatite, through experiments in synthesizing apatites under controlled conditions and through the application of different and more sophisticated analytical techniques. Several important papers were published by independent dental research groups. Marguerite Little and colleagues from the Eastman Dental Center in Rochester [52, 53] specifically addressed the issue of water in enamel. Through a combination of TGA and nuclear magnetic resonance (NMR) spectroscopy, Little and Casciani [53] identified three types of water based on their analyses: (1) Water released up to 100°C, which they called unbound, freely tumbling H_2O; (2) water released at 100-900°C, 0.8 moles H_2O based on TGA, interpreted from NMR spectra as caged but not really 'bound' in that it still was freely tumbling; and (3) water released at 900-1300°C, which they identified as 2 moles of H_2O structurally sited where others had claimed that OH resides. They also labeled the latter as caged, but freely tumbling H_2O. Little et al. [53] ultimately "speculated that enamel apatite and synthetic apatites may not be a hydroxy- but a hydro-apatite" (p. 565). It is unfortunate they did not recognize that the H_2O released above 900°C actually represented dehydroxylation of OHAp in which hydroxyl ions were released from the apatite channels in the form of H_2O. They did, however, identify 0.8 moles of actual molecular H_2O that was structurally incorporated in enamel apatite. Between the two studies referenced above came a short publication by Myers [54] on water trapped in enamel. In comparing proton NMR analyses on bioapatite samples dried at 110°C, Myers found a strong, relatively narrow peak for H_2O in enamel, even after heating it to 200°C, but only weak and very broad water peaks for dentin and bone.

Heating enamel to 500°C did eliminate the peak. Myers concluded that the persistence of such a narrow water peak to temperatures well above those that remove adsorbed moisture demonstrated this water to be " 'trapped' rather than bound in the structure" (p. 713). A larger-scale NMR study on enamel and dentine by Myrberg [55] for the most part confirmed the findings of Little and Casciani and of Myers [53, 54]. The above NMR papers are based on wide-line ^1H NMR spectroscopy.

Biophysicists Carlström et al. [56], whose work had stimulated Myers [54] to carry out his investigation, believed that their optical microscopic analyses of the (changes in) birefringence of tooth enamel under controlled conditions indicated approximately 3.3 wt% 'firmly bound' H_2O in enamel, a result that they said confirmed Little et al.'s [52] work. Dental researcher Raquel LeGeros and colleagues in 1969 published one of the first of their many important contributions to our knowledge of enamel and bone mineral [11]. This particular paper focused on new determinations for two, specific structural sites that can be filled by carbonate ions as they substitute into the OHAp structure, now known as type A and type B. The paper is noteworthy for two additional contributions. The authors combined XRD and infrared (IR) spectroscopic information on carefully synthesized carbonated hydroxylapatite samples that had been heated to specific temperatures, thereby correlating the application of three analytical techniques. In addition, they hypothesized, that in apatites precipitated from solution, a slight increase (over geological, almost pure OHAp) in the *a*-axis length of the unit cell could be due to trapped water in the structure. This hypothesis became an accepted criterion in future years for inferring the presence of water in OHAp. They also observed that the (type-A) substitution of CO_3 for OH (in the channels) only took place if the synthesis was done in the absence of water.

The geologists of the 1960s were not to be outdone in the realm of new developments. Brophy and Nash [57] also applied XRD and IR analysis to heated samples, in this case fossil bone. They recorded about 5 wt% H_2O. Their work was an interesting extension of Aeby's heat-treatment analyses of fossil bone in the mid-1870s. Dale Simpson wrote a series of papers in the 1960s about carbonated apatites and alkali-bearing apatites that he synthesized at approximately 100°C and subsequently heated to 900°C. He reported [58] that these low-temperature apatites synthesized in aqueous solutions were both Ca-deficient and water-rich. Based on a 'modified Penfield method' of analysis (see [59]), his potassium-bearing apatites yielded 4.5-6.5 wt% H_2O, the structural incorporation of which he believed was the cause of the enlarged a-axis value in their unit cell. In other sodium- and potassium-bearing carbonated apatites, Simpson [60] found that a total of 11 wt% H_2O was continuously evolved from room temperature to 900°C. Some of this water clearly must have been adsorbed, but a large proportion of it would appear to have been structurally incorporated. Simpson seems to have adopted McConnell's [61] model of bone mineral, in that he proposed the water in his samples to occur as H_3O^+ ions substituting for Ca^{2+} and $(H_4O_4)^{4-}$ ions substituting for $(PO_4)^{3-}$ groups. In characterizing the rather rare geological occurrences of hydroxylapatite, McConnell [62] stated: "On the basis of the premise that all structural positions of the apatite are filled (i.e., there are no 'holes') the structural formula becomes $[Ca_{10}[F_{0.2}(OH)_{1.5}(H_2O)_{0.3}][(PO_4)_{5.7}(H_4O_4)_{0.3}]$" (p. 425), in this case placing molecular H_2O in the channel sites after calculating full occupancy of calcium in its two sites.

Analytical chemists Paulik and colleagues [63] applied IR analysis to deproteinated natural bone in conjunction with TGA and differential thermal analyses (DTA). From their results, they concluded that "bone salt may be considered as a water- and carbonate ion-bearing non-stoichiometric hydroxylapatite" (p. 418), or more specifically, "a transitionary substance between pure hydroxylapatite, carbonateapatite [sic!], and octocalcium [sic!] phosphate" (p. 421). Their analyses, done on materials that had been surficially dehydrated in methanol, yielded 5.4 wt% of what they labeled as structural water. Paulik et al.. [63], however, seemed as confused as several chemists before them concerning the final release of 1.8 wt% H_2O at temperatures over 1000°C, still referring to this reaction as the dehydration of apatite rather than as the destabilization of OHAp as it fully dehydroxylates to form a mixture of $Ca_3(PO_4)_2$ and $Ca(OH)_2$ (the latter probably formed initially as CaO, which hydrates to $Ca(OH)_2$ very readily in air [18]).

James Elliott, a crystallographer who would focus his attention on natural and synthetic apatites for the next 40 years, summarized the decade's progress in the synthesis and structural characterization of chlor-, fluor-, and hydroxyl-apatites of sufficient size for high-precision x-ray-diffraction analysis [64]. He acknowledged that the carbonated forms of apatite were more of a challenge to grow and thus to fully characterize, but he was hopeful that more modern analytical techniques such as nuclear magnetic resonance and neutron-scattering diffraction would provide additional information even in the absence of large single crystals. The issue of water in (carbonated) hydroxylapatites arose in several different contexts. First, there is Elliot's [64] recognition that a number of chemical analytical techniques do not distinguish among the possible speciations of hydrogen within the apatite structure, such as OH^-, molecular H_2O, $(HPO_4)^{2-}$, or H_3O^+. Secondly, there is the confounding factor that the heating of samples in order to determine the concentration of lesser components in apatite, a technique that had been used since the mid-1800s, actually could cause reactions that changed the species that were retrieved. For instance, there was the well-accepted method of quantifying the amount of $(HPO_4)^{2-}$ in bone mineral or synthetic carbonated apatite by measuring the amount of pyrophosphate derived by heating the sample to 600°C:

$$2(HPO_4)^{2-} \rightarrow (P_2O_7)^{4-} + H_2O$$

Heating apatite therefore could evolve water that did not previously exist as that species. Such water might be misinterpreted as the high-temperature release of crystallographically incorporated water. Elliott [64] summarized several papers in which proton magnetic resonance had been used, especially on enamel, to determine whether water was crystallographically controlled in the apatite. As indicated in the paragraphs above, the spectra from this technique did not lend themselves to unique interpretation. Thus, the chemical affinity (including bonding) of water for the apatite and the state of mobility of water in bulk samples of enamel were not totally clear. Elliott [64] pre-saged a number of types of analytical studies that would prove important for the characterization of apatite minerals, including Rietveld analysis in x-ray diffraction. However, he was perhaps too sanguine in his final comment: "We can therefore expect in the near future a rapid increase in knowledge, and therefore understanding of these compounds" (p. 305).

The 1970s brought more detailed modeling of how and where water molecules were sited in the apatite structure, but not from the mineralogists. McConnell [65] instead wrote a rather defensive paper in which he reiterated that many apatites seem to contain appreciable structural water and that various observed anomalies/variations in the a-axis length of the unit cell could be explained by H_2O occupancy in the lattice. Although McConnell's model of incorporation of $(H_4O_4)^{4-}$ groups in carbonated apatites never had been fully substantiated, many researchers had attributed their water analyses in apatite to this explicit species. McConnell [65], overlooking this lack of documentation, was quite indignant at Elliott's [64] reminder of the lack of direct evidence for that species. Elliott and his colleagues would go on to study biological and synthetic apatites in great detail, particularly applying the vibrational spectroscopic methods that McConnell [65] indicated "should be disregarded as meaningless" (p. 1665). The times were changing, however, and the vibrational spectroscopic methods of IR and, later, Raman would become of increasing application and value in the characterization and structural-chemical understanding of biological and synthetic carbonated apatites.

In 1970, the chemists Joris and Amberg [66,67] published a pair of papers in which they investigated the mechanism behind the non-stoichiometry of uncarbonated hydroxylapatites. They documented that Ca-deficiency in the apatites was accompanied by a decrease in the concentration of hydroxyl ions in the channels. Among other observations, the increase in value of the unit-cell *a* parameter in Ca-deficient apatites suggested to Joris and Amberg [66, 67] that H_2O was in the channel sites, most likely residing in the vacancies left by OH-ions whose absence was demanded by charge balance. They provided a quantitative and diagrammatically illustrated model of how the H_2O molecules could be accommodated in the channel sites of hydroxylapatite [67]. This paper, however, does not appear to be well-cited in the modern literature.

The dental researchers once again used modern techniques of the day to document water in bioapatite. In his presentation at the 1969 international Enamel II conference (published in 1971), Casciani [68] presented his interpretation of extensive NMR analyses of enamel, dentine, and bone. He acknowledged not only general statements that he felt confident to make, but also uncertainties about several details. For instance, he firmly stated that crystallographically incorporated water exists in all types of calcified tissues and that the less crystalline of these tissues (dentin and bone) contain more water than the more crystalline (enamel). His estimate still remained at approximately 6 wt% water within enamel, and his description of how water was retained and bound in the apatite structure did not seem to have progressed beyond the interpretations of Little and Casciani [53], as described above. Moreover, Casciani [68] re-introduced confusion into the understanding of the nature of enamel itself, by suggesting that some of this 'trapped' but 'unbound' water was contained in other non-apatite phases that he thought were constituents of the 'hard parts' of mineralized tissue. Among these other known hydrated phases that he called upon were octacalcium phosphate $[Ca_8(HPO_4)_2(PO_4)_4{\cdot}5H_2O]$ and calcium hydrogen phosphate dihydrate, brushite $[Ca(HPO_4)\cdot 2H_2O]$. Dibdin [69] followed with additional NMR analyses of enamel, in which he tried to clarify some of the complications in interpreting NMR line width in terms of the mobility of H_2O and the spectral distinction between H_2O and OH^-. He showed evidence for both of those hydrogen-species in enamel, and he warned about the need to remove and exclude adsorbed water before analysis of the samples. As in the previous decade, the NMR papers referenced above from the 1970s are based on proton NMR of powdered samples (no MAS, magic-angle spinning) in which interpretations of water mobility are based on peak widths.

The published work of LeGeros in the 1960's and their own experimental syntheses and analyses of carbonated apatites led French researchers Vignoles [70] and Bonel et al. [71] (as summarized in [72]) to propose the following charge-balancing mechanism in carbonated apatites synthesized in Na-bearing solutions:

$$Ca_{10-x}\,Na_{0.67x}(PO_4)_{6-x}(CO_3)_x(H_2O)_x(OH)_{2-0.33x}$$

They not only recognized vacancies in both the Ca- and OH-sites, but they also listed H_2O as an explicit component, whose concentration was controlled by the number of vacancies in the apatite structure, most specifically by vacancies in the channels.

LeGeros et al.'s [73] frequently cited 1978 paper appears to have marked a turning point in the acknowledgment of, as the title indicates, the different "types of 'H_2O' in human enamel and in precipitated apatites." The authors used IR spectroscopy in conjunction with heating techniques to monitor both the release of water at specific temperatures from and the accompanying changes in synthetic carbonated apatites. They concluded that adsorbed water could be distinguished by its release below 200°C based on the reversibility in its re-absorption. Lattice water was released at 200-400°C, and this water could not be reversibly re-incorporated. The proof offered for water's structural incorporation was that the a-axis length (at room temperature) was smaller in the apatites after the high-temperature release of water. LeGeros et al. [73], however, once again brought up the drawback to heating a sample in order to quantify its structural water content; they demonstrated in their IR spectra that original $(HPO_4)^{2-}$ thermally decomposed into $(P_2O_7)^{4-}$ with the evolution of H_2O. Likewise, Termine et al. [74] had recognized the release of several weight percent H_2O at 600°C. They also observed, however, that the $(HPO_4)^{2-}$ proportion of bone mineral "phosphate" is substantial (15-20%) in the youngest of animals, but decreases with age. This age-dependence in hydrogen phosphate content of bone could produce additional water artifacts in the TGA data. The application of new analytical techniques in the late 1960s and throughout the 1970s, coupled with a clearer understanding of both the usefulness and pitfalls of heat-treatment techniques (e.g., TGA), would appear to have set the stage for final clarification of the structural-chemical relations of water in the apatite lattice. That great breakthrough did not materialize imminently, however, as related below.

In the 1980s, D.W. Holcomb and colleagues documented some useful modifications in their application of TGA. Holcomb and Young [75] carried out heating studies on tooth enamel, with the additional precaution of making their accompanying IR analyses in a dry nitrogen atmosphere to avoid spectral interference from adsorbed water. They agreed with LeGeros et al. [73] concerning the presence of lattice water and how its thermal release affects the XRD a-axis parameter. In a remarkably data-rich published study, Young and Holcomb [76] used multiple analytical techniques to characterize (uncarbonated and carbonated) hydroxylapatite synthesized by six different accepted methods of the day. They detected numerous differences in IR spectral features, XRD parameters, ease of deuterium exchange with the hydrogen species, etc. among samples that, in principle, were very similar. There was much evidence for structural water, especially in samples precipitated at 1 atm. from aqueous solutions, therefore at temperatures below 100°C. The authors saw a possible correlation between the amount of structural water in the latter samples and the ease with which diffusion occurred (monitored as deuterizability). Interestingly, Young and Holcomb [76] did not find a correlation in their IR data between the abundance of hydroxyl vacancies and the amount of incorporated water in their samples. The wealth of data in this paper likely could be further mined by modern analysts.

By the later 1980s, researchers were using temperature-programed analysis of apatites in conjunction with IR and XRD monitoring of changes in the solid sample. Mortier et al. [77] summarized their understanding of how the broader community viewed the calcium-phosphate phases. "Compounds with compositions intermediate between HAp and OCP are conventionally called calcium-deficient apatites (DAp)...Some authors...have proposed that they be described by the following formula $Ca_{10-z}(HPO_4)_z(PO_4)_{6-x}(OH)_{2-z} \cdot nH_2O$ ..." (p. 266). This statement suggested some degree of acceptance of water as an intrinsic component of non-stoichiometric apatite by the rapidly growing and diversifying biomaterials community.

The 1990s brought re-consideration of the accepted two mechanisms for substitution of carbonate in apatite and re-invigoration of studies into the variability of hydroxyl concentration in apatite, both of which topics are crystallochemically intertwined with the mechanism of H_2O incorporation. Analytical chemists De Maeyer et al. [78] published an insightful paper on Na- and carbonate-containing apatites that were produced through hydrolysis of the more acidic calcium hydrogen phosphate phase monetite ($CaHPO_4$). The authors did explicit chemical analyses for Ca, P, Na, and CO_3. Unfortunately, OH concentration was calculated, based on the assumption of charge balance, and H_2O subsequently was calculated by difference from an assumed 100 wt% total. Their IR spectra still showed water in samples dried under vacuum at 25°C. The carbonate concentrations of the samples ranged from approximately 8 to 21 wt%, i.e., from the level in bone upward. In this perhaps earliest of studies in which water concentration was evaluated as a function of carbonate concentration, water was *calculated* as approximately 5 wt% across the entire range of samples. Later work suggests this finding of constant water concentration to be very significant. Despite the lack of correlation between the concentration of water and that of the carbonate substituent believed to induce vacancies in the apatite, De Maeyer et al. [78] inferred that this "structural water" resides in vacancies. Christian Rey and colleagues [17], who have made numerous contributions over several decades to the understanding of nanocrystalline carbonated apatite, applied magic-angle-spinning (MAS) proton NMR to natural bone. They concluded that the concentration of OH ions in bone apatite crystallites was sufficiently low to be undetectable by that technique. One would infer thereby a high density of vacancies in the channels. As part of this study, the authors indeed detected H_2O in their samples, but they regarded this water as simply a constituent of collagen or an adsorbent on the apatite crystallites.

4.3. Heated Apatites But Unheated Discussion: 2000-2011

As most readers of this book and chapter will realize, in the past decade there has been an explosion in the literature on carbonated apatite within the biomedical (e.g., medical, dental, biological engineering), chemical, environmental, and biomaterials sciences. The chemical-structural adaptability of the mineral apatite has allowed it to be tailored in a multitude of ways for a huge range of applications. Thus, based on their specific goals and needs, individual research groups have explored the relationships among the chemistry, crystallographic structure, and properties of apatite.

In this time period in the progression of apatite research, one can ask whether there are still researchers who do not agree that there is structurally incorporated water in non-stoichiometric (carbonated) apatite and what their evidence is. Likewise, one can query those who do provide explicit evidence of the presence of structural water regarding the nature of their sample preparation and analytical techniques. The follow-on questions to the latter group concern where in the crystal structure the H_2O resides and what the mechanism is that controls its incorporation. Lastly arises the question that brought about the research for and writing of this chapter – why is water still not regarded in the broader medical-dental research community as an *essential* and relatively abundant constituent of the mineral in bone, dentin, and enamel? Why is it that modern review articles that carefully explain the incorporation of carbonate into apatite and carbonate's relation to the vacancies in hydroxyl sites do not emphasize H_2O in a like manner?

Not all the recent literature that addresses hydrogen species in bone mineral and its synthetic equivalents discusses the issue of water. Among the studies not explicitly focused on water, those that address OH-ions in bone apatite typically have also uncovered H_2O in their analyses. The present author freely admits, however, that even she ignored the issue of water in her own paper dealing with OH in non-carbonated apatite [18]. Cho et al. [20], and Kolmas and Kolodziejski [21], who applied NMR spectroscopy to bone and dental apatites, respectively, were concerned primarily with quantifying the concentration of OH-ions in bioapatites. Those two studies revealed consistent results, e.g., that bone and dentin apatite (long recognized to be very similar in properties) contain only about 20% of the concentration of OH-ions that is found in pure hydroxylapatite. Both groups also detected considerable H_2O in their NMR spectra, yet both dismissed this water as adsorbed on the crystallites or contained within the collagen.

It would appear that TGA-based studies of apatite would produce irrefutable evidence for or against the structural incorporation of water. All thermal experiments on non-stoichiometric apatites release water. Surely the temperature of release would distinguish adsorbed water from structural water. Several published papers, however, fail to make this distinction, and the reader is left wondering at how high of a release temperature the authors still would label the water as 'adsorbed'. Thus, Tas [79] followed an accepted biomaterials preparation procedure in heating to 900°C the OHAp precipitated at 37°C from synthetic body fluids. The author reported that the broad IR band for 'adsorbed water' on the initial material 'diminished' after the sample had been heated to 900°C, leaving one to wonder about the distribution of water released throughout that large temperature interval. Unexplained is why the IR spectrum of the 900°C sample in their Figure 2d (in [79]) still records the presence of some, although much less, water. Likewise, in a paper on the thermal decomposition of carbonated apatites aqueously precipitated at 90°C, Lafon et al. [80] carried out TGA studies up to 1400°C in air, N_2, or CO_2. They reported "20-600°C: departure of adsorbed water and synthesis residuals" (p. 1131). Their paper does not distinguish the amount of non-adsorbed water. Similarly, Taylor et al. [81] had reported for inelastic neutron scattering analyses of synthetic hydroxylapatite that "after heating SRM 2910 to 850°C the bands...are better resolved following the removal of the underlying water bands after heating...The absence of translational modes that could be assigned to water would imply that the water present in both samples is surface and not lattice bound water..." (p. 3143). Antonakos et al. [82] straightforwardly, but inappropriately, stated that "...sample #6 was heat-treated at 530°C to derive the adsorbed water..." (p. 3044).

In contrast, several studies based on a combination of TGA results on tooth material and synthetic apatites in conjunction with analyses of the associated solids were carried out with due attention to the temperatures of volatile release. Reyes-Gasga et al. [83] heated dentine and enamel, in addition to synthetic apatites, while noting the specific steps in the TGA traces and simultaneously monitoring changes in resistivity in the solids. Their results correlated well with the interpretation that adsorbed water was released from room temperature to 200°C, lattice water and organics (in the dentine and enamel) were removed from 200 to 600°C, and carbonate decomposed and was released as CO_2 at temperatures above 600°C [83]. An earlier related study by Shi et al. [84] compared the thermal behavior of tooth enamel and natural geologic, well crystalline fluorapatite from approximately room temperature to almost 500°C. Through IR spectroscopy, they observed essentially no water in geologic fluorapatite but were able to monitor dehydration in enamel. Although these authors did not regard the existence of lattice water with so much certainty as the previous group, they did conclude that at temperatures below about 325°C the adsorbed water and some portion of the lattice water were released from enamel. These papers are further evidence of the dental research community's ongoing research into water in enamel and dentin apatite.

There are also groups that recognize that bioapatite crystallites and their synthetic analogs possess a significant amount of water, but they attribute it to a "structured hydrated layer" (e.g., [19,85-87]), which hearkens back to the early work of Neuman et al. [49] and Neuman and Neuman [50]. The above researchers have shown that this reservoir of water is very reactive and also contains a huge ion reservoir that is important to the bulk composition and, perhaps, the maturation of the bone crystallites. Particularly important to the current discussion, as summarized by C. Rey and colleagues [19], is the development of enhanced interpretation of band positions in IR and NMR spectra by which one can distinguish water in the external, structured hydrate layer from water that is structurally (internally) incorporated in apatite.

Many papers since 2001 have acknowledged the presence of structural water in bone apatite, including several papers that were primarily focused on understanding the nature, location, and stabilization of water in several different reservoirs throughout bone. Various applications of nuclear magnetic resonance spectroscopy, particularly in conjunction with Rietveld analysis of XRD data, have proven very powerful in the quantitative analysis of non-stoichiometric apatites. Such attempts have not been without problems and inconsistencies, however. The above-mentioned studies by Cho et al. [20], and Kolmas and Kolodziejski [21] also were based on NMR analysis, but both sets of authors concluded that the spectral evidence for H_2O actually indicated it was adsorbed. The greater availability of and sophistication of analytical instruments in the past decade clearly have enabled the acquisition of much more data. The selection of the optimal analytical method for water analysis and the interpretation of the acquired data, however, remain a challenge. In the remaining summary, the present author does not always distinguish among the different NMR techniques that have been applied to apatite samples, but typically refers to them all simply as NMR. The reader should take note, however, that the specific technique and the specific isotope observed will affect the information that is derived.

In this author's opinion, the major breakthrough 'in the modern literature' for accepting structural water in non-stoichiometric apatite came in a paper by Ivanova et al. [88]. The latter studied the crystallography, especially the siting of CO_3 ions, in calcium-deficient carbonated apatites through collecting XRD data on sample powders simultaneously with their heating to 750°C. The combined TGA volatile-release data and XRD data on the associated solids had the additional benefit of clarifying "the localization and amount of lattice water that enhanced the reliability of the structural model" (p. 342). Ivanova et al. [88]

measured a 1.1 wt% loss of structural water by TGA, and inferred 1.4 wt% water in the structure based on their Rietveld refinement of XRD data. They recorded as adsorbed all water released up to 120°C and noted that all structural water had been released by 400-500°C. In their apatite formula, they assigned 0.3 moles of H_2O to the PO_4 site and 0.45 moles of H_2O to the channel site. The latter was shared with 1.65 moles of OH, for a total occupancy of 2.0 for the channel sites, i.e., full occupancy.

Other NMR papers directly addressed bone, rather than its synthetic apatite equivalents. Wehrli and Fernández [89] wrote a paper entitled 'Nuclear magnetic resonance studies of bone water'. Although the authors acknowledged water associated with the crystallites, they focused on the diffusion of intercrystalline water through pores and micro-channel systems in the bone. Kaflak-Hachulska et al. [90] investigated human bone by applying proton NMR under ultra-high-speed magic-angle spinning (MAS), carrying out two-dimensional ^{1}H-^{31}P heteronuclear correlation studies and analysis of ^{31}P rotational sidebands in conjunction with their cross-polarization kinetics. Among their results was a lack of detection of OH^- or $(HPO_4)^{2-}$ ions in bone mineral. The paper also suggested that some structural H_2O still existed in a synthetic OHAp even after it had been heated to 1073°C, which is difficult to envision. A major contribution of this well-explained article is the insight it provides into the complexity and model-dependence of the interpretation of such NMR spectra[2].

Two sets of published studies, first-authored by two different researchers named Wilson, mark major advancements in the understanding of water in apatite. The group of Erin Wilson, Michael Morris, and colleagues ([91, 92]) applied NMR analysis to bone specifically for the purpose of understanding the different types of water in this complex material. Their 2004 paper identified an ordered arrangement of H_2O molecules apparently located in the grooves/channels between collagen molecules. They also identified an 'ordered surface water layer' on the bone crystallites, reminiscent of the surface layer described by Neuman et al. [49] and Rey et al. [87]. Their paper in 2006 specifically identified three different types of water in bone and their apparent structural roles. In addition to recognizing surface (adsorbed) water, Wilson and colleagues identified structural H_2O using ^{1}H MAS NMR on bone and synthetic type-B carbonated apatite. They monitored differences in hydrogen-bonding environments of the H-bearing species and how the hydrogen-bonding changed as temperature was increased up to 225°C. Wilson et al. [92] found their results to be consistent with those published by Ivanova et al. [88].

Collaborators Rory Wilson and James Elliott published three papers ([72, 93, 94]) in which they applied Rietveld refinements to XRD data on synthetic apatites with different compositions and precipitated through different reactions. For each synthetic, non-stoichiometric apatite product, they provided data on the unit cell contents, which included molar H_2O occupancies of 1.05 (for Na-bearing OHAp with 12.5 wt% CO_3), 1.97 (for Ca-deficient OHAp samples prepared from $CaHPO_4$), and 2.29 (for Na-free OHAp with 11.2 wt% CO_3). The 2004 study applied Rietveld refinement to both X-ray and neutron diffraction data [72]. After chemical analyses of the samples, which totaled to 93-95 wt%, the OH^- ion concentration was calculated based on charge balance, and the H_2O content of the unit cell was calculated to account for the residual mass balance. By this method, 2.11 wt% H_2O (approximately 1 mole H_2O) was derived. Somewhat disconcertingly, the reader is told not to accept as reliable the IR-based OH-concentrations that are obviously lower than the

[2] See also Chapter 7 of this volume. The Editor.

calculated values. Wilson et al. [93] analyzed synthetic uncarbonated but Ca-deficient apatites by XRD, IR, and Raman spectroscopy followed by Rietveld refinement of the XRD data. Again, assumptions were made to derive specific OH^- and H_2O concentrations, which are not spectroscopically quantifiable in this situation. The samples were analyzed by TGA to 900°C, and the raw amounts of evolved water were modified to estimate 'true' H_2O occupancy based on assumptions about thermally induced reactions among original $(HPO_4)^{2-}$ groups, which generate H_2O. The authors' overall assessment is that Ca(2) vacancies surrounding the apatite channels and some proportion of OH-vacancies within the channels may be filled with H_2O molecules. This leads to the proposed 1.97 molecules of H_2O per unit cell. In their 2006 paper on Na-free carbonated apatites that were synthesized through hydrolysis of monetite, Wilson et al. [94] interpreted some resonances in their proton NMR spectra differently than in earlier literature. They attributed the resonances to H_2O that is structurally incorporated in the channels. The results of the TGA experiments and measured density of the samples led the authors to believe that H_2O is part of the unit cell. The proposed cell contents in this case are $Ca_{8.241}(PO_4)_{4.344}(CO_3)_{1.656}(OH)_{0.139}\cdot 2.29H_2O$.

More recently, Harris Mason and Brian Phillips have harnessed the efforts of several collaborative groups of geoscientists to apply NMR spectroscopy to study how H and Na are incorporated into apatites that exhibit combined type-A and -B carbonate substitution [95], the characterization of fluor-chlorapatite with a low OH-concentration [96], and the role of structural water in carbonated fluorapatite [97]. Although some information on H_2O is provided in each of those three articles, the latter is obviously most pertinent to the present discussions. In addition to NMR analysis, Mason et al. [97] applied IR spectroscopy, electron microprobe analysis (for chemical composition), and TGA to well-crystalline examples of natural, carbonated fluorapatites, like those that occur abundantly in large-scale phosphorite (phosphate rock) deposits. The authors observed a peak at 6.4 ppm that contained a set of spinning sidebands in the proton MAS NMR spectrum, consistent with the presence of rigid molecular water bound within the structure of the apatite. After the sample had been heated to 600°C (within a weight-loss step in the TGA pattern), the feature at 6.4 ppm was weaker, which indicated partial loss of water. NMR analysis of the sample after heating to 800°C indicated atomic re-ordering and more complex internal chemical reactions in the apatite. This very informative paper presents an internally consistent model for how and where the H_2O resides. The authors identified abundant water, i.e., 0.84 molecules per unit cell, which is definitely structural in nature and inferred to reside in the channels. Based on additional NMR spectra, the authors proposed that the water molecules are closely associated with and may stabilize structural vacancies.

In an effort to overcome some of the non-unique interpretations of NMR spectra of apatites, our own collaborative group synthesized carbonated hydroxyl- and fluor-apatites in D_2O, which were analyzed by solid-state ^2H NMR [98]. Deuteration during mineral formation permitted incontrovertible discrimination between structurally incorporated and adsorbed water. Each spectrum of apatite, containing widely different concentrations of carbonate, showed a single line using the quadrupole echo pulse sequence. Samples dried under several different conditions showed ^2H resonance, which was not detected once a sample had been heated to 500°C. These experiments confirmed the presence of lattice water in the carbonated apatite and its release by 500°C.

Thus, several independent research groups, representing a wide range of scientific, medical, and engineering fields, have provided what appears to be irrefutable evidence of the structural incorporation of significant amounts of water (on the order of 1 to several wt%) in vacancy-bearing hydroxylapatite and carbonated hydroxylapatite of the type found in bones and teeth. All of these groups indicate the c-axis channels of apatite as the major site for the H_2O molecules. Those channels have been recognized for decades as one of many opportune crystallographic features, including the structural flexibility of the Ca and PO_4 sites, that permit variability in the composition of apatite.

An additional crystallographic ramification of carbonate substitution in apatite appears to support the accommodation of water molecules in the structural channels. As explained in the introduction, substitution of carbonate for phosphate in OHAp induces vacancies not only in OH^- sites at the center of the channels, but also in Ca^{2+} sites. Numerous studies document or at least suggest that the Ca(2) sites, which are the calcium sites immediately around the channel (see Figure 1), are preferentially left vacant during substitutions that require additional charge-balancing [8,19,93,99].

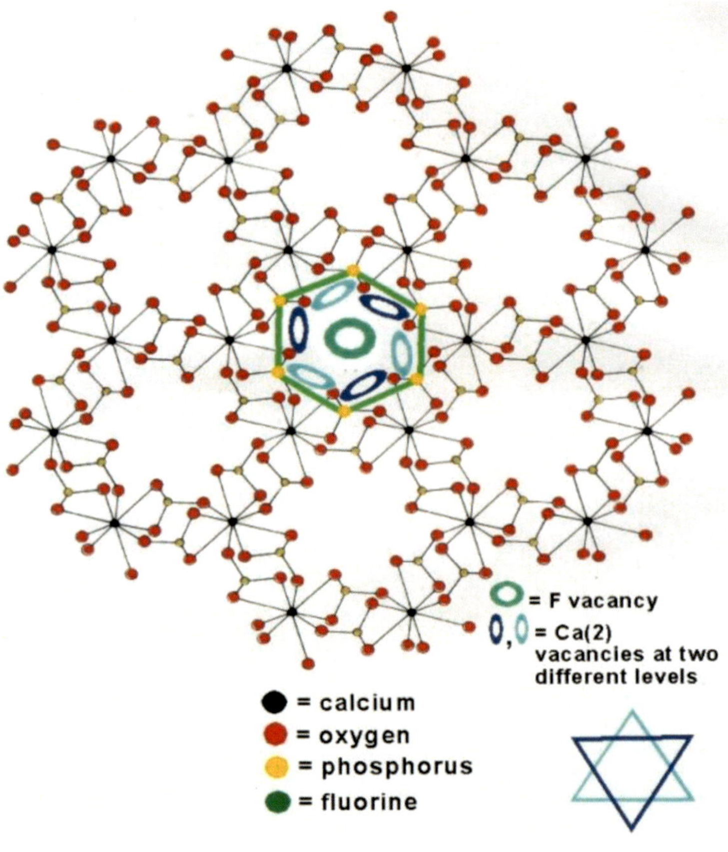

Figure 2. Apatite structure after removal of all the channel-filling anions, as well as the Ca(2) cations. The central green hexagon is the same as the outermost hexagon shown in Figure 1. A huge volume of the crystal is opened up as two types of vacancy sites have formed, i.e., absence of the central anion (shown by central thick green oval) and Ca(2) sites (shown by blue ovals) at two levels along the c-axis (distinguished by light and dark blue color). The enlarged hexagonal channels are lined by phosphate ions with negatively charged oxygens. Figure from [40] (their figure 4), with some additions. (Reproduced with kind permission of the Mineralogical Society from a paper by Beevers and McIntyre (1946)).

Creation of vacancies in all the OH sites and all the Ca(2) sites would produce remarkably large open channels in the apatite, as recognized for fluorapatite by Beevers and McIntyre [40], whose crystallographic representation is reproduced above as Figure 2 [40]. Consider that a typical bone contains carbonated OHAp with approximately 6.6 wt% CO_3. If all the carbonate in such a bioapatite demonstrated B-type substitution, its formula would be $Ca_9\square[(PO_4)_5(CO_3)](OH)\square$, where □ represents a vacancy. Hypothetically, if the Ca(2) calcium-site vacancies were three times more abundant than the Ca(1)-site vacancies, then the more specific formula would be $Ca(1)_{3.75}\square_{0.25}Ca(2)_{5.25}\square_{0.75}[(PO_4)_5(CO_3)](OH)\square$. Half the OH ions and over 12% of the Ca(2) ions would be missing, thereby making accessible for water incorporation about one-fifth of the sites left vacant in the open volume shown in Figure 2. Both the above models of vacancy-modified c-axis channels and the similarity in size between OH^- ions and H_2O molecules [100] suggest a number of ways in which H_2O molecules plausibly can be incorporated structurally in non-stoichiometric apatites.

Most proposed models for water incorporation in fact do suggest a very close relationship between the H_2O molecules and the vacancies in the OH-sites ([101] and [102], as summarized in [94]; [70] and [71], as summarized in [72]; [88,92,94,97,103,104]). The present author actually became involved in the water-in-apatite discussion while investigating the relationship between carbonate concentration and hydroxyl concentration in carbonated OHAp. If the incorporation of water is directly tied to the availability of vacancies in which to site H_2O molecules, as proposed by the authors above, then one would expect to find a robust positive correlation between the concentration of carbonate and the concentration of water in carbonated apatites. Although our initial TGA results on a suite of carbonated hydroxyl- and fluor-apatites suggested such a correlation [105], more recent unpublished analyses reveal essentially constant concentrations of water released in the interval from 200 to 500°C from hydroxylapatites of a wide range of carbonate concentrations. These data and our NMR work on deuterated carbonated apatites suggest that most of the H_2O molecules reside between the channel anions [98].

5. INTERPRETING WHY CRYSTALLOGRAPHIC INCORPORATION OF WATER WAS NOT ACCEPTED

5.1. The Issues

The literature on water in bioapatite demonstrates, as does the literature on the controversies surrounding carbonate and hydroxyl ions in bioapatite, that there are multiple reasons for the long intervals of time during which researchers did not share a community-wide view of the details of the structure and chemistry of bioapatite. In the present case, however, the pertinent question concerns why there is a hesitation to formalize the acceptance of what appears to be a well-proven contribution to our knowledge of bioapatite.

Part of the answer may come from a closer look at how well, i.e., to what level of detail, we have documented lattice water in bioapatite. The earliest evidence of lattice water (in the late 1800s) was derived by methods that we no longer acknowledge as 'cutting edge'. The evidence was only chemical data, which could not be referenced to or tested against any structural model, in that the apatite structure had not yet been determined. Then, several well-received papers presented detailed crystallographic models of fluorapatite and hydroxylapatite, providing structural constraints for future considerations of water as a component. Multiple documentations of water in carbonated hydroxylapatite subsequently were published, but the specific results did not agree. For instance, although heating experiments such as TGA reproducibly showed the evolution of water at elevated temperatures (thus, not adsorbed water), many different weight-percent values were published. Until quite recently there was not an accepted temperature interval in which to assign the released water as lattice vs. adsorbed water, although many researchers did follow the temperature intervals proposed by LeGeros et al. [73].

The sense of uncertainty of the details about water in apatite continued, however. For instance, certain knowledgeable experimentalists and analysts brought attention to the physical and chemical differences among the synthetic apatites that were under analysis [76,106-108]. The differences among the synthetic products clearly could account for much of the variation in water released during TGA experiments, but this realization only served to show how little was known about the parameters that actually control the amount of lattice water in apatite. Lastly, it became clear that the amount of water evolved from a heated non-stoichiometric apatite sample need not match the amount of lattice water that actually existed in the original sample. Unfortunately, the process of heating could induce internal reactions in the apatite that could either consume or produce molecular water. As documented above, numerous authors had recognized that acidic species such as $(HPO_4)^{2-}$ and $(HCO_3)^-$ reacted when heated so as to produce water, e.g., $2(HPO_4)^{2-} \rightarrow H_2O + (P_2O_7)^{4-}$. Such reactions would create molecular H_2O that was not originally in the apatite. In contrast, it was also known that carbonated and other non-stoichiometric apatites that inherently were depleted in OH^- would become well-crystalline, fully stoichiometric OHAp after heating to 900°C [14,15,18,23,75,79,84,109]. It was possible that the internally available lattice water in the original apatite (perhaps together with CO_3^{2-}) reacted with the latter to fully hydroxylate it during heating. If so, the TGA-evolved amount of water would be less than that in the original solid. Such variability and uncertainty do not make for a robust model of water's crystallographic incorporation.

Although thermogravimetry continued to be applied to carbonated apatite samples, NMR gradually took on increasing importance as a technique that could specify the structural positions of the isotopic species under analysis. It therefore seemed as though we might finally know exactly where the water resided in the lattice even if there were difficulties in quantifying the exact amount of water. However, the interpretation of the acquired NMR spectra was not as straightforward or objective as desired. It is clear from the literature cited above that most of the interpretations of NMR spectra with respect to lattice water are strongly model dependent. As summarized in Yoder et al. [98], several recent studies of synthetic or biological carbonated apatite have disagreed concerning lattice water even though the NMR techniques applied were quite similar. Multiple hydrogen-bearing species might exist in the same apatite sample (e.g., H_2O, OH-, HPO_4^{2-}, HCO_3^-) and thereby produce multiple peaks.

Moreover, different degrees of hydrogen bonding between these species and the rest of the structure could shift the recorded peak positions, making species identification less certain. With respect to just the one species H_2O, different 'types' of water (e.g., adsorbed, lattice-incorporated, trapped in fluid inclusions) also could complicate the differentiation among various hydrogen species. NMR, as well as IR and Raman, spectroscopy can be carried out in a dry gas environment, which at least can eliminate interferences from adsorbed water (see Figure 3). This approach has not been used in most cases, however. One therefore does not gain a sense that multiple types of studies are closing in on one, irrefutable model that accounts for exactly how much water resides in which specific crystallographic site(s) as controlled by some set of definable conditions.

It might appear that a better pathway to resolving the water question would be to focus on actual biological material. It would seem that the variability in the apatite among different samples of bone might be appreciably less than that among carbonated apatites synthesized by means of a dozen or more methodologies. Bone (also dentin) has its own problems, however. If bulk bone samples are measured by TGA or by NMR, there is interference from the abundant collagen and other organic components, which (1) give their own responses to each technique and (2) provide another medium in which non-mineral water can be stored.

It is possible to chemically dissolve the organic components from bone, leaving only the inorganic mineral. However, such aggressive chemical treatment of bone introduces the possibility that the mineral also was chemically changed. Dental enamel, on the other hand, appears to provide the ideal bioapatite for study.

Enamel has a total of only about 4 wt% combined intercrystalline water and organics [110,111], and its crystallites are on the order of 10 times wider and 1000 times longer than those of bone [34,111-113]. Enamel apatite is better crystalline as well as atomically better ordered than bone apatite.

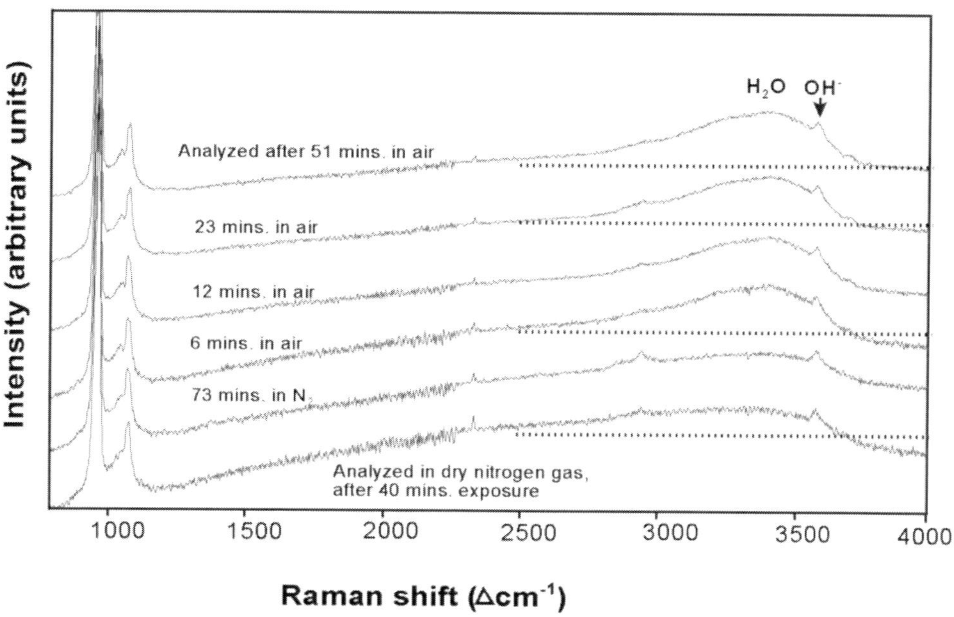

Figure 3. Raman spectra of a single aliquot of carbonated OHAp containing 6.35 wt% CO_3, synthesized aqueously at 37°C. Bottom to top sequence of spectra shows water adsorption as the apatite powder that had resided in a flow of dry nitrogen gas is exposed to room air. Analogously, viewed from the top downward, the spectra show that even after adsorbed H_2O is removed, the carbonated apatite still contains appreciable structural H_2O, as seen in the broad feature from about 2900 to 3600 Δcm^{-1} that persists even under dry gas. The O-H stretch for the hydroxyl ion at about 3570 Δcm^{-1} remains throughout the experiments. All spectra are normalized to the intensity of the P-O stretch at ~960 Δcm^{-1}. Sample kindly provided by Mitchell Sternlieb and Claude Yoder.

It is therefore no surprise that dental researchers over the decades, in many cases in conjunction with colleagues in the natural sciences, have made repeated analyses for water in apatite using increasingly sophisticated techniques. Somewhat counterintuitively, the notion of 2-3 wt% H_2O as an essential component in enamel apatite has not fully pervaded the dental research community.

5.2. Stagnation in Acceptance

The chemists in the 1890s were not particularly excited about the finding of lattice water in 'bone mineral'. They already had accepted the fact that this calcium phosphate solid was a hydrate, and it was the missing hydroxyl that ultimately interested them. These researchers were not constrained by a preconceived stoichiometry for apatite. Likewise, it has been well accepted for a long time among modern researchers that the non-crystalline (stoichiometrically less constrained) phase amorphous calcium phosphate can contain large amounts of water [114]. Chemists in the 1930s also knew about the lattice water in calcium phosphate materials known as apatitic. It was not considered startling or interesting, in that other hydrated calcium phosphate phases had been identified [7].

By the 1950s, the interesting new revelation about water and bone mineral was the recognition of an ordered hydration shell that extended over the large surface area of such small particles. Interest persists in the nature of and reactivity of that shell, e.g., as an ion reservoir (e.g., [19,115]).

Many researchers who subsequently encountered what we now would call lattice water in apatite considered it more of a nuisance than a component of significance in the mineral. Water (especially adsorbed, but also crystallographically incorporated) created an elevated background in both IR and Raman spectra, which made it difficult to make quantitative measurements for hydroxyl content, in that the narrow O-H stretch band for OH^- in hydroxylapatite is situated on the high-wave number shoulder of the broad O-H stretch band for water (see Figure 3). Similarly, the signal for water makes it difficult to define the position of the $(HPO_4)^{2-}$ resonance in the NMR spectrum [93]. As noted above, many researchers were more than willing to heat their samples to be rid of this interfering water, not always documenting the specific temperature at which they deemed the adsorbed water to have been removed. In the production of some ceramic apatitic biomaterials, the original synthetic precipitates or bone materials are heated routinely to 700°C or more [116-118].

Ivanova et al. [88] published an apatite study in which they explained the significance of the incorporation of water in the unit cell toward optimizing the Rietveld refinement of XRD data. Numerous subsequent papers have followed their lead. There is still some sense of reluctance, however, in the inclusion of water, in that its quantification remains elusive.

It appears that publications about water in apatite typically received subsequent citations, but that they did not generate a great deal of debate. Such publications were not followed by commentaries that posed pointed questions about the amounts of water that were being recorded or that demanded an internally self-consistent model of how H_2O was accommodated in the apatite structure. Presumably, medical researchers were not impressed by or concerned with a few weight percent water in inorganic mineral within a bone system that obviously has large biological reservoirs of water. The question of H_2O incorporation in bioapatite apparently has not generated either heated discussion or conflicting models (as did the notion of the incorporation of carbonate within apatite or the presence/absence of hydroxyl in bone apatite). Thus, no resolution was ever reached, because there had never been a battle. The notion of lattice water in carbonated apatite never became THE new idea that had risen through the fray in order to be incorporated into the community's now updated and expanded view of bioapatite.

5.3. Lessons for the Future

Some of the 'take-home messages' from this comparison and analysis of the literature are the importance (1) in reading the (even decades) older literature, especially focusing on papers by those who were considered to be insightful in the field, (2) in not automatically disregarding results or isolated observations simply because they do not fit the current model, and (3) in not assuming if some result is only now becoming recognized that it must not be very important, i.e., we did just fine up until now even without this information. In fact, the current information on water in apatite could help to explain discrepancies about synthetic apatite that have been reported for many years in the literature, for instance, a lack of reproducibility in experimental solubility results.

In a sense, the bone and tooth research communities have always had an identity crisis regarding the inorganic mineral component of interest. Is it hydroxylapatite or not? Well, yes, but not really... Applications of calcium *sulfate* to bone healing and in bone cements, of obvious interest to the bone, tooth, and biomaterials communities [119-122], suggest an alternative path that calcium phosphate research might have traveled. Intriguingly, calcium sulfate exists in two relatively common (geological) mineral forms, one anhydrous ($CaSO_4$) and the other hydrated ($CaSO_4 \cdot 2H_2O$). It is also easy to make the hemihydrate phase ($CaSO_4 \cdot 0.5H_2O$) that is employed in the above papers. One wonders how all these decades of bone mineral research would have played out if, very early on, two calcium phosphate phases had been accepted, one a highly crystalline unhydrated (but hydroxylated) phase OHAp, $Ca_{10}(PO_4)_6(OH)_2$, and the other a more complex hydrated phase [65] such as $Ca_{10-x}\,Na_z(PO_4)_{6-x}(HPO_4,CO_3)_x\,(OH)_{2-(x-z/2)} \cdot nH_2O$.

CONCLUSION

Scientists want to identify those ways of thinking and types of analytical techniques or approaches that will lead to progress and research success. To this end, they can benefit from the analysis of historical successes in which early hypotheses were proven to be correct and then used as springboards to new hypotheses, which, in turn, revealed important knowledge, created beneficial new techniques, and produced widely applicable new biomaterials, etc.

The above is the idealized way in which science purportedly operates: Those ideas that are shown to be wrong are rejected; repetition of this selection process culls the wrong ideas and eventually leads to documentable truths. It is disconcerting to realize, however, that there are many counter-examples to the above scenario, in which a truth or fact is discovered or a correct observation made, but this truth is either rejected or simply overlooked. Such negative-seeming stories can be equally instructive, however, if analysis of them helps scientists to avoid such mistakes in the future.

The present chapter addresses such a counter-example, the story of water in the mineral apatite. The presentation begins as an historical account, accompanied by modern interpretation of the research results. As the account nears its end, however, it becomes clear that this discussion is not only historical. The understanding of the crystallographically incorporated water in apatite and, perhaps more importantly, the determination of its biological and geological ramifications is an on-going saga. The second part of the chapter suggests possible reasons for the circuitous route to increased acceptance of water's crystallographically controlled presence in apatite, including bone mineral. Somewhat counter-intuitively, it appears that the pertinent medical communities' decades-long resistance to accepting water as an essential component in bone mineral reflects more of an avoidance than a rejection of the idea.

What is exceptional about the investigation of lattice water in apatite aside from its extremely long history? One feature of note is the absence of exciting or embittered battles in the literature that could have highlighted the need to make some major choice between competing models. The accepted formula for bone mineral, a variation on that of hydroxylapatite, did not reveal H_2O as an explicit component or even suggest where H_2O might substitute for some accepted component. Models, like mineral formulas, should provide guidance in our research, but should not obstruct new ways of thinking.

It would appear that the community is still waiting to know if and how the incorporation of a few weight percent of water in bioapatite crystallites should affect current views of inorganic bone mineral and how it responds in a dynamic, watery, organic-rich, biological system. Interested researchers in many fields should take note.

ACKNOWLEDGMENTS

The author thanks Claude Yoder for his helpful comments on parts of an earlier version of this chapter and for his on-going collaboration involving synthesis and analysis of carbonated apatites. She also benefited from earlier discussions with Brigitte Wopenka concerning the crystal-chemistry of carbonated apatite. Partial support for this work came from the Center for Materials Innovation at Washington University in St. Louis.

REFERENCES

[1] D. McConnell, *American Mineralogist* 23, 1 (1938).

[2] C. Klein and B. Dutrow, *Mineral Science*, John Wiley and Sons, Inc., New York (2007).

[3] J. M. Hughes and J. Rakovan, in *Phosphates -- Geochemical, Geobiological, and Materials Importance, 48* (Eds.: M. J. Kohn, J. Rakovan, and J. M. Hughes), Mineralogical Society of America, Chantilly, Virginia, pp. 1-12 (2002).

[4] Y. Pan and M. E. Fleet, in *Phosphates -- Geochemical, Geobiological, and Materials Importance, 48* (Eds.: M. J. Kohn, J. Rakovan, and J. M. Hughes), Mineralogical Society of America, Chantilly, Virginia, pp. 13-49 (2002).
[5] S. Gabriel, *Zeitschrift für Physiologische Chemie* 18, 257 (1894).
[6] D. McConnell, *American Journal of Science* 36, series 5, 296 (1938).
[7] J. C. Elliott, *Structure and chemistry of the apatites and other calcium orthophosphates*, Elsevier, New York (1994).
[8] D. McConnell, *Bulletin de la Societe Francaise de Mineralogie et Cristallographie* 7-9, 428 (1952).
[9] D. McConnell, *Science* 134, 213 (1961).
[10] A. S. Posner, *Science* 134, 213 (1961).
[11] R. Z. LeGeros, O. R. Trautz, E. Klein, and J. P. LeGeros, *Experientia* 25, 5 (1969).
[12] G. Penel, G. Leroy, C. Rey, and E. Bres, *Calcified Tissue International* 63, 475 (1998).
[13] J. C. Elliott, in *Phosphates -- Geochemical, Geobiological, and Materials Importance, 48* (Eds.: M. J. Kohn, J. Rakovan, and J. M. Hughes), Mineralogical Society of America, Chantilly, Virginia, pp. 427-453 (2002).
[14] R. M. Biltz and E. D. Pellegrino, *Calcified Tissue International* 36, 259 (1971).
[15] J. D. Termine and D. R. Lundy, *Calcified Tissue International* 13, 73 (1973).
[16] J. L. Meyer, *Calcified Tissue International* 27, 153 (1979).
[17] C. Rey, J. L. Miquel, L. Facchini, A. P. Legrand, and M. J. Glimcher, *Bone* 16, 583 (1995).
[18] J. D. Pasteris, B. Wopenka, J. J. Freeman, K. Rogers, E. Valsami-Jones, J. A. M. van der Houwen, and M. J. Silva, *Biomaterials* 25, 229 (2004).
[19] C. Rey, C. Combes, C. Drouet, and M. J. Glimcher, *Osteoporosis International* 20, 1013 (2009).
[20] G. Cho, Y. Wu, and J. L. Ackerman, *Science* 300, 1123 (2003).
[21] J. Kolmas and W. Kolodziejski, *Chemical Communications* 2007, 4390 (2007).
[22] R. Legros, N. Balmain, and G. Bonel, *Journal of Chemical Research (S)* 1986, 8 (1986).
[23] Z. Z. Zyman, D. V. Rokhmistrov, V. I. Glushko, and I. G. Ivanov, *Journal of Materials Science:Materials in Medicine* 20, 1389 (2009).
[24] E. von Bibra, *Chemische Untersuchungen über die Knochen und Zähne des Menschen und der Wirbeltiere*, Kunstverlag, Schweinfurt, Germany (1844).
[25] C. Aeby, *Journal für Praktische Chemie* 5, 308 (1872).
[26] C. Aeby, *Journal für Praktische Chemie* 6, 169 (1873).
[27] C. Aeby, *Journal für Praktische Chemie* 10, 408 (1875).
[28] W. E. Brown and L. C. Chow, *Annual Review of Materials Science* 6, 213 (1976).
[29] R. Maly and J. Donath, *Journal für Praktische Chemie* 7, 413 (1873).
[30] F. Wibel, *Journal für Praktische Chemie* 9, 113 (1874).
[31] C. Aeby, *Berichte der Deutschen Chemischen Gesellschaft* 7, 555 (1874).
[32] H. Watts, *A Dictionary of Chemistry and the Allied Branches of Other Sciences*, Longmans, Green, and Co., London (1875).
[33] O. Förster, *Zeitschrift für Angewandte Chemie* 5, 13 (1892).
[34] M. J. Glimcher, in *Medical Mineralogy and Geochemistry* (Eds.: N. Sahai and M. A. A. Schoonen), Mineralogical Society of America, Washington, D.C., pp. 223-282 (2006).
[35] A. Boskey, *Elements* 3, 385 (2007).

[36] W. F. de Jong, *Recueil des Travaux Chimiques des Pays-Bas* 45, 445 (1926).
[37] S. Náray-Szabó, *Zeitschrift für Kristallographie* 75, 387 (1930).
[38] M. Mehmel, *Zeitschrift für Kristallographie* 75, 323 (1930).
[39] M. Mehmel, *Zeitschrift für Physikalische Chemie [B]* 15, 223 (1931).
[40] C. A. Beevers and D. B. McIntyre, *Mineralogical Magazine* 27, 254 (1946).
[41] G. Trömel and H. Möller, *Zeitschrift für Anorganische und allgemeine Chemie* 206, 227 (1932).
[42] A. Schleede, W. Schmidt, and H. Kindt, *Zeitschrift für Elektrochemie und angewandte physikalische Chemie* 38, 633 (1932).
[43] R. Klement, *Zeitschrift für Anorganische und allgemeine Chemie* 237, 161 (1938).
[44] R. Klement, *Die Naturwissenschaften* 26, 145 (1938).
[45] S. B. Hendricks, M. E. Jefferson, and V. M. Mosley, *Zeitschrift für Kristallographie, Kristallgeometrie, Kristallphysik, Kristallchemie* 81, 352 (1932).
[46] C. Huggins, *Physiological Reviews* 17, 119 (1937).
[47] S. B. Hendricks and W. L. Hill, *Science* 96, 255 (1942).
[48] S. Eisenberger, A. Lehrman, and W. D. Turner, *Chemical Reviews* 26, 257 (1940).
[49] W. F. Neuman, T. Y. Toribara, and B. J. Mulryan, *Journal of the American Chemical Society* 75, 4239 (1953).
[50] W. F. Neuman and M. W. Neuman, *The chemical dynamics of bone mineral*, University of Chicago Press, Chicago (1958).
[51] G. W. Burnett and J. Zenewitz, *Journal of Dental Research* 37, 581 (1958).
[52] M. F. Little, E. S. Cueto, and J. Rowley, *Archives of Oral Biology* 7, 173 (1962).
[53] M. F. Little and F. S. Casciani, *Archives of Oral Biology* 11, 565 (1966).
[54] H. M. Myers, *Nature* 206, 713 (1965).
[55] N. Myrberg, *Transactions of the Royal Schools of Dentistry, Stockholm and Umea* series 2, no. 14, 1 (1968).
[56] D. Carlström, J. E. Glas, and B. Angmar, *Journal of Ultrastructural Research* 8, 24 (1963).
[57] G. P. Brophy and J. T. Nash, *American Mineralogist* 53, 445 (1968).
[58] D. R. Simpson, *American Mineralogist* 53, 432 (1968).
[59] E. F. Cruft, C. O. Ingamells, and J. Muysson, *Geochimica et Cosmochimica Acta* 29, 581 (1965).
[60] D. R. Simpson, *American Mineralogist* 49, 363 (1964).
[61] D. McConnell, *Clinical Orthopaedics* 23, 253 (1962).
[62] D. McConnell, *Archives of Oral Biology* 10, 421 (1965).
[63] F. Paulik, K. Eröss, J. Paulik, T. Farkas, and T. Vizkelety, *Hoppe-Seyler's Zeitschrift für Physiologische Chemie* 350, 418 (1969).
[64] J. C. Elliott, *Calcified Tissue International* 3, 293 (1969).
[65] D. McConnell, *American Mineralogist* 55, 1659 (1970).
[66] S. J. Joris and C. H. Amberg, *Journal of Physical Chemistry* 75, 3167 (1971).
[67] S. J. Joris and C. H. Amberg, *Journal of Physical Chemistry* 75, 3172 (1971).
[68] F. S. Casciani, in *Tooth Enamel II: Its Composition, Properties, and Fundamental Structure* (Eds.: R. W. Fearnhead and M. V. Stack), The Williams and Wilkins Company, Baltimore, pp. 14-23 (1971).
[69] G. H. Dibdin, *Archives of Oral Biology* 17, 433 (1972).

[70] M. Vignoles, Contribution á l'étude de l'influence des ions alcalins sur la carbonation dans les sites de type B des apatites phosphocalcique, Ph.D. thesis, Institut National Polytechnique de Toulouse, France, 1973.
[71] G. Bonel, J. C. Labarthe, and M. Vignoles, in *Physico-Chemie et Cristallographie des Apatites d'Interêt Biologique* (Ed.: no. 2. Colloques Internationaux du Centre National de la Recherche Scientifique), CNRS, Paris, pp. 117-125 (1973).
[72] R. M. Wilson, J. C. Elliott, S. E. P. Dowker, and R. I. Smith, *Biomaterials* 25, 2205 (2004).
[73] R. Z. LeGeros, G. Bonel, and R. Legros, *Calcified Tissue International* 26, 111 (1978).
[74] J. D. Termine, E. D. Eanes, D. J. Greenfield, M. U. Nylen, and R. A. Harper, *Calcified Tissue International* 12, 73 (1973).
[75] D. W. Holcomb and R. A. Young, *Calcified Tissue International* 31, 189 (1980).
[76] R. A. Young and D. W. Holcomb, *Calcified Tissue International* 34, S17-S32 (1982).
[77] A. Mortier, J. Lemaitre, and P. G. Rouxhet, *Thermochimica Acta* 143, 265 (1989).
[78] E. A. P. De Maeyer, R. M. H. Verbeeck, and D. E. Naessens, *Inorganic Chemistry* 32, 5709 (1993).
[79] A. C. Tas, *Biomaterials* 21, 1429 (2000).
[80] J. P. Lafon, E. Champion, D. Bernache-Assollant, R. Gibert, and A. M. Danna, *Journal of Thermal Analysis and Calorimetry* 72, 1127 (2003).
[81] M. G. Taylor, K. Simkiss, S. F. Parker, and P. C. H. Mitchell, *Physical Chemistry Chemical Physics* 1, 3141 (1999).
[82] A. Antonakos, E. Liarokapis, and T. Leventouri, *Biomaterials* 28, 3043 (2007).
[83] J. Reyes-Gasga, R. García-García, M. J. Arellano-Jiménez, E. Sanchez-Pastenes, G. E. Tiznado- Orozco, I. M. Gil-Chavarria, and G. Gómez-Gasga, *Journal of Physics D: Applied Physics* 41, 1 (2011).
[84] J. Shi, A. Klocke, M. Zhang, and U. Bismayer, *American Mineralogist* 88, 1866 (2003).
[85] S. Cazalbou, C. Combes, D. Eichert, and C. Rey, *Journal of Materials Chemistry* 14, 2148 (2004).
[86] C. Rey, C. Combes, C. Drouet, A. Lebugle, H. Sfihi, and A. Barroug, *Materialwissenschaft und Werkstofftech* 38, 996 (2007).
[87] C. Rey, C. Combes, C. Drouet, H. Sfihi, and A. Barroug, *Materials Science and Engineering C* 27, 198 (2007).
[88] T. I. Ivanova, O. V. Frank-Kamenetskaya, A. B. Kol'tsov, and V. L. Ugolkov, *Journal of Solid State Chemistry* 160, 340 (2001).
[89] F. W. Wehrli and M. A. Fernández-Seara, *Annals of Biomedical Engineering* 33, 79 (2005).
[90] A. Kaflak-Hachulska, A. Samoson, and W. Kolodziejski, *Calcified Tissue International* 73, 476 (2003).
[91] E. E. Wilson, A. Awonusi, M. D. Morris, D. H. Kohn, M. M. J. Tecklenburg, and L. W. Beck, *Journal of Bone and Mineral Research* 20, 625 (2004).
[92] E. E. Wilson, A. Awonusi, M. D. Morris, D. H. Kohn, and M. M. J. Tecklenburg, *Biophysical Journal* 90, 3722 (2006).
[93] R. M. Wilson, J. C. Elliott, S. E. P. Dowker, and L. M. Rodriguez-Lorenzo, *Biomaterials* 26, 1317 (2005).
[94] R. M. Wilson, S. E. P. Dowker, and J. C. Elliott, *Biomaterials* 27, 4682 (2006).
[95] H. E. Mason, A. Kozlowski, and B. L. Phillips, *Chemistry of Materials* 20, 294 (2008).

[96] F. M. McCubbin, H. E. Mason, H. Park, B. L. Phillips, J. B. Parise, H. Nekvasil, and D. H. Lindsley, *American Mineralogist* 93, 210 (2008).
[97] H. E. Mason, F. M. McCubbin, A. Smirnov, and B. L. Phillips, *American Mineralogist* 94, 507 (2009).
[98] C. H. Yoder, J. D. Pasteris, K. N. Worcester, and D. V. Schermerhorn, *Calcified Tissue International* 90, 60 (2012).
[99] J. Jeanjean, S. McGrellis, J. C. Rouchaud, M. Fedoroff, A. Rondeau, S. Perocheau, and A. Dubis, *Journal of Solid State Chemistry* 126, 195 (1996).
[100] G. Graziano, *Chemical Physics Letters* 396, 226 (2004).
[101] J. C. Labarthe, Contribution à l'étude de la structure et des propriétés des apatites carbonatées de type B phospho-calcique, Ph.D. thesis, Université Paul Sabatier, Toulouse, 1972.
[102] J. C. Labarthe, G. Bonel, and G. Montel, *Annales de Chimie (Paris)* 8, 289 (1973).
[103] W. Kolodziejski, *Topics in Current Chemistry* 246, 235 (2005).
[104] A. Kaflak and W. Kolodziejski, *Journal of Molecular Structure* 990, 263 (2011).
[105] J. D. Pasteris, C. H. Yoder, and B. Wopenka, *Geological Society of America Abstracts with Programs* 42, 220 (2010).
[106] D. McConnell, *Archives of Oral Biology* 3, 28 (1960).
[107] A. Ebrahimpour, M. Johnsson, C. F. Richardson, and G. H. Nancollas, *Journal of Colloid and Interface Science* 159, 158 (1993).
[108] S. Koutsopoulos, *Journal of Biomedical Materials Research* 62, 600 (2002).
[109] K. D. Rogers and P. Daniels, *Biomaterials* 23, 2577 (2002).
[110] K. Rogers and P. Zioupos, *Journal of Materials Science Letters* 18, 651 (1999).
[111] H. C. W. Skinner, *Mineralogical Magazine* 69, 621 (2005).
[112] R. Z. LeGeros and J. P. LeGeros, in *Phosphate Minerals* (Eds.: J. O. Nriagu and P. B. Moore), Springer-Verlag, New York, pp. 351-395 (1984).
[113] S. Mann, *Biomineralization: Principles and Concepts in Bioinorganic Materials*, Oxford University Press, Oxford (2001).
[114] S. V. Dorozhkin, *Acta Biomateriala* 6, 4457 (2010).
[115] L. Bertinetti, C. Drouet, C. Combes, C. Rey, A. Tampieri, S. Coluccia, and G. Martra, *Langmuir* 25, 5647 (2009).
[116] K. A. Gross and C. C. Berndt, in *Phosphates -- Geochemical, Geobiological and Material Importance, 48* (Eds.: M. J. Kohn, J. Rakovan, and J. M. Hughes), Mineralogical Society of America, Chantilly, Virginia, pp. 631-672 (2002).
[117] Y. Hong, H. Fan, B. Li, B. Guo, M. Liu, and X. Zhang, *Materials Science and Engineering R* 70, 225 (2010).
[118] M. Lombardi, P. Palmero, K. Haberko, W. Pyda, and L. Montanaro, *Journal of the European Ceramic Society* 31, 2513 (2011).
[119] R. M. Urban, *Clinical Orthopaedics and Related Research* 459, 110 (2007).
[120] L. Podaropoulos, A. A. Veis, S. Papadimitriou, and C. Alexandridis, *The Journal of Oral Implantology* 35, 28 (2009).
[121] R. Crespi, P. Capparé, and E. Gherlone, *Journal of Periodontology* 80, 210 (2009).
[122] H.-Y. Song, A. H. M. E. Raman, and B.-T. Lee, *Journal of Materials Science: Materials in Medicine* 20, 935 (2009).

In: Calcium Phosphate
Editor: Robert B. Heimann

ISBN: 978-1-62257-299-1

Chapter 4

PUTTING OXYHYDROXYAPATITE INTO PERSPECTIVE: A PATHWAY TO OXYAPATITE AND ITS APPLICATIONS

Karlis A. Gross [*] ***and Liene Pluduma***
Biomaterials Research Laboratory,
Biomaterials and Biomechanics Institute,
Riga Technical University, Riga, Latvia

ABSTRACT

Oxyhydroxyapatite and oxyapatite impact the fields of medicine, energy generation, sensing, and photonics. While oxyapatite may be stabilized with trivalent cations, oxyhydroxyapatite is inherently metastable and represents a pathway from hydroxyapatite to oxyapatite that subsequently decomposes to tricalcium phosphate and tetracalcium phosphate. Hydroxyl ions in hydroxyapatite are most mobile and upon exposure to high temperatures are the first to leave the lattice to form oxyhydroxyapatite. There have been several attempts to obtain oxyapatite by heating, but its metastability and high dehydroxylation levels promote decomposition towards tricalcium and tetracalcium phosphates. However, upon cooling the hygroscopic nature readily leads to the return of hydroxyl ions into the lattice. Phase-pure oxyapatite at room temperature has not been successfully obtained yet. The possible microstructures formed by dehydroxylation during different thermal processes will be discussed with reference to oxyhydroxyapatite solid solutions or mechanical mixtures of hydroxyaptite and oxyapatite. The largest impediment to active research in this field is the non-availability of a user-friendly approach to measure the concentration of hydroxyl ions. Different approaches will be discussed. The properties of oxyapatite are ill-defined yet, but a comment will be made on what can be expected and how this may lead to a range of different applications.

---

[*] E-mail address: kgross@rtu.lv, liene.pluduma@rtu.lv

1. Introduction

The development of oxyapatite finds its roots in the thermal processing of hydroxyapatite (HAp). Oxyapatites were initially interpreted as apatites with a depleted hydroxyl concentration replaced by oxygen ions and defects. The emphasis of this chapter is placed on hydroxyapatite and different levels of dehydroxylation as a pathway to oxyapatite. Other starting compositions may be employed as alternative routes towards oxyapatite. Regardless of the process, it will be shown that scientific studies suggest that oxyapatite (OAp) is unstable and will readily absorb water and restore the oxyhydroxyapatite (OHAp) condition.

A database search in the Web of Science reveals that oxyapatite has been a term also given to britholites or oxybritholites. These materials have silicate substituted for phosphate. Since these are thermally and chemically more stable, they are receiving more attention amongst the oxyapatites with applications for fuel cells, oxygen sensors, lasers and luminescing materials.

The third category of oxyapatites has oxygen in even greater concentrations. This is possible by introducing interstitial oxygen associated with phosphate to balance a larger cation concentration, or by placing oxygen molecules in the anion column. There are three alternatives for including molecular oxygen where the hydroxyl ions would normally be positioned in hydroxyapatite (Table 1). These are all oxygenated apatites and incorporation into the lattice includes oxygen forms such as O_2^{2-}, O_2^{-} or O_2. Very little attention has been directed to these forms as they may have not been detected or went unnoticed. This will further contribute to the variability in the properties of apatites.

This critical review will collate the information available on oxyhydroxyapatite and oxyapatite. Since processing is the essential stepping stone for producing oxyapatites, the different approaches for achieving an oxyapatite will be reviewed.

Table 1. Different types of oxygen located in the channel of oxygen-rich apatites

Oxidation state	Name of ion	Name of apatite	Reference
O^{2-}	Oxide	Oxyapatite	[1]
O_2^{2-} (O^-)	Peroxide	Peroxyapatite, oxygenated apatite	[2, 3]
O_2^-	Superoxide*	Oxygenated apatite	[4, 5]
O_2	Oxygen	Oxygenated apatite	[6]

*Also referred to as dioxide. The obsolete name is hyperoxide.

The measurement of oxyapatite with reference to the hydroxyl concentration will reveal methods for determining the level of dehydroxylation. Very limited information is available on the properties of oxyhydroxyapatites and oxyapatites, but an insight will be provided on what is presently available in the literature. As a result, very few applications are presently available, with the most activity on britholites. Some potential new applications will be discussed. This work will be presented with reference to ongoing work on manufacturing oxyapatite and hydroxyl ion measurement techniques.

2. Structure of Hydroxyapatite

Apatite is the most widely found phosphate-containing mineral. The general chemical formula of calcium apatite is $Ca_5(PO_4)_3X$, where X is an electronegative element or ion, for example a halogen ion or an OH ion. Many substitutional elements can be located in the crystal structure – metal cations (K^+, Na^+, Mn^{2+}, Ni^{2+}, Cu^{2+}, Zn^{2+}, Cd^{2+}, Fe^{3+} etc.) replace Ca^{2+} and anion complexes (AsO_4^{3-}, SO_4^{2-}, CO_3^{2-}, etc.) replace the phosphate.

Half of the periodic table elements can be incorporated in the atomic assembly of apatite. Usually, replacement of ions occurs at trace element levels, but in some cases extensive solid solutions can accommodate large concentrations of some substituting ions and complexes [7-9]. The ability to accommodate a range of substitutional elements has led to its description as isomorphic, nonstoichiometric compounds with structural defects. This modification alters the material stability, solubility and other physical properties [10].

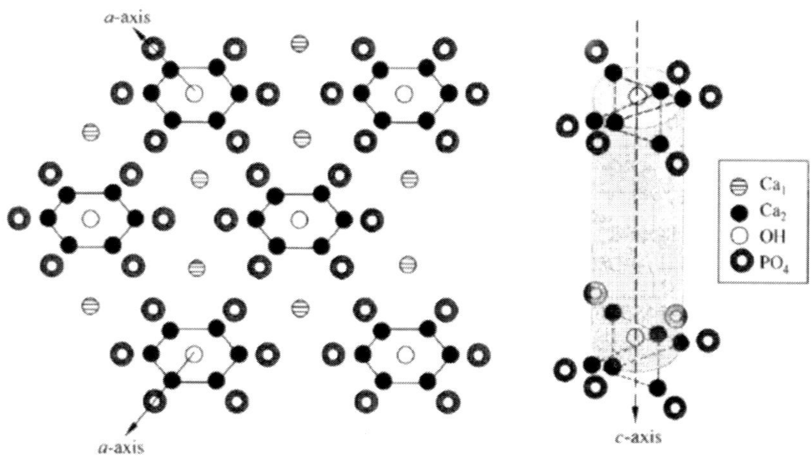

Figure 1. The positioning of Ca^{2+}, PO_4^{3-} and OH^- in the HAp structure (modified after Ref. [12]).

Biological apatite has a hexagonal crystal structure with a $P6_3/m$ space group, a = b = 9.36 - 9.64 Å, c = 6.78 - 6.90 Å, $\alpha = \beta = 90°$, $\gamma = 120°$. The crystal parameters are dependent on the type of X^- ion. To provide the lowest energy configuration for the elements in solids, the PO_4^{3-} are distributed as regular tetrahedra where P^{5+} is positioned in the centre, but O^{2-} is in the four corners.

The hexagonal unit cell consists of Ca^{2+} and PO_4^{3-} distributed within the unit cell (Figure 1). Ten calcium atoms can be divided into two different types: four Ca atoms are located in Ca(I) positions and the remaining six atoms take up Ca(II) locations in the corner of the hexagonal unit cell, surrounding the X^- ions. All six Ca(II) ions are not located within one plane. Each three Ca atoms are organized into a triangle. The X^- ion is centred inside the hexagonal unit cell. Similarly, the PO_4^{3-} are located in two triangles. The PO_4^{3-} create the base structural network that ensures stability of the apatite structure [11-13].

A wide range of calcium apatites can form in response to the type and location of the X^- ion [8]. The most familiar are fluorapatite ($Ca_5(PO_4)_3F$), chlorapatite ($Ca_5(PO_4)_3Cl$) and hydroxyapatite ($Ca_5(PO_4)_3OH$). While the fluorine atom is the smallest of the three anions and located in the centre of the Ca(II) triangle, the OH^- anion complex and the Cl^- anion are too large to remain in the plane of the Ca triangle and therefore are shifted a little above or below the plane (Cl^- is positioned further from the calcium triangle plane than OH^-) as shown in Figure 2.

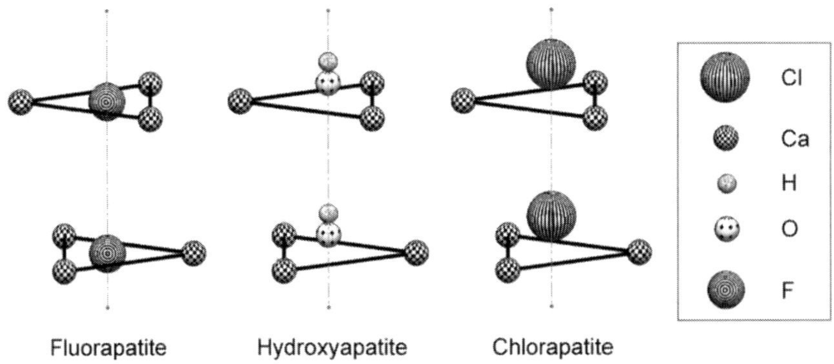

Figure 2. Location of anions in the channel for fluorapatite, hydroxyapatite and chlorapatite.

The ideal structure has a regular arrangement of OH^- in a 'head-to-tail' arrangement to form columns within the channels. The orientation of each OH^- ion may freely change between the columns [7].

It has been assumed that the characteristic crystal structure of hydroxyapatite has a space group of $P6_3/m$, but stoichiometric HAp with oriented OH^- undergoes a phase change to the monoclinic crystal structure with a $P2_1/b$ space group. While it is known that the monoclinic structure is stable at lower temperatures, there is no unified agreement on the temperature of the phase change. Henning *et al* [14] state that the conversion temperature is below 200 °C, another source suggests 200 – 250 °C [15], but a third indicates a temperature as high as 370 °C [8].

Structural water is released during thermal processing (sintering, coating production, thermal spraying, etc). The compound that displays the characteristic apatite structure and does not contain OH^- ions, but has O^{2-} ions and vacancies (V) is called oxyapatite [$Ca_{10}(PO_4)_6O_xV_x$]. In 1912, Rogers introduced the name „voelckerite", for a mineral with the oxyapatite structure, named after J.A. Voelcker, who was the first to recognize a depletion of halogen ions while studying apatites [16]. Until now, all attempts at making a pure calcium oxyapatite have appeared to be unsuccessful as the synthesis products always contained a small concentration of OH^- ions, forming a solid solution of HAp and OAp called oxyhydroxyapatite [$Ca_{10}(PO_4)_6(OH)_{2-2x}O_xV_x$]. During dehydration, a single water molecule is released from two OH^- ions. Every second calcium triangle retains an O^{2-} ion. Consequently, the vacancy in the other location causes repulsion of Ca^{2+} ions. Therefore, the Ca(II) triangles at z=1/4 and z=3/4 are no longer identical and this reduces the crystal symmetry [17]. This has been confirmed by high resolution electron microscopy that also reveals a rotation of the calcium triangles to compensate for the missing OH^- ion. As a result, oxyapatite produced by removing hydroxyl groups from the HAp structure displays a third order symmetry in the [001] direction, unlike hydroxyapatite with a $P6_3/m$ space group that has a sixth order symmetry. This experiment assumed that pure oxyapatite formed from the exposure of HAp to the electron beam. However, oxyhydroxyapatite could have been formed from the incomplete dehydroxylation throughout the thickness of the crystal observed under the electron beam [14].

The characteristic oxyapatite structure may exist not only in calcium phosphate apatites, but also in a wide variety of apatites where the electronegative species can be completely or partially replaced with O^{2-} and a vacancy. The crystal structure parameters and the resulting material properties depend on the elements that establish the structure or are substituted into the lattice [18]. There are a host of possible substitutions in oxyapatite with the following formula: $Me_5(Z0_4)_3OV$. Depending on the type of substitution, the charge balance is equilibrated in the apatite structure with the introduction of defects, that lead to nonstoichiometry. A new cation replacement will usually take place at a Ca^{2+} ion site. Gomes *et al.* [19] showed that Zn^{2+} ions did not replace Ca^{2+} but were positioned in interstitial sites and created solid solutions, $Ca_{10}Zn_x(PO_4)_6O_{2x}(OH)_{2-2x}$. To maintain charge balance, OH^- ions are replaced with O^{2-} ions within the channel to form -O-Zn-O- chains [20]. Other experiments have shown an influence of ferric ion substitution on the hydroxyapatite structure. At small ferric ion concentration, the hydroxyl ion is substituted with carbonate ($Ca^{2+} + OH^- \rightarrow Fe^{3+} + CO_3^{2-}$). However, oxygen ions are introduced in the interstitial position for higher ferric ion concentrations ($2Ca^{2+} + V \rightarrow 2Fe^{3+} + O^{2-}$). Despite the smaller size of Fe^{3+} compared to Ca^{2+}, incorporation of ferric ions into the apatite structure results in an expansion of the unit cell that suggests the inclusion of oxygen ions, whereby PO_4 tetrahedra are converted to PO_5 trigonal pyramids [20].

The apatite structure is very flexible through accommodating many substitutional elements, but these structural variations can also allow oxygen to reside in different oxidation levels. In previous situations, the oxygen is incorporated in the form of an O^{2-} ion; however, there are cases where peroxyapatite or oxygen-rich apatite can form. Table 1 shows the different forms of oxygen that further increases the complexity of the apatite composition.

Heating HAp or an A-type carbonated apatite in dry oxygen leads to incorporation of oxygen with an oxidation state of -1. Since peroxyapatite $Ca_{10}(PO_4)_6O_2$ is diamagnetic, the unit cell in the c-axis direction contains one paramagnetic peroxide ion O_2^{2-} and a vacancy, and not two paramagnetic O^- ions. An increase in the O^- ion concentration decreases the unit cell crystal parameters. The modification from an oxyapatite can be schematically shown with the following relationship: $O^{2-} + ½O_2 \rightarrow O_2^{2-}$. Peroxyapatite can be identified by Raman spectroscopy that reveals an O-O vibrational band at 750 cm^{-1}. Peroxyapatite is thermally unstable and can easily return to an oxyhydroxyapatite [16, 21].

Oxygen-rich apatite can be obtained through the hydrolysis of β-TCP in a 30% H_2O_2 solution for 12 hours. The calcium deficient apatite has a Ca/P of 1.58 and dissolves in acid to release ~ 2.1 wt% O_2. Chemical analysis shows that a large proportion of the oxygen exists in an O^- form. Spectroscopy has shown a decrease in the OH^- intensity band for samples synthesized in the presence of concentrated H_2O_2. These results suggest that oxygen replaces the OH^- ions in the apatite structure. Electron spin resonance has not shown the existence of O^- ions, but identified the superoxide (O_2^-) ion instead. Further analysis with Raman spectroscopy has revealed that oxygen rich apatites contain O_2 un O_2^{2-} ions, acidic phosphate and small concentrations of O_2^- ions. Heating these samples to 600 °C increases the O_2 content but lowers the O_2^{2-} and acidic phosphate concentration to form $P_2O_7^{4-}$ groups and water. Molecular oxygen is no longer detected at 700 °C. During heating O_2^{2-} disassociates into O_2 and O^{2-}, and O^{2-} reacts with atmospheric moisture to form OH^- ions [16].

The lanthanide-silicate apatite structure, just like the calcium phosphate apatite can have a cation replacement in the Si and La sites. Substitution of a divalent with a trivalent cation will increase the oxide content. Depending on the composition, the lanthanide-silicate apatites exhibit elevated oxygen conductivity arising from the oxygen ion mobility associated with the interstitial oxygen [22].

3. Production of Oxyhydroxyapatite and Oxyapatite

The production of oxyapatites has not been extensively reviewed, and so the following serves to cover the formation using different approaches. An initial reflection will be directed to biological apatites and then assess the formation of oxyapatites during thermal processing. A short review will be provided on apatites that may have elevated oxygen concentrations stemming from a surface treatment.

3.1. Depleted Hydroxyl Levels in Biological Apatites

Apatites formed in biological systems exhibit a range in chemistry, crystal size, crystallinity and OH^- content. The different cystalline characteristics allows these apatites to satisfy the different functional requirements in teeth and bone.

Pasteris *et al.* [23] compared apatites from different sources and noted an increase in OH^- content with samples exhibiting a higher degree of order (See also Chapter 3 of this volume. The Editor). Analyses with Raman and FTIR spectroscopy showed that nano-sized apatites display a lower hydroxyl ion concentration.

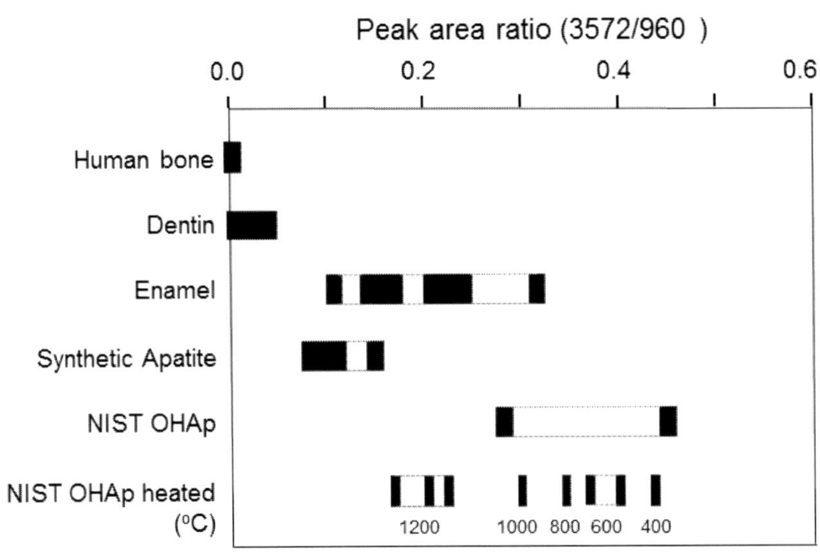

Figure 3. Hydroxyl ion concentration in biological and synthetic hydroxyapatites determined from Raman spectra. Semi-quantitative analysis of OH^- was obtained from the peak area ratio of the OH^- peak at 3572 cm^{-1} relative to the phosphate peak at 960 cm^{-1} (modified after [23]).

It is not clear yet whether OH^- is included in a different form, possibly as water molecules [24, 25], or whether the nano-dimensions make the OH^- vibrations inactive to the infra-red and laser radiation.

The crystallites in dental enamel are larger than in dentine or bone, and have a larger OH^- content as shown in Figure 3. This should be interpreted with care, since a larger CO_3^{2-} content also found in bone and dentine may influence the OH^- concentration. Both chemistry and crystal size affect the OH^- content in biologically formed apatites.

The apatite chemistry and resulting properties change in response to the source and processing history of hydroxyapatite. Hydroxyapatite, for use in the implant industry, is usually processed at high temperatures either by sintering at 1000 – 1350°C or by thermal spraying that exposes particles to significantly higher temperatures. Exposure to high temperatures can change the physical, chemical, and mechanical properties [26, 27].

3.2. Dehydroxylation

As early as 1933 it was shown that a considerable amount of constitutional water was released from hydroxyapatite upon heating to 1350°C for 7.5 hours [28]. X-ray diffraction was not able to show any marked change in the crystal unit cell parameters. Later in 1957, the release of structural water was noted to provide a contraction in the a-axis of 0.02 Å [29]. Heating will result in the following sequence of events:

- Dehydration (loss of adsorbed water);
- Dehydroxylation (loss of structural water) and formation of OHAp and then OAp
- Decomposition of OAp

Water is released from the surface and then from within the structure. Surface water is released at temperatures lower than 200 °C, but water bound in cracks or pores can be evolved up to 500 °C [30-32] (Figure 4).

Further heating will initiate structural changes and produce an oxygen ion with a water molecule [8, 14, 34]. The energetic vibration will dissociate the hydroxyl ion as the hydrogen ion bonds with the neighbouring hydroxyl ion to form a water molecule: $2OH^- \rightarrow O^{2-} + H_2O$. Gradual dehydration to OAp is an endothermic process and does not require a lot of energy (E= + 83.2 kJ/mol), therefore heating readily produces dehydroxylation [34, 35]. This process may be described by five possible kinetic processes as shown in Table 2.

Dehydroxylation is a thermokinetic process and hence occurs over a large temperature range. It depends on the hydroxyapatite characteristics (Ca/P ratio, trace elements and their concentration, crystal size, crystallinity, sample size) and processing conditions (water partial pressure, atmosphere, temperature, heating rate and time) [31, 36, 37]. While in air the process commences at 900 °C [35] in vacuum it starts already at a lower temperature of 850 °C [38]. The rate of dehydroxylation can be increased at higher temperatures or with a higher vacuum. Present work reveals that dehydroxylation may commence at even lower temperatures at higher vacuum levels.

Dehydroxylation has been prevented up to 1100 °C [31], 1200 °C [39] or even 1300 °C [35, 37] by bubbling an inert gas through water before passing it through the furnace. An atmosphere of sufficiently high water vapour partial pressure was very early proposed to maintain the hydroxyapatite composition [40].

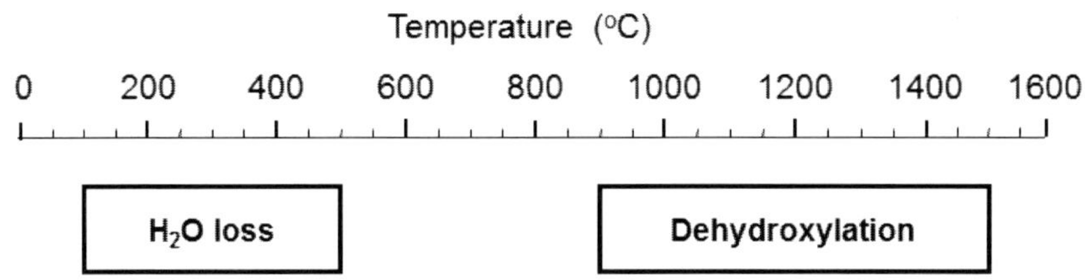

Figure 4. Temperature range of removal of adsorbed (left) and structural (right) water [33].

Table 2. The activation energy of different stages in forming oxyapatite

Kinetic process	E (kJ/mol)	Reference
OH^- liberation from the HAp lattice	530	[36]
OH^- diffusion within the lattice	190, 230	[36], [30]
$2OH^- \rightarrow H_2O + O^{2-}$	235, 251	[36], [30]
Transport of water through the lattice	?	
Lattice constitution of oxyapatite	410, 440	[36], [30]

This supports the sintering process by maintaining the species that most easily diffuse in the apatite lattice. The chemistry can thus be maintained when only processing hydroxapatite, and also solid solutions with other anions such as hydroxyfluorapatites [41] and mechanical mixtures where hydroxyapatite is just one of the phases [42]. Other reports show that dehydroxylation can be suppressed to even higher temperatures 1200 – 1450 °C [35, 43, 44]. If the presence of water vapour can maintain the hydroxyapatite composition, then the absence of moisture should facilitate the formation of oxyapatite.

3.3. Production of Oxyapatite

The production of oxyapatite has been based on anion exchange in the channels that readily introduce oxygen ions, hence the selection of hydroxyapatite or A-type carbonated apatite [16]. Even sulphate exchange has reportedly produced oxyapatite [45]. Although hydroxyapatite has been the most widely investigated, it was never completely dehydroxylated to oxyapatite. In all cases, a vacuum at high temperature has been used to encourage the removal of ions to produce water, carbon dioxide or sulphur trioxide. The easier diffusion of ions in the column will facilitate the formation of oxyapatite, but the other driving force from different curvatures around loosely contacting particles will support other diffusion paths that may lead to hydroxyl positions further from the material surface. This explains the need for long heating times. Hydroxyapatite as the starting composition will lead to oxyhydroxyapatite [34] that will transform to an oxyapatite according to the following reactions:

Hydroxyapatite to oxyapatite

$Ca_{10}(PO_4)_6(OH)_2 \rightarrow Ca_{10}(PO_4)_6(OH)_{2-2x}O_xV_x + xH_2O$ (HAp → OHAp)

$Ca_{10}(PO_4)_6(OH)_{2-2x}O_xV_x \rightarrow Ca_{10}(PO_4)_6O_xV_x + (1-x)H_2O$ (OHAp → OAp),

where V represents an OH^- vacancy in the anion column parallel to the c-axis. An equilibrium is thus established between the hydroxyl and oxygen ions such as

$2\ OH^-(s) \leftrightarrows O^{2-}(s) + V(s) + H_2O(g)$.

A-type carbonated apatite to oxyapatite

$Ca_{10}(PO_4)_6CO_3 \rightarrow Ca_{10}(PO_4)_6O + CO_2$ (CO_3-Ap → OAp)

Sulfated apatite to oxyapatite

$Ca_{10}(PO_4)_6SO_4 \rightarrow Ca_{10}(PO_4)_6O + SO_3$ (SO_4-Ap → OAp)

Complete dehydroxylation produces a row of oxygen ion and vacancy couples (x=1) [34]. X-ray diffraction studies have reported a change from the hexagonal to a monoclinic structure at 35% dehydroxylation [14]. The differences from the two structures in the XRD pattern are very slight and therefore may go unnoticed. Studies on the removal of CO_2 have revealed an immiscibility gap at 850 °C between carbonated apatite and oxyapatite [1].

Present work on the production of oxyapatite is taking a completely different approach based on the notion that anion removal is easier from a less dense structure. Heating an amorphous phase with a Ca/P ratio of 1.67 under vacuum can desorb intercluster water molecules followed by water removal from within the amorphous clusters. Water loss not only removes water from further structural rearrangements, but also provides denser calcium and phosphate ion assemblies just requiring small molecular reconfigurations for crystallization to oxyapatite. This approach has led to an oxyapatite as low as 900 °C. Ongoing experiments are revealing still lower formation temperatures.

In an article addressing the thermal stability of deuterioxyapatite [$Ca_{10}(PO_4)_6(OD)_2$], high temperature FTIR spectroscopy traditionally used to study the water in clays showed that heating produces a decrease of the intensity of the OD stretching band intensity at 2635 cm^{-1}. Based solely on the FTIR data, the loss of the OD peak at 120 °C was interpreted as the indication of formation of oxyapatite [46]. Other techniques at high-temperature Raman spectroscopy, and X-ray diffraction could now provide complimentary evidence to confirm the formation of oxyapatite. However, while the incorporation of deuterium in the reactants provides a unique approach it is costly and thus not practical.

3.4. Decomposition and Reconstitution

Heating at too high a temperature for an unnecessarily long duration will provide sufficient diffusion to create a nuclei for decomposition. Several reactions have been proposed with tricalcium phosphate present as one of the decomposition phases. The phase diagram predicts the formation of α-tricalcium phosphate and tetracalcium phosphate [40] that has been experimentally confirmed [31]. The reaction is:

$$Ca_{10}(PO_4)_6O_xV_x \rightarrow 2\ Ca_3(PO_4)_2 + Ca_4O(PO_4)_2\ (OAp \rightarrow TCP + TTCP)$$

Research findings show that it may be possible to form other decomposition phases such as pyrophosphate or calcium oxide [27, 31]. These other decomposition reactions include:

$$2\ Ca_{10}(PO_4)_6O_xV_x \rightarrow 2\ Ca_3(PO_4)_2 + Ca_2P_2O_7 + 3\ Ca_4P_2O_9$$

$$Ca_{10}(PO_4)_6O_xV_x \rightarrow 3\ Ca_3(PO_4)_2 + CaO$$

Heating a calcium-rich apatite with a Ca/P of 1.74 to 1350 °C reportedly produced a composition close to oxyapatite. An unusual suggestion indicates that the anion column may also contain excess calcium ions [47]. Decomposition of oxyapatite always produces tricalcium phosphate, but the second phase may also be a calcium rich oxyapatite.

Additional phases, combined with hydroxyapatite with the intention of improving the mechanical strength, may lead to faster decomposition. While the loss of hydroxide marks the first event before decomposition, addition of other oxides can lead to the loss of calcium first. Addition of titania will decrease the decomposition temperature to 750 °C, but yttria-stabilized zirconia will initiate decomposition only at 1100 °C [48]. Calcium stabilized zirconia was shown to prevent the decomposition of hydroxyapatite and so provides a stable apatite for thermal processing.

Heating of hydroxyapatite to 1400 °C has been found to create an 8 μm thick outer layer that transforms to tricalcium phosphate and calcium oxide, leading to cracking of the entire pellet [49]. Thin regions in scaffolds have retained their integrity with sintering returning strength to the new phase assemblage through densification.

The structural and phase changes at high temperatures may be reversed in the presence of water vapour in the atmosphere, a process known as reconstitution. A 75% dehydroxylated hydroxyapatite produced in a thermogravimetric analysis at 1000 °C has then been rehydroxylated in the presence of water vapour [16]. It is not clear whether the outside shell transforms to hydroxyapatite or whether an oxyhydroxyapatite solid solution is formed throughout the apatite. Another study has shown the reversion process or 'back diffusion through incorporation of deuterioxyl groups after exposure of a dehydroxylated hydroxyapatite to deuterium oxide vapour at 800 °C [50]. The decomposition products can also be reversibly transformed back to hydroxyapatite in the presence of water vapour [31, 35].

Cooling of oxyapatite under high vacuum has been thought to incorporate hydroxyl ions from the atmosphere [1, 43]. Recent experiments in the authors laboratory show that the vacuum level is improved over a period of several days after a thermally produced oxyapatite is stored at room temperature under vacuum conditions. It is not known which chemical species were removed from the gas phase. Further work is required to determine whether the chemical species are just adsorbed onto the surface or included in the crystal structure.

There is a large affinity for water on both surfaces of the apatite crystal. Water is attached as a molecule on the (001) surface, but dissociates to form P-OH and Ca-OH termination on the (010) surface [51]. The stronger binding energy for water on the (001) face favours inclusion into the lattice by being located at the end of the OH column. The different surface interaction will affect the properties. Electrical conduction is dictated by surface-bound moisture below 700 °C, but dictated by bulk ionic diffusion of hydroxyl ions at higher temperatures [52].

3.5. Other Processing Routes for Manufacturing Apatites

Attempts to densify hydroxyapatite using novel sintering approaches may provide faster sintering or allow processing at lower temperatures that possibly delay dehydroxylation and decomposition. Microwave sintering produces uniform heating from the induced motion of ionic species under a high frequency electromagnetic field and so provides a 10-fold reduction in sintering time. The active motion of the hydroxyl ions supports faster sintering [53], and therefore retains a higher concentration of hydroxyl ions. While spark plasma sintering can further lower the sintering temperature, early studies show that decomposition to tricalcium phosphate and tetracalcium phosphate may still occur [54]. An applied electric field establishes conditions for poling and aligning of hydroxyl ions that apparently improves the bioactivity [55]. The positioning of the oxygen ion slightly below the plane of the calcium triangle provides conditions for establishing a charge without the requirement of poling. This effect has not been investigated due to the difficulty to produce oxyapatite.

The application of lasers for processing apatites has also been investigated. Lasers can induce thermal desorption of hydroxyl groups from silica surfaces [56] and similar effects are expected on apatites. A study of rapid prototyping by laser cladding has reported complete dehydroxylation followed by decomposition to tricalcium phosphate and tetracalcium phosphate [57]. Reduced heat input and the addition of water to the atmosphere during laser ablation prevents dehydroxylation [58, 59]. Since oxyapatite readily transforms to the decomposition phases, rapid quenching is required to maintain the apatite phase to low temperatures, but this may require processing of smaller molten areas, such as in thermal spraying.

Exposure to high energy oxygen ions has been proposed to modify the surface [60]. Changes to the microstructure have been noted to increase the bioactivity, but no mention has been made about the possible incorporation of additional oxygen ions into interstitial sites or replacement of hydroxyl ions.

Surface oxygen may be activated, and care should be taken in the processing of apatites. Exposure of low temperature apatites, heated to 200 °C and then UV irradiated can create oxygen radicals on the surface [61]. Apatites heated to 1150 °C do not form oxygen radicals, O_2^-, like the low temperature apatites [62].

3.6. Oxyapatite in Thermal Spray Coatings

The rapid quenching of partially molten or fully molten particles during thermal spraying provides an opportunity to capture the oxyapatite phase. Several models have been proposed to predict the location of dehydroxylated regions in a flattened solidified droplet, a “splat” [63-66]. The first model is an update from an earlier version [65], where the outside layer becomes hydroxyl-depleted and the particle core is either solid or molten (Figure 5). To simplify the model, oxyhydroxyapatite solid solutions are not included. For the case of a fully molten particle, crystalline phase formation depends on the quenching speed. The high quenching speed will prevent the crystallization and support the formation of an amorphous phase. Slower cooling rates are expected in the upper splat layer, or in splats that are sprayed onto previously deposited splats.

The hydroxyl content within the splat dictates the ease of crystallization [67]. A melt with a hydroxyapatite composition has a lower Gibbs free energy and therefore a larger probability for crystallization.

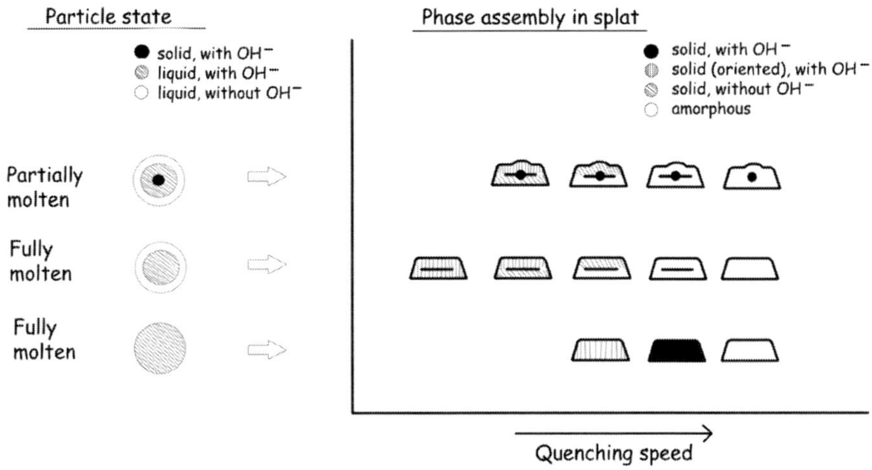

Figure 5. A model showing the formation of oxyapatite in response to the lower hydroxyl content in the molten droplet and the quenching speed of the solidifying droplet.

Since a rich hydroxyl concentration is expected to be retained in the core, the middle section of the flattened droplet will be the first to crystallize. The outer dehydroxylated layer will not have sufficient time for diffusion to form the decomposition phase, but with a lower driving force for crystallization to oxyapatite, will remain as an amorphous phase. The top of the droplet will cool more slowly, with heat dissipation through the already solidified droplet, and produce an oxyapatite in the upper part of the splat. This has been confirmed with Raman spectroscopy that has detected a lower amorphous phase content and a high oxyapatite content further from the substrate [68]. When cooling rates are even slower, the base of the splat will crystallize to an oxyapatite. At still slower quenching rates, the crystals will align with a <00*l*> orientation to provide a crystal orientation suitable for the fastest heat dissipation. The base of the splat, adjacent to the metallic substrate, will contain oxyapatite, the middle section will be composed of hydroxyapatite, but the top layer represent oxyapatite to reflect the change in composition throughout the splat.

For a partially molten particle, crystallization of the melt will occur more easily with the solid core providing a surface for heterogenous nucleation. As a consequence, a slightly higher quenching rate may support the crystallization of oxyapatite from this solid into the outer splat regions. In the last situation, where hydroxyl loss does not take place in the molten particle, crystallization will occur faster than in the dehydroxylated outer shell, but is expected to be slower than a partially molten particle, where recrystallization occurs from the unmolten core. For splats deposited onto previously formed splats, the cooling rate will be slower and so the splat types towards the left of the figure are more likely to occur.

The second model includes numerous scenarios, which also includes the undesirable effects from overheating [64]. Instead of predicting the different phases within a splat, it is assumed that the droplet solidifies before impact. Secondly, the model does not separate regions of different hydroxyl content (hydroxyapatite, oxyhydroxyapatite, oxyapatite) and places them all within the core. Furthermore, the decomposition of oxyapatite to tricalcium phosphate and tetracalcium phosphate is included in the model, but this may be prevented with the optimization of thermal spray parameters. The assumptions and the rich thermal behaviour of hydroxyapatite reflects the complexity of the resulting microstructure. Further work is required to determine formation and conditions for the different decomposition phases. The assemblage of decomposition phases has yet to be determined.

The plasma processing environment can quickly freeze high temperature phases such as oxyapatite. A water free plasma-processing environment has produced a 73% pure oxyapatite that upon contact with atmospheric moisture at room temperature is quickly reduced to 50% oxyapatite in the outer layer. Probing deeper depths within the coating with higher energy X-rays reveals that oxyapatite is retained within the core [69]. Processing in a moisture free environment, such as in low pressure plasma spraying offers the opportunity to further oxygenate the oxyapatite to produce a peroxyapatite or an apatite with molecular oxygen.

4. Measurement of Hydroxyl Concentration

The attention on oxyapatite has always been focused on the hydroxyl ions, with the understanding that an absence of hydroxyl ions corresponds to an oxyapatite structure. The reason for this arises from the many attempts trying to remove the hydroxyl ion. This section will therefore address the measurement of hydroxyl ions using a collection of complementary characterization and measurement techniques.

The varying degree of hydroxylation in hydroxyapatites and the lack of a suitable method for hydroxyl ion concentration quantification has given rise to many contradictory reports in the literature [70]. Different hydroxylation states will otherwise contribute to larger error bars when the results from many investigators are grouped together. Since the hydroxyl concentration not only influences the stoichiometry of hydroxyapatite but also the physical properties and biological response, it is important to be able to perform a quantitative analysis, that will further dictate the biological application [23]. The different techniques for this measurement have focused on

- Structure
  - X-ray diffraction (XRD)
  - Transmission electron microscopy (TEM)

- molecular bonding or local atomic arrangements
  - Fourier transform infra-red spectroscopy (FTIR)
  - Raman spectroscopy
  - Nuclear magnetic resonance spectroscopy (NMR), and

- thermal behavior
  - Thermogravimetric analysis (TGA)
  - Differential thermal analysis (DTA)

4.1. Structural Analysis

Despite the widespread use of X-ray diffraction to characterize the crystal structure and chemical phase composition, this technique does not clearly show the degree of hydroxylation. Since the replacement of two OH^- with one O^{2-} ion does not have a marked effect on the lattice parameters, XRD is not an easy method to differentiate stoichiometric hydroxyapatite from its dehydrated states. However, a shift of the (00*l*) interplanar spacings to lower angles can be used to determine the small increase in the size of the unit cell [69] as shown in Figure 6. Alternatively, the (00*l*) peak may be used to see the shift to smaller angles caused by dehydroxylation. This allows the effect of dehydroxylation to be observed in solid solutions. Easier detection is possible using mechanical mixtures of hydroxyapatite and a highly dehydroxylated phase that will produce a doublet, most easily seen at the shoulder of the (004) peak at a $2\theta = 53.1°$ with (CuK□ radiation) [69].

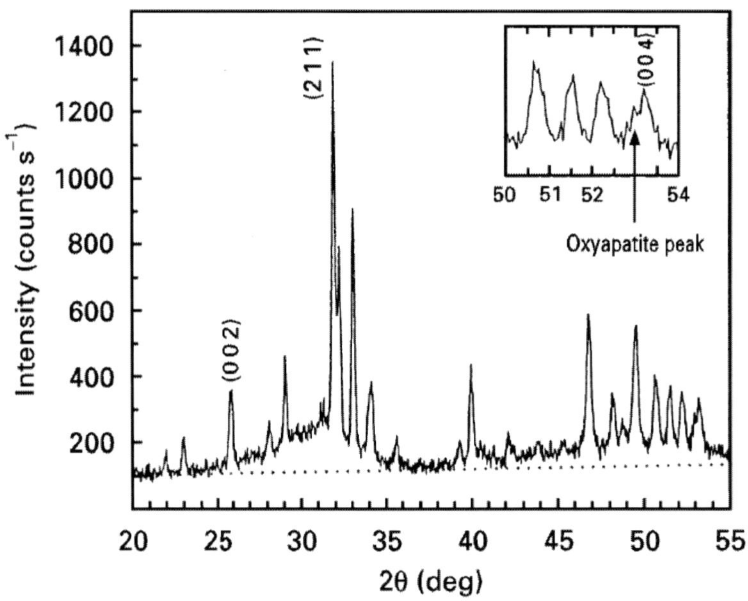

Figure 6. XRD pattern of a hydroxyapatite coating produced by air plasma spraying. The inset shows the appearance of a new phase (oxyapatite) at the shoulder of the (004) peak [69].

The (002) peak at 26.1° may also be used, however this is more complicated since the radiation from cathode-ray tubes can produce a secondary peak while residual stress can contribute to a peak shift. This requires particular attention to the source radiation and previous knowledge of the residual stress state of the sample being analysed. X-ray diffraction is useful to evaluate the crystal structure, but other methods are required to study the OH^- concentration.

Electron microscopy is also a valuable technique to determine the crystal structure, but requires a high vacuum for electron diffraction. Simultaneous exposure of hydroxyapatite to the electron beam and the high vacuum produces dehydration and consequent decomposition and thus precludes the detection of oxyapatite [14].

4.2. Spectroscopy

Both FTIR and Raman spectroscopy rely on the recognition of molecular bonding for quantitative analysis. Molecules can be identified from characteristic vibrations within the crystal structure. Vibrations are IR-active if there is a change in dipole moment, but are Raman-active if there is a change in the polarity of the molecule during vibration. Each method shows slightly different vibrational intensities, making these methods complementary. Apatite can thus be characterized by the

- absorption peak position that reflects the molecular vibration within a 3-D arrangement of surrounding ions
- peak width that reflects the crystallinity, and
- the vibrational peak height that reveals the concentration of a molecular species

Stoichiometric apatite displays only OH^- and PO_4^{3-} absorption bands. Nano-crystalline HAp exhibits a very weak OH libration at 630 cm^{-1} and a very weak vibration at 3572 cm^{-1} [23, 71]. Based on FTIR and Raman spectra only, it has been suggested that OH^- is not included in nano-sized crystals, but the actual effect of nano-sized dimensions on different spectral responses has not been explained. Heat treatment at increasingly higher temperatures decreases the absorption intensity at 630 cm^{-1}, thus reflecting the sensitivity of this peak to changes in temperature. However, the absorption peak at 3572cm^{-1} is also indicative of the OH^- concentration and becomes wider at higher temperatures as well as being more insensitive to OH^- concentration changes [71].

The other approach in quantifying the amount of oxyapatite is to focus on the oxyapatite features. Identification of OAp is possible with spectroscopy. The replacement of two monovalent ions with a vacancy and a bivalent anion removes the hydroxyl absorption bands and creates additional phosphate absorption bands. This is explained with two unequal phosphate ions – while one PO_4^{3-} is located closer to the vacancy, the other phosphate group is closer to the bivalent oxygen ion, O^{2-}. Oxyhydroxyapatite formed by heating HAp in vacuum at 800 – 1000 °C produces two additional FTIR absorption bands at 475 cm^{-1} and 433 cm^{-1}. Present work suggests that these absorption bands may move to higher wave numbers at greater levels of dehydroxylation. The intensity of these peaks depends on the degree of dehydroxylation, i.e. for dehydroxylation less than 50%, the intensity of the absorption band at 433 cm^{-1} is higher than that of the 475 cm^{-1} adsorption band [16]. Increasing levels of dehydroxylation increases the intensity of the 475 cm^{-1} band. Oxyapatite also has additional FTIR absorption peaks at 945 and 1025 cm^{-1}, but these are weaker and overlap with existing phosphate peaks [31, 72]. The Raman spectra also show a change in the phosphate absorption peaks, forming a doublet at 950 and 968 cm^{-1} [73].

The concentration of hydroxyl ions can be inferred from the intensity or area of the hydroxyl band at 3572 cm^{-1} or the phosphate band at 960 cm^{-1} [23]. Quantitative analysis requires a calibration curve made from samples with known OH^- content. This requires a sample with the maximum possible OH^- content and another sample without OH^- ions. Since ceramic processing causes dehydroxylation, the best calibration curve could then be made from solid solutions with different OH^- contents. Below it will be shown that the extremely fast heating and cooling in thermal spraying can also produce mechanical mixtures. The major obstacle until now has been the production of oxyapatite. The purest oxyapatite produced to date is 85% [74]. A further obstacle is the retention of oxyapatite, that means that mechanical mixtures could be prefered to produce calibration curves.

Raman spectroscopy offers a second approach to detect oxyapatite [75]. Virbation peaks of interest include those for OH^- and the PO_4^{3-}. Figure 7 shows the dehydroxylation of hydroxyapatite that produces a change from a single PO_4^{3-} peak at 964 cm^{-1} to a split peak with bands at 950 cm^{-1} and 972 cm^{-1} [68, 73], Figure 7.

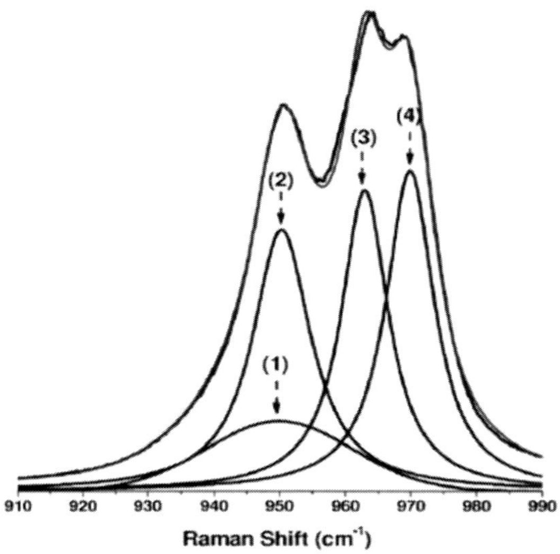

Figure 7. A Gaussian-Lorentzian peak fit for the υ_1PO_4 stretching band from a plasma sprayed coating containing an amorphous calcium phosphate (1), hydroxyapatite (3), and oxyapatite (2 and 4) [68].

No quantitative analysis has yet been performed using these peak intensities, but the peak intensities have been used to differentiate hydroxyapatite from oxyapatite and an amorphous calcium phosphate in plasma sprayed coatings [68].

4.3. Thermal Analysis

Thermogravimetric analysis (TGA) and differential thermal analysis (DTA) record real-time changes in weight and heat, and thus can follow dehydroxylation. TGA records the loss of adsorbed water and thus the progress of dehydroxylation [36]. Despite this capability, authors who have used this method have not characterized their sample to show a complete loss of hydroxyl ions [36, 70, 76]. Furthermore, there is no mention of the best parameters for the use of TGA to determine the OH content.

Complete dehydroxylation of the hydroxyapatite unit cell will release a maximum of one water molecule. The use of TGA has revealed that complete dehydroxylation and further decomposition occurs at 1450 °C, producing a weight loss of 1.79%. This can be used to measure the extent of dehydroxylation for samples heated to different processing temperatures, by measuring the amount of retained structural water. The degree of dehydroxylation (DD) is:

$$DD = \frac{1.79 - WD}{1.79} \cdot 100$$

where WD is the weight loss as a percentage upon heating to 1500 °C [70]. For this calculation, the authors have assumed that other molecules are not released within this temperature range that can contribute to the weight loss.

Another method for determining the quantity of OH^- with TGA uses a mixture of apatite and CaF_2 heated to 800 °C, whereby all the OH^- ions are replaced with F^- ions to form fluorapatite and liberate water (Figure 8). The weight loss can be used to determine the OH^- concentration assuming that water is the only species released [16]. Mass spectrometry in conjunction with TGA enables to determine the species released from the hydroxyapatite lattice during heating. This information can further be used to detect both the quantities of CO_3^{2-} and OH^- in the sample. However, thermogravimetric analysis/mass spectrometry (TGA-MS) is expensive and not readily available.

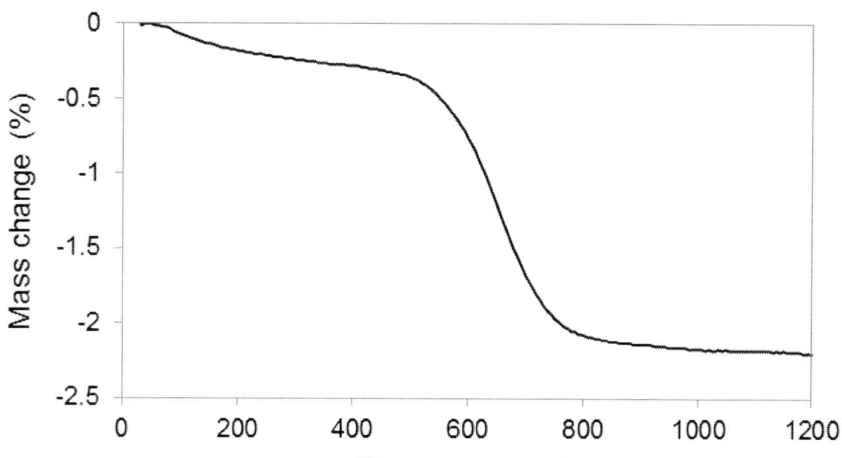

Figure 8. Thermogravimetric analysis of a HAp-CaF_2 mixture showing a weight loss corresponding to 1.7% between 420 and 800 °C.

Solid state nuclear magnetic resonance (NMR) has been used to assess the local atomic environments in thermally processed hydroxyapatite. Initial work showed a new peak in the ^{31}P spectra arising from dehydroxylation [77, 78]. Further work has compared sintered hydroxyapatite with plasma-sprayed hydroxyapatite looking at both the ^{1}H and ^{31}P spectra. Loss of hydroxyl ions creates additional peaks in the ^{1}H spectrum, but a distortion of phosphate groups produces a series of overlapping peaks in the ^{31}P spectrum [79]. (Additional information on NMR analysis of plasma-sprayed hydroxyapatite can be obtained from

Chapter 7 of this volume. The Editor). As with the other analytical techniques, further analysis relies on the availability of a pure oxyapatite sample.

5. PROPERTIES

Oxyhydroxyapatites are routinely produced during thermal processing, but are rarely described as such, usually being named as hydroxyapatites. As a result, very little data is presently available on the properties of oxyhydroxyapatites with reference to the hydroxyl concentration. The following will briefly discuss a range of different properties.

5.1. Chemical Properties

Reports on the retention of oxyapatite after cooling from high temperatures always describe the uptake of moisture [1, 46, 69, 70]. It is not known whether the outer layer of the crystal becomes partially or completely hydroxylated. Examination in a transmission electron microscope is challenging since the electron beam and high vacuum conditions may modify the hydroxyl concentration. Recent experiments by the authors on oxyapatite have revealed that it is possible to retain oxyapatite at room temperature by keeping oxyapatite under vacuum. Oxyapatite is very hygroscopic, and kept under vacuum has shown the ability to improve the vacuum by adsorbing molecular species from within the evacuated space.

5.2. Thermal Properties

The thermal stability of oxyapatite is the lowest, followed by hydroxyapatite and then chlorapatite. The decomposition is a kinetic process and so requires time for the formation of tricalcium phosphate and a calcium rich phase.

Tricalcium phosphate always appears, but the conditions determining the type of calcium rich phase is still not clear and requires further study. The distortion caused in the unit cell of oxyapatite may accommodate non-stoichiometry in the apatite.

It has been suggested that oxyapatite has a larger co-efficient of thermal expansion than hydroxyapatite [46], but this needs to be confirmed to show phase purity and bonding at high temperatures, using X-ray diffraction and Fourier transform infra-red spectroscopy, respectively.

5.3. Mechanical Properties

The mechanical properties are sensitive to composition and the microstructure (porosity and grain size). As a result, the effect of dehydroxylation on the mechanical properties is difficult to assess. Densification during sintering may be difficult since dehydroxylation can produce a vapour pressure thereby preventing pore closure [80].

A recent study showing the mechanical properties at two different temperatures gives the first clue on the effect of dehydroxylation [81]. Samples sintered at 1200 and 1250 °C show a similar density, and a negligible increase in grain size, however a decrease in elastic modulus from 108 ± 1 GPa to 100 ± 1GPa can be observed. This could have arisen from dehydroxylation. The decrease in elastic modulus with dehydroxylation is also supported from indirect evidence from another study that reports a higher co-efficient of thermal expansion for oxyapatite [46]. This follows from the observation that a higher co-efficient of thermal expansion usually is associated with a lower elastic modulus [82].

The effect of dehydroxylation on flexural strength, hardness and fracture toughness is not as pronounced as the decrease in elastic modulus. Further work is required with samples of similar density and grain size, to confirm the influence of dehydroxylation on the mechanical properties over a larger dehydroxylation range.

5.4. Biological Properties

Active research on polarized hydroxyapatite has shown greater cell attachment and bone bonding to surfaces with a negative charge arising from hydroxyl groups [83, 84].

Sintering in water vapour as opposed to air creates a six times higher charge on the polarized surfaces [15]. The separate contribution of hydroxyl content and oriented hydroxyl groups on cell response still needs to be determined. A charge may also be formed on polarized oxyapatite and peroxyapatite. Oxyapatite is unstable, but the effect of polarization in peroxyapatite on cell response has not been determined.

The next section will discuss some potential applications of oxyapatite, and then discuss some other oxyapatite compositions that are attracting attention for application in energy generation and photonics.

6. Applications

6.1. Medicine

Hydroxyapatite is routinely employed for bone bonding for all hard tissue locations, as bone filler for bone augmentation, as coatings on joint prostheses, and scaffolds for bone tissue engineering. The possibility to program the solubility through control of the crystallinity and composition provides avenues for drug release and the rich surface adsorption properties have led to interest in its use for radiation synovectomy. In all cases, oxyhydroxyapatite can form, usually unintentionally to date. Synthesis in solution enriched with oxygen ions can lead to uptake of oxygen in the lattice [85]. Inclusion of trivalent cations will usually be accompanied with an oxygen ion for charge balance.

Thermal processing without attention to the atmosphere can lead to a modification. All these situations will either produce oxygen ions in the anion column or an interstitial oxygen between the two OH^- sites in the column. The impact of this structural modification is not yet clear.

Functional oxygenated apatites have received little attention for application in biomedical situations. Both peroxide and molecular oxygen can be taken up inside the columns, with greater concentrations where Sr^{2+} substitutes for Ca^{2+} [6]. Thermal stability, however, is the greatest for calcium apatites. The release of molecular oxygen or peroxide into the body could provide an alternative means for oxygenation or antibacterial effects.

6.2. Sensing

6.2.1. Thermal Sensors

Apatites could be used as a heat sensor in thermal processes. The loss of column anions represents the first response after exposure to high temperatures and so changes in the anion concentration can possibly be used to follow the heat history of the apatite. A recently completed PhD research study revealed that plasma spraying chlorapatite produces a combination of chlorapatite and oxyapatite in the sprayed deposit [86]. The source of oxyapatite is not clear, but the results imply that oxyapatite is created even if the original apatite contains non-oxygen containing column ions. Chlorapatite as a candidate sensor material would facilitate the measurement of chloride ions in thermally unaffected zones and oxide ions for thermally exposed areas. The development of such a sensor requires quantitative analysis and mapping of the respective anion.

An indication of the cooling rate may also be possible, since oxyapatite needs a slower cooling rate for crystallization. Amorphous phases based on the oxyapatite composition have not received much attention in the literature.

Raman spectroscopy shows a lot of promise to ascertain whether the splat is amorphous or crystalline, determine the loss of anions during thermal spraying, and finally create a map of oxyapatite areas.

6.2.2. Humidity Sensors

Oxyapatite from a Y-substituted apatite composition has been suggested as a humidity sensor. Yttrium within the lattice is neutralized by a combination of oxygen ions and a point defect in the channel [17, 87]. Functionality of other sensors will depend on the anion mobility in the channels. Protons represent the main charge carrier in yttrium apatite. The electrical conductivity decreases as the OH^- concentration is lowered [17]. The spatial arrangement of hydroxyl ions and its influence on the conductivity would need to be determined before applications to such sensors.

More stable arrangements of oxygen ions in oxyapatites show promise for high reproducibility. At present, much attention is being directed to the conduction mechanisms and the structure of lanthanide-silicate apatites. There is a lot of promise for these composition materials. The oxygen ion conductivity in calcium phosphate oxyapatites has not been detected to date, but would require a protective electrolyte layer to avoid the hydroxylation.

The majority of other applications for oxyapatite are based on thermally stable britholites (REE-substituted silicate apatites), with some studies on germanates.

The silicates are more popular since they are economically more feasible and exhibit higher oxygen conductivity at lower temperatures, with only cerium gadolinium oxide with a higher conductivity [88]. A Web of Science database search on "oxyapatite" in the search string revealed that most applications include initial activity in photonics followed by a surge of interest in fuel cells and electrolytes starting from the mid 1990's. These applications will be briefly discussed.

6.3. Fuel Cells

The oxyapatites investigated for fuel cells include $M_{10}(XO_4)_6O_{2\pm y}$ where M is a rare-earth or an alkaline earth element, X is P, Si or G and y is the amount of oxygen non-stoichiometry. Lanthanoid silicates exhibit high oxygen ion conductivities [89, 90].

Conduction relying on vacancy diffusion for oxygen ion conductivity is at a similar level between 3% and 8% Y_2O_3 [91]. The improved oxygen conductivity in oxyapatites is attributed to the interstitial oxygen ions. Germanate apatites provide an alternative solution. It has been proposed that "interstitial O atoms provide a reservoir of ions that can migrate into the conducting channels of apatite, via a mechanism of inter-tunnel oxygen diffusion that transiently converts GeO_4 tetrahedra to GeO_5 distorted trigonal bipyramids" [92].

Analysis of the crystal structure with synchrotron radiation shows large oxygen ion displacement parameters that is believed to produce greater oxygen ion mobility parallel to the c-axis [93].

The activity in the area of fuel cells is being directed at understanding the defects introduced into the structure based on modifications of the composition.

Similar to calcium-rich oxyapatite providing additional oxide ions, research activities are investigating the incorporation of higher cation concentrations to increase the oxide ion concentration. There is a long list of possible substitutions, which could benefit from modeling to determine the best composition. Attention to processing will then provide the best microstructure for the highest oxide ion conductivity.

The fuel cell industry is investigating different operational conditions. Growing interest in ammonia as a fuel in solid oxide fuel cells has initiated the first assessment that shows replacement of Si in the silicon apatite structure with nitrogen from ammonia at 950 ^oC, but temperatures lower than 800 ^oC provide the required stability [94].

6.4. Photonics

Silicate oxyapatites have provided the matrix composition for producing luminescence in oxyapatites. The rich chemistry required for luminescence has led to the wide use of the sol-gel technique for incorporating transition metals. The colors reported include yellow [95], blue and green [96], red and green [97] with a promise of achieving white emission [98]. While there has been a flurry of activity to learn about the effect of nano-dimensions, a lot of work still needs to be performed to understand the cation placement in the lattice, charge compensation, defect structure, and lattice effects [99]. All these important factors affect the luminescence properties.

Early work looked at processing these compositions from the melt to grow single crystals that could then be characterized more completely. A rapid melting and solidification process could pave the way to more active research on the photonics of apatites. Luminescence is sufficiently complex and is a field that requires a lot more effort to understand the color, lifetime and quenching effects introduced by substitutional elements.

Conclusion

Oxyapatite comprises singly charged ions, negatively charged molecules, and neutral molecules at apatite structural sites or interstitial sites within the anion columns. The many attempts to produce oxyapatite have nearly attained a pure oxyapatite, retaining some OH^- ions before decomposition or have reabsorbed moisture to include OH^- ions in the structure. The best strategies have included non-traditional approaches using deuterium in place of hydrogen or a lower density arrangement such as an amorphous phase from which oxyapatite can recrystallize. Further work is required to optimize the processing conditions that will enable the development of oxyapatites.

The quantitative analyses of oxyapatite have only focused on the measurement of the hydroxyl ion content. This has only been developed for mechanical mixtures and further work needs to assess solid solutions. The successful production of oxyapatite will now support the quantification of oxyapatite as opposed to focusing on the hydroxyl ion content.

Very little information is available on the properties. This area requires much attention that will eventually reveal the change in properties upon dehydroxylation. This knowledge will have an immense impact on the implant industry where hydroxyapatite is routinely used a biomaterial.

Oxyapatites are presently used in stable forms that are based on silicate apatites with application in electrolytes, fuel cells, and photonics. Applications based on calcium phosphate apatites will need a deeper understanding of the mechanism of rehydroxylation and may need shielding from a humid atmosphere. The potential new applications hold much promise but will require attention to processing conditions and characterization.

Acknowledgments

KAG received support from the Marie Curie grant PIRG05-GA-2009-249306 and an ESF grant # 2009/0199/1DP/1.1.1.2.0/09/APIA/VIAA/090. LP has been supported by the European Social Fund 'Support for the implementation of doctoral studies at Riga Technical University'. The new approach of producing oxyapatite was initiated by Christian Rey and Karlis A Gross at the University of Toulouse and further developed in a bachelor project with Matiss Bluzma and Arnolds Ubelis at the University of Latvia.

References

[1] J. C. Trombe and G. Montel, *J. Inorg. Nucl. Chem.* 40, 15 (1978).
[2] J. C. Trombe and G. Montel, *J. Inorg. Nucl. Chem.* 40, 23 (1978).

[3] H. Zhao, X. Li, J. Wang, S. Qu, J. Weng, and X. Zhang, *J. Biomed. Mater. Res.* 52, 157 (2000).
[4] C. Rey, J. C. Trombe, and J. Dugas, *Comptes Rendus Hebdomadaires des Sciences de l'Academie des Sciences Serie C* 283, 277 (1976).
[5] H. Kanai, M. Lintuluoto, Y. Matsumura, and J.B. Moffat, *J. Mol. Catal.: Chem.* 252, 181 (2006).
[6] C. Rey, J. C. Trombe, and G. Montel, *J. Inorg. Nucl. Chem.* 40, 27 (1978).
[7] J. M. Hughes and J. Rakovan, *Rev. Mineral. Geochem.* 48, 1 (2002).
[8] L. Calderin and M. J. Stott, *Phys. Rev. B* 67, 1 (2003).
[9] K. Tonsuaadu, K. A. Gross, L. Pluduma, and M. Veiderma, *J. Therm. Anal. Calorim.* In press (2011).
[10] M. Veiderma, *Proceedings of the Estonian Academy of Sciences* (1999).
[11] B. D. Ratner, A. S. Hoffman, F. J. Schoen, and J. E. Lemons, eds., *Biomaterials Science: An Introduction to Materials in Medicine*, Elsevier Inc, Amsterdam (2004).
[12] J. Park and R. S. Lakes, eds., *Biomaterials. An Introduction*, Springer Science + Business Media, LLC (2007).
[13] T. V. Thamaraiselvi and S. Rajeswari, *Biomater. Artif. Organs* 18, (2004).
[14] P. A. Henning, A. R. Landa-Canovas, A. K. Larsson, and S. Lidin, *Acta Cryst.* B55, 170 (1999).
[15] D. Kumar, J. P. Gittings, I. G. Turner, C. R. Bowen, Bastida-Hidalgo, and S. H. Cartwell, *Acta Biomater.* 6, 1549 (2010).
[16] J. C. Elliott, ed., *Structure and chemistry of the apatites and other calcium orthophosphates*, Elsevier Inc., Amsterdam (1994).
[17] H. Owada, K. Yamashita, and T. Kanazawa, *J. Mater. Sci. Lett.* 9, 26 (1990).
[18] M. Pasero, A. R. Kampf, C. Ferraris, I. V. Pekov, J. Rakovan, and T. J. White, *Eur. J. Mineral.* 22, 163 (2010).
[19] S. Gomes, J.M. Nedelec, E. Jallot, D. Sheptyakov, and G. Renaudin, *Chem. Mater.* 23, 3072 (2011).
[20] H. R. Low, N. Phonthammachai, A. Maignan, G. A. Stewart, T. J. Bastow, L. L. Ma, and T. J. White, *Inorg. Chem.* 47, 11774 (2008).
[21] H. Yu, H. Zhang, X. Wang, Z. Gu, X. Li, and F. Deng, *J. Phys. Chem. Solids* 68, 1863 (2007).
[22] S. Guillot, S. Beaudet-Savignat, S. Lambert, P. Roussel, and R. N. Vannier, *Solid State Ionics* 185, 18 (2011).
[23] J. D. Pasteris, B. Wopenka, J. J. Freeman, K. Rogers, E. Valsami-Jones, J. A. M. van der Houwen, and M. J. Silva, *Biomaterials* 25, 229 (2004).
[24] R. Z. Le Geros, G. Bonel, and R. Legros, *Calcif. Tiss. Res.* 26, 111 (1978).
[25] J. D. Pasteris, *Structurally Incorporated Water in Bone Apatite: A Cautionary Tale*, in *Calcium phosphate: Structure, synthesis, properties and applications*, in Nova Science Publishers, Hauppauge (2012).
[26] T. Wang and A. Dorner-Reisel, *Mater. Lett.* 58, 3025 (2004).
[27] J. Chilar, A. Buchal, and M. Trunec, *J. Mater. Sci.* 34, 6121 (1999).
[28] M. A. Bredig and H. H. Frank, *Z Elektrochem* 39, 959 (1933).
[29] C. Lagergren and D. Varlstrom, *Acta Chem. Scand.* 11, 545 (1957).
[30] K. A. Gross, V. Gross, and C. C. Berndt, *J. Am. Ceram. Soc.* 81, 106 (1998).
[31] C. J. Liao, F. H. Lin, K. S. Chen, and J. S. Sun, *Biomaterials* 20, 1807 (1999).

[32] M. Koel, M. Kudrjasova, K. Tonsuaadu, M. Peld, and M. Veiderma, *Thermochim. Acta* 322, 25 (1998).
[33] L. Pluduma, *M. Eng. Sci thesis*, Riga Technical University, Riga, Latvia (2011).
[34] N. H. de Leeuw, J. R. Bowe, and J. A. L. Rabone, *Faraday Discuss. R. Soc. Chem.* 134, 195 (2007).
[35] A. A. White, I. A. Kinloch, A. H. Windle, and S. M. Best, *J. R. Soc. Interface* 7, 529 (2010).
[36] T. Wang, A. Dorner-Reisel, and E. Muller, *J. Eur. Ceram. Soc.* 24, 693 (2004).
[37] G. Muralithran and S. Ramesh, *Ceram. Int.* 26, 221 (2000).
[38] T. Nakano, K. Kaibara, Y. Umakoshi, S. Imazato, K. Ogata, A. Ehara, S. Ibisu, and M. Okazaki, *Mater. Trans.* 43, 3105 (2002).
[39] J. Zhou, X. Zhang, J. Chen, S. Zeng, and K. de Groot, *J. Mater. Sci.- Mater. Med.* 4, 83 (1993).
[40] P. V. Riboud, *Ann. Chim.* 8, 381 (1973).
[41] K. A. Gross and L. M. Rodriguez-Lorenzo, *Biomaterials* 25, 1375 (2004).
[42] K. A. Gross and K. Bhadang, *Biomaterials* 25, 1395 (2004)
[43] T. M. Sridhar, U. Kamachi-Mudali, and M. Subbaiyan, *Corros. Sci.* 45, 2337 (2003).
[44] P. Hartmann, C. Jäger, S. Barth, J. Vogel, and K. Meyer, *J. Solid State Chem.* 160, 460 (2001).
[45] M. Marraha, J. C. Heughabaert, and M. Heughebaert, *Ann. Chim. Sci. Mat.* 16, 619 (1991).
[46] M. Kikuchi, A. Yamazaki, M. Akao, and H. Aoki, *Mineral J.* 18, 79 (1996).
[47] U. Saalfeld, N. M. Meenen, and H. Saalfeld, *Neues Jahrbuch fur Mineralogie-Monatshefte* 9, 426 (1998).
[48] R. B. Heimann and T. A. Vu, *J. Mater. Sci. Lett.* 16, 437 (1997).
[49] A. Finoli, D. McKeel, J. Gerlach, and I. Nettleship, *Biomed. Mater.* 5, 015004 (2010).
[50] K. Yamashita, K. Kitagaki, and T. Umegaki, *J. Amer. Ceram. Soc.* 78, 1191 (1995).
[51] M. Corno, C. Busco, V. Bolis, S. Tosoni, and P. Ugliengo, *Langmuir* 25, (2009).
[52] J. P. Gittings, C. R. Bowen, A. C. E. Dent, I. G. Turner, F. R. Baxter, and J. B. Chaudhuri, *Acta Biomater.* 5, 743 (2009).
[53] A. Chanda, S. Dasgupta, S. Bose, and A. Bandyopadhyay, *Mater. Sci. Eng. C* 29, 1144 (2009).
[54] Y. W. Gu, N. H. Loh, K. A. Khor, S. B. Tor, and P. Cheang, *Biomaterials* 23, 37 (2002).
[55] A. Nakahira, M. Tamai, H. Aritani, S. Nakamura, and K. Yamashita, *J. Biomed. Mater. Res.* 62, 550 (2002).
[56] O. Sneh and S. M. George, *J. Phys. Chem.* 99, 4639 (1995).
[57] R. Comesana, F. Lusquinos, J. del Val, T. Malot, M. Lopez-Alvaroz, A. Riveiro, F. Quintero, M. Boutinguiza, P. Aubry, A. De Carlos, and J. Pou, *J. Europ. Ceram. Soc.* 31, 29 (2011).
[58] H. Kim, R. P. Camata, S. Lee, G. S. Rohrer, A. D. Rollett, and Y. K. Vohra, *Acta Materialia* 55, 131 (207).
[59] C. K. Wang, J. H. Chern Lin, C. P. Ju, H. C. Ong, and R. P. H. Chang, *Biomaterials* 18, 1331 (1997).

[60] E. K. Girija, S. P. Parthiban, R. V. Suganthi, K. Elayaraja, M. I. A. Joshy, R. Vani, P. Kularia, K. Asokan, D. Kanjilal, Y. Yokogawa, and S. Narayana Kalkura, *J. Ceram. Soc. Japan* 116, 320 (2008).
[61] H. Nishikawa and K. Omamiuda, *J. Mol. Catal.: Chem.* 39, 193 (2002).
[62] H. Nishikawa, *Mater. Lett.* 58, 14 (2003).
[63] S. Dyshlovenko, K. Pawlowski, B., Pateyron, I. Smurov, and J.H. Harding, *Surf. Coat. Technol.* 200, 3757 (2006).
[64] R. B. Heimann, *Surf. Coat. Technol.* 201, 2012 (2006).
[65] K. A. Gross and C. C. Berndt, *J. Biomed. Mater. Res.* 39, 580 (1998).
[66] N. Antolotti, S. Bertini, C. Fanara, X. Ranz, C. Rey, F. Rustichelli, and A. Scrivani, "Interface characterization of different apatite coatings", in *Thermal spray: Meeting the challenges of the 21st century*, C. Coddet, ed., ASM International (1998).
[67] K. A. Gross, C. C. Berndt, and H. Herman, *J. Biomed. Mater. Res.* 39, 407 (1998).
[68] I. Demnati, M. Parco, D. Grossin, I. Fagoaga, C. Drouet, G. Barykin, C. Combes, I. Braceras, S. Goncalves, and C. Rey, *Surf. Coat. Technol.* In press (2011).
[69] K. A. Gross, C. C. Berndt, P. Stephens, and R. Dinnebier, J. Mater. Sci. 33, 3985 (1998).
[70] P. A. Henning, E. Adolfsson, J. Grins, and A. Fitch, *J. Mater. Sci.* 36, 663 (2001).
[71] A. Rapacz-Kmita, C. Paluszkiewicz, A. Slosarczyk, and Z. Paszkiewicz, *J. Mol. Struct.* 744, 653 (2005).
[72] J. C. Trombe, *Ann. Chim.* 8, 251 (1973).
[73] J. C. Trombe and G. Montel, *CR Acad. Sci. Paris Ser. C* 276, 1271 (1973).
[74] X. Ranz, *PhD dissertation*, Institut National Polytechnique, Toulouse, France (1996).
[75] H. Fujimori, H. Toya, K. Ioku, S. Goto, and M. Yoshimura, *Chem. Phys. Lett.* 325, 382 (2000).
[76] S. Raynaud, E. Champion, D. Bernache-Assollant, and P. Thomas, *Biomaterials* 23, 1065 (2002).
[77] R. McPherson, N. Gane, and T. J. Bastow, *J. Mater. Sci: Mater. Med.* 6, 327 (1995).
[78] H. Sfihi, A. P. LeGrand, X. Ranz, and C. Rey, *Phosphorus Res. Bull.* 6, 13 (1996).
[79] P. Hartmann, C. Jager, S. Barth, J. Vogel, and K. Meyer, *J. Solid State Chem.* 160, 460 (2001).
[80] A. J. Ruys, M. Wei, C. C. Sorrell, M. R. Dickson, A. Brandwood, and B. K. Milthorpe, *Biomaterials* 16, 409 (1995).
[81] S. Laasri, M. Taha, A. Laghzizil, E. K. Hill, and J. Chevalier, *Mater. Res. Bulletin* 45, 1433 (2010).
[82] W.D. Callister, *Materials Science and Engineering: An introduction*, Wiley (2007).
[83] T. Kobayashi, S. Nakamura, and K. Yamashita, *J. Biomed. Mater. Res.* 57, 477 (2001).
[84] T. Kizuki, M. Ohgaki, M. Katsura, S. Nakamura, K. Hashamito, Y. Toda, S. Udagawa, and K. Yamashita, *Biomaterials* 24, 941 (2003).
[85] S. Belouafa, H. Chaair, H. Loukili, K. Digua, and B. Sallek, *Materials Research Ibero American Journal of Materials* 11, 93 (2008).
[86] I. Demnati, "Development and characterization of bioactive apatite coatings obtained by plasma spraying: aplication to biomedical implants", University of Toulouse, Touluse, France (2011).
[87] H. Owada, K. Yamashita, T. Umegaki, and T. Kanazawa, *J. Ceram. Soc. Jpn. Inter. Ed.* 97, 1076 (1989).

[88] L. Malavasi, C. A. Fisher, and M. S. Islam, *Chem. Soc. Rev.* 39, 4370 (2010).

[89] N. Takeda, Y. Itagaki, H. Aono, and Y. Sadaoka, *Sensors Actuators B* 115, 455 (2006).

[90] L. L. Reina, E. R. Losilla, M. M. Lara, S. Bruque, and M. A. G. Aranda, *J. Mater. Chem.* 14, 1142 (2004).

[91] Y. Masubuchi, M. Higuchi, S. Kikkawa, K. Kodaira, and S. Nakayama, *Solid State Ionics* 175, 357 (2004).

[92] S. S. Pramana, W. T. Klooster, and T. J. White, *Acta Cryst. B* 63, 597 (2007).

[93] R. Ali, M. Yashima, Y. Matsushita, H. Yoshioka, and F. Izumi, *J. Solid State Chem.* 182, 2846 (2009).

[94] E. Kendrick, D. Headspith, A. Orera, D. C. Apperley, R. I. Smith, M. G. Francesconi, and P. R. Slater, *J. Mater. Chem.* 19, 749 (2009).

[95] G. F. Li, D. G. Deng, X. Y. Su, Q. Wang, Y. Q. Li, Y. J. Hua, L. H. Huang, S. L. Zhao, H. P. Wang, C. X. Li, and S. Q. Xu, *Mater. Lett.* 65, 2019 (2011).

[96] G. S. R. Raju, J. Y. Park, H. C. Jung, E. Pavitra, M. K. Moon, J. H. Jeong, J. S. Yu, J. H. Kim, and H. Choi, *J. Alloys Compounds* 509, 7537 (2011).

[97] M. Yu, J. Lin, and S. B. Wang, *J. Alloys Compounds* 344, 212 (2001).

[98] G. G. Li, D. L. Geng, M. M. Shang, C. Peng, and Z. Y. Cheng, *J. Mater Chem.* 21, 13334 (2011).

[99] G. A. Waychunas, *Phosphates - Chemical, geobiological, and materials importance*, Mineralogical Society of America, Washington (2002).

In: Calcium Phosphate
Editor: Robert B. Heimann

ISBN: 978-1-62257-299-1

Chapter 5

NANO-DIMENSIONAL CALCIUM ORTHOPHOSPHATES: RECENT DEVELOPMENTS AND FUTURE APPLICATIONS

Sergey V. Dorozhkin *
Moscow, Russia

ABSTRACT

Nano-sized particles and crystals play an important role in the formation of calcified tissues of various animals. For example, nano-sized and nano-crystalline calcium orthophosphates in the form of apatites of biological origin represent the basic inorganic building blocks of bones and teeth of mammals. Namely, according the recent developments in biomineralization, tens to hundreds nano-dimensional crystals of a biological apatite are self-assembled into these complex structures. This process occurs under a strict control by bioorganic matrixes. Furthermore, both a greater viability and a better proliferation of various types of cells have been detected on smaller crystals of calcium orthophosphates. Thus, the nano-sized and nano-crystalline forms of calcium orthophosphates have a great potential to revolutionize the hard tissue-engineering field, starting from bone repair and augmentation to controlled drug delivery systems. This chapter reports on current state of the art and recent developments on the subject, starting from synthesis and characterization to biomedical and clinical applications. Furthermore, the chapter also discusses possible directions for future research and development.

1. INTRODUCTION

Living organisms can create amazing ways to produce various high-performance materials and over 60 different inorganic minerals of biological origin have already been revealed [1]. Among them, calcium orthophosphates are of special importance since they are the most important inorganic constituents of hard tissues in vertebrates [2, 3]. In the form of

* Email: sedorozhkin@yandex.ru

poorly crystalline, non-stoichiometric, ion-substituted CDHA (commonly referred to as "biological apatite"), calcium orthophosphates are present in bones, teeth, deer antlers, and tendons of mammals to give these tissues stability, hardness and function [2, 4, 5]. Although we still do not exactly know why animals use conformable calcium orthophosphates as their crucial biomineral for survival [6], current biomedical questions of persistent pathological and physiological mineralization in the body force researchers to focus on the processes, including the occurrence, formation and degradation of calcium orthophosphates in living organisms [7, 8, 9].

Biological mineralization (or biomineralization) is a process of *in vivo* formation of inorganic minerals [1, 2]. In the biomineralization processes, organized assemblies of organic macromolecules regulate nucleation, growth, morphology and assembly of inorganic crystals. *In vivo* biologically formed calcium orthophosphates (biological apatite) are always nano-dimensional and nano-crystalline. The dimensions of biological apatite crystals formed in calcified tissues are always in the range of a few to hundreds of nanometers, with the smallest building blocks on the nanometer scale [2, 4, 5, 10, 11]. For example, tens to hundreds of nanometer-sized apatite crystals in a collagen matrix are combined into self-assembled structures during bone and teeth formation [2, 4, 5]. Recent advances suggest that this is a natural selection since the nano-structured materials provide better conditions for specific interactions with proteins [12].

Due to the properties mentioned above, nano-dimensional and nano-crystalline forms of calcium orthophosphates are able to mimic both the composition and dimensions of constituent components of the calcified tissues. Thus, they can be utilized in biomineralization and as biomaterials due to their excellent biocompatibility [13, 14]. Further development of calcium orthophosphate-based biomaterials will greatly benefit from advances in nanotechnology [15] which offers unique approaches to overcome shortcomings of many conventional materials. For example, nano-sized ceramics can exhibit significant ductility before failure contributed by a grain-boundary phase. Already in 1987, Karch *et al.* reported that a nano-dimensional, brittle ceramic could tolerate a large plastic strain up to 100% [16]. In addition, nano-structured ceramics can be sintered at lower temperatures; thereby major problems associated with high temperature sintering can be avoided. Thus, nano-dimensional and nano-crystalline forms of bioceramics clearly represent a promising class of orthopedic and dental implant formulations with improved biological and biomechanical properties [17].

Many other advances have been made in the biomaterial field due to a rapid growth of nanotechnology [18]. For example, a recent theory of "aggregation-based crystal growth" [19] and a new concept of "mesocrystals" [20, 21] highlighted the roles of nano-sized particles in biological crystal engineering. In this aspect, the study of calcium orthophosphates is a specific area in nanotechnology, because they might be applied readily to repair hard skeletal tissues of mammals [22-24].

Herein, an overview of nano-dimensional and nano-crystalline apatites and other calcium orthophosphates in studies on biomineralization and biomaterials is given. The available calcium orthophosphates are listed in Table 1.

To narrow the subject of this chapter, with a few important exceptions, only non-doped and non-substituted calcium orthophosphates are considered and discussed. The reader interested in various nano-dimensional and nano-crystalline ion-substituted calcium orthophosphates [25-63] is referred to the original publications. Furthermore, details on calcium orthophosphate-based nano-dimensional biocomposites [64-85] or nano-dimensional calcium orthophosphate-based biocomposites [86-104] are available in Refs. [105,106].

Table 1. Existing calcium orthophosphates and their major properties [204, 205]

Ca/P molar ratio	Compound	Formula	Solubility @ 25 °C, -log(K_s)	Solubility @ 25 °C, g/L	pH stability range in aqueous solutions @ 25°C
0.5	Monocalcium phosphate monohydrate (MCPM)	$Ca(H_2PO_4)_2 \cdot H_2O$	1.14	~ 18	0.0 – 2.0
0.5	Monocalcium phosphate anhydrous (MCPA or MCP)	$Ca(H_2PO_4)_2$	1.14	~ 17	[c]
1.0	Dicalcium phosphate dihydrate (DCPD), mineral brushite	$CaHPO_4 \cdot 2H_2O$	6.59	~ 0.088	2.0 – 6.0
1.0	Dicalcium phosphate anhydrous (DCPA or DCP), mineral monetite	$CaHPO_4$	6.90	~ 0.048	[c]
1.33	Octacalcium phosphate (OCP)	$Ca_8(HPO_4)_2(PO_4)_4 \cdot 5H_2O$	96.6	~ 0.0081	5.5 – 7.0
1.5	α-Tricalcium phosphate (α-TCP)	α-$Ca_3(PO_4)_2$	25.5	~ 0.0025	[a]
1.5	β-Tricalcium phosphate (β-TCP)	β-$Ca_3(PO_4)_2$	28.9	~ 0.0005	[a]
1.2 – 2.2	Amorphous calcium phosphates (ACP)	$Ca_xH_y(PO_4)_z \cdot nH_2O$, n = 3 – 4.5; 15 – 20% H_2O	[b]	[b]	~ 5 – 12 [d]
1.5 – 1.67	Calcium-deficient hydroxyapatite (CDHA or Ca-def HA)[e]	$Ca_{10-x}(HPO_4)_x(PO_4)_{6-x}(OH)_{2-x}$ ($0<x<1$)	~ 85	~ 0.0094	6.5 – 9.5
1.67	Hydroxyapatite (HA, HAp or OHAp)	$Ca_{10}(PO_4)_6(OH)_2$	116.8	~ 0.0003	9.5 – 12
1.67	Fluorapatite (FA or FAp)	$Ca_{10}(PO_4)_6F_2$	120.0	~ 0.0002	7 – 12
1.67	Oxyapatite (OA, OAp or OXA)[f]	$Ca_{10}(PO_4)_6O$	~ 69	~ 0.087	[a]
2.0	Tetracalcium phosphate (TTCP or TetCP), mineral hilgenstockite	$Ca_4(PO_4)_2O$	38 – 44	~ 0.0007	[a]

[a] These compounds cannot be precipitated from aqueous solutions.

[b] Cannot be measured precisely. However, the following values were found: 25.7±0.1 (pH = 7.40), 29.9±0.1 (pH = 6.00), 32.7±0.1 (pH = 5.28) [269]. The comparative extent of dissolution in acidic buffer is: ACP >> α-TCP >> β-TCP > CDHA >> HA > FA [126].

[c] Stable at temperatures above 100°C.

[d] Always metastable.

[e] Occasionally, it is called "precipitated HA (PHA)".

[f] Existence of OA remains questionable (but see Chapters 4 and 7 of this tome).

This chapter is organized into several sections. After the brief introduction provided above, general information on the term "nano" is offered in the second section. The third section briefly compares the micron-sized and nano-dimensional calcium orthophosphates. The forth section briefly discusses the presence of nano-sized and nano-crystalline calcium orthophosphates in normal calcified tissues of mammals. The structure of nano-sized and nano-crystalline apatites is described in the fifth section. Synthesis of nano-dimensional and nano-crystalline calcium orthophosphates of various dimensions and shapes will be reviewed in the sixth section, while the biomedical applications are examined in the seventh section. Finally, in the last section a summary and reasonable future perspectives of this active research area are given.

2. General Information on the Term "Nano"

The prefix "nano" specifically means a measure of 10^{-9} units. Although it is widely accepted that the prefix "nano" specifically refers to 10^{-9} units, in the context of nano-sized and nano-crystalline materials, the units should only be those of dimensions, rather than of any other unit of the scientific measurements. Besides, for practical purposes, it appears to be unrealistic to consider the prefix "nano" to solely and precisely refer to 10^{-9} m, just as it is not considered that "micro" specifically and solely concerns something with a dimension of precisely 10^{-6}m [107]. Currently, there is a general agreement that the subject of nanoscience and nanotechnology started after the famous talk "There's plenty of room at the bottom" given by the Nobel Prize winner in physics Prof. Richard P. Feynman on December 26, 1959 at the annual meeting of the American Physical Society held at California Institute of Technology. This well-known talk has been widely published in various media (for example [108]).

In a recent extensive discussion on a framework for definitions presented to the European Commission, the term "nano-scale" has been defined as being of the order of 100 nm or less. Similarly, a nanomaterial has been defined as "any form of a material that is composed of discrete functional parts, many of which have one or more dimensions of the order of 100 nm or less" [109]. Other definitions logically follow this approach such as: a nano-crystalline material is "a material that is comprised of many crystals, the majority of which have one or more dimensions of the order of 100 nm or less" (normally without micron-sized crystals or an intergranular amorphous phase) and a nanocomposite is a "multi-phase material in which the majority of the dispersed phase components have one or more dimensions of the order of 100 nm or less" [107]. Similarly, nano-structured materials are defined as materials containing structural elements (*e.g.*, clusters, crystallites or molecules) with dimensions in the 1 to 100 nm range [110], nano-coatings represent individual layers or multilayer surface coatings of 1 – 100 nm thickness, nano-powders are extremely fine powders with an average particle size in the range of 1 – 100 nm and nano-fibers are the fibers with a diameter within 1 – 100 nm [111, 112]. It also has been proposed to extend the lower size limit to 0.1 nm [113], which would include all existing organic molecules, allowing chemists to rightly claim they have been working on nanotechnology for very many years [114].

Strictly speaking, there are serious doubts that the term "nanomaterial" has a reasonable meaning. For example, let me quote Prof. David F. Williams, the Editor-in-Chief of

Biomaterials: "*... some words which have no rational basis whatsoever become part of everyday language so rapidly, even if so illogically, that it is impossible to reverse the process and their common use has to be accepted, or perhaps, accommodated. Nanomaterial is one such word, where I have argued that it should not exist, but accept that it does through common usage and have to recognise its existence [107]. The discussion about nanomaterial provides a hint of the analysis of a biomaterial that follows, since a prefix, which is an indicator of scale, cannot specify the integer that follows (in this case a material) unless that integer can be qualified by that scale. In other words, it is very clear what a nanometre is because nano – means 10^{-9} and a metre is a measure of length. In the case of nanomaterial, what is it about the material that is 10^{-9}. Is it the dimension of a crystal within the material, or of a grain boundary, a domain, or a molecule, or is it a parameter of a surface feature of the sample, or perhaps of the resistivity or thermal conductivity of the material. Clearly this is nonsense, but one has to accept that nanomaterials are here to stay, with even some journal titles containing the word.*" [115, p. 5898, left column].

Following this logic, such terms as "nanocomposite", "nano-coatings", "nano-powders", "nano-fibers" and "nano-crystals" are also senseless and should be replaced, for example, by "composites with nano-sized (or nano-dimensional) dispersed phase(s)", "coatings of nano-sized (or nano-dimensional) thickness", "nano-sized (or nano-dimensional) powders", "fibers of nano-sized (or nano-dimensional) thickness" and "nano-sized (or nano-dimensional) crystals", respectively. At least, this has been done in this chapter.

According to their geometry, all nano-dimensional materials can be divided into three major categories: equiaxed, one-dimensional (or fibrous) and two-dimensional (or lamellar) forms. Selected examples and typical applications of each category of nano-dimensional materials and their use in biomedical applications are available in the literature [116]. It is important to note that in the literature on calcium orthophosphates there are cases, when the prefix "nano" has been applied to structures with minimum dimensions exceeding 100 nm [44, 83, 117-133].

As a rule, nano-dimensional materials can be manufactured from nearly any substance. Of crucial importance are two major characteristics conferring special properties to any nano-dimensional material. These are the quantum effects associated with the very small dimensions (currently, this is not applicable to the biomaterials field) and a large surface-to-volume ratio that is encountered at these dimensions. For instance, specific surface areas for submicron-sized particles are typically 60 – 80 m^2/g, while decreasing the particle diameters to tens of nanometers increases the specific surface area up to a factor of five, an amazing amount of surface area per mass! Furthermore, all nano-phase materials have unique surface properties when compared to micron-sized materials including an increased number of grain boundaries and surface defects, a huge surface area, and altered electronic structure [107, 134]. While less than ~ 1 % of a micron-sized particle's atoms occupy surface positions, over 10% of the atoms in a 10-nm diameter particle reside on its surface, and ~ 60 % in a 2-nm particle [135]. This very high surface-to-volume ratio of nano-dimensional materials provides a tremendous driving force for diffusion, especially at elevated temperatures, as well as causes a self-aggregation into larger particles. Besides, the solubility of many substances increases with decreasing particle size [136, 137]. What is more, nano-phase materials could have surface features (*e.g.*, a higher amount of nano-scale pores) that influence the type and amount of adsorption of selective proteins that could enhance specific osteoblast adhesion [138]. Finally and yet importantly, the nano-dimensional and nano-crystalline materials have different mechanical, electrical, magnetic and optical properties when compared to larger-grained materials of the same chemical composition [139-142].

The nano-structured materials can take the form of powders, dispersions, coatings or bulk materials. In general, nano-structured materials contain a large volume fraction (greater than 50 %) of defects such as grain boundaries, interphase boundaries and dislocations that strongly influences their chemical and physical properties. The great advantages of nano-structuring were first understood in the electronic industry with the advent of thin film deposition processes. Other application areas have followed. For example, nano-structured bioceramics were found to remedy friction and wear problems associated with joint replacement components because they are tougher and stronger than coarser-grained bioceramics [143]. Furthermore, nano-structuring has allowed chemical homogeneity and structural uniformity to an extent, which was once thought to be impossible to achieve [110]. In calcium orthophosphate bioceramics, the major target of nano-structuring is to mimic the architecture of bones and teeth [144, 145].

3. Micron- and Submicron-Sized versus Nano-Dimensional Calcium Orthophosphates

Micron-sized calcium orthophosphate-based bioceramic powders suffer from poor sinterability, mainly due to a low surface area (typically 2 – 5 m^2/g) in contrast to the specific surface area of nano-dimensional calcium orthophosphates that exceeds 100 m^2/g [146]. In addition, the resorption process of synthetic micron-sized calcium orthophosphates was found to be quite different from that of bone mineral [147].

Although the nano-dimensional and nano-crystalline features of natural calcium orthophosphates in bones and teeth had been known earlier [2, 148-153], the history of the systematic investigations of this field has started only in 1994. To confirm this statement, a careful search in scientific databases using various combinations of the keywords "nano"+"calcium phosphate", "nano"+"apatite", "nano"+"hydroxyapatite" *etc.* in the article title revealed only five papers published in 1994 [154-158]. No papers published before 1994 with the aforementioned keywords in the title were found.

Nano-dimensional (size ~ 67 nm) HA was found to have a higher surface roughness of 17 nm compared to 10 nm for the submicron-sized (~ 180 nm) HA, while the contact angles (a quantitative measure of the wetting capability of a solid by a liquid) were significantly lower for nano-sized HA (see section 6.1) when compared to submicron-sized HA (11.51). In addition, the diameters of individual pores in nano-dimensional HA compacts are several times smaller (pore diameter ~ 6.6 Å) than those of submicron grain-sized HA compacts (pore diameter range 19.8 – 31.0 Å) [159]. High surface roughness is known to enhance the osteoblast functions while a porous structure improves osteoinduction compared with smooth surfaces and nonporous structure, respectively [138]. Furthermore, nano-phase HA appeared to have ~ 11% more proteins of fetal bovine serum adsorbed per 1 cm^2 than submicron-sized HA [160]. Interfacial interactions between calcined nano-sized HA crystals and various substrates were studied and their bonding strength appeared to be influenced not only by the nature of functional groups adsorbed on the substrate but also by matching of surface roughness between the nano-sized crystals and the substrate [161]. More to the point, incorporating of nano-dimensional particles of HA into poly(acrylonitrile) fibers were found to result in rising their degree of crystallinity by about 5% [162]. In a comparative study on the effect of incorporation micron-sized and nano-sized HA particles into poly-L-lactide matrices, addition of nano-sized HA was found to influence both thermal and dynamic mechanical properties to a greater extent [163].

In general, nano-structured biomaterials [164] offer much improved performances than their larger particle-sized counterparts due to their huge surface-to-volume ratio and unusual chemical synergistic effects. Such nano-structured systems constitute a bridge between single molecules and bulk material systems [165]. For instance, powders of nano-crystalline apatites [166-172] and β-TCP [173] were found to exhibit improved sinterability and enhanced densification due to a greater surface area. This can be explained by the fact that the distances of material transport during sintering decrease for ultrafine powders with a high specific surface area, resulting in densification at lower temperature. Therefore, due to low grain growth rates, a low-temperature sintering appears to be effective to produce fine-grained apatite bioceramics [174]. Furthermore, the mechanical properties (*eg.* hardness and toughness) of HA bioceramics appear to increase as the grain size decreases from sub-micrometer to nanometer dimensions [175].

Nano-sized HA is also expected to have improved bioactivity compared to coarser crystals [176-178]. For example, Kim *et al.*, found that osteoblasts (bone-forming cells) attached to nano-sized HA/gelatin biocomposites to a significantly higher degree than to micrometer-sized analogs [179]. Increased osteoblast and decreased fibroblast (fibrous tissue-forming cells) adhesion to nano-phase ceramics [180-184] as well as to nano-crystalline HA coatings on titanium were also discovered by other researchers when compared to traditionally used plasma-sprayed HA coatings [185-187]. Scientists also observed enhanced osteoclast (bone-resorbing cells) functions that confirmed healthy remodeling of bone at the simulated implant surface [177]. Besides, the proliferation and osteogenic differentiation of periodontal ligament cells were found to be promoted by nano-phase HA when compared to dense HA bioceramics [188]. Thus, the underlying material property, responsible for this enhanced osteoblast function, appears to be the surface roughness of the nano-structured surface [18]. Interestingly, increased osteoblast adhesion was discovered on nano-sized calcium orthophosphate powders with higher Ca/P ratios [189] which points to some advantages of apatites over other calcium orthophosphates. Furthermore, a histological analysis revealed superior biocompatibility and osteointegration of bone graft substitutes when nano-sized HA was employed in biocomposites [190-192]. However, data are also available that nano-sized HA could inhibit growth of osteoblasts in a dose-dependent manner [193]. Furthermore, a cellular activity appeared to be affected by the shape and dimensions of nano-sized HA. For example, the cellular activity of L929 mouse fibroblasts on nano-sized fibers with a diameter range of 50 – 100 nm was significantly enhanced relative to that on a flat HA surface, while nano-dimensional HA needles and sheets with a diameter/thickness of less than 30 nm inhibited cellular adhesion and/or subsequent activity because cells could not form focal adhesions of sufficient size [194].

Evidently, the volume fraction of grain boundaries in nano-dimensional calcium orthophosphates is increased significantly leading to improved osteoblast adhesion, proliferation, and mineralization. Therefore, the composition of these biomaterials at the nano-scale emulates the bone's hierarchic organization, to initiate the growth of an apatite layer and to allow for a positive cellular and tissue response of bone remodeling. These examples emphasize that nano-phase materials deserve more attention when attempting to reduce orthopedic implant failure rates. However, to reduce surface energy, all nano-sized materials tend to agglomerate and, to avoid self-aggregation of calcium orthophosphate nano-sized particles [195-198], special precautions might be necessary [54, 60, 120, 199-202].

Finally yet importantly, nano-sized crystals of CDHA obtained by precipitation methods in aqueous solutions were shown to exhibit physico-chemical characteristics rather similar to those of bone apatite [203]. In particular, their chemical composition departs from stoichiometry by calcium and hydroxide ions deficiency, leading to increased solubility, and in turn the bioresorption rate *in vivo* [148, 204-206]. Nano-sized crystals of CDHA have also the property to mature in solution similar to bone apatite crystals. For example, freshly precipitated CDHA has been shown to be analogous to embryonic bone mineral crystals whereas aged precipitates resemble bone crystals of old vertebrates [203].

4. Nano-Dimensional and Nano-Crystalline Calcium Orthophosphates in Calcified Tissues of Mammals

4.1. Bones

Bone is the most typical calcified tissue of mammals and comes in all sorts of shapes and sizes designed by nature to achieve various functions to protect and mechanically support the body. The major inorganic component of bone mineral is biological apatite, a poorly crystalline, non-stoichiometric and ion-substituted CDHA [2-5, 204-207]. From a material point of view, bone can be considered an assembly of distinct levels of seven hierarchical structural units from macro- to micro- and to nano-scale (Figure 1) to meet numerous functions [2, 5, 134, 208-210].

Furthermore, all these levels permanently interact with cells and biological macromolecules. At the nano-structural level, tiny plate-like crystals of biological apatite in bone occur within discrete spaces of collagen fibrils, and grow with specific crystallographic orientation along their *c*-axes that are roughly parallel to the long axes of the collagen fibrils [211]. Type I collagen molecules are self-assembled into fibrils with a periodicity of ~ 67 nm. There are gaps of ~ 40 nm between the terminal ends of the molecules into which apatite nano-sized crystals are placed.

A biocomposite of these two constituents forms mineralized fibers. Cross-linking of fibers provides a highly dynamic system capable of modification through selection of different amino acids thus allowing for different mechanical properties for different biomaterial applications [212]. This is why bone is usually called a fiber-reinforced composite of biological origin characterized by nanometer-sized hard inclusions embedded in a soft protein matrix [213]. Although the dimensions of biological apatite crystals reported in the literature vary owing to different treatment methods and analytical techniques, they are found generally around the nanometric level with values ranging between 30 and 50 nm in length, 15 and 30 nm in width, and 2 –and 10 nm in thickness [214]. Some details on the reason of stability of such nano-dimensional apatites in bones are available in the literature [215, 216].

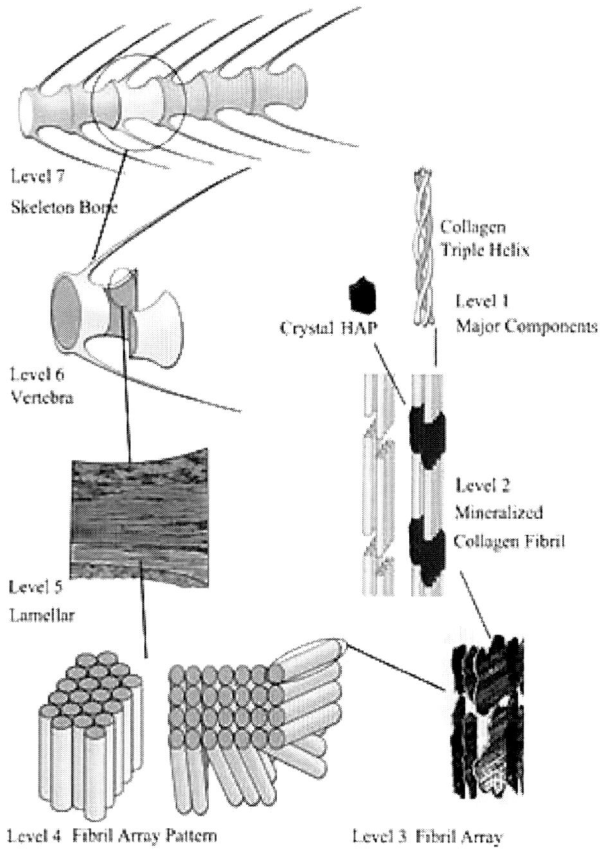

Figure 1. The seven hierarchical levels of organization of the zebrafish skeleton bone. Level 1: Isolated crystals and part of a collagen fibril with the triple helix structure. Level 2: Mineralized collagen fibrils. Level 3: The array of mineralized collagen fibrils with a cross-striation periodicity of nearly 60-70 nm. Level 4: Two fibril array patterns of organization as found in the zebrafish skeleton bone. Level 5: The lamellar structure in one vertebra. Level 6: A vertebra. Level 7: Skeleton bone. Reprinted from Ref. [208] with permission.

Why does the nanometer scale appear to be so important to bones? It was recently demonstrated that natural biocomposites exhibit a generic mechanical structure in which the nanometer sizes of mineral particles are used to ensure the optimum strength of bone, and maximum tolerance for flaws [217, 218]. Furthermore, nano-dimensional apatite has another crucial function to fulfill in organisms as it is a huge reservoir of calcium and orthophosphate ions necessary for a wide variety of metabolic functions. This reservoir offers or consumes calcium and orthophosphate ions through the so-called "remodeling" process that includes continuous resorption and formation of nano-dimensional apatite by osteoclast and osteoblast cells, respectively in a delicate equilibrium [2, 5]. Additional details on the structure, properties and composition of bones can be found in the special literature [5, 207, 219].

4.2. Teeth

Teeth are another normal calcium orthophosphate-based calcified tissue of vertebrates. Unlike bone, teeth consist of at least two different biominerals: dental enamel forming the crown, that is the part above the gum line, and dentin forming the root, that is the part below the gum line [220]. Dental enamel contains up to 98% of biological apatite, ~ 1% of bioorganic compounds and up to 2% of water. Typical structures in enamel are composed of rod-like apatite crystals measuring 25 – 100 nm and a variable length of 100 nm to 100 μm or longer, being oriented along their *c*-axes [221-223]. However, the apatite crystals in enamel were found to exhibit regular sub-domains or subunits with distinct chemical properties [224]. This subunit structure reflects an assembly mechanism for such biological crystals [225, 226]. As in bones (Figure 1), seven levels of structural hierarchy have also been discovered in human dental enamel. Moreover, the analysis of the hierarchical structures of enamel and bone suggests similarities of the scale distribution at each level [227]. In enamel, nano-sized crystals of biological apatite initially form nano-dimensional mineral fibers that always align lengthwise, aggregating into fibrils and afterwards into thicker fibers. Subsequently, prism/interprism continua are formed by the fibers. At the micro-scale, prisms are assembled into prism bands which present different arrangements across the thickness of the enamel layer. These compositional and structural characteristics impart to enamel special properties such as an anisotropic elastic modulus, effective viscoelastic properties, much higher fracture toughness, and stress-strain relationships more similar to metals than ceramics [228].

Dentin contains ~ 50% of biological apatite, ~ 30% of bioorganic compounds and ~ 20% of water. In dentin, the nano-dimensional building blocks (~ 25 nm width, ~ 4 nm thickness and ~ 35 nm length) of biological apatite are smaller than those in enamel. Dentin is analogous to bone in many aspects, for example, it has a similar composition and a hierarchical structure up to the level of the bone lamellae [204, 205]. Additional details on the structure, properties and composition of teeth can be found in the specialized literature [229].

5. Structure of Nano-Dimensional and Nano-Crystalline Apatites

Due to the apatitic structure of natural calcified tissues, apatites appear to be the best investigated compounds among the available calcium orthophosphates (Table 1). Thus, nano-dimensional and nano-crystalline apatites have been extensively studied by various physico-chemical techniques and chemical analysis methods [197, 230-242] with special attention paid to the "nano" effect (*i.e.*, the enhanced contribution of the surface compared to the volume). Unfortunately, no publications on the structure of other nano-dimensional and/or nano-crystalline calcium orthophosphates were found in the available literature.

Caused by the nano-crystalline nature, application of various diffraction techniques have not yet given much information on the fine structural details related to apatite nano-sized crystals since assemblies of nano-sized particles give only broad diffraction patterns, similar to ones from an amorphous material [230, 231]. Nevertheless, diffraction studies by electron microbeams of 35 ± 10 nm in diameter clearly indicated the crystalline character of the nano-sized particles in these assemblies. Furthermore, high-resolution transmission electron microscopy (HRTEM) results revealed that nano-sized particles of HA possess a fine single crystalline grain structure [197, 230].

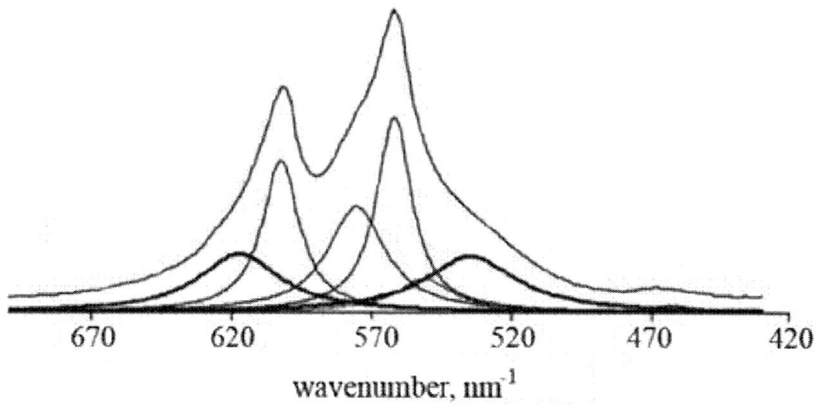

Figure 2. FTIR spectra of poorly crystalline apatites showing the non-apatitic environments of the orthophosphate ions (bold lines with peaks at 617 and 534 cm^{-1}) and the apatitic PO_4^{3-} (thin lines with peaks at 600, 575 and 560 cm^{-1}) and HPO_4^{2-} (thin line with peak at 550 cm^{-1}) in the v_4 PO_4 domain. Reprinted from Ref. [234] with permission.

Therefore, recent progress on the structure of nano-dimensional and nano-crystalline apatites has relied mainly on diverse spectroscopic methods sensitive to disturbances of the closest environments of various ions. For example, structure analysis revealed the existence of structural disorder at the particle surface which was explained by chemical interaction between the orthophosphate groups and either adsorbed water molecules or hydroxyl groups located at the surface of nano-sized apatites [232]. More to the point, infrared (FTIR) spectra of nano-crystalline apatites, in the v_4 PO_4 domain, revealed the existence of additional bands of orthophosphate ions which could not be assigned to an apatitic environment and which were not present in well-crystallized apatites (Figure 2).

These bands were assigned to non-apatitic environments of PO_4^{3-} and HPO_4^{2-} ions of the nano-sized crystals. Thus, FTIR spectra can be used to provide a sufficiently accurate evaluation of the relative contribution of such environments. Furthermore, the non-apatitic environments were found to correspond to hydrated domains of the nano-sized crystals which were distinct from the apatite domains [234]. Hence, precipitated crystals of nano-sized apatite appeared to have a hydrated surface layer containing labile ionic species which easily and rapidly could be exchanged by ions and/or macromolecules from the surrounding fluids [233, 232, 241]. In as-precipitated apatites, such a layer appears to be constituted mainly of water molecules coordinated to surface Ca^{2+} ions, approximately in a 1 : 1 ratio whereas the OH groups account only for ~ 20% of the surface hydration species. The FTIR data indicated that water molecules, located at the surface of nano-dimensional apatites, are coordinated to surface cations and experience hydrogen bonding significantly stronger than that in liquid water [240]. The hydrated surface layer is very delicate and progressively transforms into a more stable apatitic lattice upon ageing in aqueous media. Furthermore, it irreversibly altered upon drying [234]. Outgassing at temperatures up to ~ 300 °C resulted in complete surface dehydration, accompanied by a decrease of the capability to re-adsorb water. Combination of these data with rehydration tests suggested that a significant part of the surface Ca^{2+} ions, once dehydrated, could undergo a irreversible relaxation as the outgassing temperature increased [239].

Table 2. Crystallographic data of calcium orthophosphates [206]

Compound	Space group	Unit cell parameters	Z[a]	Density, g/cm^3
MCPM	triclinic $P\bar{1}$	a = 5.6261(5), b = 11.889(2), c = 6.4731(8) Å, α = 98.633(6)°, β = 118.262(6)°, γ = 83.344(6)°	2	2.23
MCPA	triclinic $P\bar{1}$	a = 7.5577(5), b = 8.2531(6), c = 5.5504(3) Å, α = 109.87(1)°, β = 93.68(1)°, γ = 109.15(1)°	2	2.58
DCPD	monoclinic Ia	a = 5.812(2), b = 15.180(3), c = 6.239(2) Å, β = 116.42(3)°	4	2.32
DCPA	triclinic $P\bar{1}$	a = 6.910(1), b = 6.627(2), c = 6.998(2) Å, α = 96.34(2)°, β = 103.82(2)°, γ = 88.33(2)°	4	2.89
OCP	triclinic $P\bar{1}$	a = 19.692(4), b = 9.523(2), c = 6.835(2) Å, α = 90.15(2)°, β = 92.54(2)°, γ = 108.65(1)°	1	2.61
α-TCP	monoclinic $P2_1/a$	a = 12.887(2), b = 27.280(4), c = 15.219(2) Å, β = 126.20(1)°	24	2.86
β-TCP	rhombohedral $R3cH$	a = b = 10.4183(5), c = 37.3464(23) Å, γ = 120°	21[b]	3.08
HA	monoclinic $P2_1/b$ or hexagonal $P6_3/m$	a = 9.84214(8), b = 2a, c = 6.8814(7) Å, γ = 120° (monoclinic) a = b = 9.4302(5), c = 6.8911(2) Å, γ = 120° (hexagonal)	4 2	3.16
FA	hexagonal $P6_3/m$	a = b = 9.367, c = 6.884 Å, γ = 120°	2	3.20
OA	hexagonal $P\bar{6}$	a = b = 9.432, c = 6.881 Å, α = 90.3°, β = 90.0°, γ = 119.9°	1	~ 3.2
TTCP	monoclinic $P2_1$	a = 7.023(1), b = 11.986(4), c = 9.473(2) Å, β = 90.90(1)°	4	3.05

[a] Number of formula units per unit cell.
[b] Per hexagonal unit cell.

In another study, elongated nano-sized crystals of CDHA, ~ 10 nm thick and of ~ 30 – 50 nm length were synthesized followed by investigations with X-ray diffraction and nuclear magnetic resonance techniques. The nano-sized crystals of CDHA were shown to consist of a crystalline core with the composition close to the stoichiometric HA and a disordered (amorphous) surface layer 1 – 2 nm thick [238, 239] with a composition close to DCPD [237]. Based on the total Ca/P ratio on the one hand, and the crystal shape on another hand, a thickness of the DCPD surface layer along the main crystal axis was estimated to be ~ 1 nm [237], close to the dimension of the unit-cell (Table 2).

A similar structure of a crystalline core with the composition of stoichiometric HA and a disordered (amorphous) surface layer was found by other researchers [243]. However, in yet another study devoted to nano-dimensional carbonate apatites [244], the model of a crystalline core and an outer amorphous layer was not confirmed. This discrepancy could perhaps be explained by the presence of carbonate ions. A lack of hydroxide in nano-dimensional apatites was detected; an extreme nanocrystallinity was found to place an upper bound on OH^- possible in apatites [245].

However, it is possible to tailor the structure of surface terminations of HA nano-sized particles to either be amorphous or crystalline by properly selecting the preparation parameters and, in particular, the temperature. In this way nano-dimensional HA without an amorphous layer at the surface has been prepared [246]. The two types of surfaces (amorphous or crystalline) of nano-dimensional HA appeared to be quite similar in terms of their first hydration layer as well as Lewis acid strength of exposed Ca^{2+} ions. Both features depend strongly on the local structure of surface sites (well probed by small molecules, such as H_2O and CO) that appeared essentially unaffected by the organization at a longer range. Interestingly, the crystalline surfaces of nano-dimensional HA treated at 573 K was found to adsorb multilayers of water to a larger extent than the amorphous ones [246].

Nevertheless, after summarizing the available data, the following statements on the structure of nano-sized crystals of apatites have been made:

(1) Nano-sized apatite crystals involve non-apatitic anionic and cationic chemical environments (“ordered and disordered HA” [238]),
(2) At least part of these environments are located at the surface of the nano-sized crystals and strongly interact with hydrated domains,
(3) Immature samples show a fine substructure of FTIR bands that is altered upon drying without leading to long-range order (LRO) modifications,
(4) The fine substructure shows striking similarities with the FTIR spectrum of OCP [235].

All these results favor a model in which nano-sized crystals of apatite are covered with a rather fragile but structured hydrated surface layer containing relatively mobile ions (mainly, bivalent anions and cations: Ca^{2+}, HPO_4^{2-}, CO_3^{2-}) at “non-apatitic” sites (Figure 3) that are supposed to be of either OCP or DCPD structure. Unfortunately, the exact structure as well as the chemical composition of this hydrated layer is still uncertain since, regrettably, the hydrated layer cannot be isolated, and it is thus not possible to standardize the methods for detailed studies [235, 237-239]. Nevertheless, it is known that the surface layer might adsorb considerable amounts of foreign compounds (molecules and ions) in the percent mass range [247]. Strictly speaking, all the aforementioned facts apply to both biological apatite of calcified tissues [248] and micron-sized apatites as well [249].

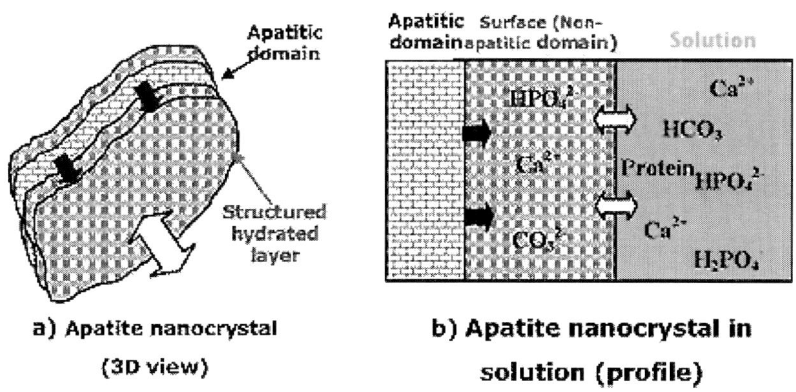

Figure 3. Schematic representation of the “surface hydrated layer model” for poorly crystalline apatite nanocrystals. Reprinted from Ref. [235] with permission.

Nonetheless, in nano-sized crystals, the composition of the hydrated surface layer contributes to the global composition in non-negligible proportion. The results of electron state spectroscopy of nano-structural HA bioceramics are available elsewhere [250, 251].

The hydrated surface layer confers unexpected properties to nano-sized apatite, is responsible for most properties of apatites, and, for example, can help to explain the regulation by biological apatites of the concentration of ions in body fluids (homoeostasis). As these properties are important for living organisms they need therefore to be considered in both material science and biotechnology [234]. The consideration of this type of surface state can help to understand and explain the behavior of biological apatites participating in homoeostasis due to their very high specific surface area thus constituting an important ion reservoir with availability depending on the maturation state. The important consequences are that the surface of nano-dimensional apatites has nothing in common with the bulk composition and that the chemistry of such materials (*e.g.*, binding of protein molecules) must be reconsidered [235, 237]. Interestingly, in response to an electrical potential, the surface of nano-sized HA bioceramics was found to exhibit dynamic changes in interfacial properties such as wettability. The wettability modification enabled both a sharp switching from hydrophilic to hydrophobic states and a microscopic wettability patterning of the HA surface which may be used to fabricate spatially arrayed HA for biological cells immobilization or gene transfer [252].

Furthermore, dry powders of nano-dimensional HA were found to contain an X-ray amorphous portion with an unspecified location [253]. After mixing of an initial nano-sized HA powder with a physiological solution (aqueous isotonic 0.9 % NaCl solution for injections), this amorphous portion was fully converted into the crystalline phase of HA. The initial average size of crystallites (~ 35 nm) was enlarged by a factor of about four within the first 100 min after mixing the powder with the physiological solution, but no further structural changes were detected during the following period [253]. In the light of the aforementioned studies, presumably, the discovered X-ray amorphous component of the initial powder was located on the surface of nano-dimensional HA.

6. Synthesis of Nano-Dimensional and Nano-Crystalline Calcium Ortho-Phosphates

6.1. General Nanotechnological Approaches

Synthesis of nano-scale materials has received considerable attention and their novel properties can find numerous applications, for example, in the biomedical field. This has encouraged the invention of chemical, physical and biomimetic methods by which such nano-sized materials can be obtained [134]. Generally, all approaches to prepare nano-dimensional and nano-crystalline materials can be categorized as "bottom-up" and "top-down" ones [142, 254]. The bottom-up approach refers to the build-up of a material from the bottom, *i.e.*, atom by atom, molecule by molecule, or cluster by cluster and then to assemble them into the final nano-structured material. An example is the production of a nano-sized powder and its compaction into the final product (*e.g.*, hot-pressed or sintered nano-structured ceramics). The top-down approach starts from a bulk material and then, via different decomminution techniques, such as milling, slicing or successive cutting, leads to the generation of nano-dimensional materials [134]. Using this approach, the novel two-dimensional carbon material graphene of just one atomic layer thickness has been prepared from bulk graphite. Furthermore, environmentally friendly methodologies of nanostructure synthesis have been summarized into a special review [255].

Concerning calcium orthophosphates, conceivably all of them (see Table 1) might be obtained in a nano-dimensional and/or a nano-crystalline state. However, not all of them (especially those with low Ca/P ionic ratios) have been prepared yet. The details on the available preparation techniques are given below.

6.2. Nano-Dimensional and Nano-Crystalline Apatites

First of all, one should stress that stoichiometric HA with well resolved X-ray diffraction pattern may be prepared mostly at temperatures exceeding ~ 700 °C either by calcining of CDHA with Ca/P molar ratio very close to 1.67 or by solid-state reactions of other calcium orthophosphates with various chemicals (*e.g.*, DCPA + CaO).

Thus, with the exception of a hydrothermal synthesis [256-258], in aqueous solutions only CDHA can be prepared [148, 204-206, 259-263]. As apatites (CDHA, HA and FA) belong to sparingly soluble compounds (Table 1), simple mixing of calcium- and orthophosphate-containing aqueous solutions at pH > 9 results in formation of extremely supersaturated solutions and, therefore, a very fast precipitation of a tremendous amount of very fine crystals [264], initially as ACP that afterwards re-crystallizes into apatites [204-206, 265-268]. The dimensions of the precipitated nano-sized crystals may be slightly increased by the Ostwald ripening approach (maturation), that is, by boiling and/or ambient aging in the mother liquid (Figure 4) [156, 169, 203, 235, 257, 267-272].

Heat treatment of ACP can be applied as well [273]. Therefore, preparation of nano-dimensional and/or nano-crystalline apatites is not a problem at all and has been known for many years [156, 157, 274-276] even though the ennobling prefix "nano" had not been used before 1994.

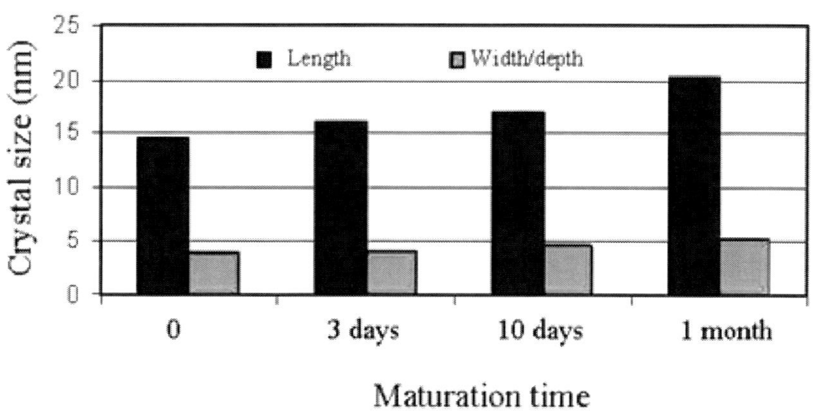

Figure 4. Variation of nanocrystalline apatite dimensions with maturation time. Reprinted from Ref. [235] with permission.

On the contrary, with the exception of thermally stable FA of which large crystals of FA can be produced by a melt-growth process [277, 278]), manufacturing of big crystals of both CDHA and HA is still a challenge.

Many different methodologies have been proposed to prepare nano-dimensional and/or nano-crystalline structures [279-286]. Prior to describing them, it is important to stress that in the vast majority of the available literature on apatites, the authors do not elaborate on the difference between CDHA and HA. Therefore, leafing through scientific papers, an attentive reader often finds statements such as: "*Because natural bone is composed of both organic components (mainly type I collagen) and inorganic components (HA), …*" [116, p. 357], "*The HA nano-rods are synthesized via a wet precipitation process ...*" [167, p. 2364], or "*… (TTCP) has been shown previously to be an essential component of self-setting calcium phosphate cements that form hydroxyapatite (HA) as the only end-product. …*" [287, abstract]. The matter of distinguishing between CDHA and HA becomes even much more complicated when researchers deal with nano-dimensional and/or nano-crystalline apatites because the assemblies of nano-sized particles give only broad diffraction patterns, similar to the ones from an amorphous material [230, 231]. While composing this review, I always tried to specify whether each quoted study dealt with CDHA or HA. Unfortunately, the necessary data were found in just a few papers. Therefore, in many cases, I was forced to refer to "apatites" without further specification. Thus, the readers are requested to be understanding of this uncertainty.

The greater part of the published reports on synthesis of nano-dimensional and/or nano-crystalline apatites is focused on the bottom-up approach. Among the available preparation techniques, wet chemical precipitation is the most popular one [74, 86, 88, 103, 119, 121, 167-169, 172, 178, 201, 203, 265, 274, 288-332]. Various authors discussed the effects of synthesis parameters, such as temperature [300-303, 322], time [301], calcium ion concentration [303], presence of surfactants [306-308], calcination [301], and use of different reagents on the morphological properties of nano-dimensional apatites. In general, the shape, stoichiometry, dimensions, and specific surface area of nano-sized apatites appears to be very sensitive to both the reaction temperature (Figure 5) and the reactant addition rate [300, 315, 322].

For example, the particle size of nano-dimensional apatites was observed to increase linearly with temperature [302, 322], an indication that sizes of nano-dimensional apatites can possibly be tailored. Furthermore, the initial pH values and reaction temperatures both play important roles in the morphology of the precipitated apatites, as well as in the phase formation and degree of crystallinity [326]. For example, significant differences in the chemical composition, morphology and amorphous character of nano-sized CDHA produced through the reaction between aqueous solutions of $Ca(NO_3)_2$ and $(NH_4)_2HPO_4$ can be induced simply by changing the pH of the reactant hydrogen phosphate solution [327]. Thus, the solvent systems, dispersant species and drying methods appear to have effects on the particle size and dispersibility. However, some conflicting results have been obtained on how certain synthesis parameters can affect the morphological properties of these nano-sized particles.

Nevertheless, it was commonly observed that nano-sized crystals of apatites synthesized through the chemical precipitation route were often highly agglomerated, forming clusters of ultra-fine primary particles [304]. The prepared nano-dimensional apatites can be consolidated to form transparent bioceramics [330].

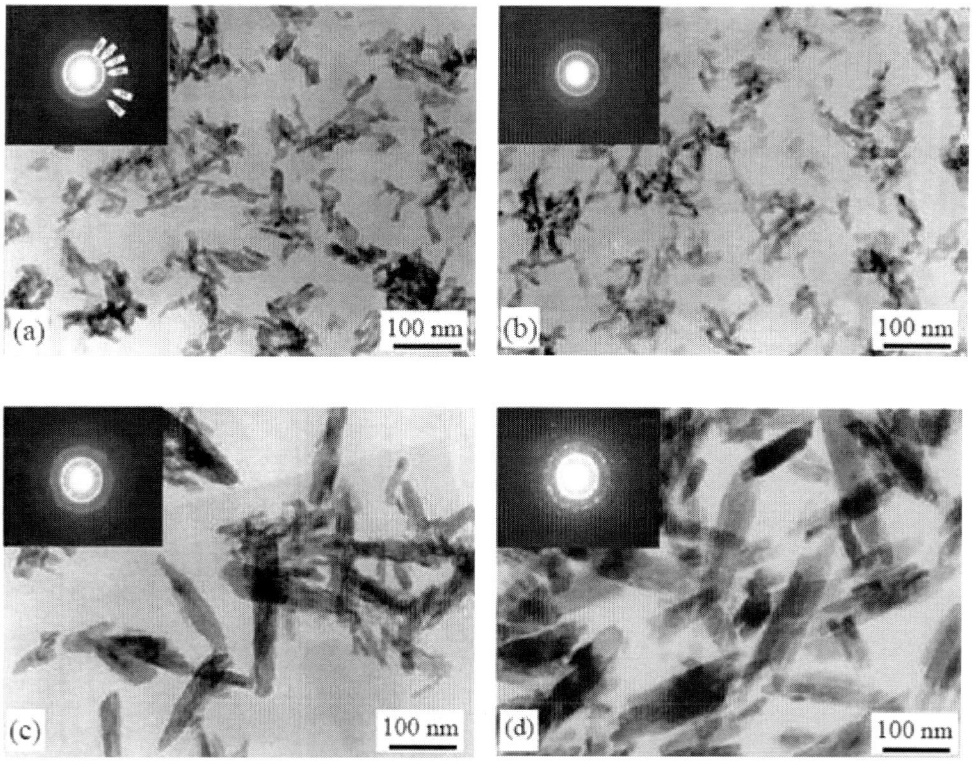

Figure 5. The influence of the reaction temperature on the crystal dimensions of precipitated CDHA: a – 25 ºC, b – 37 ºC, c – 55 ºC, d – 75 ºC.

Hydrothermal synthesis [69, 72, 156, 157, 257, 258, 288, 322, 323, 333-357] appears to be the second most popular preparation technique of nano-dimensional and/or nano-crystalline apatites. The term "hydrothermal" refers to a chemical reaction of substances in a sealed heated solution above ambient temperature and pressure [358]. This process allows synthesizing highly pure fine-grained single crystals with controlled morphology and narrow size distribution [333]. Extraneous additives such as EDTA [351], surfactants [352, 359], anionic starburst dendrimers [353] *etc.*, were utilized to modify the morphology of nano-dimensional and/or nano-crystalline apatites during their synthesis. Most of these techniques produced rod-like crystals or whiskers, whereas platy shapes were obtained in just a few cases [335, 345, 347].

Other preparation methods of nano-dimensional and/or nano-crystalline apatites of various states, shapes and sizes include sol-gel [30, 188, 231, 232, 270, 328, 360-376], co-precipitation [271, 333, 334, 377-380], mechano-chemical treatment [65, 250, 343, 348, 381-387], mechanical alloying [388, 389], ball milling [348, 383, 390-392], radio-frequency induction plasma synthesis [393, 394], vibro-milling of bones [395], flame spray pyrolysis [396], liquid-solid-solution synthesis [397], electro-crystallization [158, 398, 399], electrochemical deposition [400], microwave processing [32, 69, 288, 333, 334, 342, 356, 401-415], hydrolysis of other calcium orthophosphates [416-418], double-step stirring [419], emulsion-based [310, 349, 420-433], steam-assisted [434], sonochemical [435] and solvothermal [436] syntheses. However, still other preparation techniques are known [31, 45, 147, 154, 275, 355, 437-457]. Continuous preparation procedures are also available [200,

458]. Application of both ultrasonic [362, 459-461] and viscous systems [462] might be helpful. Furthermore, nano-dimensional HA were manufactured by a laser-induced fragmentation of HA targets in water [463-467] and in solvent-containing aqueous solutions [344, 371, 468], while dense nano-crystalline HA films could be produced by radio frequency magnetron sputtering [469, 470]. An interesting approach using sitting drop vapor diffusion technique should be mentioned as well [471]. A comparison between the sol-gel synthesis and wet chemical precipitation technique was performed and both methods were found to be suitable for synthesis of nano-dimensional apatite [328]. By means of these methods, a variety of nano-dimensional calcium orthophosphate building blocks with various structures and morphologies have been synthesized, including needle-like, spherical, fibrous and mesoporous nano-sized crystals as well as nano-sized rods, hollow spheres, layered structures and flowers as shown in Figure 6 [452].

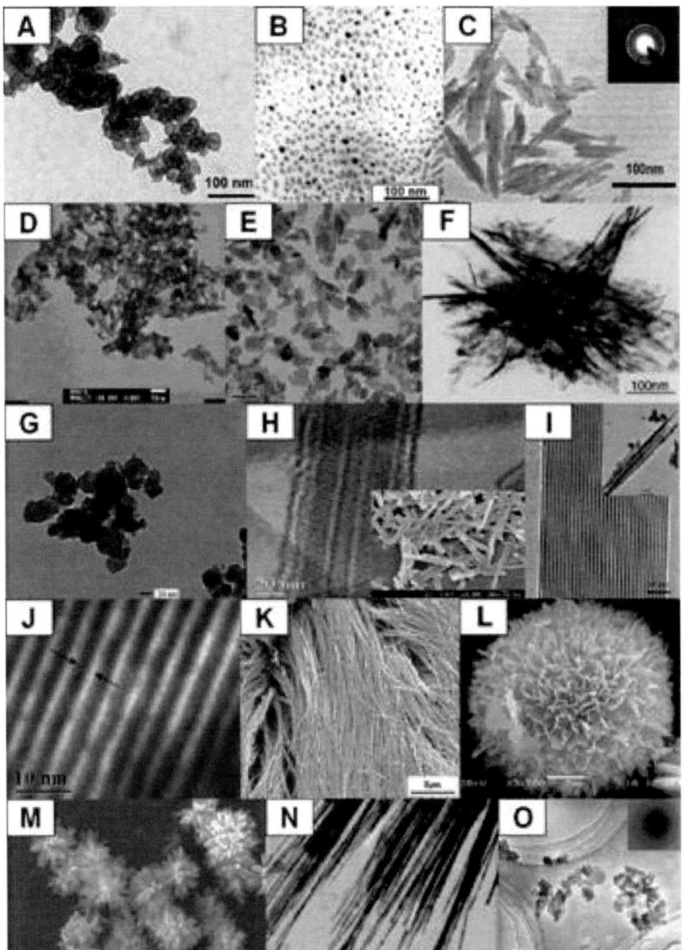

Figure 6. A variety of nano-scale calcium orthophosphates with different structures and morphologies synthesized by: (A and B) sol-gel processing, (C) co-precipitation, (D) emulsion technique, (E) hydrothermal process, (F) ultrasonic technique, (G) mechano-chemical method, (H – L) template method, (M) microwave processing, (N) emulsion-hydrothermal combination, (O) microwave-hydrothermal combination. Reprinted from Ref. [452] with permission.

Table 3. Synthesis of nanodimensional apatites – a chronological development [147]

Year	Process	Reference
1995	Synthesis of nanocrystalline HA (particle size ~ 20 nm) for the first time using calcium nitrate and diammonium hydrogen orthophosphate as precursors by solution spray dry method.	[472]
2000	Synthesis of biomimetic nanosized CDHA powders (~ 50 nm) at 37 °C and pH of 7.4 from calcium nitrate tetrahydrate and diammonium hydrogen orthophosphate salts in synthetic body fluid using a novel chemical precipitation technique.	[378]
2002	Preparation of nanosized HA particles and HA/chitosan nanocomposite.	[473]
2002	Direct precipitation from dilute calcium chloride and sodium orthophosphate solutions.	[474]
2003	Radio frequency plasma spray process employing fine spray dried HA powders (average size ~ 15 μm) as a feedstock.	[393]
2003	Sol-gel process using equimolar solutions of calcium nitrate and diammonium hydrogen orthophosphate dissolved in ethanol.	[371]
2003	Chemical precipitation through aqueous solutions of calcium chloride and ammonium hydrogen orthophosphate.	[301]
2003	Mechanochemical synthesis of nanosized HA and β-TCP powders using DCPD and CaO as starting materials.	[384]
2003	Synthesis of nano-powders via sucrose-templated sol-gel method using calcium nitrate and diammonium hydrogen orthophosphate as precursor chemicals.	[475]
2004	Hydrolysis method of DCPD and $CaCO_3$ by 2.5 M NaOH (aq).	[416]
2004	Citric acid sol-gel combustion process using calcium nitrate tetrahydrate, diammonium hydrogen orthophosphate and citric acid.	[476]

Table 3 presents the chronology of synthesis of nano-dimensional apatites for the period of 1995 – 2004 [147]. Among the methods described, the thinnest crystals of apatite (60 nm × 15 nm × 0.69 or 0.84 nm), called "two dimensional crystalline HA", have been prepared by Melikhov *et al.* [296], while the smallest ones (size between 2.1 and 2.3 nm, *i.e.* around two times the HA unit cell parameters) have been found by Biggemann *et al.* [197]. Liu *et al.* [477, 478] and Han *et al.* [476] synthesized nano-sized HA via a template-mediated and a non-template mediated sol-gel techniques, respectively. Both triethylphosphate [477, 478] and other alkylphosphates [122] can be used to produce nano-crystalline apatites. Besides, nano-dimensional ion-substituted CDHA can be precipitated from simulated [378,479] body fluids. A relatively simple sol-gel process using ethanol and/or water as a solvent to obtain the stoichiometric, nano-crystalline single phase HA has also been reported [371].

Nano-crystalline HA powder was synthesized at a low calcination temperature of 750 ºC by the citric acid sol-gel combustion method [476]. The attractive features of this method included the ability to synthesize materials with high purity, better homogeneity and high surface area in a single step [476, 480].

An array of highly ordered HA nano-sized tubes of uniform length and diameter was synthesized by a sol-gel auto-combustion method utilizing a porous anodic aluminum oxide template [370]. Varma *et al.* synthesized nano-sized HA by a polymeric combustion method and self-propagating combustion using novel body fluid solutions [481]. In another study, nano-dimensional HA was synthesized by combustion in an aqueous system containing calcium nitrate + diammonium hydrogen orthophosphate, with urea and glycine as fuels

[482]. Furthermore, nano-sized particles of both FA and β-TCP could be synthesized by simultaneous combustion of calcium carboxylate- and tributylphosphate-based precursors in a flame spray reactor [483]. Both a flame-based technique [484] and a spray drying approach [125, 485] might be applied as well. Furthermore, crystalline and phase- pure nano-sized HA and CDHA were synthesized in a continuous hydrothermal flow system using supercritical water at t < 400 °C and 24 MPa pressure [338].

Nano-dimensional powders of stoichiometric HA of ~ 20 nm particle size were synthesized by hydrolysis of a mixture of DCPD and $CaCO_3$ performed in 2.5 M aqueous solution of NaOH at 75 °C for 1 h. The only product synthesized was nano-crystalline HA, and its crystallinity was improved with increasing annealing temperature [416]. Similar results were obtained in other studies [417, 418]. Furthermore, Xu *et al.* used a radio-frequency plasma-spray process to synthesize nano-dimensional HA powders with particle size in the range of 10 – 100 nm [393]. Kuriakose *et al.* synthesized nano-crystalline HA of ~ 1.3 nm size that was thermally stable up to 1200°C [371]. Nano-crystalline platy particles of HA were directly precipitated at ambient temperature and pH ~ 7.4 from dilute aqueous solutions of calcium chloride and sodium orthophosphate. The direct precipitation of nano-sized HA was achieved by submitting the aqueous suspension to microwave irradiation immediately after mixing [474]. A simple and easy approach to synthesize thermally stable nano-structured stoichiometric HA powder under invariant pH conditions of 7.5, known as the NanoCaP process, was developed. Under these conditions, the synthesized HA not only remained in the nano-structured state but also did not exhibit any compositional fluctuations that were observed in conventional approaches to synthesize HA [12]. Other preparation techniques of nano-sized apatite might be found elsewhere [275]. Bulk bioceramics made of nano-crystalline HA with a grain size below 50 nm and a near-theoretical density can be prepared by applying high pressure (~ 3.5 GPa) through uniaxial compaction of nano-dimensional powders with subsequent sintering at 640 °C [168]. A similar approach was reported by another research group [406].

Mechano-chemical processing is another compelling method to produce nano-structured apatites in the solid state [65, 238, 343, 381-386]. For example, Yeong *et al.* [382] used appropriate amounts of DCPA and calcium oxide. The initial stage of mechanical activation resulted in a significant refinement in crystallite and particle sizes, together with a degree of amorphization in the starting powder mixture. This was followed by steady formation and subsequent growth of HA crystallites with increasing degree of mechanical activation. Finally, single-phase HA with an average particle size of ~ 25 nm, a specific surface area of ~ 76 m^2/g and high crystallinity was obtained after 20 h of mechanical activation.

The use of macromolecules as templating agents to manipulate the growth of inorganic crystals has been realized in many biological systems. For example, in the presence of biological macromolecules such as collagen, nucleation and growth of nano-crystalline apatite to form highly organized bone minerals is one of the most fascinating processes in nature. These processes might be simulated. For example, layers of nano-crystalline apatite were formed *in situ* at the surface of various chitosan and collagen films by soaking them in aqueous solutions containing calcium and orthophosphate ions. The *in situ*-synthesized particles were found to be less agglomerated which was believed to be the result of nucleation of apatite crystallites on regularly arranged side-groups located on the polymer chains [486, 487]. Another approach comprises precipitation of nano-dimensional apatites from aqueous solutions in the presence of dissolved high molecular weight poly(acrylic) acid [488, 489] acting as an inhibitor for the crystallization of apatite crystals [490, 491]. A similar inhibiting effect was found for dimethylacetamide [492], polyvinyl alcohol [291] and several other (bio)polymers [493, 494]. This type of synthesis is expected to lead to formation of nano-dimensional composites which might be structurally more comparable to bones with closely related mechanical and biological properties. Furthermore, the control of particle size of aqueous colloids of apatite nano-sized particles was described involving a presence of amino acids [495, 496]. The amino acids ensured effective growth inhibition through predominant adsorption onto the Ca-rich surfaces during the initial stages of crystallization. Thus, nano-sized particles were formed by an oriented aggregation of primary crystallite domains along the *c*-axis direction. The size of the domains was shown to be governed by interactions with the amino acid additives which restricted growth of the primary crystallites [495, 496]. Furthermore, nano-dimensional apatites can be precipitated from aqueous solutions of gelatin [70, 497]. The development of nano-sized apatite in aqueous gelatin solutions was highly influenced by the concentration of gelatin, that is, a higher concentration of gelatin induced formation of tiny (4 nm × 9 nm) nano-sized crystals while a lower concentration of gelatin contributed to the development of larger (30 nm × 70 nm) crystals. In this experiment, the higher concentration of gelatin supplied abundant reaction sites containing, for example carboxyl groups that could bind with calcium ions. This led to formation of a very large number of nuclei and the creation of a large number of tiny nano-sized crystals [70].

Although all reported approaches to produce nano-dimensional apatites have both scientific and practical relevance, only little attention has been dedicated to the physic-chemical details involved in the required careful control of the particle size distribution and particle shape. Indeed, in the case of particle size distribution, most of the reported ways to synthesize nano-dimensional apatites really produced a mixture of particles with a wide size distribution from tens to hundreds of nanometers. Moreover, the control of particle shape is another problem as these methods commonly generate needle-like or irregular particles. It is well known that bone consists of homogeneous platy crystals of biological apatite of 15 – 30 nm width and 30 – 50 nm length whereas dental enamel consists of rod-like crystals of biological apatite of 25 – 100 nm thickness and 100 nm to several microns length [2, 5, 207, 208, 210, 219, 227, 229].

The study of higher-level biomineralization and biomimetic assembly strategies involves the search for advanced methods so that the synthesis of nano-sized apatite can be accurately controlled [498]. For example, the size-controlled synthesis of materials can be achieved using limited, i.e. confined reaction spaces. For example, microemulsions have been shown to be one of the few techniques able to produce particle sizes in the range of nanometers and with minimum agglomeration [499]. Thus, microemulsions [349, 423-433], micelles [500] and reverse (inverse) micelles [339, 501-504] have been successfully applied to synthesize nano-dimensional apatites with minimal agglomeration. It was found that variation of experimental conditions including aqueous/organic phase volume ratio, pH, aging time, aging temperature, and ion concentration in the aqueous phase can affect the crystalline phase, surface area, particle size and morphology of nano-dimensional apatites.

In some cases, special polymers can be used as spatial reaction vessels for fabrication of CDHA. For example, Shchukin *et al.* employed a poly(allylamine hydrochloride)/PO_4^{3-} complex as source of orthophosphate anions to capture calcium cations and promote their reaction in the capsule volume [505]. Bose and Saha [425] synthesized spherical nano-crystalline CDHA powder with particle diameters of ~ 30 and ~ 50 nm using the emulsion route. Furthermore, nano-sized crystals of apatite could be aggregated into microspheres [438, 506].

Hexadecyl(cetyl)trimethylammonium bromide (CTAB) was selected as an efficient agent to modulate the formation of nano-sized CDHA particles [501, 502]. The particle size can be regulated feasibly well by changing the concentration of CTAB in the solution supersaturated by calcium orthophosphates. For example, three different types of spherical particles of nano-sized CDHA with average diameters of 20 ± 5, 40 ± 10 and 80 ± 12 nm were fabricated using a series of CTAB concentrations to control the particle size. The experimental results revealed that the dimensions of the prepared nano-sized CDHA were relatively uniform. In contrast, nano-sized CDHA grown in the absence of organic additives were typical, rod-like particles with lengths of hundreds of nanometers and width of tens of nanometers [507]. Colloidal formulations are known as well [54, 60, 202]. Interestingly, nano-sized apatite powders can transform from the crystalline to the amorphous state when aged for five months in 30% relative humidity [508].

To conclude this part, nano-sized particles of apatite can be functionalized and/or doped by various compounds (even by quantum dots [509, 510]) to provide new important properties [495, 511-518], *e.g.* fluorescence [27-29, 515, 516] and luminescence [54, 60, 510, 517, 518]. Both fluorescence and luminescence properties can be used to track nano-sized particles and thus to give an observable indication of agent delivery, while the particles serve to protect the agent *in vivo* until it has reached its destination.

6.3. Nano-Dimensional and Nano-Crystalline TCP

Many researchers have designed synthesis ways of nano-dimensional β-TCP. For example, Bow *et al.* synthesized β-TCP powders of ~ 50 nm particle diameter at room temperature in anhydrous methanol as a solvent [519]. With increase in aging time, a phase transformation was found to take place from initial DCPA to an intermediate ACP phase, and then to final β-TCP. The authors observed that incorporation of carbonate ions assisted in suppressing formation of ACP with apatitic structure and its transformation into poorly crystalline (almost amorphous) CDHA but favored the formation of β-TCP [519]. Nano-sized particles of both FA and β-TCP were synthesized by simultaneous combustion of calcium carboxylate- and tributylphosphate-based precursors in a flame spray reactor [483]. The same technique was used to synthesize nano-sized particles of amorphous TCP of 25 – 60 nm size [520-523] that after calcinations transformed to α-TCP or β-TCP. Nano-dimensional β-TCP powders with an average grain size of ~ 100 nm [173, 524] and below [525] were prepared by wet precipitation methods, followed by calcining at elevated temperatures. Furthermore, a sol-gel technique [526], reverse micelle-mediated synthesis [527] and a polystyrene template method [528] are also applicable. In the wet precipitation techniques, dialysis might be applied as a separation method [524]. When wet precipitation methods were used, initially nano-dimensional CDHA with a Ca/P ratio of ~ 1.50 precipitated that during calcination transformed into nano-sized β-TCP.

To synthesize nano-sized TCP, other techniques, such as milling [529, 530], high temperature flame spray pyrolysis [531] and pulsed laser ablation [532] may be employed as well. Afterwards, the nano-dimensional β-TCP powders can be compacted into 3D-specimens, followed by sintering to achieve the appropriate mechanical strength [173]. The maximal values of the bending strength, elastic modulus, Vickers hardness and compressive strength of the samples fabricated from nano-sized β-TCP powders were more than twice higher as compared to those of bioceramics obtained from micron-sized β-TCP powders. However, the degradability of bioceramics sintered from nano-dimensional powders was just about one fourth of that sintered from micron-sized powders. Thus, the degradability of β-TCP bioceramics could be additionally regulated by the particle dimensions [173].

Nano-sized whiskers of several calcium orthophosphates (HA, β-TCP and biphasic calcium phosphate BCP (HA + β-TCP)) were produced by using a novel microwave-assisted hybrid "combustion synthesis (auto ignition)/molten salt synthesis" route. Aqueous solutions containing $NaNO_3$, $Ca(NO_3)_2$ and KH_2PO_4 (with or without urea) were irradiated in a household microwave oven for 5 min at a power of 600 W. The as-synthesized precursors were then simply stirred in water at room temperature for 1 h to obtain nano-sized whiskers of the desired calcium orthophosphate [533]. Furthermore, nano-structured biphasic (HA + β-TCP) bioceramics were successfully prepared by microwave synthesis [534, 535] and a polymer matrix-mediated process [536]. Good cellular activities of the biphasic bioceramics have been reported.

Layrolle and Lebugle developed a synthesis route of nano-sized FA and other calcium orthophosphates using calcium diethoxide ($Ca(OEt)_2$) and H_3PO_4 [154] (+ NH_4F to prepare FA [537]) as the initial reagents and anhydrous ethanol as a solvent. By simple variation of the ratio of reagents, calcium orthophosphates of various chemical compositions were precipitated in ethanol. The precipitates were characterized and the results indicated that they were amorphous and nano-dimensional. Furthermore, they had large specific surface areas and possessed a high reactivity.

6.4. Other Nano-Dimensional and Nano-Crystalline Calcium Orthophosphates

Nano-sized particles of DCPD (containing some amounts of CDHA and ACP) with a relatively high monodispersity could be synthesized from aqueous solutions of calcium nitrate and H_3PO_4 in the presence of 2-carboxyethyl-phosphonic acid. They were produced in a discoid shape with a diameter of 30 – 80 nm and a height of less than ~ 5 nm. They form stable colloidal solutions with minimum agglomeration [538]. Nano-sized rods and nano-dimensional fibers of DCPD with average diameters of 25 ± 5 nm (aspect ratio ~ 6) and 76 ± 20 nm (aspect ratio ~ 40), respectively, were synthesized by a sucrose ester-based reverse microemulsion technique [539]. A similar approach was used in another study [431]. Nano-dimensional crystals of both DCPD and DCPA were prepared by an EDTA-assisted hydrothermal method [351]. An interesting approach involves precipitation of calcium orthophosphates inside nano-sized pores of another material. For example, nano-dimensional clusters of DCPD were immobilized within pores of an oxide network by immersion of this network into an acidic (pH = 2.7) calcium orthophosphate solution at 50 ºC [540]. The acid-base reaction between the calcium orthophosphate solution and the hydroxyl groups of the oxide network resulted in formation of nano-dimensional clusters of DCPD immobilized inside the oxide pores. Interestingly, the immobilized nano-dimensional clusters of DCPD were further converted into clusters of ACP and CDHA by subsequent treatment of the oxide network in alkaline solutions [540]. Hollow nano-sized shells of undisclosed calcium orthophosphates (presumably ACP) with a size distribution of (120 – 185) ± 50 nm and predictable mean shell thickness from 10 to 40 nm were prepared by crystallization onto the surface of nano-dimensional liposomes [541, 542]. Both the suspension stability and shell thickness control were achieved through introduction of carboxyethylphosphoric acid. Variation of shell thickness and stoichiometry may be a way of manipulating the dissolution kinetics of ACP coatings to control the release of encapsulated materials, necessary for drug delivery purposes [541, 542]. Other types of calcium orthophosphate shells with Ca/P ratios of 0.97 (DCPD or DCPD-like ACP) and 1.45 (CDHA or ACP) were prepared using liposome templates [543]. Roughly spherical DCPA particles ranging between 50 and 100 nm in size were synthesized via a spray-drying technique [125, 544-546] whereas ribbon-like fibers of nano-sized DCPA can be prepared through hydrolysis in urea [417]. Furthermore, nano-dimensional calcium orthophosphate powders with DCPD as the major phase have been synthesized by an inverse microemulsion system using kerosene as the oily phase, a cationic surfactant and a non-ionic surfactant [547]. Microskeletal constructions might be synthesized as well [548]. When it comes to ACP, it is nano-dimensional in the vast majority of cases. Approximately spherical nano-sized particles of ACP with a diameter of about 50 nm can be prepared by rapid precipitation from water [549] and subsequent colloidal stabilization by coating with polymers [550]. Nano-sized clusters of ACP [551] or those comprising a spherical core of 355 ± 20 DCPD units with density of 2.31 g/cm^3 and radius of 2.30 ± 0.05 nm surrounded by 49 ± 4 peptide chains with a partial specific volume of 0.7 cm^3/g, forming a tightly packed shell with an outer radius of 4.04 ± 0.15 nm were prepared by precipitation using 10 mg/ml of the 25-amino-acid N-terminal tryptic phosphopeptide of bovine β-casein as a stabilizing agent [552]. Nano-sized particles of ACP were prepared by mixing of solutions of $Ca(NO_3)_2{\cdot}4H_2O$ (450 mmol/L) in acetone and $(NH_4)_2HPO_4$ (30 mmol/L) in deionized water at pH values between 10.0 and 11.0 [553]. Furthermore, nano-dimensional particles of ACP can be prepared by electrostatic spray pyrolysis [554, 555], pulsed laser ablation [532], spray drying [125], as well as by flame spray synthesis [457]. By means of the latter technique, one can produce nano-dimensional ACPs with a broad Ca/P ratio between 0.5 and 1.5 [457].

Self-assembled shell cross-linked poly(acrylic acid-b-isoprene) micelles and/or cross-linked poly(acrylic) acid nano-sized cages in aqueous solutions may be used as templates to prepare polymer/calcium orthophosphate nano-dimensional capsules of 50 – 70 nm in diameter consisting of nano-sited spherical polymer particles enclosed within a continuous 10 – 20 nm thick surface layer of ACP [556]. Synthesis of nano-sized hollow spherical calcium orthophosphate particles using polymeric templates has also been reported by other researchers [557]. Furthermore, bundles of surfactant-coated nano-dimensional ACP filaments of ~ 2 nm in width and > 300 μm in length were synthesized in reverse micelles [558]. Bundles of the nano-dimensional filaments were stable in the reverse micelle phase up to around 5 days, thereafter being transformed to 5 nm-wide surfactant-coated CDHA rods. Discrete filaments 100 – 500 × 10 – 15 nm in size and with a linear superstructure based on the side-on stacking of surfactant-coated ACP nano-sized rods were also prepared [490]. A double reverse-micelle strategy was realized to synthesize amine, carboxylate- and polyethylene glycol surface-functionalized nano-sized calcium orthophosphate particles of an undisclosed nature [559]. Furthermore, the reverse micelle technique may also be applied to prepare nano-dimensional DCPA [501, 560]. Concerning OCP, an oriented growth of nano-dimensional belts of OCP with a clean surface has been achieved by a wet-chemical approach using cetyltrimethylammonium bromide [561]. Pulsed laser deposition technique was employed to obtain thin films of nano-crystalline OCP on pure Ti substrates [562]. The deposition was performed by a pulsed UV laser source in a flux of hot water vapor. High-resolution transmission electron microscopy (HRTEM) and grazing incidence X-ray diffraction (GIXD) investigations indicated that the coatings were composed of nano-crystalline OCP (unfortunately, the dimensions were not indicated). *In vitro* tests proved that both fibroblasts and osteoblasts adhered, reached a normal morphology, proliferated and remained vital when cultured on a nano-crystalline OCP coating, suggesting good biocompatibility and absence of cytotoxicity [562]. Nano-dimensional powders of biphasic calcium phosphates (BCP; HA + β-TCP [563-565] and HA + α-TCP [566]) have been fabricated as well. To obtain details, the interested readers are referred to the original publications. Similar to apatites (see above), nano-sized particles of TCP, ACP and other calcium orthophosphates can be functionalized and/or doped by various compounds to provide new important properties [200, 559, 566-572] such as fluorescence [569, 570], luminescence [572] or a good dispersability in organic solvents [566]. Furthermore, nano-sized calcium orthophosphates may be used as templates to manufacture nano-dimensional capsules [573].

6.5. Biomimetic Constructs Using Nano-Dimensional Particles

Morphological control of bioinorganic materials is another interested issue in biomineralization by which inorganic materials with complex morphologies can be produced. Complex forms or patterns with a hierarchical structure over several length scales are important features of biomineralization. Pattern formation in biomineralization is a process in which self-assembled organic templates are transformed by material replication into organized inorganic structures. It goes without saying that researchers try to reproduce these processes in laboratories. For example, Chen *et al.*, reported a way to create dental enamel-like structures by modifying synthetic nano-sized rods of apatite with a surfactant, bis(2-ethylhexyl)sulfosuccinate salt, that allowed the nano-sized rods to self-assemble into prism-like structures at the water/air interface [225]. A nanometer-scale rod array of apatite with preferred orientation of the *c*-axes was successfully prepared by simple soaking calcium-containing silicate glass substrates in Na_2HPO_4 aqueous solution at 80°C for various periods [574]. A biomimetic bottom-up route to obtain the first hierarchical level of bone was reported [212]. A pH-induced self-assembly of peptide-amphiphile to make a nano-structured fibrous scaffold reminiscent of extracellular bone matrix was obtained. After the cross-linking of the scaffold, the fibers were directly mineralized to CDHA forming a biocomposite, in which the crystallographic *c*-axes of the nano-sized crystals of CDHA were aligned with the long axes of the fibers. This alignment was similar to that observed between collagen fibrils and crystals of biological apatite in bones [212]. Other attempts to fabricate artificial materials having bone-like both nanostructure and chemical composition were performed and several significant achievements were obtained [575, 576].

The classical model of biomineralization considers mineral formation as an amplification process in which individual atoms or molecules are added to existing nuclei or templates [1, 2, 577]. This process occurs in the presence of various bioorganic molecules, which deterministically modify nucleation, growth and relative stability of crystallite facets. A model involving aggregation-based growth [578] recently challenged this conventional concept of crystal growth. Inorganic nano-sized crystals were found to aggregate into ordered solid phases via oriented attachment to control the reactivity of nanophase materials in nature [19, 579]. A "bricks and mortar" model was suggested to explain the biological aggregation of nano-sized apatite [580]. In this model, ACP acts as "mortar" to cement the crystalline "bricks" of nano-sized HA, with biological molecules controlling the construction process. Using nano-dimensional spheres of HA as the building blocks, highly ordered enamel-like and bone-like apatites were hierarchically constructed in the presence of glycine and glutamate, respectively. It is interesting that, during the evolution of biological apatite, the amorphous "mortar" can be eventually turned into the "brick" by phase-to-phase transformation to ensure the integrity of biominerals [580].

7. Biomedical Applications of Nano-Dimensional and Nano-Crystalline Calcium Orthophosphates

7.1. Bone Repair

Due to advances in surgical practice and rapid aging of the world population, there is continuously increasing demand for bone grafts [581]. Modern grafts should not only replace missing bones but also should be intrinsically osteoinductive by acting as scaffolds for guided bone growth. Furthermore, an ability to form a biologically active apatite layer to bond to living bone it is an essential requirement for modern biomaterials [582]. In addition, a good bone graft should provide a framework to support new blood vessels and soft tissues in forming a bridge to existing bones [581].

Calcium orthophosphate bioceramics of micron-sized dimensions have been used in dentistry, orthopedics and surgery for over 30 years because of their chemical similarity to calcified tissues of mammals and, therefore, excellent biocompatibility [148, 204-206, 259, 260]. Due to a rapid development of nanotechnology, the potential of nano-dimensional and nano-crystalline forms of calcium orthophosphates has received considerable attention [18] because they produce favorable results in repair of bone defects [583, 584]. For example, due to improved sinterability, enhanced densification and better bioactivity compared to coarser crystals, they might be chosen as major components of self-setting bone cements [14, 25, 520, 521, 585-589]. However, there is a study showing that an increase of particle and crystallite sizes of TCP did not prolong but shortened the induction time of the cement setting reaction, contrary to the common physical rules that, generally, smaller particles or crystallites should enhance reactivity [523]. Nevertheless, two general directions of the biomedical application of nano-dimensional and nano-crystalline calcium orthophosphates can be outlined: (i) as filler materials in powder form to impart bioactivity to various biocomposites and hybrid biomaterials [64-104, 179, 590], and (ii) to manufacture either dense compacts or porous scaffolds with sufficient mechanical properties [74, 96, 294, 295, 575, 576, 591, 592]. As the nano-dimensional and nano-crystalline calcium orthophosphates tend to agglomerate during heating (Figure 7) [313, 593-595], normally low-temperature [169, 371] and/or rapid consolidation [169, 258, 323, 596-602] techniques must be employed.

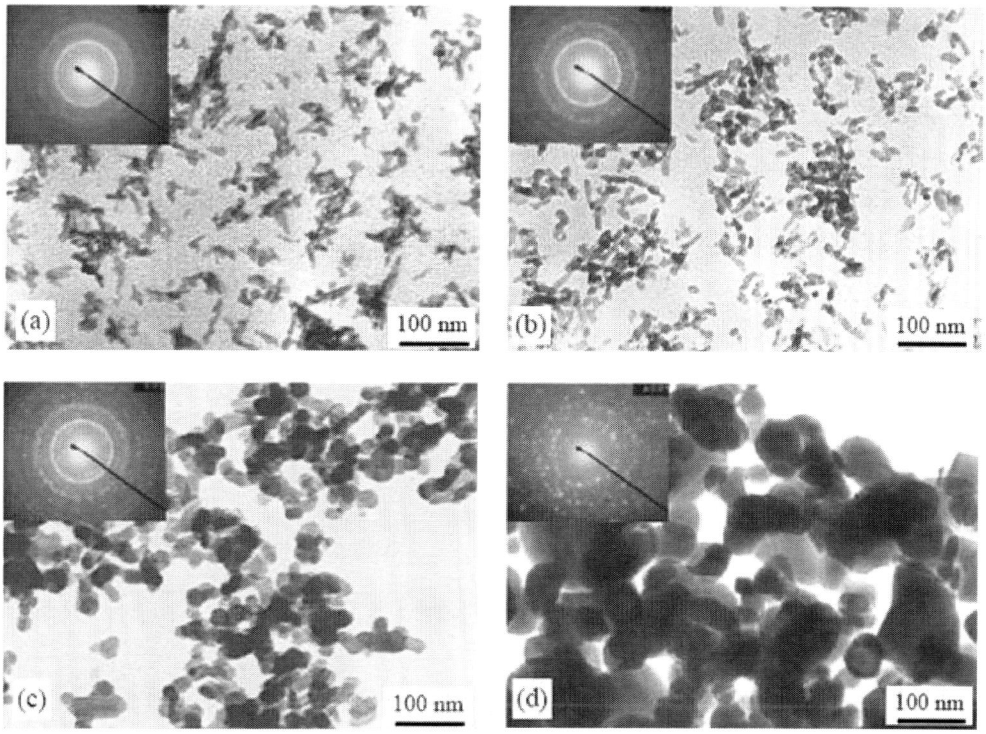

Figure 7. Particle sizes and crystallinity of HA powders after a heat treatment at various temperatures: a – 300 °C, b – 500 °C, c – 700 °C, d – 900 °C.

The low-temperature approach comprises gel hardening (at 4 °C) [371] and uni-axial pressing at 150 – 200 °C [169]. The rapid consolidation techniques comprise spark plasma sintering [169, 258, 323, 596-599], pressure sintering [597] and microwave sintering over the temperature range 1000 – 1300 °C, using a rapid sintering schedule [600-602]. Furthermore, nano-dimensional crystals of calcined HA can be fabricated by calcination at 800 °C for 1 h with an anti-sintering agent surrounding the original nano-sized CDHA particles and subsequent removal of the agent by washing after calcination [603-605]. These consolidation approaches provide a limited alteration of the initial nano-sized crystals, while the final bioceramics possess mechanical properties similar to those attained with sintered stoichiometric HA.

Already in 1990s, implants prepared from nano-dimensional apatites as well as biocomposites of nano-dimensional apatite with organic compounds were tested *in vivo* [606-608]. Cylinders made of both pure nano-dimensional apatite and organoapatite containing a synthetic peptide were analyzed 28 days after implantation into spongy bones of Chinchilla rabbits. Both implant types were well incorporated and interface events were found to be similar to those observed at human bone surfaces with regard to resorption by osteoclast-like cells and bone formation by osteoblasts. This study revealed the suitability of such materials for both bone replacement and drug release purposes [606]. Similar results were obtained in other studies [607, 608].

Among the available commercial formulations, NanOss™ bone void filler from Angstrom Medica, Inc. [609] is considered the first nanotechnological medical device receiving clearance by the US Food and Drug Administration (FDA) in 2005. It is prepared by precipitation of nano-sized calcium orthophosphates from aqueous solutions. The resulting white powder is then compressed and heated to form a dense, transparent and nano-crystalline material. NanOss™ mimics the microstructure, composition and performance of human bone as well as its mechanical strength and osteoconductivity. It is remodeled over time into human bone with applications in sports medicine, trauma treatment, spine repair, and general orthopedics applications [609].

Ostim® (Osartis GmbH and Co. KG, Obernburg, Germany) is another popular commercial formulation. This ready-to-use injectable paste received CE (Conformité Europeénne) approval in 2002. Ostim® is a suspension of synthetic needle-shaped nano-crystalline HA (average crystal dimensions: 100 × 20 × 3 nm^3; specific surface area ~ 100 m^2/g) in water, prepared by a wet chemical reaction [610]. After setting the HA content in the paste reaches ~ 35%. Ostim® does not harden when mixed with blood or spongiosa, so it is highly suitable for increasing the volume of autologous or homologous material. Simultaneously, its viscosity enables its applications to form-fit in close contact with the bone. Ostim® can be used in treatment of metaphyseal fractures and cysts, alveolar ridge augmentation, acetabulum reconstruction and periprosthetic fracture healing during hip endoprosthesis exchange operations, osteotomies, filling cages in spinal column surgery *etc.* [609-622]. Incorporated into bone, new bone formation is visible after only three months [623]. For a number of clinical applications, Ostim® can be combined with other types of calcium orthophosphate bioceramics, *e.g.*, with a HA bioceramic core (Cerabone®) [610, 624] or with biphasic (β-TCP + HA) granules (BoneSaves®) [625]. Application of such combinations of nano-crystalline Ostim® with microcrystalline calcium orthophosphate bioceramics appear to be an effective method to treat both tibia head compression fractures [610] and metaphyseal osseous volume defects in the metaphyseal spongiosa [624]. Besides, such combinations might be used for acetabular bone impaction grafting procedures [625].

Cui *et al.* developed nano-sized HA/collagen biocomposites that mimic the nanostructure of bones [208, 626]. After implantation, such biocomposites can be incorporated into bone metabolism. Due to processing difficulties and poor mechanical properties of bulk calcium orthophosphates, their applications are currently confined to non load-bearing implants and porous bodies/scaffolds. Porous 3D-biocomposites of nano-dimensional HA and collagen/polymer mimic bones in composition and microstructure and can be employed as matrix for engineering of bone tissue [89].

Owing to their low mechanical properties, the use of calcium orthophosphates in load-bearing applications is rather limited: calcium orthophosphates are too stiff and brittle for such use. While present solutions for weight-bearing applications rely mostly on bioinert metals such as cobalt-chromium alloys, titanium and its alloys, as well as stainless steel 316L, problems with stress-shielding and long-term stability can cause failures. All these metals, although nontoxic, are always bioinert and thus cannot bond directly to bone. In order to improve the biological properties of the metallic implants, nano-structured calcium orthophosphates (mainly apatite) are generally used as a coating to accelerate bone growth and enhance bone fixation [186, 187, 316, 439, 562, 627-648]. The coating techniques include thermal spraying, sputter coating, pulsed laser deposition, dynamic mixing method, dip coating, sol-gel method, electrophoretic deposition, biomimetic processimg, hot isostatic pressing and several other methods [649]. In the majority of cases, the coatings are composed of uniform nano-crystalline apatites. They are capable to assist in bone formation and promote direct osseointegration with juxtaposed bone [652-655]. For example, an enhanced new bone formation can be clearly seen on nanophase HA-coated tantalum compared to micro-scale HA-coated tantalum and non-coated tantalum (see Figure 2 in Ref. [116]). Furthermore, nano-structured calcium orthophosphates can be used as a coating material to impart surface bioactivity to other materials, *e.g.*, glasses [656] and polymers [657, 658]. Finally but yet importantly, such coatings may be patterned, *e.g.*, by direct laser writing [574] or electro-hydrodynamic atomization spraying technique [659].

7.2. Nano-Dimensional and Nano-Crystalline Calcium Orthophosphates and Bone-Related Cells

It is well known that bone-related cells, in particular osteoblasts and osteoclasts play key roles in the physiological formation of calcified tissues. Bone-related cells not only are thought to partake in the formation of biominerals and macrostructure constructions of bones, but they also continuously modulate the density, regeneration and degradation of bones. Therefore, the quest for understanding the relationship between bone-related cells and nano-sized calcium orthophosphates has been received much attention in order to elucidate the formation mechanism of bone, to prevent and cure bone-related diseases, and to design novel biomaterials. Better structural biomimetic behavior and osteoconductivity can be achieved using nano-dimensional and nano-crystalline calcium orthophosphates [178, 179, 185, 186, 660-663]. Biocompatibility of such biomaterials is the key question of their application potential for clinical use. For example, adhesion, proliferation and differentiation of mesenchymal stem cells were studied on nano-sized HA/polyamide biocomposite scaffolds. The results indicated that such biocomposites exhibit good biocompatibility and extensive osteoconductivity in contact with host bone *in vitro* and *in vivo,* and proved that nano-sized HA/polyamide scaffolds have the potential for use in orthopedic, reconstructive and maxillofacial surgery [664-666].

Most results demonstrate that nano-structured HA can improve cell attachment and mineralization *in vivo*, suggesting that nano-sized HA may be a suitable contender for clinical use in terms of bioactivity [185, 186, 190, 667-669]. The size effects of nano-dimensional HA on bone-related cells as well as the influence of crystallinity of nano-sized HA were studied [507, 670]. Different nano-sized particles of HA, typically of 20 ± 5, 40 ± 10 and 80 ± 12 nm in diameter, were prepared and their effect on the proliferation of two types of bone-related cells, bone marrow mesenchymal stem cells (MSCs) and osteosarcoma cells (U2OS and MG63) were studied. The cell culture experiments showed an improved cytophilicity of nanophase HA compared to submicron-sized HA. A greater cell viability and proliferation of MSCs were measured for nano-sized HA, in particular for 20 nm-sized particles. However, the opposite phenomenon occurred for bone tumor cells when nano-sized HA crystallites were co-cultured with cells. Nano-sized HA can inhibit proliferation of U2OS and MG63 cells and the degree of inhibition was found to be inversely proportional to the particle size, *i.e.* smaller particles possessed a greater ability to prevent cell proliferation. This suggests that nano-sized HA can exhibit favorable cell proliferation to optimize biological functionality, a process in which the particle dimensions are believed to play a key role. These *in vitro* results are of great significance for understanding the phenomena of cytophilicity and biological activity of nano-sized particles during biomineralization [507]. Furthermore, an early osteogenic signal expression of rat bone marrow stromal cells appears to be influenced by the content of nano-dimensional HA [671]. On the other hand, there is a study on early bone healing, in which an importance of nanometer thick coatings of nano-dimensional HA on titanium implants appears to be insignificant when compared to the control [672].

Studies confirmed that nano-sized ACP had an improved bioactivity when compared to nano-sized HA since better adhesion and proliferation of osteogenic cells were observed at ACP substrates [673]. However, in order to correctly understand the influence of crystallinity of the nano-sized calcium orthophosphates on the osteogenesis, it was found to be critical to use nano-sized ACP and HA of the same size distribution [670]. Thus, ACP and HA particles of ~ 20 nm size were synthesized and the effect of crystallinity were studied. The adhesion, proliferation and differentiation of MSC cells were measured on both ACP and HA films and compared at the same size scale. Surprisingly, more cells were adsorbed and proliferated on the films of the well-crystallized nano-sized HA than those on the films of nano-sized ACP. Alkaline phosphatase activity assay and RT-PCR assay were also used to evaluate the differentiation of MSC cells. The results showed that the differentiation of MSC cells from osteoblasts was promoted significantly by nano-sized HA. These experimental phenomena clearly demonstrate that the crystallized phase of HA provides a better substrate for MSC cells than ACP, when the factor of size effect is removed. This new view on the relationship between the crystallinity of calcium orthophosphates and the responses of cells emphasized the importance of both size and phase control in the application of biomedical materials [670-673].

On the other hand, the chemical composition of the samples appears to be also important. Interestingly, in spite of the fact that biological bone apatite contains substantial amounts of carbonate, among the investigated samples of nano-crystalline apatites, osteoclastic differentiation was found to be constrained on carbonate-rich samples, leading to smaller numbers of osteoclast-like cells and fewer resorption pits. Furthermore, the highest resorption rate was found for nano-dimensional HA with a low carbonate content which strongly stimulated the differentiation of osteoclast-like cells on its surface [674].

Cells are sufficiently sensitive and nano-scale alterations in topography may elicit diverse cell behavior [675-677]. How cells are capable to recognize particle size and other very small differences in the properties of nano-sized HA in these experiments remains unclear. Consequently, determining the mechanisms by which nano-sized particles of calcium orthophosphates and their size distribution exert effects on bone-related cells will require further systematic studies.

7.3. Dental Applications

Dental caries is a ubiquitous oral disease worldwide. At the initial stage of caries lesions, bacteria cause damage of dental enamel, the exterior coating of teeth that possesses remarkable hardness and resistance. As the most highly mineralized structure in vertebrates, enamel is composed of numerous needle-like apatite crystals of nano-dimensional sizes which are bundled in parallel ordered prisms to ensure unique mechanical strength and biological protection. As non-living tissue, the main constituent (~ 97 wt. %) of mature enamel being inorganic nano-dimensional apatite, enamel is scarcely self-repairing by living organisms after substantial mineral loss. Filling with artificial materials is a conventional treatment to repair damaged enamel. However, secondary caries frequently arises at the interfaces between the tooth and foreign materials [678].

Nano-dimensional HA and CDHA are often considered model compounds of dental enamel due to their chemical and phase similarities [148, 204, 205]. Therefore, enamel remineralization by using nano-dimensional apatite or other calcium orthophosphates is suggested in dental research [679]. For example, toothpastes containing nano-dimensional apatite could promote a partial remineralization of demineralized enamel [680-684], in addition to some whitening effect [685]. Furthermore, nano-sized HA can be added to methacrylate-based root canal sealers [686] as well as to a fluoride-containing mouthwash [687]. The remineralization potential of sports drink, containing nano-sized HA, was also investigated [688, 689]. A positive influence of addition of nano-dimensional β-TCP on acid demineralization and promotion of remineralization of enamel surface was detected as well [690]. Unfortunately, these chemically analogous compounds of enamel are not widely applied in clinical practices. The native structure of dental enamel is too complex to be remodeled and the synthesized apatite crystallites often have deviating dimensions, morphologies and orientations compared to the natural material, resulting in a poor adhesion and mechanical strength during dental restoration. Recent advances in biomineralization also indicate that features of smaller particles of nano-sized HA might approximate features of biological apatite more closely than features of larger HA particles that are conventionally used [13]. For example, it has been demonstrated that nano-sized HA can self-assemble to form enamel-like structures in the laboratory [225]. Therefore, a biomimetic technique is suggested as follows: the localized repair of the enamel surface can be improved by nano-sized HA (dimension of ~ 20 nm), analogues to the basic building blocks of enamel rods. Furthermore, it was found that nano-sized HA can strongly adsorb onto the enamel surface and can even be integrated into the natural enamel structure [691].

It is surprising that nano-sized HA of ~ 20 nm can significantly inhibit the mineral loss from the enamel surface [228]. Without any treatment, the demineralization of the natural enamel surface was remarkable in acidic solution (pH ~ 4.5 ± 0.1, experimental period of 2 days) and damaged sites were observed. The mass loss rate was about 0.12 ± 0.04 mg/mm^2 per day. In contrast, a layer of nano-sized HA on the treated enamel surface was almost unchanged in acidic solution. The rate of mass loss of enamel coated with nano-sized HA approached zero (< 0.02 mg/mm^2 per day), a value beyond the sensitivity of the detection method. Since the nano-sized HA coating appeared to be insensitive to dissolution, the underlying enamel surface was well protected under slightly acidic conditions. Furthermore, the enamel surface coated by ~ 20 nm-sized HA had a hardness of 4.6 ± 0.4 GPa and an elastic modulus of 95.6 ± 8.4 GPa. These data appear to be very similar to those of natural enamel samples with 4.2 ± 0.2 and 94.1 ± 5.4 GPa, respectively [228].

The similarity between ~ 20 nm-sized HA and building blocks of dental enamel yields a good fixation of artificial biomaterials to natural tissues. Moreover, the enamel structure appears to be reinforced by nano-sized HA since secondary caries formation is suppressed but hardness is retained [678, 692, 693]. This strategy may have prospective applications in dentistry as it offers an easy but effective method to reconstruct tooth enamel suffering from mineral losses. Generally, these studies also suggest that analogs of nano-dimensional building blocks of biominerals should be highlighted in the entire subject of biomineralization.

In the case of nano-dimensional DCPA, decreasing DCPA particle dimensions were found to increase the Ca- and PO_4-ions releases from DCPA-based biocomposites. Therefore, biocomposites based on nano-sized DCPA, possessing both a high strength and good release of Ca- and PO_4-ions, may provide a needed and unique combination of stress-bearing and caries-inhibiting capabilities suitable for dental applications [546].

7.4. Other Applications

Several other applications of nano-dimensional and nano-structured calcium orthophosphates are in progress, some of which are described here. For example, there is a report on a successful preparation of a multi-modal contrast agent based on nano-sized crystals of HA, engineered to show simultaneous contrast enhancement for three major molecular imaging techniques such as magnetic resonance imaging, X-ray imaging and near-infrared fluorescence imaging [694]. Furthermore, various compositions based on nano-dimensional calcium orthophosphates have been already tested for cancer treatment [48, 60, 199, 354, 570, 695-697]. Besides, nano-dimensional HA was found to be effective for proliferation inhibition of highly malignant melanoma cells [698] and human chronic myeloid leukemia K562 cells [699].

Surface modification of nano-dimensional calcium orthophosphates was performed in order to modulate their colloid stability, prevent dissolution in the case of low pH, avoid inflammation, serve as an intermediate layer to allow strong bond formation between HA and polymer matrices and potentially enhance their bioactivity, or improves its conjugation ability with special functional groups [12, 700-706]. Such surface modified nano-sized particles can be applied for oral insulin delivery [707].

In another aspect, many strategies have been employed to load various therapeutic and bioimaging agents onto nano-dimensional calcium orthophosphate (mainly apatite) surfaces [708]. In summary, these strategies can be broadly categorized into two main approaches. One approach is to load these agents during the synthesis – so called *in situ* loading. This is done by adding the desired agent(s) to the reaction mixture before the formation of a nano-dimensional calcium orthophosphate is completed. The second approach is to load the agent(s) only after a nano-dimensional calcium orthophosphate has been fully synthesized or, in other words, after the synthesis process – so called *ex situ* loading. This is mainly done through surface adsorption where the agents are adsorbed onto the surfaces of pre-synthesized nano-dimensional particles [709]. Therefore, due to established biocompatibility, ease of handling and notorious adsorption affinity, nano-sized calcium orthophosphates have been applied as non-viral carriers for drug delivery and gene therapy [133, 203, 289, 318, 325, 511, 542, 559, 600d, 695, 710-721]. After loading with genes or drugs, nano-dimensional calcium orthophosphates provide a protective environment that shields them from degradation while providing a convenient pathway for cell membrane penetration and controlled release of the genes or drugs [512]. The experimental results proved that nano-dimensional calcium orthophosphates possess a higher penetration rate into cell membranes and their transfection efficiency could be 25-fold higher than that of the micron-sized particles. Furthermore, due to the larger specific surface areas, nano-dimensional calcium orthophosphates can hold larger loads of drugs than coarser particles. These results indicate a potential of nano-sized calcium orthophosphates in gene delivery and as drug carriers [512, 722-725]. Since the charge of the particles influences their ability to pass through the cellular membrane and a positive charge is beneficial, positively charged nano-sized particles of calcium orthophosphate/polymer biocomposites were successfully applied for photodynamic therapy [726]. Furthermore, nano-dimensional calcium orthophosphates can be stably loaded with radioisotopes [318, 727].

A transfer of functional foreign nucleic acids (DNA or RNA) into nuclei of living cells (transfection) with the aim to repair missing cell function and to provide means to enhance or silence gene expression is currently used extensively in the laboratory and is fast becoming a therapeutic reality. As nucleic acids alone are unable to penetrate the cell wall, efficient carriers are required [728, 729]. Nano-dimensional calcium orthophosphates can be represented as a unique class of the non-viral vectors, which can serve as efficient and alternative DNA carriers for targeted delivery of genes [289, 696, 697, 708, 714, 730-742] and cells [567, 743-749]. The standard transfection method using calcium orthophosphates, first introduced by Graham and van der Eb in 1973 [748], is still used in biochemistry. It involves a straightforward *in situ* precipitation of calcium orthophosphate/DNA aggregates. A similar experimental approach is used to load calcium orthophosphates with drugs [719]. Interestingly, the transfection efficiency of nano-dimensional calcium orthophosphates was found to depend on the Ca/P ratio: calcium orthophosphates with Ca/P = 1.30 ratio exhibited a fourfold increase of the transfection efficiency over those with a Ca/P = 1.65 ratio [289]. This data emphasizes the importance of understanding the interaction between calcium orthophosphates and DNA to optimize the DNA uptake and its channeling to the nucleus of the cell. Besides, it has been demonstrated that surface modified particles of nano-sized calcium orthophosphates can be used *in vivo* to target genes specifically to the liver [750]. Attachment of a galactose moiety to the particle surface increased the targetability of the nano-sized particles. Furthermore, this surface modification makes it possible for site-specific gene delivery [750, 751]. Assemblies of block-copolymer/nano-sized calcium orthophosphate were prepared and used for cell transfection; a high biocompatibility of this system was emphasized [752]. Structures that are even more complex are known as well [753-755]. A schematic drawing of a functionalized nano-sized particle suitable for both imaging and drug delivery purposes is shown in Ref. [756]. Furthermore, vaccination to protect against human infectious diseases may be enhanced by using adjuvants that can selectively stimulate immuno-regulatory responses and nano-sized particles of calcium orthophosphates were found to be suitable for such purposes [757, 758].

In all these new applications of nano-sized calcium orthophosphates, knowledge of the exact internalization pathway into the cells represents the first necessary step towards the detailed investigation and optimization of the functional mechanism. The main groups of pathways into the cell are diffusion, passive and active transport, as well as a number of endocytic mechanisms [695]. Larger particles far above 10 nm are internalized by eukaryotic cells through endocytic pathways including phagocytosis, macropinocytosis, clathrin-mediated endocytosis and non clathrin-mediated endocytosis such as internalization via caveolae. To date, the exact internalization pathway of nano-sized calcium orthophosphates into cells has not been determined and there are many questions that remain to be answered, particularly, concerning possible interactions of calcium orthophosphates with nucleic acids.

Furthermore, the mechanisms of cellular uptake and transport to the cell nucleus of calcium orthophosphate/DNA nano-dimensional complexes remain unclear either. Therefore, there is a need to conduct a focused study on the synthesis of various forms of nano-sized calcium orthophosphates that could elucidate the mechanisms of binding, transport and release of attached plasmid DNA for understanding the gene delivery method. Research is also warranted to understand the tracking of DNA intracellularly [745] to understand the release and transport of DNA into cellular nuclei.

Concerning the healing abilities of nano-sized calcium orthophosphates, an *in vitro* inhibiting effect and even apoptotic action of un-functionalized nano-sized HA of about 50 nm diameter on a hepatoma cell line in the concentration range of 50 – 200 mg/l was reported [759]. A similar inhibiting effect was discovered for discrete nano-sized particles of HA, which appear to cause apoptosis of leukemia P388 cells [119] and rat macrophages [760]. This effect might be due to a harmful increase in the intracellular calcium concentration. However, the correlation between the particle dimensions and the apoptotic action of nano-sized calcium orthophosphates appears not to be straightforward. For example, the apoptosis efficacy of nano-dimensional particles of HA of various sizes was found to decrease in the order of 45 nm > 26 nm > 78 nm > 175 nm [761].

Hollow nano-sized spheres are extremely attractive constructions because they can greatly enhance the load quantity. Although these novel biomaterials can improve the total intake of drugs, they also bring new problems, *e.g.*, uncontrolled release kinetics and unreasonable metabolic pathway of the carriers [762]. In order to solve these problems, calcium orthophosphates were selected as suitable biomaterials to construct nano-dimensional hollow spheres [198, 541, 542, 717, 763, 764].

Such spheres with dimensions ranging from 110 to 180 nm were synthesized by an ultrasound-assisted wet chemical reaction in the presence of a modifier [765].

In addition, they may be prepared through emulsions [766]. Transmission electron microscopy investigations revealed that uniform nano-dimensional spheres were formed, well dispersed in the solutions. The wall thickness of the shells was about 45 nm; thus, they always had ~ 60 nm-sized internal cavities that could be used to load drugs. The hollow spheres appeared to be stable in both air and aqueous solutions without ultrasonic application. However, when an ultrasonic treatment (40 kHz, 150 W) was applied, the hollow structures were destroyed to form needle-like nano-sized crystals of calcium orthophosphates [765]. During this transformation, the encapsulated drugs and chemicals were released [559, 765]. Different from a free and slow diffusion of encapsulated drugs from the cavity through the shells [199], the release kinetics in this system was triggered and controlled by ultrasound. Furthermore, the power density of ultrasound can conveniently regulate the release dynamics. Besides, the needle-like nano-sized crystals of calcium orthophosphates formed behaved similarly as the biological apatite of bones. Thus, a combination of hollow calcium orthophosphate nanospheres and ultrasonic treatment could provide a good system for drug delivery and release [765].

8. Other Applications of Nano-Dimensional and Nano-Crystalline Calcium Orthophosphates

Only a few publications are available on non-biomedical applications of nano-dimensional and nano-crystalline calcium orthophosphates. For example, nano-sized particles of calcium orthophosphates with a mean size of 150 ± 20 nm filled with a solution containing luminol, haematin and fluorescein were found to improve the ease and accuracy of H_2O_2

sensing [767]. Besides, nano-dimensional HA particles were tested as a component of a green slow-release fertilizer composition [768]. Also, addition of nano-dimensional HA remarkably inhibits desorption of heavy metals from soils, which increases their geochemical stability in metal contaminated soils [769]. Furthermore, nano-dimensional HA was found to hold a great potential to remove cationic heavy metal species from industrial wastewater [770-774]. Finally but important, nano-dimensional and nano-crystalline calcium orthophosphates occasionally might possess catalytic activity [775, 776]. See also Chapters 15 to 19 of this volume.

SUMMARY AND PERSPECTIVES

As the basic building blocks of calcified tissues of mammals, nano-sized calcium orthophosphates with apatite structure play an important role in the reconstruction of these hard tissues. Therefore, they appear to be an almost ideal biomaterial due to their good biocompatibility and bioresorbability. Even more enhanced applications are expected in drug delivery systems [777]. However, there is still an unanswered question concerning their structure: are the majority of nano-dimensional calcium orthophosphates that appear to be amorphous according to numerous results of X-ray diffraction studies actually well-crystallized structures that only appear to be amorphous due to their nanoscopic size, or are they really amorphous with only a short-range order at the scale of a few atomic neighbors? A valuable attempt to discuss this topic is available in the literature [778].

In the future, the ability to functionalize surfaces with different molecules of varying nature and dimensions by means of their attachment to cells will enable them to act selectively on biological species such as proteins and peptides. The capability to synthesize and process nano-dimensional and nano-crystalline calcium orthophosphates with controlled structures and topographies to simulate the basic units of bones and teeth, will offer a chance to design novel proactive bioceramics required for enhanced repair efficacy. The various positive initial results on biocompatibility and biomimetic behavior of novel nano-structured bioceramics merit further studies and require confirmation. For example, much work remains to be undertaken to address the following key challenges and critical issues of nano-dimensional and nano-crystalline calcium orthophosphates [779]:

- Consistency of the processing technologies;
- Optimization of the structure and properties mimicking bones;
- Matching the strength of nano-dimensional and nano-crystalline constructs with those of bones in order to provide a uniform distribution of stresses (load sharing);
- Optimizing bioresorption without comprising the mechanical properties;
- Assessing the inflammatory response to validate their biosafety.

Furthermore, substantial research efforts are required to analyze cells and their different behavior with regard to their interactions with nano-dimensional and nano-crystalline calcium orthophosphates [779]. An important but still unsolved question is how the cells can recognize the particle dimensions and crystallinity of nano-sized calcium orthophosphates. What is the signal for nano-dimensional biomaterials to promote cell proliferation and differentiation and how can their pathways be found out? According to the experimental results on transfection, nano-sized particles can enter into cells readily but many details of this process remain unclear. For example, the pathways for nano-sized particles to enter the cells through the membranes should be revealed [780]. An increased influence of the hydrated surface layer with labile ionic species of smaller particles and crystals (see section 5 for details) may be another possible option, to be confirmed experimentally. Then, it is important to examine the metabolic process of nano-sized calcium orthophosphates inside cells, so that the existing forms of these particles during the biological processes could be understood. Further, a critical step will be the investigation of possible changes of gene or protein expression in the absence and presence of various nano-sized calcium orthophosphates that may directly be related to cell proliferation and differentiation [13].

Understanding the interactions between nano-sized particles and living cells is still a great challenge [779]. For example, elucidating the mechanisms by which cells internalize and process nano-dimensional particles is of great importance to understand their potential toxicity and to improve the targeted delivery of nano-dimensional particles for biomedical applications. Already, some data are available showing that clathrin-mediated endocytosis might be responsible for the uptake of nano-sized HA [695]. In another study, nano-dimensional particles of HA were sequestered within a specialized membrane-bound surface-connected compartment, directly connected to the extracellular matrix [781]. Future studies will likely focus on: (1) the detailed interfacial structure of nano-dimensional calcium orthophosphates and the specific adsorption of proteins or other matrices, (2) the uptake processes of the nano-sized particles by cells, and (3) the metabolism of nano-sized calcium orthophosphates inside the cells and its possible interference with physiological reactions. Another important topic is the biological safety of nano-sized particles in general [144, 145, 782, 783] and that of calcium orthophosphates in particular [178, 376, 784]. For example, toxicity of nano-sized HA was found to vary considerably, related to their physico-chemical properties. Furthermore, cell death correlate strongly with the load of nano-sized particles. Hence, the biological effect on human monocyte-derived macrophages of rod-shaped apatite, 50 – 80 nm in length, was investigated [178]. High concentrations of apatite (200 nano-sized particles per cell) were incubated for 24 hours with macrophages under both serum and serum-free conditions. This induced high release levels of lactate dehydrogenase, an indicator of cellular damage. However, lower concentrations (20 and 2 nano-sized particles per cell) of the rod-shaped apatite did not affect the cell viability relative to the control group without nano-sized apatite [178]. Similarly, investigation of intracellular dissolution of nano-sized HA as a function of time suggested that increased cytoplasmic calcium load is likely to be the cause of cell death [784]. Furthermore, nano-sized calcium orthophosphates were found to interfere with the cell cycle of cultured human ovarian granulosa cells thus increasing cell apoptosis [785]. This pilot study suggested further that the effect of nano-sized particles on ovarian function should be extensively investigated. Additional examples of cytotoxicity experiments with nano-dimensional calcium orthophosphates were well described in a recent review [709].

To finalize this topic, one should stress that *in vivo* evaluation of nano-sized particles includes the particle's activity, biodistribution and pharmacokinetic properties [786]. Ultimately, all these properties will be determined by particle dimensions, surface charge, morphology, and surface chemistry. Furthermore, nano-sized particles penetrate and leave biological organisms more readily using a number of pathways. For example, very small (<

10 nm) particles are generally eliminated from the body via renal clearance, *i.e.* being filtered through the kidneys and eliminated through urine, while nano-sized particles of larger dimensions are phagocytized by tissue macrophages of the reticuloendothelial system in the liver and spleen [713]. For example, intravenously administered nano-dimensional (~ 40 nm and ~ 200 nm) rod-shaped crystals of apatite showed clearance from the bloodstream within two hours, with ~ 90 % of them being cleared in the first 10 minutes post injection. Nano-dimensional crystals of apatite were observed primarily in the liver with a minority seen in the spleen [318]. These results indicate that bloodstream clearance occurs rapidly for a wide range of nano-dimensional sizes. The accumulation of nano-dimensional (50 – 100 nm in size) apatite in the liver was also noted in another study [510].

Thus, understanding the biological influence of nano-sized and nano-crystalline calcium orthophosphates is essential for the future development of bionanotechnology [787]. This interdisciplinary approach is very complex and the effective collaboration among scientists from different disciplines is the key to success [13].

CONCLUSION

With a high surface area, un-agglomerated nano-dimensional and nano-crystalline bioceramic particles are of interest for many applications including injectable or controlled setting bone cements, high strength porous or non-porous synthetic bone grafts and the reinforcing phase in biocomposites that attempt to mimic both the complex structure and superior mechanical properties of bone. Therefore, nano-sized and nano-crystalline calcium orthophosphates have already gained much regard in the biomedical field due to their superior biocompatibility and biomechanical properties. This can easily be seen from the continuous increase of the number of publications on this topic. At present, apatites (HA and CDHA) and β-TCP are the major calcium orthophosphates used in clinical applications. Currently, nano-dimensional apatites are used primarily as bioactive coatings on bioinert materials such as titanium and its alloys, in bone tissue repairs and implants, as well as for drug delivery purposes. Nano-sized β-TCP exhibits a significant biological affinity and activity, and responds very well to the physiological environment. Much research is expected on enhanced applications of nano-dimensional and nano-crystalline calcium orthophosphates for both drug delivery systems and as resorbable scaffolds that can be replaced by endogenous hard tissues with the passage of time [147, 787].

Although the nano-structured biomaterials may have many potential advantages in the context of promoting bone cell responses [534-536, 677], it is important to remember that studies on nanophase materials have only just begun; there are still many other issues regarding human health that must be answered. Since particles of very low size have higher reactivity and effectiveness, a rapid technical development of nanometer-scaled particles in the biomedical field leads to concerns regarding the unknown health risks of such materials [782, 783]. Nano-sized particles might induce inflammatory reactions, cytotoxicity, oxidative stresses or thrombogenesis when injected for drug delivery purposes since they can enter the human body through pores and may accumulate in the cells of the respiratory or other organ systems (when becoming dislodged through wear debris), and consequently their health effects are still largely unknown. These negative health effects could happen during commercial-scale processing of the nano-sized particles as well as during use of these materials as implants [790]. Besides, nano-sized particles might be an agent the existence of which has not been recognized by the living body's defense system [18, 144, 145]. Up to now, only a small number of short-term and small-scale health effects of single nano-dimensional materials have been examined in toxicological studies, usually concerning the lungs [783]. Therefore, prior to clinical applications, any toxicity concerns of the nanophase materials [791-796] need to be overcome.

In summary, despite the challenges that lie ahead, significant evidence now exists confirming that nanophasic biomaterials represent an important growing area of research that may improve bonding between the implants and the surrounding tissues. It has proven to be a versatile approach that can increase bone cell functions on a wide range of orthopedic implant chemistries. Even if nano-dimensional and nano-crystalline calcium orthophosphates do not provide the ultimate answer to increasing bone cell responses owing to some potential problems mentioned above, researchers have learned a tremendous amount of information concerning bone cell recognition by nano-structured surfaces that will most certainly aid in improving orthopedic implant efficacy [144, 145].

Post-Conclusion Remarks

According to Prof. D. F. Williams [115], the term "nanomaterial" should not be used because it is senseless (see section 2). Following his logic, the term "nanoapatite" is senseless as well. However, it is contained in the titles of several publications, for example in Refs. [606, 629, 653, 667]. In a slightly modified form, the term "nano-apatite" is presented in the titles of several other publications, for example in Refs. [89, 100, 226, 511, 631, 735, 789]. Furthermore, similar terms such as "nano-HA" [103, 575, 608, 659, 681, 692], "nano-hydroxyapatite" [42, 87, 91, 92, 96, 97, 102, 119, 190, 191, 294, 295, 314, 338, 451, 459, 486, 492, 578, 583, 606, 627, 628, 641, 664, 666, 677, 682, 685, 687, 693, 698, 701, 703, 705, 770, 773], nano-fluorapatite [104, 684] and "nanohydroxyapatite" [86, 93, 99, 162, 201, 239, 240, 242, 246, 253, 274, 320, 329, 362, 462, 576, 656, 664, 764] are present in the titles of still other publications. Presumably, it is wiser not to use these terms anymore.

References

[1] Mann, S. Biomineralization principles and concepts in bioinorganic materials chemistry. Oxford University Press: New York, USA, 2001; pp. 216.

[2] Lowenstam, H.A., Weiner, S. On biomineralization. Oxford University Press: New York, USA, 1989; pp. 324.

[3] Vallet-Regí, M., González-Calbet, J.M. Calcium phosphates as substitution of bone tissues. *Prog. Solid State Chem.* 2004, 32, 1-31.

[4] Weiner, S., Addadi, L. Design strategies in mineralized biological materials. *J. Mater. Chem.* 1997, 7, 689-702.

[5] Weiner, S. Wagner, H.D. The material bone: structure-mechanical function relations. *Ann. Rev. Mater. Sci.* 1998, 28, 271-298.

[6] Pasteris, J.D., Wopenka, B., Valsami-Jones, E. Bone and tooth mineralization: why apatite? *Elements* 2008, 4, 97-104.
[7] Giachelli, C.M. Ectopic calcification: gathering hard facts about soft tissue mineralization. *Am. J. Pathol.* 1999, 154, 671-675.
[8] Kirsch, T. Determinants of pathological mineralization: crystal deposition diseases. *Curr. Opin. Rheumatol.* 2006, 18, 174-180.
[9] Christian, R.C., Fitzpatrick, L.A. Vascular calcification. *Curr. Opin. Nephrol. Hypertens.* 1999, 8, 443-448.
[10] Boskey, A. Bone mineral crystal size. Osteoporosis Int. 2003, 14, Suppl. 5, S16-S20; discussion S20-S21.
[11] Alivisatos, A.P. Enhanced naturally aligned nanocrystals. *Science* 2000, 289, 736-737.
[12] Narayan, R.J., Kumta, P.N., Sfeir, C., Lee, D.H., Choi, D., Olton, D. Nanostructured ceramics in medical devices: applications and prospects. *JOM* 2004, 56, 38-43.
[13] Cai, Y., Tang, R. Calcium phosphate nanoparticles in biomineralization and biomaterials. *J. Mater. Chem.* 2008, 18, 3775-3787.
[14] Ginebra, M.P., Driessens, F.C.M., Planell, J.A. Effect of the particle size on the micro and nanostructural features of a calcium phosphate cement: a kinetic analysis. *Biomaterials* 2004, 25, 3453-3462.
[15] Nanotechnology is an application of science and engineering at the nanoscale.
[16] Karch, J., Birringer, R., Gleiter, H. Ceramics ductile at low temperature. *Nature* 1987, 330, 556-558.
[17] Webster, T.J. Nanophase ceramics: the future of orthopedic and dental implant material. In: Nanostructured materials. Ying, J.Y. Ed., Academic Press, New York, USA, 2001; pp. 125-166.
[18] Tasker, L.H., Sparey-Taylor, G.J., Nokes, L.D. Applications of nanotechnology in orthopaedics. *Clin. Orthop. Relat. Res.* 2007, 456, 243-249.
[19] Banfield, J.F., Welch, S.A., Zhang, H., Ebert, T.T., Penn, R.L. Aggregation-based crystal growth and microstructure development in natural iron oxyhydroxide biomineralization products. *Science* 2000, 289, 751-754.
[20] Cölfen, H. Bio-inspired mineralization using hydrophilic polymers. *Top. Curr. Chem.* 2007, 271, 1-77.
[21] Oaki, Y., Imai, H. Nanoengineering in echinoderms: the emergence of morphology from nanobricks. *Small* 2005, 2, 66-70.
[22] Lee, S.H., Shin, H. Matrices and scaffolds for delivery of bioactive molecules in bone and cartilage tissue engineering. *Adv. Drug Delivery Rev.* 2007, 59, 339-359.
[23] Ben-Nissan, B. Nanoceramics in biomedical applications. *MRS B*ulletin 2004, 29, 28-32.
[24] Rehman, I. Nano bioceramics for biomedical and other applications. *Mater. Technol.* 2004, 19, 224-233.
[25] Driessens, F.C.M., Boltong, M.G., de Maeyer, E.A.P., Wenz, R., Nies, B., Planell, J.A. The Ca/P range of nanoapatitic calcium phosphate cements. *Biomaterials* 2002, 23, 4011-4017.
[26] Doat, A., Fanjul, M., Pellé, F. Hollande, E., Lebugle, A. Europium-doped bioapatite: a new photostable biological probe, internalizable by human cells. *Biomaterials* 2003, 24, 3365-3371.

[27] Doat, A., Pellé, F., Gardant, N., Lebugle, A. Synthesis of luminescent bioapatite nanoparticles for utilization as a biological probe. *J. Solid State Chem.* 2004, 177, 1179-1187.

[28] Lebugle, A., Pellé, F., Charvillat, C., Rousselot, I., Chane-Ching, J.Y. Colloidal and monocrystalline Ln^{3+} doped apatite calcium phosphate as biocompatible fluorescent probes. *Chem. Commun.* 2006, 606-608.

[29] Mondejar, S.P., Kovtun, A., Epple, M. Lanthanide-doped calcium phosphate nanoparticles with high internal crystallinity and with a shell of DNA as fluorescent probes in cell experiments. *J. Mater. Chem.* 2007, 17, 4153-4159.

[30] Kalita, S.J., Bhatt, H.A. Nanocrystalline hydroxyapatite doped with magnesium and zinc: synthesis and characterization. *Mater. Sci. Eng. C* 2007, 27, 837-848.

[31] Huang, J., Jayasinghe, S.N., Best, S.M., Edirisinghe, M.J., Brooks, R.A., Rushton, N., Bonfield, W. Novel deposition of nanosized silicon substituted hydroxyapatite by electrostatic spraying. *J. Mater. Sci.* Mater. Med. 2005, 16, 1137-1142.

[32] Pon-On, W., Meejoo, S., Tang, I.M. Incorporation of iron into nano hydroxyapatite particles synthesized by the microwave process. *Int. J. Nanosci.* 2007, 6, 9-16.

[33] Predoi, D., Barsan, M., Andronescu, E., Vatasescu-Balcan, R.A., Costache, M. Hydroxyapatite – iron oxide bioceramic prepared using nano-size powders. *J. Optoelectronics Adv. Mater.* 2007, 9, 3609-3613.

[34] Bakunova, N.V., Fomin, A. S., Fadeeva, I.V., Barinov, S.M., Shvorneva, L.I. Silicon-containing hydroxylapatite nanopowders. *Russ. J. Inorg. Chem.* 2007, 52, 1492-1497.

[35] Miao, S., Weng, W., Cheng, K., Du, P., Shen, G., Han, G. Preparation of nano-sized strontium containing tricalcium phosphate particles. *Key Eng. Mater.* 2007, 330-332, 263-266.

[36] Liu, Y., Zhou, R., Mo, A., Chen, Z., Wu, H. Synthesis and characterization of yttrium/hydroxyapatite nanoparticles. *Key Eng. Mater.* 2007, 330-332, 295-298.

[37] Wu, H.C., Wang, T.W., Sun, J.S., Wang, W.H., Lin, F.H. A novel biomagnetic nanoparticle based on hydroxyapatite. *Nanotechnology* 2007, 18, 165601 (9 pages).

[38] Rameshbabu, N., Kumar, T.S.S., Prabhakar, T.G., Sastry, V.S., Murty, K.V.G.K., Rao, K.P. Antibacterial nanosized silver substituted hydroxyapatite: synthesis and characterization. *J. Biomed. Mater. Res. A* 2007, 80A, 581-591.

[39] Fujii, E., Ohkubo, M., Tsuru, K., Hayakawa, S., Osaka, A., Kawabata, K., Bonhomme, C., Babonneau, F. Selective protein adsorption property and characterization of nano-crystalline zinc-containing hydroxyapatite. *Acta Biomater.* 2006, 2, 69-74.

[40] Chowdhury, E.H., Akaike, T. Fibronectin-coated nano-precipitates of calcium-magnesium phosphate for integrin-targeted gene delivery. *J. Control. Release* 2006, 116, e68–e69.

[41] Low, H.R., Phonthammachai, N., Maignan, A., Stewart, G.A., Bastow, T.J., Ma, L.L., White, T.J. The crystal chemistry of ferric oxyhydroxyapatite. *Inorg. Chem.* 2008, 47, 11774-11782.

[42] Zhang, S.M., Hu, W., Zhou, W., Li, J., Liu, Y.H., Qiu, Z.Y. Dialysis preparation of zinc-substituted nano-hydroxyapatite and its characterization. *Key Eng. Mater.* 2007, 330-332, 219-222.

[43] Pon-On, W., Meejoo, S., Tang, I.M. Substitution of manganese and iron into hydroxyapatite: core/shell nanoparticles. *Mater. Res. Bull.* 2008, 43, 2137-2144.

[44] Zou, C., Weng, W., Cheng, K., Du, P., Shen, G., Han, G. Preparation of nanosized β-tricalcium phosphate particles with Zn substitution. *J. Mater. Sci. Mater. Med.* 2008, 19, 1133-1136.

[45] Hwang, K.S., Hwangbo, S., Kim, J.T. Silver-doped calcium phosphate nanopowders prepared by electrostatic spraying. *J. Nanoparticle Res.* 2008, 10, 1337-1341.

[46] Lee, D., Sfeir, C., Kumta, P.N. Novel in-situ synthesis and characterization of nanostructured magnesium substituted β-tricalcium phosphate (β-TCMP). *Mater. Sci. Eng.* C 2009, 29, 69-77.

[47] Petchsang, N., Pon-On, W., Hodak, J.H., Tang, I.M. Magnetic properties of Co-ferrite-doped hydroxyapatite nanoparticles having a core/shell structure. *J. Magnetism Magnetic Mater.* 2009, 321, 1990-1995.

[48] Hou, C.H., Hou, S.M., Hsueh, Y.S., Lin, J., Wu, H.C., Lin, F.H. The *in vivo* performance of biomagnetic hydroxyapatite nanoparticles in cancer hyperthermia therapy. *Biomaterials* 2009, 30, 3956-3960.

[49] Chen, F., Huang, P., Zhu, Y.J., Wu, J., Zhang, C.L., Cui, D.X. The photoluminescence, drug delivery and imaging properties of multifunctional Eu^{3+}/Gd^{3+} dual-doped hydroxyapatite nanorods. *Biomaterials* 2011, 32, 9031-9039.

[50] Cacciotti, I., Bianco, A., Lombardi, M., Montanaro, L. Mg-substituted hydroxyapatite nanopowders: synthesis, thermal stability and sintering behaviour. *J. Eur. Ceram. Soc.* 2009, 29, 2969-2978.

[51] Bianco, A., Cacciotti, I., Lombardi, M., Montanaro, L. Si-substituted hydroxyapatite nanopowders: synthesis, thermal stability and sinterability. *Mater. Res. Bull.* 2009, 44, 345-354.

[52] Capuccini, C., Torricelli, P., Boanini, E., Gazzano, M., Giardino, R., Bigi, A. Interaction of Sr-doped hydroxyapatite nanocrystals with osteoclast and osteoblast-like cells. *J. Biomed. Mater. Res. A* 2009, 89A, 594-600.

[53] Jiang, H., Li, Y., Zuo, Y., Yang, W., Zhang, L., Li, J., Wang, L., Zou, Q., Cheng, L., Li, J. Physical and chemical properties of superparamagnetic Fe-incorporated nano hydroxyapatite. *J. Nanosci. Nanotechnol.* 2009, 9, 6844-6850.

[54] Al-Kattan, A., Dufour, P., Dexpert-Ghys, J., Drouet, C., Preparation and physicochemical characteristics of luminescent apatite-based colloids. *J. Phys. Chem. C* 2010, 114, 2918-2924.

[55] Hou, C.H., Chen, C.W., Hou, S.M., Li, Y.T., Lin, F.H. The fabrication and characterization of dicalcium phosphate dihydrate-modified magnetic nanoparticles and their performance in hyperthermia processes *in vitro*. *Biomaterials* 2009, 30, 4700-4707.

[56] Hanifi, A., Fathi, M.H., Sadeghi, H.M.M., Varshosaz, J. Mg^{2+} substituted calcium phosphate nano particles synthesis for non viral gene delivery application. *J. Mater. Sci. Mater. Med.* 2010, 21, 2393-2401.

[57] Stojanović, Z., Veselinović, L., Marković, S., Ignjatović, N., Uskoković, D. Hydrothermal synthesis of nanosize pure and cobalt-exchanged hydroxyapatite. *Mater. Manuf. Process* 2009, 24, 1096-1103.

[58] Veselinović, L., Karanović, L., Stojanović, Z., Bračko, I., Marković, S., Ignjatović, N., Uskoković, D. Crystal structure of cobalt-substituted calcium hydroxyapatite nanopowders prepared by hydrothermal processing. *J. Appl. Crystallogr.* 2010, 43, 320-327.

[59] Evis, Z., Webster, T.J. Nanosize hydroxyapatite: doping with various ions. *Adv. Appl. Ceram.* 2011, 110, 311-320.

[60] Al-Kattan, A., Girod-Fullana, S., Charvillat, C., Ternet-Fontebasso, H., Dufour, P., Dexpert-Ghys, J., Santran, V., Bordère, J., Pipy, B., Bernad, J., Drouet, C. Biomimetic nanocrystalline apatites: emerging perspectives in cancer diagnosis and treatment. *Int. J. Pharm.* 2012, 423, 26-36.

[61] Kaflak, A., Kolodziejski, W. Complementary information on water and hydroxyl groups in nanocrystalline carbonated hydroxyapatites from TGA, NMR and IR measurements. *J. Mol. Struct.* 2011, 990, 263-270.

[62] Kaflak, A., Ślósarczyk, A., Kolodziejski, W. A comparative study of carbonate bands from nanocrystalline carbonated hydroxyapatites using FT-IR spectroscopy in the transmission and photoacoustic modes. *J. Mol. Struct.* 2011, 997, 7-14.

[63] Li, Y., Widodo, J., Lim, S., Ooi, C.P. Synthesis and cytocompatibility of manganese (II) and iron (III) substituted hydroxyapatite nanoparticles. *J. Mater. Sci.* 2012, 47, 754-763.

[64] Li, W., Gao, L. Fabrication of Hap-ZrO_2 (3Y) nano-composite by SPS. *Biomaterials* 2003, 24, 937-940.

[65] Wang, L., Nemoto, R., Senna, M. Microstructure and chemical states of hydroxyapatite/silk fibroin nanocomposites synthesized via a wet-mechanochemical route. *J. Nanopart. Res.* 2002, 4, 535-540.

[66] Nemoto, R., Wang, L., Ikoma, T., Tanaka, J., Senna, M. Preferential alignment of hydroxyapatite crystallites in nanocomposites with chemically disintegrated silk fibroin. *J. Nanopart. Res.* 2004, 6, 259-265.

[67] Fang, L.M., Leng, Y., Gao, P. Processing and mechanical properties of HA/UHMWPE nanocomposites. *Biomaterials* 2006, 27, 3701-3707.

[68] Sugawara, A., Yamane, S., Akiyoshi, K. Nanogel-templated mineralization: polymer-calcium phosphate hybrid nanomaterials. *Macromol. Rapid Commun.* 2006, 27, 441-446.

[69] Pushpakanth, S., Srinivasan, B., Sreedhar, B., Sastry, T.P. An *in situ* approach to prepare nanorods of titania – hydroxyapatite (TiO_2 – HAp) nanocomposite by microwave hydrothermal technique. *Mater. Chem. Phys.* 2008, 107, 492-498.

[70] Chang, M.C., Ko, C.C., Douglas, W.H. Preparation of hydroxyapatite-gelatin nanocomposite. *Biomaterials* 2003, 24, 2853-2862.

[71] Hao, J., Liu, Y., Zhou, S., Li, Z., Deng, X. Investigation of nanocomposites based on semi-interpenetrating network of [L-poly(ε-caprolactone)]/[net-poly(ε-caprolactone)] and hydroxyapatite nanocrystals. *Biomaterials* 2003, 24, 1531-1539.

[72] Deng, X.M., Hao, J.Y., Wang, C.S. Preparation and mechanical properties of nanocomposites of poly(D,L-lactide) with Ca-deficient hydroxyapatite nanocrystals. *Biomaterials* 2001, 22, 2867-2873.

[73] Hong, Z., Zhang, P., He, C., Qiu, X., Liu, A., Chen, L., Chena, X., Jing, X. Nanocomposite of poly(L-lactide) and surface grafted hydroxyapatite: mechanical properties and biocompatibility. *Biomaterials* 2005, 26, 6296-6304.

[74] Ramay, H.R.R., Zhang, M. Biphasic calcium phosphate nanocomposite porous scaffolds for load-bearing bone tissue engineering. *Biomaterials* 2004, 25, 5171-5180.

[75] Cross, K.J., Huq, N.L., Palamara, J.E., Perich, J.W., Reynolds, E.C. Physicochemical characterization of casein phosphopeptide – amorphous calcium phosphate nanocomplexes. *J. Biol. Chem.* 2005, 280, 15362-15369.

[76] Murugan, R., Ramakrishna, S. Development of nanocomposites for bone grafting. *Comp. Sci. Tech.* 2005, 65, 2385-2406.

[77] Liou, S.C., Chen, S.Y., Liu, D.M. Phase development and structural characterization of calcium phosphate ceramics – polyacrylic acid nanocomposites at room temperature in water-methanol mixtures. *J. Mater. Sci. Mater. Med.* 2004, 15, 1261-1266.

[78] Sung, Y.M., Shin, Y.K., Ryu, J.J. Preparation of hydroxyapatite/zirconia bioceramic nanocomposites for orthopaedic and dental prosthesis applications. *Nanotechnology* 2007, 18, 065602 (6 pages).

[79] Sreedhar, B., Aparna, Y., Sairam, M., Hebalkar, N. Preparation and characterization of HAP / carboxymethyl chitosan nanocomposites. *J. Appl. Polym. Sci.* 2007, 105, 928-934.

[80] Pramanik, N., Biswas, S.K., Pramanik, P. Synthesis and characterization of hydroxyapatite/poly(vinyl alcohol phosphate) nanocomposite biomaterials. *Int. J. Appl. Ceram. Technol.* 2008, 5, 20-28.

[81] Jevtić, M., Radulović, A., Ignjatović, N., Mitrić, M., Uskoković, D. Controlled assembly of poly(D,L-lactide-*co*-glycolide)/hydroxyapatite core-shell nanospheres under ultrasonic irradiation. *Acta Biomater.* 2009, 5, 208-218.

[82] Li, X., Chang, J. Preparation of bone-like apatite – collagen nanocomposites by a biomimetic process with phosphorylated collagen. *J. Biomed. Mater. Res. A* 2008, 85A, 293-300.

[83] Ohsawa, H., Ito, A., Sogo, Y., Yamazaki, A., Ohno, T. Synthesis of albumin/DCP nano-composite particles. *Key Eng. Mater.* 2007, 330-332, 239-242.

[84] Wilberforce, S.I., Finlayson, C.E., Best, S.M., Cameron, R.E. The influence of the compounding process and testing conditions on the compressive mechanical properties of poly(D,L-lactide-co-glycolide)/α-tricalcium phosphate nanocomposites. *J. Mech. Behav. Biomed. Mater.* 2011, 4, 1081-1089.

[85] Wilberforce, S.I., Finlayson, C.E., Best, S.M., Cameron, R.E. A comparative study of the thermal and dynamic mechanical behaviour of quenched and annealed bioresorbable poly-L-lactide/α-tricalcium phosphate nanocomposites. *Acta Biomater.* 2011, 7, 2176-2184.

[86] Degirmenbasi, N., Kalyon, D.M., Birinci, E. Biocomposites of nanohydroxyapatite with collagen and poly(vinyl alcohol). *Colloids Surf. B* 2006, 48, 42-49.

[87] Zhang, X., Li, Y.B., Zuo, Y., Lv, G.Y; Mu, Y.H., Li, H. Morphology, hydrogen-bonding and crystallinity of nano-hydroxyapatite/polyamide 66 biocomposites. *Composites A* 2007, 38, 843-848.

[88] Wei, J., Li, Y.B., Lau, K.T. Preparation and characterization of a nano apatite/polyamide$_6$ bioactive composite. *Composites B* 2007, 38, 301-305.

[89] Wei, J., Li, Y.B. Tissue engineering scaffold material of nano-apatite crystals and polyamide composite. *Eur. Polym. J.* 2004, 40, 509-515.

[90] Rauchmann, M.A., Wichelhaus, T.A., Stirnal, V., Dingeldein, E., Zichner, L., Schnettler, R., Alt, V. Nanocrystalline hydroxyapatite and calcium sulphate as biodegradable composite carrier material for local delivery of antibiotics in bone infections. *Biomaterials* 2005, 26, 2677-2684.

[91] Szaraniec, B., Rosół, P., Chłopek, J. Carbon composite material and polysulfone modified by nano-hydroxyapatite. *e-Polymers* 2005, no. 030.

[92] Pramanik, N., Mohapatra, S., Pramanik, P. Processing and properties of nano-hydroxyapatite (n-HAp) / poly(ethylene-co-acrylic acid) (EAA) composite using a phosphonic acid coupling agent for orthopedic applications. *J. Am. Ceram. Soc.* 2007, 90, 369-375.

[93] Ren, Y.J., Sun, X.D., Cui, F.Z., Wei, Y.T., Cheng, Z.J., Kong, X.D. Preparation and characterization of Antheraea pernyi silk fibroin based nanohydroxyapatite composites. *J. Bioact. Compat. Polym.* 2007, 22, 465-474.

[94] Xu, H.H.K., Sun, L., Weir, M.D., Takagi, S., Chow, L.C., Hockey, B. Effects of incorporating nanosized calcium phosphate particles on properties of whisker-reinforced dental composites. *J. Biomed. Mater. Res. B (Appl. Biomater.)* 2007, 81B, 116-125.

[95] Zhou, G., Li, Y., Zhang, L., Zuo, Y., Jansen, J.A. Preparation and characterization of nano-hydroxyapatite/chitosan/konjac glucomannan composite. *J. Biomed. Mater. Res. A* 2007, 83A, 931-939.

[96] Liu, L., Liu, J., Wang, M., Min, S., Cai, Y., Zhu, L., Yao, J. Preparation and characterization of nano-hydroxyapatite/silk fibroin porous scaffolds. *J. Biomater. Sci. Polymer Edn.* 2008, 19, 325-338.

[97] Xu, F., Li, Y.B., Deng, Y., Xiong, J. Porous nano-hydroxyapatite/poly(vinyl alcohol) composite hydrogel as artificial cornea fringe: characterization and evaluation *in vitro*. *J. Biomater. Sci. Polymer Edn.* 2008, 19, 431-439.

[98] Huang, J., Lin, Y.W., Fu, X.W., Best, S.M., Brooks, R.A., Rushton, N., Bonfield, W. Development of nanosized hydroxyapatite reinforced composites for tissue engineering scaffolds. *J. Mater. Sci. Mater. Med.* 2007, 18, 2151-2157.

[99] Yusong, P., Dangsheng, X., Xiaolin, C. Mechanical properties of nanohydroxyapatite reinforced poly(vinyl alcohol) gel composites as biomaterial. *J. Mater. Sci.* 2007, 42, 5129-5134.

[100] Deng, C., Weng, J., Lu, X., Zhou, S.B., Wan, J.X., Qu, S.X., Feng, B., Li, X.H., Cheng, Q.Y. Mechanism of ultrahigh elongation rate of poly(D,L-lactide)-matrix composite biomaterial containing nano-apatite fillers. *Mater. Lett.* 2008, 62, 607-610.

[101] Sundaraseelan, J., Sastry, T.P. Fabrication of a biomimetic compound containing nano hydroxyapatite – demineralised bone matrix. *J. Biomed. Nanotechnol.* 2007, 3, 401-405.

[102] Teng, S., Chen, L., Guo, Y., Shi, J. Formation of nano-hydroxyapatite in gelatin droplets and the resulting porous composite microspheres. *J. Inorg. Biochem.* 2007, 101, 686-691.

[103] Meng, Y.H., Tang, C.Y., Tsui, C.P., Chen, D.Z. Fabrication and characterization of needle-like nano-HA and HA/MWNT composites. *J. Mater. Sci. Mater. Med.* 2008, 19, 75-81.

[104] Lin, J., Zhu, J., Gu, X., Wen, W., Li, Q., Fischer-Brandies, H., Wang, H., Mehl, C. Effects of incorporation of nano-fluorapatite or nano-fluorohydroxyapatite on a resin-modified glass ionomer cement. *Acta Biomater.* 2011, 7, 1346-1353.

[105] Dorozhkin, S.V. Calcium orthophosphate-based biocomposites and hybrid biomaterials. *J. Mater. Sci.* 2009, 44, 2343-2387.

[106] Dorozhkin, S.V. Biocomposites and hybrid biomaterials based on calcium orthophosphates. *Biomater.* 2011, 1, 3-56.

[107] Williams, D.F. The relationship between biomaterials and nanotechnology. *Biomaterials* 2008, 29, 1737-1738.

[108] Feynman, R.P. There's plenty of room at the bottom. *J. Microelectromechanical Systems* 1992, 1, 60-66.

[109] European Commission, Scientific Committee on Emerging and Newly Identified Health Risks (SCENIHR). Opinion on "the scientific aspects of the existing and proposed definitions relating to products of nanoscience and nanotechnologies". Adopted Brussels: European Commission; 29 November 2007.

[110] Moriarty, P. Nanostructured materials. *Rep. Prog. Phys.* 2001, 64, 297-381.

[111] Webster, T.J., Ahn, E.S. Nanostructured biomaterials for tissue engineering bone. *Adv. Biochem. Eng. Biotechnol.* 2006, 103, 275-308.

[112] Streicher, R.M., Schmidt, M., Fiorito, S. Nanosurfaces and nanostructures for artificial orthopedic implants. *Nanomedicine* 2007, 2, 861-874.

[113] Havancsak, K. Nanotechnology at present and its promises in the future. *Mater. Sci. Forum* 2003, 414-415, 85-94.

[114] Duncan, R. Nanomedicines in action. *Pharm. J.* 2004, 273, 485-488.

[115] Williams, D.F. On the nature of biomaterials. *Biomaterials* 2009, 30, 5897-5909.

[116] Liu, H., Webster, T.J. Nanomedicine for implants: a review of studies and necessary experimental tools. *Biomaterials* 2007, 28, 354-369.

[117] Murugan, R., Ramakrishna, S. Bioresorbable composite bone paste using polysaccharide based nano hydroxyapatite. *Biomaterials* 2004, 25, 3829-3835.

[118] Murugan, R., Ramakrishna, S. Aqueous mediated synthesis of bioresorbable nanocrystalline hydroxyapatite. *J. Cryst. Growth* 2005, 274, 209-213.

[119] Li, G., Huang, J., Li, Y., Zhang, R., Deng, B., Zhang, J., Aoki, H. *In vitro* study on influence of a discrete nano-hydroxyapatite on leukemia P388 cell behavior. *Biomed. Mater. Eng. 2007*, 17, 321-327.

[120] Ganesan, K., Kovtun, A., Neumann, S., Heumann, R., Epple, M. Calcium phosphate nanoparticles: colloidally stabilized and made fluorescent by a phosphate-functionalized porphyrin. *J. Mater. Chem.* 2008, 18, 3655-3661.

[121] Kim, H.W., Kim, H.E. Nanofiber generation of hydroxyapatite and fluor-hydroxyapatite bioceramics. *J. Biomed. Mater. Res. B (Appl. Biomater.)* 2005, 77B, 323-328.

[122] Cihlar, J., Castkova, K. Direct synthesis of nanocrystalline hydroxyapatite by hydrothermal hydrolysis of alkylphosphates. *Monatshefte für Chemie* 2002, 133, 761-771.

[123] Lak, A., Mazloumi, M., Mohajerani, M., Kajbafvala, A., Zanganeh, S., Arami, H., Sadrnezhaad, S.K. Self-assembly of dandelion-like hydroxyapatite nanostructures via hydrothermal method. *J. Am. Ceram. Soc.* 2008, 91, 3292-3297.

[124] Mukesh, U., Kulkarni, V., Tushar, R., Murthy, R.S.R. Methotrexate loaded self stabilized calcium phosphate nanoparticles: a novel inorganic carrier for intracellular drug delivery. *J. Biomed. Nanotechnol.* 2009, 5, 99-105.

[125] Sun, L., Chow, L.C., Frukhtbeyn, S.A., Bonevich, J.E. Preparation and properties of nanoparticles of calcium phosphates with various Ca/P ratios. *J. Res. Natl. Inst. Stand. Technol.* 2010, 115, 243-255.

[126] Sylvie, J., Sylvie, T.D., Pascal, P.M., Fabienne, P., Hassane, O., Guy, C. Effect of hydroxyapatite and β-tricalcium phosphate nanoparticles on promonocytic U937 cells. *J. Biomed. Nanotechnol.* 2010, 6, 158-165.
[127] Sokolova, V., Knuschke, T., Kovtun, A., Buer, J., Epple, M., Westendorf, A.M. The use of calcium phosphate nanoparticles encapsulating Toll-like receptor ligands and the antigen hemagglutinin to induce dendritic cell maturation and T cell activation. *Biomaterials.* 2010, 31, 5627-5633.
[128] Wu, H.C., Wang, T.W., Bohn, M.C., Lin. F.H., Spector, M. Novel magnetic hydroxyapatite nanoparticles as non-viral vectors for the glial cell line-derived neurotrophic factor gene. *Adv. Funct. Mater.* 2010, 20, 67-77.
[129] Gergely, G., Wéber, F., Lukács, I., Illés, L., Tóth, A.L., Horváth, Z.E., Mihály, J., Balázsi, C. Nano-hydroxyapatite preparation from biogenic raw materials. *Cent. Eur. J. Chem.* 2010, 8, 375-381.
[130] Ergun, C., Evis, Z., Webster, T.J., Sahin, F.C. Synthesis and microstructural characterization of nano-size calcium phosphates with different stoichiometry. *Ceram. Int.* 2011, 37, 971-977.
[131] Ge, X., Leng, Y., Ren, F., Lu, X. Integrity and zeta potential of fluoridated hydroxyapatite nanothick coatings for biomedical applications. *J. Mech. Behav. Biomed. Mater.* 2011, 4, 1046-1056.
[132] Wang, J., Chen, X., Yang, X., Xu, S., Zhang, X., Gou, Z. A facile pollutant-free approach toward a series of nutritionally effective calcium phosphate nanomaterials for food and drink additives. *J. Nanopart. Res.* 2011, 13, 1039-1048.
[133] Sokolova, V., Knuschke, T., Buer, J., Westendorf, A.M., Epple, M. Quantitative determination of the composition of multi-shell calcium phosphate-oligonucleotide nanoparticles and their application for the activation of dendritic cells. *Acta Biomater.* 2011, 7, 4029-4036.
[134] Traykova, T., Aparicio, C., Ginebra, M.P., Planell, J.A. Bioceramics as nanomaterials. *Nanomedicine* 2006, 1, 91-106.
[135] Grainger, D.W., Castner, D.G. Nanobiomaterials and nanoanalysis: opportunities for improving the science to benefit biomedical technologies. *Adv. Mater.* 2008, 20, 867-877.
[136] Nelson, K.G. The Kelvin equation and solubility of small particles. *J. Pharmac. Sci.* 1972, 61, 479-480.
[137] Fan, C., Chen, J., Chen, Y., Ji, J., Teng, H.H. Relationship between solubility and solubility product: the roles of crystal sizes and crystallographic directions. *Geochim. Cosmochim. Acta* 2006, 70, 3820-3829.
[138] Sato, M., Webster, T.J. Nanobiotechnology: implications for the future of nanotechnology in orthopedic applications. *Expert Rev. Med. Dev.* 2004, 1, 105-114.
[139] Hahn, H. Unique features and properties of nanostructured materials. *Adv. Eng. Mater.* 2003, 5, 277-284.
[140] Aronov, D., Karlov, A., Rosenman, G. Hydroxyapatite nanoceramics: basic physical properties and biointerface modification. *J. Eur. Ceram. Soc.* 2007, 27, 4181-4186.
[141] Ramsden, J.J., Freeman, *J. The nanoscale. Nanotechnol. Percept.* 2009, 5, 3-25.
[142] Rempel, A.A. Nanotechnologies. Properties and applications of nanostructured materials. *Russ. Chem. Rev.* 2007, 76, 435-461.

[143] Catledge, S.A., Fries, M.D., Vohra, Y.K., Lacefield, W.R., Lemons, J.E., Woodard, S., Venugopalan, R. Nanostructured ceramics for biomedical implants. *J. Nanosci. Nanotechnol.* 2002, 2, 1-20.

[144] Balasundarama, G., Webster, T.J. A perspective on nanophase materials for orthopedic implant applications. *J. Mater. Chem.* 2006, 16, 3737-3745.

[145] Balasundarama, G., Webster, T.J. Nanotechnology and biomaterials for orthopedic medical applications. *Nanomedicine* 2006, 1, 169-176.

[146] Padilla, S., Izquierdo-Barba, I., Vallet-Regí, M. High specific surface area in nanometric carbonated hydroxyapatite. *Chem. Mater.* 2008, 20, 5942-5944.

[147] Kalita, S.J., Bhardwaj, A., Bhatt, H.A. Nanocrystalline calcium phosphate ceramics in biomedical engineering. *Mater. Sci. Eng.* C 2007, 27, 441-449.

[148] LeGeros, R.Z. Calcium phosphates in oral biology and medicine. Karger: Basel, Switzerland, 1991; pp. 210.

[149] Mann, S. The study of biominerals by high resolution transmission electron microscopy. *Scan. Electron. Microsc.* 1986, Pt. 2, 393-413.

[150] Katsura, N. Nanospace theory for biomineralization. *Dent. Jpn.* (Tokyo) 1990, 27, 57-63.

[151] Cuisinier, F.J.G., Voegel, J.C., Yacaman, J., Frank, R.M. Structure of initial crystals formed during human amelogenesis. *J. Cryst. Growth* 1992, 116, 314-318.

[152] Cuisinier, F.J.G., Steuer, P., Senger, B., Voegel, J.C., Frank, R.M. Human amelogenesis: high resolution electron microscopy of nanometer-sized particles. *Cell Tissue Res.* 1993, 273, 175-182.

[153] Brès, E.F., Moebus, G., Kleebe, H.J., Pourroy, G., Werkmann, J., Ehret, G. High resolution electron microscopy study of amorphous calcium phosphate. *J. Cryst. Growth* 1993, 129, 149-162.

[154] Layrolle, P., Lebugle, A. Characterization and reactivity of nanosized calcium phosphate prepared in anhydrous ethanol. *Chem. Mater.* 1994, 6, 1996-2004.

[155] Cui, F.Z., Wen, H.B., Zhang, H.B., Ma, C.L., Li, H.D. Nanophase hydroxyapatite-like crystallites in natural ivory. *J. Mater. Sci. Lett.* 1994, 13, 1042-1044.

[156] Li, Y.B., de Wijn, J., Klein, C.P.A.T., de Meer, S.V., de Groot, K. Preparation and characterization of nanograde osteoapatite-like rod crystals. *J. Mater. Sci. Mater. Med.* 1994, 5, 252-255.

[157] Li, Y.B., de Groot, K., de Wijn, J., Klein, C.P.A.T., de Meer, S.V. Morphology and composition of nanograde calcium phosphate needle-like crystals formed by simple hydrothermal treatment. *J. Mater. Sci. Mater. Med.* 1994, 5, 326-331.

[158] Shirkhanzadeh, M. X-ray diffraction and Fourier transform infrared analysis of nanophase apatite coatings prepared by electrocrystallization. *Nanostruct. Mater.* 1994, 4, 677-684.

[159] Webster, T.J., Ergun, C., Doremus, R.H., Siegel, R.W., Bizios, R. Specific proteins mediate enhanced osteoblast adhesion on nanophase ceramics. *J. Biomed. Mater. Res.* 2000, 51, 475-483.

[160] Chan, C.K., Kumar, T.S.S., Liao, S., Murugan, R., Ngiam, M., Ramakrishnan, S. Biomimetic nanocomposites for bone graft applications. *Nanomedicine* 2006, 1, 177-188.

[161] Okada, M., Furukawa, K., Serizawa, T., Yanagisawa, Y., Tanaka, H., Kawai, T., Furuzono, T. Interfacial interactions between calcined hydroxyapatite nanocrystals and substrates. *Langmuir* 2009, 25, 6300-6306.

[162] Mikołajczyk, T., Rabiej, S., Bogun, M. Analysis of the structural parameters of polyacrylonitrile fibers containing nanohydroxyapatite. *J. Appl. Polym. Sci.* 2006, 101, 760-765.

[163] Wilberforce, S.I.J., Finlayson, C.E., Best, S.M., Cameron, R.E. The influence of hydroxyapatite (HA) microparticles (m) and nanoparticles (n) on the thermal and dynamic mechanical properties of poly-L-lactide. *Polymer* 2011, 52, 2883-2890.

[164] There are both nano-sized biomaterials and nanostructured biomaterials, which should be differentiated from each other. Nano-sized biomaterials refer to individual molecular level biomaterials such as single proteins (are not considered in this review), while nanostructured biomaterials refer to any biomaterials whose structure or morphology can be engineered to get features with nanometer-scale dimensions [165]. This review is limited to calcium orthophosphate-based nanostructured biomaterials only.

[165] Thomas, V., Dean, D.R., Vohra, Y.K. Nanostructured biomaterials for regenerative medicine. *Curr. Nanosci.* 2006, 2, 155-177.

[166] LeGeros, R.Z. Biodegradation and bioresorption of calcium phosphate ceramics. *Clin. Mater.* 1993, 14, 65-88.

[167] Wang, J., Shaw, L.L. Morphology-enhanced low-temperature sintering of nanocrystalline hydroxyapatite. *Adv. Mater.* 2007, 19, 2364-2369.

[168] Fomin, A.S., Barinov, S.M., Ievlev, V.M., Smirnov, V.V., Mikhailov, B.P., Belonogov, E.K., Drozdova, N.A. Nanocrystalline hydroxyapatite ceramics produced by low-temperature sintering after high-pressure treatment. *Dokl. Chem.* 2008, 418, 22-25.

[169] Drouet, C., Bosc, F., Banu, M., Largeot, C., Combes, C., Dechambre, G., Estournes, C., Raimbeaux, G., Rey, C. Nanocrystalline apatites: from powders to biomaterials. *Powder Technol.* 2009, 190, 118-122.

[170] Ramesh, S., Tan, C.Y., Bhaduri, S.B., Teng, W.D., Sopyan, I. Densification behaviour of nanocrystalline hydroxyapatite bioceramics. *J. Mater. Process. Technol.* 2008, 206, 221-230.

[171] Skorokhod, V.V., Solonin, S.M., Dubok, V.A., Kolomiets, L.L., Katashinskii, V.P., Shinkaruk, A.V. Pressing and sintering of nanosized hydroxyapatite powders. *Powder Metall. Metal Ceram.* 2008, 47, 518-524.

[172] Sung, Y.M., Lee, J.C., Yang, J.W. Crystallization and sintering characteristics of chemically precipitated hydroxyapatite nanopowder. *J. Cryst. Growth* 2004, 262, 467-472.

[173] Lin, K., Chang, J., Lu, J., Wu, W., Zeng, Y. Properties of β-$Ca_3(PO_4)_2$ bioceramics prepared using nanosized powders. *Ceram. Int.* 2007, 33, 979-985.

[174] Tanaka, Y., Hirata, Y., Yoshinaka, R. Synthesis and characteristics of ultra-fine hydroxyapatite particles. *J. Ceram. Proc. Res.* 2003, 4, 197-201.

[175] Wang, J., Shaw, L.L. Nanocrystalline hydroxyapatite with simultaneous enhancements in hardness and toughness. *Biomaterials* 2009, 30, 6565-6572.

[176] Stupp, S.I., Ciegler, G.W. Organoapatites: materials for artificial bone. I. Synthesis and microstructure. *J. Biomed. Mater.* Res. 1992, 26, 169-183.

[177] Webster, T.J., Ergun, C., Doremus, R.H., Siegel, R.W., Bizios, R. Enhanced osteoclast-like cell functions on nanophase ceramics. *Biomaterials* 2001, 22, 1327-1333.

[178] Huang, J., Best, S.M., Bonfield, W., Brooks, R.A., Rushton, N., Jayasinghe, S.N., Edirisinghe, M.J. *In vitro* assessment of the biological response to nanosized hydroxyapatite. *J. Mater. Sci. Mater. Med.* 2004, 15, 441-445.
[179] Kim, H.W., Kim, H.E., Salih, V. Stimulation of osteoblast responses to biomimetic nanocomposites of gelatin-hydroxyapatite for tissue engineering scaffolds. *Biomaterials* 2005, 26, 5221-5230.
[180] Webster, T.J., Siegel, R.W., Bizios, R. Osteoblast adhesion on nanophase ceramics. *Biomaterials* 1999, 20, 1221-1227.
[181] Webster, T.J., Ergun, C., Doremus, R.H., Siegel, R.W., Bizios, R. Enhanced functions of osteoblast on nanophase ceramics. *Biomaterials* 2000, 21, 1803-1810.
[182] Smith, I.O., McCabe, L.R., Baumann, M.J. MC3T3-E1 osteoblast attachment and proliferation on porous hydroxyapatite scaffolds fabricated with nanophase powder. *Int. J. Nanomed.* 2006, 1, 189-194.
[183] Nelson, M., Balasundaram, G., Webster, T.J. Increased osteoblast adhesion on nanoparticulate crystalline hydroxyapatite functionalized with KRSR. *Int. J. Nanomed.* 2006, 1, 339-349.
[184] Liu, H., Yazici, H., Ergun, C., Webster, T.J., Bermek, H. An *in vitro* evaluation of the Ca/P ratio for the cytocompatibility of nano-to-micron particulate calcium phosphates for bone regeneration. *Acta Biomater.* 2008, 4, 1472-1479.
[185] Sato, M., Sambito, M.A., Aslani, A., Kalkhoran, N.M., Slamovich, E.B., Webster, T.J. Increased osteoblast functions on undoped and yttrium-doped nanocrystalline hydroxyapatite coatings on titanium. *Biomaterials* 2006, 27, 2358-2369.
[186] Thian, E.S., Huang, J., Best, S.M. Barber, Z.H., Brooks, R.A., Rushton, N., Bonfield, W. The response of osteoblasts to nanocrystalline silicon-substituted hydroxyapatite thin films. *Biomaterials* 2006, 27, 2692-2698.
[187] Palin, E., Liu, H., Webster, T.J. Mimicking the nanofeatures of bone increases bone-forming cell adhesion and proliferation. *Nanotechnology* 2005, 16, 1828-1835.
[188] Sun, W., Chu, C., Wang, J., Zhao, H. Comparison of periodontal ligament cells responses to dense and nanophase hydroxyapatite. *J. Mater. Sci. Mater. Med.* 2007, 18, 677-683.
[189] Ergun, C., Liu, H., Webster, T.J., Olcay, E., Yılmaz, Ş., Sahin, F.C. Increased osteoblast adhesion on nanoparticulate calcium phosphates with higher Ca/P ratios. *J. Biomed. Mater. Res. A* 2008, 85A, 236-241.
[190] Lewandrowski, K.U., Bondre, S.P., Wise, D.L., Trantolo, D.J. Enhanced bioactivity of a poly(propylene fumarate) bone graft substitute by augmentation with nano-hydroxyapatite. *Biomed. Mater. Eng.* 2003, 13, 115-124.
[191] Zhou, D.S., Zhao, K.B., Li, Y., Cui, F.Z., Lee, I.S. Repair of segmental defects with nano-hydroxyapatite / collagen / PLA composite combined with mesenchymal stem cells. *J. Bioactive Compat. Polym.* 2006, 21, 373-384.
[192] Khanna, R., Katti, K.S., Katti, D.R. Bone nodules on chitosan-polygalacturonic acid-hydroxyapatite nanocomposite films mimic hierarchy of natural bone. *Acta Biomater.* 2011, 7, 1173-1183.
[193] Xu, Z., Sun, J., Changsheng, L., Jie, W. Effect of hydroxyapatite nanoparticles of different concentrations on rat osteoblast. *Mater. Sci. Forum* 2009, 610-613, 1364-1369.

[194] Okada, S., Nagai, A., Oaki, Y., Komotori, J., Imai, H. Control of cellular activity of fibroblasts on size-tuned fibrous hydroxyapatite nanocrystals. *Acta Biomater.* 2011, 7, 1290-1297.

[195] Krut'ko, V.K., Kulak, A.I., Lesnikovich, L.A., Trofimova, I.V., Musskaya, O.N., Zhavnerko, G.K., Paribok, I.V. Influence of the dehydration procedure on the physicochemical properties of nanocrystalline hydroxylapatite xerogel. *Russ. J. General Chem.* 2007, 77, 336-342.

[196] Severin, A.V., Komarov, V.F., Bozhevol'nov, V.E., Melikhov, I.V. Morphological selection in suspensions of nanocrystalline hydroxylapatite leading to spheroidal aggregates. *Russ. J. Inorg. Chem.* 2005, 50, 72-77.

[197] Biggemann, D., da Silva, M.H.P., Rossi, A.M., Ramirez, A.J. High-resolution transmission electron microscopy study of nanostructured hydroxyapatite. *Microsc. Microanal.* 2008, 14, 433-438.

[198] Hagmeyer, D., Ganesan, K., Ruesing, J., Schunk, D., Mayer, C., Dey, A., Sommerdijk, N.A.J.M., Epple, M. Self-assembly of calcium phosphate nanoparticles into hollow spheres induced by dissolved amino acids. *J. Mater. Chem.* 2011, 21, 9219-9223.

[199] Kester, M., Heakal, Y., Fox, T., Sharma, A., Robertson, G.P., Morgan, T.T., Altinoğlu, E.I., Tabaković, A., Parette, M.R., Rouse, S.M., Ruiz-Velasco, V., Adair, J.H. Calcium phosphate nanocomposite particles for *in vitro* imaging and encapsulated chemotherapeutic drug delivery to cancer cells. *Nano Lett.* 2008, 8, 4116-4121.

[200] Welzel, T., Meyer-Zaika, W., Epple, M. Continuous preparation of functionalised calcium phosphate nanoparticles with adjustable crystallinity. *Chem. Commun.* 2004, 1204-1205.

[201] Nichols, H.L., Zhang, N., Zhang, J., Shi, D., Bhaduri, S., Wen, X. Coating nanothickness degradable films on nanocrystalline hydroxyapatite particles to improve the bonding strength between nanohydroxyapatite and degradable polymer matrix. *J. Biomed. Mater. Res. A* 2007, 82A, 373-382.

[202] Bouladjine, A., Al-Kattan, A., Dufour, P., Drouet, C., New advances in nanocrystalline apatite colloids intended for cellular drug delivery. *Langmuir* 2009, 25, 12256-12265.

[203] Rey, C., Hina, A., Tofighi, A., Glimcher, M.J. Maturation of poorly crystalline apatites: chemical and structural aspects *in vivo* and *in vitro*. *Cell Mater.* 1995, 5, 345-356.

[204] Dorozhkin, S.V. Calcium orthophosphates. *J. Mater. Sci.* 2007, 42, 1061-1095.

[205] Dorozhkin, S.V. Calcium orthophosphates in nature, biology and medicine. *Materials* 2009, 2, 399-498.

[206] Elliott, J.C. Structure and chemistry of the apatites and other calcium orthophosphates; Elsevier: Amsterdam, Holland, 1994; pp. 404.

[207] Olszta, M.J., Cheng, X., Jee, S.S., Kumar, R., Kim, Y.Y., Kaufmane, M.J., Douglas, E.P., Gower, L.B. Bone structure and formation: a new perspective. *Mater. Sci. Eng.* R 2007, 58, 77-116.

[208] Cui, F.Z., Li, Y., Ge, J. Self-assembly of mineralized collagen composites. *Mater. Sci. Eng.* R 2007, 57, 1-27.

[209] Meyers, M.A., Chen, P.Y., Lin, A.Y.M., Seki, Y. Biological materials: structure and mechanical properties. *Prog. Mater. Sci.* 2008, 53, 1-206.

[210] Currey, J.D. Hierarchies in biomineral structures. *Science* 2005, 309, 253-254.

[211] Rubin, M.A., Jasiuk, I., Taylor, J., Rubin, J., Ganey, T., Apkarian, R.P. TEM analysis of the nanostructure of normal and osteoporotic human trabecular bone. *Bone* 2003, 33, 270-282.

[212] Hartgerink, J.D., Beniash, E., Stupp, S.I. Self-assembly and mineralization of peptide-amphiphile nanofibers. *Science* 2001, 294, 1684-1688.

[213] Ji, B., Gao, H. Elastic properties of nanocomposite structure of bone. *Compos. Sci. Technol.* 2006, 66, 1212-1218.

[214] Wang, L., Nancollas, G.H., Henneman, Z.J., Klein, E., Weiner, S. Nanosized particles in bone and dissolution insensitivity of bone mineral. *Biointerphases* 2006, 1, 106-111.

[215] Xie, B., Nancollas, G.H. How to control the size and morphology of apatite nanocrystals in bone. *Proc. Natl. Acad. Sci. USA* 2011, 107, 22369-22370.

[216] Hu, Y.Y., Rawal, A., Schmidt-Rohr, K. Strongly bound citrate stabilizes the apatite nanocrystals in bone. *Proc. Natl. Acad. Sci. USA* 2011, 107, 22425-22429.

[217] Gao, H., Ji, B., Jager, I.L., Arz, E., Fratzl, P. Materials become insensitive to flaws at nanoscale: lessons from nature. *Proc. Natl. Acad. Sci. USA* 2003, 100, 5597-5660.

[218] Gupta, H.S., Seto, J., Wagermaier, W., Zaslansky, P., Boesecke, P., Fratzl, P. Cooperative deformation of mineral and collagen in bone at the nanoscale. *Proc. Natl. Acad. Sci. U. S. A.* 2006, 103, 17741-17746.

[219] Currey, J.D. Bones: structure and mechanics. Princeton University Press: Princeton, USA, 2006; pp. 456.

[220] Porter, A.E., Nalla, R.K., Minor, A., Jinschek, J.R., Kisielowski, C., Radmilovic, V., Kinney, J.H., Tomsia, A.P., Ritchie, R.O. A transmission electron microscopy study of mineralization in age-induced transparent dentin. *Biomaterials* 2005, 26, 7650-7660.

[221] Kirkham, J., Brookes, S.J., Shore, R.C., Wood, S.R., Smith, D.A., Zhang, J., Chen, H., Robinson, C. Physico-chemical properties of crystal surfaces in matrix-mineral interactions during mammalian biomineralisation. *Curr. Opin. Colloid Interf. Sci.* 2002, 7, 124-132.

[222] Daculsi, G., Mentanteau, J., Kerebel, L.M., Mitre, D. Length and shape of enamel crystals. *Calcif. Tissue Int.* 1984, 36, 550-555.

[223] Robinson, C., Connell, S., Kirkham, J., Shorea, R., Smith, A. Dental enamel – a biological ceramic: regular substructures in enamel hydroxyapatite crystals revealed by atomic force microscopy. *J. Mater. Chem.* 2004, 14, 2242-2248.

[224] Chen, H., Tang, Z., Liu, J., Sun, K., Chang, S.R., Peters, M.C., Mansfield, J.F., Czajka-Jakubowska, A., Clarkson, B.H. Acellular synthesis of a human enamel-like microstructure. *Adv. Mater.* 2006, 18, 1846-1851.

[225] Chen, H., Clarkson, B.H., Sun, K., Mansfield, J.F. Self-assembly of synthetic hydroxyapatite nanorods into an enamel prism-like structure. *J. Colloid Interf. Sci.* 2005, 288, 97-103.

[226] Robinson, C. Self-oriented assembly of nano-apatite particles: a subunit mechanism for building biological mineral crystals. *J. Dental Res.* 2007, 86, 677-679.

[227] Cui, F.Z., Ge, J. New observations of the hierarchical structure of human enamel, from nanoscale to microscale. *J. Tissue Eng. Regen. Med.* 2007, 1, 185-191.

[228] He, L.H., Swain, M.V. Enamel – a "metallic-like" deformable biocomposite. *J. Dent.* 2007, 35, 431-437.

[229] Nelson, S.J. Wheeler's dental anatomy, physiology and occlusion. 9th Ed., W. B. Saunders: Philadelphia, USA. 2009; pp. 368.

[230] Suvorova E.I., Buffat P.A. Electron diffraction from micro- and nanoparticles of hydroxyapatite. *J. Microscopy* 1999, 196, 46-58.

[231] Panda, R.N., Hsieh, M.F., Chung, R.J., Chin, T.S. X-ray diffractometry and X-ray photoelectron spectroscopy investigations of nanocrytalline hydroxyapatite synthesized by a hydroxide gel technique. *Jpn. J. Appl. Phys.* 2001, 40, 5030-5035.

[232] Panda, R.N., Hsieh, M.F., Chung, R.J., Chin, T.S. FTIR, XRD, SEM and solid state NMR investigations of carbonate-containing hydroxyapatite nano-particles synthesized by hydroxide-gel technique. *J. Phys. Chem. Solids* 2003, 64, 193-199.

[233] Eichert, D. Sfihi, H., Combes, C., Rey, C. Specific characteristics of wet nanocrystalline apatites. Consequences on biomaterials and bone tissue. *Key Eng. Mater.* 2004, 254-256, 927-930.

[234] Rey, C., Combes, C., Drouet, C., Sfihi, H., Barroug, A. Physico-chemical properties of nanocrystalline apatites: implications for biominerals and biomaterials. *Mater. Sci. Eng. C* 2007, 27, 198-205.

[235] Eichert, D., Drouet, C., Sfihi, H., Rey, C., Combes, C. Nanocrystalline apatite-based biomaterials: synthesis, processing and characterization. In: Biomaterials research advances. Kendall J.B. Ed., Nova Science Publishers, Inc., USA, 2007; Chapter 5, pp. 93-143.

[236] Aronov, D., Rosenman, G. Trap state spectroscopy studies and wettability modification of hydroxyapatite nanobioceramics. *J. Appl. Phys.* 2007, 101, 034701 (5 pages).

[237] Jäger, C., Welzel, T., Meyer-Zaika, W., Epple, M. A solid-state NMR investigation of the structure of nanocrystalline hydroxyapatite. *Magn. Reson. Chem.* 2006, 44, 573-580.

[238] Isobe, T., Nakamura, S., Nemoto, R., Senna, M., Sfihi, H. Solid-state double nuclear magnetic resonance of calcium phosphate nanoparticules synthesized by wet-mechanochemical reaction. *J. Phys. Chem.* B 2002, 106, 5169-5176.

[239] Bertinetti, L., Tampieri, A., Landi, E., Ducati, C., Midgley, P.A., Coluccia, S., Martra, G. Surface structure, hydration, and cationic sites of nanohydroxyapatite: UHR-TEM, IR, and microgravimetric studies. *J. Phys. Chem. C* 2007, 111, 4027-4035.

[240] Bertinetti, L., Tampieri, A., Landi, E., Bolis, V., Busco, C., Martra, G. Surface structure, hydration and cationic sites of nanohydroxyapatite. *Key Eng. Mater.* 2008, 361-363, 87-90.

[241] Bertinetti, L., Drouet, C., Combes, C., Rey, C., Tampieri, A., Coluccia, S., Martra, G. Surface characteristics of nanocrystalline apatites: effect of Mg surface enrichment on morphology, surface hydration species, and cationic environments. *Langmuir* 2009, 25, 5647-5654.

[242] Gopi, D., Indira, J., Prakash, V.C.A., Kavitha, L. Spectroscopic characterization of porous nanohydroxyapatite synthesized by a novel amino acid soft solution freezing method. *Spectrochim. Acta A* 2009, 74A, 282-284.

[243] Rossi, A.M., da Silva, M.H.P., Ramirez, A.J., Biggemann, D., Caraballo, M.M., Mascarenhas, Y.P., Eon, J.G., Moure, G.T. Structural properties of hydroxyapatite with particle size less than 10 nanometers. *Key Eng. Mater.* 2007, 330-332, 255-258.

[244] Ramirez, C.A.O., Costa, A.M., Bettini, J., Ramirez, A.J., da Silva, M.H.P., Rossi, A.M. Structural properties of nanostructured carbonate apatites. *Key Eng. Mater.* 2009, 396-398, 611-614.

[245] Pasteris, J.D., Wopenka, B., Freeman, J.J., Rogers, K., Valsami-Jones, E., van der Houten, J.A.M., Silva, M.J. Lack of OH in nanocrystalline apatite as a function of degree of atomic order: implications for bone and biomaterials. *Biomaterials* 2004, 25, 229-238.
[246] Sakhno, Y., Bertinetti, L., Iafisco, M., Tampieri, A., Roveri, N., Martra, G. Surface hydration and cationic sites of nanohydroxyapatites with amorphous or crystalline surfaces: a comparative study. *J. Phys. Chem. C* 2010, 114, 16640-16648.
[247] Zyman, Z.Z., Epple, M., Rokhmistrov, D., Glushko, V. On impurities and the internal structure in precipitates occurring during the precipitation of nanocrystalline calcium phosphate. *Mat.-Wiss. u. Werkstofftech.* 2009, 40, 297-301.
[248] Cazalbou, S., Combes, C., Eichert, D., Rey, C. Adaptative physico-chemistry of bio-related calcium phosphates. *J. Mater. Chem.* 2004, 14, 2148-2153.
[249] Eichert, D., Salomé, M., Banu, M., Susini, J., Rey, C. Preliminary characterization of calcium chemical environment in apatitic and non-apatitic calcium phosphates of biological interest by X-ray absorption spectroscopy. *Spectrochim. Acta B* 2005, 60B, 850-858.
[250] Rosenman, G., Aronov, D., Oster, L., Haddad, J., Mezinskis, G., Pavlovska, I., Chaikina, M., Karlov, A. Photoluminescence and surface photovoltage spectroscopy studies of hydroxyapatite nano-bio-ceramics. *J. Luminescence* 2007, 122-123, 936-938.
[251] Melikhov, I.V., Teterin, Y.A., Rudin, V.N., Teterin, A.Y., Maslakov, K.I., Severin, A.V. An X-ray electron study of nanodisperse hydroxyapatite. *Russ. J. Phys. Chem. A* 2009, 83, 91-97.
[252] Aronov, D., Rosenman, G., Karlov, A., Shashkin, A. Wettability patterning of hydroxyapatite nanobioceramics induced by surface potential modification. *Appl. Phys. Lett.* 2006, 88, 163902 (3 pages).
[253] Rau, J.V., Generosi, A., Ferro, D., Minozzi, F., Paci, B., Albertini, V.R., Dolci, G., Barinov, S.M. *In situ* time-resolved X-ray diffraction study of evolution of nanohydroxyapatite particles in physiological solution. *Mater. Sci. Eng.* C 2009, 29, 1140-1143.
[254] Arora, A. Ceramics in nanotech revolution. *Adv. Eng. Mater.* 2004, 6, 244-247.
[255] Mao, Y., Park, T.J., Zhang, F., Zhou, H., Wong, S.S. Environmentally friendly methodologies of nanostructure synthesis. *Small* 2007, 3, 1122-1139.
[256] Ioku, K., Yoshimura, M. Stochiometric apatite fine single crystals by hydrothermal synthesis. *Phosphorus Res. Bull.* 1991, 1, 15-20.
[257] Chen, J.D., Wang, Y.J., Wei, K., Zhang, S.H., Shi, X.T. Self-organization of hydroxyapatite nanorods through oriented attachment. *Biomaterials* 2007, 28, 2275-2280.
[258] Guo, X., Xiao, P., Liu, J., Shen, Z. Fabrication of nanostructured hydroxyapatite via hydrothermal synthesis and spark plasma sintering. *J. Am. Ceram. Soc.* 2004, 88, 1026-1029.
[259] Brown, P.W., Constantz B., Eds., Hydroxyapatite and related materials. CRC Press: Boca Raton, FL, USA, 1994; pp. 343.
[260] Amjad, Z., Ed., Calcium phosphates in biological and industrial systems. Kluwer Academic Publishers: Boston, MA, USA, 1997; pp. 529.

[261] Hughes, J.M., Kohn, M., Rakovan, J., Eds., Phosphates: geochemical, geobiological and materials importance, Series: Reviews in Mineralogy and Geochemistry. Vol. 48; Mineralogical Society of America: Washington, D.C., USA, 2002; pp. 742.

[262] Chow, L.C., Eanes, E.D., Eds., Octacalcium phosphate. Karger: Basel, Switzerland, 2001; pp. 168.

[263] Brès, E., Hardouin, P., Eds., Les matériaux en phosphate de calcium. Aspects fondamentaux. / Calcium phosphate materials. Fundamentals. Sauramps Medical: Montpellier, France, 1998; pp. 176.

[264] Komarov, V.F., Kibalchitz, V. Precipitation of apatite through highly saturated solutions. *Moscow Univ. Bull. Chem. Dic.* 1979, 2680-2685.

[265] Prakash, K.H., Kumar, R., Ooi, C.P., Cheang, P., Khor, K.A. Conductometric study of precursor compound formation during wet-chemical synthesis of nanocrystalline hydroxyapatite. *J. Phys. Chem. B* 2006, 110, 24457-24462.

[266] Tao, J., Pan, H., Wang, J., Wu, J., Wang, B., Xu, X., Tang, R. Evolution of amorphous calcium phosphate to hydroxyapatite probed by gold nanoparticles. *J. Phys. Chem. C* 2008, 112, 14929-14933.

[267] Chane-Ching, J.Y., Lebugle, A., Rousselot, I., Pourpoint, A., Pelle, F. Colloidal synthesis and characterization of monocrystalline apatite nanophosphors. *J. Mater. Chem.* 2007, 17, 2904-2913.

[268] Zyman, Z.Z., Rokhmistrov, D.V., Glushko, V.I. Structural and compositional features of amorphous calcium phosphate at the early stage of precipitation. *J. Mater. Sci. Mater. Med.* 2010, 21, 123-130.

[269] Wei, M., Ruys, A.J., Milthorpe, B.K., Sorrell, C.C. Solution ripening of hydroxyapatite nanoparticles: effects on electrophoretic deposition. *J. Biomed. Mater. Res.* 1999, 45, 11-19.

[270] Zhu, X., Eibl, O., Berthold, C., Scheideler, L., Geis-Gerstorfer, J. Structural characterization of nanocrystalline hydroxyapatite and adhesion of pre-osteoblast cells. *Nanotechnology* 2006, 17, 2711-2721.

[271] Rusu, V.M., Ng, C.H., Wilke, M., Tiersch, B., Fratzl, P., Peter, M.G. Size-controlled hydroxyapatite nanoparticles as self-organized organic – inorganic composite materials. *Biomaterials* 2005, 26, 5414-5426.

[272] Wang, Y.J., Lai, C., Wei, K., Tang, S.Q. Influence of temperature, ripening time and cosurfactant on solvothermal synthesis of calcium phosphate nanobelts. *Mater. Lett.* 2005, 59, 1098-1104.

[273] Li, Y.B., Li, D., Weng, W. Preparation of nano carbonate-substituted hydroxyapatite from an amorphous precursor. *Int. J. Appl. Ceram. Technol.* 2008, 5, 442-448.

[274] Zhang, S., Gonsalves, K.E. Preparation and characterization of thermally stable nanohydroxyapatite. J. Mater. Sci. Mater. Med. 1997, 8, 25-28.

[275] Ferraz, M.P., Monteiro, F.J., Manuel, C.M. Hydroxyapatite nanoparticles: a review of preparation methodologies. *J. Appl. Biomater. Biomech.* 2004, 2, 74-80.

[276] Ahn, E.S., Gleason, N.J., Nakahira, A., Ying, J.Y. Nanostructure processing of hydroxyapatite-based bioceramics. *Nano Lett.* 2001, 1, 149-153.

[277] Mazelsky, R., Hopkins, R.H., Kramer, W.E. Czochralski-growth of calcium fluorophosphates. *J. Cryst. Growth* 1968, 3-4, 260-264.

[278] Loutts, G.B., Chai, B.H.T. Growth of high-quality single crystals of FAP ($Ca_5(PO_4)_3F$) and its isomorphs. *Proc. SPIE – Int. Soc. Optical Eng.* 1993, 1863, 31-34.

[279] Siegel, R.W. Creating nanophase materials. *Sci. Am.* 1996, 275, 42-47.
[280] Hu, J., Odom, T.W., Lieber, C.M. Chemistry and physics in one dimension: synthesis and properties of nanowires and nanotubes. *Acc. Chem. Res.* 1999, 32, 435-445.
[281] Schmidt, H.K. Nanoparticles for ceramic and nanocomposite processing. *Mol. Cryst. Liq. Cryst.* 2000, 353, 165-179.
[282] Cushing, B.L., Kolesnichenko, V.L., O'Connor, C.J. Recent advances in the liquid-phase syntheses of inorganic nanoparticles. *Chem. Rev.* 2004, 104, 3893-3946.
[283] Wang, X., Zhuang, J., Peng, Q., Li, Y. A general strategy for nanocrystal synthesis. *Nature* 2005, 437, 121-124.
[284] Yin, Y., Alivisatos, A.P. Colloidal nanocrystal synthesis and the organic-inorganic interface. *Nature* 2005, 437, 664-670.
[285] Mao, Y., Park, T.J., Zhang, F., Zhou, H., Wong, S.S. Environmentally friendly methodologies of nanostructure synthesis. *Small* 2007, 3, 1122-1139.
[286] Ma, M.G., Zhu, J.F. Recent progress on fabrication of calcium-based inorganic biodegradable nanomaterials. Recent Patents on *Nanotechnology* 2010, 4, 164-170.
[287] Takagi, S., Chow, L.C., Ishikawa, K. Formation of hydroxyapatite in new calcium phosphate cements. *Biomaterials* 1998, 19, 1593-1599.
[288] Meejoo, S., Maneeprakorn, W., Winotai, P. Phase and thermal stability of nanocrystalline hydroxyapatite prepared via microwave heating. *Thermochim. Acta* 2006, 447, 115-120.
[289] Kumta, P., Sfeir, C., Lee, D.H., Olton, D., Choi, D. Nanostructured calcium phosphates for biomedical applications: novel synthesis and characterization. *Acta Biomater.* 2005, 1, 65-83.
[290] Liou, S.C., Chen, S.Y., Lee, H.Y., Bow, J.S. Structural characterization of nanosized calcium deficient apatite powders. *Biomaterials* 2004, 25, 189-196.
[291] Mollazadeh, S., Javadpour, J., Khavandi, A. *In situ* synthesis and characterization of nanosized hydroxyapatite in poly(vinyl alcohol) matrix. *Ceram. Int.* 2007, 33, 1579-1583.
[292] Bigi, A., Boanini, E., Gazzano, M., Rubini, K., Torricelli, P. Nanocrystalline hydroxyapatite – polyaspartate composites. *Biomed. Mater. Eng.* 2004, 14, 573-579.
[293] Chen, H., Sun, K., Tang, Z., Law, R.V., Mansfield, J.F., Czajka-Jakubowska, A., Clarkson, B.H. Synthesis of fluorapatite nanorods and nanowires by direct precipitation from solution. *Cryst. Growth Des.* 2006, 6, 1504-1508.
[294] Kong, L., Gao, Y., Cao, W., Gong, Y., Zhao, N., Zhang, X. Preparation and characterization of nano-hydroxyapatite/chitosan composite scaffolds. *J. Biomed. Mater. Res. A* 2005, 75A, 275-282.
[295] Kong, L., Gao, Y., G. Lu, Gong, Y., Zhao, N., Zhang, X. A study on the bioactivity of chitosan/nano-hydroxyapatite composite scaffolds for bone tissue engineering. *Eur. Polym. J.* 2006, 42, 3171-3179.
[296] Melikhov, I.V., Komarov, V.F., Severin, A.V., Bozhevol'nov, V.E., Rudin, V.N. Two-dimensional crystalline hydroxyapatite. *Dokl. Phys. Chem.* 2000, 373, 355-358.
[297] Zhao, Y., Zhang, Y., Ning, F., Guo, D., Xu, Z. Synthesis and cellular biocompatibility of two kinds of HAP with different nanocrystal morphology. *J. Biomed. Mater. Res. B (Appl. Biomater.)* 2007, 83B, 121-126.
[298] Ganesan, K., Epple, M. Calcium phosphate nanoparticles as nuclei for the preparation of colloidal calcium phytate. *New J. Chem.* 2008, 32, 1326-1330.

[299] Zhang, Y., Lu, J. A simple method to tailor spherical nanocrystal hydroxyapatite at low temperature. *J. Nanopart. Res.* 2007, 9, 589-594.

[300] Bouyer, E., Gitzhofer, F., Boulos, M.I. Morphological study of hydroxyapatite nanocrystal suspension. *J. Mater. Sci. Mater. Med.* 2000, 11, 523-531.

[301] Pang, Y.X., Bao, X. Influence of temperature, ripening time and calcination on the morphology and crystallinity of hydroxyapatite nanoparticles. *J. Eur. Ceram. Soc.* 2003, 23, 1697-1704.

[302] Kumar, R., Prakash, K.H., Cheang, P. Khor, K.A. Temperature driven morphological changes of chemically precipitated hydroxyapatite nanoparticles. *Langmuir* 2004, 20, 5196-5200.

[303] Li-yun, C., Chuan-bo, Z., Jian-feng, H. Influence of temperature, [Ca^{2+}], Ca/P ratio and ultrasonic power on the crystallinity and morphology of hydroxyapatite nanoparticles prepared with a novel ultrasonic precipitation method. *Mater. Lett.* 2005, 59, 1902-1906.

[304] Afshar, A., Ghorbani, M., Ehsani, N., Saeri, M.R., Sorrell, C.C. Some important factors in the wet precipitation process of hydroxyapatite. *Mater. Des.* 2003, 24, 197-202.

[305] Wei, M., Ruys, A.J., Milthorpe, B.K., Sorrell, C.C. Precipitation of hydroxyapatite nanoparticles: effects of precipitation method on electrophoretic deposition. *J. Mater. Sci. Mater. Med.* 2005, 16, 319-324.

[306] Liu, Y., Hou, D., Wang, G. A simple wet chemical synthesis and characterization of hydroxyapatite nanorods. *Mater. Chem. Phys.* 2004, 86, 69-73.

[307] Saha, S.K., Banerjee, A., Banerjee, S., Bose, S., Synthesis of nanocrystalline hydroxyapatite using surfactant template systems: role of templates in controlling morphology. *Mater. Sci. Eng. C* 2009, 29, 2294-2301.

[308] Shanthi, P.M.S.L., Ashok, M., Balasubramanian, T., Riyasdeen, A., Akbarsha, M.A. Synthesis and characterization of nano-hydroxyapatite at ambient temperature using cationic surfactant. *Mater. Lett.* 2009, 63, 2123-2125.

[309] Mobasherpour, I., Heshajin, M.S., Kazemzadeh, A., Zakeri, M. Synthesis of nanocrystalline hydroxyapatite by using precipitation method. *J. Alloys Compd.* 2007, 430, 330-333.

[310] Phillips, M.J., Darr, J.A., Luklinska, Z.B., Rehman, I. Synthesis and characterization of nanobiomaterials with potential osteological applications. *J. Mater. Sci. Mater. Med.* 2003, 14, 875-882.

[311] Lee, S.J., Yoon, Y.S., Lee, M.H., Oh, N.S. Nanosized hydroxyapatite powder synthesized from eggshell and phosphoric acid. *J. Nanosci. Nanotechnol.* 2007, 7, 4061-4064.

[312] Monmaturapoj, N. Nanosize hydroxyapatite powders preparation by wet-chemical precipitation route. *J. Metals Mater. Miner.* 2008, 18, 15-20.

[313] Ramesh, S., Tan, C.Y., Sopyan, I., Hamdi, M., Teng, W.D. Consolidation of nanocrystalline hydroxyapatite powder. *Sci. Technol. Adv. Mater.* 2007, 8, 124-130.

[314] Zhou, W., Zhang, S.M., Hu, W., Qiu, Z.Y., Liu, Y.H. Dialysis efficiency in rapid synthesis of nano-hydroxyapatite. *Key Eng. Mater.* 2007, 330-332, 211-214.

[315] Shi, H.B., Zhong, H., Liu, Y., Gu, J.Y., Yang, C.S. Effect of precipitation method on stoichiometry and morphology of hydroxyapatite nanoparticles. *Key Eng. Mater.* 2007, 330-332, 271-274.

[316] Monkawa, A., Ikoma, T., Yunoki, S., Ohta, K., Tanaka, J. Electrophoretic deposition of hydroxyapatite nanocrystal. *Key Eng. Mater.* 2006, 309-311, 643-646.

[317] Fujii, S., Okada, M., Furuzono, T. Hydroxyapatite nanoparticles as stimulus-responsive particulate emulsifiers and building block for porous materials. *J. Coll. Interf. Sci.* 2007, 315, 287-296.

[318] Ong, H.T., Loo, J.S.C., Boey, F.Y.C., Russell, S.J., Ma, J., Peng, K.W. Exploiting the high-affinity phosphonate – hydroxyapatite nanoparticle interaction for delivery of radiation and drugs. *J. Nanopart. Res.* 2008, 10, 141-150.

[319] Silva, G.W.C., Ma, L., Hemmers, O., Lindle, D. Micro-structural characterization of precipitation-synthesized fluorapatite nano-material by transmission electron microscopy using different sample preparation techniques. *Micron* 2008, 39, 269-274.

[320] Poinern, G.E., Brundavanam, R.K., Mondinos, N., Jiang, Z.T. Synthesis and characterisation of nanohydroxyapatite using an ultrasound assisted method. *Ultrason. Sonochem.* 2009, 16, 469-474.

[321] Doğan, Ö., Öner, M. The influence of polymer architecture on nanosized hydroxyapatite precipitation. *J. Nanosci. Nanotechnol.* 2008, 8, 667-674.

[322] Loo, S.C.J., Siew, Y.E., Ho, S., Boey, F.Y.C., Ma, J. Synthesis and hydrothermal treatment of nanostructured hydroxyapatite of controllable sizes. *J. Mater. Sci. Mater. Med.* 2008, 19, 1389-1397.

[323] Guo, X., Gough, J.E., Xiao, P., Liu, J., Shen, Z. Fabrication of nanostructured hydroxyapatite and analysis of human osteoblastic cellular response. *J. Biomed. Mater. Res. A* 2007, 82A, 1022-1032.

[324] Safronova, T.V., Putlyaev, V.I., Sergeeva, A.I., Kunenkov, E.V., Tret'yakov, Y.D. Synthesis of nanocrystalline calcium hydroxyapatite from calcium saccharates and ammonium hydrogen phosphate. *Dokl. Chem.* 2009, 426, 118-123.

[325] Iafisco, M., Palazzo, B., Marchetti, M., Margiotta, N., Ostuni, R., Natile, G., Morpurgo, M., Gandin, V., Marzano, C., Roveri, N. Smart delivery of antitumoral platinum complexes from biomimetic hydroxyapatite nanocrystals. *J. Mater. Chem.* 2009, 19, 8385-8392.

[326] Wang, P., Li, C., Gong, H., Jiang, X., Wang, H., Li, K. Effects of synthesis conditions on the morphology of hydroxyapatite nanoparticles produced by wet chemical process. *Powder Technol.* 2010, 203, 315-321.

[327] Leskiv, M., Lagoa, A.L.C., Urch, H., Schwiertz, J., da Piedade, M.E.M., Epple, M. Energetics of calcium phosphate nanoparticle formation by the reaction of $Ca(NO_3)_2$ with $(NH_4)_2HPO_4$. *J. Phys. Chem. C* 2009, 113, 5478-5484.

[328] Rodrigues, L.R., Motisuke, M., Zavaglia, C.A.C. Synthesis of nanostructured hydroxyapatite: a comparative study between sol-gel and aqueous solution precipitation. *Key Eng. Mater.* 2009, 396-398, 623-626.

[329] Medvecky, L., Sopcak, T., Durisin, J., Briancin, J. Nanohydroxyapatite prepared from non-toxic organic Ca^{2+} compounds by precipitation in aqueous solution. *Mater. Lett.* 2011, 65, 3566-3569.

[330] Okada, M., Furuzono, T. Low-temperature synthesis of nanoparticle-assembled, transparent, and low-crystallized hydroxyapatite blocks. *J. Coll. Interf. Sci.* 2011, 360, 457-462.

[331] Sheykhan, M., Heydari, A., Ma'mani, L., Badiei, A. The synthesis and spectroscopic characterization of nano calcium fluorapatite using tetra-butylammonium fluoride. *Spectrochim. Acta A* 2011, 83, 379-783.
[332] Kazemzadeh, R., Behnamghader, A., Hesaraki, S. Effect of synthesis temperature on phase and morphological characteristics of hydroxyapatite nanoparticles. *Adv. Mater. Res.* 2011, 264-265, 1329-1333.
[333] López-Macipe, A., Gómez-Morales, J., Rodríguez-Clemente, R. Nanosized hydroxyapatite precipitation from homogeneous calcium/citrate/phosphate solutions using microwave and conventional heating. *Adv. Mater.* 1998, 10, 49-53.
[334] Siddharthan, A., Seshadri, S.K., Kumar, T.S.S. Rapid synthesis of calcium deficient hydroxyapatite nanoparticles by microwave irradiation. *Trends Biomater. Artif. Organs* 2005, 18, 110-113.
[335] Ioku, K., Yamauchi, S., Fujimori, H., Goto, S., Yoshimura, M. Hydrothermal preparation of fibrous apatite and apatite sheet. *Solid State Ionics* 2002, 151, 147-150.
[336] Kasahara, H., Ogata, N., Ogihara, T. Effect of starting solution on the formation of calcium phosphate nano particles by hydrothermal process. *J. Ceram. Soc. Jpn.* 2004, 112, 650-654.
[337] Lemos, A.F., Rocha, J.H.G., Quaresma, S.S.F., Kannana, S., Oktar, F.N., Agathopoulos, S., Ferreira, J.M.F. Hydroxyapatite nano-powders produced hydrothermally from nacreous material. *J. Eur. Ceram. Soc.* 2006, 26, 3639-3646.
[338] Chaudhry, A.A., Haque, S., Kellici, S., Boldrin, P., Rehman, I., Khalid, F.A., Darr, J.A. Instant nano-hydroxyapatite: a continuous and rapid hydrothermal synthesis. *Chem. Commun.* 2006, 2286-2288.
[339] Cao, M., Wang, Y., Guo, C., Qi, Y., Hu, C. Preparation of ultrahigh-aspect-ratio hydroxyapatite nanofibers in reverse micelles under hydrothermal conditions. *Langmuir* 2004, 20, 4784-4786.
[340] Jinlong, N. Hydrothermal synthesis of nano-crystalline hydroxyapatite. *Key Eng. Mater.* 2007, 330-332, 247-250.
[341] Ryu, I.Y., Kim, D.J., Han, J.S., Lee, M.H. Influence of two-step sintering variables on phase stability of hydrothermally prepared HAp nano powders. *Key Eng. Mater.* 2008, 361-363, 91-94.
[342] Han, J.K., Song, H.Y., Saito, F., Lee, B.T. Synthesis of high purity nanosized hydroxyapatite powder by microwave-hydrothermal method. *Mater. Chem. Phys.* 2006, 99, 235-239.
[343] Suchanek, W.L., Shuk, P., Byrappa, K., Riman, R.E., TenHuisen, K.S., Janas, V.F. Mechanochemical-hydrothermal synthesis of carbonated apatite powders at room temperature. *Biomaterials* 2002, 23, 699-710.
[344] Guo, X., Xiao, P. Effects of solvents on properties of nanocrystalline hydroxyapatite produced from hydrothermal process. *J. Eur. Ceram. Soc.* 2006, 26, 3383-3391.
[345] Xin, R., Yu, K. Ultrastructure characterization of hydroxyapatite nanoparticles synthesized by EDTA-assisted hydrothermal method. *J. Mater. Sci.* 2009, 44, 4205-4209.
[346] Zhang, C., Yang, J., Quan, Z., Yang, P., Li, C., Hou, Z., Lin, J. Hydroxyapatite nano- and microcrystals with multiform morphologies: controllable synthesis and luminescence properties. *Cryst. Growth Des.* 2009, 9, 2725-2733.

[347] Zhang, H.B., Zhou, K.C., Li, Z.Y., Huang, S.P. Plate-like hydroxyapatite nanoparticles synthesized by the hydrothermal method. *J. Phys. Chem. Solids* 2009, 70, 243-248.
[348] Abdel-Aal, E.A., El-Midany, A.A., El-Shall, H. Mechanochemical-hydrothermal preparation of nano-crystallite hydroxyapatite using statistical design. *Mater. Chem. Phys.* 2008, 112, 202-207.
[349] Sun, Y., Guo, G., Tao, D., Wang, Z. Reverse microemulsion-directed synthesis of hydroxyapatite nanoparticles under hydrothermal conditions. *J. Phys. Chem. Solids* 2007, 68, 373-377.
[350] Du, X., Chu, Y., Xing, S., Dong, L. Hydrothermal synthesis of calcium hydroxyapatite nanorods in the presence of PVP. *J. Mater. Sci.* 2009, 44, 6273-6279.
[351] Xin, R., Ren, F., Leng, Y. Synthesis and characterization of nano-crystalline calcium phosphates with EDTA-assisted hydrothermal method. *Mater. Des.* 2010, 31, 1691-1694.
[352] Yan, L., Li, Y., Deng, Z., Zhuang, J., Sun, X. Surfactant-assisted hydrothermal synthesis of hydroxyapatite nanorods. *Int. J. Inorg. Mater.* 2001, 3, 633-637.
[353] Zhang, F., Zhou, Z., Yang, S., Mao, L., Chen, H., Yu, X. Hydrothermal synthesis of hydroxyapatite nanorods in the presence of anionic starburst dendrimer. *Mater. Lett.* 2005, 59, 1422-1425.
[354] Pathi, S.P., Lin, D.D., Dorvee, J.R., Estroff, L.A., Fischbach, C. Hydroxyapatite nanoparticle-containing scaffolds for the study of breast cancer bone metastasis. *Biomaterials* 2011, 32, 5112-5122.
[355] Zhu, A., Lu, Y., Si, Y., Dai, S. Frabicating hydroxyapatite nanorods using a biomacromolecule template. *Appl. Surf. Sci.* 2011, 257, 3174-3179.
[356] Wang, Y.Z., Fu, Y. Microwave-hydrothermal synthesis and characterization of hydroxyapatite nanocrystallites. *Mater. Lett.* 2011, 65, 3388-3390.
[357] Manafi, S., Rahimipour, M.R. Synthesis of nanocrystalline hydroxyapatite nanorods via hydrothermal conditions. *Chem. Engin. Technol.* 2011, 34, 972-976.
[358] Byrappa, K., Haber, M. Handbook of hydrothermal technology: a technology for crystal growth and materials processing. Noyes Publications: New Jersey, USA, 2002, 893 pp.
[359] Yu-Song, P. Surface modification of nanocrystalline hydroxyapatite. *Micro and Nano Lett.* 2011, 6, 129-132.
[360] Chai, C.S., Ben-Nissan, B. Bioactive nanocrystalline sol-gel hydroxyapatite coatings. *J. Mater. Sci. Mater. Med.* 1999, 10, 465-469.
[361] Ben-Nissan, B., Green, D.D., Kannangara, G.S.K., Chai, C.S., Milev, A. ^{31}P NMR studies of diethyl phosphite derived nanocrystalline hydroxyapatite. *J. Sol-Gel Sci. Technol.* 2001, 21, 27-37.
[362] Gopi, D., Govindaraju, K.M., Victor, C.A.P., Kavitha, L., Rajendiran, N. Spectroscopic investigations of nanohydroxyapatite powders synthesized by conventional and ultrasonic coupled sol–gel routes. *Spectrochim. Acta A: Mol. Biomol. Spectrosc.* 2008, 70, 1243-1245.
[363] Natarajan, V.U., Rajeswari, S. Influence of calcium precursors on the morphology and crystallinity of sol–gel-derived hydroxyapatite nanoparticles. *J. Cryst. Growth* 2008, 310, 4601-4611.
[364] Ben-Nissan, B., Choi, A.H. Sol-gel production of bioactive nanocoatings for medical applications. *Part 1: an introduction. Nanomedicine* 2006, 1, 311-319.

[365] Choi, A.H., Ben-Nissan, B. Sol-gel production of bioactive nanocoatings for medical applications. *Part 2: current research and development. Nanomedicine* 2007, 2, 51-61.
[366] Kim, T.S., Kumta, P.N. Sol-gel synthesis and characterization of nanostructured hydroxyapatite powder. *Mater. Sci. Eng*. B 2004, 111, 232-236.
[367] Rajabi-Zamani, A.H., Behnamghader, A., Kazemzadeh, A. Synthesis of nanocrystalline carbonated hydroxyapatite powder via nonalkoxide sol-gel method. *Mater. Sci. Eng*. C 2008, 28, 1326-1329.
[368] Sopyan, I., Toibah, A.R., Natasha, A.N. Nanosized bioceramic hydroxyapatite powders via sol-gel method. *Int. J. Mech. Mater. Eng*. 2008, 3, 133-138.
[369] Padmanabhan, S.K., Balakrishnan, A., Chu, M.C., Lee, Y.J., Kim, T.N., Cho, S.J. Sol–gel synthesis and characterization of hydroxyapatite nanorods. *Particuology* 2009, 7, 466-470.
[370] Yuan, Y., Liu, C., Zhang, Y., Shan, X. Sol-gel auto-combustion synthesis of hydroxyapatite nanotubes array in porous alumina template. *Mater. Chem. Phys*. 2008, 112, 275-280.
[371] Kuriakose, T.A., Kalkura, S.N., Palanichamy, M., Arivuoli, D., Dierks, K., Bocelli, G., Betzel, C. Synthesis of stoichiometric nano crystalline hydroxyapatite by ethanol-based sol-gel technique at low temperature. *J. Cryst. Growth* 2004, 263, 517-523.
[372] Jahandideh, R., Behnamghader, A., Rangie, M., Youzbashi, A., Joughehdoust, S., Tolouei, R. Sol-gel synthesis of FHA nanoparticles and CDHA agglomerates from a mixture with a nonstochiometric Ca/P ratio. *Key Eng. Mater*. 2009, 396-398, 607-610.
[373] Pang, X., Zeng, H., Liu, J., Wei, S., Zheng, Y. The properties of nanohydroxyapatite materials and its biological effects. *Mater. Sci. Applications* 2010, 1, 81-90.
[374] Sanosh, K.P., Chu, M.C., Balakrishnan, A., Lee, Y.J., Kim, T.N., Cho, S.J. Synthesis of nano hydroxyapatite powder that simulate teeth particle morphology and composition. *Curr. Appl. Phys*. 2009, 9, 1459-1462.
[375] Darroudi, M., Eshtiagh-Hosseini, H., Housaindokht, M.R., Youssefi, A. Preparation and characterization of fluorohydroxyapatite nanopowders by nonalkoxide sol-gel method. *Digest J. Nanomater. Biostruct*. 2010, 5, 29-33.
[376] Montazeri, N., Jahandideh, R., Biazar, E. Synthesis of fluorapatite-hydroxyapatite nanoparticles and toxicity investigations. *Int. J. Nanomedicine* 2011, 6, 197-201.
[377] Li, B., Wang, X.L., Guo, B., Xiao, Y.M., Fan, H.S., Zhang, X.D. Preparation and characterization of nano hydroxyapatite. *Key Eng. Mater*. 2007, 330-332, 235-238.
[378] Tas, A.C. Synthesis of biomimetic Ca-hydroxyapatite powders at 37°C in synthetic body fluids. *Biomaterials* 2000, 21, 1429-1438.
[379] Wu, Y.S., Lee, Y.H., Chang, H.C. Preparation and characteristics of nanosized carbonated apatite by urea addition with coprecipitation method. *Mater. Sci. Eng*. C 2009, 29, 237-241.
[380] Swain, S.K., Sarkar, D. A comparative study: hydroxyapatite spherical nanopowders and elongated nanorods. *Ceram. Int*. 2011, 37, 2927-2930.
[381] Rameshbabu, N., Kumar, T.S.S., Murugan, R., Rao, K.P. Mechanochemical synthesis of nanocrystalline fluorinated hydroxyapatite. *Int. J. Nanosci*. 2005, 4, 643-649.
[382] Yeong, K.C.B., Wang, J., Ng, S.C. Mechanochemical synthesis of nanocrystalline hydroxyapatite from CaO and $CaHPO_4$. *Biomaterials* 2001, 22, 2705-2712.

[383] Coreno, J.A., Coreno, O.A., Cruz, R.J.J., Rodriguez, C.C. Mechanochemical synthesis of nanocrystalline carbonate-substituted hydroxyapatite. *Optical Mater.* 2005, 27, 1281-1285.
[384] el Briak-Ben Abdeslam, H., Mochales, C., Ginebra, M.P., Nurit, J., Planell J.A., Boudeville, P. Dry mechanochemical synthesis of hydroxyapatites from dicalcium phosphate dihydrate and calcium oxide: a kinetic study. *J. Biomed. Mater. Res, A* 2003, 67A, 927-937.
[385] Nakamura, S., Isobe, T., Senna, M. Hydroxyapatite nano sol prepared via a mechanochemical route. *J. Nanopart. Res.* 2001, 3, 57-61.
[386] Nasiri-Tabrizi, B., Honarmandi, P., Ebrahimi-Kahrizsangi, R., Honarmandi, P. Synthesis of nanosize single-crystal hydroxyapatite via mechanochemical method. *Mater. Lett.* 2009, 63, 543-546.
[387] Sharifah, A., Iis, S., Mohd, H., Singh, R. Mechanochemical synthesis of nanosized hydroxyapatite powder and its conversion to dense bodies. *Mater. Sci. Forum* 2011, 694, 118-122.
[388] Fathi, M.H., Zahrani, E.M. Fabrication and characterization of fluoridated hydroxyapatite nanopowders via mechanical alloying. *J. Alloys Compd.* 2009, 475, 408-414.
[389] Fathi, M.H., Zahrani, E.M. Mechanical alloying synthesis and bioactivity evaluation of nanocrystalline fluoridated hydroxyapatite. *J. Cryst. Growth* 2009, 311, 1392-1403.
[390] Silva, C.C., Graça, M.P.F., Valente, M.A., Sombra, A.S.B. Crystallite size study of nanocrystalline hydroxyapatite and ceramic system with titanium oxide obtained by dry ball milling. *J. Mater. Sci.* 2007, 42, 3851-3855.
[391] Zahrani, E.M., Fathi, M.H. The effect of high-energy ball milling parameters on the preparation and characterization of fluorapatite nanocrystalline powder. *Ceram. Int.* 2009, 35, 2311-2323.
[392] Mochales, C., Wilson, R.M., Dowker, S.E.P., Ginebra, M.P. Dry mechanosynthesis of nanocrystalline calcium deficient hydroxyapatite: structural characterization. *J. Alloys Compounds* 2011, 509, 7389-7394.
[393] Xu, J.L., Khor, K.A., Dong, Z.L., Gu, Y.W., Kumar, R., Cheang, P. Preparation and characterization of nanosized hydroxyapatite powders produced in a radio frequency (rf) thermal plasma. *Mater. Sci. Eng. A* 2004, 374, 101-108.
[394] Xu, J.L., Khor, K.A., Kumar, R., Cheang, P. RF induction plasma synthesized calcium phosphate nanoparticles. *Key Eng. Mater.* 2006, 309-311, 511-514.
[395] Ruksudjarit, A., Pengpat, K., Rujijanagul, G., Tunkasiri, T. Synthesis and characterization of nanocrystalline hydroxyapatite from natural bovine bone. *Curr. Appl. Phys.* 2008, 8, 270-272.
[396] Cho, J.S., Kang, Y.C. Nano-sized hydroxyapatite powders prepared by flame spray pyrolysis. *J. Alloys Compd.* 2008, 464, 282-287.
[397] Wang, X., Zhuang, J., Peng, Q., Li, Y. Liquid-solid-solution synthesis of biomedical hydroxyapatite nanorods. *Adv. Mater.* 2006, 18, 2031-2034.
[398] Shirkhanzadeh, M. Direct formation of nanophase hydroxyapatite on cathodically polarized electrodes. *J. Mater. Sci. Mater. Med.* 1998, 9, 67-72.
[399] Montalbert-Smith, R., Palma, C.A., Arias, J.D., Montero, M.L. Formation of hydroxyapatite nanosized and other apatites by electrolysis process. *Key Eng. Mater.* 2009, 396-398, 579-582.

[400] Gao, J.H., Guan, S.K., Chen, J., Wang, L.G., Zhu, S.J., Hu, J.H., Ren, Z.W. Fabrication and characterization of rod-like nano-hydroxyapatite on MAO coating supported on Mg–Zn–Ca alloy. *Appl. Surf. Sci.* 2011, 257, 2231-2237.

[401] Liu, J., Li, K., Wang, H., Zhu, M., Xu, H., Yan, H. Self-assembly of hydroxyapatite nanostructures by microwave irradiation. *Nanotechnology* 2005, 16, 82-87.

[402] Rameshbabu, N., Rao, K.P., Kumar, T.S.S. Accelerated microwave processing of nanocrystalline hydroxyapatite. *J. Mater. Sci.* 2005, 40, 6319-6323.

[403] Ran, X., Chen, J., Ran, J., Gou, L., Zhang, X. Synthesis of nanosized carbonated hydroxyapatite under microwave irradiation. *Key Eng. Mater.* 2007, 330-332, 303-306.

[404] Siddharthan, A., Seshadri, S.K., Kumar, T.S.S. Microwave accelerated synthesis of nanosized calcium deficient hydroxyapatite. *J. Mater. Sci. Mater. Med.* 2004, 15, 1279-1284.

[405] Liu, J., Li, K., Wang, H., Zhu, M., Yan, H. Rapid formation of hydroxyapatite nanostructures by microwave irradiation. *Chem. Phys. Lett.* 2004, 396, 429-432.

[406] Krishna, D.S.R., Siddharthan, A., Seshadri, S.K., Kumar, T.S.S. A novel route for synthesis of nanocrystalline hydroxyapatite from eggshell waste. *J. Mater. Sci. Mater. Med.* 2007, 18, 1735-1743.

[407] Seo, D.S., Hwang, K.H., Lee, J.K. Nanostructured hydroxyapatite by microwave sintering. *J. Nanosci. Nanotechnol.* 2008, 8, 944-948.

[408] Arami, H., Mohajerani, M., Mazloumi, M., Khalifehzadeh, R., Lak, A., Sadrnezhaad, S.K. Rapid formation of hydroxyapatite nanostrips via microwave irradiation. *J. Alloys Compd.* 2009, 469, 391-394.

[409] Lak, A., Mazloumi, M., Mohajerani, M.S., Zanganeh, S., Shayegh, M.R., Kajbafvala, A., Arami, H., Sadrnezhaad, S.K. Rapid formation of mono-dispersed hydroxyapatite nanorods with narrow-size distribution via microwave irradiation. *J. Am. Ceram. Soc.* 2008, 91, 3580-3584.

[410] Rameshbabu, N., Kumar, T.S.S., Rao, K.P. Synthesis of nanocrystalline fluorinated hydroxyapatite by microwave processing and its *in vitro* dissolution study. *Bull. Mater. Sci.* 2006, 29, 611-615.

[411] Kumar, A.R., Kalainathan, S., Saral, A.M. Microwave assisted synthesis of hydroxyapatite nano strips. *Cryst. Res. Technol.* 2010, 45, 776-778.

[412] Kalita, S.J., Verma, S. Nanocrystalline hydroxyapatite bioceramic using microwave radiation: synthesis and characterization. *Mater. Sci. Eng. C* 2010, 30, 295-303.

[413] Vani, R., Raja, S.B., Sridevi, T.S., Savithri, K., Devaraj, S.N., Girija, E.K., Thamizhavel, A., Kalkura, S.N. Surfactant free rapid synthesis of hydroxyapatite nanorods by a microwave irradiation method for the treatment of bone infection. *Nanotechnology* 2011, 22, 285701.

[414] Cabrera, J.L., Velázquez-Castillo, R., Rivera-Muñoz, E.M. Synthesis of hydroxyapatite nanostructures using microwave heating. *J. Nanosci. Nanotechnol.* 2011, 11, 5555-5561.

[415] Zyman, Z., Goncharenko, A., Rokhmistrov, D., Epple, M. Nanocrystalline calcium-deficient hydroxyapatite prepared by a microwave-assisted solvent-free reaction. *Mat.-Wiss. u. Werkstofftech.* 2011, 42, 154-157.

[416] Shih, W.J., Chen, Y.F., Wang, M.C., Hon, M.H. Crystal growth and morphology of the nanosized hydroxyapatite powders synthesized from $CaHPO_4{\cdot}2H_2O$ and $CaCO_3$ by hydrolysis method. *J. Cryst. Growth* 2004, 270, 211-218.

[417] Zhang, Y., Lu, J. The transformation of single-crystal calcium phosphate ribbon-like fibres to hydroxyapatite spheres assembled from nanorods. *Nanotechnology* 2008, 19, 155608 (10 pages).
[418] Furuichi, K., Oaki, Y., Imai, H. Preparation of nanotextured and nanofibrous hydroxyapatite through dicalcium phosphate with gelatin. *Chem. Mater.* 2006, 18, 229-234.
[419] Yoruç, A.B.H., Koca, Y. Double step stirring: a novel method for precipitation of nano-sized hydroxyapatite powder. *Digest J. Nanomater. Biostructures* 2009, 4, 73-81.
[420] Jarudilokkul, S., Tanthapanichakoon, W., Boonamnuayvittaya, V. Synthesis of hydroxyapatite nanoparticles using an emulsion liquid membrane system. *Colloids Surf. A* 2007, 296, 149-153.
[421] Lim, G.K., Wang, J., Ng, S.C., Gan, L.M. Nanosized hydroxyapatite powders from microemulsions and emulsions stabilized by a biodegradable surfactant. *J. Mater. Chem.* 1999, 9, 1635-1639.
[422] Guo, G., Sun, Y., Wang, Z., Guo, H. Preparation of hydroxyapatite nanoparticles by reverse microemulsion. *Ceram. Int.* 2005, 31, 869-872.
[423] Lim, G.K., Wang, J., Ng, S.C., Gan, L.M. Formation of nanocrystalline hydroxyapatite in nonionic surfactant emulsions. *Langmuir* 1999, 15, 7472-7477.
[424] Sun, Y., Guo, G., Wang, Z., Guo, H. Synthesis of single-crystal HAP nanorods. *Ceram. Int.* 2006, 32, 951-954.
[425] Bose, S., Saha, S.K. Synthesis and characterization of hydroxyapatite nanopowders by emulsion technique. *Chem. Mater.* 2003, 15, 4464-4469.
[426] Lai, C., Tang, S.Q., Wang, Y.J., Wei, K. Formation of calcium phosphate nanoparticles in reverse microemulsions. *Mater. Lett.* 2005, 59, 210-214.
[427] Jiang, F.X., Lu, X.Y., Zhang, M.L., Weng, J. Regulating size, morphology and dispersion of nano-crystallites of hydroxyapatite by pH value and temperature in microemulsion system. *Key Eng. Mater.* 2008, 361-363, 195-198.
[428] Sato, K., Hotta, Y., Nagaoka, T., Yasuoka, M., Watari, K. Agglomeration control of hydroxyapatite nano-crystals grown in phase-separated microenvironments. *J. Mater. Sci.* 2006, 41, 5424-5428.
[429] Li, H., Zhu, M.Y., Li, L.H., Zhou, C.R. Processing of nanocrystalline hydroxyapatite particles via reverse microemulsions. *J. Mater. Sci.* 2008, 43, 384-389.
[430] Koetz, J., Baier, J., Kosmella, S. Formation of zinc sulfide and hydroxylapatite nanoparticles in polyelectrolyte-modified microemulsions. *Colloid Polym. Sci.* 2007, 285, 1719-1726.
[431] Lim, H.N., Kassim, A., Huang, N.M. Preparation and characterization of calcium phosphate nanorods using reverse microemulsion and hydrothermal processing routes. *Sains Malaysiana* 2010, 39, 267-273.
[432] Furuzono, T., Walsh, D., Sato, K., Sonoda, K., Tanaka, J. Effect of reaction temperature on the morphology and size of hydroxyapatite nanoparticles in an emulsion system. *J. Mater. Sci. Lett.* 2001, 2, 111-114.
[433] Sadjadi, M.A.S., Akhavan, K., Zare, K. Preparation of hydroxyapatite nanoparticles by reverse microemulsions and polyelectrolyte-modified microemulsions. *Res. J. Chem. Environment* 2011, 15, 959-962.

[434] Shen, S.C., Chia, L., Ng, W.K., Dong, Y.C., Tan, R.B.H. Solid-phase steam-assisted synthesis of hydroxyapatite nanorods and nanoparticles. *J. Mater. Sci.* 2010, 45, 6059-6067.

[435] Jevtić, M., Mitrić, M., Škapin, S., Jančar, B., Ignjatović, N., Uskoković, D. Crystal structure of hydroxyapatite nano-rods synthesized by sonochemical homogenous precipitation. *Cryst. Growth Des.* 2008, 8, 2217-2222.

[436] Wang, Y.J., Lai, C., Wei, K., Chen, X., Ding, Y., Wang, Z.L. Investigations on the formation mechanism of hydroxyapatite synthesized by the solvothermal method. *Nanotechnology* 2006, 17, 4405-4412.

[437] Cao, L.Y., Zhang, C.B., Huang, J.F. Synthesis of hydroxyapatite nanoparticles in ultrasonic precipitation. *Ceram. Int.* 2005, 31, 1041-1044.

[438] Liu, J., Wu, Q., Ding, Y. Self-assembly and fluorescent modification of hydroxyapatite nanoribbon spherulites. *Eur. J. Inorg. Chem.* 2005, 20, 4145-4149.

[439] Huang, J., Jayasinghe, S.N., Su, X., Ahmad, Z., Best, S.M., Edirisinghe, M.J., Brooks, R.A., Rushton, N., Bonfield, W. Electrostatic atomisation spraying: a novel deposition method for nano-sized hydroxyapatite. *Key Eng. Mater.* 2006, 309-311, 635-638.

[440] Hwang, K.S., Kim, B.H. Preparation of calcium phosphate nano-powders prepared by sol-gel assisted-electrostatic spraying method. *J. Mater. Sci.* 2005, 40, 4665-4666.

[441] Uota, M., Arakawa, H., Kitamura, N., Yoshimura, T., Tanaka, J., Kijima, T. Synthesis of high surface area hydroxyapatite nanoparticles by mixed surfactant-mediated approach. *Langmuir* 2005, 21, 4724-4728.

[442] Chu, M., Liu, G. Preparation and characterization of hydroxyapatite/liposome core – shell nanocomposites. *Nanotechnology* 2005, 16, 1208-1212.

[443] Huang, F., Shen, Y., Xie, A., Zhu, J., Zhang, C., Li, S., Zhu, J. Study on synthesis and properties of hydroxyapatite nanorods and its complex containing biopolymer. *J. Mater. Sci.* 2007, 42, 8599-8605.

[444] Wang, A., Liu, D., Yin, H., Wu, H., Wada, Y., Ren, M., Jiang, T., Cheng, X., Xu, Y. Size-controlled synthesis of hydroxyapatite nanorods by chemical precipitation in the presence of organic modifiers. *Mater. Sci. Eng. C* 2007, 27, 865-869.

[445] Ye, F., Guo, H., Zhang, H. Biomimetic synthesis of oriented hydroxyapatite mediated by nonionic surfactants. *Nanotechnology* 2008, 19, 245605 (7 pages).

[446] Han, Y., Wang, X., Li, S. A simple route to prepare stable hydroxyapatite nanoparticles suspension. *J. Nanoparticle Res.* 2009, 11, 1235-1240.

[447] Tseng, Y.H., Kuo, C.S., Li, Y.Y., Huang, C.P. Polymer-assisted synthesis of hydroxyapatite nanoparticle. *Mater. Sci. Eng. C* 2009, 29, 819-822.

[448] Klinkaewnarong, J., Swatsitang, E., Maensiri, S. Nanocrystalline hydroxyapatite powders by a chitosan-polymer complex solution route: synthesis and characterization. *Solid State Sci.* 2009, 11, 1023-1027.

[449] Li, Y., Li, D., Xu, Z. Synthesis of hydroxyapatite nanorods assisted by Pluronics. *J. Mater. Sci.* 2009, 44, 1258-1263.

[450] Nayar, S., Sinha, M.K., Basu, D., Sinha, A. Synthesis and sintering of biomimetic hydroxyapatite nanoparticles for biomedical applications. *J. Mater. Sci. Mater. Med.* 2006, 17, 1063-1068.

[451] Yao, X., Yao, H., Li, G., Li, Y. Biomimetic synthesis of needle-like nano-hydroxyapatite templated by double-hydrophilic block copolymer. *J. Mater. Sci.* 2010, 45, 1930-1936.

[452] Hong, Y., Fan, H., Li, B., Guo, B., Liu, M., Zhang, X. Fabrication, biological effects, and medical applications of calcium phosphate nanoceramics. *Mater. Sci. Eng. R* 2010, 70, 225-242.

[453] Mostaghaci, B., Fathi, M.H., Sheikh-Zeinoddin, M., Soleimanian-Zad, S. Bacterial synthesis of nanostructured hydroxyapatite using *Serratia marcescens* PTCC 1187. *Int. J. Nanotechnol.* 2009, 6, 1015-1030.

[454] Nathanael, A.J., Hong, S.I., Mangalaraj, D., Chen, P.C. Large scale synthesis of hydroxyapatite nanospheres by high gravity method. *Chem. Eng. J.* 2011, 173, 846-854.

[455] Parisi, M., Stoller, M., Chianese, A. Production of nanoparticles of hydroxy apatite by using a rotating disk reactor. *Chem. Eng. Trans.* 2011, 24, 211-216.

[456] Yuan, J., Wu, Y., Zheng, Q., Xie, X. Synthesis and characterization of nano hydroxylapatite by reaction precipitation in impinging streams. *Adv. Mater. Res.* 2011, 160-162, 1301-1308.

[457] Mohn, D., Doebelin, N., Tadier, S., Bernabei, R.E., Luechinger, N.A., Stark, W.J., Bohner, M. Reactivity of calcium phosphate nanoparticles prepared by flame spray synthesis as precursors for calcium phosphate cements. *J. Mater. Chem.* 2011, 21, 13963-13972.

[458] Tadic, D., Veresov, A., Putlayev, V.I., Epple, M. *In-vitro* preparation of nanocrystalline calcium phosphates as bone substitution materials in surgery. *Mat.-Wiss. u. Werkstofftech.* 2003, 34, 1048-1051.

[459] Qiu, Y., Xia, H., Jiang, H. Fabrication of nano-hydroxyapatite using a novel ultrasonic atomization precipitation method. *J. Nanosci. Nanotechnol.* 2010, 10, 2213-2218.

[460] Rouhani, P., Taghavinia, N., Rouhani, S. Rapid growth of hydroxyapatite nanoparticles using ultrasonic irradiation. *Ultrasonics Sonochemistry* 2010, 17, 853-856.

[461] Giardina, M.A., Fanovich, M.A. Synthesis of nanocrystalline hydroxyapatite from $Ca(OH)_2$ and H_3PO_4 assisted by ultrasonic irradiation. *Ceram. Int.* 2010, 36, 1961-1969.

[462] Sadjadi, M.S., Meskinfam, M., Sadeghi, B., Jazdarreh, H., Zare, K. *In situ* biomimetic synthesis, characterization and *in vitro* investigation of bone-like nanohydroxyapatite in starch matrix. *Mater. Chem. Phys.* 2010, 124, 217-222.

[463] Mhin, S.W., Ryu, J.H., Kim, K.M., Park, G.S., Ryu, H.W., Shim, K.B., Sasaki, T., Koshizaki, N. Simple synthetic route for hydroxyapatite colloidal nanoparticles via a Nd:YAG laser ablation in liquid medium. *Appl. Phys. A* 2009, 96A, 435-440.

[464] Musaev, O.R., Dusevich, V., Wieliczka, D.M., Wrobel, J.M., Kruger, M.B. Nanoparticle fabrication of hydroxyapatite by laser ablation in water. *J. Appl. Phys.* 2008, 104, 084316 (5 pages).

[465] Boutinguiza, M., Lusquiños, F., Riveiro, A., Comesaña, R., Pou, J. Hydroxylapatite nanoparticles obtained by fiber laser-induced fracture. *Appl. Surf. Sci.* 2009, 255, 5382-5385.

[466] Boutinguiza, M., Pou, J., Lusquiños, F., Comesaña, R., Riveiro, A. Production of calcium phosphate nanoparticles by laser ablation in liquid. *Physics Procedia* 2011, 12, 54-59.

[467] Boutinguiza, M., Comesaña, R., Lusquiños, F., Riveiro, A., Pou, J. Production of nanoparticles from natural hydroxylapatite by laser ablation. *Nanoscale Res.* Lett. 2011, 6, 1-5.

[468] Zuo, Y., Li, Y.B., Wei, J., Yan, Y. Influence of ethylene glycol on the formation of calcium phosphate nanocrystals. *J. Mater. Sci. Technol.* 2003, 19, 628-630.

[469] Barinov, S.M., Belonogov, E.K., Ievlev, V.M., Kostyuchenko, A.V., Putlyaev, V.I., Tret'yakov, Y.D., Smirnov, V.V., Fadeeva, I.V. Synthesis of dense nanocrystalline hydroxyapatite films. *Dokl. Phys. Chem.* 2007, 412, 15-18.

[470] Mello, A., Mavropoulos, E., Hong, Z., Ketterson, J.B., Rossi, A.M. Nanometer coatings of hydroxyapatite characterized by glancing-incidence X-ray diffraction. *Key Eng. Mater.* 2009, 396-398, 369-372.

[471] Iafisco, M., Morales, J.G., Hernández-Hernández, M.A., García-Ruiz, J.M., Roveri, N. Biomimetic carbonate-hydroxyapatite nanocrystals prepared by vapor diffusion. *Adv. Eng. Mater.* 2010, 12, B218-B223.

[472] Luo, P., Nieh, T.G. Synthesis of ultrafine hydroxyapatite particles by a spray dry method. *Mater. Sci. Eng.* C 1995, 3, 75-78.

[473] Chen, F., Wang, Z.C., Chang, J.L. Preparation and characterization of nanosized hydroxyapatite particles and hydroxyapatite / chitosan nano-composite for use in biomedical materials. *Mater. Lett.* 2002, 57, 858-861.

[474] Sarig, S., Kahana, F. Rapid formation of nanocrystalline apatite. *J. Cryst. Growth* 2002, 237-239, 55-59.

[475] Bose, S., Saha, S.K. Synthesis of hydroxyapatite nanopowders via sucrose-templated sol-gel method. *J. Am. Ceram. Soc.* 2003, 86, 1055-1057.

[476] Han, Y., Li, S., Wang, X., Chen, X. Synthesis and sintering of nanocrystalline hydroxyapatite powders by citric acid sol-gel combustion method. *Mater. Res. Bull.* 2004, 39, 25-32.

[477] Liu, D.M., Yang, Q., Troczynski, T., Tseng, W.J. Structural evolution of sol-gel-derived hydroxyapatite. *Biomaterials* 2002, 23, 1679-1687.

[478] Liu, D.M., Troczynski, T., Tseng, W.J. Water-based sol-gel synthesis of hydroxyapatite: process development. *Biomaterials* 2001, 22, 1721-1730.

[479] Wang, F., Li, M.S., Lu, Y.P., Ge, S.S. Synthesis of nanocrystalline hydroxyapatite powders in stimulated body fluid. *J. Mater. Sci.* 2005, 40, 2073-2076.

[480] Wang, J., Shaw, L.L. Synthesis of high purity hydroxyapatite nanopowder via sol-gel combustion process. *J. Mater. Sci. Mater. Med.* 2009, 20, 1223-1227.

[481] Varma, H.K., Kalkura, S.N., Sivakumar, R. Polymeric precursor route for the preparation of calcium phosphate compounds. *Ceram. Int.* 1998, 24, 467-470.

[482] Ghosh, S.K., Roy, S.K., Kundu, B., Datta, S., Basu, D. Synthesis of nano-sized hydroxyapatite powders through solution combustion route under different reaction conditions. *Mater. Sci. Eng. B* 2011, 176, 14-21.

[483] Loher, S., Stark, W.J., Maciejewski, M., Baiker, A., Pratsinis, S.E., Reichardt, D., Maspero, F., Krumeich, F., Günther, D. Fluoro-apatite and calcium phosphate nanoparticles by flame synthesis. *Chem. Mater.* 2005, 17, 36-42.

[484] Trommer, R.M., Santos, L.A., Bergmann, C.P. Nanostructured hydroxyapatite powders produced by a flame-based technique. *Mater. Sci. Eng. C* 2009, 29, 1770-1775.

[485] Chow, L.C., Sun, L., Hockey, B. Properties of nanostructured hydroxyapatite prepared by a spray drying technique. *J. Res. Natl. Inst. Stand. Technol.* 2004, 109, 543-551.

[486] Li, J., Chen, Y.P., Yin, Y., Yao, F., Yao, K. Modulation of nano-hydroxyapatite size via formation on chitosan-gelatin network film *in situ*. *Biomaterials* 2007, 28, 781-790.

[487] Zhai, Y., Cui, F.Z., Wang, Y. Formation of nano hydroxyapatite on recombinant human like collagen fibrils. *Curr. Appl. Phys.* 2005, 5, 429-432.

[488] Liou, S.C., Chen, S.Y., Liu, D.M. Synthesis and characterization of needlelike apatitic nanocomposite with controlled aspect ratios. *Biomaterials* 2003, 24, 3981-3988.
[489] Liou, S.C., Chen, S.Y., Liu, D.M. Manipulation of nanoneedle and nanosphere apatite/poly(acrylic acid) nanocomposites. *J. Biomed. Mater. Res. B* (Appl. Biomater.) 2005, 73B, 117-122.
[490] Amjad, Z. Performance of polymeric additives as HA crystal growth inhibitors. *Phosphorus Res. Bull.* 1995, 5, 1-12.
[491] Kamitahara, M., Kawashita, M., Kokubo, T., Nakamura, T. Effect of polyacrylic acid on the apatite formation of a bioactive ceramic in a simulated body fluid: fundamental examination of the possibility of obtaining bioactive glass-ionomer cements for orthopedic use. *Biomaterials* 2001, 22, 3191-3196.
[492] Wang, X., Li, Y., Wei, J., de Groot, K. Development of biomimetic nano-hydroxyapatite/poly(hexamethylene adipamide) composites. *Biomaterials* 2002, 23, 4787-4791.
[493] Sinha, A., Nayar, S., Agrawak, A.C. Synthesis of nanosized and microporous precipitated hydroxyapatite in synthetic polymers and biopolymers. *J. Am. Ceram. Soc.* 2003, 86, 357-359.
[494] Liao, S., Watari, F., Zhu, Y., Uo, M., Akasaka, T., Wang, W., Xu, G., Cui, F. The degradation of the three layered nano-carbonated hydroxyapatite/collagen/PLGA composite membrane *in vitro*. *Dental Mater.* 2007, 23, 1120-1128.
[495] Gonzalez-McQuire, R., Chane-Ching, J.Y., Vignaud, E., Lebugle, A., Mann, S. Synthesis and characterization of amino acid-functionalized hydroxyapatite nanorods. *J. Mater. Chem.* 2004, 14, 2277-2281.
[496] Rosseeva, E.V., Golovanova, O.A., Frank-Kamenetskaya, O.V. The influence of amino acids on the formation of nanocrystalline hydroxyapatite. *Glass Phys. Chem.* 2007, 33, 283-286.
[497] Zhan, J., Tseng, Y.H., Chan, J.C.C., Mou, C.Y. Biomimetic formation of hydroxyapatite nanorods by a single-crystal-to-single-crystal transformation. *Adv. Funct. Mater.* 2005, 15, 2005-2010.
[498] Xu, A.W., Ma, Y., Cölfen, H. Biomimetic mineralization. *J. Mater. Chem.* 2007, 17, 415-449.
[499] Pileni, M. The role of soft colloidal templates in controlling the size and shape of inorganic nanocrystals. *Nature Mater.* 2003, 12, 145-150.
[500] Wu, Y., Bose, S. Nanocrystalline hydroxyapatite: micelle templated synthesis and characterization. *Langmuir* 2005, 21, 3232-3234.
[501] Wei, K., Lai, C., Wang, Y. Solvothermal synthesis of calcium phosphate nanowires under different pH conditions. *J. Macromolec. Sci. A* 2006, 43A, 1531-1540.
[502] Lai, C., Tang, S.Q., Wang, Y.J., Wei, K., Zhang, S.Y. Insight into shape control mechanism of calcium phosphate nanopartiles in reverse micelles solution. *Synth. React. Inorg. Met. Met. Org. Nano-Metal Chem.* 2005, 35, 717-725.
[503] Banerjee, A., Bandyopadhyay, A., Bose, S. Hydroxyapatite nanopowders: synthesis, densification and cell-materials interaction. *Mater. Sci. Eng. C* 2007, 27, 729-735.
[504] Han, J.Y., Tan, T.T.Y., Loo, J.S.C. Utilizing inverse micelles to synthesize calcium phosphate nanoparticles as nano-carriers. *J. Nanoparticle Res.* 2011, 13, 3441-3454.

[505] Shchukin, D.G., Sukhorukov, G.B., Möhwald, H. Biomimetic fabrication of nanoengineered hydroxyapatite/polyelectrolyte composite shell. *Chem. Mater. 2003*, 15, 3947-3950.
[506] Mateus, A.Y.P., Ferraz, M.P., Monteiro, F.J. Microspheres based on hydroxyapatite nanoparticles aggregates for bone regeneration. *Key Eng. Mater.* 2007, 330-332, 243-246.
[507] Cai, Y., Liu, Y., Yan, W., Hu, Q., Tao, J., Zhang, M., Shi, Z., Tang, R. Role of hydroxyapatite nanoparticle size in bone cell proliferation. *J. Mater. Chem.* 2007, 17, 3780-3787.
[508] Mossaad, C., Tan, M.C., Starr, M., Payzant, E.A., Howe, J.Y., Riman, R.E. Size-dependent crystalline to amorphous uphill phase transformation of hydroxyapatite nanoparticles. *Cryst. Growth Des.* 2011, 11, 45-52.
[509] Quantum dots, first developed in the early 1980's, are crystalline semi-conducting nanodimensional particles comprised of a metalloid crystalline core and a "cap" or "shell" that shields the core and renders the dots biologically compatible. They are used or being developed for use in electronics, biomedical imaging and surveillance.
[510] Guo, Y., Shi, D., Lian, J., Dong, Z., Wang, W., Cho, H., Liu, G., Wang, L., Ewing, R.C. Quantum dot conjugated hydroxylapatite nanoparticles for *in vivo* imaging. *Nanotechnology* 2008, 19, 175102 (6 pages).
[511] Liu, Q., de Wijn, J.R., de Groot, K., van Blitterswijk, C.A. Surface modification of nano-apatite by grafting organic polymer. *Biomaterials* 1998, 19, 1067-1072.
[512] Palazzo, B., Iafisco, M., Laforgia, M., Margiotta, N., Natile, G., Bianchi, C.L., Walsh, D., Mann, S., Roveri, N. Biomimetic hydroxyapatite-drug nanocrystals as potential bone substitutes with antitumor drug delivery properties. *Adv. Funct. Mater.* 2007, 17, 2180-2188.
[513] Lee, H.J., Choi, H.W., Kim, K.J., Lee, S.C. Modification of hydroxyapatite nanosurfaces for enhanced colloidal stability and improved interfacial adhesion in nanocomposites. *Chem. Mater.* 2006, 18, 5111-5118.
[514] Lee, S.C., Choi, H.W., Lee, H.J., Kim, K.J., Chang, J.H., Kim, S.Y., Choi, J., Oh, K.S., Jeong, Y.K. *In-situ* synthesis of reactive hydroxyapatite nanocrystals for a novel approach of surface grafting polymerization. *J. Mater. Chem.* 2007, 17, 174-180.
[515] Li, L., Liu, Y.K., Tao, J.H., Zhang, M., Pan, H.H., Xu, X.R., Tang, R.K. Surface modification of hydroxyapatite nanocrystallite by a small amount of terbium provides a biocompatible fluorescent probe. *J. Phys. Chem. C* 2008, 112, 12219-12224.
[516] Neumeier, M., Hails, L.A., Davis, S.A., Mann, S., Epple, M. Synthesis of fluorescent core-shell hydroxyapatite nanoparticles. *J. Mater. Chem.* 2011, 21, 1250-1254.
[517] Wang, W., Shi, D., Lian, J., Guo, Y., Liu, G., Wang, L., Ewing, R.C. Luminescent hydroxylapatite nanoparticles by surface functionalization. *Appl. Phys. Lett.* 2006, 89, 183106 (3 pages).
[518] Liu, H., Xi, P., Xie, G., Chen, F., Li, Z., Bai, D., Zeng, Z. Biocompatible hydroxyapatite nanoparticles as a redox luminescence switch. *J. Biol. Inorg. Chem.* 2011;16, 1135-1140.
[519] Bow, J.S., Liou, S.C., Chen, S.Y. Structural characterization of room-temperature synthesized nanosized β-tricalcium phosphate. *Biomaterials* 2004, 25, 3155-3161.

[520] Brunner, T.J., Bohner, M., Dora, C., Gerber, C., Stark, W.J. Comparison of amorphous TCP nanoparticles to micron-sized α-TCP as starting materials for calcium phosphate cements. *J. Biomed. Mater. Res. B (Appl. Biomater.)* 2007, 83B, 400-407.

[521] Brunner, T.J., Grass, R.N., Bohner, M., Stark, W.J. Effect of particle size, crystal phase and crystallinity on the reactivity of tricalcium phosphate cements for bone reconstruction. *J. Mater. Chem.* 2007, 17, 4072-4078.

[522] Döbelin, N., Brunner, T.J., Stark, W.J., Eggimann, M., Fisch, M., Bohner, M. Phase evolution of thermally treated amorphous tricalcium phosphate nanoparticles. *Key Eng. Mater.* 2009, 396-398, 595-598.

[523] Bohner, M., Brunner, T.J., Döbelin, N., Tang, R., Stark, W.J. Effect of thermal treatments on the reactivity of nanosized tricalcium phosphate powders. *J. Mater. Chem.* 2008, 18, 4460-4467.

[524] Liu, Y.H., Zhang, S.M., Liu, L., Zhou, W., Hu, W., Li, J., Qiu, Z.Y. Rapid wet synthesis of nano-sized β-TCP by using dialysis. *Key Eng. Mater.* 2007, 330-332, 199-202.

[525] Abdel-Fattah, W.I., Reicha, F.M., Elkhooly, T.A. Nano-beta-tricalcium phosphates synthesis and biodegradation: 1. Effect of microwave and SO_4^{2-} ions on β-TCP synthesis and its characterization. *Biomed. Mater.* 2008, 3, 034121 (13 pages).

[526] Sanosh, K.P., Chu, M.C., Balakrishnan, A., Kim, T.N., Cho, S.J. Sol-gel synthesis of pure nano sized β-tricalcium phosphate crystalline powders. *Curr. Appl. Phys.* 2010, 10, 68-71.

[527] Dasgupta, S., Bandyopadhyay, A., Bose, S. Reverse micelle-mediated synthesis of calcium phosphate nanocarriers for controlled release of bovine serum albumin. *Acta Biomater.* 2009, 5, 3112-3121.

[528] Xia, C., Deng, X., Lin, Y.H., Nan, C.W. Preparation and characterisation of nano-sized beta-tricalcium phosphate with a ps template method. Int. *J. Mater. Product Technol.* 2010, 37, 257-262.

[529] Choi, D., Kumta, P.N. Mechano-chemical synthesis and characterization of nanostructured β-TCP powder. *Mater. Sci. Eng.* C 2007, 27, 377-381.

[530] Nikcević, I., Maravić, D., Ignjatović, N., Mitrić, M., Makoveć, D., Uskoković, D. The formation and characterization of nanocrystalline phases by mechanical milling of biphasic calcium phosphate/poly-L-lactide biocomposite. *Mater. Transact.* 2006, 47, 2980-2986.

[531] Cho, J.S., Jung, D.S., Han, J.M., Kang, Y.C. Nano-sized α and β-TCP powders prepared by high temperature flame spray pyrolysis. *Mater. Sci. Eng. C* 2009, 29, 1288-1292.

[532] Boutinguiza, M., Pou, J., Lusquiños, F., Comesaña, R., Riveiro, A. Laser-assisted production of tricalcium phosphate nanoparticles from biological and synthetic hydroxyapatite in aqueous medium. *Appl. Surf. Sci.* 2011, 257, 5195-5199.

[533] Jalota, S., Tas, A.C., Bhaduri, S.B. Microwave-assisted synthesis of calcium phosphate nanowhiskers. *J. Mater. Res.* 2004, 19, 1876-1881.

[534] Rameshbabu, N., Rao, K.P. Microwave synthesis, characterization and in-vitro evaluation of nanostructured biphasic calcium phosphates. *Curr. Appl. Phys.* 2009, 9, S29-S31.

[535] Li, B., Chen, X., Guo, B., Wang, X., Fan, H., Zhang, X. Fabrication and cellular biocompatibility of porous carbonated biphasic calcium phosphate ceramics with a nanostructure. *Acta Biomater.* 2009, 5, 134-143.
[536] Guha, A.K., Singh, S., Kumaresan, R., Nayar, S., Sinha, A. Mesenchymal cell response to nanosized biphasic calcium phosphate composites. *Colloids Surf.* B 2009, 73, 146-151.
[537] Layrolle, P., Lebugle, A. Synthesis in pure ethanol and characterization of nanosized calcium phosphate fluoroapatite. *Chem. Mater.* 1996, 8, 134-144.
[538] Andres, C., Sinani, V., Lee, D., Gun'ko, Y., Kotov, N. Anisotropic calcium phosphate nanoparticles coated with 2-carboxyethylphosphonic acid. *J. Mater. Chem.* 2006, 16, 3964-3968.
[539] Lim, H., Kassim, A., Huang, N., Hashim, R., Radiman, S., Khiew, P., Chiu, W. Fabrication and characterization of 1D brushite nanomaterials via sucrose ester reverse microemulsion. *Ceram. Int.* 2009, 35, 2891-2897.
[540] Shirkhanzadeh, M., Sims, S. Immobilization of calcium phosphate nano-clusters into alkoxy-derived porous TiO_2 coatings. *J. Mater. Sci. Mater. Med.* 1997, 8, 595-601.
[541] Schmidt, H.T., Ostafin, A.E. Liposome directed growth of calcium phosphate nanoshells. *Adv. Mater.* 2002, 14, 532-535.
[542] Schmidt, H.T., Gray, B.L., Wingert, P.A., Ostafin, A.E. Assembly of aqueous-cored calcium phosphate nanoparticles for drug delivery. *Chem. Mater.* 2004, 16, 4942-4947.
[543] Yeo, C.H., Zein, S.H.S., Ahmad, A.L., McPhail, D.S. Comparison of DOPA and DPPA liposome templates for the synthesis of calcium phosphate nanoshells. *Ceram. Int.* 2012, 38, 561-570.
[544] Xu, H.H.K., Sun, L., Weir, M.D., Antonucci, J.M., Takagi, S., Chow, L.C., Peltz, M. Nano DCPA – whisker composites with high strength and Ca and PO_4 release. *J. Dent. Res.* 2006, 85, 722-727.
[545] Xu, H.H.K., Weir, M.D., Sun, L., Takagi, S., Chow, L.C. Effects of calcium phosphate nanoparticles on Ca-PO_4 composite. *J. Dent. Res.* 2007, 86, 378-383.
[546] Xu, H.H.K., Weir, M.D., Sun, L. Nanocomposites with Ca and PO_4 release: effects of reinforcement, dicalcium phosphate particle size and silanization. *Dental Mater.* 2007, 23, 1482-1491.
[547] Singh, S., Bhardwaj, P., Singh, V., Aggarwal, S., Mandal, U.K. Synthesis of nanocrystalline calcium phosphate in microemulsion – effect of nature of surfactants. *J. Colloid Interf. Sci.* 2008, 319, 322-329.
[548] Wals, D., Mann, S. Chemical synthesis of microskeletal calcium phosphate in bicontinuous microemulsions. *Chem. Mater.* 1996, 8, 1944-1953.
[549] Ma, Z., Chen, F., Zhu, Y.J., Cui, T., Liu, X.Y. Amorphous calcium phosphate/poly(D,L-lactic acid) composite nanofibers: electrospinning preparation and biomineralization. *J. Coll. Interf. Sci.* 2011, 15, 371-379.
[550] Urch, H., Vallet-Regí, M., Ruiz, L., Gonzalez-Calbet, J.M., Epple, M. Calcium phosphate nanoparticles with adjustable dispersability and crystallinity. *J. Mater. Chem.* 2009, 19, 2166-2171.
[551] Holt, C., Wahlgren, N.M., Drakenberg, T. Ability of a β-casein phosphopeptide to modulate the precipitation of calcium phosphate by forming amorphous dicalcium phosphate nanoclusters. *Biochem. J.* 1996, 314, 1035-1039.

[552] Holt, C., Timmins, P.A., Errington, N., Leaver, J. A core-shell model of calcium phosphate nanoclusters stabilized by β-casein phosphopeptides, derived from sedimentation equilibrium and small-angle X-ray and neutron-scattering measurements. *Eur. J. Biochem.* 1998, 252, 73-78.
[553] Duan, B., Wang, M., Zhou, W.Y., Cheung, W.L. Synthesis of Ca-P nanoparticles and fabrication of Ca-P/PHBV nanocomposite microspheres for bone tissue engineering applications. *Appl. Surf. Sci.* 2008, 255, 529-533.
[554] Hwang, K.S., Jeon, K.O., Jeon, Y.S., Kim, B.H. Hydroxyapatite forming ability of electrostatic spray pyrolysis derived calcium phosphate nano powder. *J. Mater. Sci.* 2006, 41, 4159-4162.
[555] Hwang, K.S., Jeon, K.O., Jeon, Y.S., Kim, B.H. Hydroxyapatite forming ability of electrostatic spray pyrolysis derived calcium phosphate nano powder. *J. Mater. Sci. Mater. Med.* 2007, 18, 619-622.
[556] Perkin, K.K., Turner, J.L., Wooley, K.L., Mann, S. Fabrication of hybrid nanocapsules by calcium phosphate mineralization of shell cross-linked polymer micelles and nanocages. *Nano Lett.* 2005, 5, 1457-1461.
[557] Tjandra, W., Ravi, P., Yao, J., Tam, K.C. Synthesis of hollow spherical calcium phosphate nanoparticles using polymeric nanotemplates. *Nanotechnology* 2006, 17, 5988-5994.
[558] Sadasivan, S., Khushalani, D., Mann, S. Synthesis of calcium phosphate nanofilaments in reverse micelles. *Chem. Mater.* 2005, 17, 2765-2770.
[559] Morgan, T.T., Muddana, H.S., Altinoglu, E.I., Rouse, S.M., Tabakovic, A., Tabouillot, T., Russin, T.J., Butler, P.J., Eklund, P., Yun, J.K., Kester, M., Adair, J.H. Encapsulation of organic molecules in calcium phosphate nanocomposite particles for intracellular imaging and drug delivery. *Nano Lett.* 2008, 8, 4108-4115.
[560] Lai, C., Wang, Y.J., Wei, K. Nucleation kinetics of calcium phosphate nanoparticles in reverse micelle solution. *Colloids Surf. A* 2008, 315, 268-274.
[561] Yang, X., Gao, X., Gan, Y., Gao, C., Zhang, X., Ting, K., Wu, B.M., Gou, Z. Facile synthesis of octacalcium phosphate nanobelts: growth mechanism and surface adsorption properties. *J. Phys. Chem.* C 2010, 114, 6265-6271.
[562] Socol, G., Torricelli, P., Bracci, B., Iliescu, M., Miroiu, F., Bigi, A., Werckmann, J., Mihailescu, I.N. Biocompatible nanocrystalline octacalcium phosphate thin films obtained by pulsed laser deposition. *Biomaterials* 2004, 25, 2539-2545.
[563] Cho, J.S., Ko, Y.N., Koo, H.Y., Kang, Y.C. Synthesis of nano-sized biphasic calcium phosphate ceramics with spherical shape by flame spray pyrolysis. *J. Mater. Sci. Mater. Med.* 2010, 21, 1143-1149.
[564] Farzadi, A., Solati-Hashjin, M., Tahmasebi-Birgani, Z., Aminian, A. Microwave-assisted synthesis and characterization of biphasic calcium phosphate nanopowders. *Ceram. Transact.* 2010, 218, 59-65.
[565] Farzadi, A., Solati-Hashjin, M., Bakhshi, F., Aminian, A. Synthesis and characterization of hydroxyapatite/β-tricalcium phosphate nanocomposites using microwave irradiation. *Ceram. Int.* 2011, 37, 65-71.
[566] Pan, L., Li, Y., Zou, C., Weng, W., Cheng, K., Song, C., Du, P., Zhao, G., Shen, G., Wang, J., Han, G. Surface modification of nanosized biphasic α-TCP/HA powders. *Key Eng. Mater.* 2007, 330-332, 223-226.

[567] Urch, H., Franzka, S., Dahlhaus, D., Hartmann, N., Hasselbrink, E., Epple, M. Preparation of two-dimensionally patterned layers of functionalised calcium phosphate nanoparticles by laser direct writing. *J. Mater. Chem.* 2006, 16, 1798-1802.
[568] Sokolova, V., Prymak, O., Meyer-Zaika, W., Cölfen, H., Rehage, H., Shukla, A., Epple, M. Synthesis and characterization of DNA functionalized calcium phosphate nanoparticles. *Mat.-Wiss. u. Werkstofftech.* 2006, 37, 441-445.
[569] Muddana, H.S., Morgan, T.T., Adair, J.H., Butler, P.J. Photophysics of Cy3-encapsulated calcium phosphate nanoparticles. *Nano Lett.* 2009, 9, 1559-1566.
[570] Altinoğlu, E.I., Russin, T.J., Kaiser, J.M., Barth, B.M., Eklund, P.C., Kester, M., Adair, J.H. Near-infrared emitting fluorophore-doped calcium phosphate nanoparticles for *in vivo* imaging of human breast cancer. *ACS Nano* 2008, 2, 2075-2084.
[571] Schwiertz, J., Wiehe, A., Gräfe, S., Gitter, B., Epple, M. Calcium phosphate nanoparticles as efficient carriers for photodynamic therapy against cells and bacteria. *Biomaterials* 2009, 30, 3324-3331.
[572] Chen, F., Zhu, Y.J., Zhang, K.H., Wu, J., Wang, K.W., Tang, Q.L., Mo, X.M. Europium-doped amorphous calcium phosphate porous nanospheres: preparation and application as luminescent drug carriers. *Nanoscale Res. Lett.* 2011, 6, 1-9.
[573] Schwiertz, J., Meyer-Zaika, W., Ruiz-Gonzalez, L., González-Calbet, J.M., Vallet-Regí, M., Epple, M. Calcium phosphate nanoparticles as templates for nanocapsules prepared by the layer-by-layer technique. *J. Mater. Chem.* 2008, 18, 3831-3834.
[574] Hayakawa, S., Li, Y., Tsuru, K., Osaka, A., Fujii, E., Kawabata, K. Preparation of nanometer-scale rod array of hydroxyapatite crystal. *Acta Biomater.* 2009, 5, 2152-2160.
[575] Liao, S.S., Cui, F.Z., Zhang, W., Feng, Q.L. Hierarchically biomimetic bone scaffold materials: nano-HA/collagen/PLA composite. *J. Biomed. Mater. Res. B (Appl. Biomater.)* 2004, 69B, 158-165.
[576] Thomas, V., Dean, D.R., Jose, M.V., Mathew, B., Chowdhury, S., Vohra, Y.K. Nanostructured biocomposite scaffolds based on collagen co-electrospun with nanohydroxyapatite. *Biomacromolecules* 2007, 8, 631-637.
[577] de Yoreo, J.J., Vekilov, P.G. Principles of crystal nucleation and growth. *Rev. Mineral. Geochem.* 2003, 54, 57-93.
[578] Liao, S., Xu, G., Wang, W., Watari, F., Cui, F., Ramakrishna, S., Chan, C.K. Self-assembly of nano-hydroxyapatite on multi-walled carbon nanotubes. *Acta Biomater.* 2007, 3, 669-675.
[579] Penn, R.L., Banfield, J.F. Imperfect oriented attachment: dislocation generation in defect-free nanocrystals. *Science* 1998, 281, 969-971.
[580] Tao, J., Pan, H., Zeng, Y., Xu, X., Tang, R. Roles of amorphous calcium phosphate and biological additives in the assembly of hydroxyapatite nanoparticles. *J. Phys. Chem. B* 2007, 111, 13410-13418.
[581] Hing, K.A. Bone repair in the twenty-first century: biology, chemistry or engineering? *Philos. Trans. R. Soc. Lond. A* 2004, 362, 2821-2850.
[582] Kokubo, T., Kim, H.M., Kawashita, M. Novel bioactive materials with different mechanical properties. *Biomaterials* 2003, 24, 2161-2175.
[583] Fu, J.M., Miao, B., Jia, L.H., Lü, K.L. Nano-hydroxyapatite for repair of rabbit jaw bone defect: bone mineral density analysis. *J. Clin. Rehabil. Tissue Eng. Res.* 2009, 13, 2387-2390.

[584] Zhou, H., Lee, J. Nanoscale hydroxyapatite particles for bone tissue engineering. *Acta Biomater.* 2011, 7, 2769-2781.
[585] Barralet, J.E., Lilley, K.J., Grover, L.M., Farrar, D.F., Ansell, C., Gbureck, U. Cements from nanocrystalline hydroxyapatite. *J. Mater. Sci. Mater. Med.* 2004, 15, 407-411.
[586] Lilley, K.J., Gbureck, U., Wright, A.J., Farrar, D.F., Barralet, J.E. Cement from nanocrystalline hydroxyapatite: effect of calcium phosphate ratio. *J. Mater. Sci. Mater. Med.* 2005, 16, 1185-1190.
[587] Neira, I.S., Kolen'ko, Y.V., Lebedev, O.I., van Tendeloo, G., Gupta, H.S., Matsushita, N., Yoshimura, M., Guitián, F. Rational synthesis of a nanocrystalline calcium phosphate cement exhibiting rapid conversion to hydroxyapatite. *Mater. Sci. Eng. C* 2009, 29, 2124-2132.
[588] Dorozhkin, S.V. Calcium orthophosphate cements for biomedical application. *J. Mater. Sci.* 2008, 43, 3028-3057.
[589] Dorozhkin, S.V. Calcium orthophosphate cements and concretes. *Materials* 2009, 2, 221-291.
[590] Fu, Q., Zhou, N., Huang, W., Wang, D., Zhang, L., Li, H. Effects of nano HAP on biological and structural properties of glass bone cement. *J. Biomed. Mater. Res. A* 2005, 74A, 156-163.
[591] Strnadova, M., Protivinsky, J., Strnad, J., Vejsicka, Z. Preparation of porous synthetic nanostructured HA scaffold. *Key Eng. Mater.* 2008, 361-363, 211-214.
[592] Kim, J.Y., Lee, J.W., Lee, S.J., Park, E.K., S.Y. Kim, Cho, D.W. Development of a bone scaffold using HA nanopowder and micro-stereolithography technology. *Microelectronic Engineering* 2007, 84, 1762-1765.
[593] Severin, A.V., Komarov, V.F., Bozhevol'nov, V.E., Melikhov, I.V. Morphological selection in suspensions of nanocrystalline hydroxylapatite leading to spheroidal aggregates. *Russ. J. Inorg. Chem.* 2005, 50, 72-77.
[594] Krylova, I.V., Ivanov, L.N., Bozhevol'nov, V.E., Severin, A.V. Self-organization processes and phase transitions in nanocrystalline hydroxyapatite according to exoemission data. *Russ. J. Phys. Chem. A* 2007, 81, 241-245.
[595] Veljovic, D., Jokic, B., Jankovic-Castvan, I., Smiciklas, I., Petrovic, R., Janackovic, D. Sintering behaviour of nanosized HAP powder. *Key Eng. Mater.* 2007, 330-332, 259-262.
[596] Zhang, F., Lin, K., Chang, J., Lu, J., Ning, C. Spark plasma sintering of macroporous calcium phosphate scaffolds from nanocrystalline powders. *J. Eur. Ceram. Soc.* 2008, 28, 539-545.
[597] Grossin, D., Banu, M., Sarda, S., Martinet-Rollin, S., Drouet, C., Estournès, C., Champion, E., Rossignol, F., Combes, C., Rey, C. Low temperature consolidation of nanocrystalline apatites toward a new generation of calcium phosphate ceramics. *Ceram. Eng. Sci. Proc.* 2010, 30, 113-126.
[598] Chaudhry, A.A., Yan, H., Gong, K., Inam, F., Viola, G., Reece, M.J., Goodall, J.B.M., ur Rehman, I., McNeil-Watson, F.K., Corbett, J.C.W., Knowles, J.C, Darr, J.A. High-strength nanograined and translucent hydroxyapatite monoliths via continuous hydrothermal synthesis and optimized spark plasma sintering. *Acta Biomater.* 2011, 7, 791-799.

[599] Eriksson, M., Liu, Y., Hu, J., Gao, L., Nygren, M., Shen, Z. Transparent hydroxyapatite ceramics with nanograin structure prepared by high pressure spark plasma sintering at the minimized sintering temperature. *J. Eur. Ceram. Soc.* 2011, 31, 1533-1540.

[600] Kutty, M.G., Loertscher, J., Bhaduri, S., Bhaduri, S.B., Tinga, W.R. Microwave sintering of nanocrystalline hydroxyapatite. *Ceram. Eng. Sci. Proc.* 2001, 22, 3-10.

[601] Vijayan, S., Varma, H. Microwave sintering of nanosized hydroxyapatite powder compacts. *Mater. Lett.* 2002, 56, 827-831.

[602] Ramesh, S., Tan, C.Y., Bhaduri, S.B., Teng, W.D. Rapid densification of nanocrystalline hydroxyapatite for biomedical applications. *Ceram. Int.* 2007, 33, 1363-1367.

[603] Okada, M., Furuzono, T. Fabrication of high-dispersibility nanocrystals of calcined hydroxyapatite. *J. Mater. Sci. 2006*, 41, 6134-6137.

[604] Okada, M., Furuzono, T. Nanosized ceramic particles of hydroxyapatite calcined with an anti-sintering agent. *J. Nanosci. Nanotechnol.* 2007, 7, 848-851.

[605] Okada, M., Furuzono, T. Calcination of rod-like hydroxyapatite nanocrystals with an anti-sintering agent surrounding the crystals. *J. Nanopart. Res.* 2007, 9, 807-815.

[606] Müller-Mai, C.M., Stupp, S.I., Voigt, C., Gross, U. Nanoapatite and organoapatite implants in bone: histology and ultrastructure of the interface. *J. Biomed. Mater. Res.* 1995, 29, 9-18.

[607] Du, C., Cui, F.Z., Feng, Q.L., Zhu, X.D., de Groot, K. Tissue response to nano-hydroxyapatite/collagen composite implants in marrow cavity. *J. Biomed. Mater. Res.* 1998, 42, 540-548.

[608] Du, C., Cui, F.Z., Zhu, X.D., de Groot, K. Three-dimensional nano-HAp/collagen matrix loading with osteogenic cells in organ culture. *J. Biomed. Mater. Res.* 1999, 44, 407-415.

[609] Paul, W., Sharma, C.P. Nanoceramic matrices: biomedical applications. *Am. J. Biochem. Biotechnol.* 2006, 2, 41-48.

[610] Huber, F.X., McArthur, N., Hillmeier, J., Kock, H.J., Baier, M., Diwo, M., Berger, I., Meeder, P.J. Void filling of tibia compression fracture zones using a novel resorbable nanocrystalline hydroxyapatite paste in combination with a hydroxyapatite ceramic core: first clinical results. *Arch. Orthop. Trauma Surg.* 2006, 126, 533-540.

[611] Smeets, R., Jelitte, G., Heiland, M., Kasaj, A., Grosjean, M., Riediger, D., Yildirim, M., Spiekermann, H., Maciejewski, O. Hydroxylapatit-Knochenersatzmaterial (Ostim®) bei der Sinusbodenelevation. *Schweiz Monatsschr. Zahnmed.* 2008, 118, 203-208.

[612] Gerlach, K.L., Niehues, D. Die Behandlung der Kieferzysten mit einem neuartigen nanopartikulären Hydroxylapatit. *Mund Kiefer GesichtsChir.* 2007, 11, 131-137.

[613] Schwarz, F., Bieling, K., Latz, T., Nuesry, E., Becker, J. Healing of intrabony periimplantitis defects following application of a nanocristalline hydroxyapatite (Ostim™) or a bovine-derived xenograft (Bio-Oss™) in combination with a collagen membrane (Bio-Gide™). A case series. *J. Clin. Periodontol.* 2006, 33, 491-499.

[614] Strietzel, F.P., Reichart, P.A., Graf, H.L. Lateral alveolar ridge augmentation using a synthetic nano-crystalline hydroxyapatite bone substitution material (Ostim®). Preliminary clinical and histological results. *Clin. Oral Implants Res.* 2007, 18, 743-751.

[615] Spies, C., Schnürer, S., Gotterbarm, T., Breusch, S. Tierexperimentelle Untersuchung des Knochenersatzstoffs Ostim™ im knöchernen Lager des Göttinger Miniaturschweins. *Z. Orthop. Unfall.* 2008, 146, 64-69.
[616] Thorwarth, M., Schultze-Mosgau, S., Kessler, P., Wiltfang, J., Schlegel, K.A. Bone regeneration in osseous defects using a resorbable nanoparticular hydroxyapatite. *J. Oral Maxillofac. Surg.* 2005, 63, 1626-1633.
[617] Brandt, J., Henning, S., Michler, G., Schulz, M., Bernstein, A. Nanocrystalline hydroxyapatite for bone repair. *Key Eng. Mater.* 2008, 361-363, 35-38.
[618] Huber, F.X., Hillmeier, J., Herzog, L., McArthur, N., Kock, H.J., Meeder, P.J. Open reduction and palmar plate-osteosynthesis in combination with a nanocrystalline hydroxyapatite spacer in the treatment of comminuted fractures of the distal radius. *J. Hand Surg. (Brit.)* 2006, 31B, 298-303.
[619] Huber, F.X., Hillmeier, J., McArthur, N., Kock, H.J., Meeder, P.J. The use of nanocrystalline hydroxyapatite for the reconstruction of calcaneal fractures: preliminary results. *J. Foot Ankle Surg.* 2006, 45, 322-328.
[620] Laschke, M.W., Witt, K., Pohlemann, T., Menger, M.D. Injectable nanocrystalline hydroxyapatite paste for bone substitution: *in vivo* analysis of biocompatibility and vascularization. *J. Biomed. Mater. Res. B (Appl. Biomater.)* 2007, 82B, 494-505.
[621] Spies, C.K.G., Schnürer, S., Gotterbarm, T., Breusch, S. The efficacy of Biobon™ and Ostim™ within metaphyseal defects using the Göttinger Minipig. *Arch. Orthop. Trauma Surg.* 2009, 129, 979-988.
[622] Chitsazi, M.T., Shirmohammadi, A., Faramarzie, M., Pourabbas, R., Rostamzadeh, A.N. A clinical comparison of nano-crystalline hydroxyapatite (Ostim) and autogenous bone graft in the treatment of periodontal intrabony defects. Medicina Oral, Patologia Oral y Cirugia Bucal 2011, 16, 448-453.
[623] Huber, F.X., Belyaev, O., Hillmeier, J., Kock, H.J., Huber, C., Meeder, P.J., Berger, I. First histological observations on the incorporation of a novel nanocrystalline hydroxyapatite paste OSTIM® in human cancellous bone. BMC Musculoskelett. Disord. 2006, 7, 50 (14 pages).
[624] Huber, F.X., Berger, I., McArthur, N., Huber, C., Kock, H.P., Hillmeier, J., Meeder, P.J. Evaluation of a novel nanocrystalline hydroxyapatite paste and a solid hydroxyapatite ceramic for the treatment of critical size bone defects (CSD) in rabbits. *J. Mater. Sci. Mater. Med.* 2008, 19, 33-38.
[625] Arts, J.J.C., Verdonschot, N., Schreurs, B.W., Buma, P. The use of a bioresorbable nano-crystalline hydroxyapatite paste in acetabular bone impaction grafting. *Biomaterials* 2006, 27, 1110-1118.
[626] Zhang, W., Liao, S.S., Cui, F.Z. Hierarchical self-assembly of nano-fibrils in mineralized collagen. *Chem. Mater.* 2003, 15, 3221-3226.
[627] Li, X., Huang, J., Edirisinghe, M.J. Development of nano-hydroxyapatite coating by electrohydrodynamic atomization spraying. *J. Mater. Sci. Mater. Med.* 2008, 19, 1545-1551.
[628] Guo, L., Li, H. Fabrication and characterization of thin nano-hydroxyapatite coatings on titanium. *Surf. Coat. Technol.* 2004, 185, 268-274.
[629] Thian, E.S., Ahmad, Z., Huang, J., Edirisinghe, M.J., Jayasinghe, S.N., Ireland, D.C., Brooks, R.A., Rushton, N., Bonfield, W., Best, S.M. Electrosprayed nanoapatite: a new generation of bioactive material. *Key Eng. Mater.* 2008, 361-363, 597-600.

[630] Han, Y., Xu, K., Montay, G., Fu, T., Lu, J. Evaluation of nanostructured carbonated hydroxyapatite coatings formed by a hybrid process of plasma spraying and hydrothermal synthesis. *J. Biomed. Mater. Res.* 2002, 60, 511-516.

[631] Li, P. Biomimetic nano-apatite coating capable of promoting bone ingrowth. *J. Biomed. Mater. Res. A* 2003, 66A, 79-85.

[632] Mendes, V.C., Moineddin, R., Davies, J.E. The effect of discrete calcium phosphate nanocrystals on bone-bonding to titanium surfaces. *Biomaterials* 2007, 28, 4748-4755.

[633] Oh, S.H., Finõnes, R.R., Daraio, C., Chen, L.H., Jin, S. Growth of nano-scale hydroxyapatite using chemically treated titanium oxide nanotubes. *Biomaterials* 2005, 26, 4938-4943.

[634] Ma, J., Wong, H., Kong, L.B., Peng, K.W. Biomimetic processing of nanocrystallite bioactive apatite coating on titanium. *Nanotechnology* 2003, 14, 619-623.

[635] Gu, Y.W., Tay, B.Y., Lim, C.S., Yong, M.S. Nanocrystallite apatite formation and its growth kinetics on chemically treated porous NiTi. *Nanotechnology* 2006, 17, 2212-2218.

[636] Hu, R., Lin, C.J., Shi, H.Y. A novel ordered nano hydroxyapatite coating electrochemically deposited on titanium substrate. *J. Biomed. Mater. Res. A* 2007, 80A, 687-692.

[637] Bigi, A., Boanini, E., Bracci, B., Facchini, A., Panzavolta, S., Segatti, F., Sturba, L. Nanocrystalline hydroxyapatite coatings on titanium: a new fast biomimetic method. *Biomaterials* 2005, 26, 4085-4089.

[638] Narayanan, R., Seshadri, S.K., Kwon, T.Y., Kim, K.H. Electrochemical nano-grained calcium phosphate coatings on Ti-6Al-4V for biomaterial applications. *Scripta Mater.* 2007, 56, 229-232.

[639] Thian, E.S., Huang, J., Best, S.M., Barber, Z.H., Bonfield, W. Nanostructured apatite coatings for rapid bone repair. *Key Eng. Mater.* 2006, 309-311, 519-522.

[640] Cai, X., Gong, P., Man, Y., Chen, Z., He, G. The construction and characterization of nano-FHA bioceramic coating on titanium surface. *Key Eng. Mater.* 2007, 330-332, 333-336.

[641] Citterio, H., Jakani, S., Benmarouane, A., Millet, P., Lodini, A. Nano-hydroxyapatite coatings on titanium substrates. Finite element analysis of process and experimental plasma thermal sprayed coatings. *Key Eng. Mater.* 2008, 361-363, 745-748.

[642] Lee, S.H., Kim, H.E., Kim, H.W. Nanosized hydroxyapatite coatings on Ti substrate with TiO_2 buffer layer by e-beam deposition. *J. Am. Ceram. Soc.* 2007, 90, 50-56.

[643] Nishimura, I., Huang, Y., Butz, F., Ogawa, T., Lin, A., Wang, C.J. Discrete deposition of hydroxyapatite nanoparticles on a titanium implant with predisposing substrate microtopography accelerated osseointegration. *Nanotechnology* 2007, 18, 245101 (9 pages).

[644] Narayanan, R., Kwon, T.Y., Kim, K.H. Preparation and characteristics of nano-grained calcium phosphate coatings on titanium from ultrasonated bath at acidic pH. *J. Biomed. Mater. Res. B (Appl. Biomater.)* 2008, 85B, 231-239.

[645] Hahn, B.D., Park, D.S., Choi, J.J., Ryu, J., Yoon, W.H., Kim, K.H., Park, C., Kim, H.E. Dense nanostructured hydroxyapatite coating on titanium by aerosol deposition. *J. Am. Ceram. Soc.* 2009, 92, 683-687.

[646] Narayanan, R., Kwon, T.Y., Kim, K.H. Direct nanocrystalline hydroxyapatite formation on titanium from ultrasonated electrochemical bath at physiological pH. *Mater. Sci. Eng*. C 2008, 28, 1265-1270.
[647] Yousefpour, M., Afshar, A., Yang, X., Li, X., Yang, B., Wu, Y., Chen, J., Zhang, X. Nano-crystalline growth of electrochemically deposited apatite coating on pure titanium. *J. Electroanal. Chem*. 2006, 589, 96-105.
[648] Mendes, V.C., Moineddin, R., Davies, J.E. Discrete calcium phosphate nanocrystalline deposition enhances osteoconduction on titanium-based implant surfaces. *J. Biomed. Mater. Res. A* 2009, 90A, 577-585.
[649] Yang, Y., Kim, K.H., Ong, J.L. A review on calcium phosphate coatings produced using a sputtering process – an alternative to plasma spraying. *Biomaterials* 2005, 26, 327-337.
[650] Nies, B., Rößler, S., Reinstorf, A. Formation of nano hydroxyapatite – a straightforward way to bioactivate bone implant surfaces. *Int. J. Mat. Res. (formerly Z. Metallkd.)* 2007, 98, 630-636.
[651] Jalota, S., Bhaduri, S.B., Tas, A.C. Effect of carbonate content and buffer type on calcium phosphate formation in SBF solutions. *J. Mater. Sci. Mater. Med.* 2006, 17, 697-707.
[652] Chen, F., Lam, W.M., Lin, C.J., Qiu, G.X., Wu, Z.H., Luk, K.D.K., Lu, W.W. Biocompatibility of electrophoretical deposition of nanostructured hydroxyapatite coating on roughen titanium surface: *in vitro* evaluation using mesenchymal stem cells. *J. Biomed. Mater. Res. B (Appl. Biomater.)* 2007, 82B, 183-191.
[653] Thian, E.S., Ahmad, Z., Huang, J., Edirisinghe, M.J., Jayasinghe, S.N., Ireland, D.C., Brooks, R.A., Rushton, N., Bonfield, W., Best, S.M. The role of electrosprayed nanoapatites in guiding osteoblast behaviour. *Biomaterials* 2008, 29, 1833-1843.
[654] Bigi, A., Nicoli-Aldini, N., Bracci, B., Zavan, B., Boanini, E., Sbaiz, F., Panzavolta, S., Zorzato, G., Giardino, R., Facchini, A., Abatangelo, G., Cortivo, R. *In vitro* culture of mesenchymal cells onto nanocrystalline hydroxyapatite coated Ti13Nb13Zr alloy. *J. Biomed. Mater. Res. A* 2007, 82A, 213-221.
[655] Bigi, A., Fini, M., Bracci, B., Boanini, E., Torricelli, P., Giavaresi, G., Aldini, N.N., Facchini, A., Sbaiz, F., Giardino, R. The response of bone to nanocrystalline hydroxyapatite-coated Ti13Nb11Zr alloy in an animal model. *Biomaterials* 2008, 29, 1730-1736.
[656] Thian, E.S., Huang, J., Ahmad, Z., Edirisinghe, M.J., Jayasinghe, S.N., Ireland, D.C., Brooks, R.A., Rushton, N., Best, S.M., Bonfield, W. Influence of nanohydroxyapatite patterns deposited by electrohydrodynamic spraying on osteoblast response. *J. Biomed. Mater. Res. A* 2008, 85A, 188-194.
[657] Furuzono, T., Masuda, M., Okada, M., Yasuda, S., Kadono, H., Tanaka, R., Miyatake, K. Increase in cell adhesiveness on a poly(ethylene terephthalate) fabric by sintered hydroxyapatite nanocrystal coating in the development of an artificial blood vessel. *ASAIO J.* 2006, 52, 315-320.
[658] Yanagida, H., Okada, M., Masuda, M., Ueki, M., Narama, I., Kitao, S., Koyama, Y., Furuzono, T., Takakuda, K. Cell adhesion and tissue response to hydroxyapatite nanocrystal-coated poly(L-lactic acid) fabric. *J. Biosci. Bioeng*. 2009, 108, 235-243.

[659] Li, X., Huang, J., Edirisinghe, M.J. Development of template-assisted electrohydrodynamic atomization spraying for nanoHA patterning. *Key Eng. Mater.* 2008, 361-363, 585-588.

[660] Shi, Z.L., Huang, X., Cai, Y.R., Tang, R.K., Yang, D.S. Size effect of hydroxyapatite nanoparticles on proliferation and apoptosis of osteoblast-like cells. *Acta Biomater.* 2009, 5, 338-345.

[661] Liu, Y., Wang, G., Cai, Y., Ji, H., Zhou, G., Zhao, X., Tang, R., Zhang, M. *In vitro* effects of nanophase hydroxyapatite particles on proliferation and osteogenic differentiation of bone marrow-derived mesenchymal stem cells. *J. Biomed. Mater. Res. A* 2009, 15, 1083-1091.

[662] Zhu, X., Eibl, O., Scheideler, L., Geis-Gerstorfer, J. Characterization of nano hydroxyapatite/collagen surfaces and cellular behaviors. *J. Biomed. Mater. Res. A* 2006, 79A, 114-127.

[663] Zhu, W., Zhang, X., Wang, D., Lu, W., Ou, Y., Han, Y., Zhou, K., Liu, H., Fen, W., Peng, L., He, C., Zeng, Y. Experimental study on the conduction function of nano-hydroxyapatite artificial bone. *Micro Nano Lett.* 2010, 5, 19-27.

[664] Wang, H., Li, Y., Zuo, Y., Li, J., Ma, S., Cheng, L. Biocompatibility and osteogenesis of biomimetic nano-hydroxyapatite/polyamide composite scaffolds for bone tissue engineering. *Biomaterials* 2007, 28, 3338-3348.

[665] Zhang, Y.F., Cheng, X.R., Chen, Y., Shi, B., Chen, X.H., Xu, D.X., Ke, J. Three-dimensional nanohydroxyapatite / chitosan scaffolds as potential tissue engineered periodontal tissue. *J. Biomater. Appl.* 2007, 21, 333-349.

[666] Huang, Y.X., Ren, J., Chen, C., Ren, T.B., Zhou, X.Y. Preparation and properties of poly(lactide-co-glycolide) (PLGA) / nano-hydroxyapatite (NHA) scaffolds by thermally induced phase separation and rabbit mscs culture on scaffolds. *J. Biomater. Appl.* 2008, 22, 409-432.

[667] Thian, E.S., Ahmad, Z., Huang, J., Edirisinghe, M.J., Jayasinghe, S.N., Ireland, D.C., Brooks, R.A., Rushton, N., Bonfield, W., Best, S.M. Bioactivity of nanoapatite produced by electrohydrodynamic atomization. *J. Bionanosci.* 2007, 1, 60-63.

[668] Pezzatini, S., Solito, R., Morbidelli, L., Lamponi, S., Boanini, E., Bigi, A., Ziche, M. The effect of hydroxyapatite nanocrystals on microvascular endothelial cell viability and functions. *J. Biomed. Mater. Res. A* 2006, 76A, 656-663.

[669] Pezzatini, S., Morbidelli, L., Solito, R., Paccagnini, E., Boanini, E., Bigi, A., Ziche, M. Nanostructured HA crystals up-regulate FGF-2 expression and activity in microvascular endothelium promoting angiogenesis. *Bone* 2007, 41, 523-534.

[670] Hu, Q., Tan, Z., Liu, Y., Tao, J., Cai, Y., Zhang, M., Pan, H., Xu, X., Tang, R. Effect of crystallinity of calcium phosphate nanoparticles on adhesion, proliferation, and differentiation of bone marrow mesenchymal stem cells. *J. Mater. Chem.* 2007, 17, 4690-4698.

[671] Kim, K., Dean, D., Lu, A., Mikos, A.G., Fisher, J.P. Early osteogenic signal expression of rat bone marrow stromal cells is influenced by both hydroxyapatite nanoparticle content and initial cell seeding density in biodegradable nanocomposite scaffolds. *Acta Biomater.* 2011, 7, 1249-1264.

[672] Svanborg, L.M., Hoffman, M., Andersson, M., Currie, F., Kjellin, P., Wennerberg, A. The effect of hydroxyapatite nanocrystals on early bone formation surrounding dental implants. *Int. J. Oral Maxillofac. Surg.* 2011, 40, 308-315.

[673] Balasundaram, G., Sato, M., Webster, T.J. Using hydroxyapatite nanoparticles and decreased crystallinity to promote osteoblast adhesion similar to functionalizing with RGD. *Biomaterials* 2006, 27, 2798-2805.

[674] Detsch, R., Hagmeyer, D., Neumann, M., Schaefer, S., Vortkamp, A., Wuelling, M., Ziegler, G., Epple, M. The resorption of nanocrystalline calcium phosphates by osteoclast-like cells. *Acta Biomater.* 2010, 6, 3223-3233.

[675] Stevens, M.M., George, J.H. Exploring and engineering the cell surface interface. *Science* 2005, 310, 1135-1138.

[676] Martínez, E., Engel, E., Planell, J.A., Samitier, J. Effects of artificial micro- and nano-structured surfaces on cell behaviour. *Annals Anat.* 2009, 191, 126-135.

[677] Lee, D.H., Han, J.S., Yang, J.H., Lee, J.B. MC3T3-E1 cell response to pure titanium, zirconia and nano-hydroxyapatite. *Int. J. Modern Phys. B* 2009, 23, 1535-1540.

[678] Onuma, K., Yamagishi, K., Oyane, A. Nucleation and growth of hydroxyapatite nanocrystals for nondestructive repair of early caries lesions. *J. Cryst. Growth* 2005, 282, 199-207.

[679] Huang, S., Gao, S., Cheng, L., Yu, H. Remineralization potential of nano-hydroxyapatite on initial enamel lesions: an *in vitro* study. *Caries Res.* 2011, 45, 460-468.

[680] Roveri, N., Battistella, E., Bianchi, C.L., Foltran, I., Foresti, E., Iafisco, M., Lelli, M., Naldoni, A., Palazzo, B., Rimondini, L. Surface enamel remineralization: biomimetic apatite nanocrystals and fluoride ions different effects. *J. Nanomater.* 2009, 746383 (9 pages).

[681] Lv, K., Zhang, J., Meng, X., Li, X. Remineralization effect of the nano-HA toothpaste on artificial caries. *Key Eng. Mater.* 2007, 330-332, 267-270.

[682] Jeong, S.H., Jang, S.O., Kim, K.N., Kwon, H.K., Park, Y.D., Kim, B.I. Remineralization potential of new toothpaste containing nano-hydroxyapatite. *Key Eng. Mater.* 2006, 309-311, 537-540.

[683] Tschoppe, P., Zandim, D.L., Martus, P., Kielbassa, A.M. Enamel and dentine remineralization by nano-hydroxyapatite toothpastes. *J. Dent.* 2011, 39, 430-437.

[684] Wang, C.J., Zhang, Y.F., Wei, J., Wei, S.C. Repair of artificial enamel lesions by nano fluorapatite paste containing fluorin. *J. Clin. Rehabil. Tiss. Eng. Res.* 2011, 15, 6346-6350.

[685] Kim, B.I., Jeong, S.H., Jang, S.O., Kim, K.N., Kwon, H.K., Park, Y.D. Tooth whitening effect of toothpastes containing nano-hydroxyapatite. *Key Eng. Mater.* 2006, 309-311, 541-544.

[686] Collares, F.M., Leitune, V.C.B., Rostirolla, F.V., Trommer, R.M., Bergmann, C.P., Samuel, S.M.W. Nanostructured hydroxyapatite as filler for methacrylate-based root canal sealers. *Int. Endodontic J.* 2012, 45, 63-67.

[687] Kim, M.Y., Kwon, H.K., Choi, C.H., Kim, B.I. Combined effects of nano-hydroxyapatite and NaF on remineralization of early caries lesion. *Key Eng. Mater.* 2007, 330-332, 1347-1350.

[688] Lee, H.J., Min, J.H., Choi, C.H., Kwon, H.G., Kim, B.I. Remineralization potential of sports drink containing nano-sized hydroxyapatite. *Key Eng. Mater.* 2007, 330-332, 275-278.

[689] Min, J.H., Kwon, H.K., Kim, B.I. The addition of nano-sized hydroxyapatite to a sports drink to inhibit dental erosion – *in vitro* study using bovine enamel. *J. Dent.* 2011, 39, 629-635.

[690] Hong, Y.W., Kim, J.H., Lee, B.H., Lee, Y.K., Choi, B.J., Lee, J.H., Choi, H.J. The effect of nano-sized β-tricalcium phosphate on remineralization in glass ionomer dental luting cement. *Key Eng. Mater.* 2008, 361-363, 861-864.

[691] Li, L., Pan, H.H., Tao, J.H., Xu, X.R., Mao, C.Y., Gu, X.H., Tang, R.K. Repair of enamel by using hydroxyapatite nanoparticles as the building blocks. *J. Mater. Chem.* 2008, 18, 4079-4084.

[692] Meng, X., Lv, K., Zhang, J., Qu, D. Caries inhibitory activity of the nano-HA *in vitro*. *Key Eng. Mater.* 2007, 330-332, 251-254.

[693] Li, B.G., Wang, J.P., Zhao, Z.Y., Sui, Y.F., Zhang, Y.X. Mineralizing of nano-hydroxyapatite powders on artificial caries. *Rare Metal. Mat. Eng.* 2007, 36, 128-130.

[694] Ashokan, A., Menon, D., Nair, S., Koyakutty, M. A molecular receptor targeted, hydroxyapatite nanocrystal based multi-modal contrast agent. *Biomaterials* 2010, 31, 2606-2616.

[695] Bauer, I.W., Li, S.P., Han, Y.C., Yuan, L., Yin, M.Z. Internalization of hydroxyapatite nanoparticles in liver cancer cells. *J. Mater. Sci. Mater. Med.* 2008, 19, 1091-1095.

[696] Liu, T., Tang, A., Zhang, G.Y., Chen, Y.X., Zhang, J.Y., Peng, S.S., Cai, Z.M. Calcium phosphate nanoparticles as a novel nonviral vector for efficient transfection of DNA in cancer gene therapy. *Cancer Biother. Radiopharm.* 2005, 20, 141-149.

[697] Czupryna, J., Tsourkas, A. Suicide gene delivery by calcium phosphate nanoparticles. A novel method of targeted therapy for gastric cancer. *Cancer Biol. Ther.* 2006, 5, 1691-1692.

[698] Li, B., Guo, B., Fan, H., Zhang, X. Preparation of nano-hydroxyapatite particles with different morphology and their response to highly malignant melanoma cells *in vitro*. *Appl. Surf. Sci.* 2008, 255, 357-360.

[699] Dai, H., Pei, C., Han, Y., Xinyu, W., Li, S. Inhibitory effect of hydroxyapatite nanoparticles on K562 cells. *Mater. Sci. Forum* 2011, 685, 352-356.

[700] Borum, L., Wilson, O.C. Surface modification of hydroxyapatite. Part II. Silica. *Biomaterials* 2003, 24, 3681-3688.

[701] Lee, H.J., Kim, S.E., Choi, H.W., Kim, C.W., Kim, K.J., Lee, S.C. The effect of surface-modified nano-hydroxyapatite on biocompatibility of poly(ε-caprolactone)/hydroxyapatite nanocomposites. *Eur. Polym. J.* 2007, 43, 1602-1608.

[702] Wilson, O.C., Hull, J.R. Surface modification of nanophase hydroxyapatite with chitosan. *Mater. Sci. Eng. C* 2008, 28, 434-437.

[703] Liao, J.G., Wang, X.J., Zuo, Y., Zhang, L., Wen, J.Q., Li, Y.B. Surface modification of nano-hydroxyapatite with silane agent. *J. Inorg. Mater.* 2008, 23, 145-149.

[704] Wang, Y., Xiao, Y., Huang, X., Lang, M. Preparation of poly(methyl methacrylate) grafted hydroxyapatite nanoparticles via reverse ATRP. *J. Coll. Interf. Sci.* 2011, 15, 415-421.

[705] Deng, C., Xiao, X., Yao, N., Yang, X.B., Weng, J. Effect of surface modification of nano-hydroxyapatite particles on *in vitro* biocompatibility of poly (ε-caprolactone)-matrix composite biomaterials. *Int. J. Polym. Mater.* 2011, 60, 969-978.

[706] Jensen, T., Baas, J., Dolathshahi-Pirouz, A., Jacobsen, T., Singh, G., Nygaard, J.V., Foss, M., Bechtold, J., Bünger, C., Besenbacher, F., Søballe, K. Osteopontin

functionalization of hydroxyapatite nanoparticles in a PDLLA matrix promotes bone formation. *J. Biomed. Mater. Res. A* 2011, 99A, 94-101.

[707] Ramachandran, R., Paul, W., Sharma, C.P. Synthesis and characterization of PEGylated calcium phosphate nanoparticles for oral insulin delivery. *J. Biomed. Mater. Res. B (Appl. Biomater.)* 2009, 88B, 41-48.

[708] Uskoković, V., Uskoković, D.P. Nanosized hydroxyapatite and other calcium phosphates: chemistry of formation and application as drug and gene delivery agents. *J. Biomed. Mater. Res. B (Appl. Biomater.)* 2010, 96B, 152-191.

[709] Loo, S.C., Moore, T., Banik, B., Alexis, F. Biomedical applications of hydroxyapatite nanoparticles. *Curr. Pharm. Biotechnol.* 2010, 11, 333-342.

[710] Fu, H., Hu, Y., McNelis, T., Hollinger, J.O. A calcium phosphate-based gene delivery system. *J. Biomed. Mater. Res. A* 2005, 74A, 40-48.

[711] Liu, T.Y., Chen, S.Y., Liu, D.M., Liou, S.C. On the study of BSA-loaded calcium-deficient hydroxyapatite nano-carriers for controlled drug delivery. *J. Control. Release* 2005, 107, 112-121.

[712] Barroug, A., Kuhn, L.T., Gerstenfeld, L.C., Glimcher, M.J. Interactions of cisplatin with calcium phosphate nanoparticles: *in vitro* controlled adsorption and release. *J. Orthop. Res.* 2004, 22, 703-708.

[713] Cheng, X.G., Kuhn, L.T. Chemotherapy drug delivery from calcium phosphate nanoparticles. *Int. J. Nanomed.* 2007, 2, 667-674.

[714] Maitra, A. Calcium phosphate nanoparticles: second-generation nonviral vectors in gene therapy. *Expert Rev. Mol. Diagn.* 2005, 5, 893-905.

[715] Yang, X.C., Walboomers, X.F., van den Dolder, J., Yang, F., Bian, Z., Fan, M.W., Jansen, J.A. Non-viral bone morphogenetic protein 2 transfection of rat dental pulp stem cells using calcium phosphate nanoparticles as carriers. *Tissue Eng. A* 2008, 14, 71-81.

[716] Altinoğlu, E.I., Adair, J.H. Calcium phosphate nanocomposite particles: a safer and more effective alternative to conventional chemotherapy? *Future Oncology* 2009, 5, 279-281.

[717] Joyappa, D.H., Kumar, C.A., Banumathi, N., Reddy, G.R., Suryanarayana, V.V.S. Calcium phosphate nanoparticle prepared with foot and mouth disease virus P1-3CD gene construct protects mice and guinea pigs against the challenge virus. *Veter. Microbiol.* 2009, 139, 58-66.

[718] Dreesen, I.A.J., Lüchinger, N.A., Stark, W.J., Fussenegger, M. Tricalcium phosphate nanoparticles enable rapid purification, increase transduction kinetics, and modify the tropism of mammalian viruses. *Biotechnol. Bioeng.* 2009, 102, 1197-1208.

[719] Tang, Q.L., Zhu, Y.J., Wu, J., Chen, F., Cao, S.W. Calcium phosphate drug nanocarriers with ultrahigh and adjustable drug-loading capacity: one-step synthesis, in situ drug loading and prolonged drug release. *Nanomedicine* 2011, 7, 428-434.

[720] Pittella, F., Zhang, M., Lee, Y., Kim, H.J., Tockary, T., Osada, K., Ishii, T., Miyata, K., Nishiyama, N., Kataoka, K. Enhanced endosomal escape of siRNA-incorporating hybrid nanoparticles from calcium phosphate and PEG-block charge-conversional polymer for efficient gene knockdown with negligible cytotoxicity. *Biomaterials* 2011, 32, 3106-3114.

[721] Behera, T., Swain, P. Antigen adsorbed calcium phosphate nanoparticles stimulate both innate and adaptive immune response in fish, *Labeo rohita* H. *Cell. Immunol.* 2011, 271, 350-359.

[722] Chu, T.C., He, Q., Potter, D.E. Biodegradable calcium phosphate nanoparticlesas a new vehicle for delivery of a potential ocular hypotensive agent. *J. Ocular Pharmacol. Therapeutics* 2002, 18, 507-514.

[723] Paul, W., Sharma, C.P. Porous hydroxyapatite nanoparticles for intestinal delivery of insulin. *Trends Biomater. Artif. Organs* 2001, 14, 37-38.

[724] Victor, S.P., Kumar, T.S.S. Tailoring calcium-deficient hydroxyapatite nanocarriers for enhanced release of antibiotics. *J. Biomed. Nanotechnol.* 2008, 4, 203-209.

[725] Kilian, O., Alt, V., Heiss, C., Jonuleit, T., Dingeldein, E., Flesch, I., Fidorra, U., Wenisch, S., Schnettler, R. New blood vessel formation and expression of VEGF receptors after implantation of platelet growth factor-enriched biodegradable nanocrystalline hydroxyapatite. *Growth Factors* 2005, 23, 125-133.

[726] Klesing, J., Wiehe, A., Gitter, B., Grafe, S., Epple, M. Positively charged calcium phosphate/polymer nanoparticles fro photodynamic therapy. *J. Mater. Sci. Mater. Med.* 2010, 21, 887-892.

[727] Ling, J.Y., Loo, S.C., Phung, N.T., Boey, F., Ma, J. Controlled size and morphology of EDTMP-doped hydroxyapatite nanoparticles as model for ^{153}Samarium-EDTMP doping. *J. Mater. Sci. Mater. Med.* 2008, 19, 2993-3003.

[728] Jordan, M., Schallhorn, A., Wurm, F.M. Transfecting mammalian cells: optimization of critical parameters affecting calcium-phosphate precipitate formation. *Nucleic Acids Res.* 1996, 24, 596-601.

[729] Sokolova, V.V., Epple, M. Inorganic nanoparticles as carriers of nucleic acids into cells. *Angew. Chem. Int. Ed.* 2008, 47, 1382-1395.

[730] Olton, D., Li, J., Wilson, M.E., Rogers, T., Close, J., Huang, L., Kumta, P.N., Sfeir, C. Nanostructured calcium phosphates (NanoCaPs) for non-viral gene delivery: influence of the synthesis parameters on transfection efficiency. *Biomaterials* 2007, 28, 1267-1279.

[731] Bisht, S., Bhakta, G., Mitra, S., Maitra, A. pDNA loaded calcium phosphate nanoparticles: highly efficient non-viral vector for gene delivery. *Int. J. Pharm.* 2005, 288, 157-168.

[732] Chowdhury, E.H., Akaike, T. A bio-recognition device developed onto nano-crystals of carbonate apatite for cell-targeted gene delivery. *Biotechnol. Bioeng.* 2005, 90, 414-421.

[733] Bisht, S., Chattopadhyay, D., Maitra, A. Intraperitoneal administration of calcium phosphate nanoparticles encapsulating pSVβgal elicits immune response to encoded protein. *J. Biomed. Nanotechnol.* 2006, 2, 229-238.

[734] Zhu, S.H., Huang, B.Y., Zhou, K.C., Huang, S.P., Liu, F., Li, Y.M., Xue, Z.G., Long, Z.G. Hydroxyapatite nanoparticles as a novel gene carrier. *J. Nanopart. Res.* 2004, 6, 307-311.

[735] Chowdhury, E.H., Kutsuzawa, K., Akaike, T. Designing smart nano-apatite composites: the emerging era of non-viral gene delivery. *Gene Ther. Mol. Biol.* 2005, 9, 301-316.

[736] Chowdhury, E.H., Maruyama, A., Kano, A., Nagaoka, M., Kotaka, M., Hirose, S., Kunou, M., Akaike, T. pH-sensing nano-crystals of carbonate apatite: effects on

intracellular delivery and release of DNA for efficient expression into mammalian cells. *Gene* 2006, 376, 87-94.

[737] Chowdhury, E.H. pH-sensitive nano-crystals of carbonate apatite for smart and cell-specific transgene delivery. *Expert Opin. Drug Deliv.* 2007, 4, 193-196.

[738] Chowdhury, E.H., Akaike, T. High performance DNA nano-carriers of carbonate apatite: multiple factors in regulation of particle synthesis and transfection efficiency. *Int. J. Nanomed.* 2007, 2, 101-106.

[739] Pedraza, C.E., Bassett, D.C., McKee, M.D., Nelea, V., Gbureck, U., Barralet, J.E. The importance of particle size and DNA condensation salt for calcium phosphate nanoparticle transfection. *Biomaterials* 2008, 29, 3384-3392.

[740] Zhou, C., Yu, B., Yang, X., Huo, T., Lee, L.J., Barth, R.F., Lee, R.J. Lipid-coated nano-calcium-phosphate (LNCP) for gene delivery. *Int. J. Pharm.* 2010, 392, 201-208.

[741] Giger, E.V., Puigmartí-Luis, J., Schlatter, R., Castagner, B., Dittrich, P.S., Leroux, J.C. Gene delivery with bisphosphonate-stabilized calcium phosphate nanoparticles. *J. Control. Release* 2011, 28, 87-93.

[742] Olton, D.Y., Close, J.M., Sfeir, C.S., Kumta, P.N. Intracellular trafficking pathways involved in the gene transfer of nano-structured calcium phosphate-DNA particles. *Biomaterials*. 2011, 32, 7662-7670.

[743] Welzel, T., Radtke, I., Meyer-Zaika, W., Heumann, R., Epple, M. Transfection of cells with custom-made calcium phosphate nanoparticles coated with DNA. *J. Mater. Chem.* 2004, 14, 2213-2217.

[744] Sokolova, V.V., Radtke, I., Heumann, R., Epple, M. Effective transfection of cells with multi-shell calcium phosphate-DNA nanoparticles. *Biomaterials* 2006, 27, 3147-3153.

[745] Sokolova, V.V., Kovtun, A., Heumann, R., Epple, M. Tracking the pathway of calcium phosphate/DNA nanoparticles during cell transfection by incorporation of red-fluorescing tetramethylrhodamine isothiocyanate-bovine serum albumin into these nanoparticles. *J. Biol. Inorg. Chem.* 2007, 12, 174-179.

[746] Sokolova, V.V., Kovtun, A., Prymak, O., Meyer-Zaika, W., Kubareva, E.A., Romanova, E.A., Oretskaya, T.S., Heumann, R., Epple, M. Functionalisation of calcium phosphate nanoparticles by oligonucleotides and their application for gene silencing. *J. Mater. Chem.* 2007, 17, 721-727.

[747] Neumann, S., Kovtun, A., Dietzel, I.D., Epple, M., Heumann, R. The use of size-defined DNA-functionalized calcium phosphate nanoparticles to minimise intracellular calcium disturbance during transfection. *Biomaterials* 2009, 30, 6794-6802.

[748] Graham, F.L., van der Eb, A.J. A new technique for the assay of infectivity of human adenovirus 5 DNA. *Virology* 1973, 52, 456-467.

[749] Kovtun, A., Heumann, R., Epple, M. Calcium phosphate nanoparticles for the transfection of cells. *Bio-Med. Mater. Eng.* 2009, 19, 241-247.

[750] Roy, I., Mitra, S., Maitra, A., Mozumdar, S. Calcium phosphate nanoparticles as novel non-viral vectors for targeted gene delivery. *Int. J. Pharm.* 2003, 250, 25-33.

[751] Li, J., Chen, Y.C., Tseng, Y.C., Mozumdar, S., Huang, L. Biodegradable calcium phosphate nanoparticle with lipid coating for systemic siRNA delivery. *J. Controlled Release* 2010, 142, 416-421.

[752] Kakizawa, Y., Kataoka, K. Block copolymer self-assembly into monodispersive nanoparticles with hybrid core of antisense DNA and calcium phosphate. *Langmuir* 2002, 18, 4539-4543.

[753] Wang, K.W., Zhou, L.Z., Sun, Y., Wu, G.J., Gu, H.C., Duan, Y.R., Chen, F., Zhu, Y.J. Calcium phosphate/PLGA-mPEG hybrid porous nanospheres: a promising vector with ultrahigh gene loading and transfection efficiency. *J. Mater. Chem.* 2010, 20, 1161-1166.

[754] Epple, M., Kovtun, A. Functionalized calcium phosphate nanoparticles for biomedical application. Key Eng. Mater. 2010, 441, 299-305.

[755] Sokolova, V., Neumann, S., Kovtun, A., Chernousova, S., Heumann, R., Epple, M. An outer shell of positively charged poly(ethyleneimine) strongly increases the transfection efficiency of calcium phosphate/DNA nanoparticles. *J. Mater. Sci.* 2010, 45, 4952-4957.

[756] Epple, M., Ganesan, K., Heumann, R., Klesing, J., Kovtun, A., Neumann, S., Sokolova, V. Application of calcium phosphate nanoparticles in biomedicine. *J. Mater. Chem.* 2010, 20, 18-23.

[757] He, Q., Mitchell, A.R., Johnson, S.L., Wagner-Bartak, C., Morcol, T., Bell, S.J.D. Calcium phosphate nanoparticle adjuvant. *Clin. Diagn. Lab. Immunol.* 2000, 7, 899-903.

[758] He, Q., Mitchell, A.R., Morcol, T., Bell, S.J.D. Calcium phosphate nanoparticles induce mucosal immunity and protection against herpes simplex virus type 2. Clin. Diagn. Lab. Immunol. 2002, 9, 1021-1024.

[759] Liu, Z.S., Tang, S.L., Ai, Z.L. Effects of hydroxyapatite nanoparticles on proliferation and apoptosis of human hepatoma BEL-7402 cells. *World J. Gastroenterol.* 2003, 9, 1968-1971.

[760] Sun, J., Ding, T. P53 reaction to apoptosis induced by hydroxyapatite nanoparticles in rat macrophages. *J. Biomed. Mater. Res. A* 2009, 88A, 673-679.

[761] Yuan, Y., Liu, C., Qian, J., Wang, J., Zhang, Y. Size-mediated cytotoxicity and apoptosis of hydroxyapatite nanoparticles in human hepatoma HepG2 cells. *Biomaterials* 2010, 31, 730-740.

[762] Allen, T.M., Cullis, P.R. Drug delivery systems: entering the mainstream. *Science* 2004, 303, 1818-1822.

[763] Schmidt, H.T., Kroczynski, M., Maddox, J., Chen, Y., Josephs, R., Ostafin, A.E.J. Antibody-conjugated soybean oil-filled calcium phosphate nanoshells for targeted delivery of hydrophobic molecules. *Microencapsulation* 2006, 23, 769-781.

[764] Ferraz, M.P., Mateus, A.Y., Sousa, J.C., Monteiro, F.J. Nanohydroxyapatite microspheres as delivery system for antibiotics: release kinetics, antimicrobial activity, and interaction with osteoblasts. *J. Biomed. Mater. Res. A* 2007, 81A, 994-1004.

[765] Cai, Y., Pan, H., Xu, X., Hu, Q., Li, L., Tang, R. Ultrasonic controlled morphology transformation of hollow calcium phosphate nanospheres: a smart and biocompatible drug release system. *Chem. Mater.* 2007, 19, 3081-3083.

[766] Zhou, W.Y., Wang, M., Cheung, W.L., Guo, B.C., Jia, D.M. Synthesis of carbonated hydroxyapatite nanospheres through nanoemulsion. *J. Mater. Sci. Mater. Med.* 2008, 19, 103-110.

[767] Wingert, P.A., Mizukami, H., Ostafin, A.E. Enhanced chemiluminescent resonance energy transfer in hollow calcium phosphate nanoreactors and the detection of hydrogen peroxide. *Nanotechnology* 2007, 18, 295707 (7 pages).

[768] Kottegoda, N., Munaweera, I., Madusanka, N., Karunaratne, V. A green slow-release fertilizer composition based on urea-modified hydroxyapatite nanoparticles encapsulated wood. *Curr. Sci.* 2011, 101, 73-78.

[769] Chen, J.H., Wang, Y.J., Zhou, D.M., Cui, Y.X., Wang, S.Q., Chen, Y.C. Adsorption and desorption of Cu(II), Zn(II), Pb(II), and Cd(II) on the soils amended with nanoscale hydroxyapatite. *Environ. Prog. Sustain. Energy* 2010, 29, 233-241.

[770] Wang, D., Chu, L., Paradelo, M., Peijnenburg, W.J., Wang, Y., Zhou, D. Transport behavior of humic acid-modified nano-hydroxyapatite in saturated packed column: effects of Cu, ionic strength, and ionic composition. *J. Coll. Interf. Sci.* 2011, 15, 398-407.

[771] Mobasherpour, I., Salahi, E., Pazouki, M. Comparative of the removal of Pb^{2+}, Cd^{2+} and Ni^{2+} by nano crystallite hydroxyapatite from aqueous solutions: adsorption isotherm study. *Arab. J. Chem.* 2012, (early view).

[772] Handley-Sidhu, S., Renshaw, J.C., Yong, P., Kerley, R., Macaskie, L.E. Nano-crystalline hydroxyapatite bio-mineral for the treatment of strontium from aqueous solutions. Biotechnol. Lett. 2011, 33, 79-87.

[773] Gandhi, R.M., Kousalya, G.N., Meenakshi, S. Removal of copper(II) using chitin/chitosan nano-hydroxyapatite composite. *Int. J. Biol. Macromolecules* 2011, 48, 119-124.

[774] Manocha, L.M., Disher, I.A., Manocha, S. Sorption of cadmium ions on (AB-type) carbonated hydroxyapatite nanoparticles. *Adv. Sci. Lett.* 2011, 4, 44-50.

[775] Ma'mani, L., Heydari, A., Shiroodi, R.K. Nanohydroxyapatite microspheres as a biocompatible and recoverable catalyst for synthesis of carbon -phosphorous bond formation. *Curr. Org. Chem.* 2009, 13, 758-762.

[776] Liu, Y., Zhong, H., Li, L., Zhang, C. Temperature dependence of magnetic property and photocatalytic activity of Fe_3O_4/hydroxyapatite nanoparticles. *Mater. Res. Bull.* 2010, 45, 2036-2039.

[777] Yih, T.C., Al-Fandi, M. Engineered nanoparticles as precise drug delivery systems. *J. Cell. Biochem.* 2006, 97, 1184-1190.

[778] Celotti, G., Tampieri, A., Sprio, S., Landi, E., Bertinetti, L., Martra, G., Ducati, C. Crystallinity in apatites: how can a truly disordered fraction be distinguished from nanosize crystalline domains? *J. Mater. Sci. Mater. Med.* 2006, 17, 1079-1087.

[779] Christenson, E.M., Anseth, K.S., van den Beucken, J.J.J.P., Chan, C.K., Ercan, B., Jansen, J.A., Laurencin, C.T., Li, W.J., Murugan, R., Nair, L.S., Ramakrishna, S., Tuan, R.S., Webster, T.J., Mikos, A.G. Nanobiomaterial applications in orthopedics. *J. Orthop. Res.* 2007, 25, 11-22.

[780] Schmidt, S.M., Moran, K.A., Kent, A.M.T., Slosar, J.L., Webber, M.J., McCready, M.J., Deering, C., Veranth, J.M., Ostafin, A. Uptake of calcium phosphate nanoshells by osteoblasts and their effect on growth and differentiation. *J. Biomed. Mater. Res. A* 2008, 87A, 418-428.

[781] Motskin, M., Müller, K.H., Genoud, C., Monteith, A.G., Skepper, J.N. The sequestration of hydroxyapatite nanoparticles by human monocyte-macrophages in a compartment that allows free diffusion with the extracellular environment. *Biomaterials* 2011, 32, 9470-9482.

[782] Powell, M.C., Kanarek, M.S. Nanomaterials health effects – Part 1: background and current knowledge. *Wisconsin Med. J.* 2006, 105, 16-20.

[783] Powell, M.C., Kanarek, M.S. Nanomaterials health effects – Part 2: uncertainties and recommendations for the future. *Wisconsin Med. J.* 2006, 105, 18-23.
[784] Motskin, M., Wright, D.M., Muller, K., Kyle, N., Gard, T.G., Porter, A.E., Skepper, J.N. Hydroxyapatite nano and microparticles: correlation of particle properties with cytotoxicity and biostability. *Biomaterials* 2009, 30, 3307-3317.
[785] Liu, X., Qin, D., Cui, Y., Chen, L., Li, H., Chen, Z., Gao, L., Li, Y., Liu, J. The effect of calcium phosphate nanoparticles on hormone production and apoptosis in human granulosa cells. *Reproductive Biology and Endocrinology* 2010, 8, 32 (8 pages).
[786] Li, S., Huang, L. Pharmacokinetics and biodistribution of nanoparticles. *Mol. Pharm.* 2008, 5, 496-504.
[787] Bionanotechnologies are modeled after biological substances and structures, or combine nanomaterials with biological substances. They include materials such as biochips, drug release systems, nanofibers, hybrid nanobiodevices, molecular electronics and biomimetics (synthetic genes, proteins and viruses) [788].
[788] Moghimi, S.J., Hunter, A.C., Murray, J.C. Nanomedicine: current status and future prospects. *FASEB J.* 2005, 19, 311-330.
[789] Xu, H.H.K., Weir, M.D., Simon, C.G., Jr. Injectable and strong nano-apatite scaffolds for cell/growth factor delivery and bone regeneration. *Dental Mater.* 2008, 24, 1212-1222.
[790] Watari, F., Abe, S., Tamura, K., Uo, M., Yokoyama, A., Totsuka, Y. Internal diffusion of micro/nanoparticles inside body. *Key Eng. Mater.* 2008, 361-363, 95-98.
[791] Oberdorster, G., Oberdorster, E., Oberdorster, J. Nanotoxicology: an emerging discipline evolving from studies of ultrafine particles. *Environ. Health Perspect.* 2005, 113, 823-839.
[792] Nel, A., Xia, T., Mädler, L., Li, N. Toxic potential of materials at the nanolevel. *Science* 2006, 311, 622-627.
[793] Jahnen-Dechent, W., Simon, U. Function follows form: shape complementarity and nanoparticle toxicity. *Nanomedicine* 2008, 3, 601-603.
[794] Singh, N., Manshian, B., Jenkins, G.J.S., Griffiths, S.M., Williams, P.M., Maffeis, T.G.G., Wright, C.J., Doak, S.H. NanoGenotoxicology: the DNA damaging potential of engineered nanomaterials. *Biomaterials* 2009, 30, 3891-3914.
[795] Dhawan, A., Sharma, V., Parmar, D. Nanomaterials: a challenge for toxicologists. *Nanotoxicology* 2009, 3, 1-9.
[796] Dwivedi, P.D., Misra, A., Shanker, R., Das, M. Are nanomaterials a threat to the immune system? *Nanotoxicology* 2009, 3, 19-26.

In: Calcium Phosphate
Editor: Robert B. Heimann

ISBN: 978-1-62257-299-1

Chapter 6

CONTROLLED PRECIPITATION OF CALCIUM PHOSPHATES IN GLASS-CERAMICS

C. Ritzberger**, *M. Schweiger, V. M. Rheinberger and W. Höland
Ivoclar Vivadent AG, Research and Development,
Schaan, Principality of Liechtenstein

ABSTRACT

The controlled phase formation of calcium phosphate crystals is an important process in the development of biomaterials. In dentistry, glass-ceramics of this type are mainly used as restorative biomaterials for the fabrication of inlays, crowns and bridges. Fluoroapatite is a crystal phase that controls the optical properties of glass-ceramics. As a result, it enables the translucency of the material to be adapted to that of the natural tooth structure. Therefore, fluoroapatite glass-ceramics are popular veneering materials in dentistry. This study examines the synthesis, microstructure, properties and main application areas of these materials. Furthermore, calcium phosphates are known to be the main components of bioactive materials, such as apatite or metaphosphate-containing glass-ceramics. The microstructures and properties of these types of biomaterials are also discussed.

1. INTRODUCTION

Glass-ceramics are composed of at least one glass phase and at least one crystal phase. They are fabricated from one base glass according to controlled nucleation and crystallization processes [1,2]. As a result, these materials allow a variety of different properties to be combined: for example, the characteristics of glasses (e.g. high light transmission) and those of ceramics (e.g. high strength). Both of these properties, that is, strength and optical characteristics, are extremely important in the fabrication and application of restorative dental

* E-mail address: ch.ritzberger@ivoclarvivadent.com

biomaterials. These products are expected to simulate the characteristics of natural teeth and, in some cases, even surpass them. The attainment of high strength and toughness in biomaterials has been documented elsewhere [1-3]. The current chapter examines the possibilities available for tailoring the optical properties of glass-ceramics by forming phosphate crystals in these materials. For this purpose, the precipitation of fluoroapatite is discussed. The joining of these materials with high-strength glass-ceramics or sintered ceramics is also presented.

Furthermore, crystalline calcium phosphate is of considerable importance in bioactive glass-ceramics as mentioned by Dorozhkin [4].[1] Therefore, apatite-containing glass-ceramics and calcium metaphosphate glass-ceramics functioning as bioactive materials, are also introduced.

Due to the different chemical compositions, properties and applications of these two types of materials, the authors of this chapter have decided to divide each section into a subsection on glass-ceramics for restorative dental applications and one on glass-ceramics for use as bioactive materials.

2. Synthesis and Microstructure

2.1. Glass-Ceramics for Restorative Dental Applications

Fundamental research for the development of glass-ceramics containing ($Ca_5(PO_4)_3F$) crystals was conducted in the 1990s, and needle-like apatite was developed in glass-ceramics for the first time [5]. Furthermore, Hill and Clifford [6] developed apatite-mullite glass ceramics, and Miosescu et al. [7] introduced the controlled crystallization of apatite in glass-ceramics.

The goal of developing a dental restorative material which combines high chemical durability with favourable processing and optical properties was fulfilled with the SiO_2 – Al_2O_3 – Na_2O – K_2O – CaO – P_2O_5 – F glass forming system. A typical composition range is 45-70 wt% SiO_2, 5-22 Al_2O_3, 4-13 Na_2O, 3-9 K_2O, 1.5-11 CaO, 0.5-6 P_2O_5 and 0.1-2.5 F [1, 2, 8-10]. Glasses in this composition range show a tendency for glass-in-glass phase separation. The phenomenon of phase separation has been studied in depth by Vogel [11, 12]. This type of phase separation is characterized by a phosphate-containing droplet phase in which the fluoride ions form a reservoir. At the same time, this phase may be enriched with alkali and alkaline earth ions. In the composition range examined in this study, the droplet phase is enriched with alkaline earth ions in the form of Ca^{2+}. The glass matrix is rich in silica. Phase separation represents a preliminary phase of nucleation and enables the nano-phase separation of calcium phosphates. This glass composition with contents of approx. 1.5-11 wt% CaO and 0.5-6 P_2O_5 (see composition range above) allows fluoroapatite crystals to develop in their unique needle-like shape [13]. Studies by Müller et al. [14] have shown that crystal growth proceeds according to the mechanism of Ostwald ripening. That is, large crystals grow at the expense of small ones during isothermal processing at 1050 °C, for example [1,2]. A typical microstructure of a glass-ceramic with needle-shaped fluoroapatite crystals is shown in Figure 1.

[1] See also Chapter 5 of this volume. The Editor.

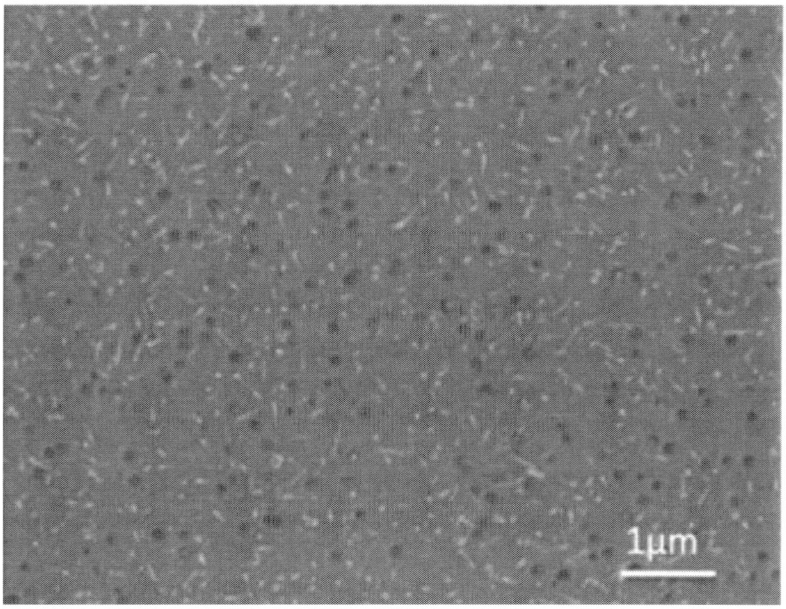

Figure 1. Volume crystallization of nano-scaled fluoroapatite based glass-ceramics (fracture surface etched for 10 s with 3% HF acid / SEM).

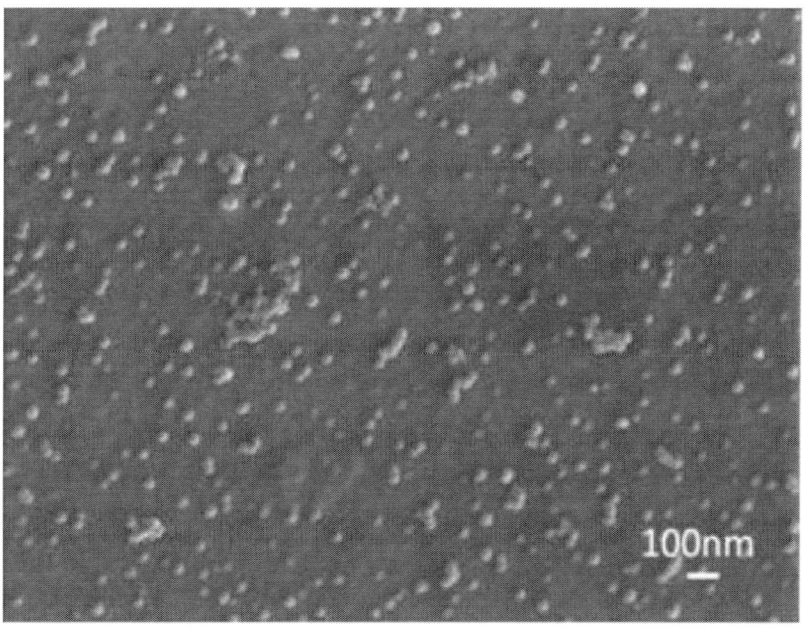

Figure 2. Volume crystallization and sintering to process powder compacts of fluoroapatite-based glass-ceramics (fracture surface etched with 3% HF acid / SEM).

It is important to note that the crystallization process can be conducted in such a way that the crystals touch each other. In other words, the crystallization process can be controlled. This aspect is of utmost importance for attaining the desired properties in a material.

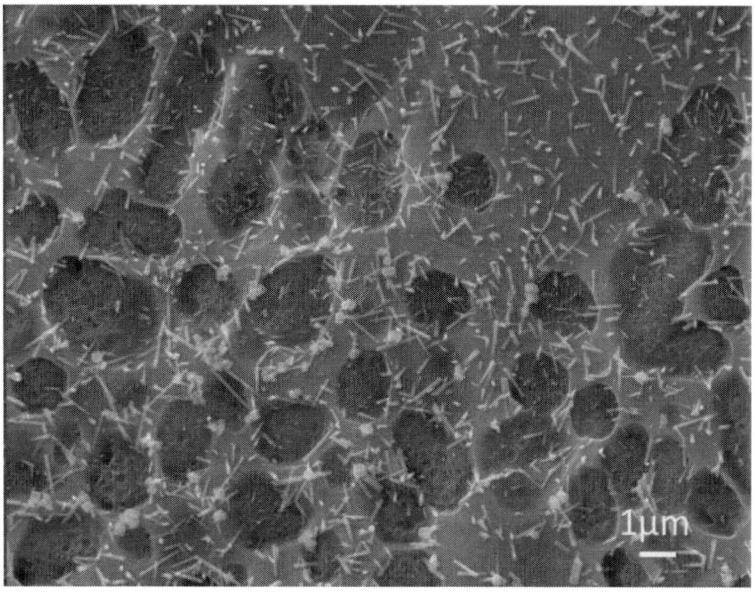

Figure 3. Volume crystallization and sintering to process powder compacts of fluoroapatite-based glass-ceramics (fracture surface etched for 10 s with 3% HF acid / SEM). The holes represent the volume of the leucite crystals etched away during preparation.

If a particularly narrow composition range is selected, such as 57.6-62 wt% SiO_2, 12-16 Al_2O_3, 5-8 Na_2O, 9-13 K_2O, 0.6-2.4 CaO, 0.4-2.5 P_2O_5 and 0.3-1.5 F, fluoroapatite ($Ca_5(PO_4)_3F$) can be precipitated in the glass-ceramic in nanoscopic dimensions, in other words, smaller than 100 nm. The microstructure of such a glass-ceramic is shown in Figure 2. The base glass was crystallized at a temperature of 600°C.

Apart from these two types of glass-ceramics containing nanoapatite or needle-like fluoroapatite, glass-ceramics with two crystal phases, that is, leucite ($KAlSi_2O_6$) and fluoroapatite have been developed using a method of double controlled crystallization [15-18]. The characteristic composition range of these glass-ceramics is 49-58 wt% SiO_2, 11-21 Al_2O_3, 1-10 Na_2O, 3-9 K_2O, 2.5-12 CaO, 0.5-6 P_2O_5 and 0.2-2.5 F. The way in which the crystal phases form during thermal treatment of the base glass is quite unusual.

A total of three solid state reactions take place in parallel: the formation and growth of fluoroapatite by volume crystallization, the formation and growth of the secondary rhenanite ($NaCaPO_4$) phase, and the nucleation and growth of leucite by surface crystallization. The typical microstructure of a leucite-apatite glass-ceramic is shown in Figure 3.

2.2. Bioactive Glass-Ceramics

Bioactive inorganic materials can be used as bone replacements in human medicine and as additives in toothpaste. BIOGLASS® is a prime example of these materials [19]. This specialized glass is heat treated to form a glass-ceramic.

A different type of a bioactive glass-ceramic represents the apatite-wollastonite ($Ca_5(PO_4)_3F$-$CaSiO_3$) glass-ceramics [20-21]. The product is called A-W glass-ceramic. The characteristic chemistry of this glass-ceramic is within the system SiO_2-CaO-MgO-P_2O_5-F.

The base glass for this material is processed in the form of granules. During thermal treatment, apatite and wollastonite crystals form within the glass matrix.

A different bioactive glass-ceramic was developed by controlling the crystallization of two types of crystals in monolithic glasses. Both fluoroapatite and mica could be formed in glasses of the SiO_2-CaO-MgO-K_2O-P_2O_5-F system, resulting in a machinable bioactive glass-ceramic [1, 2, 22, 23]. Chen et al. [22] investigated the structure and properties of a fluoroapatite/mica glass-ceramic in the SiO_2-MgO-Al_2O_3-K_2O-TiO_2-F system useful for machinable processing. Faeghi-Nia et al. [24] were successful in developing a apatite-wollastonite-mica glass-ceramic. The material was produced as powder compacts.

In addition to the bioactive glass-ceramics containing apatite, glass-ceramics featuring calcium phosphate in the form of metaphosphate were developed. Abe [25, 26], for example, managed to crystallize calcium metaphosphate $Ca(PO_3)_2$ in glass-ceramics. This significant research represented the first directionally-controlled crystallization of glass-ceramics. Subsequently, this materials system was further developed and Kasuga et al. [27-30] used it as a basis for producing a very successful biomaterials system.

Bioactive glass-ceramics based on rhenanite ($NaCaPO_4$)-type of crystals were also developed by Höland et al. [31] and Sitarz et al. [32]. Knabe et al. [33] reported on bioresorbable Ca-phosphate glass-ceramics.

3. Properties

3.1. Glass-Ceramics for Restorative Dental Applications

Apatite-containing glass-ceramics feature properties that are highly sought-after in restorative dentistry. These properties include high chemical durability and the ability to control the translucency of the material and its coefficient of thermal expansion (CTE). The material's chemical durability in acid, evaluated according to the dental standard ISO 6872, is rated as very good (weight loss < 50 $\mu g/cm^2$). The translucency of the material, expressed as the contrast ratio (CR) (according to British Standard BS 5612), can be adjusted between 0.3 and 0.9 (corresponding to 30 and 90 % transmission), depending on the crystallization degree of the glass-ceramics. Under the conditions posed by the envisaged application, the CTE was established to be $10.5 \cdot 10^{-6}$ K^{-1} m/m. Apart from these parameters mentioned, the glass-ceramics demonstrate excellent processing properties when processed using two different techniques: sintering of powder compacts and moulding with the aid of viscous flow phenomena. Leucite-apatite glass-ceramics in the form of powder compacts exhibit a CTE of $11.6 - 13.1 \cdot 10^{-6}$ K^{-1} m/m and the same outstanding chemical durability as that of apatite glass-ceramics. The light transmission of the materials can be adjusted to between 10 and 90 %. The materials can also be either sintered or moulded (viscous flow).

3.2. Bioactive Glass-Ceramics

Apatite-wollastonite glass-ceramics are characterized by high bioactivity combined with favourable mechanical properties. The bioactivity of the material complies with ISO 23317.

In other words, hydroxyapatite forms on the surface of the glass-ceramic in simulated body fluid. The *in vivo* performance corresponds to the results of the *in vitro* examination, in which a bond was formed to human bone without any connective tissue capsule. Furthermore, the material exhibits useful mechanical properties including a bending strength of 220 MPa, a compressive strength of 1080 MPa, Young's modulus of 118 GPa, Vickers hardness of 680 MPa and fracture toughness (K_{IC}) of 2.0 MPa · $m^{0.5}$ [34]. The mica-apatite glass-ceramics can be machined and demonstrate a flexural strength of 140 to 180 MPa. As the materials can be shaped with tungsten carbide tools, diamond tools are not required.

Metaphosphate glass-ceramics such as the bioactive materials mentioned, are known to be stable, yet also resorbable products suitable for the replacement of hard tissue.

4. Application

4.1. Glass-Ceramics for Restorative Dental Applications

Apatite glass-ceramics containing needle-shaped crystals have been specifically developed to further improving the optical properties of metal-free dental restorations. The aim of these developments was to imitate the appearance of natural dental tissue. As a result, fluoroapatite glass-ceramics exhibit a suitable interplay with natural white light to achieve a natural-looking appearance.

Therefore, these biomaterials are preferably used for veneering ZrO_2 ceramic frameworks (dental crown copings and multi-unit bridge structures), crowns and small bridges, as well as lithium-disilicate-based glass-ceramic crowns. Figure 4 shows the application of this type glass-ceramic as a three-unit metal-free dental bridge.

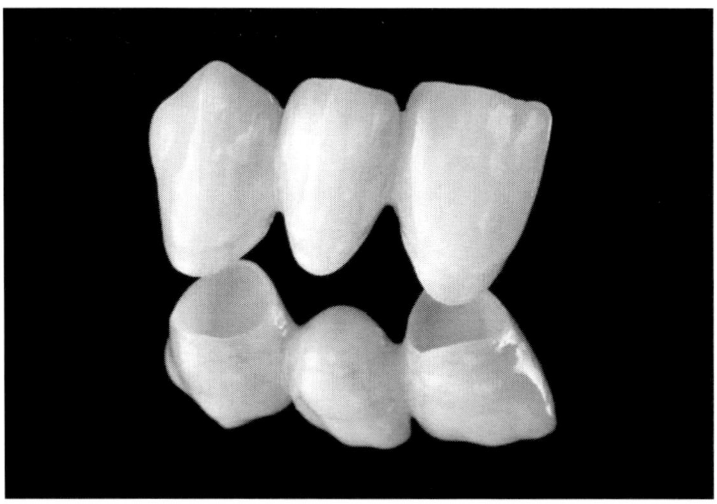

Figure 4. Three-unit dental bridge positioned on a mirror. The framework consists of a lithium disilicate glass-ceramic. The veneering biomaterial is an apatite-containing glass-ceramics. The fluoroapatite crystals show a needle-like morphology. Reproduced from W. Höland, Glaskeramik, vdf Zürich, UTB Stuttgart, 2006; all rights reserved.

The apatite glass-ceramic is joined to the sintered ZrO_2 ceramic or the lithium disilicate glass-ceramic by sintering or by moulding, taking advantage of the viscous flow phenomenon in the latter case.

Because of their favourable coefficient of thermal expansion, leucite-apatite glass-ceramics are popular materials for covering metal frameworks in restorative dentistry. They are particularly suitable for long-span, multi-unit bridges. Nevertheless, before the glass-ceramics can be applied, the metal framework has to be coated with a more opaque material. Subsequently, the glass-ceramic material is either sintered to or moulded on the framework.

4.2. Bioactive Glass-Ceramics

Bioactive A-W glass-ceramic has been extensively used in orthopaedic human medicine as a spacer, a vertebra substitute, an intervertebral spacer, an iliac crest prosthesis, and as a filler [34,35]. Machinable bioactive glass-ceramics, however, are popularly used in ENT surgery, for example, in the replacement of the tympanic bone. Metaphosphate glass-ceramics have also been tested for replacing bone. This biomaterial presents the possibility of being completely resorbed.

Conclusion

Calcium phosphate and, in particular fluoroapatite can be formed in glass-ceramics according to the methods of controlled nucleation and crystallization. The needle-like shape of the hexagonal crystals formed in this way is particularly advantageous. Their special properties allow this type of glass-ceramics to be used as a biomaterial in restorative dentistry. Furthermore, apatite glass-ceramics can be used as bioactive materials for replacement of bone in human medicine.

References

[1] W. Höland and G.H. Beall, *Glass-ceramic technology,* The Am. Ceram. Society and Wiley, USA, (2002 and 2006).

[2] W. Höland and G.H. Beall, *Glass-ceramic technology,* Wiley, USA, (2012) in the press.

[3] W. Höland, M. Schweiger, V.M. Rheinberger, and H. Kappert, *Adv. Appl. Ceram.* 108, 373, (2009).

[4] S.V. Dorozhkin, *J. Mater. Sci.*, 42, 1061 (2007).

[5] W. Höland, M. Frank, M. Schweiger, and V. Rheinberger, *Glastech. Ber. Glass Sci. Technol.* 67C, 117, (1994).

[6] Clifford and R. Hill, *J. Non-Cryst. Sol.* 196, 346, (1996).

[7] Moisescu, G. Carl, and C. Rüssel, *Phosphorus Res. Bul.* 10, 515, (1999).

[8] M. Schweiger, *Report Ivoclar Vivadent* 25, (2006).

[9] W. Höland, V. Rheinberger, S. Wegner, and M. Frank, *J. MATER. SCI.-MATER. M.* 11, 1, (2000).

[10] W. Höland, V. Rheinberger, and M. Schweiger, *Philos. T. Roy. Soc. A* 361, 575, (2003).
[11] W. Vogel, *Structure and Crystallization of Glasses*, Edition Leipzig, Pergamon Press, Oxford (1971).
[12] W. Vogel, *Chemistry of Glass*, The Am. Ceram. Society, USA, (1985).
[13] F. Krumeich, R. Nesper, C. Ritzberger, E. Apel, V. Rheinberger, and W. Höland , *Microsc. Microanal.* , 13, 3, (2007).
[14] R. Müller, L.A. Abu-Hilal, S. Reinisch, and W. Höland, *J. Mater. Sci.* 34, 65, (1999).
[15] W. Höland, *Glaskeramik*, UTB, vdf, Stuttgart, Germany and Zürich, Switzerland (2006).
[16] W. Höland, *J. Non-Cryst. Solids*, 219, 192 (1997).
[17] W. Höland, V. Rheinberger, M. Frank, *J. Non-Cryst. Solids*, 253, 170 (1999).
[18] W. Höland, M. Schweiger, S. Cramer von Clausbruch, and V. Rheinberger, *Glastech. Ber. Glass Sci. Technol.*, 73, C1, 12, (2000).
[19] L.L. Hench, D.E. Day, W. Höland, and V.M. Rheinberger, *Int. J. Appl. Glass Sci.* 1, 104, (2010).
[20] T. Kokubo, S. Ito, S. Sakka, and T. Yamamuro, *J. Mater. Sci.*, 21, 536 (1986).
[21] C. Ohtsuki, *Bioceramics and their clinical application*. Woodhead Publ. Ltd, Cambridge, England, Ed. T. Kokubo, 501, (2008).
[22] X. Chen, H. Zhang, Y. Meng, Y. Li, and C. Lin, *Phys. Chem. Glasses*: *Eur. J. Glass Sci. Technol. B*, 52, 3, 133 (2011).
[23] G. Schneider, K. Blechschmidt, D. Linde, P. Litschko, T. Körbs, and E. Beleites, *J. Mater. Sci.: Mater. Med.*, 21, 2853-2859, (2010).
[24] Faeghi-Nia, V.K. Marghussian, E. Taheri-Nassaj, M.J. Pascual, and A. Durán,, *J. Am. Ceram. Soc.*, 92, 7, 1514, (2009).
[25] Y. Abe, *Nature* 282, 55, (1979).
[26] Y. Abe, T. Kasuga, H. Hosono, and K. de Groot, *J. Am. Ceram. Soc.* 67, C142, (1984).
[27] T. Kasuga, A. Ichino, and Y. Abe, *J. Ceram. Soc. Japan* 100, 1089 (1992).
[28] T. Kasuga, H. Hosono, and Y. Abe, *Phosphorus Sulfur* 76, 247 (1993).
[29] T. Kasuga, Y. Ota, K. Tsuji, and Y. Abe, *J. Am. Ceram. Soc.* 79, 1821 (1996).
[30] T. Kasuga and M. Norigami, T. MRS Japan 29, 2933, (2004).
[31] M. Höland, A. Dommann, W. Höland, E. Apel, and V. Rheinberger, *Glass Sci. Technol.*, 78, 153 (2005).
[32] M. Sitarz, K. Bulat, D. Suka, *Phys. Chem. Glasses: Eur. J. Glass Sci. Technol. B*, 52, 3,115, (2011).
[33] C. Knabe and P. Ducheyne, *Bioceramics and their clinical application*, Woodhead Publ. Ltd, Cambridge, England, Ed. T. Kokubo, 133, (2008).
[34] T. Kokubo, *Bioceramics and their clinical application*. Woodhead Publ. Ltd, Cambridge, England, Ed. T. Kokubo, (2008).
[35] T. Yamamuro, *Bioceramics and their clinical application*. Woodhead Publ. Ltd, Cambridge, England, Ed. T. Kokubo, 583-605, (2008).

In: Calcium Phosphate
Editor: Robert B. Heimann

ISBN: 978-1-62257-299-1

Chapter 7

LASER-RAMAN, NUCLEAR MAGNETIC RESONANCE (NMR) AND ELECTRON ENERGY LOSS (EEL) SPECTROSCOPY STUDIES OF PLASMA-SPRAYED HYDROXYAPATITE COATINGS

Robert B. Heimann *

Applied Mineralogy and Materials Science, Görlitz, Germany

ABSTRACT

The phase composition of HAp coatings deposited onto Ti6Al4V coupons by atmospheric plasma spraying (APS) was studied by laser-Raman, nuclear magnetic resonance (NMR), and electron energy loss (EEL) spectroscopy. The samples investigated comprised as-sprayed HAp coatings as well as coatings incubated for different lengths of time (up to 12 weeks) in simulated body fluid (r-SBF) under physiological conditions. Incubation in SBF further advanced the proportion of crystalline HAp since the disordered structures, the thermal decomposition products, and ACP exhibit substantially higher solubility.

1. INTRODUCTION

Atmospheric plasma spraying (APS) is still the method of choice to deposit osteoconductive HAp coatings on biomedical implants despite many efforts to develop alternate ways that would circumvent the drawback of thermal decomposition of the target material in the extremely hot plasma jet. However, it is well known that a certain degree of decomposition is beneficial to the biomedical performance of a coating as the easily soluble decomposition products of HAp will provide the means of *p*H-dependent precipitation of nano- to microcrystalline 'bone-like' apatite from body fluid (for example [1], [2]) owing to the high level of supersaturation immediately adjacent to the implant-bone interface.

* Email: robert.heimann@ocean-gate.de

Consequently, the parameters of plasma spraying have to be carefully controlled (for example [3]) to prevent excessive overheating of the hydroxyapatite powder injected into the plasma jet. The mechanism of thermal decomposition of hydroxyapatite during contact with the extremely hot plasma jet can be investigated by several surface analytical techniques including Raman and NMR spectroscopy. Results of such studies will be presented below.

2. Materials and Methods

Ti6Al4V coupons of dimensions 50x20x 2mm were grit-blasted with alumina grit (0.5-1mm) using pressurized air of 4 bars at an angle of 75°, cleaned ultrasonically in a 7.5% Tickopur® solution at 40°C for 15 minutes, and air dried. Hydroxyapatite (HAp) powder was obtained from DePuy Bioland, Toulouse, France and plasma-sprayed onto the coupons using an atmospheric plasma spray system PM-M1000® (Plasma-Technik, Wohlen, Switzerland) equipped with an F4 plasmatron supplied by Sulzer Metco (Deutschland) GmbH. The plasma spray parameters applied to deposit the HAp are shown in Table 1.

The incubation experiments were performed in r-SBF (revised simulated body fluid, [2]) under physiological conditions at 37 ± 0.5°C and *p*H 7.40 for 1 and 12 weeks. The composition of the r-SBF is shown in Table 2.

Table 1. Plasma spray parameters

Plasma power (kW)	32
Plasma current (A)	560
Argon flow rate (slpm)*	45
Hydrogen flow rate (slpm)	6.5
Powder carrier gas flow rate (slpm)	5.5
Powder feeder disk rotation (% of maximum)	20
Powder feeder stirrer rotation (% of maximum)	40
Spray stand-off distance (mm)	90

*standard liter per minute.

Table 2. Composition of r-SBF in mmol, stabilized with HEPES [2-(4-(2-hydroxyethyl)-1-piperazinyl)ethane] sulfonic acid + 1 M NaOH [2]

Na^+	142.0	Cl^-	103.0
K^+	5.0	HCO_3^-	27.0
Mg^{2+}	1.5	HPO_4^{2-}	1.0
Ca^{2+}	2.5	SO_4^{2-}	0.5

Laser-Raman spectra were recorded with a Raman microprobe (T 64000, Jobin Yvon) using a monochromatic Ar^+ laser at 514.5 nm wavelength with a focus diameter of about 3 μm.^{1}H- and ^{31}P-MAS- NMR spectra as well as 2D-HETCOR-CP-NMR spectra were acquired with BRUKER AMX400 and TECMAG (300 MHz) spectrometers.[1]

3. Results and Discussion[2]

3.1. Thermal Decomposition of Hydroxyapatite

Figure 1 shows the binary phase diagram of the system CaO-P_2O_5 highlighting the region of particular interest around the composition of apatite.

As shown in the phase diagram hydroxyapatite melts incongruently in the absence of water at 1843 K to form a mixture of α'-$Ca_3(PO_4)_2$ (high-temperature tricalcium phosphate, α'-C_3P[3]) and $Ca_4O(PO_4)_2$ (tetracalcium phosphate, C_4P) [4].

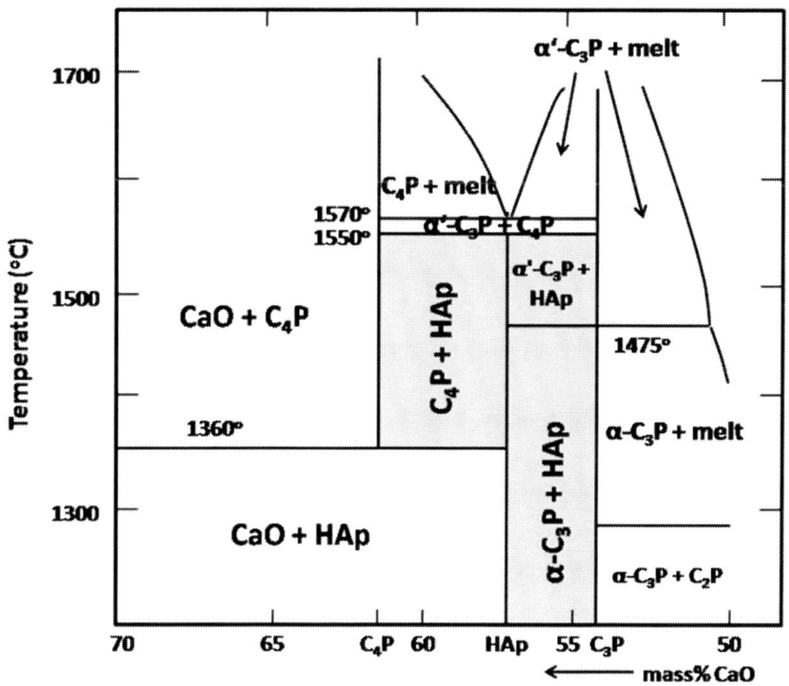

Figure 1. High temperature portion of the phase diagram of the quasi-binary system CaO-P_2O_5-H_2O at a water partial pressure of 65.5 kPa [4]. Note that at 1570°C, incongruent melting of hydroxyapatite (HAp) occurs under decomposition into α'-C_3P (α'-TCP) and C_4P (TetrCP). The shaded areas indicate the regions of relative stability of hydroxyapatite.

[1] The author is indebted to Dr. Klaus Irmer, Technische Universität Bergakademie Freiberg, Germany for the Raman measurements and to Dr. Peter Hartmann, formerly at the Institute of Optoelectronics, Universität Jena, Germany for the NMR measurements.

[2] The text is a reworked and shortened excerpt from the chapter "Novel approaches towards design and biofunctionality of plasma-sprayed osteoconductive calcium phosphate coatings for biomedical implants: The concept of bond coats", published in *Trends in Biomaterials Research*, P.J. Pannone (ed.), Nova Science Publishers, Inc., New York, 2007 [34].

[3] Cement chemistry notation: C = CaO, P = P_2O_5

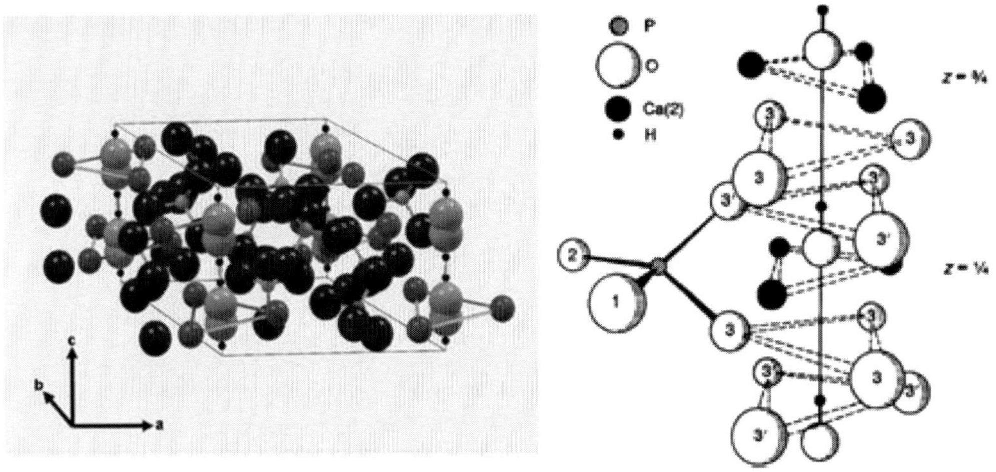

Figure 2. Left: Crystallographic structure of hydroxyapatite showing the orthohexagonal unit cell with a = 943.2 pm, b = a√3, c = 688.14 pm [11]. Right: Detail of the arrangement of ions around the 6_3 screw axis [12].

The actual melting temperature depends on the water partial pressure: at a water partial pressure $p_{H2O} = 6.7 \cdot 10^4$ Pa hydroxyapatite is stable up to 1633 K and its melting point increases to 1823 K at $p_{H2O} = 1.33 \cdot 10^2$ Pa. Since the rate of phase transition of C_3P or C_4P to HAp also increases with p_{H2O}, during sintering of HAp under low water conditions a kinetic barrier exists that makes it difficult to predict the volume fraction of high temperature phases and their relative stability in the quenched product at room temperature [5].

The thermal decomposition a HAp powder particle injected into the hot plasma jet of an atmospheric plasma spray (APS) system by a blast of an inert carrier gas suffers during its flight along a trajectory is to a large extent determined by the grain size, thermal conductivity, density and velocity of the particle as well as the kinematic viscosity of the plasma and the plasma gas density [6, 7]. Since HAp melts incongruently the oxyapatite (OAp) formed by dehydroxylation of oxyhydroxyapatite (OHAp) (step 2) decomposes into tri- and tetracalcium phosphates (step 3) and the entire transformation sequence can be expressed in four steps [8, 9]:

Step 1: $Ca_{10}[(OH)_2/(PO_4)_6] \rightarrow Ca_{10}[(OH)_{2-2x}O_x\square_x/(PO_4)_6] + xH_2O$

Step 2: $Ca_{10}[(OH)_{2-2x}O_x\square_x/(PO_4)_6] \rightarrow Ca_{10}[O_x\square_x/(PO_4)_6] + (1-x)H_2O$

Step 3: $Ca_{10}[O_x\square_x/(PO_4)_6] \rightarrow 2\ Ca_3(PO_4)_2 + Ca_4O(PO_4)_2$

Step 4a: $Ca_3(PO_4)_2 \rightarrow 3\ CaO + P_2O_5 \uparrow$

Step 4b: $Ca_4O(PO_4)_2 \rightarrow 4\ CaO + P_2O_5 \uparrow$

The squares † in the formulae of oxyhydroxyapatite and oxyapatite refer to lattice vacancies in the OH positions along the crystallographic *c*-axis in the structure of hydroxyapatite (Figure 2). In the completely dehydroxylated oxyapatite there exists a linear chain of O^{2-} ions parallel to the *c*-axis each one followed by a vacancy (x = 1 in step 2). Two broad lines centered at 1.5 (band D) and 5.6 (band A) ppm observed in the ^{31}P-MAS-NMR spectrum suggest the presence of different distortions of the $(PO_4)^{3-}$ tetrahedra depending on their position relative to the Ca ions in the central chains (Figure 8). While OAp appears to be stable in the absence of moisture it will readily transform to hydroxyapatite in the presence of water [10] according to the equation

$$O^{2-}(s) + \square(s) + H_2O(g) \rightarrow 2OH^{-}(s).$$

The equilibrium temperature of step 3 is determined by the temperature of incongruent melting of hydroxyapatite at 1843 K [4].

Based on this decomposition sequence a model was developed [13, 8, 14] showing the in-flight evolution of individual calcium phosphate phases. Figure 3, left shows that during the short residence time of the HAp particle in the plasma jet the innermost core is still at a temperature below 1823 K owing to its low thermal conductivity, consisting of coexisting HAp, OHAp and OAp as stable phases (steps 1 and 2). The existence of OHAp and OAp was postulated earlier [15] and was subject to controversy for many years but has now been well established thanks to modern analytical techniques. A more recent study [16] revealed that (i) a small amount of OH^- ions is always present in the structure of OAp and (ii) even a loss of 75% of the chemically bound water maintains the apatite channel structure.

The limiting composition is then $Ca_{10}(PO_4)_6(OH)_{0..5}O_{0.75}\square_{0.75}$ corresponding to a water loss of 1.34% [17]. Hence it is suggested to abandon the term oxyapatite in favour of the more general term oxyhydroxyapatite that could be described either as a non-stoichiometric defect hydroxyapatite with a partially distorted structure [18, 19] or a solid solution of stoichiometric HAp and oxyapatite [20]. Vibrational spectroscopy showed a loss of the intensity of the OH stretching vibration band at 3571 cm^{-1} (Figure 4, right) and the OH librational band at 630 cm^{-1} [21, 22]. Since OHAp (and OAp) show only small differences in lattice parameters compared to HAp, x-ray diffraction is not the most suitable tool to distinguish among these phases. However, the small shift of the (002) peak position of OAp to lower diffraction angles compared to HAp may be used for this purpose [23].

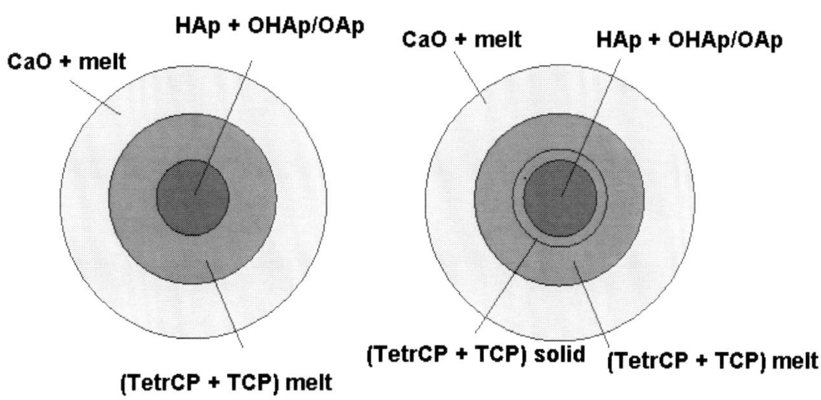

Figure 3. Schematic model of the thermal decomposition of a spherical hydroxyapatite particle subjected to high temperature in a plasma jet, left at a high water partial pressure of 65.5 kPa [4], right at a low water partial pressure of 1.3 kPa [24].

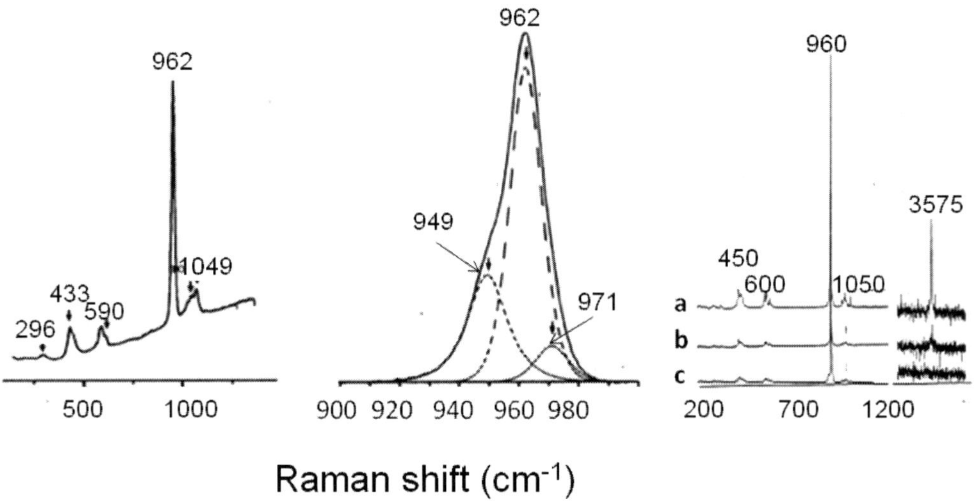

Figure 4. Laser-Raman spectra of as-sprayed hydroxyapatite showing the four normal vibrational modes of the PO_4^{3-} tetrahedron (left) and the ν_1 Raman mode deconvoluted by a Gaussian-Lorentzian fit (center). The right spectrum shows the principal Raman specta of as-received HAp powder (a), HAp powder annealed at 1300°C for 3 hours (b) and an APS HAp coating on a Ti6Al4V substrate (c). The OH stretching vibrations at 3575 cm^{-1} are also shown [31].

Recently this simple model was modified [24] to include the product of the solid state dehydration transformation of HAp into tri- and tetracalcium phosphates between 1633 and 1843 K that has been experimentally [16, 25] found to occur at very low water partial pressure (Figure 3, right). Hence the second shell of the spherical model has been heated to a temperature of 1633<T<1843 K, just below the incongruent melting point of HAp that underwent solid state decomposition to a mixture of α'-TCP and TetrCP. The third shell was heated to temperatures above 1843 K and consists of a melt with a Ca/P ratio of 1.67 (step 3, [9]). The outermost shell is composed of CaO and a melt whose Ca/P ratio shifts by continuous evaporation of P_2O_5 along the liquidus of the binary phase diagram CaO-P_2O_5 towards CaO-richer phases, *i.e.* TetrCP (step 4). On impact with the target surface this molten phase solidifies quickly to produce ACP with various Ca/P ratios [9, 26, 27].

3.2. Raman Spectroscopy

The thermal dehydroxylation of hydroxyapatite towards oxy/oxyhydroxyapatite can be investigated by Raman and nuclear magnetic resonance spectroscopy as well as infrared spectroscopy. Figure 4, left shows a laser-Raman spectrum (λ = 514.6 nm; 5mW) of the surface of a plasma-sprayed hydroxyapatite coating in the full spectral range between 200 and 1500 cm^{-1} with the four principal vibrational modes of the PO_4^{3-} tetrahedron: the large intensity symmetric stretching mode of the P-O bonds ν_1 at 962 cm^{-1}, the doubly degenerate O-P-O bending mode ν_2 at 433 cm^{-1}, the triply degenerate asymmetric P-O stretching mode ν_3

at 1049 cm^{-1} as well as the triply degenerate O-P-O bending mode ν_4 at 590 cm^{-1} [22, 28][4]. IR investigations on plasma-sprayed HAp coatings showed a noticeable difference between spectra obtained from the surface and the bulk of the coating manifest in a higher degree of splitting of the asymmetric P-O stretching vibration mode (ν_3) in the surface spectra [29]. This has been attributed to an increased perturbed PO_4^{3-} environment due to dehydroxylation of HAp and/or an increased disorder of hydroxyl ions in the crystalline network.

Figure 4, center shows the ν_1 mode composed of a main contribution at 962 cm^{-1} typical for well-ordered HAp [28] and weaker signals centered around 949 and 971 cm^{-1} obtained by fitting the composite Raman peak with a Gaussian-Lorentzian function.

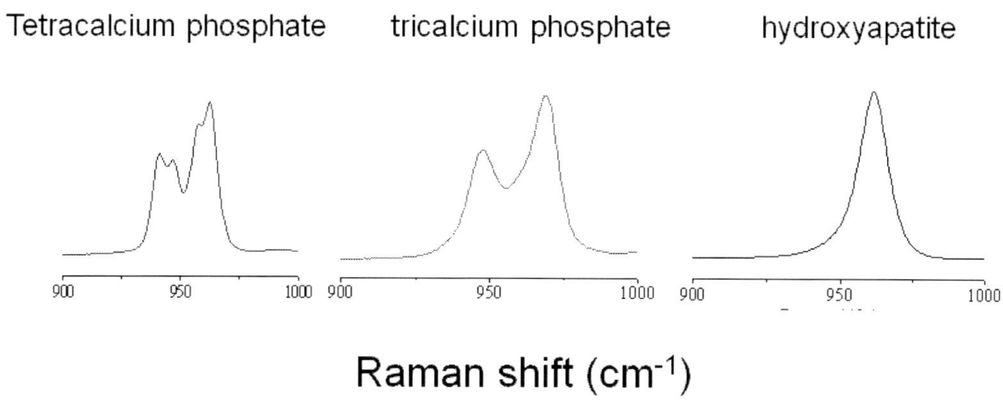

Figure 5. Laser-Raman standard spectra of TetrCP (left), ß-TCP (center) and HAp (right) [19].

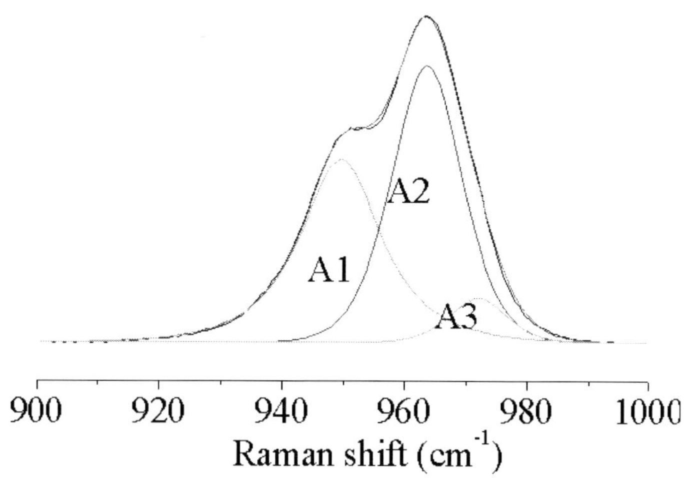

Figure 6. Gaussian-Lorentzian deconvolution of a laser-Raman spectrum of an as-sprayed HAp coating. A1: TetrCP+ß-TCP+ACP+OHAp, A2: HAp, A3: ß-TCP+ACP+ OHAp. The major contribution compositions are shown in bold [19].

[4] These Raman shifts deviate from the values given by Cusco *et al.* [28] that were obtained from measurements on phosphate ions in aqueous solution. In crystals such as hydroxyapatite the crystal field induces distortion of the PO_4^{3-} tetrahedra, giving rise to shifts and splitting of the Raman signals of the normal modes as shown in Fig. 4, left.

Table 3. Quantitative composition in mass% of HAp coatings after incubation in r-SBF for 1 and 12 weeks (based on Raman data)

Peak position	As-sprayed	1 week incubation	12 weeks incubation
A1	44	35	12
A2	43	53	75
A3	13	11	5

The signal at 949 cm^{-1} can be associated with ACP [30] as well as to changes of the environment around the orthophosphate tetrahedra caused by OH^- vacancies as typical for oxyhydroxyapatite [22], and the vibration at 971 cm^{-1} may be related to tricalcium phosphate formed during thermal decomposition of oxyapatite. Consequently, the intensity of the ν_1 vibration, together with the relative intensities of the side peaks, will give information on the degree of ordering of the structure of crystalline hydroxyapatite. The decrease of the intensity of the OH stretching vibration at 3575 cm^{-1} from as-received HAp powder (a) to powder annealed at 1300°C (b) to a plasma-sprayed coating (c) is indicative of the consecutive loss of OH ions with increased heating (Figure 4, right).

The acquired spectra were compared to standards obtained from pure tetracalcium phosphate (TetrCP), ß-tricalcium phosphate (ß-TCP) and hydroxyapatite (HAp) (Figure 5). TetrCP shows a bimodal spectrum with doublets of 942 – 947 and 957 – 963 cm^{-1} (Figure 5, left).

The spectrum of ß-TCP is characterized by two Raman bands centred around 947 and 970 cm^{-1} (Figure 5, center), The HAp spectrum shows only a single band centred at 962 cm^{-1} (Figure 5, right). The deconvoluted Raman spectrum of an as-sprayed HAp coating is shown in Figure 6. From the appearance of the spectra it is evident that plasma-sprayed HAp coatings consist of well-crystallized HAp (A_2) with concurrent strong contribution of TetrCP+ß-TCP+ACP+OHAp (band A1 centred around 950 cm^{-1}) and small amounts of ß-TCP+ACP+ OHAp (band A3 centred around 972 cm^{-1}).

A quantitative evaluation of the as-sprayed coatings and coatings incubated in r-SBF for 1 and 12 weeks is shown in Table 3. The proportion of HAp in the coatings increases strongly from 43 mass% in as-sprayed coatings to 75 mass% after incubation for 12 weeks [19]. In parallel the proportions of dehydration products such as oxyhydroxyapatite (OHAp) and oxyapatite (OAp), respectively as well as thermal decomposition products such as tri- and tetracalcium phosphates (ß-TCP, TetrCP) and ACP decrease substantially owing to the much higher solubility of these phases in r-SBF.

3.3. Nuclear Magnetic Resonance Spectroscopy

Even more information on the degree of disorder introduced by thermal treatment of hydroxyapatite during plasma spraying can be obtained by nuclear magnetic resonance (NMR) spectroscopy. The positions and shift of ^{1}H-MAS and ^{31}P-MAS NMR peaks are indicative of the environment of the PO_4^{3-} tetrahedra in calcium orthophosphates and allow distinguishing between dehydroxylation (oxyhydroxyapatite) and decomposition (tricalcium phosphate, tetracalcium phosphate) products of hydroxyapatite. Figures 7 and 8 show ^{1}H- and ^{31}P NMR spectra, respectively of as-sprayed hydroxyapatite [32, 19, 22].

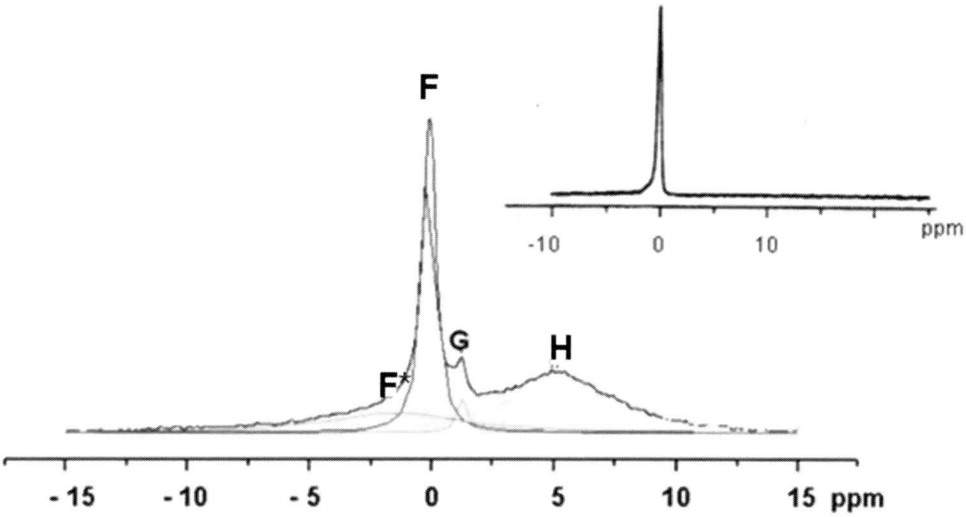

Figure 7. ^{1}H-MAS NMR spectrum of as-sprayed hydroxyapatite [32]. The inset shows the spectrum of a completely ordered, highly crystalline hydroxyapatite [18].

While the ^{1}H-MAS NMR spectrum of stoichiometric, highly crystalline hydroxyapatite shows only one single band position at -0.1 ± 0.1 ppm (Figure 7, inset; band F) in plasma-sprayed hydroxyapatite coatings additional bands can be detected centred at -1.3 ± 0.3 ppm (band F*) and +5.2 ± 0.2 ppm (band H). The main band F represents the proton band position of crystalline stoichiometric hydroxyapatite whereas the weak isotropically shifted band F* can be attributed to protons in distorted HAp short range ordered (SRO) structures such as oxyhydroxyapatite. The band H relates to OH^- positions missing only one neighbouring OH^- ion and hence indicates the existence of isolated pairs of strongly coupled protons in the channel structure of HAp [18]. The band G at about +1.3 ppm possibly belongs to free water. The features of the ^{31}P-MAS NMR spectrum of plasma-sprayed hydroxyapatite are even more complex and hence open to discussion. Figure 8, inset shows the single band position C at +2.3 ± 0.1 ppm of the PO_4^{3-} tetrahedra in well-ordered, highly crystalline hydroxyapatite. In plasma-sprayed HAp coatings several additional band positions can be discerned neither of which can be unequivocally attributed to the known chemical shifts of thermal decomposition phases such as tricalcium phosphate or tetracalcium phosphate.

Instead, they were related to variously distorted PO_4^{3-} groups as follows (Figure 8). Band A at +5.0 ± 0.2 ppm may belong to very strongly distorted PO_4^{3-} groups without associated OH^- groups, band B at +3.0 ± 0.2 ppm is thought to be associated with distorted PO_4^{3-} groups with single or paired OH^- groups, band C at +2.3 ± 0.1 ppm is characteristic of well-ordered HAp with undistorted PO_4^{3-} groups, and finally band D at +1.5 ± 0.2 ppm has been assigned to a strongly distorted PO_4^{3-} environment without neighbouring OH^- [32, 22].

2D-^{1}H/^{31}P-CP MAS HETCOR NMR spectroscopy is a useful tool to investigate the changes in phase composition introduced during incubation of APS hydroxyapatite coatings in simulated body fluid. Additional work using 2D-double quantum ^{1}H/^{31}P cross polarisation (CP) hetero-correlation (HETCOR) NMR spectroscopy suggested that the phosphorus position D represents an oxyapatite, $Ca_{10}(PO_4)_6O$ short-range ordered (SRO) structure [18].

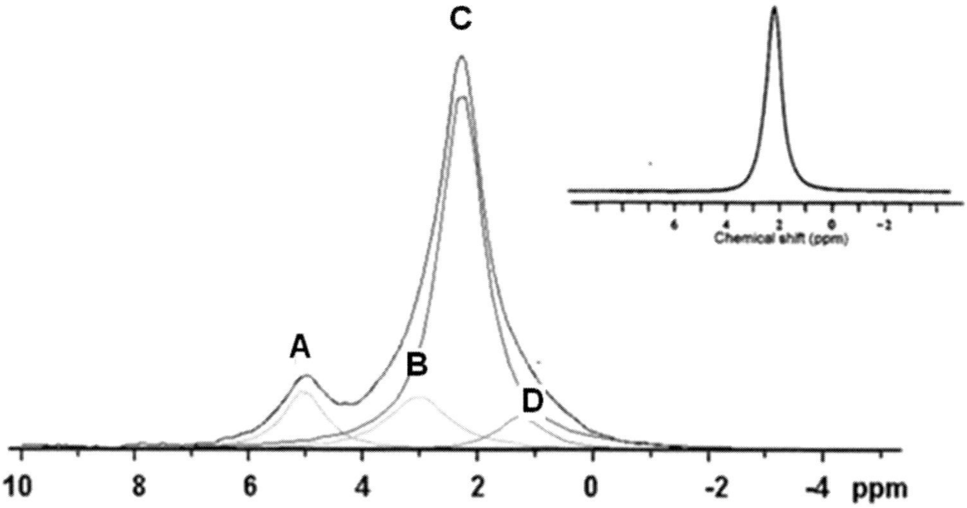

Figure 8. ^{31}P-MAS NMR spectrum of as-sprayed hydroxyapatite [32]. The inset shows the spectrum of a completely ordered, highly crystalline hydroxyapatite [18].

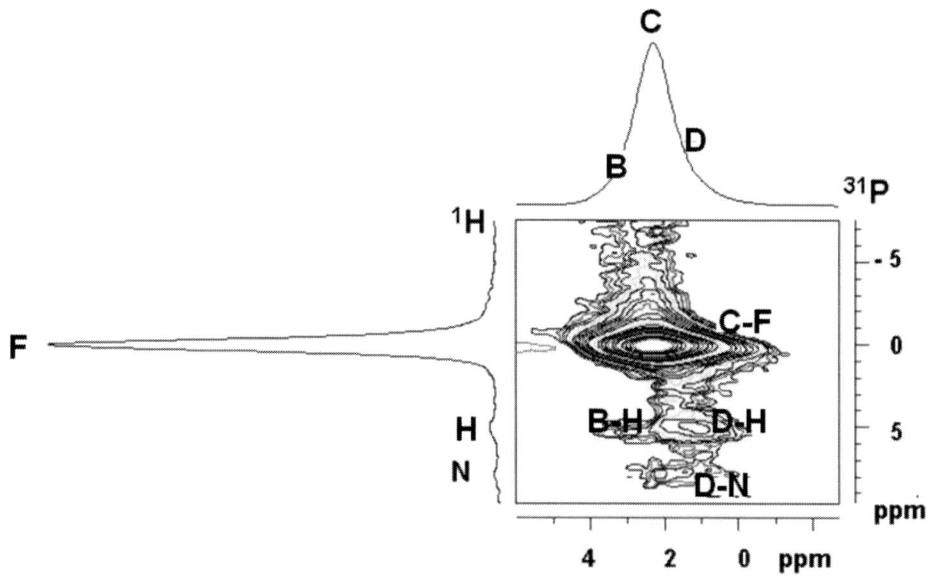

Figure 9. 2D-^{1}H/^{31}P-CP-HETCOR NMR spectrum of a plasma-sprayed hydroxyapatite coating [19,22].

Since this chemical shift is identical to that of crystalline tetracalcium phosphate it can be concluded that with increasing dehydroxylation of oxyhydroxyapatite the Ca^{2+} and O^{2-} coordinations of the PO_4^{3-} tetrahedra as represented by the ^{31}P D-band eventually approach those of tetracalcium phosphate, *i.e.* there exist a structural succession HAp → OHAp → OAp → TetrCP (+ TCP) as postulated earlier.

Figure 9 shows a 2D-^{1}H/^{31}P-CP-HETCOR NMR spectrum of a plasma-sprayed hydroxyapatite coating [19, 22]. The main contribution of the cross sectional peak is an C-F correlation peak (crystalline and highly stoichiometric, well-ordered hydroxyapatite) with some minor additions of B-H (distorted PO_4^{3-} with single or paired OH^-) and D-H (strongly distorted PO_4^{3-} without neighboring OH) that can be associated with a completely dehydroxylated apatite structure, *i.e.* oxyapatite *sensu lato* and Ca-deficient amorphous calcium phosphate (ACP), respectively.

The week and broad N band in the ^1H spectrum at about +7.5 ppm was attributed by Heimann *et al.* [19] to isolated OH^- groups in the z channel of apatite, signaling an advanced dehydroxylation state, i.e. the presence of OAp. These results suggest that stoichiometric hydroxyapatite and oxyapatite as well as ACP are present in the coating investigated. It is remarkable that some disordered hydroxapatite remains in the coating even after incubation in r-SBF for 12 weeks (Figure 10) despite its supposedly higher solubility. On the other hand, there is a remaining contribution of the D-H band and structures involving the B-band may also be present. This means that oxyapatite and tetracalcium phosphate, respectively have not been dissolved away completely.

The NMR spectrum changes during incubation of plasma-sprayed HAp coatings in r-SBF (Figure 10). The weak peak B-H disappears during incubation already after 1 week exposure time and may thus be assigned to TetrCP as confirmed by the data of Table 3.

Table 4. Quantitative composition in mass% of HAp coatings after incubation in r-SBF for 1 and 12 weeks (based on NMR data). Peak assignment of ^{31}P-MAS NMR spectrum as in Figure 8 (A: very strongly distorted PO_4 environment without OH, *i.e.* TetrCP and ß-TCP. B: distorted PO_4 groups with single or paired OH (+ TetrCP/TCP). C: well-ordered HAp. D: strongly distorted PO_4 environment without nearby OH, *i.e.* OAp and/or ACP [19]

Peak position	As-sprayed	1 week incubation	12 weeks incubation	Assignment
A	18	11	11	TetrCP + ß-TCP
B	21	14	11	TetrCP/ß-TCP
C	46	68	74	HAp
D	15	7	5	OAp/ACP

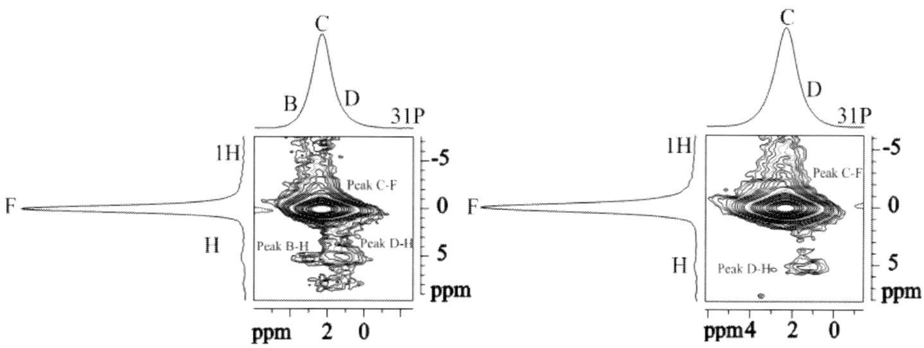

Figure 10. 2D-^{31}P (x-axis)-^{1}H (y-axis) HETCOR-CP-NMR spectra of as-sprayed HAp coatings (left) and incubated for 12 weeks in r-SBF (right).

Even after 12 weeks incubation (Figure 10, right) the coating still retains substantial amounts of non-HAp phases that render the coating essentially unstable under physiological conditions. Quantitative evaluation of the coating compositions based on the NMR results are summarized in Table 4.

3.4. Cathodoluminescence (CL), Scanning Transmission Electron Microscopy (STEM), and Electron Energy Loss Spectroscopy (EELS)

As shown in Tables 3 and 4 and the related NMR spectra a sizeable proportion of amorphous calcium phosphate (ACP) exists in the plasma-sprayed coatings that on contact with simulated body fluid transforms back to crystalline hydroxyapatite.

When the molten HAp particles impinge at the metal surface the idealised and neat phase separation shown in Figure 3 above will be completely lost. The result is an inhomogeneous calcium phosphate layer in which HAp, OHAp/OAp, ß-TCP (whitlockite), TetrCP (hilgardite), CaO (oldhamite), and ACP of various compositions are interspersed on a nano- to microcrystalline scale [35]. At the immediate interface to the metal substrate a thin layer of ACP exists formed by rapid quenching of the outermost shell with heat transfer rates beyond 10^6 K/s. Patches of ACP are also being scattered throughout the crystalline calcium phosphate phases as shown in Figure 11, right in an SEM micrograph in cathodoluminescence (CL) mode [35]. The thin ACP layer at the coating interface takes on a specific significance as its high solubility *in vivo* and pronounced cracking by shrinking on transformation to rehydroxylized HAp [23] are considered leading causes of coating delamination *in vivo* that is frequently observed in clinical application. Hence much research is directed towards deposition of well-crystallised hydroxyapatite layers with a minimum content of ACP. However, there is a caveat: too high a crystalline content may compromise coating performance in hip and dental prostheses owing to reduced adhesion strength and enhanced dissolution through a crack network [23].

Scanning transmission electron microscopy (STEM) studies on Ti6Al4V coupons coated with hydroxyapatite and incubated in protein-free r-SBF show very instructively how the as-deposited layer of ACP gradually transforms to nano-crystalline HAp on contact with the fluid. Figure 12, left reveals at high magnification that a transformation front sweeps through the volume of the ACP from top right to bottom left (direction of arrow) forming HAp nano-crystals at its trailing edge and leaving in its wake a rather porous coarse crystalline HAp [27,34].

Electron diffraction images show interplanar spacings of HAp at 0.336 nm (00.2), 0.306 nm (21.0) and 0.277 (11.2) as well as weak diffraction rings presumably related to TCP and TetrCP. Larger single crystals of HAp reveal in HRTEM micrographs (Figure 12, right) diffraction fringes spaced 0.39 nm apart, corresponding to (111) of HAp at d = 0.389 nm.

Study of the characteristics of amorphous calcium phosphate (ACP) by electron energy loss spectroscopy (EELS) confirms that in as-sprayed hydroxyapatite coatings two types of oxygen with different binding energies exist [27]. Figure 13, top shows an EEL spectrum of ACP. Oxygen ions bound to Ca and P are responsible for the main EEL peak of the O-K edge at 535 eV. In addition, there are oxygen ions in the structure of ACP carrying an additional negative charge gained from hydrogen thus turning the latter into a proton.

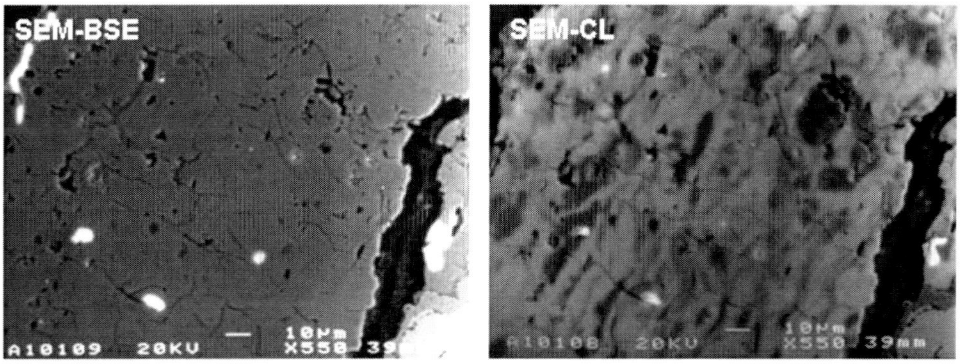

Figure 11. SEM micrographs in backscattered electron (BSE) mode (left) and cathodoluminescence (CL) mode (right) of a plasma-sprayed hydroxyapatite coating [35]. The dark regions in the CL image were identified as amorphous calcium phosphate (ACP).

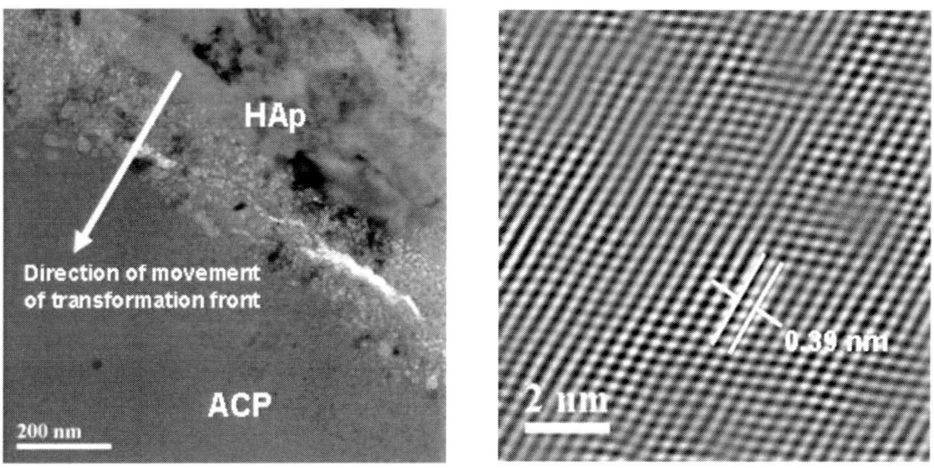

Figure 12. Left: STEM micrograph of ACP during transformation to crystalline HAp on contact with protein-free r-SBF for 24 weeks. Right: HRTEM diffraction fringes showing a spacing of d = 0.389 nm corresponding to (111) of hydroxyapatite [27, 34].

This causes increased shielding of the negative oxygen nuclear charge. In turn, the excitation of the 1s electrons of oxygen requires less energy than in the case of the undistorted orthophosphate group. This process is expressed by a small pre-peak (arrow in Figure 13, top) at the low energy slope of the O-K edge at 525 eV [33], predicting the existence of isolated protons and OH^{-}-ions in the ACP structure as suggested by the NMR data above. On crystallisation of the amorphous phase these protons will be lost and consequently the EEL spectrum of more or less stoichiometric HAp shows only the oxygen K-edge at 535 eV (Figure 13, bottom). It is realised that this information on the presence of water in ACP just scratches the surface of the problem. Much more information on the crucial role played by structural water incorporated into bone apatite is being provided by Pasteris in Chapter 3 of this volume [36].

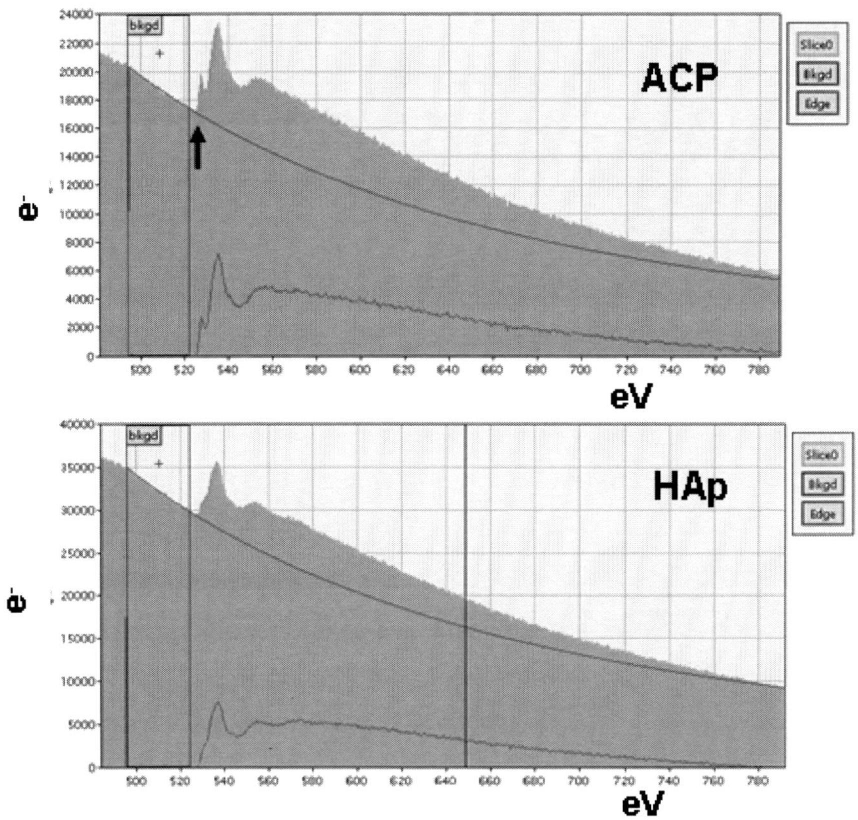

Figure 13. Electron energy loss spectra (EELS) of ACP (top) and crystalline HAp (bottom) in an as-sprayed hydroxyapatite coating [27]. The ACP structure is characterised by an additional peak (arrow) at the low energy slope of the oxygen K-edge at 535 eV.

Conclusion

The complex reaction sequence of the thermal dehydroxylation/ decomposition of hydroxyapatite subjected to a hot plasma jet during deposition on biomedical implants can be reasonably interpreted by laser-Raman and Nuclear Magnetic Resonance (NMR) spectroscopy. The first dehydroxylation steps lead to intermediate structures such as oxyhydroxyapatite/oxyapatite by successive abstraction of hydroxyl ions under maintaining the hexagonal lattice symmetry of the starting phase. The missing OH^- positions are occupied by mobile O^{2-} ions in oxyhydroxyapatite or by lattice vacancies in oxyapatite. They are of vital importance for understanding the kinetics of the dehydroxylation reactions.

With increasing temperature incongruent melting of oxyapatite occurs, forming α'-tricalcium phosphate that a still higher temperature decompose into CaO and P_2O_5. The extremely high temperature of the plasma jet shifts the Ca/P ratio by continuous evaporation of P_2O_5 towards CaO-richer phases such as tetracalcium phosphate. On impact with the much cooler substrate surface the melt droplets will be quenched rapidly and solidify with a high proportion of amorphous calcium phosphate (ACP) with various Ca/P ratios. Contact with body fluid *in vivo* after implantation of the coated device leads to partial recovery of the hydroxyapatite structure through biomimetic reaction paths.

REFERENCES

[1] A.L.Boskey and A.S. Posner, *J. Phys. Chem.* 77(19), 2313 (1973).

[2] H.M. Kim, T. Miyazaki, T. Kokubo, T. Nakamura, *Bioceramics*, 13, 192 (2001).

[3] R.B. Heimann, *J. Thermal Spray Technol.*, 8(4), 597 (1999).

[4] P.V. Riboud, *Ann. Chim.*, 8, 381 (1973).

[5] L.L. Hench, *J. Am. Ceram. Soc.*, 74(7), 1487 (1991).

[6] L. Pawlowski, *The Science and Engineering of Thermal Spray Coatings*, 2nd ed.,Wiley and Sons, Ltd., 2008.

[7] R.B. Heimann, *Plasma Spray Coating: Principles and Applications*, 2nd ed., Wiley-VCH, Weinheim, 2008.

[8] O. Graßmann and R.B. Heimann,. *J. Biomed. Mater. Res.*, 53(6), 685 (2000).

[9] M.T. Carayon and J.L. Lacout,. *J. Solid State Chem.*, 172, 339 (2003).

[10] G. Montel, G. Bonel, J.C. Trombe, J.C. Heughebaert and C. Rey, *Pure Appl. Chem.*, 52, 973 (1980).

[11] A.S. Posner, A. Perloff and A.F. Diorio, *Acta Cryst.*, 11, 308 (1958).

[12] J.C. Elliot, *Structure and Chemistry of the Apatites and other Calcium Orthophosphates,* Elsevier: Amsterdam, London, New York, Tokyo, 1994.

[13] N. Antolotti, S. Bertini, C. Fanaro, X. Ranz, C. Rey, F. Rusticchelli and A. Scrivani, *In*: *Thermal Spray. Meeting the Challenges of the 21st Century,* C.Coddet (ed.). Proc. 15th ITSC, May 25-29, 1998, Nice, France, p. 1121-1126.

[14] S. Dyshlovenko, B. Pateyron, L. Pawlowski and D. Murano, *Surf. Coat. Technol.*, 179, 110 (2004).

[15] J.C. Trombe and G. Montel,. *J. Inorg.Nucl. Chem.*, 40, 15-21, 27 (1978).

[16] C.J. Liao, F.H. Lin, K.S. Chen and J.S. Sun, *Biomaterials*, 20, 1807 (1999).

[17] J.C. Trombe and G. Montel, *Comp. Rend. Acad. Sci. Paris*, 273, 462 (1971).

[18] P. Hartmann, S. Barth, J. Vogel and Ch. Jäger,. In: *Applied Mineralogy in Research, Economy, Technology, Ecology and Culture*, Proc. ICAM 2000, Göttingen, D. Rammlmair et al. (eds.), A.A.Balkema, Rotterdam, Brookfield, Vol. 1, pp. 147-150, 2000.

[19] R.B. Heimann, H.V. Tran and P. Hartmann, *Mat.-wiss. u. Werkstofftechn.*, 34(12), 1163 (2003).

[20] T. Kijima and M. Tsutsumi, *J. Am. Ceram. Soc.*, 62 (9/10), 455 (1979).

[21] E. Park, R.A. Condrate, D. Lee, K. Kociba and P.K. Gallagher, *J. Mater. Sci.: Mater. Med.*, 13, 211(2002).

[22] H.V. Tran, *Investigation into the thermal dehydroxylation and decompo-sition of hydroxyapatite during atmospheric plasma spraying: NMR and Raman spectroscopic study of as-sprayed coatings and coatings incubated in simulated body fluid.* Ph.D. Thesis, Department of Mineralogy, Technische Universität Bergakademie Freiberg, D-09596 Freiberg, Germany, 2004.

[23] K.A. Gross, V. Gross and C.C. Berndt, *J. Am. Ceram. Soc.*, 81(1), 106 (1998).

[24] S. Dyshlovenko, B. Pateyron, L. Pawlowski and D. Murano, *Surf. Coat. Technol.*, 187, 408 (2004).
[25] C. Santos, A. Pazo and F. Guitian, In: *Materials in Clinical Applications*, P. Vincenzini (ed.), Techna, Faenza, Italy, pp. 11-18, 1995.
[26] K.A. Gross, C.C. Berndt and H. Herman, *J. Biomed. Mater. Res.*, 39(3), 407 (1998).
[27] R.B. Heimann and R. Wirth, *Biomaterials*, 27, 823 (2006).
[28] R. Cusco, F. Guitian, S. de Aza and L. Artus, *J. Eur. Ceram. Soc.*, 18, 1301 (1998).
[29] E. Park, R.A. Condrate and D. Lee, *Mater. Letters*, 36(1-4), 38 (1998).
[30] M. Weinlaender, J. Beumer III, E.B. Kenney, P.K. Moy and F. Adar, *J. Mater. Sci.: Mater. Med.*, 3, 397 (1992).
[31] R.B. Heimann and T.A. Vu, *J. Thermal Spray Technol.*, 6, 145 (1997).
[32] P. Hartmann, Ch. Jäger, S. Barth, J. Vogel and K. Meyer, *J. Solid State Chem.*, 160, 460 (2001).
[33] R. Wirth, *Phys. Chem. Min.*, 24, 561 (1996).
[34] R.B. Heimann, In: *Trends in Biomaterials Research*, P.J. Pannone (ed.), Nova Science Publishers, Inc., New York, 2007, pp 1-80.
[35] J. Götze, R.B. Heimann, H. Hildebrandt and U. Gburek, *Mat.-wiss. u. Werkstofftechn.*, 32, 130 (2001).
[36] J.D. Pasteris, In: *Calcium Phosphates: Structure, Synthesis, Properties and Applications,* R.B. Heimann (ed.), Nova Science Publishers, Inc., New York, Chapter 3. In print.

In: Calcium Phosphate
Editor: Robert B. Heimann

ISBN: 978-1-62257-299-1

Chapter 8

ELECTROCHEMICAL DEPOSITION OF CALCIUM PHOSPHATE COATINGS ON A PROSTHETIC TITANIUM ALLOY SUBSTRATE

Richard Drevet*[*] *and Hicham Benhayoune
INSERM UMR-S 926, URCA, France

ABSTRACT

Calcium phosphate coatings synthesized by electrodeposition on titanium alloy substrates (Ti6Al4V) are presented in this chapter. This process occurs at low temperature and allows obtaining thin coatings with a controlled thickness. Reaction mechanisms of the process are here described and the influence of the experimental parameters is investigated. Among them, the current density is used to control the chemical composition of the coating, but its value is limited because high values lead to a mechanically weak coating. To solve this problem, pulsed electrodeposition is proposed to obtain uniform coatings without the holes and the craters usually observed with the direct current electrodeposition. Higher current density can be employed, but its value is still limited and the chemical composition of the coating restricted to a calcium-deficient hydroxyapatite. Thus, hydrogen peroxide is added to the electrolytic solution to obtain monophasic coatings made of stoichiometric hydroxyapatite. *In vitro* behaviour of the electrodeposited coatings is tested in physiological medium. The results show that the calcium phosphate coating protects the titanium alloy against corrosion and promotes the precipitation of a bone-like apatite on its surface indicating the bioactivity of this prosthetic material.

1. INTRODUCTION

The electrodeposition of calcium phosphate coatings was first proposed in 1991 by Shirkhanzadeh [1]. This chemical process occurs at low temperature and the coating thickness is controlled.

[*] E-mail address: richard.drevet@univ-reims.fr

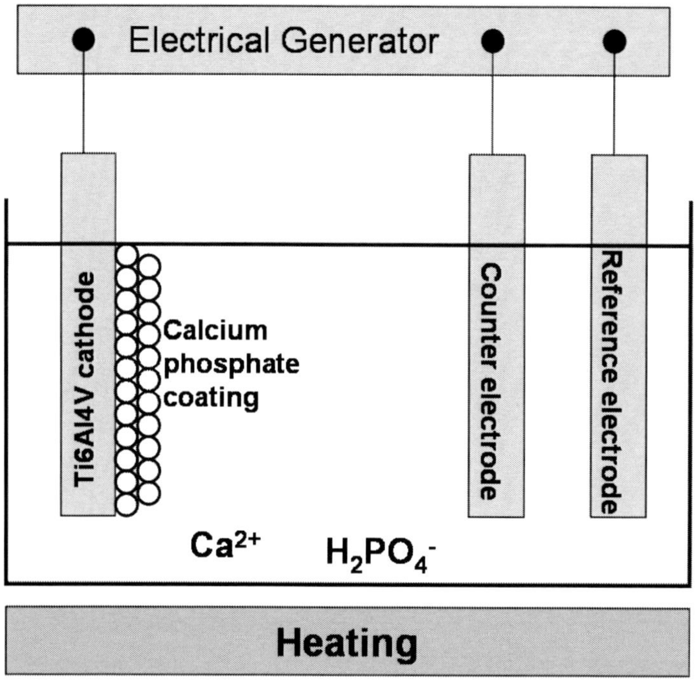

Figure 1. Schematic view of the experimental setup.

The technique uses a three-electrode system immersed in an electrolytic solution and connected to an electrical generator [2]. The electrodes comprise a cathode, also called the working electrode consisting of titanium alloy Ti6Al4V (i.e. the prosthesis), a platinum counter electrode, and a reference electrode. The electrolytic solution consists of calcium ions and dihydrogen phosphate ions obtained by salts dissolution in aqueous media.

The electrical generator provides the necessary energy to produce electrochemical reactions to the electrode surface, a stream of ions in the electrolytic solution, and a stream of electrons in the external circuit. This method allows controlling the chemical parameters related to the solution (salt concentrations, temperature, and pH) and the electrical parameters (voltage between the electrodes and current density through the circuit).

There are two ways of using the electrical generator. The first is the potentiostatic mode [3-5] in which the potential of the working electrode is kept constant compared to the reference electrode. The second is the galvanostatic mode (or intensiostatic) in which the current density value is kept constant during the deposition. The latter has the advantage of being able to control the energy provided at the electrode surface and thus the resulting electrochemical reactions [6].

2. Reaction Mechanisms for Electrodeposition of Calcium Phosphate Coatings

The electrolytic solution is prepared by dissolving a calcium salt and a phosphate salt in an aqueous medium. Generally, calcium nitrate tetrahydrate ($Ca(NO_3)_2.4H_2O$) and ammonium dihydrogen phosphate ($NH_4(H_2PO_4)$) are used. They dissolve in aqueous media as follow:

$$Ca(NO_3)_2.4H_2O \xrightarrow{H_2O} Ca^{2+} + 2NO_3^- \quad (1)$$

$$NH_4(H_2PO_4) \xrightarrow{H_2O} NH_4^+ + H_2PO_4^- \quad (2)$$

When an electric current is applied to the circuit, chemical reactions occur at the electrode-electrolyte interfaces: oxidation at the anode and reduction at the cathode. These reactions lead to local variations of the pH value of the solution which, by intermediate acid-base reactions, cause the precipitation of a calcium phosphate coating at the cathode surface. Therefore, three consecutive steps are necessary to obtain a calcium phosphate coating by electrodeposition: electrochemical reactions, acid-base reactions, and precipitation.

2.1. Electrochemical Reactions

These are redox reactions occurring at the electrode-electrolyte interface. The main reactions involve the water solvent, the anodic oxidation of which is:

$$2H_2O \rightarrow O_2\uparrow + 4H^+ + 4\bar{e} \quad (3)$$

and the cathodic reduction is:

$$2H_2O + 2\bar{e} \rightarrow H_2\uparrow + 2OH^- \quad (4)$$

If the pH of the solution is acidic, reduction of the proton can also take place:

$$2H^+ + 2\bar{e} \rightarrow H_2\uparrow \quad (5)$$

These reduction reactions locally increase the pH value of the solution at the cathode-electrolyte interface and then produce acid-base reactions. Nevertheless, some authors also describe other electrochemical reactions leading to a local variation of the pH value [7-10] according to:

$$O_2 + 2H_2O + 4\bar{e} \rightarrow 4OH^- \quad (6)$$

$$O_2 + 2H_2O + 2\bar{e} \rightarrow 2OH^- + H_2O_2 \quad (7)$$

$$NO_3^- + 2H^+ + 2\bar{e} \rightarrow NO_2^- + H_2O \quad (8)$$

$$NO_3^- + 10H^+ + 8\bar{e} \rightarrow NH_4^+ + 3H_2O \tag{9}$$

$$NO_3^- + H_2O + 2\bar{e} \rightarrow NO_2^- + 2OH^- \tag{10}$$

$$NO_3^- + 7H_2O + 8\bar{e} \rightarrow NH_4^+ + 10OH^- \tag{11}$$

$$NO_3^- + 6H_2O + 8\bar{e} \rightarrow NH_3 + 9OH^- \tag{12}$$

$$H_2PO_4^- + H_2O + 2\bar{e} \rightarrow H_2PO_3^- + 2OH^- \tag{13}$$

However, the amounts of oxygen, nitrate ions and dihydrogen phosphate ions are extremely low compared to the amount of water, the solution solvent. The reduction of water (or proton) is considered as the major cathodic reaction in this electrolytic process.

Thus, the pH values generated in the vicinity of the cathode lead to the dissociation of dihydrogen phosphate ions according to an acid-base process.

2.2. Acid-Base Reactions

The pH variations caused by the reduction of water dissociate the dihydrogen phosphate ions in the solution. The probability of occurrence of the phosphate ion moieties as a function of the pH value is shown in Figure 2 as proposed by Lynn and Bonfield [11].

When the electrolytic solution is prepared, its pH value is typically between 4 and 5, and dihydrogen phosphate ions ($H_2PO_4^-$) predominate.

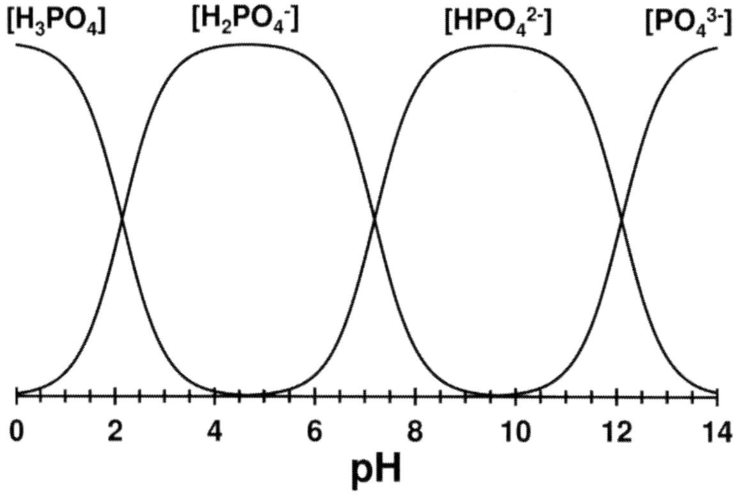

Figure 2. Distribution of the phosphate species as a function of the pH value.

If the local pH value at the cathode is between 7.2 and 11.8, hydrogen phosphate ions (HPO_4^{2-}) predominate according to reaction (14):

$$H_2PO_4^- \rightarrow HPO_4^{2-} + H^+ \qquad (14)$$

For pH values above 11.8, phosphate ions (PO_4^{3-}) predominate according to reaction (15):

$$HPO_4^{2-} \rightarrow PO_4^{3-} + H^+ \qquad (15)$$

2.3. Calcium Phosphate Phase Precipitation

The reactions (14) and (15) lead to a local ionic supersaturation which causes precipitation of calcium phosphate phases with low solubility product and high thermodynamic stability. As a function of the pH values obtained near the cathode, various compounds can precipitate [12-13]:

- Brushite (Ca/P = 1.00):

$$Ca^{2+} + HPO_4^{2-} + 2H_2O \rightarrow CaHPO_4.2H_2O \qquad (16)$$

- Octacalcium phosphate (Ca/P = 1.33):

$$8Ca^{2+} + 2HPO_4^{2-} + 4PO_4^{3-} + 5H_2O \rightarrow Ca_8(HPO_4)_2(PO_4)_4.5H_2O \qquad (17)$$

- Hydroxyapatite (Ca/P = 1.67):

$$10Ca^{2+} + 6PO_4^{3-} + 2OH^- \rightarrow Ca_{10}(PO_4)_6(OH)_2 \qquad (18)$$

- Calcium-deficient apatite (1.33 < Ca/P < 1.67):

$$(10-x)Ca^{2+} + xHPO_4^{2-} + (6-x)PO_4^{3-} + (2-x)OH^- \rightarrow Ca_{10-x}(HPO_4)_x(PO_4)_{6-x}OH_{2-x}$$

with $0 < x < 2$ (19)

When the calcium-deficient apatitic phase obtained by reaction (19) is thermally treated above 800°C, it crystallizes into a mixture of phases whose proportions depend on its stoichiometry (hence its atomic Ca/P ratio). This phase mixture can be composed of hydroxyapatite or tricalcium phosphate (TCP). Tricalcium phosphate is a calcium-deficient apatite with x = 1 and its atomic Ca/P ratio is exactly 1.5. During thermal treatment it crystallizes into β-TCP above 800°C and α–TCP above 1100°C [14].

$$Ca_9(HPO_4)(PO_4)_5(OH) \xrightarrow{thermal \quad treatment} 3Ca_3(PO_4)_2 + H_2O \qquad (20)$$

TCP β-TCP, then α-TCP

The composition of the calcium phosphate coating depends on the pH value near the cathode, which is linked to the current density imposed by the generator. However, other experimental parameters such as the electrolyte temperature or the ionic concentrations may also change the chemical composition of the coating.

3. Influence of the Experimental Parameters on Calcium Phosphate Coating Properties

3.1. Ionic Concentrations

The electrolytic solution is commonly prepared with a molar ratio of 1.67 between calcium ions and dihydrogen phosphate ions. Huelin *et al.* established that concentrations greater than 0.050 mol·L^{-1} for calcium ions and 0.030 mol·L^{-1} for dihydrogen phosphate ions promote brushite synthesis. In fact, an increase of the dihydrogen phosphate concentration hinders the formation of hydroxide ions and thus the increase of the pH value near the cathode [15].

3.2. Temperature

An increase of the electrolytic solution temperature is essentially used in order for degassing the carbon dioxide (CO_2) dissolved in water which is able to form carbonic acid (H_2CO_3). In an alkaline medium, bicarbonate ions (HCO_3^-) are formed followed by carbonate ions (CO_3^{2-}).

$$CO_2 + H_2O \rightarrow H_2CO_3 \qquad (21)$$

$$H_2CO_3 \rightarrow HCO_3^- + H^+ \qquad (22)$$

$$HCO_3^- \rightarrow CO_3^{2-} + H^+ \qquad (23)$$

Bicarbonate and carbonate ions can react with calcium ions in the electrolyte solution to form calcium carbonate which may precipitate at the cathode, thus decreasing the concentration of calcium ions in the electrolyte solution.

$$2HCO_3^- + Ca^{2+} \rightarrow CaCO_3 + H_2CO_3 \qquad (24)$$

$$CO_3^{2-} + Ca^{2+} \rightarrow CaCO_3 \quad (25)$$

Ban and Maruno noted that an increase of temperature leads to a higher rate of deposition of calcium phosphate crystals with larger size [16]. They also demonstrated that a temperature above 70°C is favourable to the precipitation of octacalcium phosphate rather than an apatitic phase [17].

3.3. Current Density

The reduction of water (production of hydroxide ions and release of hydrogen gas H_2) increases with the current density applied, thereby increasing the local pH value in the vicinity of the cathode. Zhang *et al.* [18] studied the variation of the pH value in the vicinity of the cathode as a function of the applied current density. For example, with a current density of 5 mA·cm^{-2}, the pH reaches a value slightly higher than 8. For this pH value, hydrogen phosphate ions are predominant (see Figure 2).

Several authors have characterized the phases obtained as a function of the applied current density. Zhang *et al.* have obtained a mixture of brushite and octacalcium phosphate at a current density of 1 mA·cm^{-2}, but tricalcium phosphate at a current density of 3 mA·cm^{-2}. Ban *et al.* got a coating of octacalcium phosphate at a current density of 1.3 mA·cm^{-2} [17]. Abdel-Aal *et al.* have obtained brushite at current densities ranging between 0.375 and 6 mA·cm^{-2} [19]. Dumelié *et al.* synthesized octacalcium phosphate by applying a current density of 2 mA·cm^{-2} but at a current density up to 10 mA·cm^{-2} a calcium-deficient apatitic phase with a tricalcium phosphate, stoichiometry [20].

The current density most frequently used and reported in the scientific literature is 10 mA·cm^{-2}. We can observe at high magnification that the electrodeposited coating obtained under this condition is composed of acicular crystals about 1 μm length and 0.1 μm width (Figure 3).

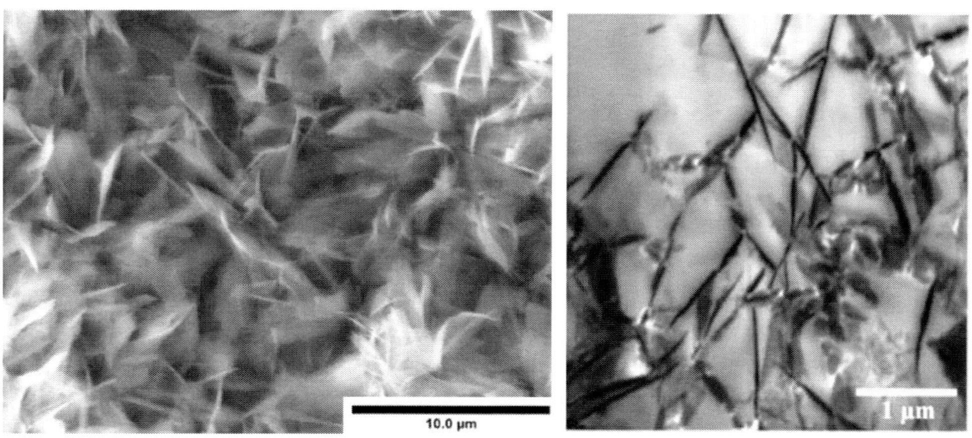

Figure 3. Calcium phosphate coating obtained by direct current electrodeposition (10 mA·cm^{-2}): (a) SEM image, (b) STEM image.

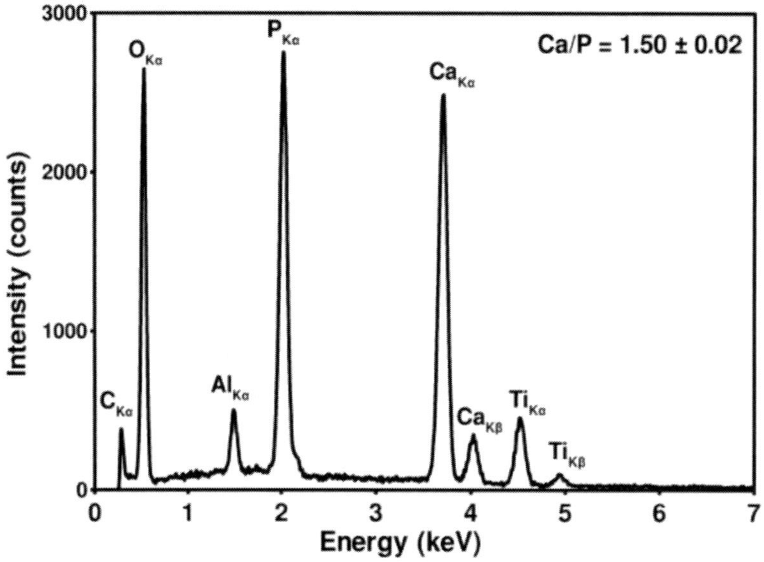

Figure 4. X-ray spectrum of the calcium phosphate coating obtained by direct current electrodeposition (10 mA·cm^{-2}).

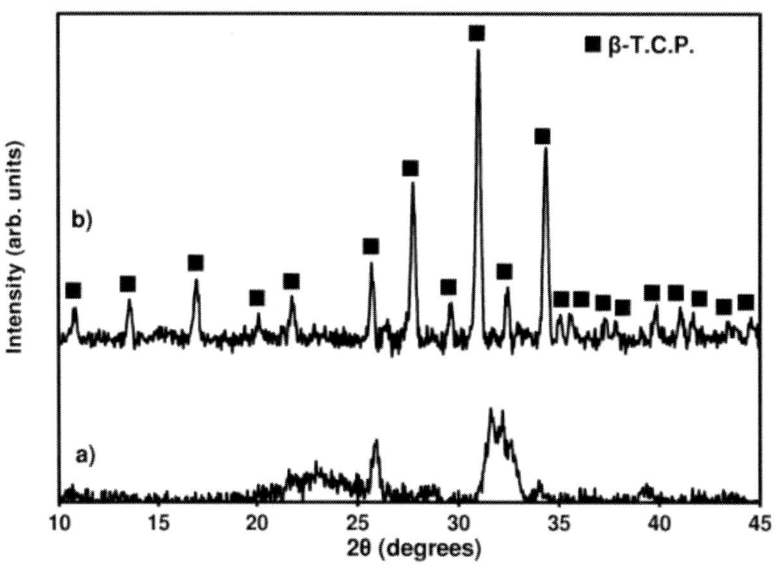

Figure 5. X-ray diffraction of the calcium phosphate coating obtained by direct current electrodeposition (10 mA·cm^{-2}): (a) before and (b) after thermal treatment.

X-ray microanalysis from the X-ray spectrum of this coating indicates an atomic Ca/P ratio equal to 1.50. This ratio corresponds to a specific calcium-deficient hydroxyapatite called tricalcium phosphate (TCP) which is crystallized into β-tricalcium phosphate after a thermal treatment above 800°C [21]:

$$Ca_9(HPO_4)(PO_4)_5(OH) \xrightarrow{thermal \quad treatment} 3Ca_3(PO_4)_2 + H_2O$$

TCP β-TCP

3.4. Problem with the Process

The current density is limited to 10 mA·cm^{-2} to avoid excessive production of hydrogen bubbles generated by the reduction of water. The resulting pH value in the vicinity of the cathode is thus limited. When a calcium phosphate coating is electrodeposited with a current density exceeding 10 mA·cm^{-2}, the amount of hydrogen bubbles generated is so large that orderly precipitation of the coating is disturbed at the cathode, causing numerous and large craters which weaken the cohesion of the coating [22].

As the current density is limited (and thus the pH value at the vicinity of the cathode), the chemical composition of the calcium phosphate coating is limited too, with an atomic Ca/P ratio equal to 1.50. It is a surprising result as all the users of the process prepare an electrolytic solution containing calcium ions and dihydrogen phosphate ions with a Ca/P ratio equal to 1.67 which is the characteristic ratio of the stoichiometric hydroxyapatite. In order to solve this problem, some authors propose to employ pulsed currents [23, 24].

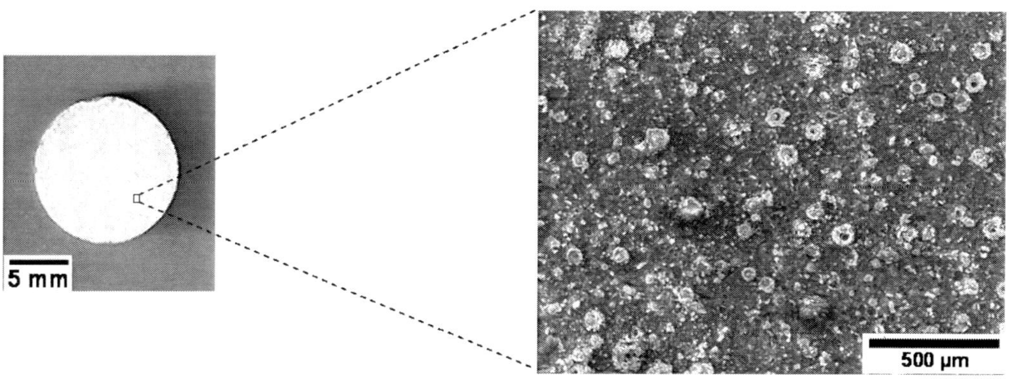

Figure 6. Photograph and SEM image of the electrodeposited calcium phosphate coating at 15 mA·cm^{-2}.

3.5. Pulsed Current Electrodeposition

A pulsed current for electrodeposition of calcium phosphate coatings is employed to reduce the problems associated with the accumulation of hydrogen bubbles at the surface of the cathode by creating current intermissions during which the bubbles can leave the substrate surface. Optimization of the intermission (t_b = 2 min.) and the deposition times (t_d = 1 min.) was proposed by Benhayoune *et al.* [25]. Then, the process can be performed with higher current densities up to 15 mA·cm^{-2}.

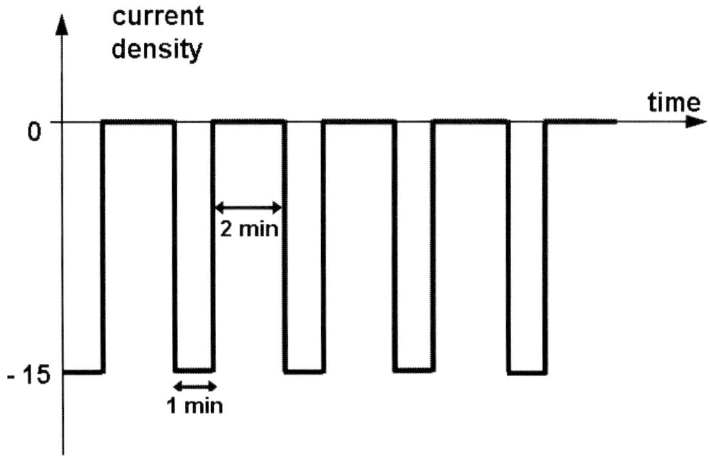

Figure 7. Diagram of pulsed current.

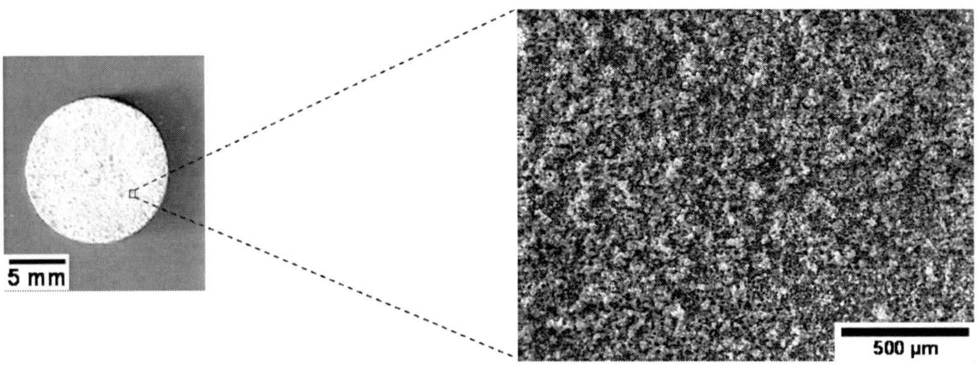

Figure 8. Photograph and SEM image of an electrodeposited calcium phosphate coating using a pulsed current of 15 $mA \cdot cm^{-2}$.

Figure 9. Calcium phosphate coating electrodeposited by pulsed current: (a) SEM image, (b) STEM image.

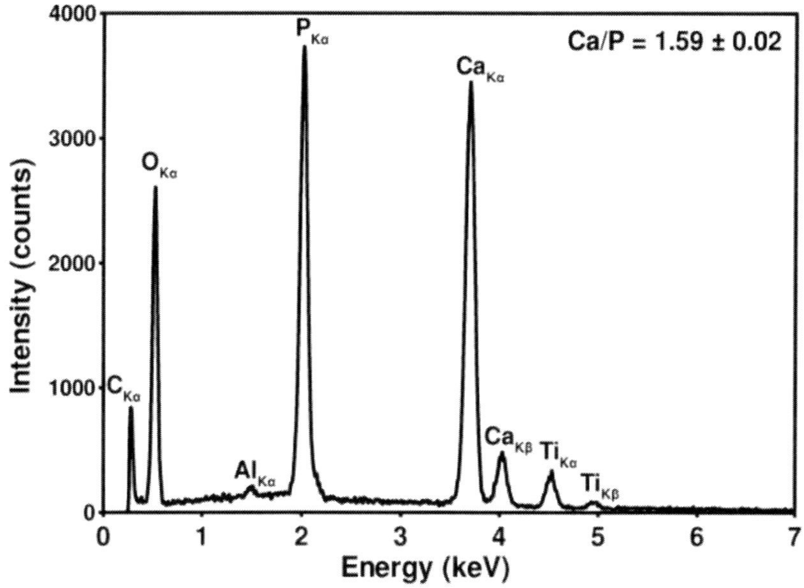

Figure 10. X-ray spectrum of an electrodeposited calcium phosphate coating obtained by pulsed current mode.

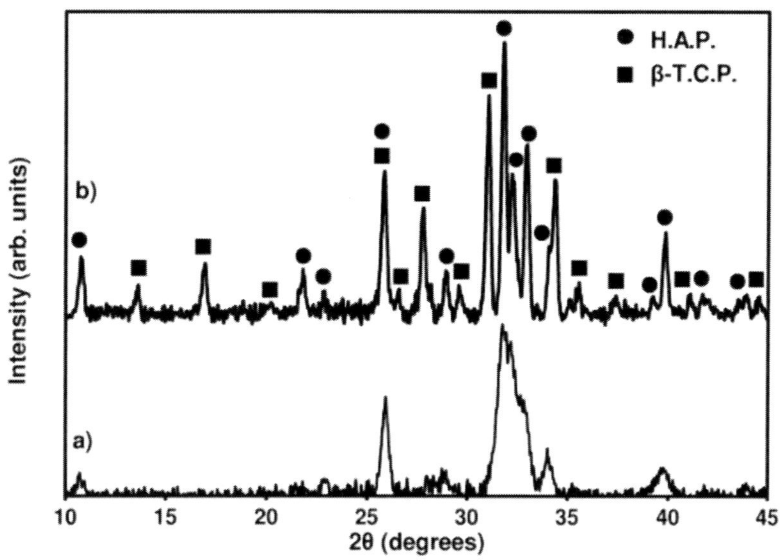

Figure 11. X-ray diffraction of the calcium phosphate coating obtained by pulsed current electrodeposition: (a) before and (b) after thermal treatment.

Figure 8 shows a scanning electron micrograph of an electrodeposited calcium phosphate coating produced using a pulsed current. The homogeneity of the coating and the absence of holes and craters indicate that a current density of 15 mA·cm^{-2} has successfully been applied in the pulsed regime.

Higher magnification images obtained by SEM and STEM are presented in Figure 9 and indicate that the coating is composed of agglomerates formed by needles finer than those obtained using the direct current mode.

The corresponding X-ray spectrum is presented in Figure 10 indicating that this coating has an atomic Ca/P ratio of 1.59.

This Ca/P ratio corresponds to calcium-deficient hydroxyapatite which leads to a mixture of hydroxyapatite (HAP) and β-tricalcium phosphate (β-TCP) after thermal treatment [26, 27]. This is confirms by X-ray diffraction (Figure 11) performed before and after thermal treatment. By using the method described in the standard ISO 13779-2, the phase's proportion can be quantified [28]. In the present case, the biphasic compound contains 52% in hydroxyapatite and 48% in β-tricalcium phosphate, that corresponds to an atomic ratio Ca / P = 1.58. This result confirms the atomic ratio determined by X-ray microanalysis.

According to reaction (19), precipitation of such a calcium-deficient hydroxyapatite phase can occur only if hydrogen phosphate ions are prevalent in the solution. Therefore it indicates that the pH value reached near the cathode is not sufficiently high to obtain stoichiometric hydroxyapatite in spite of using a current density of 15 $mA \cdot cm^{-2}$ although it appears to be the maximum value accessible by the pulsed current mode. In order to increase the pH value at the vicinity of the cathode, incorporation of hydrogen peroxide into the electrolytic solution has been proposed in combination with the pulsed current [29].

3.6. Electrodeposition with Hydrogen Peroxide (H_2O_2)

Incorporation of hydrogen peroxide in the electrolytic solution has been often used to electrodeposit ceramic coatings, mainly by Zhitomirsky [30, 31]. In the particular case of calcium phosphate coatings, only few studies have been conducted [32]. Hydrogen peroxide is expected to react and produce hydroxide ions in order to increase the pH value at the cathode. As a stronger oxidative reagent than water, it is first reduced at the cathode, producing only hydroxide ions by the following reaction:

$$H_2O_2 \quad + \quad 2\bar{e} \quad \rightarrow \quad 2OH^- \qquad (26)$$

Hydrogen peroxide incorporation into the electrolytic solution is limited to 9 % in volume as described by Zhitomirsky [33]: *"When the rate of* OH^- *generation is faster than the rate of* OH^- *consumption by hydrolysis reactions at the electrode, a fraction of the* OH^- *ions generated at the cathode is transported away by the electric current and diffusion. In this case, the high pH boundary moves away from the electrode surface, resulting in lower adhesion of the deposits."*

We can observe at high magnification in Figure 12 that the electrodeposited coating obtained under this condition is more compact, less porous and composed of thin needles agglomerates which form spheroids that are about 1 μm in diameter. The corresponding X-ray spectrum is presented in Figure 13, indicating that this coating has an atomic Ca/P ratio of 1.67.

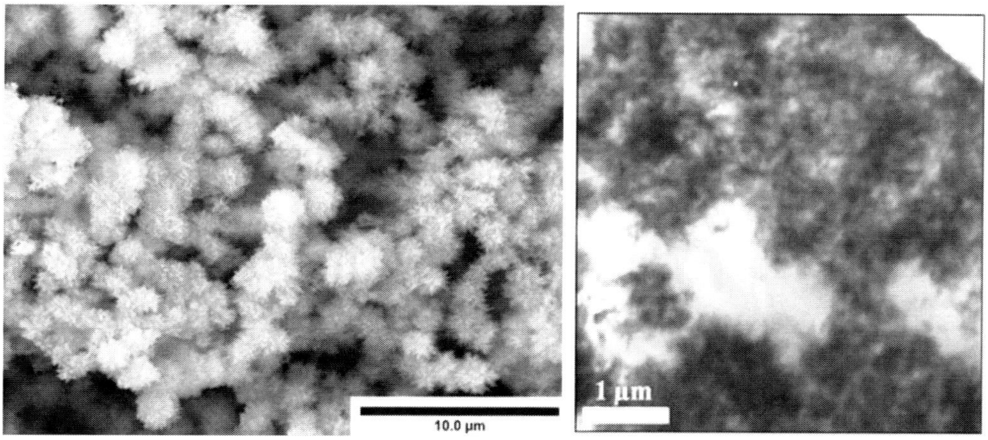

Figure 12. Electrodeposited calcium phosphate coating by pulsed current from 9 vol% H_2O_2 electrolyte: (a) SEM image, (b) STEM image. Reprinted from Ref. [29] with kind permission from Elsevier Publication.

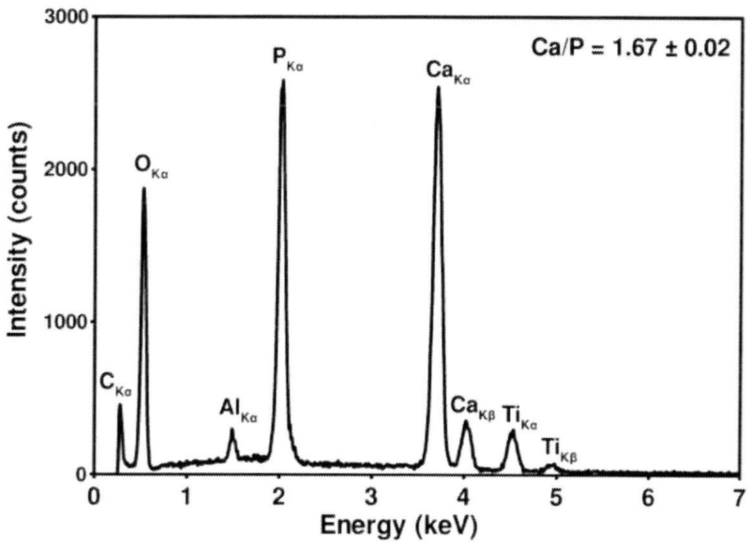

Figure 13. X-ray spectrum of electrodeposited calcium phosphate coating by pulsed current from 9 vol% H_2O_2 electrolyte.

The diffractograms of Figure 14 indicate that the coating is only composed of stoichiometric hydroxyapatite, fully crystallized after the thermal treatment. The corresponding atomic ratio is Ca/P=1.67.

Variation of the hydrogen peroxide amount introduced into the electrolytic solution until 9 vol% can be used to control the stoichiometry of the calcium phosphate coating electrodeposited by pulsed current. Without hydrogen peroxide, the electrodeposition process leads to a calcium-deficient hydroxyapatite coating, whereas a combination of pulsed current electrodeposition and 9 vol% of hydrogen peroxide leads to stoichiometric hydroxyapatite. The modification of the stoichiometry implies a modulation of the behaviour of the compound in physiological medium.

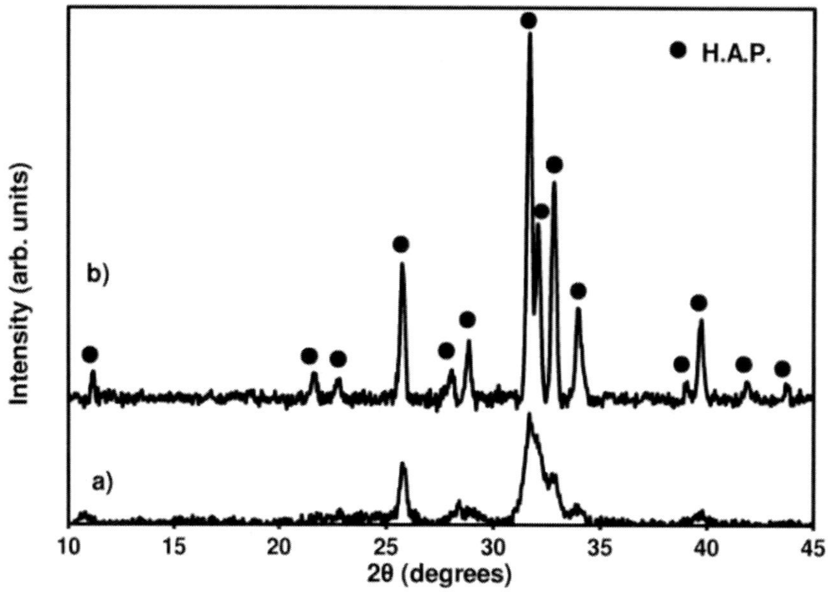

Figure 14. X-ray diffraction of the calcium phosphate coating obtained by pulsed current electrodeposition from 9 vol% H_2O_2 electrolyte: (a) before and (b) after thermal treatment.

4. *In Vitro* Behaviour

When a prosthetic material is implanted, its surface is the location of chemical interactions with the surrounding environment. Body fluids contain anions, cations and dissolved oxygen that provides a very corrosive environment [34]. Mu *et al.* have studied the corrosion of a titanium alloy implant immersed in physiological medium and observed that it produces a continuous ionic release up to 28 days of immersion [35]. Guibert *et al.* have noted *in vivo*, the presence of titanium, aluminium and vanadium up to several millimetres deep into the tissues surrounding a metallic Ti6Al4V implant [36]. Eisenbarth *et al.* have assessed the cytotoxicity of these ionic releases and showed that they cause a decrease in cell proliferation in conjunction with an increased inflammatory response [37, 38]. However, when a metallic alloy is coated with calcium phosphate, a corrosion mitigation is generally observed [39, 40].

Moreover, when they are immersed in physiological medium, calcium phosphates undergo partial dissolution. The resulting local ionic supersaturation leads to the precipitation of bone-like apatite whose role is to create a continuous link with the surrounding bone tissue [41-44]. This behaviour defines the bioactivity of an implant material according to the definition proposed by Hench [45].

4.1. Corrosion Study

The corrosion behaviours of the uncoated Ti6Al4V, the Ca-def HAP/Ti6Al4V system and the HAP/Ti6Al4V system, were studied in Dulbecco's Modified Eagle Medium

(D.M.E.M.), a physiological medium composed of ions and physiological elements as glucose, amino acids and vitamins. The samples were immersed in the physiological medium, the temperature of which was maintained at 37°C and their corrosion behaviour was studied by a stationary method, the Tafel representation of the polarisation curves (potential variation as a function of the current density). The polarisation curves are used to reveal the corrosion potential ($E_{corr.}$) and the corrosion current density ($i_{corr.}$) at the solid / liquid interface.

Figure 15 presents the polarisation curves of the HAP/Ti6Al4V system and the Ca-def HAP/Ti6Al4V system compared to those of the uncoated Ti6Al4V alloy. The electrodeposited coating shifts the corrosion potential $E_{corr.}$ to nobler values. This shift is more pronounced in the case of HAP/Ti6Al4V system and is accompanied by a decrease of the corrosion current density $i_{corr.}$ (see Table 1).

These results indicate that the electrodeposited calcium phosphate coatings act as a protective barrier for the titanium alloy against corrosion under physiological conditions, particularly in the case of the stoichiometric hydroxyapatite coating which is more compact and less porous than the calcium-deficient hydroxyapatite coating.

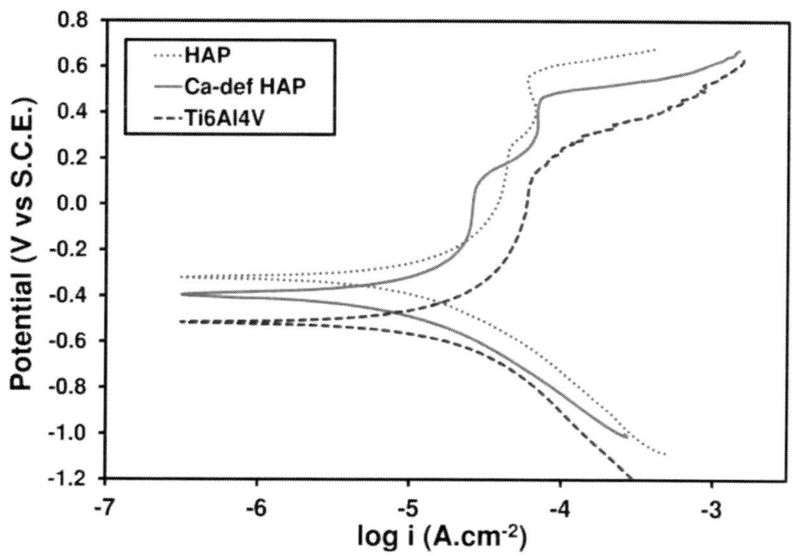

Figure 15. Polarisation curves of uncoated Ti6Al4V substrate, Ca-def HAP coating/Ti6Al4V and HAP coating/Ti6Al4V.

Table 1. Corrosion parameters extracted from the polarisation curves

Sample	E_{corr} (mV / ECS)	i_{corr} ($\mu A.cm^{-2}$)
HAP/Ti6Al4V	- 321	10
Ca-def HAP/Ti6Al4V	- 394	21
Ti6Al4V	- 514	47

4.2. Dissolution-Precipitation Reactions

During immersion in the physiological medium, calcium phosphate ceramics undergo dissolution-precipitation reactions the kinetic of which depends on their chemical composition. These reactions modify the concentrations of calcium and phosphate ions in the physiological medium and thus are characteristic of the biological activity of the implant. To highlight the behaviour of the calcium phosphate coatings electrodeposited by pulsed current mode, three systems (uncoated Ti6Al4V substrate, Ca-def HAP/Ti6Al4V and HAP/Ti6Al4V) were immersed in triplicate in DMEM maintained at 37°C for different time periods : 1, 6 and 12 hours, and 1, 2, 4, 7, 14, 21 and 28 days. For each period and each sample, physiological media were collected and analyzed in triplicate by Induced Coupled Plasma–Atomic Emission Spectroscopy (ICP-AES) to monitor the evolution of the calcium and phosphorus concentrations in the physiological medium.

Figure 16 shows the evolution of the calcium and phosphorus concentrations in the physiological medium as a function of the immersion period for the three systems (uncoated Ti6Al4V substrate, Ca-def HAP/Ti6Al4V and HAP/Ti6Al4V). First, the results indicate that immersion of the uncoated alloy does not change the composition of the solution during the 28 days of the study.

However, the study of the HAP / Ti6Al4V and Ca-def HAP / Ti6Al4V systems reveals a significant variation of calcium and phosphorus concentrations in DMEM. A significant decrease is observed between 0 and 7 days, suggesting that the precipitation of a calcium phosphate compound occurred at the surface of the electrodeposited coating. This decrease is larger and faster for the Ca-def HAP coating than that observed for the HAP coating. Indeed, in the case of the Ca-def HAP coating this decrease is observed from 12 hours of immersion and reached 50% of the initial value. In the case of the HAP coating, this decrease is observed from 4 days and reached 25% of the initial value.

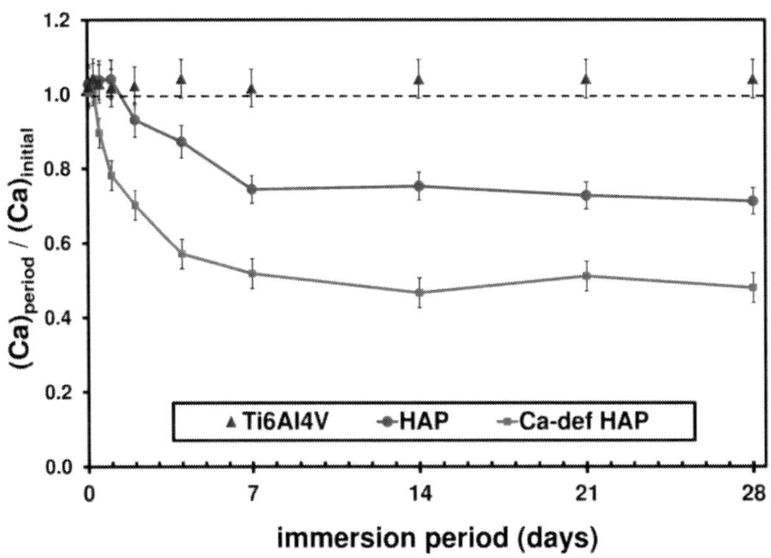

Figure 16a. Calcium evolution in DMEM as a function of the immersion period.

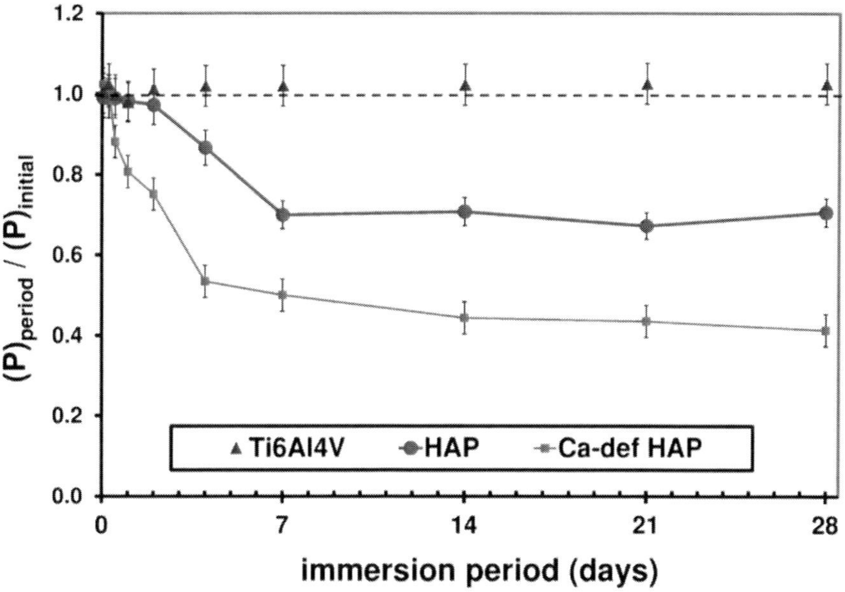

Figure 16b. Phosphorus evolution in DMEM as a function of the immersion period.

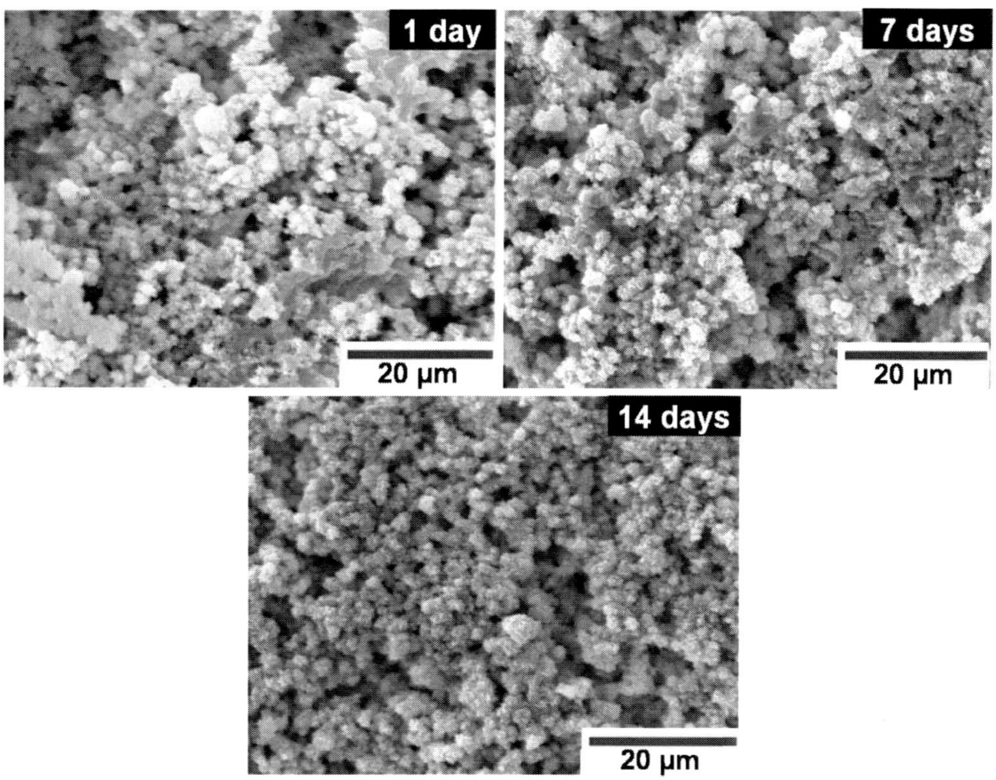

Figure 17. SEM micrographs of the coating showing the modification of the morphology after 1, 7 and 14 days of immersion time in DMEM.

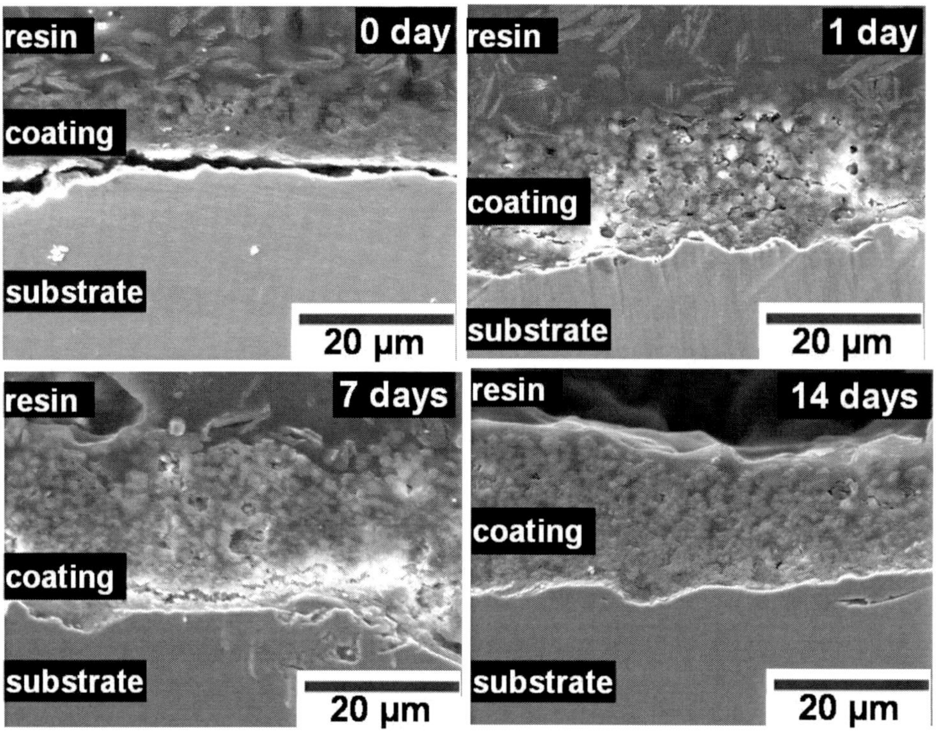

Figure 18. Coating/substrate interface evolution as a function of immersion time in DMEM.

Finally, there are no more significant variations of the calcium and phosphorus concentrations after 7 days of immersion. These changes in the composition of the physiological medium indicate that the dissolution / precipitation reactions are more pronounced for the Ca-def HAP coating the solubility of which is higher than that of the HAP coating [46]. The precipitation of the new apatite layer on the surface of the implant modifies its morphology, especially in the case of the Ca-def HAP coating. In order to observe this modification, SEM images of Figure 17 present the surface of the coated Ti6Al4V implant after immersion in DMEM during 1, 7 and 14 days.

After one day of immersion, the coating reveals needle-like crystals agglomerated into spheres linked by a new layer with porosity still apparent. After 7 and 14 days of immersion, the needles agglomerated into spheres become thinner and the pores of the coating are less important, suggesting that the precipitation of the apatite phase has partially filled the pores. These modifications are also observed by SEM cross-sections presented on Figure 18. After immersion, the coating thickness is enhanced reaching about 25 µm instead of about 10 µm for the as-deposited coatings.

X-ray emission spectra shown in Figure 19 are used to obtain more information on the new apatite layer formed in the physiological medium. The characteristic K_α peaks of titanium, aluminium and vanadium decrease significantly when the immersion time increases. This result is consistent with the thickness increase and the porosity decrease observed by SEM. After immersion, the presence of characteristic K_α peaks of sodium, magnesium, chlorine and potassium suggests that the precipitation of the new apatite layer has taken up the ions of the physiological medium to form a "bone-like" apatite [47].

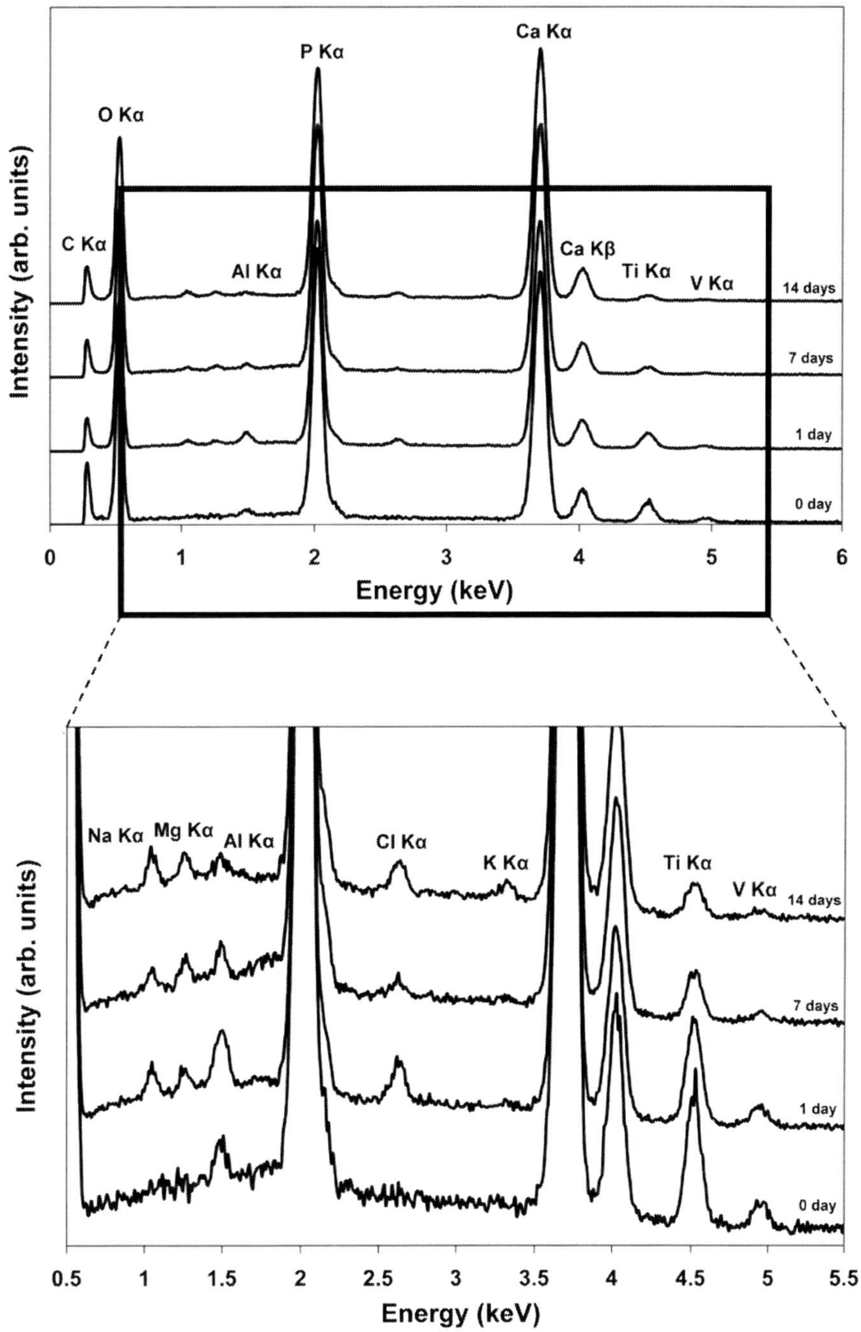

Figure 19. EDXS spectra as a function of immersion time in DMEM.

CONCLUSION

In this chapter, electrodeposition of calcium phosphate coatings was presented by evaluating the three steps comprising the process: electrochemical reaction, acid-base reaction, and precipitation reaction. The mechanisms occurring during these three steps have been described. The influence of the chemical and electrical parameters on the composition of the electrodeposited calcium phosphate coating was studied in order to understand the limits of the process. When a current density of 15 $mA \cdot cm^{-2}$ is used in direct current mode, the electrodeposited coating is heterogeneous and thus fragile. Consequently, in the direct current mode, calcium phosphate coating electrodeposition is limited to 10 $mA \cdot cm^{-2}$ that constrains the coating composition. To solve this problem two combined methods have been proposed and studied: the use of pulsed currents and the incorporation of hydrogen peroxide (H_2O_2) in the electrolyte solution.

Pulsed currents have been employed to apply the required current density of 15 $mA \cdot cm^{-2}$ and to obtain a homogenous Ca-def HAP coating that lead to a biphasic compound made of 52% HAP and 48% β-TCP after an appropriate thermal treatment. Moreover, the addition of 9 vol% hydrogen peroxide to the electrolyte solution has also modified the chemical composition of the electrodeposited prosthetic coating. Under these conditions it is composed of stoichiometric HAP fully crystallized after an appropriate thermal treatment.

The combination of the two methods allows obtaining homogenous calcium phosphate coating with a controlled stoichiometry. The behaviour of the coatings was studied in DMEM physiological solution. Polarisation measurements have shown that the synthesized coating protect the titanium alloy against corrosion, especially in the case of the HAP coating. Dissolution/precipitation reactions occurring during immersion of the coating were studied and revealed that the process occurred during the first 7 days and that it was more pronounced in the case of the Ca-def HAP coating the solubility of which is higher than that of a HAP coating. This precipitation of a new apatite layer caused surface modifications of the implant that have been observed by SEM. The chemical composition of this layer has been identified to be a 'bone-like' apatite, confirming and highlighting the bioactivity of electrodeposited calcium phosphate coatings synthesized by application of a pulsed current.

REFERENCES

[1] M. Shirkhanzadeh, *J. Mater. Sci. Lett.* 10, 1415 (1991).

[2] R. Drevet, F. Velard, S. Potiron, D. Laurent-Maquin, H. Benhayoune, *J. Mater. Sci.: Mater. Med.* 22, 753 (2011).

[3] M. Shirkhanzadeh, *J. Mater. Sci.: Mater. Med.* 9, 67 (1998).

[4] M. Manso, C. Jiménez, C. Morant, P. Herrero and J. M. Martinez-Duart, *Biomaterials* 21, 1755 (2000).

[5] C. Montero-Ocampo, D. Villegas and L.Veleva, *J. Electrochem. Soc.* 152, 692 (2005).

[6] J. Redepenning and J. P. MacIsaac, *Chem. Mater.* 2, 625 (1990).

[7] R. Narayanan, T. Y. Kwon and K. H. Kim, *Mater. Sci. Eng. C* 28, 1265 (2008).

[8] G. H. A. Therese and P. V. Kamath, *Chem. Mater.* 12, 1195 (2000).

[9] N. Eliaz and T. M. Sridhar, *Cryst. Growth Des.* 8, 3965 (2008).

[10] N. Eliaz and M. Eliyahu, *J. Biomed. Mater. Res. A* 80, 621 (2007).

[11] K. Lynn and W. Bonfield, *Acc. Chem. Res.* 38, 202 (2005).

[12] Y. Han, K. Xu and J. Lu, *J. Mater. Sci.: Mater. Med.* 10, 243 (1999).

[13] Rakngarm and Y. Mutoh, *Mater. Sci. Eng. C* 29, 275 (2009).

[14] S. Raynaud, E. Champion, D. Bernache-Assollant and P. Thomas, *Biomaterials* 23, 1065 (2002).
[15] S. D. Huelin, H. R. Baker, E. F. Merschrod and K. M. Poduska, *Cryst. Growth Des.* 6, 2634 (2006).
[16] S. Ban and S. Maruno, *Biomaterials* 16, 977 (1995).
[17] S. Ban and S. Maruno, *Biomaterials* 19, 1245 (1998).
[18] J. M. Zhang, C. J. Lin, Z. D. Feng and Z. W. Tian, *J. Electroanal. Chem.* 452, 235 (1998).
[19] E. A. Abdel-Aal, D. Dietrich, S. Steinhaeuser and B. Wielage, *Surf. Coat. Technol.* 202, 5895 (2008).
[20] N. Dumelié, H. Benhayoune, C. Rousse-Bertrand, S. Bouthors, A. Perchet, L. Wortham, J. Douglade, D. Laurent-Maquin and G. Balossier, *Thin Solid Films* 492, 131 (2005).
[21] S. Raynaud, E. Champion and D. Bernache-Assolant, *Biomaterials* 23, 1073 (2002).
[22] S. Rössler, A. Sewing, M. Stölzel, R. Born, D. Scharnweber, M. Dard and H. Worch, *J. Biomed. Mater. Res. A* 64, 655 (2003).
[23] S. Lin, R. Z. LeGeros and J. P. LeGeros, *J. Biomed. Mater. Res. A*.66, 819 (2003).
[24] J. P. LeGeros, S. Lin, D. Mijares, F. Dimaano and R. Z. LeGeros, *Key Eng. Mater.* 284, 247 (2005).
[25] H. Benhayoune, R. Drevet, J. Faure, S. Potiron, T. Gloriant, H. Oudadesse and D. Laurent-Maquin, *Adv. Eng. Mater.* 12, B193 (2010).
[26] S. Raynaud, E. Champion, D. Bernache-Assolant and J. P. Laval, *J. Am. Ceram. Soc.* 84, 359 (2001).
[27] S. C. Liou and S. Y. Chen, *Biomaterials* 23, 4541 (2002).
[28] Standard for Implants for surgery -- Hydroxyapatite -- Part 2: Coatings of hydroxyapatite ISO 13779-2 (2008).
[29] R. Drevet, H. Benhayoune, L. Wortham, S. Potiron, J. Douglade and D. Laurent-Maquin, *Mater. Charact.* 61, 786 (2010).
[30] Zhitomirsky, *J. Eur. Ceram. Soc.* 18, 849 (1998).
[31] Zhitomirsky, *J. Eur. Ceram. Soc.* 19, 2581 (1999).
[32] Wang, Y. Chao, Q. Wan, Z. Zhu and H. Yu, *Acta Biomater.* 5, 1798 (2009).
[33] Zhitomirsky, *Adv. Colloid Interface Sci.* 97, 279 (2002).
[34] R. Singh and N. Dahotre, *J. Mater. Sci.: Mater. Med.* 18, 725 (2007).
[35] Y. Mu, T. Kobayashi, M. Sumita, A. Yamamoto and T. Hanawa, *J. Biomed. Mater. Res. B : Appl. Biomater.* 49, 238 (2000).
[36] G. Guibert, J. L. Irigaray, P. Moretto, T. Sauvage, J. L. Kemeny, A. Cazenave and E. Jallot, *Nucl. Instr. Meth. Phys. Res. B* 251, 246 (2006).
[37] E. Eisenbarth, D. Velten, M. Müller, R. Thull and J. Breme, *Biomaterials* 25, 5705 (2004).
[38] E. Eisenbarth, D. Velten, K. Schenk-Meuser, P. Linez, V. Biehl, H. Duschner, J. Breme and H. Hildebrand, *Biomol. Eng.* 19, 243 (2002).
[39] R. M. Souto, M. M. Laz and R. L. Reis, *Biomaterials* 24, 4213 (2003).
[40] P. G. Coelho, S. L. de Assis, I. Costa and V. P. Thompson, *J. Mater. Sci.: Mater. Med.* 20, 215 (2009).
[41] F. Barrere, T. A. Mahmood, K. De Groot and C. A. Van Blitterswijk, *Mater. Sci. Eng. R.* 59, 38 (2008).

[42] P. Ducheyne and Q. Qiu, *Biomaterials* 20, 2287 (1999).
[43] Q. Zhang, J. Chen, J. Feng, Y. Cao, C. Deng and X. Zhang, *Biomaterials* 24, 4741 (2003).
[44] P. Ducheyne, S. Radin, L. King, *J. Biomed. Mater. Res.* 27, 25 (1993).
[45] L. Hench, *J. Am. Ceram. Soc.* 81, 1705 (1998).
[46] W. Cao and L. L. Hench, *Ceram. Int.* 22, 493 (1996).
[47] H. Wang, N. Eliaz, Z. Xiang, H. P. Hsu, M. Spector and L. W. Hobbs, *Biomaterials* 27, 4192 (2006).

In: Calcium Phosphate
Editor: Robert B. Heimann

ISBN: 978-1-62257-299-1

Chapter 9

NANOSTRUCTURED HYDROXYAPATITE COATINGS FOR MEDICAL IMPLANTS DEPOSITED BY RADIO-FREQUENCY MAGNETRON SPUTTERING: PROPERTIES AND BIOMEDICAL APPLICATION

Roman A. Surmenev[1], Maria A. Surmeneva[1], Vladimir F. Pichugin[1] and Igor A. Khlusov[1,2]

[1]Tomsk Polytechnic University (TPU), Scientific Educational Centre 'Biocompatible Materials and Bioengineering' at TPU, Tomsk, Russia
[2]Siberian State Medical University, Tomsk, Russia

ABSTRACT

This chapter focuses on current achievements in the field of radio-frequency (RF) magnetron sputter (RFMS) deposition of biocompatible CaP-based coatings for medical implants with an emphasis on their composition, structure, mechanical and biological performance. The chapter shows that RFMS allows preparing dense, homogeneous, pore-free and highly adherent biocompatible coatings able to prevent leaching of toxic ions from metal to the surrounding tissues, and capable of stimulating osteogenesis of a new bone. The main advantages and limitations of this technique are presented as well as the most important challenges and critical issues are highlighted.

1. INTRODUCTION

Metallic biomaterials such as stainless steel (SS), cobalt-based alloys, and titanium (Ti) and its alloys are widely used as artificial hip joints, bone plates and dental implants due to their excellent mechanical properties and endurance [1]. However, there are some problems associated with metallic implants including corrosion and wear in biological environments resulting in ion release and accumulation of wear debris, poor implant fixation owing to the lack of osteoconductivity and osteoinductivity, and infections due to bacterial adhesion and colonization at the implantation site. Because of poor osteoconductivity, metallic implants often become encapsulated by fibrous tissue thus prolonging the healing time [2]. To overcome these surface-related problems, a variety of surface modification techniques have been used on metallic implants, including chemical treatment, physical, and biological methods [3]. Over the last decade, biomaterials research has been focused on the improvement of implant design features in an attempt to accelerate bone healing at early implantation times. Considering that the implant surface is the first part of the implant that interacts with the host, surface modification has been essential in enhancing biocompatible and osseoconductive properties of the implant [3,4]. Hydroxyapatite (HA, $Ca_{10}(PO_4)_6(OH)_2$) is the main chemical constituent of bone tissue (~70%) and has been widely employed in non-load-bearing applications, as a bioactive material with a desirable bone response [5]. However, due to the very brittle nature of bulk HA ceramics, it cannot be used in orthopedic devices that must withstand substantial forces during their expected lifetimes. Still, as a coating HA provides significant improvement of the metallic implants osteoconductivity as it is chemically very similar to inorganic bone matrix [6,7].

It has been established that HA coating must meet the minimal requirements as described by the U.S. Food and Drug Administration (FDA) and the International Organization for Standardization (ISO) [8-11].

A variety of techniques have been employed to obtain CaP-based, most frequently HA, coatings with desirable features including powder as well as suspension or liquid plasma spraying (PS) [12-16], radio-frequency magnetron sputtering (RFMS) [17-21], direct current (DC) magnetron sputtering [22], pulsed laser deposition (PLD) [23-24], ion implantation [25], ion beam sputtering [26,27], ion beam dynamic mixing (IBDM) [28,29], and ion beam-assisted deposition (IBAD) [30-33], electrophoretic deposition [34,35], sol-gel [36,37], micro-arc oxidation [38], biomimetic coating techniques [39,40], hot isostatic pressing [41], etc. Additional information on the methodologies for CaP ceramic coatings preparation and investigation of their performance can be found elsewhere [7,42].

According to the opinion of researchers in medical material science and bioengineering, the main processes of histogenesis take place at the interface of artificial material (implant) and biological tissue [1].

The biological response to artificial materials is determined by certain factors including their surface topography, chemical composition, degradation rate, dissolution product type, and mechanical characteristics of the implant. Therefore, at the present time there are developing approaches to use the material with technological and biological characteristics for specific diseases and patients determining the concepts of 'intelligent implants' and 'personalized therapy'.

Plasma processes allow adjusting surface properties of solids with negligible effect on their bulk [43]. In addition, it has recently been shown that plasma-based technology is very promising as it can modify surface properties, even creating a nano-sized surface topography [44,45]. Moreover, RFMS is regarded an advanced technology for biomedical applications [42]. Nevertheless, a common problem for actual and developing implants is their surface optimization for enhancing bone cells growth and bone tissue formation.

2. PROPERTIES OF RFMS-DEPOSITED CAP COATINGS

2.1. Concept of RFMS

Since 1952, sputtering is a reported physical phenomenon and has now developed into a thin film deposition technique [46,47]. The process operates at room temperature but in a vacuum environment. In general, sputtering involves the surface erosion of target materials. The erosion of target material is largely driven by momentum exchange as a result of collisions between high energy gaseous ions generated by plasma or ion guns, and atoms from the target surface. The high energy gaseous ions, accelerated by high voltage, bombard and collide with atoms on the target surface. As a result, the atoms or occasionally entire molecules of the target material are then ejected and transported to the substrate surface, subsequently resulting in a thin film that forms by mechanical interlocking and chemical bonding with diverse substrates [7,47-49].

There are several modes of sputtering that are generally classified into RFMS [18,45], RF diode sputtering [50,51], DC magnetron sputtering [52], and ion-beam sputtering [53,54]. As HA is an electrically insulating material, DC sputtering cannot be used as a deposition process. This is because the HA target serves as a cathode, attracting positively charged ions, and the insulating nature of HA determines the accumulation of these charges on the HA surfaces. Consequently, fluxes of ions and electrons to the surface are balanced, regardless of the negative potential applied to the cathode. As this equalization occurs, the sheath voltage across the HA target falls, and no voltage remains to drive the discharge. In contrast to a DC source, RF sputtering utilizes an alternating current (AC) source, which allows to sputter ceramic targets, particularly HA [47].

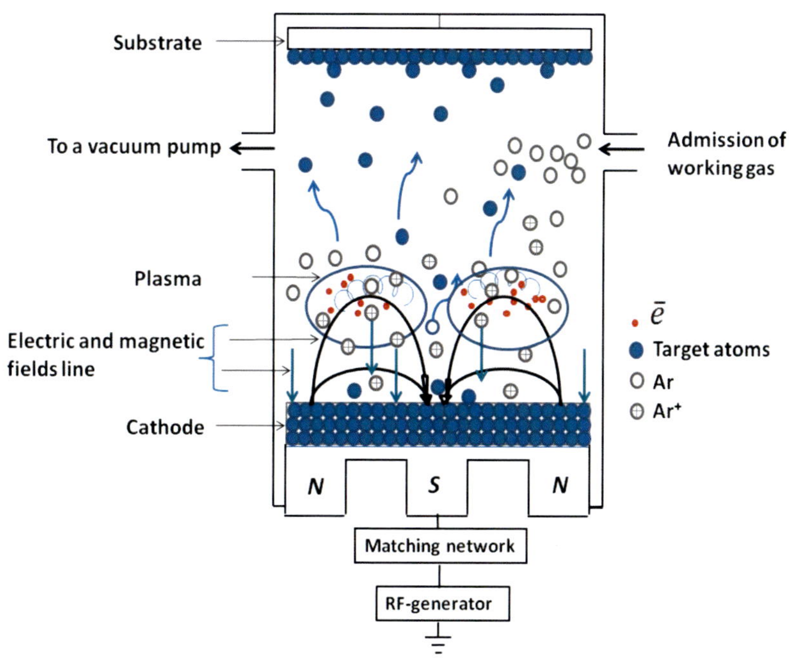

Figure 1. Schematic diagram of a typical RFMS facility.

Table 1. The comparison between the properties of RF magnetron sputter deposited coatings and that prepared by other most-frequently used plasma-assisted techniques

Method	Thickness*, μm	Main advantages	Main limitations
RF magnetron sputtering	~0.04-3.5	Uniform coating thickness; dense pore-free coating; ability to coat heat-sensitive substrates; high-purity films; ability to control the coating structure (amorphous or crystalline) and the Ca/P ratio; bonding strength of the coating exceeds 30 MPa	Line-of-sight technique; expensive; time consuming due to low deposition rate; in most cases Ca/P ratio of the coating is higher than that of synthetic HA due to P_2O_5 evaporation and coating resputtering effect
Plasma spraying	~30-300	High deposition rates; regarded as the most efficient and economical technique; coatings usually have micro-rough surface and porosity to enhance the adhesion of osteoblast cells and their subsequent proliferation; in most cases tensile adhesion strength above 15 MPa	Line-of-sight technique; high temperatures induce decomposition; rapid cooling produces cracks in coatings; poor control of chemical and physical coating parameters; lack of uniformity of the coating thickness; poor control of biodegradation
Laser ablation (or pulsed laser deposition)	~0.05-10.00	Coatings by crystalline and amorphous phases; dense and porous with controlled phases and Ca/P ratios; MAPLE further increases the variety of molecules and macromolecules, in particular heat-labile compounds which can be co-deposited with CaPs	Line-of-sight technique; expensive; high substrate temperature (usually 350-750 °C) is necessary to prepare a crystalline HA coating

* - variations are possible due to different deposition parameters.

RFMS was used to prepare HA coating by Yamashita [55], Jansen, Wolke et al. [18,19,21,56,57], and Yoshinari et al. [58]. The typical setup for the preparation of CaP-based thin films (Figure 1) includes a vacuum chamber, an RF generator, a matching network, a magnetron, and a cooling system.

Typically, RFMS employs a sinusoidal wave generator operating at 13.56 MHz [59-64], 5.28 MHz [45,65] or 1.78 MHz [66,67]. The use of the matching network and a controller allows the RF supply to match the generator output impedance to the impedance of cathode and anode. The parameters that directly affect the quality and integrity of the CaP coating include discharge power, gas flow rate, working pressure and atmosphere (argon, oxygen, water vapor or a mixture thereof), substrate temperature, deposition time, post-heat treatment, negative substrate bias, substrate position relative to the target erosion profile [6,21,45,60,68-76]. The substrate is kept at a constant temperature during the deposition process (e.g. 100 °C) [69,77], or subjected to plasma irradiation without any additional external heating [45,65,78,79]. Different types of CaP-based coatings are deposited, such as pure HA [6,45,67,80], dicalcium pyrophosphate (DCPP) [81,82], pyrophosphate (β-$Ca_2P_2O_7$) [83,84], silicon-containing HA (Si-HA) [85-88], carbonated HA [89,90], antibacterial silver-containing HA [91], fluoridated HA/carbon nanotubes (CNTs) [92], and Zn, Mg, and Al-doped CaPs [93]. As substrates different materials are used such as NiTi [17,79], Ti [65,91,94], Ti6Al4V [95-97], Ti-5Al-2.5Fe [66,67], 316 L stainless steel [68], polyethylene

(PE) and polytetrafluoroethylene (PTFE) [71,98,99], and polymethylmetacrylate (PMMA) [78]. The substrates can be abraded with SiC [58,100], grit-blasted with TiO_2 particles [101,102], or Al_2O_3 [83,103 to increase the substrates surface roughness before CaP film deposition.

A sputter target can be produced by sintering according to the procedures described in the literature [45,104]. The target material used in the sputtering process can also be a copper disk coated with plasma-sprayed HA [105,106]. A cold isostatic pressing method is also used to prepare a target for sputtering [59]. The most important characteristics of RFMS-deposited coatings in comparison to other available methods are summarized in Table 1. Further experimental results for the comparison between the CaP coatings properties deposited using RFMS and other most frequently used techniques is reported in the studies [3,6,42].

2.2. Surface Morphology of Coatings

It is well-established that surface morphology plays a crucial role in the overall success of an implant [107]. Either rough or flat surface morphologies of implants are needed in any particular clinical case. RFMS technique allows preparing highly dense, uniform, continuous coatings without any microcracks, surface and bulk porosity. The technique allows maintaining the initial substrate surface roughness, rendering the coatings beneficial in clinical trials where the initial surface roughness of an implant should not be significantly influenced [6,45,76,79,90,108]. It is very important in the process of coating of porous materials that no surface porosity should be concealed for further bone ingrowth.

The typical patterns of the surface morphology of RFMS-deposited films are presented in Figure 2. The surface morphology of the coatings can be significantly influenced by the variation of deposition parameters, e.g. smooth (Figure 2a) or nano/micro patterned (Figure 2b) surface morphology was obtained on the surface of polished mirror-like silicon substrates.

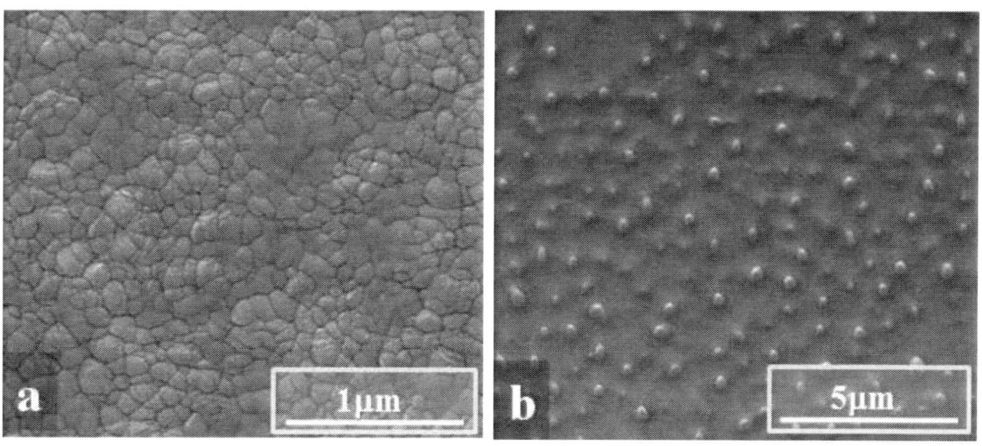

Figure 2. Surface morphology a typical RFMS-deposited HA coatings: (a,b) RF power – 290 W, deposition time 180 min, working atmosphere – argon. The substrate holder was grounded (a), and DC-bias of -100 V was applied (b). Nano/micro-patterned surface morphology obtained with DC-substrate bias, while other deposition control parameters were kept constant.

Moreover, the annealing of the as-sputtered coatings was reported to result in a change in surface morphology, and depending on the annealing temperature and time, the change in surface morphology was attributed to changes in the crystal structure and residual stress [42,109].

2.3. Influence of Deposition Parameters

2.3.1. Chemical Composition and Ca/P Ratio

The most important parameters that affect the RFMS-deposited CaP coatings properties are RF power level, working pressure and type of the atmosphere (argon, oxygen, water vapor and a mixture thereof) in the vacuum chamber, negative substrate bias, substrate temperature, and target degradation.

The chemical composition of the outermost coating surface is important because it will be in direct contact with bone tissue and dissolves first at the initial stage of implantation. X-ray photoelectron spectroscopy (XPS), X-ray fluorescence (XRF), Rutherford backscattering spectrometry (RBS), and energy dispersive X-ray (EDX) analyses of CaP-based coatings typically reveal the presence of O, C, Ca, and P [6,42]. The EDX-mapping of RFMS-deposited films showed a homogeneous distribution of Ca and P on the film surface, proving the absence of segregation phenomena which are usually responsible for the formation of undesired second phases [78]. The Ca/P ratio is one of the main characteristics of a biocompatible film that depends on the applied deposition parameters [21,45,82,106,110]. An overview of Ca/P ratio of RFMS-deposited coatings is presented in Table 2.

It can be observed that CaP films with different Ca/P ratios, including those close to Ca/P=1.67 typical for stoichiometric HA, can be prepared using RFMS [42,45,68,70]. The reasons that can lead to a deviation from the coatings stoichiometry and subsequently to different resorption rates *in vivo* and *in vitro* are presented below [5].

Sputtering of a multicomponent target such as HA often resulted in a coating stoichiometry different from the bulk target (Table 2). As an example, Ca/P ratio from 1.51±0.02 to 1.82±0.02 revealed the formation of $Ca_3(PO_4)_2$ and CaO phases [111]. The components of the target such as hydroxide, oxygen, and hydrogen may not be completely transferred to the substrates owing to the low environmental pressure in the chamber. As a result, the composition of the coating may differ from that of the target material, depending on the type of sputtering system and the deposition parameters [6]. The difference in Ca/P ratio between the HA target and the sputtered coatings is attributed to the preferential deposition of Ca ions and a possible significant loss of phosphorus ions, which may be pumped away before reaching the substrate or be re-sputtered out of the growing film by incoming ions or neutrals [6,20,70,77,112,113]. Different sputtering rates for calcium and phosphorus at the target can also affect Ca/P ratio of the coatings. It could depend on the history of target processing that alters the target stoichiometry if sputtering rates are not identical for the different components [75,114]. Increase in Ca/P is observed when annealing temperature is raised [61, 115, 116]. Boyd et al. reported that the same coatings that were annealed at 500°C with a soak time of 2 h had Ca/P ratios much higher than the corresponding as-deposited surfaces [68]. This was most likely due to the evaporation of P_2O_5 species from the surface of these coatings during the annealing [42].

Table 2. The typical Ca/P ratios of the RF magnetron sputter deposited coatings

Ca/P ratio	Deposition conditions or type of exposure	Reference
1.4	as-deposited HA in an inert atmosphere	[78]
1.55	as-deposited HA in an inert atmosphere with 20% O_2	
1.73±0.08	as-deposited HA	[93]
1.6-2.6	as-deposited HA	[6]
1.69±0.02	as-deposited HA after annealing in air at 550 °C for 1 h	[127]
1.8	as-deposited B-type carbonated HA in Ar atmosphere	[90]
1.53-3.88	as-deposited CaP and HA	[45]
2.1	as-deposited CaP	[20]
1.9-2.5	as-deposited CaP	[21]
1.8-2.2	as-deposited CaP	[56]
~1.66-2.3	as-deposited CaP at different oxygen content	[70]
1.5-2	as-deposited CaP	[18]
1.8-2	heat-treated CaP at 500 °C for 2 h	[106]
0.55-2.1	as-deposited CaP	[83]
1.73-1.78	as-deposited CaP	[67, 120]
1.78±0.07	as-deposited CaP	[61]
2.17±0.06	as-deposited CaP after annealing at 400 °C	
~1.6-3.7	as-deposited CaP at grounded substrate and negative substrate bias of up to 200 V	[60]
~1.6-4.6	as-deposited CaP at grounded substrate and negative substrate bias of up to 100 V	[64]
~1.15	as-deposited CaP in Ar atmosphere	[77]

Van Dijk et al. studied the influence of Ar pressure on the Ca/P ratio of the sputtered layers [57]. Both the deposition rate and Ca/P ratio tended to increase with increasing Ar pressure [57, 68]. However, Lee et al. [112] and Xu et al. [64] revealed that Ca/P ratio decreased with increasing Ar pressure. This contradiction can be explained by the difference in deposition rates. In the Ar pressure range used by Ozeki et al (0.5-5.0 Pa), the deposition rate decreased with increasing Ar pressure, whereas for van Dijk et al., the rate increased [74]. Van Dijk et al. found that a better stoichiometry of the films with respect to HA was obtained by intentional introduction of oxygen in the sputtering gas [70]. It was explained by the fact that O atoms were pumped away, and that the bonding strength of P atoms arriving at the growing film layer was weak, so that P atoms were more easily re-sputtered at a higher discharge power [20, 70, 74]. Van Dijk et al. revealed that Ca/P ratio of the films increased with the discharge power [21, 57]. It is also reported that a negative substrate bias tend the Ca/P ratio to increase [45, 60, 64, 70]. This phenomenon can be connected with the decreasing of the current density of negative charged phosphorus-containing species or preferable re-sputtering from film surface phosphorus atoms by the charged particles accelerated in anode dark sheath.

2.3.2. Phase Composition and Structure

The phase composition and structure, as well as Ca/P ratio of the RF magnetron sputter-deposited coatings depend on the deposition parameters. This method allows the preparation of CaP coatings of either amorphous or crystalline structure of a definite phase composition that along with Ca/P ratio influences the coating behavior *in vitro* and *in vivo* [3,82]. Despite the utilization of a partially crystalline target, the RFMS process usually produces amorphous coatings or CaP compounds [6,45,56,79,84,100,105,117,118]. It is generally known that the amorphous phase in CaP coatings easily dissolves and thus loses long-term stability in clinical use [42,119]. The CaP coatings deposited at low power using RFMS can undergo a decreased dissolution rate under physiological conditions when subjected to post-deposition annealing [68].

Since it is critical to maintain the integrity of HA coatings, controlling heat-treatment temperatures and heating environment (air, water vapor) results in conversion of amorphous coatings to HA ones of different degrees of crystallinity [6,61,83,109]. Annealing *in situ* at 300-600°C [114], 550°C [64,66,120], 300-530°C [45,65,79] or after deposition usually at the temperature of 300-700ºC in an inert gas, water vapor, air or vacuum enhances the film crystallinity [18,20,56,57,66,100,109,121,122]. Crystalline HA coatings were obtained during *in situ* annealing at 500 and 600°C, as well as hybrid HA/TiO_2 films at 700°C [115]. With identical heating time, the coatings heated in the presence of water vapor had a significantly higher degree of crystallinity compared with the coatings heated without water vapor. This suggests that the presence of water vapor during heat treatments plays an important role in increasing CaP crystallinity [123]. Studies by Van Dijk et al. [20] and Yang et al. [94] revealed that 600°C is probably the optimum annealing temperature to change an amorphous CaP coating into a crystalline coating, which was also found in the case of Si-HA coatings [87]. Annealing results in coating cracks, owing to the differences in the thermal expansion coefficients of HA and the underlying metal substrate. This manifests itself in a weakened bond strength [20,54].

A rapid infrared treatment at 300-700ºC was used to enhance the crystallinity of thin HA films to obtain suitable dissolution properties without cracking and/or detachment of the coatings [58,83,84]. The amorphous coatings were instable in simulated body fluid (SBF) and dissolved partially within four weeks of incubation. All coatings infrared heat-treated at 550°C appeared to be stable in SBF [56, 84]. The CaP films were crystallized under hydrothermal conditions in distilled water at 100-170ºC and different pH 5.0, 7.0, 9.0 [76,117].

The crystallinity of the film increased with treatment time and process temperature [42, 76]. Moreover, it is known that the configuration of the confining magnetic field strongly affects the sputtering characteristics of a magnetron [124]. Therefore, the geometry of the samples position is of crucial importance for the properties of the RFMS-deposited CaP coating [45, 124, 125]. This non-uniformity is responsible for the existence of the cathode erosion profile and the distribution of plasma density. So, as-deposited HA coatings of a sufficient crystallinity were prepared by RFMS [42, 45, 65, 79, 126]. The RFMS-deposited coating (Figure 3) structure is close to that of HA with a preferable (002) crystallographic orientation [45, 65]. Peaks at 25.8° (002), 53.1° (004), 31.8° (211), 32.2° (112), and 32.9° (300) appear, but the last three peaks overlap [60, 61, 65]. Besides the peaks assigned to crystalline HA, no peaks were found that could be attributed to other CaPs like tricalcium phosphate (TCP), tetracalcium phosphate (TTCP) or to CaO [65]. According to the Scherrer's equation the average crystallite size in direction (002) for the coatings after 180 minutes of deposition was 48 nm at Position 1 («within» the target erosion profile), and 20 nm at

Position 2 («out of» the target erosion profile). The reason explaining different crystallite sizes can be the temperature gradient along the substrate that is caused by a limited size of the target erosion zone.

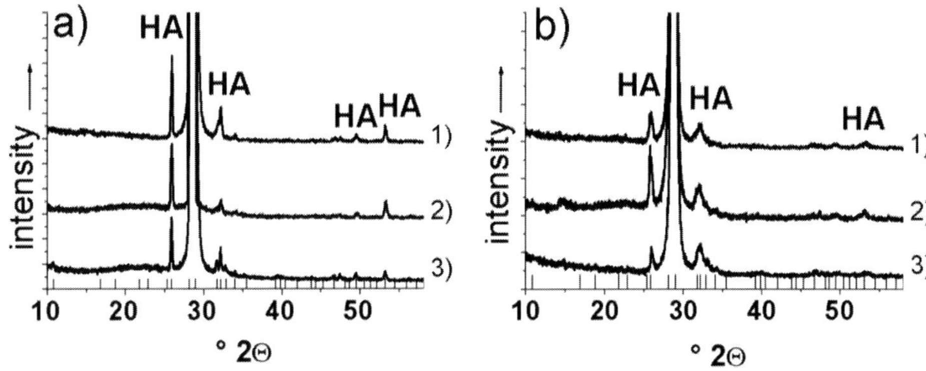

Figure 3. XRD patterns of CaP coating deposited at 290 W for 180 min. (a) Position 1, (b) Position 2. Substrate bias: 1) -100 V, 2) -50 V, 3) grounded substrate. "HA" denotes the peaks attributed to HA. The vertical lines show the peaks obtained for the HA target. The substrate is a silicon wafer.

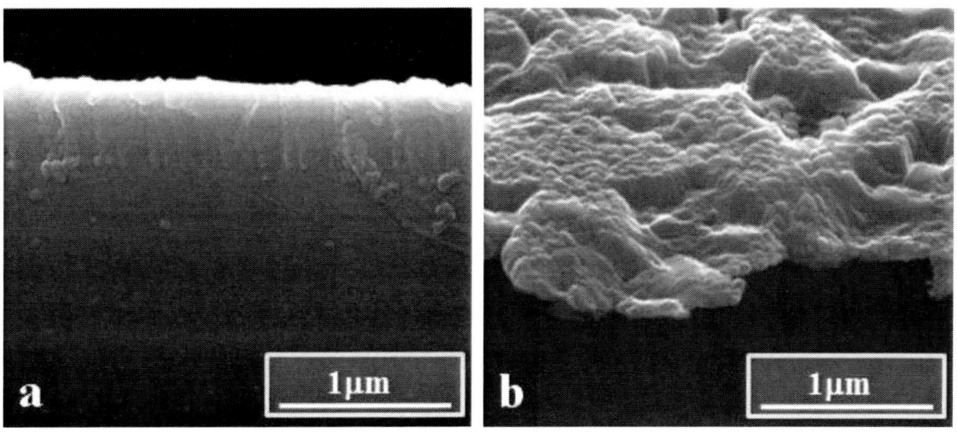

Figure 4. Cross-section patterns of RFMS-deposited CaP film before a) and after etching b) in 0.1 mol HCl solution. Figure a) unpublished results, reprinted with the authorization of Prof. M. Epple (University of Duisburg-Essen).

The increased influx of ions resulted in a larger crystallite size and a more crystalline coating [42]. The (002) texture is a common phenomenon when RFMS is used [45, 60, 63-66, 68, 76, 94, 106, 120, 127], and is also found in case of PLD [128, 129] and PS deposited films [13, 130]. It is also reported that this preferable crystallographic orientation is not changed after 1 and 3 weeks of implantation into the tibia of a goat [106].

A typical cross-section pattern of RFMS-deposited coating is shown in Figure 4. The coatings structure before (Figure 4a) and after (Figure 4b) etching in 0.1 mol HCl solution is seen. The preferential orientation of grains (002), which is perpendicular to the substrate surface (Figure 4b) can be observed.

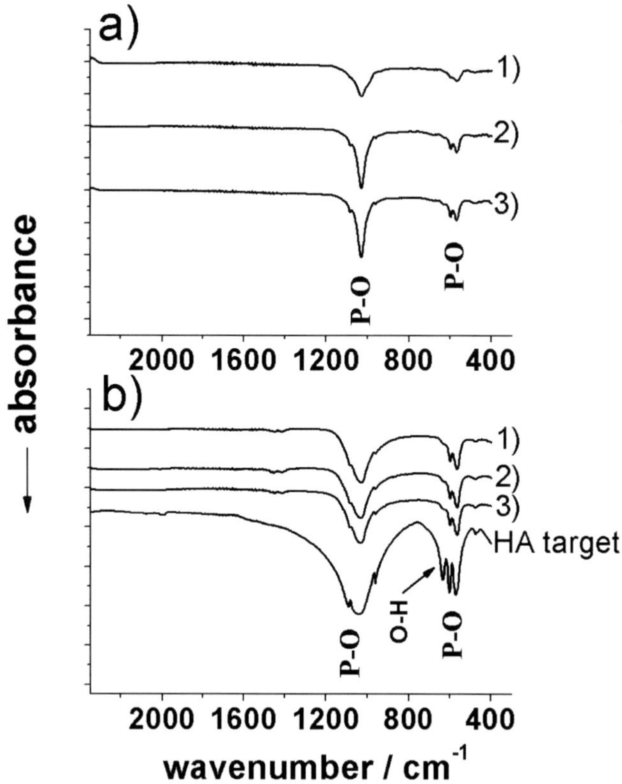

Figure 5. IR spectra of CaP coatings with the deposition time of 180 min in the position 1. RF power: (a) 30 W, (b) 290 W. Substrate bias: 1) -100 V, 2) -50 V, 3) grounded substrate. The substrates were KBr single crystals. IR spectrum of HA target (KBr pellet) is shown.

The FTIR analysis showed the presence of absorption bands typical for HA (Figure 5). However, an alteration of the band characteristics (amplitude, width at half height) occurred, which was caused by changes in the surrounding of ions PO_4^{3-} in the lattice of HA and reorientation of the crystallites within the coating. In IR-spectra, the bands assigned to the valence and deformative vibrations of PO_4^{3-} ions at 570 cm^{-1} (ν_4), 601 cm^{-1} (ν_4) and 1031 cm^{-1} (ν_3) [21, 54, 70, 131] were found. The bands of O-H ions in the deposited coatings were only weakly resolved at 631 and 3571 cm^{-1}, therefore a part of the ions OH^- was lost during the target sputtering that affects the coatings structural integrity [132]. It was reported that the OH^- group variation indicated different degrees of dehydroxylation as a result of HA decomposition during the coating or treatment process [123, 133].

The coating structure in this case corresponds to the partly dehydroxylated HA with the chemical formula $Ca_{10}(PO_4)_6(OH)_{2-2x}O_xV_x$, where V denotes a vacancy ($0<x<1$) [45, 65, 77, 123, 133, 134], which was found in PS [132] and IBAD films [30]. An OH^- peak at 3572 cm^{-1} was observed for CaP coatings heat-treated for 2 and 4 h in the presence and absence of water vapor [123].

Thin highly crystalline nanometer thick HA films were also prepared using a right-angle RFMS approach [80, 93, 135]. These coatings showed some advantages over conventionally sputtered films in which ambient temperature deposition yields nano-crystalline and nearly stoichiometric HA coatings under appropriate conditions, thereby avoiding the need for *in-situ* or *ex-situ* annealing.

The rotation of the substrate during RFMS plays a crucial role in the films structure. Films sputtered with a discharge power level lower than 800 W and all films sputtered with rotating substrate holder appeared to have an amorphous structure [21]. Wolke et al. [18] demonstrated that stationary deposited sputter coatings with the thickness of 2.5 μm are crystalline with a preferred (001) crystallographic orientation, whereas the rotated sputtered coatings with a thickness of 10 μm were amorphous. This phenomenon can be attributed to differences in substrate temperature under various experimental conditions. A rotating substrate holder leads to a decrease in the sputtering rate and energy, equivalent to a low substrate temperature. Consequently, the grain size of the deposited coating will be small, resulting in an amorphous appearance [18, 136-138].

The target degradation should be taken into account, thus it affects the quality of RFMS-deposited films [21, 75]. The changes in the target phase composition after several deposition runs were found in studies [74, 75]. The annulus region of the target surface, where sputtering under RF magnetron conditions normally occurs, showed severe surface degradation after only one deposition run, as indicated by significant PO_4^{3-} and OH^- depletion [75]. This deterioration continued after each subsequent deposition cycle but to a much lesser extent than that observed in the initial sputtering period.

2.4. Mechanical Characteristics

2.4.1. Adhesion Strength

The mechanical properties of CaP-based coatings determine the overall success of an implant [65]. The mechanical properties of the coatings depend on many parameters such as density, phase and chemical composition, and crystallinity [111]. An overview of RFMS-deposited CaP coatings adhesion strength is presented in Table 3. Investigation of the coating bond strength is typically performed according to ISO 4624 (pull-off test) [139], ASTM C633 (tensile adhesion test) [97, 129, 140], ASTM F-1147-05 (tensile adhesion test) [141] or DIN 50,161 (shear strength) [142]. Another technique to study adhesion strength of RFMS-deposited films is the scratch test [65, 69, 143]. The scratch test is performed with reference to ISO 20502:2005(E) [144, 145]. Laser spallation technique was used to measure the tensile strength of interfaces [95].

The ISO 13779-2:2008 standard requirement for tensile adhesion strength of HA coatings in the surgical implant application is 15 MPa [10, 42, 140, 146]. The data presented in Table 3 revealed that RFMS-deposited coatings meet the requirements of the standard.

Many authors reported that adhesion strength could still be enhanced using different interlayers such as TiO_2 [66, 112, 115, 116, 147], Ti [63, 118], TiN [120], fluoridated HA/CNT composite [92], CHA_xTi_{1-x} (x=0-1) [90], and HA/(ZrO_2+Y_2O_3) [96].

These interlayers can reduce the mismatch of thermal expansion coefficients between HA and substrates or increase the surface area of the material, wettability and heat conductivity. The graded HA/Ti coatings with enhanced adhesion strength compared with monolithic HA films were also deposited [97, 148].

Table 3. A summary of RF magnetron sputter deposited CaP-based coatings adhesion strength

Adhesion strength, MPa		Type of exposure	Reference
Pull-off	Tensile		
	~35	as-sputtered graded HA/Ti and annealed at 600 °C	[97]
>60		as-sputtered ACP, and OAp and air heat-treated	[122]
~40		heat treated OAp in vacuum	
68.5±3.74		as-deposited carbonated HA	[90]
84.8±0.78		annealed carbonated HA at 550ºC/1 h	
~30		as-deposited HA PMMA	[78]
	64.6±10.7	as-deposited HA	[63]
	69.4±8.5	as-deposited HA after 14 weeks in SBF	
~30		as-deposited HA	[67]
59.9±12.4		as-deposited HA	[62]
59.5±6.5		as-deposited 95HA/5Ti	
64.8±6.2		as-deposited 85HA/15Ti	
66.1±7.8		as-deposited 85HA/15Ti after 14 weeks in SBF	
	59.6	heat-treated HA(+ZrO_2+Y_2O_3) at 600 ºC for 4 hours	[96]
59.9±12.4		as-deposited HA	[148]
49.6±12.4		as-deposited HA after 30 days in a high humidity environment (>95%)	
59.5±6.5		as-deposited 95HA/5Ti	

The bond strength of HA(+ZrO_2+Y_2O_3)/Ti6Al4V composite coatings is enhanced with the increase of sputtering power and the addition of (ZrO_2+Y_2O_3) [96]. The bonding strength of fluoridated HA/CNTs composite coating composites was 83.1 MPa, which indicated significant improvement compared with 71.40 MPa of non-CNTs ones [92].

Xu et al. reported that calcium titanate ($CaTiO_3$) at the substrate coating interface promoted a higher interfacial bonding strength [64]. Whilst on the one hand, $CaTiO_3$ is beneficial in ensuring a better interfacial stability of the bioimplant coating, on the other hand, it can potentially reduce the coating's biomimetic response. Pichugin et al. using the scratch-test observed that the as-sputtered coatings with the thickness less than 1.6 μm on Ti did not delaminate until the indenter reached the substrate-coating interface, whereas thicker coatings collapsed by exfoliation, splits and chips along the scratching direction. According to the results of acoustic emission and friction coefficient tests all coatings with a thickness up to 2.7 μm had a high adhesion strength to Ti and NiTi, which is in good agreement with the data reported by other authors (Table 3).

The reason for a high adhesion strength is a thin oxide layer of TiO_2 that promotes a formation of a strong covalent bond between the substrate and CaP coating with the energy above 0.5 eV [65]. When CaP films are deposited by RFMS in argon atmosphere, there is only a scarce quantity of oxygen atoms available that limit the substrate oxidation. Berezhnaya et al. reported that 0.2 μm thick TiO_2 interlayer in HA/Ti system inhibited Ti oxidation, and the reaction intermediates formed during high temperature heat treatment (900-1000 °C) enhance the adhesion between the Ti and HA [116]. Thus, the experimental data summarized in this section reveal that RFMS deposition technique allows us to prepare CaP coatings with the adhesion strength that meets requirements of the standard [8, 10, 42].

2.4.2. Nanohardness and Young's Modulus

Nanoindentation technique is typically used to study the nanohardness (*H*) and Young´s modulus (*E*) of CaP-based coatings according to Oliver and Pharr´s method [149]. Hardness is a property related to the yield strength and can be useful to predict the strength of the coating, stress state within the coating, and the abrasion resistance on the coating surface [150, 151]. An overview of thin films nanohardness and Young´s modulus is presented in Table 4.

It should be pointed out that the RFMS process can cause a shift of the Bragg peaks of the HA. This is most probably attributed to the lattice compression of HA due to the presence of residual stress.

The thermal expansion coefficient of sintered HA is larger than that of either Ti6Al4V alloy or pure Ti ($\sim 14 \cdot 10^{-6}$ K^{-1} [152, 153], $(8.4\text{-}8.9) \cdot 10^{-6}$ K^{-1} [152-154], $8.5 \cdot 10^{-6}$ K^{-1} [155], respectively), hence HA will be in a tensile stress state after being cooled from the elevated temperature. The presence of compressive stresses in the crystalline sputter coating was found [56].

The data for amorphous coating heat-treated at 650°C indicated an almost stress free film. The change in apatite composition due to dehydroxylation will also modify the resulting stress condition [156]. Stress conditions can cause the CaP coatings delamination in the experiments *in vivo* or *in vitro* and be a reason for implant failure shortly after surgery; therefore, they should be avoided.

The mentioned above different interlayers between CaP-based coating and the substrate can reduce the mismatch of thermal expansion coefficients, thus avoiding stress state formation. Other reliable and inexpensive ways are still a challenge for the research community for the nearest future.

The values of the Young's modulus for the cortical bone and trabecular (cancellous) bone are 7-30 GPa and 0.05-0.5 GPa, respectively, as reported by Hench [157]. In Ref. [158] it is shown that bone nanohardness ranges from 0.63±0.10 GPa for cancellous bone to 0.88±0.16 in its cortical counterpart. From the comparison of these data with that presented in Table 4 it can be concluded that the Young's modulus of RFMS-deposited films is higher than that of bone, which was also found in case of other plasma-assisted techniques [42].

Table 4. The nanohardness and the Young's modulus of CaP-based coatings

Hardness, GPa	Young's modulus, GPa	Type of the coating	Reference
~10	~110	as-deposited HA	[65]
-	~131-163	HA after annealing at 550 ºC at 1 hour	[111]
4.5-5.2	130-155	TiN/HA composite	[67]
5	130	TiN/HA composite	[66]
10±1	147±10	HA after annealing at 550 ºC at 1 hour	[127]
~3.4-4	~125	as-deposited HA	[120]
~5-5.2	~140	as-deposited HA with TiN buffer layer	
10-12	-	as-deposited HA	[104]

Hence, a stress shielding effect can be expected, therefore attempts to make the Young's modulus of HA coatings to be closer to that of bone is a challenge for the research community for the nearest future.

3. Biomedical Characteristics of Magnetron Sputter Deposited Coatings

3.1. Toxicological Properties and Biocompatibility of RFMS Coatings *In Vitro*

According to Taranobu et al. [159] the increasing dissolution rate of CaP ceramics in the organism stimulates the osteoinductive activity, but decreases the biocompatibility of the biomaterial [159]. Amorphous (soluble) CaPs may either be cytotoxic [160, 161] or decrease the functional activity of cells [162].

Titanium samples of 1×1 cm^2 size with RFMS-deposited HA or Si-HA coatings were prepared for comparative analysis in vitro. The cytotoxicity tests were performed with frozen BALB/c mouse cells. Bone marrow cells taken from thigh bones were thawed according to the standard procedure, washed twice to remove the kryo-preserving agent DMSO, and brought to a concentration of up to 10^7 viable karyocytes in 1 ml of DMEM medium. The samples tested were immersed in the cells suspension and cultivated in sterile tubes with 2 ml culture medium for 3.5 hours at 37 °C. Then the cells were homogeneously pipetted in the medium volume, and the cytotoxicity of the samples was measured with 0.4 % trypan blue according to the ISO 10993-5 specifications. The control agent was the cells suspension without the samples.

A culture of prenatal stromal human lung cells (Stem Cell Bank Ltd., Tomsk) was used to study the cells response *in vitro* in direct contact with the surface. These cells represent a stromal stem cells pool which preserves a stable karyotype *ex vivo* and is oncogenically safe. The cells are free from extraneous viral (HIV, hepatitis, herpes etc.) and bacterial (syphilis, mycoplasma, chlamydiae etc.) agents. After thawing the cell vitality was estimated to be 93% as determined by the standard ISO 10993-5 specifications in the 0.4 % trypan blue test The samples studied were deposited into the wells of 24-well plates (Costar) and subsequently a cell suspension with a concentration of $5 \cdot 10^4$ viable karyocytes was added into 1 ml of nutritional medium with a composition of 80% DMEM/F12 medium (GIBKO, Great Britain), 20% cow embryo blood serum (Sigma-Aldrich, USA), 280 mg/l of l-glutamine ("Biolot", Russia), 50 mg/l of gentamycini sulfas ("Darnitsa", Ukraine), 10 mM of beta-glycerophosphate (Sigma-Aldrich, USA), 50 μg/l of ascorbic acid ("Dalkhimfarm", RF), and 10^{-6} M of dexamethasone (KPKA, Slovenia). The control growth agent was a fibroblast-like cells culture on plastic wells. After four days the implants were removed and air-cured. Samples with surface-adhesive cells were prepared for scanning electron microscopy (SEM). Surface-adhesive cells were fixed in 2.5 % glutaric dialdehyde phosphate-buffered solution for 30 minutes, followed by 1% osmium tetroxide solution for 30-40 minutes with further washing in double phosphate buffer (pH=7.2-7.4). Then the cells were dehydrated in aqueous ethanol solutions with the rising concentration of 30, 50, 70, 90, 100 % for 15 minutes in each solution and twice in 100 % acetone. Then the samples were studied by SEM (Phillips SEM 515).

The RFMS-deposited CaP-based coatings on titanium scaffolds did not reveal any statistically significant toxic effect (Table 5) that could have been identified through dead cells stained with 0.4 % trypan blue (necrosis indicator). It is known that cells death processes may be used to determine the implant's biocompatibility and performance (survival/failure) in the organism [163]. Thus, RFMS-deposited Si-HA coatings do not cause necrotic changes of cells despite their higher solubility compared to that of pure HA.

Table 5. Average toxicity indicators of RF magnetron sputter deposited CaP coatings after exposure to the DMEM medium during 3.5 hours, X±SD(m), n=15

Indicator	1 Medium toxicity control	2 HA coating	3 Si-HA coating
% of dead cells in culture	9.45±11.14 (4,98)	8.88±9.61 (2.48)	14.65±9.28 (2.40)

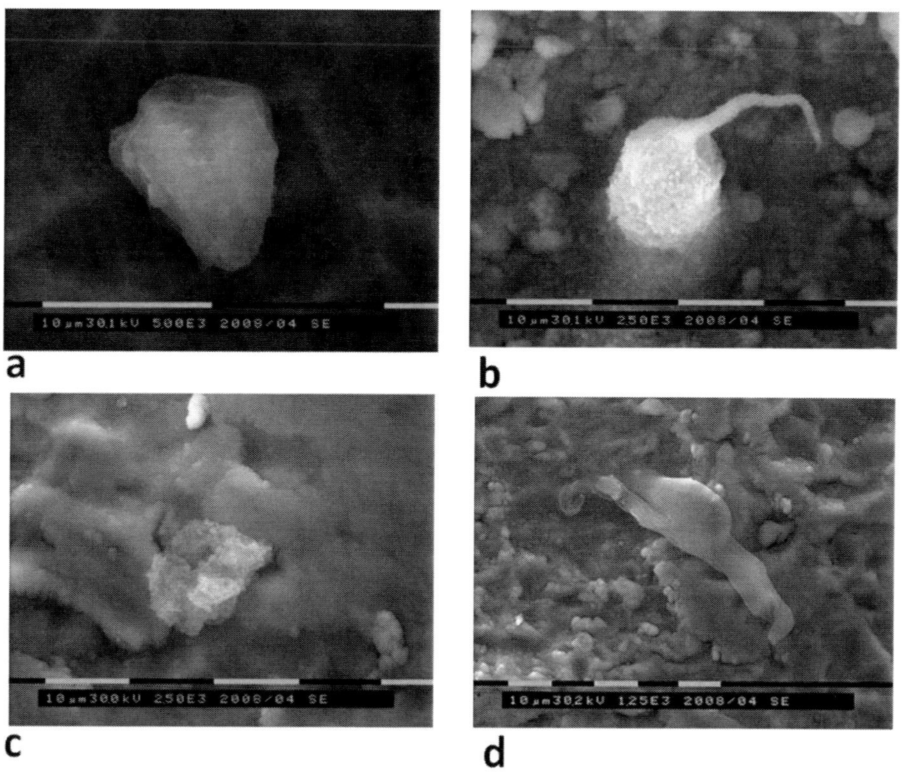

Figure 6. (a) SEM images of (a) a typical stromal cell with ellipsoidal shape on RFMS-deposited CaP coating (magnification ×5000, bar 10 μm), (b) a typical stromal cell with pseudopodia (magnification ×2500, bar 10 μm), (c) a spreading stromal cell (magnification ×2500, bar 10 μm), (d) a typical fibroblast-like stromal cell (magnification ×2500, bar 10 μm). The images were kindly provided by Ms. E.V. Legostayeva in cooperation with Materials Science Centre at Tomsk State University (Tomsk, Russia).

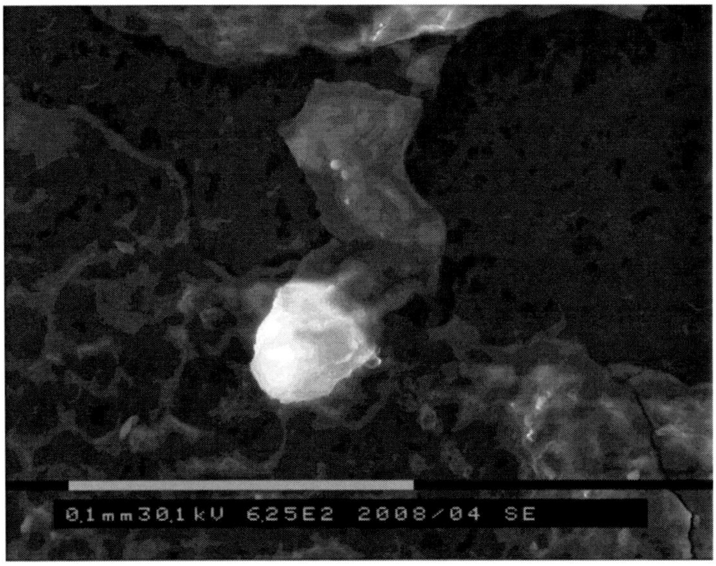

Figure 7. Typical SEM image of a stromal cell lodged in the place of a marginal defect of an RFMS-deposited CaP coating (magnification ×625, bar 100 μm).

Analysis of the direct interaction between fibroblast-like lung cell cultures and RF magnetron sputter-deposited CaP coatings carried out by SEM revealed that the artificial surface topography of the samples affected the morphology of stromal cells. During short-term 4-days contact *in vitro* in 85-95% of the cases the cells were situated on submicroscopic irregularities of relief and revealed a rounded or oval morphology (Figure 6a), yielded pseudopodia (Figure 6b) if they were able to find appropriate peculiarities of the surface micro-architecture.

In the cavities of RFMS-deposited CaP coatings the cells were spread flatly (Figure 6c), at the cavities and crest edges they obtained fibroblast-like morphology (Figure 6d). If the coatings integrity was damaged the cells preferably adhered to the CaP surface rather than the surface of titanium substrate (Figure 7).

It is known that the cells morphology changes reveal their multifunctional maturation process [164]. Thus, the fibroblast-like cells morphological maturation was taking an active place at the interface with the artificial surface of RFMS-deposited CaP coatings.

3.2. Osteogenous Properties (Specific Activity) of RFMS-Deposited HA Coatings *In Vitro* and *In Vivo*

Investigations of niches for stem cells was begun with R. Schofield's study published in 1978, one of the most important contributions to the cellular and molecular biology [165]. In 1964, Curtis and Varde [166] proposed a crucial influence of the surface topography in determination of cells behavior. Moreover, the significance of the structural and functional organization of microenvironment to reproduce physiological conditions for stem cells *ex vivo* were specified [167, 168]. It was also mentioned that the existence of hierarchical niches was possible, specialized as stem cells microenvironment for self-maintaining and differentiating of hematopoietic cells [167] as well as for 'silent' and actively proliferating hematopoietic stem cells (HSCs) [169].

The functioning of endothelial (perivascular, vascular) and osteoblast (endosteal) niches for HSCs was analyzed [170]. It was suggested that osteoblast (hypoxic) niches support HSCs in the G_0 phase (inactive, 'silent' state), whereas vascular microterritories promoted the mitotic cycle of stem cells due to a high oxygen tension [170]. Other functional and metabolic (intra- and intercellular) signals keeping stem cells inactive were also considered [169].

It was suggested that the direct contact with multipotent mesenchymal stromal cells (MMSCs) and osteoblasts impedes the HSCs activation [168, 171]. There are several types of osteoblasts. The differentiation stage of hemopoietically active osteoblasts is still unknown [172]. Moreover, a part of them promotes HSCs pool self-maintenance, the others, presumably, regulate HSCs proliferation and differentiation.

The probable mechanism of controlling pluripotent and committed HSCs is assumed to be the size of hematopoietic niches [170], and the elasticity and toughness of extracellular matrix [173]. Moreover, on one hand, osteoblasts contact with hematopoietic elements, on the other hand, they contact with the bone surface, the main non-organic component of which is HA and other CaPs.

However, the properties of bone as a specialized extracellular substance representative were almost overlooked by the researchers. Some authors pointed at particular anatomic localization of endosteal niches with a different function [174]. The physical and chemical differences of bone tissue in its different parts are known. The cortical bone tissue is thinner, denser and more calcareous [174], its remodeling rate is significantly lower than that of the trabecular area [175]. At the same time the bone surface in any state is a natural scaffold for stromal and hematopoietic cells.

Three structural components of endosteal niches for HSCs can be distinguished: mesenchymal cells, extracellular matrix, and mineral part of bone. However, the impact of mineral bone substance on the hematopoietic niche functioning is estimated only by the extracellular calcium concentration [176]. Some attempts have been made to model 'artificial niches' for HSCs *in vitro* [177]. Moreover, first experimental data have been obtained about the existence of artificial osteogenous niches for stromal stem cells, and their close connection with quantitative parameters of the mineral bone substance relief [178]. The quantitative parameters of artificial endosteal niche allow controlling MMSC maturation and their osteogenous differentiation.

The osteogenous properties of HA and TCP, which are the bases of the bone mineral matrix, have been convincingly proven by the phenomenon of ectopic bone formation whereby the bone tissue is formed at the surface of implanted materials. An adequate approach to determine possible osteogenous properties of CaP materials is an ectopic bone formation experiment when an artificial sample is implanted subcutaneously or intramuscularly without using growth factors [179].

We obtained reproducible results on subcutaneous growth of bone tissue from syngenic bone marrow that was preliminarily put on the surface of rough RFMS-deposited CaP coatings on titanium [162, 180]. Therefore, modeling of different relief peculiarities and mineral matrix parameters with the implants coated with HA coating is of obvious interest for studying the structural and functional states of stromal stem cells *in vitro* and remodeling of the 'bone/bone marrow' system *in vivo*.

In experiments, discs were used of pure titanium (diameter 12 mm, thickness 1 mm) completely coated with either smooth HA or Si-HA coatings as a source of the model mineral bone matrix. The smooth ($R_a < 1$ μm) HA and Si-HA coatings were deposited on titanium by RFMS [45]. Nano-sized powders (particle diameter 80 nm) of either pure HA or Si-HA with silicon contents of 4.9 mass % ($Ca_{10}(PO_4)_{4.28}(SiO_4)_{1.72}(OH)_{0.28}$), synthesized by mechanochemical activation, were used to prepare the targets for our RFMS set-up. Morphology and chemical composition of the coatings were investigated by Quanta 200 ESEM FEG (FEI). Coating surface roughness was measured by profilometry (Talysurf 5-120; resolution 10 nm).

Sterile extracts of coated samples were obtained according to the ISO 10993-5 after 7-days dissolution at 37 °C in 4 ml of isotonic (0.9 %) sodium chloride solution. To reproduce the conditions of osteosynthesis in a clinical situation (acidosis, antibiotics injection), 30 mg/l gentamycini sulfas was added to the solution.

A short-term (1-, 4- or 8-day) *in vitro* co-culturing of prenatal stromal human lung cells (PSHLCs) with tested discs was done in a 24-well plate (Costar) with a concentration of viable karyocyte cells of 3×10^4 in 1 ml of osteogenous medium. The content of the medium, methods for estimatation of the cytochemistry and cultivated cells morphology, alkaline (ALP) and acid phosphatases (ACP) activity, osteocalcin (OC) concentration in supernatants, optical microscopy and SEM of samples were applied according to the protocols previously described in the study [178].

The free calcium and non-organic phosphorus concentrations in the intercellular environment were measured by the colorimetric standard method with the automated biochemical analyzer 'Hitachi 911'. Collagen was revealed according to van Gieson's picrofuchsine staining.

The assessment of tumor necrosis factor-alpha (TNFα), interleukins-2 and 4 (IL-2, IL-4) concentrations in cell culture supernatants was done using an enzyme immunodetection set produced by 'Protein contour' ('Proteinovy kontur', Saint-Petersburg) according to the manufacturer's instructions. The stain intensity was measured and recorded using the calibration curve and photometer 'Multiscan EX' (USA) at a wavelength 450 nm. The absolute cytokine concentration was calculated according to the calibration curve.

For heterotopic (ectopic) bone formation experiments *in vivo* 40 BALB/c male mice were used from the collection fund of the Research Institute for Pharmacology (Siberian Branch of the Russian Academy, Tomsk). In one variation of the ectopic test the artificial sample was implanted subcutaneously or intramuscularly [179]. Reproducible results were obtained on mice subcutaneous osteogenesis from syngenic bone marrow preliminary applied on the CaP surface [180]. The bone marrow was the source of stromal stem cells and growth factors. At a separate implantation of either scaffolds or bone marrow fragments no tissue plates formation was observed.

In a current study two implants were implanted under etherization for each of 20 BALB/c mice used in the experiment (one implant at each side of the abdomen's centre line) with a bone marrow column (the marrow average area was 7.5 mm^2) taken from the thigh bone under aseptic conditions. For cell adhesion the bone marrow column was cultivated on the scaffold at 37°C for 45 minutes in a culture medium containing 95% of the DMEM/F12 medium (GIBKO, Great Britain) and 5% of cow embryo blood serum (Sigma-Aldrich, USA).

After 45 days, the animals were sacrificed with ether narcosis and the implants removed. Then the tissue plates were removed from the disc surface, decalcified, and embedded in paraffin. Several thin cuts made vertically at the implant's surface were stained with hematoxylin and eosin for histological examinations. The growth of bone marrow and/or a bone on the implant surface was considered a positive result, the development of connective, muscle or adipose tissue a the negative one. The bone and bone marrow areas were measured using computer morphometry as described in the study [180].

The experimental results obtained were processed using the statistical description techniques as well as methods for testing of statistical hypotheses provided by the standard program package Statistica (version 6.0). The normal-theory test (Kolmogorov-Smirnov's test) was used to analyze the data sampling provided. Typically, in data sampling a distribution of indicators was observed that differed from a normal distribution. Therefore, the Mann-Whitney's distribution-free test was implemented to estimate the statistical significance of sampling differences. The results were represented as the arithmetic average (X), the median (Me), 25 % (Q1) and 75 % (Q3) quartiles or the statistical deviation (SD) and standard error of mean (m). The Spearman's correlation analysis was performed to reveal the relationship between the indicators studied. The differences were considered statistically significant at the significance level $p < 0.05$.

Designs of artificial matrices able to biomimetically reproduce the cell and tissue microenvironments are considered promising bioengineering pathways. Moreover, it is believed that the extracellular matrix peculiarities are able to regulate the MMSCs lineage [181]. In turn, the MMSCs represent a heterogeneous multicellular pool [182]. Osteoblasts are distinguished by their functional capacity [175] and topographical arrangement (trabecular or cortical). The cortical bone tissue is thinner, denser, and more calcified [174], and its remodeling rate is significantly lower than that of the trabecular area [175]. Meanwhile the bone surface in any of its state is a natural scaffold for stromal and hematopoietic cells.

The investigations performed earlier by our group showed that RFMS-deposited CaP coatings represented a thin (< 1-2 μm), nonporous, smooth (R_a < 1 μm) apatite layer with a Ca/P ratio typically higher than that of stoichiometric HA (> 1.67). From our point of view, similar artificial surfaces imitate to some extent a bone mineral matrix in a quiescent state out of the remodeling area.

The PSHLCs were used by us in experiments *in vitro*, allowing to obtain information about the early stages of development and regeneration of bone tissue since it is known that embryonic stem cells (ESCs) are able to organize themselves the niches for their development [183]. On the other hand, micro-environmental factors including the physico-chemical ones regulate self-maintenance/differentiation of ESCs in the same way as for tissue-specific stem cells [167]. To understand the results obtained for the stromal cells we used some analogies with the niches for HSCs which are the most studied micro-territories closely connected with stromal cells of hemopoietic microenvironment.

According to the results of our previous experiments, the PSHLCs interacting with CaP artificial surfaces gain an advantage in revealing a morpho-functional activity in comparison with the culture of cells on plastics [184]. In our studies they were specified by various forms in case of a short-term *in vitro* contact with smooth HA coatings. The morphology of PSHLCs in 85-95% of all cases remained rounded or oval, and they were situated on submicroscopic irregularities of the relief. If they found appropriate peculiarities of the surface micro-architecture, they gained fibroblast-like morphology. In a few artificial cavities on the surface the cells were spread flatly. Thus it was possible to measure the area of micro-territories (niches) they occupied.

It happened that exactly those cells that were located in the cavities occupied a significant surface area and were stained with alkaline phosphatase (ALP). However, the optical density of the stained cells was minimal (Table 6).

A weak cytochemical ALP activity (Figure 8a), a marker of maturation and osteogenous differentiation of stromal stem cells [175, 185, 186] was revealed in discreet areas of cells contact with the smooth CaP matrix, which imitated the surface of an intact (out-of modeling phase) lamellar or trabecular bone.

Table 6. Computer morphometric results of prenatal stromal human lung cells stained with the alkaline phosphatase (ALP) after cultivation for 4-days on the surface of CaP coatings, X ± SD(m)

Groups of coatings	Part of stained cells out of seeded cells on 1 mm^2 surface, %	D, c.u.o.d.	Cell staining area (S), μm^2	S of the territory occupied by the cell, μm^2	Cell staining S/Territory S, %
Growth control of cells on plastics, n_2=12	-	5.19±1.57 (0.45)	145.11±5.18 (1.49)	-	-
Smooth coatings, the average Ra=0,94 μm (n=10), n_2=12	12 n_1=20	-4.36 ±18.17 (4.06)	111.47±218.65 (48.89)	461.58±744.60 (166.50)	23.61±30.08 (6.73)
Smooth structured coatings, the average Ra=0,87 μm (n=10), n_2=13	10 n_1=13	-0.78 ±14.34 (4.14)	283.26±336.91 (97.26)	822±1180 (340.82)	34.46±30.82 (8.90)

Note: here and in the tables 7 - 8: n is the number of artificial surface roughness measurements; n_1 is the number of counted optical pictures (visual fields); n_2 is the number of counted cells in visual fields, c.u.o.d. is an abbreviation for conventional units of optical density.

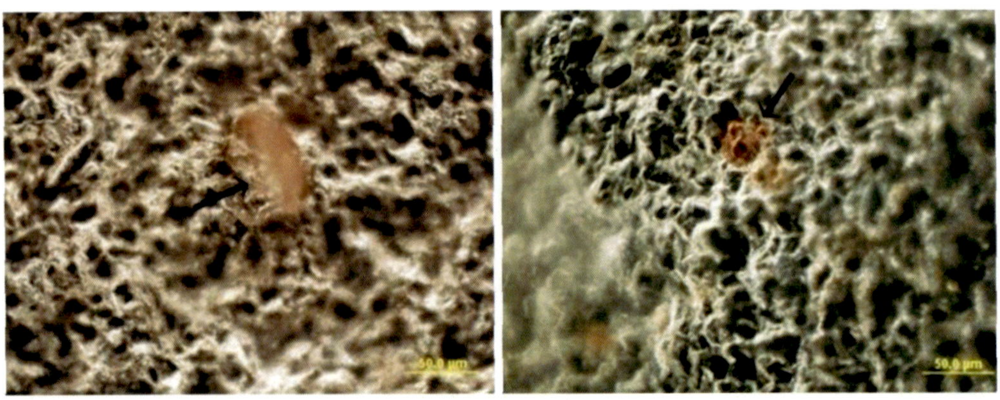

Figure 8. Optical microscopy images of stromal cells stained with (a) alkaline phosphatase and (b) acid phosphatase (marked with black arrows) on RFMS-deposited CaP coatings (dark field) (magnification ×500).

Table 7. Mineral and biochemical composition of supernatants of the prenatal stromal human lung cells (growth control) after culturing for 4-days in presence of model implants with CaP surface, X ± SD(m), p

Analyzed groups	Average Ra, μm, n=10	Calcium, mM	Phosphate ions, mM	ALP activity, E/l	ACP activity, E/l	Osteocalcin, ng/ml
Culture medium	-	1.51	9.19	44.15	0.02	0.43
Cells growth control on plastics, n_l=5	-	1.53 ±0.06 (0.03)	8.71±0.17 (0.08)	46.64 ±1.16 (0.52)	0.04±0.02 (0.009)	0.74±0.14 (0.08)
Smooth coatings, n_l=6	0.94	1.44 ±0.17 (0.10)	8.06±0.46# (0.27) <0.046	47.03 ±5.10 (2.94)	0.073±0.012# (0.007) <0.041	0.80±0.22 (0.13)
Smooth structured coatings, n_l=5	1.14	1.61 ±0.08 (0.05)	8.35±0.08 (0.05)	51.30 ±2.09# (1.21) <0.006	0.097±0.031# (0.018) <0.018	0.77±0.10 (0.06)

Note: (#) is statistically significant difference with the cells growth control on plastics according to Mann-Whitney's U-criterion at p<0.05; n is the number of roughness measurements of artificial surfaces; n_l is the number of analyzed plate wells.

Table 8. Levels (% of the culture medium background) of secretable factors after culturing of fibroblast-like cells for 8-days, X ± SD (m), p

Analyzed group (scaffold name)	Alkaline phosphatase	Osteocalcin	Acid phosphatase
Cells on plastics (control), n_l=3	119.32±6.89 (3.95)	119.49±7.11 (4.17)	104±4.55 (2.26)
Cells in presence of smooth coating discs with Ra=1,04 μm (n=10), n_l=4	90.53±4.37# (2.16) <0.001	96.94±12.25# (6.13) <0.04	113.64±4.55 (2.26)

Note: (#) is statistically significant difference with the cells growth control on plastics according to Mann-Whitney's U-criterion at p<0.05; n is the number of roughness measurements of artificial surfaces; n_l is the number of analyzed plate wells.

A specific function of osteoblasts is the synthesis of the components of bone marrow. Nevertheless, the ALP activity and OC concentration, which are the markers of metabolically active osteoblasts in supernatants of cell culture contacting with the prototype of a smooth bone did not differ from that of the cells on plastics (Table 7) [175, 182].

Moreover, the PSHLCs osteogenous activity in the presence of the smooth HA surface was even decreased with time in comparison with the cells growth control on plastics (Table 8). So, on the eighth day of cultivating a statistically significant decrease of the ALP activity and the OC level was marked in cell culture supernatants (by 29% and 23%, respectively). In other words, within a short-term cultivation period on smooth surfaces it is a very unlikely event for PSHLCs to mature into osteoblasts, which then form the mineral bone matrix.

Table 9. Computer morphometric results of prenatal stromal human lung cells stained within the acid phosphatase (ACP) after cultivating on the surface of CaP coatings for 4-days, X (Q1- Me-Q3), p

Groups of coatings	Average Ra, μm, n=10	Part of stained cells out of the seeded cells for 1 mm² of the surface, %	D, c.u.o.d.	Cell staining area (S), μm²	Cells shape, %	
					Regular shape	Irregular shape
Smooth coatings, n_1=12	0.75	9.86 (6.75-10.13)	24.08 (3.56-7.78-13.25) n_2=35	209.89 (78-216.5) n_2=35	85.71	14.29
Smooth structured coatings, n_1=10	1.05	39.04* (23.64-40.52) <0.011	13.63 (10.23-17.30) n_2=104	148.78 (73.37-160.5) n_2=104	95.19	4.81

Note: (*) is statistically significant difference with the smooth coatings group according to Mann-Whitney's U-criterion.

On the other hand, the percentage of cells stained with the ACP on smooth HA coatings did not differ from the part of ALP-positive elements (within 10-12%). However, ACP-positive PSHLCs had a high optical density, a significant stained area (at the average 15×15 μm^2, Figure 8b), and a regular shape (Table 9).

The concentration of phosphate ions in the intercellular liquid for a 4-days cell culture contacting with the lamellar bone prototype decreased in comparison with the culture on the surface of plastics, whereas there was observed a 83% growth of ACP activity (Table 7). Apparently, RFMS-deposited CaP coatings may serve as initiators of secretion of an active ferment into the intercellular liquid. Meanwhile, ACP has both a hydrolytic (phosphate radical detaching from a molecule of phosphorous acid) and a transferase function (free phosphate radical shifting to the acceptor) [175].

In connection with that, immature fibroblast precursors from the pool of MMSCs [186] and monocytes/macrophages [175, 187] may be a source of ACP. The increase of TNFα secretion the classical producers of which are the cells of the mononuclear phagocyte system (Table 10), counts in favor of the latter. The presence of hematopoietic cells in the PSHLCs culture cannot be omitted due to the presence of non-bone ALP isoform in the supernatants [184]. On the other hand, osteoblasts [188] and human bone marrow stromal cell lines [189] are also able to secrete TNFα. However, in contrast to lymphocytes and macrophages [190], non-stimulated human MMSCs [182] do not produce interleukin-4 (IL-4). The absence of cytokine secretion by the cells culture as a response to the contact with artificial materials was stated as a result of our experiments (Table 10). Van Gieson's stain for collagen gave a positive reaction for 25% of cells (Table 11, Figure 9) adhering to smooth CaP coatings. Based on the results published elsewhere [164], we suggest an initial phase of collagen precipitation on cells surface which represent a precursor matrix for building intercellular substance of the connective tissue and its derivatives. The facts presented allow concluding that in case of colonization of smooth HA coatings imitating the mineral bone matrix in a quiescent state (out-of remodeling area), the PSHLCs are slowly differentiating and maturing probably towards collagen-producing fibroblast-like cells.

Table 10. Cytokines concentrations (pg/ml) in supernatants in 24 hours culture of prenatal stromal human lung cells (growth control) in presence of model implants with smooth CaP surface, X ± SD(m), p

Analyzed groups, n_1=3	Average Ra, μm, n=10	TNFα	IL-2	IL-4
Cells on plastic surface (control)	-	46.52 ± 1.16 (0.67)	78.01 ± 0.77 (0.44)	66.33 ± 2.21 (1.28)
Cells in presence of discs with smooth surface	0.17	62.26 ± 6.0 (3.47)# <0.011	77.32 ± 2.75 (1.59)	65.86 ± 5.76 (3.33)

Note: (#) is statistically significant difference with the cells growth control on plastics according to Mann-Whitney's U-criterion at $p<0.05$; n is the number of roughness measurements of artificial surfaces; n_1 is the number of analyzed plate wells.

Table 11. Computer morphometric results of prenatal stromal human lung cells stained for collagen after cultivating on the surface of CaP coatings for 4-days, X (Q1- Q3), p

Groups of coatings	Average Ra, μm, n=10	Part of collagen-positive cells out of the seeded cells for 1 mm^2 of the surface, %	Cell shape, %	
			Regular	Irregular
Smooth coatings, n_1=10	0.46	24.85 (16.88-30.39) n_2=83	91.36	8.64
Smooth structured coatings, n_1=10	0.75	9.22* (6.75-10.13) n_2=30 <0.00002	96.67	3.33

Note: n is the number of roughness measurements of artificial surfaces; n_1 is the number of counted optical pictures (visual fields); n_2 is the number of counted cells in visual fields.

It is considered that endosteal niches formed by inactive cells are serving to maintain a slowly re-populated HSCs taht are mainly in a quiescent state [188, 191]. The function of the endosteal niches in various bone tissue areas is unknown [176], but a close interaction of endochondral bone formation and hemopoiesis initiation processes can be assumed [192].

Thus, based on experimental modeling of the smooth (two-dimensional) surface of bone marrow, it is supposed that 'silent' niches for HSCs are connected with the trabecular and cortical endosteum apart from the remodeling and vascularization area. Consequently, topographically those hematopoietic niches should remain over a long period of time mainly on a slowly renewing cortical endosteum. Moreover, stromal stem cells are a considerable component of those 2D micro-territories. Furthermore, Wilson and Trumpp [168] supposed that the interaction between the bone-marrow MMSCs and HSCs facilitates the maintenance of the latter apart from the cells cycle both *in vitro* and *in vivo*. The knowledge gathered in cells biology *in vitro* needs to be validated *in vivo* [193], including implementation of the ectopic (heterotopic) osteogenesis test, when bone structures are formed while artificial materials are implanted either subcutaneously or intramuscularly [179, 180].

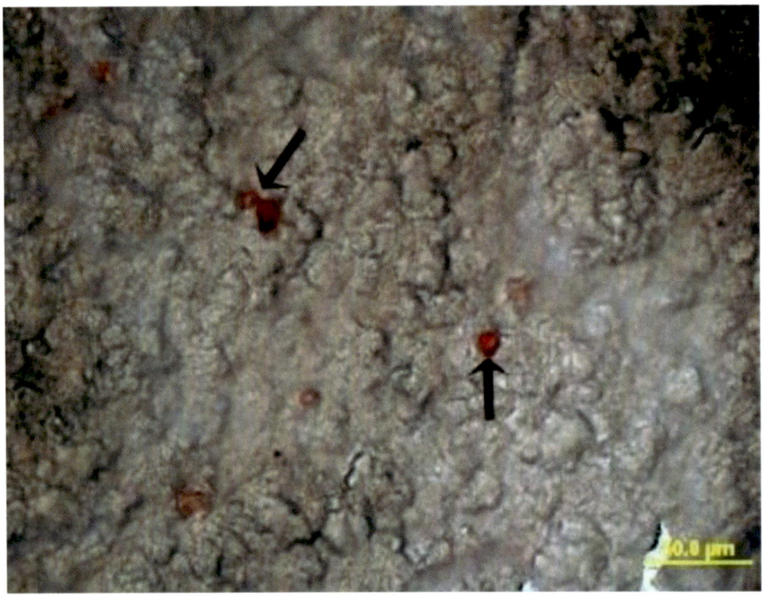

Figure 9. Optical microscopy image of stromal cells stained for collagen (marked with a black arrow) on RFMS-deposited CaP coating (bright field) (magnification ×500).

The application of materials with optimal physico-chemical characteristics allows us to culture subcutaneously the bone marrow surrounded with a fibrous (trabecular) bone tissue plate from the bone marrow column applied previously on an implant *in vitro* [180].

The bone marrow heterotopic metaplasia takes place through the activation of donor MMSCs pool, which are differentiated into chondro-/osteoblast precursors and descendants [194]. In its turn, the endochondral ossification is a key condition for the development of hematopoietic micro-environment, bone-marrow lacunes and niches for HSCs of a recipient [192].

Previously, we revealed [180] that there was a direct correlation ($r = 0.77$; $p < 0.05$) between bone formation in the ectopic osteogenesis test and Ca/P ratio of the coatings deposited of an implant surface. At the same time, the results of the present study showed that if Ca/P ratio of the coatings with a low surface roughness was higher than that of stoichiometric HA, RFMS-deposited coatings do not possess the capacity to induce an ectopic development of the mouse bone/bone marrow system. The probability of tissue plate formation, bone growth and formation of hemopoiesis loci on smooth HA coatings equals to zero (10 out of 10 implantations did not have positive results) despite their high biocompatibility. Thus, 2D-CaP coatings imitating the state of a smooth, poorly renewing mineral matrix, do not stimulate the development of functional activity of osteogenous and hematopoietic cells.

Depending on physico-chemical properties (degree of crystallinity and porosity, solubility, surface roughness, elemental and phase composition etc.) different CaP materials have a different capacity to maintain the bone formation process [195]. The artificial CaP matrix dissolution is a significant part of its osteogenous properties intensification [196]. It is connected, first of all, with the amplification of extracellular calcium concentration which plays an important role in the regulation of the stromal and hematopoietic pool of stem cells [170].

In connection with that we decided to examine the influence of the model mineral matrix dissolution on its bone-forming activity. Si-HA is considered to be one of prospective biodegradable materials for biomedical application. Since silicate anions (SiO_4^{4-}) are a natural part of the intertissue liquid [197] their presence in glass-ceramics and HA compounds accelerates the material degradation and osteointegration [198]. The mechanism of this Si-HA biomaterial activity may be the appearance of TCP in their phase structure [142] and Ca/P ratio decrease.

Actually, incorporation of silicate-ions into the composition of the smooth RFMS-deposited HA coating increased by 2.5 times (Pu<0.001) the release of calcium *in vitro* into the electrolyte (0.9 % sodium chloride solution) (Table 12) at the same HA and Si-HA coating roughness ($R_a < 1$ μm).

Table 12. The Ca^{2+} concentration after 7-days of dissolution of samples with RF magnetron sputter deposited coatings in the 0.9 % sodium chloride solution at 37°C with 30 mg/l of gentamycini sulfas, X ± m, p

Number of group	Samples, n=6	Ca^{2+} release into solution, % of control
1	Magnetron sputtering on Ti scaffold, HA coating	354 ± 68
2	Magnetron sputtering on Ti scaffold, Si-HA coating	947 ± 74* <0.001

Note: as controls there were extracts of titanium scaffolds without RF magnetron sputter deposited coatings; *) is the statistically significant discrepancy according to Student's *t*-criterion.

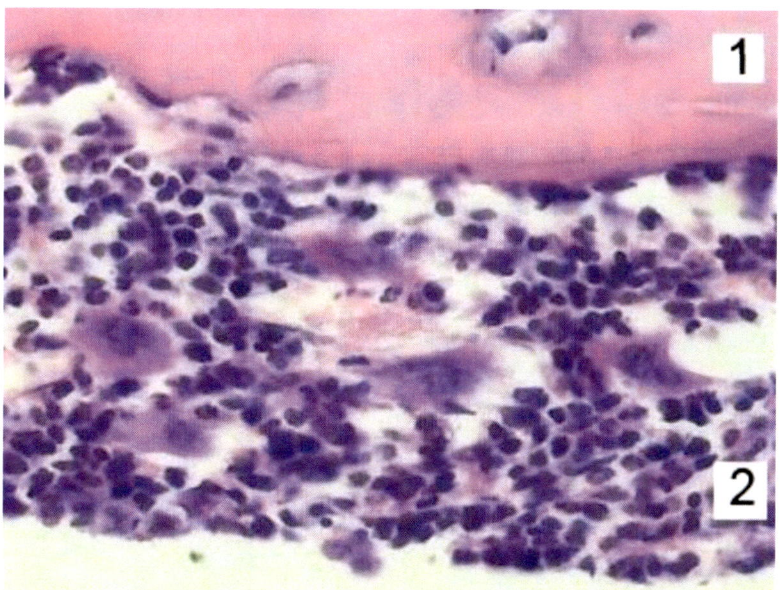

Figure 10. Cross section of tissue plate grown on textured implants with RFMS-deposited CaP coating in heterotopic test with mice. 1 - membrane reticulated bone tissue; 2 - bone marrow. Stained with hematoxylin and eosin (magnification ×200).

Moreover, the tissue plate formation increased from 0 % up to 50 % (3 out of 6 implants promoted ectopic osteogenesis) the histological composition of which corresponded to the rough fibrous bone tissue with lacunes filled with red bone marrow (Figure 10).

The results obtained confirmed [176] the significance of the mineral bone substance as an extracellular calcium source for cell cooperation. Thus, it was found a calcium-sensing receptor on stromal element [199] and the HSCs [200] membranes are necessary for their endosteum homing and, apparently, for the endosteal niche formation. Under the modeled conditions the increased dissolution of a smooth HA coating, first of all, might promote better bone marrow stroma adhesion to the substrate if it was previously cultivated *in vitro* before being implanted into mice.

It is obvious that the microenvironment's control of stem cell fate includes surface topography [201] and various extracellular matrix components (cell products, ions, cytokines, integrins, fibrous proteins etc.) [193]. Nevertheless, significant differences were observed between 2D and 3D cell cultures as well as their behavior *in vivo* [202]. To regulate the vital activity of stem cells and outcome the intercellular matrix structure is of considerable importance as well as the matrix three-dimensionality [203] and appearing mechanical forces [204].

Significant biomechanical shear loads were observed that may lead to a complete or partial deprivation of the cells population and heterotopic osteogenesis efficiency decrease. It was also revealed for bone marrow organoid culture situated on subcutaneous implanted samples. Hence for further experiments titanium substrates were mechanically macrotextured before preparation of the smooth RFMS-deposited HA coatings in such a way as to allow us to form parallel grooves with the depth and width of up to 1 mm while keeping the surface roughness index the same ($R_a < 1$ μm).

Then, after 45 days of subcutaneous growth on 3D-macrotextured implant surfaces tissue plates were observed that corresponded histologically to the positive results of the subcutaneous osteogenesis test. The probability to form the system bone/bone marrow on textured CaP coatings increased by 83.3 % (5 out of 6 implants promoted the ectopic osteogenesis). This result is in good agreement with data previously obtained by us for rough ($R_a > 2$ μm) RFMS-deposited CaP surfaces [180]. Moreover, similar structuring of Si-HA coatings allowed us to obtain the best results in a heterotopic test (4 out of 4 implants promoted the bone formation and hemopoiesis).

According to a biomimetic approach, it is suggested that in a certain approximation the macrostructured RFMS-deposited CaP surfaces imitate the physico-chemical properties of trabecular bone. Moreover, during experiments *in vitro* a similar coating relief strengthens the cytochemical and secretory activity of the PSHLCs culture mainly towards ACP-positive cells.

Hence macrotexturing (groove formation) of a smooth RFMS-deposited HA coatings promoted significant growth of ACP-positive cells (from 9.9 % up to 39.0 %, $P < 0.08$; $n = 19$) at a small part of 150 μm^2 area. Thus, the PSHLCs culture on the surface of macrotextured implants was remodeled more actively, gaining gradually osteogenetic properties in comparison with analogues on smooth HA coatings.

Osteoclasts and their precursors (first of all, monocytes/macrophages) play a versatile role in remodeling of the hematopoietic microenvironment. Like cellular components of the hematopoietic niche [174, 191] they promote the mobilization of hematopoietic precursors into the circulating blood [205]. For a limited number of osteoclasts, lacune formation is violated at the loci of extramedullary hematopoiesis that impedes the formation of active hematopoietic micro-territories of resident bone-marrow stroma and migrating of HSCs [188].

The question now arising is: how can cell molecular reactions of stromal and hematopoietic stem cells on scaffolds be controlled?

According to general principles of hematopoietic micro-environment formation [206], the endosteal niche regulates the HSCs through cell contacts and humoral signals [203]. Much data indicated the dual nature of a regulatory stimulus in osteoblast niches. Thus secretory factors of positive and negative stromal regulation of HSCs were described [174]. Most likely, the direct contact with the MMSCs and osteoblasts impedes the HSCs activation [168, 171]. In turn, it is considered that two aspects of HSCs pool self-maintaining can be distinguished, either cell proliferation or limitation of their differentiation [207].

In our studies, the TNFα secretion during contact of PSHLCs with the smooth bone matrix model was increased by 34 % as compared with the culture of cells on plastics (Table 10). In this context the TNFα may be one of multifunctional molecular factors regulating the self-maintenance/commitment of stromal stem cells, which increase the MMSCs survival rate by autocrine and paracrine regulation, but impede their differentiation into osteoblasts [208].

Also, TNFα and its receptors are the molecular components of the osteoblast niche for HSCs [188] that inhibit the proliferation of early hematopoietic precursors [209]. Then, the TNFα stimulates osteoclasts to remodel the bone and an implant that is inevitably accompanied by changes in morpho-functional parameters of both osteogenous and hematopoietic niches [163].

The current studies demonstrated (Table 9) that the macrostructured HA surfaces stimulate the ACP-positive cells *in vitro* that remodel the bone matrix under natural conditions [175]. At the same time, both the surface area and ALP-positive cells staining slightly increased (Table 6). For *in vitro* characterization of the artificial surfaces with osteogenous niches a structural-functional index was found that represents the ratio of the area of cells (osteoblast) for ALP-staining to the area of artificial surface cavities occupied by the stained cells (S_{ALP}/S_{niche}, %) [184]. In the current studies the S_{ALP}/S_{niche} index was also increased from 23% up to 35% on structured titanium substrates with CaP coatings (Table 6), accompanied by an increase in the efficiency of the bone/bone-marrow system remodeled in the heterotopic test *in vivo*.

These results allow suggesting that the function of the endosteal niche (silent/activated; [174]) depends on the structural-functional state of the bone matrix. The niche size [170], elasticity and stiffness of the extracellular matrix [173], as well as the calcium release from the bone matrix [176] may be probable mechanisms to control the behavior of pluripotent and committed HSCs.

At the same time, osteoclast-like cells that change bone and implant topography apparently promote switching on the morphofunctional parameters of osteogenous and hematopoietic niches from their quiescent state into an active one [163].

Thus, the weak osteogenous commitment of stromal stem cells on the 'smooth' ($R_a < 1$ μm) RFMS-deposited HA coating *in vitro* is associated with the average osteogenous niche index less than 35%, insignificant surface dissolution, and increased TNFα secretion in cell cultures. Moreover, it is accompanied by the decay of ectopic bone formation and hematopoiesis *in vivo*. A similar experimental bone state model allows imitating and examining 'silent' endosteal niches for osteogenous and hematopoeic stem cells.

Furthermore, it allows introducimg and testing physico-chemical and biological components that convert the niches from their quiescent state into an active one.

It is known that the manipulation of structural characteristics of HA ceramics improves their osteogenous properties [210]. In this case the physical (surface macrostructuring) and chemical (substitution of silicate ions into the structure of the coating) modification of HA coatings promote the appearance and intensification of their osteogenous potential.

However, all types of RFMS-deposited HA coatings studied stimulated the activity of ACP-positive cells. For successful osteoinduction an implant has to activate osteoblasts. Excessive osteoclast stimulation leads to the resorption of artificial material and bone tissue, deterioration of osteointegration with an expected implant failure [163]. Thus, the experimental results described above showed that there exist materials and coatings with an indirect osteoinductive effect through predominant activation of ACP-positive cells functions. In this case, for bioactive materials and coatings designed for bone tissue engineering the existence of a paradoxical group of 'resorptive osteogenous implants' should be emphasized which are able to stimulate remodeling of the system bone/bone marrow in an experiment. These implants can find a clinical application due to certain physico-chemical properties such as low thickness, mechanical resistance, and regulated dissolution that can be varied during the technological stage.

4. Challenges and Future Trends

Future research should concentrate on both the improvements of the RFMS process and the investigation of the characteristics of CaP-based coatings. RFMS is a line-of-sight technique, therefore, additional measures e.g. substrate rotation and movement should be taken to prepare uniform coatings on implants with complex geometry. Moreover RFMS can cause dehydroxylation of HA during biocompatible film deposition [45, 65]. Therefore, if a coating of pure HA is to be deposited the addition of water vapor into the deposition system should become commonplace [42].

Although it is generally accepted that HA and CaP coatings deposited by RFMS improve bone strength and the initial osseointegration rate, the optimal coating properties required to achieve the maximum bone response are still to be determined. A better understanding in structure-property relationship would provide good guidelines for implant designs [211]. Moreover, there are only a limited number of *in vivo* studies on RFMS-deposited coatings available in the literature [122, 212]. The *in vivo* testing should demonstrate stability in biological environment for up to 1 month, which corresponds to the initial healing phase of implants [213].

The influence of surface chemistry modifications and topography at a nano/micro scale level on biological response, as well as hydrophilicity, surface charge, and wettability should be studied more carefully [214-217].

Very few articles were devoted to the modeling of the growth of RFMS-deposited CaP-based coatings [45, 218, 219]. The proposed mechanisms of CaP-based growth via RF magnetron sputtering is discussed in some studies [45, 64]. Particle ejection from the sputtering target, particle collisions with the background gas, charge state of the sputtered particles, and re-sputtering of the deposited film influence the composition and the structure of the film [218, 219]. This subject still needs particular investigations to prepare coatings with tailored characteristics.

To conclude, future investigations of RFMS deposition of CaP-based coatings should address [6, 33, 42, 220-222] i) the effect of coating crystallinity (including the importance of low crystallinity), chemical composition and surface topography on the cell differentiation, ii) therapeutic capabilities of the coating (templates for the *in situ* delivery of drugs and osseoinductive agents (peptides, hormones and growth factors) at the required times), and iii) long-term *in vivo* clinical trials to optimize coating properties and to continue the preclinical research to provide additional understanding of bone responses to coated implant surfaces of different properties.

ACKNOWLEDGMENTS

The authors would like to thank Prof. Dr. Matthias Epple (University of Duisburg-Essen) and Prof. Dr. M.V. Chaikina (Institute of Solid State Chemistry and Mechanochemistry, Siberian Branch of the Russian Academy of Sciences) for fruitful discussions that undoubtedly ensured a significant improvement of the quality of this chapter.

The research has been accomplished with support from the federal and international special-purpose programs 'Researches and development within priority guidelines of Russian scientific and technological complex for the years 2007-2013' (Government contract No 16.512.11.2087), 'Academic and academic and teaching staff of innovative Russia' for the years 2009-2013 (Government contracts П861 from 25.05.2010, 16.740.11.0205 from 24.09.2010), the analytical departmental special-purpose program (ADSPP), 'Development of higher school scientific potential for the years 2009-2011' (project registration number 2.1.1/14204), the RAS presidium grant No 5FNM-27 'Fundamental sciences into medicine', the integration project No 126 SB RAS and grant No 09-04-00287a of the Russian Foundation for Basic Research, DAAD, BMBF (Rus 10/020), Russian President Grants (MK-6360.2010.8, MK-1149.2012.8) and Tomsk Polytechnic University (No 11-03/2012).

REFERENCES

[1] J. B. Brunski, in: B. D. Ratner, A. S. Hoffman, F. J. Schoen, J. E. Lemons, Biomaterials Science: An Introduction to Materials in Medicine, Elsevier Inc., New York, 2004, pp. 137-153.

[2] M. Roy, A. Bandyopadhyay, and S. Bose, *Surface and Coatings Technology*, 205, 2785 (2011).

[3] L. Sun, C. C. Berndt, K. A. Gross, and A. Kucuk, *J. Biomedical Materials Research,* 58, 570 (2001).

[4] C. C. G. Moura, M. A. Souza, P. Dechichi, D. Zanetta-Barbosa, C. C. Teixeira, and P. G. Coelho, *J Biomedical Materials Research,* 94A, 103 (2010).

[5] S. V. Dorozhkin, and M. Epple, *Angewandte Chemie Int. Ed. in English*, 41, 3130 (2002).

[6] B. León, and J. A. Jansen, *Calcium Phosphate Coatings for Medical Implants*, Springer, New York (2009).
[7] S. R. Paital, and N. B. Dahotre, *Materials Science Engineering R*, 66, 1 (2009).
[8] *ISO 13779-4:2002.* Implants for surgery -- Hydroxyapatite -- Part 4: Determination of coating adhesion strength. www.iso.org.
[9] *ISO 13779-3:2008.* Implants for surgery - Hydroxyapatite - Part 3: Chemical analysis and characterization of crystallinity and phase purity. http://www.iso.org.
[10] *ISO 13779-2:2008.* Implants for surgery - Hydroxyapatite - Part 2: Coatings of hydroxyapatite, http://www.iso.org/.
[11] *F.D.A., (Food and Drug Administration, Washington, DC, 1992) p. 1.* http://www.fda.gov/.
[12] R. B. Heimann, *CMU. Journal*, 1, 23 (2002).
[13] K. A. Gross, and S. Saber-Samandari, *Surface and Coatings Technology*, 203, 2995 (2009).
[14] R. Jaworski, C. Pierlot, L. Pawlowski, M. Bigan, and M. Martel, *Surface and Coatings Technology*, 203, 2092 (2009).
[15] L. Łatka, L. Pawlowski, D. Chicot, C. Pierlot, and F. Petit, *Surface and Coatings Technology*, 205, 954 (2010).
[16] Y. Huang, L. Song, T. Huang, X. Liu, Y. Xiao, Y. Wu, F. Wu, and Z. Gu, *Biomedical Materials*, 5, 054113 (2010).
[17] K. Ozeki, T. Yuhta, H. Aoki, and Y. Fukui, *Bio-Medical Materials and Engineering*, 13, 271 (2003).
[18] J. G. C. Wolke, K. van Dijk, H. G. Schaeken, K. de Groot, and J. A. Jansen, *J. Biomedical Materials Research*, 28, 1477 (1994).
[19] J. A. Jansen, J. G. C. Wolke, S. Swann, J. P. C. M. van der Waerden, and K. de Groot, *Clinical Oral Implants Research*, 4, 28 (1993).
[20] K. van Dijk, H. G. Schaeken, J. G. C. Wolke, and J. A. Jansen, *Biomaterials*, 17, 405 (1996).
[21] K. van Dijk, H. G. Schaeken, J. G. C. Wolke, C. H. M. Maree, F. H. P. M. Habraken, J. Verhoeven, and J. A. Jansen, *J. Biomedical Materials Research*, 29, 269 (1995).
[22] D. V. Shtansky, N. A. Gloushankova, A. N. Sheveiko, Ph. V. Kiryukhantsev-Korneev, I. A. Bashkova, B. N. Mavrin, S. G. Ignatov, S. Yu. Filippovich, and C. Rojas, *Surface and Coatings Technology*, 205, 728 (2010).
[23] B. León, in: B. León, J. A. Jansen (Eds.), *Pulsed laser deposition of thin calcium phosphate coatings*, Springer, New York, 2009, pp. 101-155.
[24] M. Jelínek, M. Weiserová, T. Kocourek, M. Zezulová, and J. Strnad, *Laser Physics*, 21, 1265 (2011).
[25] T. R. Rautray, R. Narayanan, T.-Y. Kwon, and K.-H. Kim, *Thin Solid Films*, 518, 3160 (2010).
[26] S. Sobieszczyk, and A. Zieliński, *Advanced Materials Science*, 8, 35 (2008).
[27] H. L. Barthell, T. A. Archuleta, and R. Kossowsky, *Mater. Res. Soc. Symp. Proc.*, 110, 709 (1989).
[28] M. Yoshinari, Y. Watanabe, Y. Ohtsuka, and T. Derand, *J. Dental Research*, 76, 1485 (1997).
[29] M. Yoshinari, in B. Leon, J. A. Jansen (Eds.), *Thin Calcium Phosphate Coatings for Medical Implants*, Springer, New York, 2009, pp. 157-174.

[30] X. Bai, S. Sandukas, M. R. Appleford, J. L. Ong, and A. Rabiei, *Acta Biomaterialia*, 5, 3563 (2009).
[31] X. Bai, *Processing and Characterization of Functionally Graded Hydroxyapatite Coatings for Biomedical Implants*, Ph.D thesis, North Carolina State University, 2009, 3430563, 239 p.
[32] P. G. Coelho, and J. E. Lemons, *J. Biomedical Materials Research*, 90A, 351 (2009).
[33] T. R. Rautray, R. Narayanan, T.-Y. Kwon, and K.-H. Kim, *J. Biomedical Materials Research B: Applied Biomaterials*, 93B, 581 (2010).
[34] I. Zhitomirsky, and L. Gal-Or, *J. Materials Science: Materials in Medicine*, 8, 213 (1997).
[35] J. Ma, C. Wang, and K. W. Peng, *Biomaterials*, 24, 3505 (2003).
[36] D. Liu, Q. Yang, and T. Troczynski, *Biomaterials,* 23, 691 (2002).
[37] K. Hwang, and Y. Lim, *Surface and Coatings Technology*, 115, 172 (1999).
[38] J. Sun, Y. Han, and X. Huang, *Surface and Coatings Technology*, 201, 5655 (2007).
[39] E. Zhang, C. Zou, and S. Zeng, *Surface and Coatings Technology*, 203, 1075 (2009).
[40] Y. Liu, E. B. Hunziker, *Biomimetic Coatings and Their Biological Functionalization*, in B. Leon, J. A. Jansen (Eds.), Thin Calcium Phosphate Coatings for Medical Implants, Springer, New York, 2009, pp. 303-305.
[41] H. Wie, H. Hero, and T. Solheim, *Int. J. Oral Maxillofac. Implants*, 13, 837 (1998).
[42] R.A. Surmenev, *Surface and Coatings Technology,* 206, 2035 (2012).
[43] L. Russo, N. Shaikh, L. Cipolla, F. Nicotra, C. Riccardi, and S. Zanini, *J. Biotechnology*, 150, 202 (2010).
[44] P. K. Chu, *Surface and Coatings Technology*, 203, 2793 (2009).
[45] R. A. Surmenev, M. A. Surmeneva, K. E. Evdokimov, V. F. Pichugin, T. Peitsch, and M. Epple, *Surface and Coatings Technology*, 205, 3600 (2011).
[46] A. W. Warner, and C. D. Stockbridge, *J. Applied Physics*, 34, 437 (1963).
[47] J. L. Ong, Y. Yang, S. Oh, M. Appleford, W. Chen, Y. Liu, K.-H. Kim, S. Park, J. Bumgardner, W. Haggard, C. Mauli Agrawal, D.L. Carner, and N. Oh, *Calcium Phosphate Coating Produced by a Sputter Deposition Process*, in B. Leon, J.A. Jansen (Eds.), Thin Calcium Phosphate Coatings for Medical Implants, Springer, New York, 2009, pp. 175-198.
[48] D. M. Mattox, *Handbook of Physical Vapor Deposition (PVD) Processing (Second Edition)*, Elsevier (2010).
[49] P. M. Martin, *Handbook of Deposition Technologies for Films and Coatings (Third Edition)*, Elsevier (2010).
[50] G. Sobe, H. Schreiber, G. Weise, and A. Heinrich, *Contributions to Plasma Physics,* 33, 325 (2006).
[51] J. V. Anguita, M. Thwaites, B. Holton, P. Hockley, S. Rand, and S. Haughton, *Plasma Processes Polymers*, 4, 48 (2007).
[52] K. P. Mohanchandra, K. K. Ho, and G. P. Carman, *J. Intelligent Mater. Systems and Structures*, 15, 387 (2004).
[53] C. Wang, Z. Chen, L. Guan, Z. Liu, P. Wang, S. Zheng, and X. Liao, *Surface and Coatings Technology*, 130, 39 (2000).
[54] J. L. Ong, and L. C. Lucas, *Biomaterials*, 15, 337 (1994).
[55] K. Yamashita, T. Arashi, K. Kitagaki, S. Yamasa, and T. Umegaki, *J. American Ceramic Society*, 77, 2401 (1994).

[56] J. G. C. Wolke, J. P. C. M. van der Waerden, K. de Groot, and J. A. Jansen, *Biomaterials*, 18, 483 (1997).

[57] K. van Dijk, H. G. Schaeken, C. H. M. Marée, J. Verhoeven, J. C. G. Wolke, F. H. P. M. Habraken, and J. A. Jansen, *Surface and Coatings Technology*, 76-77, 206 (1995).

[58] M. Yoshinari, T. Hayakawa, J. G. C. Wolke, K. Nemoto, and J. A. Jansen, *J. Biomedical Materials Research*, 37, 60 (1996).

[59] T. Hayakawa, M. Yoshinari, K. Takahashi, and K. Nemoto, *J. Oral Tissue Engineering*, 1, 41 (2004).

[60] J. Long, L. Sim, S. Xu, and K. Ostrikov, *Chemical Vapor Deposition*, 13, 299 (2007).

[61] A. R. Boyd, B. J. Meenan, and N. S. Leyland, *Surface and Coatings Technology*, 200, 6002 (2006).

[62] S.-J. Ding, C.-P. Ju, and J.-H. Chern Lin, *J. Biomedical Materials Research*, 47, 551 (1999).

[63] S.-J. Ding, *Biomaterials*, 24, 4233 (2003).

[64] S. Xu, J. Long, L. Sim, C. H. Diong, and K. Ostrikov, *Plasma Processes Polymers*, 2, 373 (2005).

[65] V. F. Pichugin, R. A. Surmenev, E. V. Shesterikov, M. A. Ryabtseva, E. V. Eshenko, S. I. Tverdokhlebov, O. Prymak, and M. Epple, *Surface and Coatings Technology*, 202, 3913 (2008).

[66] V. Nelea, C. Morosanu, M. Bercu, and I. N. Mihailescu, *J. Materials Science: Materials in Medicine*, 18, 2347 (2007).

[67] V. Nelea, C. Morosanu, M. Iliescu, and I. N. Mihailescu, *Applied Surface Science*, 228, 346 (2004).

[68] A. R. Boyd, H. Duffy, R. McCann, M. L. Cairns, and B. J. Meenan, *Nuclear Instruments and Methods in Physics Research B*, 258, 421 (2007).

[69] M. Hamdi, J. A. Toque, and A. Ide-Ektessabi, *Key Engineering Materials*, 443, 469 (2010).

[70] K. van Dijk, J. Verhoeven, C. H. M. Marée, F. H. P. M. Habraken, and J. A. Jansen, *Thin Solid Films*, 304, 191 (1997).

[71] B. Feddes, A. M. Vredenberg, J. G. C. Wolke, and J. A. Jansen, *Surface and Coatings Technology*, 185, 346 (2004).

[72] K. van Dijk, C. H. M. Marée, J. Verhoeven, F. H. P. M. Habraken, and J. A. Jansen, *J. Biomedical Materials Research*, 42, 266 (1998).

[73] K. Ozeki, J. M. Janurudin, H. Aoki, and Y. Fukui, *Applied Surface Science*, 253, 3397 (2007).

[74] K. Ozeki, T. Yuhta, Y. Fukui, and H. Aoki, *Surface and Coatings Technology*, 160, 54 (2002).

[75] A. Boyd, M. Akay, and B. J. Meenan, *Surface and Interface Analysis,* 35, 188 (2003).

[76] K. Ozeki, H. Aoki, and T. Masuzawa, *Applied Surface Science*, 256, 7027 (2010).

[77] J. A. Toque, M. Hamdi, A. Ide-Ektessabi, and I. Sopyan, *Int. J. Modern Physics B*, 23, 5811 (2009).

[78] G. Socol, A. M. Macovei, F. Miroiu, N. Stefan, L. Duta, G. Dorcioman, I. N. Mihailescu, S. M. Petrescu, G. E. Stan, D. A. Marcov, A. Chiriac, and I. Poeata, *Materials Science Engineering B*, 169, 159 (2010).

[79] R. A. Surmenev, M. A. Ryabtseva, E. V. Shesterikov, V. F. Pichugin, T. Peitsch, and M. Epple, *J. Materials Science: Materials in Medicine*, 21, 1233 (2010).

[80] A. Mello, E. Mavropoulos, Z. Hong, J. B. Ketterson, and A. Rossi, *Key Engineering Materials*, 396 - 398, 369 (2009).

[81] Y. Yan, J. G. C. Wolke, A. De Ruijter, Li Yubao, and J. A. Jansen, *J. Biomedical Materials Research*, 78A, 42 (2006).

[82] Y. Yonggang, J. G. C. Wolke, L. Yubao, and J. A. Jansen, Clinical Oral Implants Research, 18, 345 (2007).

[83] Y. Yonggang, J. G. C. Wolke, L. Yubao, and J. A. Jansen, *J. Materials Science: Materials in Medicine*, 18, 1061 (2007).

[84] Y. Yonggang, J. G. C. Wolke, L. Yubao, and J. A. Jansen, *J. Biomedical Materials Research*, 76A, 744 (2006).

[85] V. F. Pichugin, M. A. Surmeneva, R. A. Surmenev, I. A. Khlusov, and M. Epple, Journal of Surface Investigation. X-ray, Synchrotron and Neutron Techniques, 5, 863 (2011).

[86] M. A. Surmeneva, R.A. Surmenev, V.F. Pichugin, S.S. Chernousova, and M. Epple, Journal of Surface Investigation. X-ray, Synchrotron and Neutron Techniques, 5, 1202 (2011).

[87] E. S. Thian, J. Huang, S. M. Best, Z. H. Barber, and W. Bonfield, *J. Biomedical Materials Research*, 78A, 121 (2006).

[88] E. S. Thian, J. Huang, S. M. Best, Z. H. Barber, and W. Bonfield, *Biomaterials*, 26, 2947 (2005).

[89] K. Yamashita, T. Yagi, and T. Umegaki, *J. American Ceramic Society*, 79, 3313 (1996).

[90] L. E. Sima, G. E. Stan, C. O. Morosanu, A. Melinescu, A. Ianculescu, R. Melinte, J. Neamtu, and S. M. Petrescu, *J. Biomedical Materials Research*, 95A, 1203 (2010).

[91] W. Chen, Y. Liu, H. S. Courtney, M. Bettenga, C. M. Agrawal, J. D. Bumgardner, and J. L. Ong, *Biomaterials*, 27, 5512 (2006).

[92] L. Ding, Y. Zheng, Q. B. Wan, X. B. Pei, and S. Y. Chen, *Materials Science Forum*, 675-677, 869 (2011).

[93] Z. Hong, A. Mello, T. Yoshida, L. Luan, P. H. Stern, A. Rossi, D. E. Ellis, and J. B. Ketterson, *J. Biomedical Materials Research*, 93A, 878, (2010).

[94] Y. Yang, J. D. Bumgardner, R. Cavin, D. L. Carnes, and J. L. Ong, *J. Dental Research*, 82, 449 (2003).

[95] J. E. G. Hulshoff, K. van Dijk, J. E. de Ruijter, F. J. R. Rietveld, L. A. Ginsel, and J. A. Jansen, *J. Biomedical Materials Research*, 40, 464 (1998).

[96] Y. T. Zhao, Z. Zhang, Q. X. Dai, D. Y. Lin, and S. M. Li, *Surface and Coatings Technology*, 200, 5354 (2006).

[97] M. Chen, D. Liu, C. You, X. Yang, and Z. Cui, *Surface and Coatings Technology*, 201, 5688 (2007).

[98] B. Feddes, J. G. C. Wolke, A. M. Vredenberg, and J. A. Jansen, *Biomaterials*, 25, 633 (2004).

[99] B. Feddes, J. G. C. Wolke, A. M. Vredenberg, and J. A. Jansen, *Surface and Coatings Technology*, 184, 247 (2004).

[100] E. S. Thian, J. Huang, S. M. Best, Z. H. Barber, and W. Bonfield, *J. Materials Science*, 16, 411 (2005).

[101] J. G. Wolke, J. P. van der Waerden, H. G. Schaeken, and J. A. Jansen, *Biomaterials*, 24, 2623 (2003).

[102] R. Torensama, P. J. ter Brugge, J. A. Jansen, and C. G. Figdor, *Clinical Oral Implants Research*, 14, 569 (2003).
[103] T. Hayakawa, M. Yoshinari, K. Nemoto, J.G. Wolke, nd J.A. Jansen, *Clinical Oral Implants Research*, 11, 296 (2000).
[104] V. M. Ievlev, E. P. Domashevskaya, V. I. Putlyaev, Yu. D. Tret'yakov, S. M. Barinov, E. K. Belonogov, A. V. Kostyuchenko, M. I. Petrzhik, and F. V. Kiryukhantsev-Korneev, *Glass Physics and Chemistry*, 34, 608 (2008).
[105] X. Lu, and Y. Leng, J. Biomedical Materials Research, 66A, 677 (2003).
[106] J. G. C. Wolke, K. de Groot, and J. A. Jansen, *J. Biomedical Materials Research (Applied Biomaterials)*, 43, 270 (1998).
[107] F. Variola, J. B. Brunski, G. Orsini, P. Tambasco de Oliveira, R. Wazen, and A. Nanci, *Nanoscale*, 3, 335 (2011).
[108] P. G. Coelho, G. Cardaropoli, and M. Suzuki, *Clinical Implants Dentistry Related. Research*, 11, 292 (2009).
[109] Y. Yang, K.-H. Kim, and J. L. Ong, *Biomaterials*, 26, 327 (2005).
[110] K. Ozeki, Y. Fukui, and H. Aoki, *Applied Surface Science*, 253, 5040 (2007).
[111] R. Snyders, E. Bousser, D. Music, J. Jensen, S. Hocquet, and J.M. Schneider, *Plasma Processes Polymers*, 5, 168 (2008).
[112] B.-H. Lee, and N. Koshizaki, *Nanotechnology*, 19, 415303 (2008).
[113] J. L. Ong, G. N. Raikar, and T. M. Smoot, *Biomaterials*, 18, 1271 (1997).
[114] V. M. Pantojas, E. Velez, D. Hernández, and W. Otaño, *PR Health Sciences Journal*, 28, 258 (2009).
[115] A. R. Boyd, G. A. Burke, H. Duffy, M. Holmberg, C. O' Kane, B. J. Meenan, and P. Kingshott, *J. Materials Science: Materials in Medicine*, 22, 71 (2011).
[116] A. Y. Berezhnaya, V. O. Mittova, E. V. Kukueva, and I. Y. Mittova, *Inorganic Materials*, 46, 971 (2010).
[117] K. Ozeki, H. Aoki, and Y. Fukui, *J. Materials Science*, 40, 2837 (2005).
[118] J.Z. Shi, C.Z. Chen, H.J. Yu, and S.J. Zhang, *Bulletin Materials Science*, 31, 877 (2008).
[119] C.-Y. Yang, T.-M. Lee, Y.-Z. Lu, and C.-W. Yang, *J. Med. Biol. Eng.*, 30, 91 (2009).
[120] V. Nelea, C. Morosanu, M. Iliescu, and I.N. Mihailescu, *Surface and Coatings Technology*, 173, 315 (2003).
[121] K. Yamashita, M. Matsuda, T. Arashi, and T. Umegaki, *Biomaterials*, 19, 1239 (1998).
[122] K. Ueda, T. Kawasaki, T. Goto, J. Kurihara, H. Nakagawa, H. Kawamura, and M. Taira, *J. Biomechanical Science Engineering*, 4, 392 (2009).
[123] Y. Yang, K. H. Kim, C. M. Agrawal, and J. L. Ong, *J. Dental Research*, 82, 833 (2003).
[124] J. Musschoot, D. Depla, G. Buyle, J. Haemers, and R. De Gryse, *J. Physics D: Applied Physics,* 39, 3989 (2006).
[125] S. Ido, and K. Nakamura, *Jpn. J. Applied Physics*, 35, 2302 (1996).
[126] X. Lu, and Y. Leng, *J. Biomedical Materials Research B*, 90, 438 (2009).
[127] R. Snyders, D. Music, D. Sigumonrong, B. Schelnberger, J. Jensen, and J. M. Schneider, Appl. Phys. Lett. 90 (2007) 3 pages.
[128] G. P. Dinda, J. Shin, and J. Mazumder, Acta Biomaterialia, 5, 1821 (2009).
[129] H. C. Man, K. Y. Chiu, F. T. Cheng, and K. H. Wong, *Thin Solid Films*, 517, 5496 (2009).

[130] F. J. García-Sanz, M. B. Mayor, J. L. Arias, J. Pou, B. León, and M. Pérez-Amor, *J. Materials Science: Materials in Medicine*, 8, 861 (1997).
[131] J. Weng, X.G. Liu, X.D. Li, and X.D. Zhang, *Biomaterials*, 16, 39 (1995).
[132] Q. Tang, R. Brooks, N. Rushton, and S. Best, *J. Matererials Science: Materials in Medicine*, 21, 173 (2010).
[133] E. E. Berry, and C. B. Baddiel, *Spectrochim. Acta A*, 23, 1781 (1976).
[134] Y. C. Tsui, C. Doyle, and T. W. Clyne, *Biomaterials*, 19, 2015 (1998).
[135] Z. Hong, L. Luan, S. B. Paik, B. Deng, D. E. Ellis, J. B. Ketterson, A. Mello, J. G. Eon, J. Terra, and A. Rossi, *Thin Solid Films*, 515, 6773 (2007).
[136] T. J. Kistenmacher, W. W. Bryden, J. S. Morgan, D. Dayan, R. Fainchtein, and T.O. Poehler, *J. Materials Research*, 6, 1300 (1991).
[137] C. M. Cotell, *Applied Surface Science*, 69, 140 (1993).
[138] C. M. Cotell, D. B. Chrisey, K. S. Grabowsky, and J. A. Sprague, *J. Applied Biomaterials*, 3, 87 (1992).
[139] R. Gadow, A. Killinger, and N. Stiegler, *Surface and Coatings Technology*, 205, 1157 (2010).
[140] Y. Huang, L. Song, X. Liu, Y. Xiao, Y. Wu, J. Chen, F. Wu, and Z. Gu, *Biofabrication*, 2, 045003 (2010).
[141] G. Singh, S. Singh, and S. Prakash, *Surface and Coatings Technology*, 205, 4814 (2011).
[142] P. S. Gomes, C. Botelho, M. A. Lopes, J. D. Santos, and M. H. Fernandes, *J. Biomedical Materials Research B*, 94, 337 (2010).
[143] J. A. Toque, M. K. Herliansyah, M. Hamdi, A. Ide-Ektessabi, and I. Sopyan, *J. Mechanical Behavior Biomedical Materials*, 3, 324 (2010).
[144] ISO 20502:2005(E). Fine ceramics (advanced ceramics, advanced technical ceramics)-determination of adhesion of ceramic coatings by scratch testing. Geneva, Switzerland: International organization for standardization ISO.
[145] P. Rajesh, C.V. Muraleedharan, M. Komath, and H. Varma, *J. Materials Science: Materials in Medicicne*, 22, 497 (2011).
[146] W. Xue, H. L. Hosick, A. Bandyopadhyay, S. Bose, C. Ding, K. D. K. Luk, K. M. C. Cheung, and W. W. Lu, *Surface and Coatings Technology*, 201, 4685 (2007).
[147] E. van der Wal, A. M. Vredenberg, P. J. Ter Brugge, J. G. C. Wolke, and J. A. Jansen, *Biomaterials*, 27, 1333 (2006).
[148] S.-J. Ding, T.-L. Lee, and Y.-H. Chu, *J. Materials Science Letters*, 22, 479 (2003).
[149] W. S. Oliver, and G. M. Pharr, *J. Materials Research*, 7, 1564 (1992).
[150] K. A. Gross, S. Saber-Samandari, and K. S. Heemann, *J. Biomedical Materials Research* B, 93, 1 (2010).
[151] S. Saber-Samandari, and K. A. Gross, *J. European Ceramic Society*, 29, 2641 (2009).
[152] S. Saber-Samandari, C. C. Berndt, and K. A. Gross, *Acta Biomaterialia*, 7, 874 (2011).
[153] A. Carradò, *Applied Materials Interface,* 2, 561 (2010).
[154] C. Wu, Y. Ramaswamy, D. Gale, W. Yang, K. Xiao, L. Zhang, Y. Yin, and H. Zreiqat, *Acta Biomaterialia*, 4, 569 (2008).
[155] D. R. Lide, *CRC Handbook of Chemistry and Physics*, Internet Version (2005).
[156] S. Saber-Samandari, and K. A. Gross, *Surface and Coatings Technology*, 203, 1660 (2009).
[157] L. L. Hench, *J. American Ceramic Society,* 81, 1705 (1998).

[158] G. X. Ni, Y. S. Choy, W. W. Lu, A. H. W. Ngan, K. Y. Chiu, Z. Y. Li, B. Tang and K. D. K. Luk, *Biomaterials*, 27, 1963 (2006).
[159] O. Taranobu, J. Nailo, N. Takahashi et al, "Oral. Implantol. And Biomater" (1989). Proc.3rd Jnt. Congr.Implantos. and Biomater. Stomatol., Ap. 27-29 1988 (Osaka) pp. 233-238.
[160] R. O. C. Oreffo, F. C. M. Driessens, J. A. Planell and J. T. Triffitt, *Biomaterials,* 19, 1845 (1998).
[161] H. Katsufumi, Y. Takao, K. Yoshihiko et al., *J. Biomedical Materials Research,* 23, 1049 (1989).
[162] A. V. Karlov, I. A. Khlusov, N. S. Pozhen'ko and M. V. Chaikina, *Bull. Exp. Biol. Med.* 138, 316 (2004).
[163] B. D. Ratner, A. S. Hoffman, F. J. Schoen, J. E. Lemons, *Biomaterials Science: An Introduction to Materials in Medicine*, Elsevier Inc., New York, 2004.
[164] V. V. Serov and A. B. Shekhter, *A connective tissue (functional morphology and general pathology)*, Meditsina, Moscow (1981) (in Russian).
[165] R. Schofield, *Blood Cells,* 4, 7 (1978).
[166] A. S. Curtis and M. Varde, *J. Natl. Cancer Inst.* 33, 15 (1964).
[167] S. M. Dellatore, A. S. Garsia and W. M. Miller, *Curr. Opin. Biotechnol.,* 19, 534 (2008).
[168] A. Wilson and A. Trumpp, *Nat. Rev. Immunol.*, 6, 93 (2006).
[169] F. Arai, T. Suda. Preprint, 2008 at http://www.stembook.org, doi/10.3824/stembook.1.6.1.
[170] T. Yin, and L. Li, *Journal of clinical investigation* 116, 1195 (2006).
[171] A. P. D. N. De Barros, C. M. Takiya, L. R. Garzoni et al., *PLoS One* 5, e9093 (2010).
[172] R. L. Porter, and L.M. Calvi, *Arch. Biochem. Biophys.*, 473, 193 (2008).
[173] S. Eshghi, and D. V. Schaffer. Preprint, 2008 at http://www.stembook.org, doi/10.3824/stembook.1.5.1.
[174] L L. E. Purton, and D. T. Scadden. Preprint, 2008 at http://www.stembook.org, doi/10.3824/stembook.1.28.1.
[175] B. L. Riggs and L. J. Melton III, *Osteoporosis. Etiology, Diagnosis, and Management. Second Edition,* Lippincott-Raven Publishers, Philadelphia-New York (1996).
[176] D. T. Scadden, *Best Pract. Res. Clin. Haematol.,* 20, 19 (2007).
[177] M. P. Lutolf, R. Doyonnas, K. Havenstrite, K. Koleckar and H. M. Blau, *Integr. Biol. (Camb).,* 1, 59 (2009).
[178] I. A. Khlusov, M. Yu. Khlusova, K. V. Zaitsev, T. D. Kolokol'tsova, Yu. P. Sharkeev, V. F. Pichugin, E.V. Legostaeva, I. E. Trofimova, A. S. Klimov and A. I. Zhdanova, *Bul. Exp. Biol. Med.*, 150, 535 (2011).
[179] C. Klein, K. De Groot, W. Chen et. al., *Biomaterials,* 15, 31 (1994).
[180] I. A. Khlusov, A. V. Karlov, Yu. P. Sharkeev et al., *Bull. Exp. Biol. Med.*, 140, 144 (2005).
[181] C. M. Kolf, E. Cho, and R. S. Tuan, Arthritis Res. Ther., 9, 204 (2007).
[182] F. Aerts and G. Wagemaker, *Mesenchymal stem cell engineering and transplantation,* J.A. Nolta "Genetic Engineering of Mesenchymal Stem Cells", Springer (2006), pp. 1–44.
[183] R. Peerani, B. M. Rao, C. Bauwens, T. Yin, G. A. Wood, A. Nagy, E. Kumacheva and P. W. Zandstra, *EMBO* 26, 4744 (2007).

[184] I. A. Khlusov, G. B. Slepchenko, G. T. Dambaev, L. V. Zagrebin, S. S. Shestov, S. A. Antipov, T. A. Feduschak, M. Yu. Khlusova, O. V. Kokorev, A. Ye. Yermakov, M. A. Uymin and A. M. Nekrasova, *Trace Elements and Nanoparticles: Certain Approaches to Cancer Targeted Diagnostics and Therapy,* L.M. Gutiérrez, "Neuro-Oncology and Cancer Targeted Therapy", Nova Science Publishers Inc., New York (2010), pp.1-64.

[185] A. J. Friedenstein, *Osteogenic stem cells in bone marrow*, J.N.M.Heershe, J.A.Kanis "Bone and Mineral Research", Elsevier Science Publishers, Amsterdam, The Netherlands (1990), pp. 243-272.

[186] Q. He, C. Wan and G. Li, *Stem cells,* 25, 69 (2007).

[187] F. G. J. Hayhoe and D. Quaglino, Haematological cytochemistry, Churchill Livingstone, Edinburgh, London and New York (1980).

[188] R. S. Taichman, M. J. Reilly, R. S. Verma et al., *Br. J. Haematol.*, 112, 428 (2001).

[189] K. Thalmeier, P. Meissner, G. Reisbach et al., *Exp.Hematol.*, 24, 1 (1996).

[190] R. J. Tushinski and J. J. Mule, Biology of cytokines: interkeukins, V. T. DeVita, S. Hellman, S. A. Rosenberg "Biologic therapy of cancer: second edition. Translated from English", Meditsina, Moscow (2002), pp. 90-105 (in Russian).

[191] B. J. Frisch, R. L. Porter and L. M. Calvi, *Curr. Opin. Support. Palliat. Care,* 2, 211 (2008).

[192] C. K. F. Chan, C. C. Chen, C. A. Luppen et al., *Nature,* 457, 490 (2009).

[193] M. P. Lutolf, P. M. Gilbert and H. M. Blau, *Nature,* 462, 433 (2009).

[194] A. Ya. Friedenstein and E. A. Louriya, *Cellular bases of hemopoietic microenvironments*, Meditsina, Moscow (1980) (in Russian).

[195] J. D. De Bruijn, *Calcium phosphate biomaterials: Bone-bonding and biodegradation properties*, Leiden (1993).

[196] C. J. Damien, J. L. Ricci, P. Christel, H. Alexander and J. L. Patat, *Calcif. Tissue Int.,* **55,** 151 (1994).

[197] M. F. Morks, *J. Mechanical Behavior Biomedical Materials,* S1, 105 (2008).

[198] I. R. Gibson, S. M. Best and W. Bonfield, *J.Bio. Mater. Res. Symp.* 44, 422 (1999).

[199] S. E. McNeil, S. A. Hobson, V. Nipper and K.D. Rodland, *J. Bio. Chem.*, 273, 1114 (1998).

[200] G. B. Adams, K. T. Chabner, I. R. Alley, D. P. Olson, Z. M. Szczepiorkowski, M. C. Poznansky, C. H. Kos, M. R. Pollak, E. M. Brown and D. T. Scadden, *Nature,* 439, 599 (2006).

[201] N. J. Sniadecki, R. A. Desai, S. A. Ruiz and C.S. Chen, *Annals of Biomedical Engineering,* 34, 59 (2006).

[202] A. Birgersdotter, R. Sandberg and I. Ernberg, *Semin. Cancer Biol.*, 15, 405 (2005).

[203] D. Jing, A.-V. Fonseca, N. Alakel, F.A. Fierro, K. Muller, M. Bornhauser, G. Ehninger, D. Corbeil and R. Ordemann, *Haematologica,* 95, 542 (2010).

[204] D. E. Discher, D. J. Mooney and P. W. Zandstra, *Science,* 324, 1673 (2009).

[205] O. Kollet, A. Dar, S. Shivtiel, A. Kalinkovich, K. Lapid and Y. Sztainberg, *Nat. Med.*, 12, 657 (2006).

[206] J. J. Trentin, *American J. Pathol.,* 65, 621 (1971).

[207] A. W. Duncan, F. M. Rattis, L. N. DiMascio, K. L. Congdon, G. Pazianos, C. Zhao, K. Yoon, J. M. Cook, K. Willert, N. Gaiano and T. A. Reya, *Nat.Immunol.* 6, 314 (2005).

[208] W. Li, B. Yu, M. Li et al., *Biochem Biophys Res. Commun.*, 391, 1228 (2010).

[209] *Clinical oncohematology: Guide for physicians*, M. A. Volkova, Medicine, Moscow (2001) (in Russian).
[210] M. Marcacci, E. Kon, S. Zaffagnini, A. Vascellari, M.P. Neri and F. Iacono, *Chir. Org. Mov*. LXXXVIII, 33 (2003).
[211] G. M. Wu, W. D. Hsiao, and S. F. Kung, *Surface and Coatings Technology*, 203, 2755 (2009).
[212] A. Fügl, C. Ulm, S. Tangl, C. Vasak, R. Gruber, and G. Watzek, *Clinical Oral Implants Research*, 20, 183 (2009).
[213] P. Pezeshki, S. Lugowski, and J.E. Davies, *J. Biomedical Materials Research*, 94A, 660 (2010).
[214] A. Murakami, T. Arimoto, D. Suzuki, M. Iwai-Yoshida, F. Otsuka, Y. Shibata, T. Igarashi, R. Kamijo, and T. Miyazaki, *Nanomedicine: Nanotechnology, Biology, and Medicine,* in press (2011), doi:10.1016/j.nano.2011.07.001.
[215] S. R. Paital, and N. B. Dahotre, *Acta Biomaterialia*, 5, 2763 (2009).
[216] D. M. Dohan Ehrenfest, P. G. Coelho, B.-S. Kang, Y.-T. Sul, and T. Albrektsson, *Trends in Biotechnology*, 28, 198 (2010).
[217] V. Bucci-Sabattini, C. Cassinelli, P. G. Coelho, A. Minnici, A. Trani, and D. M. Dohan Ehrenfest, *Oral Surg. Oral Med. Oral Pathol. Oral Radiol. Endod.*, 109, 217 (2010).
[218] B. Feddes, J. G. C. Wolke, J. Jansen, and A. M. Vredenberg, *J. Applied Physics*, 93, 9503 (2003).
[219] B. Feddes, J. G. C. Wolke, J. A. Jansen, and A. M. Vredenberg, *J. Applied Physics*, 93, 662 (2003).
[220] K. J. Jeong, and D. S. Kohane, *Therapeutic Delivery*, 2, 737 (2011).
[221] V. Mouriño, and A. R. Boccaccini, *J. R. Soc. Interface*, 7, 209 (2010).
[222] A. P. Tomsia, M. E. Launey, J. S. Lee, M. H. Mankani, U. G. K. Wegst, and E. Saiz, *Int. J. Oral Maxillofac. Impl.*, 26 (Suppl. 1), 25 (2011).

In: Calcium Phosphate
Editor: Robert B. Heimann

ISBN: 978-1-62257-299-1

Chapter 10

EFFECT OF ß-TCP CERAMICS ON OSTEOGENIC CELLS AND BONE HEALING

Eliezer Gilsohn and Hanna Rapaport [*]
Department of Biotechnology Engineering,
Ben-Gurion University, Beer-Sheva, Israel

ABSTRACT

Healing of bone fractures and defects may be enhanced by a variety of bone grafting materials. Synthetic scaffolds which may replace the traditional autogenous bone graft treatment have become more common in recent years due to ease of use, improvement in patient comfort and the ability to treat large defects. The most commonly used synthetic scaffolds are calcium phosphate ceramics, mainly β-tricalcium phosphate (β-TCP). Numerous studies have shown that β-TCP is osteoconductive and supports bone healing *in vivo*. Various types of β-TCP ceramics are in commercial use and in development, exhibiting different physical parameters such as shape, size and porosity. In this chapter we aim at elucidating how these parameters affect cellular activity en-route to bone regeneration. This chapter will analyze the data presented by various studies that characterized the interaction between β-TCP and osteogenic cells, either *in vitro* or *in vivo*, the effect of β-TCP on tissue response to the implanted material and the bone healing with respect to gene expression, cell localization and β-TCP resorption. This chapter will contribute to better understanding of bone defect treatment and analyze trends in further development of β-TCP and other calcium phosphate synthetic scaffolds for bone regeneration.

1. INTRODUCTION

Healing of bone fractures and defects may be expedited by bone grafting which facilitates tissue regeneration. It is currently accepted that there are several essential elements that are required in order for regeneration to occur: a scaffold which should be osteoconductive;

[*] E-mail address: hannarap@bgu.ac.il

signaling molecules; and osteogenic cells that may respond to the signals to form new bone tissue. Therefore, for many years the gold standard for bone grafting has been autogenous bone transplant, which satisfies these requirements.

Using autogenous bone, however, has several major drawbacks. Harvesting of autogenous bone requires an additional surgical procedure, which causes discomfort to the patient and increases morbidity risks, chances for infection, and pain. Moreover, the amount of graft material required for clinical use is often too large and thus autografting becomes infeasible, and finally the additional surgical procedure involved increases the overall expenses of the treatment. Of note, due to prolonged ischemia, and subsequent insufficient revascularization, the free block autografts may undergo necrosis and/or resorption [1-3].

For these reasons, researchers have been developing various grafts based on allogeneic tissue, xenografts harvested from animal cadavers, or synthetic scaffolds such as minerals and biopolymers. Of the many types of scaffolds that have been developed, the most widely used materials are calcium phosphate ceramics and, in particular, β-tricalcium phosphate (β-TCP) and hydroxyapatite (HA). Both materials are osteoconductive, so that they provide a substrate on which osteoblasts can deposit bone [4-7]. HA and β-TCP differ in several aspects, the main one being resorption rate. HA is a crystalline calcium phosphate mineral similar to that present in bone. However, HA will be resorbed at a much slower rate than β-TCP [4, 8], limiting the ability of the bone to naturally remodel, hence preventing it from reaching its natural mechanical and biological properties. Moreover, β-TCP has been shown to be more osteoconductive than HA [9]. Therefore, the use of β-TCP is considered by many to be more suitable for bone grafting. One process which often utilizes β-TCP is alveolar ridge augmentation which is required to increase bone volume prior to insertion of dental implants [10-13].

One of the most important parameters to consider when examining biomaterials is the tissue's response to the implanted material. In recent years there have been many studies that characterized the interactions of several osteogenic cell types with β-TCP, either using *in-situ* hybridization (staining osteogenic-specific mRNA) or microscopy to follow cell localization and interaction with β-TCP particles [12, 14, 15]. The function of osteoclast cells in β-TCP resorption has also been examined [12, 16]. Moreover, the characteristics of β-TCP, such as its porosity, granule size and morphology have been studied in relation to β-TCP remodeling activity [17, 18]. The aim of this chapter is, therefore, to summarize how cells interact with β-TCP, both *in vitro* and *in vivo* and how these interactions affect bone regeneration.

2. Activation of Osteoblasts Following β-TCP Implantation

Several studies have examined how the osteoconductivity of β-TCP is manifest, i.e. whether or not osteoblasts are activated at the surface of β-TCP particles, and what the timeframe is of this activation. In a paper by Neo et al.. [19], temporal and spatial osteoblast activation was examined around β-TCP particles. The authors drilled 3 mm holes in rat *tibiae*, filled the holes with β-TCP particles (150-300 μm, 1.5% porosity) and, using a digoxygenin-stained RNA probe for procollagen α1, followed osteoblast localization in thin-sections over 28 days. After three days, the spaces between β-TCP particles were filled with blood cells and fibrin clots, with very few osteoblasts. By day 7 healing had started to occur centripetally (from the periphery to the center of the defect) and new osteoblast-lined *trabeculae* were formed between β-TCP particles. The osteoblasts were not directly on the surface of β-TCP particles (there was a layer of procollagen α1-negative cells between β-TCP and the osteoblasts) but within young bone matrix deposited in between the particles. By day 14 all particles were surrounded by new bone and the number of osteoblasts had decreased dramatically. A similar study by Ohsawa et al. [15] further showed, using in-situ staining, that by day 7 there was expression of osteocalcin and osteonectin in procollagen α1 positive cells on newly formed bone and in blood clots in control animals. However, expression of osteopontin in the control animals was restricted to procollagen α1 positive osteoblasts on the woven bone of the mineralization front and was weak in the region where mineralization had terminated.

In animals treated with β-TCP, however, osteopontin positive cells were seen mainly on particle surfaces and some of them were procollagen α1 negative cells. The authors speculated the osteopontin-expressing procollagen α1 negative cells to be macrophages that are part of the foreign-body response caused by the β-TCP particles. It was therefore shown that osteopontin is expressed mostly in early stages of osteoblast differentiation and bone formation.

The additional finding that osteopontin-positive cells were mainly seen on the particles implies that the β-TCP particles may be important for early osteoblast differentiation, maybe partially accounting for the β-TCP osteoconductivity attributes. This finding was further demonstrated by studies that showed that osteopontin-positive cells are found mainly on the surface of β-TCP particles and that filling of defects with β-TCP results in elevated osteopontin mRNA levels until day 10, after which osteopontin in both the control and experiment group decreases, again indicating the role of osteopontin in early differentiation [20, 21]. These studies also showed the ability of β-TCP particles to promote faster bone regeneration.

The observations that further osteoblast activation and bone-matrix formation do not necessarily occur on the surface of β-TCP particles were supported by an earlier study by Neo et al.. [22] which showed, using transmission electron microscopy, that new bone formed in regions non-adjacent to the β-TCP particles and subsequently approached the particles and bonded to them.

A recent study by Lu et al.. [23], using isolated human primary osteoblasts (HOBs) cultured on β-TCP, suggested that β-TCP exerts its ability to induce osteoblast differentiation through autocrine BMP2-dependent activation of osteoblasts. The BMP2 inhibitor Noggin prevented early osteoblast differentiation (indicated by decreased mRNA levels of runx2 and osteopontin), but did not change mRNA levels of later differentiation genes (osteocalcin and bone sialoprotein).

This observation supports previously described findings that implied that attachment to β-TCP particle surface is only required for initial differentiation. The study also showed that the BMP2-dependent activation requires the α2β1 integrin receptor and mitogen-activated protein kinase (MAPK), as using α2β1 integrin or MAPK inhibitors attenuated expression of BMP2 and its receptors [23, 24]. The authors proposed a model by which β-TCP binds to integrin receptors on the osteoblasts, which in turn activates the BMP2 autocrine loop, leading to osteoblast differentiation.

3. Interaction between B-TCP and Osteoclasts

One of the desirable characteristics of an artificial scaffold is its ability to be resorbed, remodeled and ultimately replaced by new bone, a process which involves the activity of the multi-nuclear osteoclast cells [25]. Under normal conditions, bone resorption by osteoclasts is balanced with bone formation by osteoblasts. Thus, it is crucial to characterize the interaction between the resident osteoclasts and the artificial implant.

In a study by Zerbo et al.. [12], osteoclast localization was followed in histological sections taken from human biopsies following treatment with β-TCP for maxillary sinus floor elevation. Cells were identified by staining for Tartrate Resistant Acid Phosphatase (TRAP) activity. The authors did not find any multi-nuclear osteoclasts in or on β-TCP particles. The number of mono- or bi-nucleate TRAP-positive cells found on the β-TCP particles decreased as the particles were further away from residual bone. These findings led the authors to hypothesize that osteoclasts play only a minor role in β-TCP resorption and that chemical dissolution probably accounts for most of the resorption.

These results were supported by a study by Detsch et al.. [26]. Here the authors showed that the surface of hydroxyapatite (HA) enhanced the formation of giant multi-nuclear cells out of osteoclast-like cells derived from a human leukome monocytic lineage. However, osteoclast-like cells on β-TCP formed smaller cells, smaller lacunae and smaller actin rings. The release of calcium was quicker in β-TCP than in HA, resulting in a higher calcium concentration around β-TCP particles. The authors reasoned that the high calcium level around β-TCP particles prevented osteoclast migration and function. In this study, the *in vitro* osteoclastogenesis was observed starting from day 14, which may explain results of a study that claimed HA not to be resorbed by osteoclasts, but examined resorption only after two days [27].

Further evidence of weaker osteoclast activation on β-TCP than on HA comes from work by Lange et al.. [16]. This work examined, using ELISA, expressions of cytokines important for dendritic cells and osteoclast differentiation in human peripheral blood mononuclear cells that were *in vitro* grown on HA or β-TCP. The authors found that the expression of Granulocyte Macrophage Colony Stimulating factor (GM-CSF) and Receptor activator of nuclear factor kappa-B ligand (RANKL), known for promoting dendritic cell activation and osteoclastogenesis, respectively, was lower in cells grown on β-TCP as opposed to cells grown on HA. Expression of pro-inflammatory cytokines such as IL-8 and IL-1β was similar for both treatments. Therefore, although HA and β-TCP both induce similar expression of pro-inflammatory cytokines, β-TCP leads to reduced osteoclastogenesis.

Most *in vitro* studies dealing with osteoclasts were performed on mononuclear precursor cells that are differentiated into osteoclasts using various methods. This is due to the difficulty in isolating osteoclast cells as they are very scarce under physiological conditions (around 1% of the bony surface) [28] and the difficulty in handling calcified tissues. This could be a reason for contradicting results reported for various *in vitro* studies on osteoclast activity. A recent work examined the effect of β-TCP granule size, shape and porosity on osteoclast activity *in vivo*, by implanting different particles subcutaneously in Wistar rats and following osteoclast activity using histological measurements [29]. In this work, the authors showed that β-TCP has the potential to induce formation of multi-nucleated giant osteoclasts. The induction was independent of particle porosity but appeared to be greater in particles of polygonal shape and mean size of 150-500 μm. Bigger and less porous particles induced formation of relatively fewer TRAP-positive multi-nucleated cells. The authors thus concluded using an *in vivo* model that the shape, size and porosity of β-TCP particles may highly affect their osteoclast-based resorption. This implies that previously reported contradicting results of *in vitro* studies may arise from differences in β-TCP particle size that were not accounted for.

4. The Effect of B-TCP Microstructure on Cell Behavior

The microstructure of β-TCP particles, and in particular particle porosity, has a pronounced effect on cell behavior. As mentioned above, different porosity levels may affect osteoclast differentiation and function to a great extent. Porosity level has an effect on bone formation and osteoblasts as well.

In a study by Knabe et al.. [18], a sinus floor augmentation procedure was performed on humans using autogenous bone chips combined with either low porosity β-TCP particles (35% porosity) or high porosity particles (65% porosity). Six months post procedure, biopsies were taken, immune-stained for various bone markers, and healing as well as particle degradation were evaluated histomorphometrically. Patients that were treated with higher porosity β-TCP showed greater bone regeneration and particle degradation.

The observation that higher porosity β-TCP induces better bone regeneration was further examined in a study by Kasten et al.., that observed growth of mesenchymal stem cells (MSCs) on β-TCP of 25%, 65% or 75% porosity, both *in vivo* and *in vitro* [30]. When cultured *in vitro*, MSCs proliferated better on scaffolds of 65% or 75% porosity, as indicated by total protein levels. Levels of alkaline phosphatase (ALP) activity, however, were similar among all scaffolds, suggesting that porosity may not affect cell differentiation. When scaffolds were loaded with MSCs and implanted in SCID mice, higher levels of calcification on scaffolds of 65% and 75% were found in histological sections, supporting the role of larger pore size in aiding bone regeneration. Noticeably, *in vivo* ALP levels were higher in the 65% scaffolds compared to both the 75% and the 25% scaffolds, suggesting that not only the pore size but also other microstructure properties (such as pore shape) may affect cell behavior.

Interestingly, a study with contradictory results was conducted by Isaac et al.. [31]. In this study cell adhesion, viability and differentiation of bone marrow stromal cells (BMSCs) was examined on β-TCP of 0%, 25% or 45% porosity. Here it was found that cell viability and differentiation decrease with increasing porosity, as indicated by alkaline phosphatase activity and metabolic assays. Nevertheless, as the study by Kasten et al.. [30] described above examined the effect of porosity *in vivo*, it is therefore more likely that higher porosity positively affects cell differentiation. Moreover, the pore size in the work of Isaac et al.. [31] was significantly lower than pore sizes in the studies by Kasten et al.. [30] and Knabe et al.. [18] (approximately 1μm).

Indeed, in addition to overall porosity, pore size may also affect cell behavior.

A recent study by Feng et al.. [32] examined the effect of pore size on cell behavior *in vivo*, by implanting β-TCP cylinders of similar porosity but different pore sizes into rabbits and examining fibrous tissue and blood vessels within the cylinders. The authors showed that smaller pore size enabled the growth of significantly more fibrous material inside the β-TCP scaffold (with highest fibrous growth observed in cylinders with pores of 300-400 μm), presumably due to the higher surface area enabling growth of more cells. This result is supported by previous studies which indicated that smaller pore size induces faster bone ingrowth into scaffolds [8, 33]. The study by Feng et al.. [32] further showed that as opposed to the positive effect on fibrous growth, neo-vascularization was hindered, and new vessel size was smaller in cylinders with small pore sizes (smaller than 400 μm).

Yuan et al.. [34] reported similar results regarding hydroxyapatite. This study examined the osteogenesis induced by two types of hydroxyapatite having different porous structures (J-HA: Ø 3.5 mm x 8.5 mm, 70% overall porosity, average pore size 200 μm and S-HA: Ø 5 mm x 6 mm, overall 60-70% porosity and average pore size 400 μm) with the most significant difference being the pore size that was larger in the S-HA sample. Ceramics were implanted in dorsal muscles of dogs and examined by histology after 3 and 6 months. Bone was found in all S-HA implants after 3 months as well as after 6 months, whereas no bone formation was detected in the J-HA samples. These results indicate that the microstructure architecture of the calcium phosphate ceramic, and pore size in particular, influences osteoinductive capacitance in ectopic bone formation in soft tissue.

5. Modifications of Β-TCP

In the light of the information accumulated over the years regarding the use of β-TCP as a synthetic scaffold, various attempts have been made to improve the osteoconductivity of β-TCP. One such method is using biphasic calcium-phophate (BCP), which contains an HA phase and a β-TCP phase.

It has been shown that BCP is more osteoinductive than HA alone, since bone formation was faster and more significant when BCP was implanted into dog femoral bone or muscle than when HA was implanted [35]. Notable, the pore sizes of both the BCP and HA in this study were similar, so it is therefore unlikely that the advantage of BCP in osteoinduction is due to larger pore size. However, the authors noted that BCP has a higher specific surface area than HA, which could lead to concentration of more growth factors that in turn render BCP more osteoinductive. The difference in surface area may also affect the dissolution rate of the calcium phosphate in BCP thus leading to higher osteoinductive potential. Resorption of either BCP or HA was hardly observed at any time point in the described work. It was postulated that calcium-deficient BCP used in this study behaved similarly to previously reported calcium-deficient TCP that was also not resorbed *in vivo* after two and a half years. The formation of bone on the surface of the ceramics was suggested as another reason for the lack of resorption. Bone formed on BCP surface could have slowed down the chemical dissolution and decreased the material's surface area exposed to osteoclast activity.

Another study showed that β-TCP alone induces faster bone formation than BCP, whereas BCP induces faster bone formation compared to HA [36]. This study, which examined bone healing of minipig mandible defects treated with either autogenous bone, β-TCP, BCP or HA, showed that while after 24 weeks all defects were almost equally filled with lamellar bone, examination at earlier time points revealed that β-TCP or autogenous bone induced faster healing than BCP and HA. In addition, treatment with BCP resulted in faster healing than treatment with HA. After 24 weeks, almost all of the β-TCP was resorbed, while most of the BCP and HA were still visible. Another study from the same group [37] examined the bone healing capacity and resorption of several BCP compounds with different HA/β-TCP ratios. This study showed that a higher amount of β-TCP in the BCP results in higher resorption and faster bone formation.

A study by Lindgren et al.. [38] examined BCP resorption by implanting BCP microimplants into human patients that underwent sinus floor augmentation, extracting the implants after 8 months and examining the implant after extraction. This study showed, using energy-dispersive X-ray spectroscopy (EDS) measurements, that when BCP is resorbed, there is a reduction in the Ca/P ratio, which the authors suggest stems from the β-TCP component of the BCP being gradually replaced by calcium-deficient hydroxyapatite. This is supported by an observation by Jensen et al.. [36] in which toluidine blue dye was shown to penetrate a few microns into the outer surface of BCP particles, which according to the authors is due to dissolution of the β-TCP component at the surface.

Interestingly, a work by Fellah et al.. [39] showed that BCP better supports bone formation than autogenous bone. In this paper, the osteoconductivity of BCP granules and autologous bone chips was compared by examining bone healing in a critical-size femoral epiphysis model in goats. Resorption of autogenous bone chips started approximately 6 weeks post procedure, while the BCP granules remained stable. However, more bone tissue was seen in defects implanted with BCP than with autogenous bone at 12 weeks post procedure. The authors suggested that these results may arise from the use of polytetrafluoroethylene (PTFE) membranes as a barrier between the examined material and the defect perimeter. Taken together, all the results described show that BCP combines the high osteoconductivity of β-TCP with the slow dissolution properties of HA. Since the osteoconductivity/resorption of BCP is dependent on the β-TCP/HA ratio, it stands to reason that different BCP would be used for different fracture locations (bearing vs. non-bearing bone, for example).

Another approach to improve the characteristics of β-TCP is doping of β-TCP with various materials. One such method applied by Cai et al.. [40] is reinforcing β-TCP with $45P_2O_5$–22CaO–$25Na_2O$–8MgO bioglass. This bioglass-reinforced β-TCP showed significantly higher compression strength compared to β-TCP of similar porosity. Scanning electron microscopy (SEM) showed apatite crystal formation on the bioglass-reinforced β-TCP but not on pure β-TCP. In addition, the bioglass-reinforced β-TCP better supported adhesion and proliferation of MC3T3-E1 osteoblast cells, manifested both by number of cells eluted from the matrices and by SEM analysis showing cell adhesion.

In addition to doping β-TCP with bioglass, it is currently widely accepted that silicon-substituted calcium phosphates exhibit better biological properties compared to pure calcium phosphates. Silicon has been shown to improve the osteoconductivity of BCP. In the study by Byun et. al [41], silicone-doped and non-doped BCP particles were compared *in vitro* in terms of affecting MG-63 osteoblast cell differentiation, adhesion, proliferation, and gene-expression (using RT-PCR). The study showed that silicone-doped BCP particles better supported cell viability and cell proliferation when compared to non-doped particles. Moreover, SEM analysis revealed that MG-63 cells on silicone-doped BCP display stronger adhesion than cells on non-doped BCP. Cell differentiation as measured by Alizarin Red staining and expression levels of the differentiation genes osteonectin and osteopontin were also shown to be enhanced on silicone-doped BCP.

Numerous articles have confirmed the metabolic effect of Si on bone, as was presented in a review by Pietak et al.. [42]. For example, research by Patel et al.. [43] showed that granules of Si-substituted HA lead to increase in bone ingrowth and bone–implant coverage compared to pure HA mineral implanted in a rabbit model. Similar results were obtained by Hing et al.. [44] using porous scaffolds. Moreover, a study by Mastrogiacoma et al.. [45] showed in a long-term *in vivo* experiment in sheep, that Si-TCP scaffolds were resorbed much faster than HA scaffolds of similar porosity, and were replaced by newly formed highly mineralized laminar bone tissue within two years.

However, in a recently published review by Bohner [46] the advantage of silicon-substituted calcium phosphates was revisited, arguing that there is a lack in experimental evidence to link the improved biological performance of Si-substituted calcium phosphates with Si release. Bohner further claims that there are not enough studies comparing the *in vivo* effect of Si-substituted calcium phosphates with non-substituted calcium phosphates and stresses that the 'superior biological performance' of the Si-substituted materials should be correlated with measured silicon levels *in vivo* and related to therapeutic concentrations.

A different approach involves the addition of biological elements to β-TCP in order to increase osteoconductivity and accelerate bone healing. Bone Morphogenic Protein 2 (BMP-2) has been known for many years to be able to induce differentiation of mesenchymal stem cells into osteoblasts and thus is able to drive osteoinduction [47, 48]. The release of recombinant human BMP-2 (rhBMP-2) from different scaffolds has been examined. Release from soft matrices such as collagen, resulting in bone formation, occurs in a burst rather than a gradual release [49]. The release of rhBMP-2 from harder scaffolds such as β-TCP has therefore been examined in several studies in order to combine their osteoinductive/osteoconductive properties and induce sustained release of rhBMP-2. For example, in a study by Sohier et al.. [50] β-TCP was chemically etched (to increase the contact surface) and loaded with rhBMP-2. Loaded and non-loaded β-TCPs were implanted into the back muscles of rabbits to examine osteoinduction, but into rabbit fermoral bone defects to examine osteoconduction. Histological sections taken four weeks post-procedure showed that the rhBMP-2-loaded β-TCP was able to induce bone formation within the back muscle, while non-loaded β-TCP was not. Furthermore, loaded β-TCP showed a 1.3-fold increased new bone formation in bone defects, as measured via histomorphometry.

Another biological element that can be added to β-TCP in order to improve bone healing are osteogenic cells. In a study by Seebach et al.. [51] mesenchymal stem cells (MSCs) and/or endothelial progenitor cells (EPCs) were seeded onto β-TCP scaffolds that were transplanted into a femoral critical size defect of immune-suppressed rats. Using immuno-histochemistry, micro-CT analysis and mechanical testing, the vascularization and bone formation were determined 1, 4 and 8 weeks post procedure. Both vascularization and new bone formation were significantly increased in defects that were treated with both MSCs and EPCs, in comparison with defects treated with only one or none of the cell types. In another study, Janicki et al.. [52] used subcutaneous transplantation into immune-compromised mice in order to examine the effect of MSCs addition to β-TCP scaffolds. This study showed that MSCs-loaded β-TCP granules smaller than 0.7 mm showed a 3.8-fold increase in bone formation compared to granules of 0.7-1.4 mm. Furthermore, β-TCP loaded with MSCs supported more bone formation than BCP loaded with MSCs. When MCSs-loaded β-TCP particles were incubated with chondrogenic differentiation medium five weeks prior to transplantation, a collagen type II + proteoglycan-rich cartilage-like tissue formed and underwent endochondral ossification after subcutaneous transplantation. This led to formation of bone with marrow and a haematopoietic environment, which was not achieved when MSCs or β-TCP alone were transplanted.

CONCLUSION

The understanding how β-TCP affects cell behavior and bone healing is crucial in order to be able to treat patients in an efficient way and in order to design bone regenerative materials that best support bone healing. The studies we have presented in this chapter shed light on the cell behavior and β-TCP characteristics that affect bone healing.

Works from recent years have shown the importance of β-TCP osteoconductivity, especially at an early cell-differentiation stage. The conclusions that arise from the research presented support a model in which early cell differentiation, characterized by expression of genes such as osteopontin and runx2, is achieved by adhesion of cells to the surface of the β-TCP particles. Adhesion through integrin receptors activates a BMP-2 autocrine loop that, in turn, activates differentiation. Later differentiation, however, is achieved in the spaces between β-TCP particles, as fully differentiated osteoblasts are not observed in direct contact with β-TCP particles, and bone formation centers start in the spaces between particles. This observation stresses the importance of inter-particle spaces for bone healing and suggests that the use of a bone-filler material that combines the osteoconductivity of β-TCP with an inter-particle matrix that supports later cell-differentiation, may lead to improved bone healing. An example of such an approach can be obtained from research in our laboratory (Amosi et al., submitted for publication), which applied a peptide-based hydrogel in combination with β-TCP to heal a rat femoral bone defect. These results clearly showed that the combination of a hydrogel and the β-TCP particles resulted in higher bone formation *in vivo*. This was supported by *in vitro* tests that showed that osteoblasts in culture better differentiate on the hydrogel/ β-TCP composite.

In terms of β-TCP resorption, recent data points to enhanced calcium release from β-TCP in comparison to HA, mainly due to chemical dissolution. Osteoclast activation is much less enhanced on β-TCP than on HA, and is dependent on particle size and shape. It therefore appears to be important to design β-TCP particles to be of the right shape and size in order to control resorption kinetics.

Indeed, designing physical properties of β-TCP particles is also crucial to enhance resulting bone healing. The studies described herein showed that higher porosity in β-TCP particles induces better bone healing and resorption. Particles with smaller pore size induce better bone formation but lesser vascularization. Thus the design of pore-size in implanted β-TCP should be done in context of the treated bone – if a large defect is treated, vascularization is more critical and thus a larger pore size should be used, but if a smaller defect is treated, particles with smaller pore sizes should be used to induce faster healing (e.g. in sinus floor augmentation). In addition, if a larger defect is treated and hence enhanced physical strength/lower resorption is desired, β-TCP can be augmented by combining it with HA to form BCP that is osteoconductive but less resorbable. The ratio between β-TCP and HA in BCP will control the desired efficacy vs. resorption which is suitable for the specific treated defect.

To increase β-TCP efficacy, it can be further augmented by biological elements such as rhBMP-2 or osteogenic cells. The disadvantage of adding biological elements to synthetic scaffolds is in the final product price, product stability and, in the case of addition of cells, the need for a further procedure to extract cells from the patient.

In summary, in this chapter we have shown that there are many elements influencing the biological response of cells and tissue growth to β-TCP particles. We have demonstrated the interplay between parameters that induce enhanced cell differentiation/bone formation and parameters that induce slower resorption/confer stronger physical characteristics. When developing novel β-TCP-based materials for bone healing, it is crucial to consider the context of the bone to be treated, in order to develop a scaffold with the optimal characteristics in terms of porosity, pore size, pore shape, pore inter-connectivity and particle shape.

Acknowledgments

The writing of this chapter was supported by B.G. Negev, the Horowitz Foundation, and in part by BIO-DISC, a cooperation in biotechnology research between Germany (BMBF) and Israel (The Ministry of Industry, Trade and Labor, and The Ministry of Science and Technology). We thank Dr. Fabian Peters (Curasan AG, Germany) for commenting on the manuscript.

References

[1] Albrektsson, T. and Linder, L., *Arch. Orthop. Trauma Surg.* 98, 189 (1981).
[2] Boyne, P. J., *J Periodontol.* 45, 378 (1974).
[3] Nystrom, E., Kahnberg, K. E. and Albrektsson, T., *Int. J. Oral Maxillofac. Implants.* 8, 167 (1993).
[4] Holmes, R. E., Bucholz, R. W. and Mooney, V., *J Orthop Res.* 5, 114 (1987).
[5] Jarcho, M., *Clin. Orthop. Relat. Res.* 259 (1981).
[6] Metsger, D. S., Driskell, T. D. and Paulsrud, J. R., *J. Am. Dent Assoc.* 105, 1035 (1982).
[7] Yaszemski, M. J., Payne, R. G., Hayes, W. C., Langer, R. and Mikos, A. G., *Biomaterials.* 17, 175 (1996).
[8] Eggli, P. S., Muller, W. and Schenk, R. K., *Clin. Orthop. Relat. Res.* 127 (1988).
[9] Lu, J. X., Gallur, A., Flautre, B., Anselme, K., Descamps, M., Thierry, B. and Hardouin, P., *J. Biomed. Mater Res.* 42, 357 (1998).
[10] Suba, Z., Takacs, D., Matusovits, D., Barabas, J., Fazekas, A. and Szabo, G., *Clin. Oral Implants Res.* 17, 102 (2006).
[11] Szabo, G., Huys, L., Coulthard, P., Maiorana, C., Garagiola, U., Barabas, J., Nemeth, Z., Hrabak, K. and Suba, Z., *Int. J. Oral Maxillofac. Implants.* 20, 371 (2005).
[12] Zerbo, I. R., Bronckers, A. L. J. J., de Lange, G. and Burger, E. H., *Biomaterials.* 26, 1445 (2005).
[13] Zerbo, I. R., Zijderveld, S. A., de Boer, A., Bronckers, A. L. J. J., de Lange, G., ten Bruggenkate, C. M. and Burger, E. H., *Clin. Oral Implants Res.* 15, 724 (2004).

[14] Beuvelot, J., Pascaretti-Grizon, F., Filmon, R., Moreau, M.-F., Basle, M. F. and Chappard, D., *J. Biomed. Mater Res. A*. 96, 108.
[15] Ohsawa, K., Neo, M., Matsuoka, H., Akiyama, H., Ito, H., Kohno, H. and Nakamura, T., *Journal of biomedical materials research*. 52, 460 (2000).
[16] Lange, T., Schilling, A. F., Peters, F., Haag, F., Morlock, M. M., Rueger, J. M. and Amling, M., *Biomaterials*. 30, 5312 (2009).
[17] Ghanaati, S., Barbeck, M., Orth, C., Willershausen, I., Thimm, B. W., Hoffmann, C., Rasic, A., Sader, R. A., Unger, R. E., Peters, F. and Kirkpatrick, C. J., *Acta Biomater*. 6, 4476.
[18] Knabe, C., Koch, C., Rack, A. and Stiller, M., *Biomaterials*. 29, 2249 (2008).
[19] Neo, M., Herbst, H., Voigt, C. F. and Gross, U. M., *J. Biomed. Mater Res*. 39, 71 (1998).
[20] Neo, M., Voigt, C. F., Herbst, H. and Gross, U. M., *Journal of biomedical materials research*. 39, 1 (1998).
[21] Shiratori, K., Matsuzaka, K., Koike, Y., Murakami, S., Shimono, M. and Inoue, T., *Biomed Res*. 26, 51 (2005).
[22] Neo, M., Nakamura, T., Ohtsuki, C., Kokubo, T. and Yamamuro, T., *J. Biomed. Mater Res*. 27, 999 (1993).
[23] Lu, Z. and Zreiqat, H., *Tissue Eng. Part A*. 16, 3075.
[24] Lu, Z. and Zreiqat, H., *Biochem. Biophys. Res. Commun*. 394, 323.
[25] Teitelbaum, S. L., Cao, X., Mimura, H., Chiba, M. and Ross, F. P., *Miner Electrolyte Metab*. 21, 193 (1995).
[26] Detsch, R., Mayr, H. and Ziegler, G., *Acta Biomater*. 4, 139 (2008).
[27] Yamada, S., Heymann, D., Bouler, J. M. and Daculsi, G., *Biomaterials*. 18, 1037 (1997).
[28] Parfitt, A. M., Drezner, M. K., Glorieux, F. H., Kanis, J. A., Malluche, H., Meunier, P. J., Ott, S. M. and Recker, R. R., *J. Bone Miner Res*. 2, 595 (1987).
[29] Ghanaati, S., Barbeck, M., Orth, C., Willershausen, I., Thimm, B. W., Hoffmann, C., Rasic, A., Sader, R. A., Unger, R. E., Peters, F. and Kirkpatrick, C. J., *Acta Biomater*. 6, 4476.
[30] Kasten, P., Beyen, I., Niemeyer, P., Luginbuhl, R., Bohner, M. and Richter, W., *Acta Biomater*. 4, 1904 (2008).
[31] Isaac, J., Hornez, J.-C., Jian, D., Descamps, M., Hardouin, P. and Magne, D., *J. Biomed. Mater Res. A*. 86, 386 (2008).
[32] Feng, B., Jinkang, Z., Zhen, W., Jianxi, L., Jiang, C., Jian, L., Guolin, M. and Xin, D., *Biomed. Mater*. 6, 015007.
[33] Grynpas, M. D., Pilliar, R. M., Kandel, R. A., Renlund, R., Filiaggi, M. and Dumitriu, M., *Biomaterials*. 23, 2063 (2002).
[34] Yuan, H., Kurashina, K., de Bruijn, J. D., Li, Y., de Groot, K. and Zhang, X., *Biomaterials*. 20, 1799 (1999).
[35] Yuan, H., van Blitterswijk, C. A., de Groot, K. and de Bruijn, J. D., *J. Biomed. Mater Res. A*. 78, 139 (2006).
[36] Jensen, S. S., Yeo, A., Dard, M., Hunziker, E., Schenk, R. and Buser, D., *Clin Oral Implants Res*. 18, 752 (2007).
[37] Jensen, S. S., Bornstein, M. M., Dard, M., Bosshardt, D. D. and Buser, D., *J. Biomed. Mater Res. B. Appl. Biomater*. 90, 171 (2009).

[38] Lindgren, C., Hallman, M., Sennerby, L. and Sammons, R., *Clin. Oral Implants Res.* 21, 924.

[39] Fellah, B. H., Gauthier, O., Weiss, P., Chappard, D. and Layrolle, P., *Biomaterials.* 29, 1177 (2008).

[40] Cai, S., Xu, G. H., Yu, X. Z., Zhang, W. J., Xiao, Z. Y. and Yao, K. D., *J. Mater Sci. Mater Med.* 20, 351 (2009).

[41] Byun, I.-S., Sarkar, S. K., Anirban Jyoti, M., Min, Y.-K., Seo, H.-S., Lee, B.-T. and Song, H.-Y., *J. Mater Sci. Mater Med.* 21, 1937.

[42] Pietak, A. M., Reid, J. W., Stott, M. J. and Sayer, M., *Biomaterials.* 28, 4023 (2007).

[43] Patel, N., Best, S. M., Bonfield, W., Gibson, I. R., Hing, K. A., Damien, E. and Revell, P. A., *J. Mater Sci. Mater Med.* 13, 1199 (2002).

[44] Hing, K. A., Revell, P. A., Smith, N. and Buckland, T., *Biomaterials.* 27, 5014 (2006).

[45] Mastrogiacomo, M., Corsi, A., Francioso, E., Di Comite, M., Monetti, F., Scaglione, S., Favia, A., Crovace, A., Bianco, P. and Cancedda, R., *Tissue Eng.* 12, 1261 (2006).

[46] Bohner, M., *Biomaterials.* 30, 6403 (2009).

[47] Sampath, T. K. and Reddi, A. H., *Proc. Natl. Acad. Sci. USA.* 78, 7599 (1981).

[48] Urist, M. R. and Strates, B. S., *J. Dent Res.* 50, 1392 (1971).

[49] Liu, Y., Wu, G. and de Groot, K., *J. R. Soc. Interface.* 7 Suppl 5, S631.

[50] Sohier, J., Daculsi, G., Sourice, S., de Groot, K. and Layrolle, P., *J. Biomed. Mater Res. A.* 92, 1105.

[51] Seebach, C., Henrich, D., Kahling, C., Wilhelm, K., Tami, A. E., Alini, M. and Marzi, I., *Tissue Eng. Part A.* 16, 1961.

[52] Janicki, P., Kasten, P., Kleinschmidt, K., Luginbuehl, R. and Richter, W., *Acta Biomater.* 6, 3292.

In: Calcium Phosphate
Editor: Robert B. Heimann
ISBN: 978-1-62257-299-1

Chapter 11

CALCIUM PHOSPHATE BONE CEMENTS

Sandra Pina*[*] and José M. F. Ferreira
Department of Materials Engineering and Ceramics,
University of Aveiro, CICECO, Portugal

ABSTRACT

Calcium phosphate cements (CPC) have been in clinical use for the last 15 years. Their most salient features include good biocompatibility and resorbability, excellent bioactivity, non-cytotoxicity, development of osteoconductive pathways and sufficient compressive strength for a number of applications. The major advantages of the CPC include a fast setting time, excellent mouldability, outstanding biocompatibility, and easy manipulation; therefore, the cements are commonly used in filling bone defects and trauma surgeries as mouldable paste-like bone substitute materials. Besides, like any other bioceramics, CPC provide the opportunity for bone grafting using alloplastic materials, which are unlimited in quantity and provide no risk of infectious diseases. At present, there are two types of CPC depending on the end-product of the reaction: apatite (AP) cements and dicalcium phosphate dihydrate (DCPD or brushite) cements. Apatite is formed above pH 4.2, while brushite is preferentially formed in CPC when the pH value of the paste is < 4.2, although it may grow even up to pH 6.5. Brushite cements have raised special interest because they are resorbed *in vivo* much faster than apatite ones. Moreover, brushite is metastable under physiological conditions and brushite-based cements possess shorter setting times.

In recent years, ionic incorporations in CaP (for example, α- and β-TCP) bioceramics, such as Mg, Sr, Zn, Na, K, F and carbonates, have been the subject of great interest owing to the critical role these ions play in the biological processes after implantation. The presence of foreign ions in the structure of synthetic apatites can alter a number of structural, physico-chemical and biological properties of CaP, such as lattice parameters, crystallinity, solubility, dissolution, resorption, and bone bonding capability. The studies performed so far form a stable base from which minor modifications should provide improved functional bone substitutes. However, there remains a need for improving the overall performance of CPC, especially in terms of injectability and mechanical properties aiming at specific clinical applications such as vertebroplasty and

[*] Email: sandra.pina@ua.pt

kyphoplasty. This requires driving the research towards identifying the key factors and better understanding their roles in determining the most relevant properties of CPC. Future developments will enable the commercialization of better and more differentiated products that should improve the clinical outcome and hence the quality of life of the patients.

1. Introduction

During the last century, new materials and surgical techniques have drastically changed the lives of millions of patients. Biomaterials have made an important contribution to modern health care and will expand further as musculoskeletal disorders, especially osteoporosis and fragility fractures, will increase among an aging population. Each biomaterial considered for potential clinical applications has specific chemical, physical and mechanical properties, which might originate variations in host/material response. However, the biological characteristics and the anatomic site of implantation are also important for the behaviour and outcome of the implant. In surgery for fragility fractures, biomaterials are still sparsely used for bone repair. In about 10% of all reconstructive operations caused by traumatic, resectional or congenital defects, bone transplants and bone substitute materials are necessary.

The discovery in early 1980s of self-setting cements based on calcium phosphates (CaP) has opened up a new era in the medical applications of bone grafting. Synthetic bone substitutes made of calcium phosphates have chemical compositions similar to those of bone and dentine, and can be fitted with bone defects. They are also particularly suited for minimally invasive surgical techniques, will harden *in situ* and provide stability for any defect geometry. Calcium phosphate cements (CPC) are resorbable, biocompatible, and osteoconductive.

Nevertheless, the commercially available formulations still suffer from several shortcomings related to unsatisfactory levels of some properties and justify the continuous research effort that has been devoted to synthetic bone substitutes as cements.

The purpose of this chapter is to provide a review on the chemical, physical and mechanical properties and clinical applications of calcium phosphate cements and to report the latest developments and tendencies.

2. Calcium Phosphate Cements

In 1832, Ostermann prepared a CaP biomaterial in the form of a paste that set *in situ* to form a solid material [1]. However, it took more than one and a half centuries until Brown and Chow in 1986 [2] first presented this new form of CaP, currently known as calcium phosphate cements (CPCs).

CPCs are made of an aqueous solution and of one or several CaPs, which upon mixing, dissolve and precipitate into a less soluble CaP and sets by the entanglement of the growing crystals, providing mechanical rigidity to the cement. When the paste becomes sufficiently stiff, it can be placed into a defect as a substitute for the damaged part of bone. Subsequently it hardens *in situ* while still within the operating theatre. It hardens in generally < 20 min at body temperature (37°C) and then displays limited solubility. The hardened cements have porosities of 40-60%, with pore sizes ranging typically from 0.1 to 10 μm [3]. The relative stability and solubility of various CaPs is the major driving force for the setting reactions that occur in this type of cements. CPCs are mechanically much stronger in compression than in tension or shear, because entangled crystals are not well bonded. Compressive strength values are typically 5–10 times larger than that of tensile strength.

CPCs are resorbable, osteoconductive, non-cytotoxic, create chemical bonds to the host bones, restore contour, and in general have chemical composition and x-ray diffraction pattern similar to those of natural bone [4-7]. The major advantages of the CPCs include fast setting time, excellent mouldability, outstanding biocompatibility, and easy manipulation [7, 8]. Hence, the cements are more versatile in handling characteristics than prefabricated CaP granules or blocks. Besides, as any other bioceramics, CPCs provide the opportunity for bone grafting using alloplastic materials, which are unlimited in quantity and provide no risk of infectious diseases [5]. There are a number of CPCs products available in the market that are summarized in Table 1.

Chemical reactions taking place during setting of CPCs depend on chemical composition. Nevertheless, it can be stated that only two major chemical types of the setting reaction are possible. The first type is an acid-base reaction that depends on the hydration rates of the acid and basic salts, followed by the neutralization of the by-products. Examples are:

$$2Ca_4(PO_4)_2O + 2CaHPO_4 \rightarrow Ca_{10}(PO_4)_6(OH)_2 \quad (1)$$

or

$$\beta - Ca_3(PO_4)_2 + Ca(H_2PO_4)_2H_2O + 7H_2O \rightarrow 4CaHPO_4.2H_2O \quad (2)$$

In equation (1) the basic tetracalcium phosphate (TTCP) reacts with the acidic DCPA/DCPD resulting in the formation of hydroxyapatite (HA) with a neutral pH. A steady state in terms of ion concentration in the solution is kept as long as the rate of dissolution of DCPA/DCPD and TTCP exceeds the rate of HA formation [18, 19]. Finally, the hardening of the cement results from the interlocking of HA crystals [6]. Several deviations from the chemical equation (1) have been studied in details [20].

The second type of the setting reaction might be defined as hydrolysis of a metastable CaP in an aqueous medium, e.g. when the initial and final CaPs have the same Ca/P molar ratio. Examples are cements made of α, β-tricalcium phosphate (TCP), amorphous calcium phosphate (ACP), TTCP plus aqueous solution, where they re-crystallize to calcium deficient HA (CDHA) upon contact with water. An example is shown by equation (3):

$$3\alpha - Ca_3(PO_4)_2 + H_2O \rightarrow Ca_9(HPO_4)(PO_4)_5OH \quad (3)$$

Experimental details on the above mentioned hydrolysis reactions are available elsewhere [21, 22]. The hydration process of CPCs is characterized by a slow exothermic reaction (thus preventing the attainment of high curing temperatures) during which the cement does not shrink. Previous results showed that the temperature peaks at a highest value of 37°C after 3 hours [23].

Table 1. Marketed calcium phosphate cements formulations [9-17]

Company	Product	Composition: Powder	Composition: Liquid	End product	Resorbable
ETEX (US)	α-BSM	ACP, DCPD	UASS	Apatite	yes
	CarriGen	CaP, SCC, sodium bicarbonate, SC	n.d.*		
Shanghai Rebone Biomaterials Co (CN)	Rebone	TTCP, DCP	H_2O	Apatite	n.d.*
Skeletal Kinetics (US)	Callos™	α-TCP, $CaCO_3$, MCPM	sodium silicate	CAP	yes
	Callos Inject™ and Callos Impact™	α-TCP (close to Callos™)	n.d.*		
	OsteoVation EX Inject	close to Callos™	n.d.*		
Lorenz Surgical (GER)	Mimix	α-TCP, TTCP, HA, CA	H_2O, CA	Apatite	n.d.*
Teknimed (FR)	Cementek®	TTCP, α-TCP, MCPM	H_2O, $Ca(OH)_2$, H_3PO_4	Apatite	yes
	Cementek® LV	α-TCP, TTCP, SG, dimethylsiloxane			
Berkeley Advanced Biomaterials (US)	Cem-Ostetic™	n.d.*	H_2O	Apatite	n.d.*
	Tri-Ostetic™				
Calcitec (US)	Osteofix	CaP, CaO	phosphate buffer	Apatite	n.d.*
Graftys (FR)	Graftys® HBS	TCP, ACP, BCP	PBS	Apatite	yes
	Graftys® Quickset	CaP salts, HPMC	n.d.*		
Biomatlante (FR)	MCPC	α-TCP, ACP, BCP	PBS	Apatite	n.d.*
Mitsubishi Materials (J)	Biopex®	α-TCP, TTCP, DCPD	H_2O, Sodium succinate, sodium chondroitin sulphate	CAP	yes
	Biopex®-R	α-TCP, TTCP, DCPD, HA, $Mg_3(PO_4)_2$, $NaHSO_3$			
Stryker (US)	BoneSource	TTCP, DCP	H_2O, mixture of Na_2HPO_4 and NaH_2PO_4	Apatite	yes

Company	Product	Composition		End product	Resorbable
		Powder	Liquid		
	Hydroset™	DCPD, TTCP, TC	H_2O, PVP, sodium phosphate	Apatite	n.d.*
Biomet (US)	Calcibon®	α-TCP, DCP, CC, PHA	H_2O, Na_2HPO_4	CAP	n.d.*
	Biocement D	α-TCP, DCP, CC, PHA		CAP	yes
Synthes (US)	Norian SRS®	α-TCP, CC, MCPM	H_2O, Na_2HPO_4	CAP	yes
	Norian CRS®				
	ChronosInject®	β-TCP, MCPM, MP, SHPy, MS, SH	H_2O, SH	Brushite	
	Norian® SRS Fast Set Putty	n.d.* (close to Norian SRS/CRS)		CAP	yes
	Norian® CRS Fast Set Putty	n.d.*			
	Norian Drillable	CaP powder, bioresorbable fibers	SH	CAP	n.d.*
Kasios (FR)	Jectos Eurobone®	β-TCP, PA, TTSPy, SA	H_2O, H_3PO_4, H_2SO_4	Brushite	n.d.*
	Jectos+	n.d.*	n.d.*		n.d.*
Produits Dentaires SA (CH)	PD VitalOs®	β-TCP, MCPM, PA, SHPy	$CaSO_4 \cdot 2H_2O$, H_2O, H_3PO_4	Brushite	yes

*n.d. = not documented.

ACP: amorphous calcium phosphate; BCP: biphasic calcium phosphate; CA: citric acid; CAP: carbonated apatite; CC: calcium carbonate; DCP: dicalcium phosphate; DCPD: dicalcium phosphate dehydrate; MCPM: monocalcium phosphate monohydrate; MP: dimagnesium phosphate trihydrate; MS: magnesium sulphate; PA: phosphoric acid; PBS: phosphate buffered solution; PHA: precipitated hydroxyapatite; PVP: polyvynilpyrrolidone; SC: sodium carbonate; SCC: sodium carboxymethyl cellulose; SHPy: sodium dihydrogen phosphate; SG: sodium glycerophosphate; SH: sodium hyaluronate; TC: trisodium citrate; TCP: tricalcium phosphate; TTCP: tetracalcium phosphate; TTSPy: tetrasodium pyrophosphate; UASS: Unbuffered aqueous saline solution.

The liquid phases involved in the setting reaction of CPCs are aqueous solutions. The systems of brushite cements, i.e. TTCP + DCPD, harden in water, whereas other systems usually require the use of sodium phosphate solution (approximately 0.25 mol L^{-1}) as the liquid [24]. For applications requiring longer working times, glycerine or polyethylene glycol can be used [25].

Depending upon the pH value of a cement paste after setting, the CPCs can be classified according to their end product into apatite (AP) cements and dicalcium phosphate dehydrate (DCPD or brushite) cements. AP is formed above pH 4.2, while brushite is preferentially formed in CPCs when pH value of the paste is < 4.2 [26], although it may grow even up to pH 6.5, due to kinetics reasons [26].

As CDHA (calcium-deficient hydroxyapatite) is similar to the chemical composition of bone, AP cements have been more extensively investigated and produced by several companies (Table 1). Nonetheless, brushite cements have raised interest because they are more soluble and resorbed *in vivo* much faster than AP cements [4]. Although AP cements show higher mechanical strength, they have slow *in vivo* resorption rates that interfere with the bone regeneration process [27, 28].

Since water is not a reactant in the setting reaction of the majority of AP cements, the quantity of water actually needed for setting of AP cements is very small [29, 30]. Conversely, for brushite cements, water always participates in the chemical transformations because it is necessary for DCPD formation. Hence, brushite cements are always hydraulic, while usually this term is not associated with AP cements [4]. Moreover, brushite is metastable in physiological conditions and brushite-based cements possess faster setting reactions [26, 31-33].

2.1. Apatite Cements

Apatite cement formulations contain precipitated HA and/or CDHA as the end product of the reaction (eq.1), although some have carbonates present and form carbonated apatites ($Ca_{8.8}(HPO_4)_{0.7}(PO_4)_{4.5}(CO_3)_{0.7}(OH)_{1.3}$) [34]. They are more degradable than HA but still degrade slowly.

AP cements usually contain TCP and/or TTCP phases in the powder constituents [35], although a K- and Na-contained CDHA is also available [36]. TCP tends to be more resorbable than HA and occurs in two crystallographic forms, β and α. α-TCP is biocompatible and more biodegradable than β-TCP, but less stable than the β-phase [38]. Thus, α-TCP is more reactive in aqueous systems, has a higher specific energy and it can be hydrolyzed to a mixture of other calcium phosphates. It never occurs in biological ossifications but, due to its hydraulic power, is used in CPCs [38, 39].

The setting process of this type of cements occurs according to one reaction (see eq.1) at near the physiological pH, except for Norian SRS® and Cementek® that set according two chemical reactions: precipitation of DCPD, followed by the precipitation of CDHA or CAP [29]:

$$\alpha - Ca_3(PO_4)_2 + Ca(H_2PO_4)_2H_2O + 7H_2O \rightarrow 4CaHPO_4.2H_2O \tag{4}$$

$$5.2CaHPO_4 \cdot 2H_2O + 3.6CaCO_3 \rightarrow$$
$$Ca_{8.8}(HPO_4)_{0.7}(PO_4)_{4.5}(CO_3)_{0.7}(OH)_{1.3} + 2.9CO_2 + 12H_2O \quad (5)$$

The initial chemical reaction (eq.4) is very fast, and provokes DCPD formation and setting of the cement paste within seconds. The formation of CAP (eq.5) is a result of the reaction between DCPD with remaining α-$Ca_3(PO_4)_2$ and $CaCO_3$, thus resulting in the cement hardening. A similar two-step hardening mechanism was reported for cement consisting of MCPM and CaO, to form CDHA [40]. The solubility of AP cements in aqueous solutions is comparable to the mineral part of bone, which means that they are insoluble at neutral pH and increasingly soluble as the pH decreases.

The mechanical strength of AP cements can vary with the composition and with the liquid-to-powder ratio (LPR). When the LPR is low, the mechanical strength is high due to the low porosity of the hardened cement. Further, the initial mechanical properties of AP cements can vary with implantation time. Previous studies showed that the mechanical properties of AP cements tend to increase continually, in contrast to brushite cements, which decrease initially and increase when bone grows [41, 42].

The biocompatibility of AP cements is very good and the biodegradability is larger than that of HA, but still slow [43].

2.2. Brushite Cements

Brushite cements were introduced in 1987 by Mirtchi and Lemaitre [44], whereby DCPD is the major end product of the setting reaction (eq.2). Other formulations have been already proposed including β-TCP + H_3PO_4 and TTCP + MCPM + CaO. It is important to notice that β-TCP + H_3PO_4 formulations have several advantages over β-TCP + MCPM formulations, namely: (i) easier and faster preparation, (ii) better control of the chemical composition and reactivity, and (iii) improved physico-chemical properties such as longer setting times and larger tensile strengths due to a higher homogeneity. However, the use of H_3PO_4 might weaken the biocompatibility of the cement formulation, owing to low pH values during setting [45].

Brushite cements are acidic during setting, since DCPD can only precipitate from solutions at pH values below 6; hence the reaction is very rapid corresponding to the setting stage. Despite this initial high reactivity, the hardening stage of brushite cements typically lasts one day until completion, due to an increase of the pH of the paste at the end of the setting reaction. In order to control the onset of the setting reaction, inhibitors of crystal nucleation and growth, and less soluble reagents (for example, HA instead of β-TCP) can be used. Moreover, the use of monodisperse and fine powder particles is vital to provide an overall setting reaction.

Brushite cements are biocompatible and bioresorbable. They are faster resorbed *in vivo* and bear a rapid decrease in strength even though strength of the healing bone increases as bone ingrowth occurs. Conversely, low mechanical strength, short setting times, and consequently a rapid increase in viscosity to enable injection through hypodermic needles, prevent brushite cements from a broader clinical application. Use of chondroitin 4-sulfate

[46] and glycolic acid [33] as setting retardants is an option to get more workable and less viscous pastes of brushite cements.

Brushite cements have a fast and linear degradation rate of 0.25 mm/week [47], which may lead to formation of an immature bone. Adding β-TCP granules to the cement paste could solve this problem because they act as bone anchors and encourage formation of mature bone [47, 48].

The biocompatibility of brushite-based cements has been tested in various experimental applications, compositions and *in vivo* [27, 28, 49]. Brushite cements are generally well tolerated by the bone and soft tissue environment *in vivo*, such that cement resorption was closely followed by new bone formation. Histological measurements indicated good biocompatibility of brushite cements, with almost complete absence of inflammatory cells [50]. Moreover, bone conduction over an intact cortical surface was obtained with brushite cements, indicating that they may be used in minimally invasive vertical bone augmentation procedures without the need for stabilizing devices.

2.3. Ionic Co-Substitutions

The introduction of foreign ions into the structure of synthetic calcium phosphate phases used as starting powders in the formulation of CPCs can alter a number of structural, physico-chemical and biological properties, such as lattice parameters, crystallinity, solubility in the setting liquid, resorption and bone bonding capability [51-54]. Under this perspective, a number of research results have been reported so far on the trace elemental incorporation into α, β-TCP or HA structures with the different ions including Mg, Sr, Zn, Na, K, F and carbonates [51-66]. For example, carbonate substitution causes the formation of smaller and more soluble apatite, while fluoride incorporation has the opposite effects [67-69]. Mg incorporation in apatite is limited but causes reduction in crystal size and increase in solubility [70]. Likewise, Sr causes an increase in solubility [10, 71]. It is also interesting to note that chemical elements not found in natural bones can be intentionally incorporated into CaP biomaterials to obtain special properties. Examples are addition of Ag and Cu [72, 73] that have been used to impart antimicrobial effects, while radioactive Sm-153 and Re-186 isotopes have been incorporated into HA microspheres and injected into knee joints to treat rheumatoid joint synovitis [74].

Among CaP ceramics, β-TCP shows high biocompatibility and resorbability in bone tissue [54]. While β-TCP implants are as osteoconductive as HA ceramic implants they are simultaneously replaced by new bone tissue after some time. Therefore, β-TCP combines biodegradability with osteoconductivity in such a sense that the process taking care of its biodegradation is exactly the process that also induces the formation of new bone. As a result, β-TCP is an adequate ion carrier.

Incorporation of ions into α, β-TCP structures has been experimentally proven through quantitative phase analysis and structural refinement of the powders performed by x-ray diffraction and Rietveld refinement techniques [51, 53, 56, 57, 75-78]. It has been found empirically that extensive ion substitutions can occur in a crystal if the valence of the replacing ion is within one unit of that of the ion being replaced and if the radii of the two ions are similar. Moreover, the accommodation of a foreign ion into the atomic site purely depends on the prevailing geometric conditions at the crystallographic site and also on the nature and size of the ion.

3. Calcium Phosphate Cement Properties

From a clinical point of view, the desirable relevant properties of CPCs are adequate flow properties, short setting times, easy injectability, mechanical properties comparable to those of trabecular bone, and a resorption rate that is neither too fast nor too slow.

3.1. Dynamics of Flow

Rheological properties have not been the subject of extensive investigations in the past. These properties are crucial in gaining understanding of the fundamentals of the dynamics of flow of an injectable CPC through the delivery system (cannula) and its subsequent interdigitation into the cancellous bone. This knowledge can also help in optimizing the design of the cannula, establishing the optimum time for injection and the optimum viscosity at the time of injection, and minimizing the risk for cement extravasations [5]. Besides, the first initial setting period, when the cements lose gradually their plasticity, can be accessed by rheological measurements as well.

The rheological behaviour of a system depends on many intrinsic factors such as the concentration of solid particles in a suspension, particle size and particle size distribution, particle shape, the pH value of the suspending medium, type and amount of dispersing agents of other processing additives (binders, plasticizers, lubricants), stabilization mechanism; or on the concentration of polymer molecules in solution, their molecular weight and molecular weight distribution; as well as on outside influences such as the ambient pressure, the strength of a magnetic or an electric field, and other testing conditions including the type of load, the degree the duration of load, the temperature, etc.

The rheological behaviour can be classified as Newtonian and non-Newtonian. The shear viscosity η of a Newtonian fluid is dependent on the intensity of the shear load. Examples of ideal viscous materials include: low molecular liquids such as water, solvents, mineral oils (without polymer additives), standard oils, pure and clean bitumen (without associative superstructures, i.e. at sufficiently high temperatures) and blood plasma. In Non-Newtonian systems the evolution of viscosity as a function of shear rate γ can show a decreasing (shear-thinning flow behaviour or 'structural viscosity') or an increasing trend (shear-thickening flow behaviour or 'dilatancy'). The overall relation between the shear stress τ and the shear rate $d\gamma/dt = \dot{\gamma}$ for systems without an (apparent) yield point can be expressed by $\tau = \dot{\gamma}^n$. For samples displaying Newtonian flow behaviour ($n \equiv 1$), the shear viscosity is independent of the degree of shear, and the plots τ versus γ are straight lines passing through the origin (Figure 1.a), while the plots η versus γ are straight horizontal lines (Figure 1b).

Figures 1a and b, also show typical plots of shear-thinning ($n < 1$), and shear-thickening ($n > 1$) flow behaviour. In Figure 1a, the diagram is represented with linear scale, while in Figure 1b a logarithmic scale was used. These last scales are recommended when the course of the curves should be made visible at very low τ and γ values.

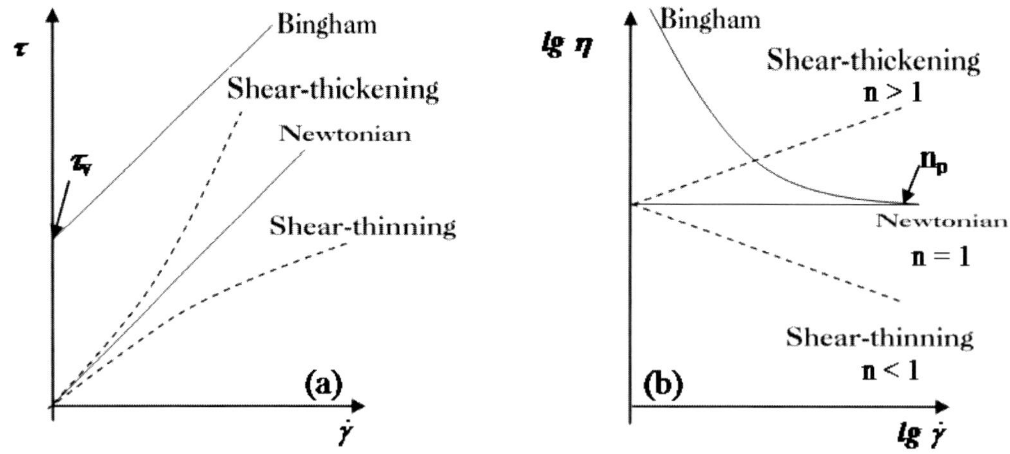

Figure 1. Comparison of the (a) flow and (b) viscosity curves.

Examples of shear-thickening materials include dispersions with a high concentration of solids or polymers such as ceramic suspensions, starch dispersions etc. The flow curve shows an increasing curve slope (Figure 1a), i.e. the viscosity η increases with increasing shear rate dγ/dt.

Plastic systems exhibit a yield point and "plastic flow" is the deformation process which can only occur above the yield point; below this point no or only elastic deformation occurs. The simplest or ideal plastic behaviour has been proposed by Eugen C. Bingham [79]. At the yield point the loaded sample starts to be deformed and the rate of deformation increases linearly with the shear rate. Inhomogeneous plastic behaviour cannot be described unequivocally using mathematics. It can only be represented using the results of empirical tests or appropriate rheological models.

The terms plastic, ideal plastic, viscoplastic or elastoplastic are often used to mean different things. It is useful to understand what these terms might mean, but their use should be avoided when performing scientific rheological tests. In most cases, samples can be characterized as viscoelastic in a limited deformation range. However, when a material cannot be sheared homogeneously it is necessary to use special relative measuring systems. In this case, it is better to work with the measured raw data values e.g. the torque, deformation speed and deflection angle instead of the rheological parameters e.g. shear stress, shear rate, deformation, viscosity and shear modulus.

Bone cements in general are considered viscoelastic materials as they change from having primarily liquid-like properties immediately after mixing to having primarily solid-like properties once cured. As an example, flow curves for Sr-, Mg- and Zn-substituted cement pastes exhibiting similar shear-thinning characteristics are presented in Figure 2.

In general, the storage modulus (G') corresponds to the elastic behaviour of the material and determines its inherent rigidity It thus depends on the ability of the material to store mechanical energy [80]. Conversely, the loss modulus (G") corresponds to the viscous behaviour of the material and has a strong influence on its toughness and is dependent on the material's ability to absorb and dissipate mechanical energy [80].

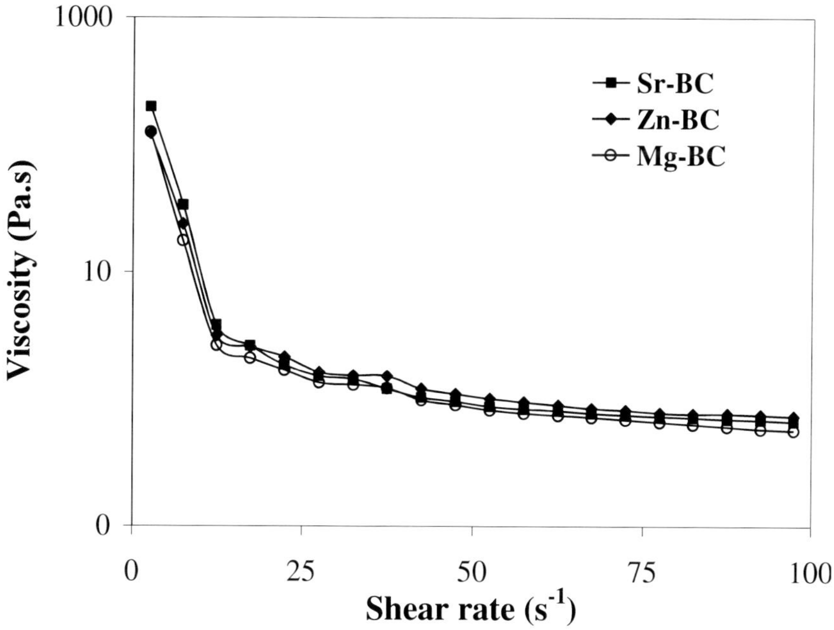

Figure 2. Viscosity vs shear rate curves of cement pastes prepared with LPR of 0.35 mL/g. Sr-BC: Sr-substituted bone cement; Zn-BC: Zn-substituted bone cement; Mg-BC: Mg-substituted bone cement (adapted from [85]).

One of the first studies related to CaP paste rheology by Bujake [81] investigated the viscosity of toothpaste-like suspensions containing DCPD, glycerine and water. This author observed an increase of the viscosity with a decrease of the particle size. Similar results were obtained in other studies [80-84]. Rao and Kannan [82] examined the yield stress and viscosity of HA suspensions. They observed a yield stress of 5 Pa, and a shear-thinning followed by shear-thickening behaviour. Generally, shear-thickening appears to occur at high particle loading [80, 84].

The study of the rheological behaviour of α–TCP slurries using a viscometer allowed looking at the effect on cement viscosity of various parameters during the initial setting period, such as the particle size of the α–TCP powder, the LPR and the addition of dispersants [86]. Also, Baroud et al. [87] studied the effects of liquid-to-powder ratio (LPR), milling of the powder particles and the presence of additives on the rheological properties of a non-reactive system. The study also demonstrated the importance of determining the rheological properties of the pastes in order to fully understand the injectability data. In addition, the authors recommended increasing the viscosity of the mixing liquid, decreasing the particle interactions, and decreasing the particle size of the powder.

In an attempt to improve the rheological properties of CPCs, the influence of several additives to the liquid phase such as lactic acid, glycerol, chitosan, citric acid or soluble polymers, on the injectability of CPCs have been studied by several authors [88-90]. These additives increase the viscosity of the liquid phase up to several orders of magnitude and make the cement pastes putty-like, since they improve the wettability and increase the surface charge of the cement particles due to ions adsorption at the particles' surfaces [91].

3.2. Setting Time and Reaction Rate

The setting of bone cements is a continuous process that involves: (1) dissolution of reactants, (2) nucleation of new crystals, and (3) growth of crystals. After these processes, newly formed crystals entangle, the cement loses its viscofluid properties, and transforms into a solid body [92]. Commonly, CPC should set slowly to provide enough time to a surgeon to perform implantation by moulding or injection, but harden fast to prevent delaying the operation, allowing the surgeon to close the defects shortly after cement placement. The setting and hardening process of CPCs have been determined with two standardized methods, namely the Vicat needle method (ASTM C191-92) and the Gilmore needle method (ASTM C266-89). Basically, both methods rely on visual inspection of CPCs surfaces, whether or not the test specimen accepts the needles without leaving any mark on the surface. Whereas the Vicat method only uses one needle to determine the final setting time (FST), the Gilmore method uses two different needles to measure the initial (IST) and FST setting times of CPCs. IST is measured with a light and thick needle (113 ± 0.5 g, Ø = 2.12 ± 0.05 mm, excerting a static pressure of 0.3 MPa), while FST is measured with a heavy and thin needle (453.6 ± 0.5 g, Ø = 1.06 ± 0.05 mm, excerting a static pressure of 5 MPa) [93]. IST indicates the time during which the material must not be manipulated to avoid serious damage to the cement structure, whereas FST defines the time when the material has hardened, but maximum strength for some materials has not yet achieved. As far as clinical applications are concerned, the cement should be implanted before IST and the desired range of the setting times (in minutes) may be: $3 \leq IST < 8$ and $FST \leq 15$. These parameters are represented schematically in Figure 3. To guarantee that the surgeon has at least 1 min time to apply and mould the paste, the cohesion time (CT), also called swelling time, must be at least 1 min before IST. As the mixing in a mortar takes about 1 min, the shortest CT that can be allowed is about 2 min [89]. However, still there is not a consensus on what the ideal values of IST and FST should be. Ginebra et al. [94] have suggested that IST should be between 4 and 8 min, while FST should be between 10 and 15 min. On the other hand, Driessens et al. [95] have suggested that the cement should be injected before IST, while the wound should be closed after FST, making sure that the cement is not deformed between IST and FST since otherwise cracks could be induced. In contrast, Sarda et al. [96] have suggested that the cement probably reaches a sufficient strength to withstand the pressure during wound closing well before FST.

The Vicat and Gilmore methods have been used since the discovery of CPCs. However, these methods have not escaped criticism because of their inherent subjectivity. For this reason, some studies have been performed to record the setting process in a continuous manner and in real time by non-destructive methods, such as isothermal calorimetry [31, 97, 98], pulse-echo ultrasound technique [99, 100] and AC impedance spectroscopy [101]. In the isothermal calorimetry method, the thermal reactions occurring during CPC setting reactions were followed as a function of time and composition. In addition, other methods have been proposed. For example, Hofmann et al. [32] followed the setting kinetics with Fourier-transform infrared (FTIR) while Grover et al. [102] used x-ray diffraction (XRD).

Several factors can control the setting time and setting reaction of CPC pastes, such as ion modification, LPR, particle size, particle shape, and additives (accelerators and retarders). The presence of Sr ions in Sr-substituted α, or β- TCP can extend the setting time of the cement [55-59].

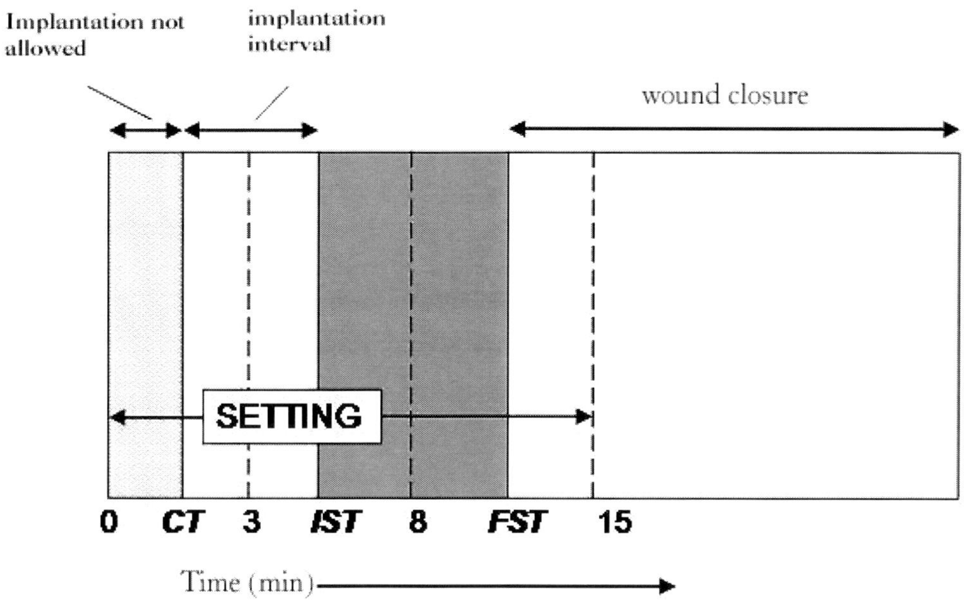

Figure 3. Diagram of the setting parameters relevant for CPC. CT: cohesion time; IST: initial setting time; FST: final setting time (adapted from [95]).

Conversely, the addition of $SrCl_2$ to brushite cements results in a shorter setting time [103]. Mg ions tend to delay the setting process in comparison to pure α-TCP cements [56, 62], and can have a strong inhibitory effect on the growth of brushite crystals [104]. Si-substituted β-TCP brushite-based cements show higher setting times in comparison with silica-free cements [55].

Additives such as gelling agents can also be used to control the setting reaction. For example, a delay of the setting process was observed in the referred studies [62] when poly(ethyleneglycol) (PEG) and hydroxyl propyl methylcellulose (HPMC) were used as gelling agents, being less pronounced in the case of PEG in comparison to HPMC. These agents are hydrophilic polymers that form a network structure with water molecules, which then become less free to spontaneously react with inorganic CaPs. On the other hand, the polymeric species might be adsorbed at the surface of the CaP particles thus delaying the dissolution/precipitation processes. Consequently, the cement network formed may integrate with the polymer structure. This phenomenon is more pronounced for high molecular-weight polymers such as HPMC, which could explain the more accentuated decrease of setting time in the presence of this additive in comparison to cement pastes with added PEG.

Another example was shown by Bohner et al. [105] that heat treatment at 500ºC could extend the initial part of the setting reaction from a few minutes to a few hours providing a potential approach to better control of the setting reaction. Brunner et al. [106] demonstrated that nano-sized amorphous CaP particles could be used to produce cements reacting within an hour.

More detailed discussion about the factors used to control the setting reaction is presented in section 4.

3.3. Injectability and Cohesion

The viscous and cohesive properties are of major interest in the case of injectable cements. Injectability and cohesion are required for applications with limited accessibility and narrow cavities, and when there is a need for precise placement of the paste to conform to a defect area such as encountered in periodontal bone repair and tooth root canal fillings.

As there are no standard procedures to assess injectability property, indigenous test methods were developed for the purpose [35, 107, 108] as shown in Figure 4. According to this empirical method, injectability is reported as the percentage by weight of the cement paste that could be extruded without demixing from a standard syringe, through the small hole of a long needle (usually, Ø = 2 mm and 10 cm in length), pressing the plunger manually or applying an equivalent external force (~ 100 N) [35]. Frequently, CPCs pastes show liquid-phase separation, the so-called filter-pressing effect, provoked by the pressure applied to the cement paste after a certain injection pressure or after a certain injection time [109, 110].

In the case of demixing, the exact composition of the extruded part of the paste becomes unknown. For example, in 4.5 mL Norian SRS® cement paste, only ~ 3 mL of paste is injectable, while up to 1.5 mL of the cement remains in the syringe [95].

A good cohesion of the cement pastes during mixing is essential to avoid the possible occurrence of inflammatory reactions. Cohesion was reached when no disintegration of the cement paste was observed in the fluid, which can be obtained by keeping a high viscosity or using cohesion promoters, e.g., 1% aqueous solution of sodium alginate or other chemicals [11, 38, 111].

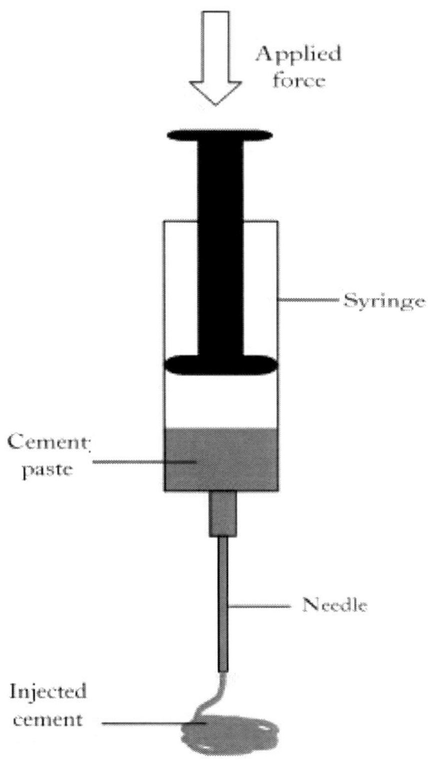

Figure 4. Schematic representation of the experimental setup used to measure the injectability of CPCs.

Due to the setting process, the rheological properties of CPCs are transient and constantly changing. Combined with this is the complication that the injection process may destroy the evolving structure and thereby affect the measurements. There are different ways to improve injectability and cohesion, such as increasing the extrusion velocity, using a small syringe with short cannulas and large diameter, increasing the LPR, decreasing the powder particle size, modifying particle size distribution and particle–particle interactions, and using additives in the mixing liquids [35]. Setting retarders, such as gelling agents can improve injectability by an increase of the viscosity of the liquid. The particle–particle interactions can be reduced by using additives such as carboxylic acids and by ion modification of the cement reagents. For example, the presence of sodium citrate or citric acid in the mixing liquid increases the injectability [112]. Besides, adding Mg ions to α-TCP or Si ions to β-TCP has been found to increase DCP cement injectability [55, 56], while doping α-TCP with Sr decreases the cement injectability [56]. Although injectability of cements is enhanced with increasing LPR [89], this approach often has a detrimental effect on the mechanical properties [113]. In addition, α-BSM® is well injectable due to the small particle size of the powder. However, a small particle size requires a larger amount of mixing liquid, which decreases cohesion and mechanical strength [4, 97]. An indirect approach to improve injectability is to add CaP particles that act as spacers between other particles, as is the case when DCPA is added to Biocement D [4]. Another approach to improve injectability is by adding polymers to increase the viscosity and prevent the filter-pressing effect [90, 91, 114]. Sarda et al. [114] found that the injectability of α-TCP/water mixtures was increased by 50-100% with the addition of citrate ions. Similar results were obtained for brushite and AP cements [91, 112].

The influence of all the above mentioned factors is discussed in detail in section 4 below.

3.4. Mechanical Behaviour

As bone cements are meant to be used as bone substitutes, and because one of bone's first functions is to support the body weight, the mechanical behaviour of the cements is of a paramount importance. The mechanical strength required for cements must be at least as high as that of trabecular bone which is close to 10 MPa. There are two mechanical tests particularly important for implanted CPCs: compression strength (CS) (Figure 5A) that requires sample dimensions of 6.0 mm diameter x 12.0 mm height, whereas in the diametral compression strength (Figure 5 B) cylindrical samples with dimensions of 6.0 mm diameter x 3.0 mm height are used. Diametral compression strength (frequently denominated by several authors, as diametral tensile strength, DTS), also named the Brazilian test or indirect tension test, is routinely used for brittle materials as ceramics [115]. Direct tensile tests are rarely applied to brittle materials since such materials do not deform until rupture, and misalignment of the specimens may induce a tangential stress component that may lower the measured strength of the materials. Therefore, the DTS test allows measuring tensile strength of materials by applying a compressive load in an experimental setup similar to the one used for axial compression, to keep simple sample shape, and to avoid the problem of mounting specimens on a universal testing machine. The advantage of this type of test is that fracture starts within the specimen and the measured value does not depend on the surface state.

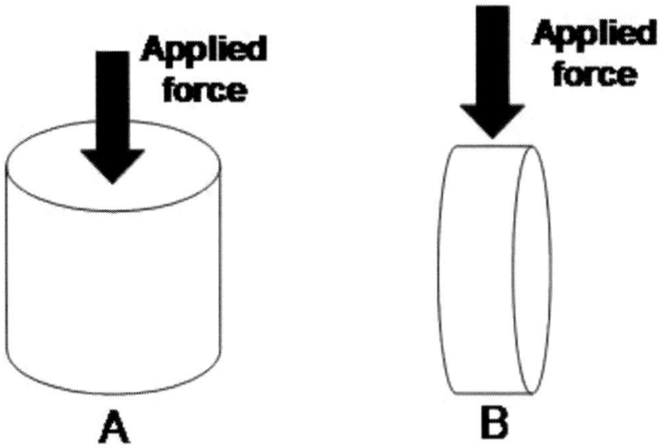

Figure 5. Schematic representation of the experimental setups used to measure the (A) compression and (B) diametral compression strengths.

CPCs are strong enough at compression with values in the range of 10-100 MPa, whereas they possess a low tensile strength (1-10 MPa) [113]. The poor mechanical properties noticeably restrict a wider clinical application of CPCs; however after 12 weeks of implantation, CS values higher than those of normal bone (60-70 MPa) have been reported [41]. One must bear in mind that the value found for mechanical properties depends, not only on the product, but also on the storage conditions. In particular, the condition of soaking for at least 1 day in water, Ringer's solution or PBS solution at 37ºC is thought to be relevant for evaluating the suitability of these cements as biomaterials. As shown by Driessens [116], the strength of brushite cements obtained by mixing α-TCP or TTCP with acidic solutions depends critically on their storage conditions.

Varying the crystallinity or the particle size of the materials used in the solid phase may also alter the CS. Since CPC are relatively insoluble at neutral and alkaline pHs, their porosity is related to the LPR used in the starting mixture. Obviously, high porosity cement would be expected to exhibit low CS.

These mechanical properties may also vary with implantation time, and animal studies have shown that mechanical properties of AP cements tend to increase continually, in contrast to those of brushite cements, which initially decrease and then increase when bone grows [41, 42]. This is the result of differences in porosity and bioresorption between AP and brushite cements.

To improve the mechanical properties of CPCs, incorporation of water-soluble polymers into the pastes has been attempted [117, 118]. It was observed that the strength increased mainly due to the polymers' capacity to entangle. A research group reported similar results using sodium alginate and sodium polyacrylate [119]. Studies show that composites of α-BSM® with polycations (polyethylenimine and polyallylamine hydrochloride) exhibited compressive strengths up to 6 times those of pure α-BSM® material [120]. Similarly, pyrophosphate salts and α-hydroxyl carboxylic acids such as citric, glycolic and tartaric acid and their salts, allow easier mixing of the cement, allowing for an decreased LPR, thus resulting in improved mechanical properties [33, 130, 112, 121].

Mechanical properties of CPCs were found to decrease exponentially with increasing porosity since the presence of pores leads to cracks in the hardened cement. Pressure can be applied to reduce the porosity of CPCs [113, 115, 122]. Another way to control the mechanical strength is by varying LPR in the hardening mixture. Elevated CS would be applicable in cranioplasty for regions requiring significant soft-tissue support. For smaller bone defects, such as root canal fillings, low CS cements must be used [30]. An empirical relationship between strength and porosity is reported in the literature [123]. As porosity is mainly due to an excess of water used in CPCs formulations, efforts to reduce the amount of water were made. Also, a decrease in water content leads to a strong increase in viscosity of the CPC pastes. As CPCs set with almost constant volume, the final porosity can be predicted from the initial composition [4, 124].

Other factors affecting strength are the reagents used in the solid phase, particle sizes, incorporation of fillers into the solid phase and various liquid phases [125]. The ionic exchange must also be taken into account. To date, Lilley et al. [61] have shown that the presence of Mg had a strong effect on cement composition and strength, namely by increasing the proportion of brushite and decreasing the CS. According to these authors [61], Mg could be used to adjust the composition and rate of hydration of the cement. Mg has been also reported to have a stabilising role of non-crystalline CaPs, preventing crystallisation into other more stable CaP phases [126]. Pina et al. [62] showed that Mg-containing brushite-forming cement appeared slightly stronger in comparison to Mg-free cements, due to a more favourable crystalline phase composition. Moreover, Sr-containing cement specimens exhibited higher CS in comparison to Sr-free cements [57]. In a later report [56], the role of Mg- and Sr-substituted cements relating to CS was evaluated. It was demonstrated that Sr-substituted cements were mechanically stronger than with Mg-substituted ones, demonstrating that other benefits can be obtained by properly designing the bone cement composition.

On the other hand, adding metallic ions such as $SrCl_2$ to the cement setting reaction has no beneficial effect on the mechanical properties of the cement [127], although β-TCP doped with Si ions has an increased CS when compared with silica-free cements [55].

3.5. Biocompatibility and Bioresorbability

Biocompatibility, osteoconductivity, and bioresorbability are all features of CPCs that make them attractive bone substitute candidates.

Biocompatibility expresses the ability of a biomaterial to perform with appropriate reactions in applications planned for a particular objective. Biocompatibility and bone-bonding ability of a material have been evaluated by *in vitro* and *in vivo* tests.

In vitro biocompatibility assays are usually preceded by mineralization tests in simulated body fluid (SBF) with ion concentrations nearly equal to those of human blood plasma [128-130] with the formation of apatite onto the materials surface. The biocompatibility screening tests include cell cultures [131] available on the market. The bone-bonding ability of a material is often evaluated. Moreover, there is a wide range of repeatable and reproducible methods, which are regulated by national and international standards for commercial use and for the scientific development of new materials and products. Detailed analysis of the surface apatite formed in SBF by thin film x-ray diffraction, Fourier transform infrared spectroscopy (FTIR), scanning electron microscopy (SEM) and transmission electron microscopy (TEM) showed that it was similar to bone mineral in its composition and structure [131]. As a result, it was speculated that osteoblasts might preferentially proliferate and differentiate to produce apatite and collagen on its surface. Thus secondary apatite formed might bond to the surface apatite as well as to the surrounding bone. Consequently, a tight chemical bond is generated between the biomaterial and the living bone through the apatite layer. An example is that β-TCP bonds in either SBF [133, 134] or *in vivo* [135-137] environments to living bone due to its high resorbability.

In vivo assessment of the biomaterials biocompatibility is a critical step in their development and implementation. While *in vitro* studies yield information on important fundamental interactions with biomaterials, they cannot replace *in vivo* evaluation. An animal model is an appropriate way to identify a material with *in vivo* bone bioactivity, with special attention being paid to the similarities and differences in the trabecular bone composition and architecture of humans. In this regard, pigs and sheep are suitable models because their bone structure is comparable to those of humans. CPC revealed extensive bone formation immediately after implantation without any inflammatory tissue response [112]. Bone colonization occurred much earlier and faster in CPC than in CaP ceramics [107]. Polymers [88], gelatin [138], and collagen [111] can be added to improve the biocompatibility of CPCs. Adding chitosan and citric acid improved the biocompatibility and decreased the initial inflammatory responses of self-curing CPCs [11]. Addition of 20% citric acid decreased the initial inflammatory response and enabled good bone bonding to be established.

One of the most appealing characteristics of CPCs is their resorbability *in vivo*. Upon implantation, these materials act as osteoconductive scaffolds but, with time, degrade and are replaced by bone during the remodelling process [139, 140]. The products of the cements' degradation are Ca^{2+} and PO_4^{3-} ions, which are easily excreted or recycled by the body. CPCs degradation occurs by the combination of two processes: (i) *in vivo* dissolution, which is strictly related to their composition and particle size [141], and (ii) cell-mediated resorption mainly by osteoclasts. Cell-mediated resorption is advantageous since it mimics the natural process of bone turn-over, in which osteoclasts resorb bone and osteoblasts subsequently secrete bone matrix [142, 143]. The bioresorption process of bone substitutes obviously depends on their chemical nature, being closely related with CaPs solubility in aqueous solutions.

A variety of characteristics may influence the resorption rate. Among them are the age, sex, and general metabolic health of the host. Taking these variables into consideration, the complete resorption of CPCs may take from 3 to 36 months. Additional studies are needed in order to describe the degradability of these materials in clinical models [144].

Brushite and AP cements have an excellent bioresorbability and ability to form new bone, with AP being less bioresorbable than brushite [31]. Moreover, when the crystal size increases or the porosity decreases, the bioresorption will take longer. Norian SRS® and α-BSM® are therefore expected to bioresorb faster than BoneSource®, Biopex and Cementek. Norian SRS® was evaluated in canine tibial and femoral metaphyseal defects.

Ionic exchange phenomena occurring with CPCs are associated with reactivity towards bone bonding. The adverse effect of trace mineral deprivation on bone metabolism in animals has been recognized for many years and is known to be related to specific defects in organic bone matrix synthesis. To date, Mg acts as a surrogate for Ca in transport and mineralization processes [145, 146], but it also exerts a large number of other actions, including enzyme co-factor function and modulation of the action of hormones, growth factors, and cytokines.

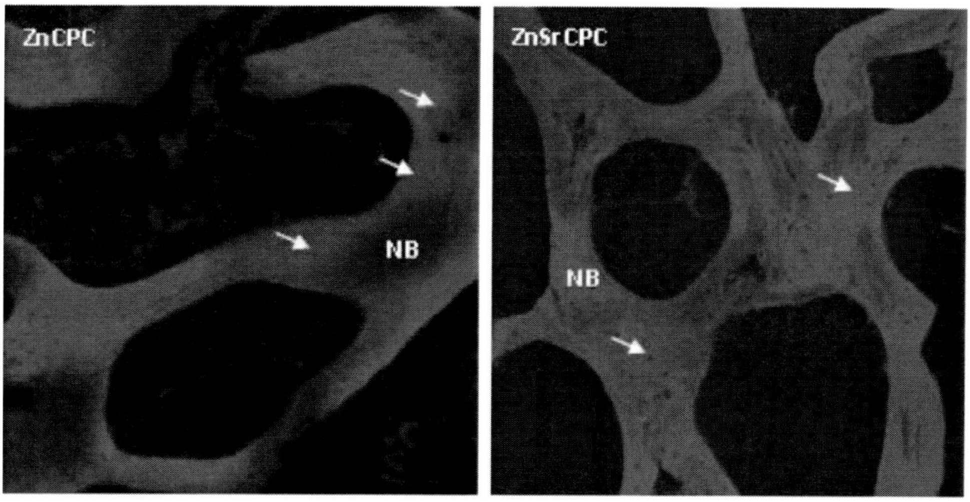

Figure 6. Confocal fluorescent micrographs of histological HandE stained sections showing the tissue osteoconductive response to Zn-substituted brushite cements (ZnCPC) and ZnSr-substituted brushite cements (ZnSrCPC), after 2 months of implantation. NB: new bone; arrows: osteoclast-like cells (adapted from [85]).

Mg also has a direct effect on the bone formation processes of resorption and mineral aggregation [147]. Moreover, adding Mg ions to the cement paste can retard the transformation of brushite into HA and thus reduces the possibility of inflammation [5] while Sr increases osteoclast apoptosis and enhances preosteoblastic cell proliferation and collagen synthesis [148]. Zn deficiency causes a reduction in osteoblastic activity, collagen and chondroitin sulfate synthesis, and alkaline phosphatase (ALP) activity [75, 149].

Zinc is essential for life and reproduction, and is a component of the cell nucleus, mitochondria, cytoplasm, cell membranes, and cell walls [150]. Zinc is a constituent of about 300 enzymes, and Zn ions are located in the catalytic site as well as in the structural site of the enzyme complex [151, 152]. Recent studies on Zn- and ZnSr-substituted brushite cements injected into trabecular bone defects in pigs proved that Zn and Sr are good inductors of osteoprogenitor cell proliferation and differentiation [58]. Results indicate that the investigated cements are biocompatible, osteoconductive, and good candidate materials for use as bone substitutes. The pattern of trabecular bone that was formed in the implantation area was qualitatively similar to the adjacent old trabecular bone in the sections' periphery. Also, results indicated that Sr is a more potent inhibitor of osteoclastic activity than Zn, as much fewer osteoclast-like cells could be found in ZnSr-containing implants (Figure 6), in good agreement with previous studies indicating that Sr-containing ionic cements are more osteoconductive than Zn-containing ionic cements [153].

4. Influence of Factors on Calcium Phosphate Cement Properties

The properties of CaP bone cements are influenced by a wide spectrum of factors, such as, the presence of additives, the powder particle size, the liquid-to-powder ratio (LPR), and the method used to mix the powder and liquid constituents of CPCs.

4.1. Additives

Additives that have been studied are fluidificants, gelling agents, reinforcing agents, and antibiotics [154]. Further, radiopacifiers might be used [155]. The main requirements for additives are non-toxicity as well as the absence of inhibition of the cement setting reaction.

Fluidificants added to the liquid are aimed at enhancing the flowability of the cement pastes. An example is citric acid, which decreases the CS during initial setting of the cement but increases its CS at saturation [114]. Moreover, citric acid also acts as setting stimulator and improves the mechanical properties of the hardened cements [156]. For several cements, mixing with sodium citrate or citric acid has resulted in some effects on the IST [91, 114], while for other cements the effect was insignificant [112].

The main role of *gelling agents* is to improve the workability of the pastes. These kind of agents significantly increase the injectability as well as the setting times. Two examples of workability-improving agents are polysaccharides and gelatin, which lead to a reduction in total porosity and achievement of a more compact microstructure. Hydroxylpropylmethyl-cellulose (HPMC) is another gelling agent capable of improving the pastes cohesiveness, since HPMC is a polysaccharide able to develop hydrogen bonds with water and to form a viscous solution [157, 158].

Reinforcing agents can be added to enhance the mechanical properties of CPCs. Bohner et al. [159] used citrate, pyrophosphate and sulphate to modify the properties of CPCs. Moreover, in order to decrease CPCs brittleness, reinforcing fibres and polymers were added to the cement matrix as a part of a periodontal disease model, yielding promising results [160]. The incorporation of biodegradable poly(lactic-co-glycolic) acid (PLGA) microspheres into CPC could provide the early strength needed, and would allow the generation of macropores for bone ingrowth as they degrade. A study reported that a CPC containing 30 wt % PLGA microspheres showed that this material exhibited excellent injectability, high early strength and satisfactory setting time and *in situ* macropore-formation [161]. Moreover, the addition of sodium citrate in CPC containing PLGA microspheres can reduce the viscosity and the yield stress of the paste, as well as the compressive strength can be improved.

Chitosan is also used as an additive in CPC with the purpose of reducing the periodontal pockets [162, 163]. Addition of chitosan oligosaccharide and collagen polypeptide to salmon-calcitonin loaded CPC was used to adjust its physical and releasing properties [164].

Studies have reported on the influence of additives in final CPCs properties. Wang et al. [165] showed that the addition of disodium phosphate, poly(ethyleneglycol) (PEG 200), glycerin and citric acid influenced the injectability and the setting time of an ACP + DCPD bone substitute material. The study suggested that injectability and the setting time could be balanced and the cement could be prepared by optimizing the additives and their concentrations according to different clinical requirements. Leroux et al. [90] found that glycerol greatly improved the injectability and increased the setting time but decreased the mechanical properties. Lactic acid reduced the setting time, increased the material toughness, but limited the dissolution rate. After injection, the cement did not present any disintegration. The effect of lactic acid was correlated with the formation of calcium complexes. Chitosan alone improved the injectability, increased the setting time, and limited the evolution of the cement by maintaining the CPC phase.

The main role of an *antibiotic* is to reduce infections in the body [166, 167]. However, its presence affects cement properties, as the addition of flomoxef sodium, for instance, leads to an increase in cement consistency, a decrease in DTS and an increase in the set mass porosity [168]. Both setting time and CS of a brushite CPC increased when gentamicin sulphate was added either as a powder or as an aqueous solution.

4.2. Powder Particle Characteristics

Powder particle characteristics include the average particle size, the particle size distribution, and the specific surface area. Particle size distribution and surface area are interdependent characteristics which influence the behaviour of bone substitutes.

A reduction in particle size produces a significant decrease in both initial and final setting times, an acceleration of the rate of hardening but an unclear trend with regards to mechanical properties [8]. In a study described by Liu et al. [19], a steeper increase in pH and acceleration of the dissolution of DCPA was observed by reducing the particle size of TTCP. For instance, the commercial reagent DCPA presents a median particle size of about 1 μm, while Chow et al. [169] reported that TTCP should be ground to a median particle size of approximately 15 μm.

4.3. Liquid-to-Powder Ratio

The properties of CPCs are influenced by LPR since a higher amount of mixing liquid markedly increases both IST and FST, and decreases mechanical properties [35, 110]. On the other hand, decreasing the LPR and adding reinforcing fibres decrease the cements' porosity and increase their mechanical properties [110], while injectability was found not to be affected by LPR within the range of 0.22 – 0.26 mL g^{-1}, but drops by nearly 100% between 0.2 and 0.22 ml g^{-1} [91]. In a study reported by Baroud et al. [87] it was found that changing LPR and the milling time significantly affected the rheological properties.

4.4. Mixing Procedure

Mixing methods of bone cement are categorized as manual (or hand) mixing, centrifugation, vacuum mixing, and combined mechanical mixing.

During *manual mixing*, the powder component is added to the liquid (which may or may not have been chilled to a temperature that is usually between 15°C and 16°C) in a polymeric (usually polypropylene or polyethylene) bowl. Then, these components are stirred with a polypropylene or polyethylene spatula with a speed of 1 or 2 Hz for 45–120 s. To be injected, the cement paste must be then transferred into a syringe. Two commercial products, Norian SRS® and α-BSM®, are sold with a reactant pack containing the powder and the mixing liquid, and require the use of a mixing machine and an injection gun [4]. The electrically powered mixing machine of Norian SRS® mixes the cement paste within 70-80 s, and enables a fast and consistent filling of the application syringe. In spite of this kind of mixing system being very simple, it has some drawbacks. As the mixing procedure depends on the user and as the setting time of the cement is temperature dependent, the setting time varies from user to user.

In *centrifugation mixing*, the hand-mixed dough is immediately poured into a syringe (from which the nozzle is detached) that is then promptly placed in a centrifuge and spun with a speed of 2,300–4,000 rpm for 30–180 s [138, 170].

Applying *vacuum* to self-setting acrylic cements was first described in dentistry, but Lidgren et al. [171] studied vacuum-mixed cements as bone cements in 1984. They analysed the mechanical strength of high- and low-viscosity gentamicin-containing cements by using three different mixing procedures: hand, vibration, and vacuum stirring. They reported that vacuum mixing improved the flexural and compression strength especially for high-viscosity cement. In another study by Lidgren et al. [172] in 1987 it was shown that vacuum mixing improved fracture strength, maximal deflection, modulus of elasticity, and hardness when compared with hand mixing. The fatigue life was ten times longer after vacuum mixing. Vacuum mixing delayed the setting time by 1 min, and also decreased the peak temperature. Radiographic analysis showed that vacuum mixing mainly reduced not only the microporosity, but also the macroporosity. The mechanical properties deteriorated slightly after a storage time of two months in Ringer's solution, but the differences between the mixing procedures remained unchanged [29].

5. Clinical Application of CPCs

CPCs are non-toxic and resorbable materials with osteoconduction, excellent biocompatibility, low exothermicity of setting reactions, but their mechanical properties are not ideal, limiting their clinical use. The main purpose of CPCs is to fill small bone defects or to increase bone volume in metaphyseal bone, thus reducing the need for bone graft. The combination of a self-setting nature, mouldability, biocompatibility and a great potential for being replaced by bone, make these materials to be used by bone remodelling cells for reconstruction of damaged parts of bones [139, 140, 173]. Besides, some formulations were found to possess an antimicrobial activity, as well as promote osteoblast cell adhesion and gene expression *in vitro* [22, 49].

CPCs have successfully been used for treatment of the distal radius fracture. Moreover, other successful attempts have been made to use these materials for calcaneal fractures, hip fractures, augmentation of osteoporotic vertebral bodies, tibial plateau fractures, restoration of pedicle screw fixation, reinforcement of both thoracolumbar burst fractures, cancellous bone screws, in wrist arthrodesis and for fixation of titanium implants [174, 175]. They can also be applied to reinforce osteoporotic vertebral bodies [124, 176]. Stankewich et al. [177] and Goodman et al. [178] showed augmentation of femoral neck fracture with CPC, which improved the initial stability and failure strength of the fractures. Furthermore, cadaver studies revealed that brushite cement is able to increase the bone mineral density of osteoporotic vertebrae, the vertebral stiffness, and the maximum force. However, while the cement had no significant effect on healthy vertebra, in 12% of the cases cement extrusion to the spinal canal occurred [179].

The use of CPCs for craniofacial and maxillofacial applications has been another valuable application. TTCP + DCPA cement that showed to be efficacious in animal studies was evaluated in several medical centres for cranial defect repair as part of a U.S. FDA-approved study in human subjects [180]. Sliindo et al. [181] reported that BoneSource^{TM} was employed to increase the supraorbital ridge in dogs, as well as in a variety of skull base defects. It was also used in 24 patients to increase or obliterate the frontal and ethmoid sinus regions and mastoid cavities. Cranial base reconstruction with CPCs has been successful for translabyrinthine, middle cranial fossa, and suboccipital craniectomy defects, as well as for extensive temporal bone fractures [182].

In addition, several clinical studies have shown that brushite cements are capable of regenerating oral and maxillofacial bone in atrophic areas, buccal dehiscence defects, and maxillary sinuses [183, 184].

CPCs have been employed for dental applications too. These materials were tried as root canal fillers and for pulp capping [185, 186]. A previous study showed that a hydraulic CPC was injected as a bone filler for gaps around oral implants placed on the medial femoral condyles of 6 goats and found excellent bone formation around the graft material. However, the degradation rate of the cement appeared to be very slow and no resorption was observed [187]. In another study, researchers used a CPC for direct pulp capping and found its capability of producing secondary dentin after 24 weeks [188].

Due to their outstanding properties (injectability, biodegradation, setting at room temperature, large surface area, and near neutral pHs [6, 7]), CPCs are very attractive candidates to be used as a delivery system for therapeutic peptides, antibiotics, anticancer drugs, anti-inflammatory drugs, and bone morphogenetic protein.

The antibiotics and antiseptics that have been assayed for drug release from brushite cements include vancomycin, doxycycline, tetracycline, ciprofloxacin, gentamicin and chlorhexidine, while the growth factors that have been tested include PDGF, VEGF, RANKL, bone morphogenetic protein-2 and heparin. These drugs will be slowly released through set cement pores, after their incorporation into a liquid and a powder phase of the cements. However, it ought to be taken into account that incorporation of drugs in the cements should not affect the final CPCs properties, such as the setting and hardening processes, rheological behaviour and injectability. Thus, the clinical performance of the drug-delivery system must be evaluated through *in vivo* tests, before application [6, 7].

The abovementioned possibility offers an attractive and efficient solution for the treatment of bone diseases such as tumours, osteoporosis or osteomyelitis [189].

In a previous study growth-factor cement (GFC) was reported [190]. A combination of bone morphogenetic protein-2, transforming growth factor β, platelet-derived growth factor, and a basic fibroblast growth factor were used in a CPC for treatment of peri-implant defects in a dog model [190]. Results indicated a significant effect of GFC on increased bone-to-implant contact and amount of bone per surface area. Further examples of CPCs application as drug-delivery systems are given in the literature [6, 7, 95].

Conclusion and Future Trends

Calcium phosphate cements have unique characteristics for bone substitution compared with other biomaterials. Their excellence relies on good biocompatibility, excellent bioactivity, self-setting characteristics, low setting temperature, adequate stiffness, and easy shaping for any complicated geometry. In addition, CPC resemble the composition of bone mineral and thus induce a biological response similar to the one generated during bone remodelling.

The results obtained so far form a stable base from which minor modifications should provide improved functional bone substitutes. The main shortcomings of these cements are still related to their poor injectability and insufficient mechanical properties, especially for load bearing clinical applications. Therefore, further research and development work should focus on the improvement of the flow properties of cement pastes by decreasing the LPR used in its preparation, while avoiding the filter-pressing effects. The enhancement of the mechanical strength as well as the biological properties of CPCs is also desired. The incorporation of foreign ions into the formulation of CPCs is another important trend to improve their biological behaviour.

REFERENCES

[1] E. Fernández, M. Ginebra, M.G. Boltong, F. Driessens, J. Ginebra, E.A. De Maeyer, R.M. Verbeeck, J.A. Planell, *J. Biomed. Mater. Res.* 32, 367 (1996).
[2] W. E. Brown and L. C. Chow, *J. Dent. Res.* 62, 672 (1983).
[3] K. L. Low, S. Tan, S. Zein, J. Roether, V. Mourin, A. Boccaccini, *J. Biomed. Mater. Res. B: Appl. Biomater.* 94, 273 (2010).
[4] M. Bohner, *Inj.-Int. J. Care Inj.* 31, 37 (2000).
[5] S. V. Dorozhkin, *J. Mater. Sci: Mater. Med.* 43, 3028 (2008).
[6] M. P. Ginebra, T. Traykova, J. A. Planell, *J. Cont.Rel.* 113, 102 (2006).
[7] M. P. Ginebra, T. Traykova, J. A. Planell, *Biomater.* 27, 2171 (2006).
[8] M. Bohner, *J. Mater. Chem.* 17, 3980 (2007).
[9] J. P. Schmitz, J. O. Hollinger, S. B. Milan, *J. Oral Maxillof. Surg.* 57, 1122 (1999).
[10] R. Z. LeGeros, in *Monographs in Oral Science*, Karger, New York (1991).
[11] I. Khairoun, F. C. M. Driessens, M. G. Boltong, J. A. Planell, R. Wenz, *Biomater.* 20, 393 (1999).
[12] P. Frayssinet *et al.*, *J. Mater. Sci. Mater. Med.* 11, 811 (2000).
[13] P. Frayssinet, L. Gineste, P. Conte, J. Fages, N. Rouquet, *Biomater.* 19, 971 (1998).
[14] J. Lemaitre, C. Pittet, D. Brendlen. United Stated Patent No. 10/491,189, 2004.
[15] D. Brendlen, J. Lemaître, paper presented at the 13th GRIBOI Meeting, Baltimore, MD (2003).
[16] M. Bohner, *Eur. Cell.Mater.* 20, 1 (2010).
[17] S. Tanaka, T. Kishi, R. Shimogoryo, S. Matsuya, K. Ishikawa, *Dent. Mater. J.* 22, 301 (2003).
[18] I. C. Chow, *Mater. Res.: Soc. Symp. Proc.* 599, 27 (2000).
[19] C. S. Liu, H. F. Shao, F. Y. Chen, H. Y. Zheng, *Biomater.* 24, 4103 (2003).
[20] K. Ishikawa, S. Takagi, L. C. Chow, K. Suzuki, *J. Biomed. Mater. Res.* 46, 504 (1999).
[21] K. S. TenHuisen, P. W. Brown, *Biomater.* 19, 2209 (1998).
[22] C. H. Tsai, R. M. Lin, C. P. Ju, J. H. C. Lin, *Biomater.* 29, 984 (Mar, 2008).
[23] C. Liu, Y. Huang, J. Chen, *J. Biomed. Mater. Res. B* 69, 73 (2004).
[24] S. Takagi, L. C. Chow, K. Ishikawa, *Biomater.* 19, 1593 (1998).

[25] H. H. K. Xu, L. E. Carey, C. G. Simon, S. Takagi, L. C. Chow, *Dent. Mater.* 23, 433 (2007).
[26] U. Gbureck, S. Dembski, R. Thull, J. E. Barralet, *Biomater.* 26, 3691 (2005).
[27] B. B. Constantz BR, Ison IC, Fulmer MT, Baker J, McKinney LA, Goodman SB, Gunasekaren S, Delaney DC, Ross J, Poser R., *J. Biomed. Mater. Res. B* 43, 451 (1998).
[28] A.Gisep, R.Wieling, M. Bohner, S. Matter, E. Schneider, B. Rahn, *J. Biomed. Mater. Res. A* 66, 532 (2003).
[29] J. L. Lacout, E. Mejdoubi, M. Hamad. , *J. Mater. Sci. – Mater. Med.* 7, 371 (1996).
[30] D. D. Weiss, M. A. Sachs, C. R. Woodard, *J. Long. Term. Eff. Med. Impl.* 13, 41 (2003).
[31] M. Bohner, U. Gbureck, *J. Biomed. Mater. Res. B-Appl. Biomater.* 84B, 375 (2008).
[32] M. P. Hofmann, A. M. Young, U. Gbureck, S. N. Nazhat, J. E. Barralet, *J. Mater. Chem.* 16, 3199 (2006).
[33] F. T. Marino, J. Torres, M. Hamdan, C. R. Rodriguez, E. L. Cabarcos, *J. Biomed. Mater. Res. B – Appl. Biomater.* 83, 571 (2007).
[34] B. Constantz, I. Ison, M. Fulmer, *Science* 267, 1796 (1995).
[35] M. Bohner, G. Baroud, *Biomater.* 26, 1553 (2005).
[36] A. C. Tas, F. Aldinger, *J. Mater. Sci.-Mater. Med.* 16, 167 (2005).
[37] X. Yin, M. J. Stott, A. Rubio, *Phys. Rev. B – Condens. Matter. Mater. Phys.* 68, 205 (2003).
[38] H. L. R. Alves, L. A. dos Santos, C. P. Bergmann, *J. Mater. Sci.-Mater. Med.* 19, 2241 (2008).
[39] H. W. Kim, Y. H. Koh, Y. M. Kong, J. G. Kang, H. E. Kim, *J. Mater. Sci.-Mater. Med.* 15, 1129 (2004).
[40] P. Boudeville, S. Serraj, J.M. Leloup, J. Margerit, B.Pauvert, A. Terol, *J. Mater. Sci.-Mater. Med.* 10, 99 (1999).
[41] M. Ikenaga, P. Hardouin, J. Lemaitre, H. Andrianjatovo, B. Flautre, *J. Biomed. Mater. Res. – Appl. Biomater.* 40, 139 (1998).
[42] Y. Miyamoto *et al.*, *Biomaterials* 16, 855 (1995).
[43] S. Young, M. Holde, S. Gunasekaran, R. Poser, B. Constantz, in *44th Annual Meeting, Orthopaedic Research Society, New Orleans, USA* (1998).
[44] J. Lemaitre, A. Mirtchi, A. Mortier, *Silicates Ind.* 10, 141 (1987).
[45] M. Bohner, P. vanLanduyt, H. P. Merkle, J. Lemaitre, *J. Mater. Sci. – Mater. Med.* 8, 675 (1997).
[46] F. T. Mariño, J. Mastio, C. Rueda, L. Blanco, E.L. Cabarcos, *J. Mater. Sci. – Mater. Med.* 18, 1195 (2007).
[47] K. Ohura, M. Bohner, P. Hardouin, J. Lemaitre, G. Pasquier, B. Flautre, *J. Biomed. Mater. Res. B* 30 193 (1996).
[48] B. Flautre, C. Maynou, J. Lemaitre, P. Van Landuyt, P. Hardouin, *J. Biomed. Mater. Res.B – Appl. Biomater.* 63, 413 (2002).
[49] M. Bohner, F. Theiss, D. Apelt, W. Hirsiger, R. Houriet, G. Rizzoli, E. Gnos, C. Frei, J. Auer, B. von Rechenberg, *Biomaterials* 24, 3463 (2003).
[50] F. Tamimi, B. Kumarasami, C. Doillon, U. Gbureck, D. Nihouannen, E. L. Cabarcos, J. Barralet, *Acta Biomaterialia* 4, 1315 (2008).
[51] S. Kannan, S. Pina, J. M. F. Ferreira, *J. Amer. Ceram. Soc.* 89, 3277 (2006).

[52] S. Kannan, J. M. F. Ferreira, *Chem. Mater.* 18, 198 (2006).
[53] S. Kannan, F. Goetz-Neunhoeffer, J. Neubauer, J. M. F. Ferreira, *J. Amer. Ceram. Soc.* 91, 1 (2008).
[54] S. Kannan, I. A. F. Lemos, J. H. G. Rocha, J. M. F. Ferreira, *J. Sol. St. Chem.* 178, 3190 (2005).
[55] Z. G. Huan, J. Chang, *Acta Biomater.* 5, 1253 (2009).
[56] S. Pina, P. M. C. Torres, J. M. F. Ferreira, *J. Mater. Sci. – Mater. Med.* 21, 431 (2010).
[57] S. Pina, P. M. C. Torres, F. Goetz-Neunhoeffer, J. Neubauer, J. M. F. Ferreira, *Acta Biomater.* 6, 928 (2010).
[58] S. Pina, S.I. Vieira, P. Rego, P.M.C. Torres, O.A.B. da Cruz e Silva, E.F. da Cruz da Silva, J.M.F. Ferreira, *Eur. Cell. Mater.* 20, 162 (2010).
[59] S. Pina, S.I. Vieira, P.M.C. Torres, F.Goetz-Neunhoeffer, J. Neubauer, O.A.B. da Cruz e Silva, E.F. da Cruz da Silva, J.M.F. Ferreira, *J. Biomed. Mater. Res. B* 94, 414 (2010).
[60] U. Klammert, T. Reuther, M. Blank, I. Reske, J.E. Barralet, L. Grover, *Acta Biomater.* 6, 1529 (2010).
[61] K. J. Lilley, U. Gbureck, J. C. Knowles, D. F. Farrar, J. E. Barralet, *J. Mater. Sci.-Mater. Med.* 16, 455 (2005).
[62] S. Pina, S. M. Olhero, S. Gheduzzi, A. W. Miles, J. M. F. Ferreira, *Acta Biomater.* 5, 1233 (2009).
[63] S. Kannan, A. Rebelo, J. M. F. Ferreira, *J. Inorg. Biochem.* 100, 1692 (2006).
[64] S. Kannan, J. H. G. Rocha, J. M. F. Ferreira, *J. Mater. Chem.* 16, 286 (2006).
[65] S. Kannan, J. M. Ventura, J. M. F. Ferreira, *Chem. Mater.* 17, 3065 (2005).
[66] F. Guo, B. Li, *Sheng Wu Yi Xue Gong Cheng Xue Za Zhi* 27, 328 (2010).
[67] R. Z. Legeros, C. Frondoza, S. Lin, R. Rohanizadeh, J. P. Legeros, *J. Dent. Res.* 82, B36 (2003).
[68] R. Z. LeGeros, R. Kijkowska, C. Bautista, M. Retino, J. P. LeGeros, *J. Dent. Res.* 75, 60 (1996).
[69] R. Z. Legeros, M. Retino, D. Mijares, K. Hoshino, S. Lin, J.P. Legeros, *J. Dent. Res.* 77, 916 (1998).
[70] R. Z. LeGeros, A. M. Gatti, R. Kijkowska, D. Q. Mijares, J. P. LeGeros, *Bioceram. 16* 254-2, 127 (2004).
[71] R. LeGeros, S. Lin, J. LeGeros, S. Cazalbou, C. Combes, I. Dupin-Roger, C. Rey, *Calc. Tiss. Inter.* 74, S95 (2004).
[72] T. N. Kim, Q. L. Feng, J. Kim, J. Wu, H. Wang, G. Chen, F. Cui, *J. Mater. Sci. – Mater. Med.* 9, 129 (1998).
[73] J. Li, Y. Li, Y. Zuo, *J. Funct. Mater.* 37, 635 (2006).
[74] M. Chinol, S. Vallabhajosula, S. J. Goldsmith, M. J. Klein, *J. Nucl. Med.* 34, 1536 (1993).
[75] X. Li, Y. Sogo, A. Ito, H. Mutsuzaki, N. Ochiai, T. Kobayashi, *Mater. Sci .Eng.* 29, 969 (2008).
[76] S. Kannan, F. Goetz-Neunhoeffer, J. Neubauer, J. M. F. Ferreira, *J. Amer. Ceram. Soc.* 92, 1592 (2009).
[77] M. H. Alkhraisat, C. Moseke, L. Blanco, J.E. Barralet, E. Lopez-Cabarcos, U. Gbureck, *Biomater.* 29, 4691 (2008).
[78] A. Ito, K. Ojima, H. Naito, N. Ichinose, T. Tateishi, *J. Biomed. Mater. Res.* 50, 178 (2000).

[79] E.C. Bingham, *U.S. Bureau Stand. Bull.* 13, 309 (1916).
[80] S. M. Olhero, J. M. F. Ferreira, *Powd. Techn.* 139, 69 (2004).
[81] J. E. J. Bujake, *J. Pharm. Sci.* 54, 1599 (1965).
[82] R. R. Rao, T. S. Kannan, *J. Amer. Ceram. Soc.* 84, 1710 (2001).
[83] J. M. Tian, Y. Zhang, X. M. Guo, L. M. Dong, *Ceram. Int.* 28, 299 (2002).
[84] H. H. Xu, D. S. Keller, Montreal, QC, Canada, June 16-19 2003.
[85] S. Pina, J. Ferreira, *Materials* 3, 519 (2010).
[86] J. Friberg, E. Fernandez, S. Sarda, M. Nilson, M.P. Ginebra, S. Martinez, *Key. Eng. Mater.* 192-195, 777 (2001).
[87] G. Baroud, E. Cayer, M. Bohner, *Acta Biomater.* 1, 357 (2005).
[88] M. P. Ginebra, A. Rilliard, E. Fernandez, C. Elvira, J. San Roman, J. Planell, *J. Biomed. Mater. Res.* 57, 113 (2001).
[89] I. Khairoun, M. G. Boltong, F. C. M. Driessens, J. A. Planell, *J. Mater. Sci.-Mater. Med.* 9, 425 (1998).
[90] L. Leroux, Z. Hatim, M. Freche, J. L. Lacout, *Bone* 25, 31S (1999).
[91] U. Gbureck, J. E. Barralet, K. Spatz, L. M. Grover, R. Thull, *Biomater.* 25, 2187 (2004).
[92] M. P. Ginebra, E. Fernandez, E. A. De Maeyer, R.M. Verbeeck, M.G. Boltong, J. Ginebra, *J. Dent. Res.* 76, 905 (1997).
[93] U. Gbureck, O. Grolms, J. E. Barralet, L. M. Grover, R. Thull, *Biomater.* 24, 4123 (2003).
[94] M. P. Ginebra, E. Férnandez, M.G. Boltong, O. Bermudez, J.A. Planell, F. Driessens, *Clin. Mater.* 17, 99 (1994).
[95] F. C. M. Driessens, J. A. Planell, M. G. Boltong, I. Khairoun, M. P. Ginebra, *Proc. Inst. Mech. Eng. Part H – J. Eng. Med.* 212, 427 (1998).
[96] S. Sarda, E. Fernandez, J. Llorens, S. Martinez, M. Nilsson, J.A. Planell, *J. Mater. Sci.-Mater. Med.* 12, 905 (2001).
[97] M. Bohner, A. K. Malsy, C. L. Camire, U. Gbureck, *Acta Biomater.* 2, 343 (2006).
[98] M. P. Hofmann, S. N. Nazhat, U. Gbureck, J. E. Barralet, *J. Biomed. Mater. Res. B-Appl. Biomater.* 79, 360 (2006).
[99] J. Carlson, M. Nilsson, E. Fernandez, J. A. Planell, *Biomater.* 24, 71 (2003).
[100] M. Nilsson, Lund University, Faculty of Medicine, Lund, Sweden (2003).
[101] C. Liu, Y. Huang, H. Zheng, *J. Amer. Ceram. Soc.* 82, 1052 (1999).
[102] L. M. Grover, U. Gbureck, A. M. Young, A. J. Wright, J. E. Barralet, *J. Mater. Chem.* 15, 4955 (2005).
[103] M. H. Alkhraisat, F. T. Marino, J. R. Retama, L. B. Jerez, E. Lopez-Cabarcos, *J. Biomed. Mater. Res. A* 84, 710 (Mar 1, 2008).
[104] J. L. Giocondi, B. S. El-Dasher, G. H. Nancollas, C. A. Orme, *Philos. Trans. R. Soc. A Math. Phys. Eng. Sci.* 368, 1937 (2010).
[105] M. Bohner, R. Luginbühl, C. Reber, N. Doebelin, G. Baroud, E. Conforto, *Acta Biomater.* 5, 3524 (2009).
[106] T. J. Brunner, M. Bohner, C. Dora, C. Gerber, W. J. Stark, *J. Biomed. Mater. Res. B-Appl. Biomater.* 83, 400 (2007).
[107] E. F. Burguera, H. H. K. Xu, L. M. Sun, *J. Biomed. Mater. Res. B-Appl. Biomater.* 84, 493 (2008).

[108] E. F. Burguera, H. H. K. Xu, M. D. Weir, *J. Biomed. Mater. Res. B-Appl. Biomater.* 77, 126 (2006).
[109] M. Habib, G. Baroud, F. Gitzhofer, M. Bohner, *Acta Biomater.* 4, 1465 (2008).
[110] M. Bohner, U. Gbureck, J. E. Barralet, *Biomater.* 26, 6423 (2005).
[111] S. del Valle, N. Miño, F. Muñoz, A. González, J.A. Planell, M.P. Ginebra, *J. Mater. Sci. – Mater. Med.* 18, 353 (2007).
[112] J. E. Barralet, L. M. Grover, U. Gbureck, *Biomater.* 25, 2197 (2004).
[113] K. Ishikawa and K. Asoaka, *J. Biomed. Mater. Res.* 29, 1537 (1995).
[114] S. Sarda, E. Férnandez, M. Nilson, M. Balcells, J. A. Planell, *J. Biomed. Mater. Res.* 61, 652 (2002).
[115] L. C. Chow, S. Hirayama, S. Takagi, E. Parry, *J. Biomed. Mater. Res.* 53, 511 (2000).
[116] F. C. M. Driessens, M. G. Boltong, E. A. P. De Maeyer, R. M. H. Verbeeck, R. Wenz, *J. Mater. Sci.-Mater. Med.* 11, 453 (2000).
[117] K. Miyazaki, T. Horibe, J. M. Anntonucci, S. Takagi, C. L. C., *Dent. Mater.* 9, 46 (1993).
[118] K. Miyazaki, T. Horibe, J. M. Anntonucci, S. Takagi, L. C. Chow, *Dent. Mater.* 9, 41 (1993).
[119] L. Santos, L.C. Oliveira, E.C. Rigo, R.G. Carrodeguas, A.O. Boschi, A.C. De Arruda, *Bone* 25, 99 (1999).
[120] R. Mickiewicz, A. Mayes, D. Knaack, *J. Biomed. Mater. Res.* 61, 581 (2002).
[121] M. Bohner, H. P. Merkle, P. V. Landuyt, G. Trophardy, J. Lemaitre, *J. Mater. Sci. – Mater. Med.* 11, 111 (1999).
[122] E. Barralet, T. Gaunt, A. J. Wright, I. R. Gibson, J. C. Knowles, *J. Biomed. Mater. Res. B* 63, 1 (2002).
[123] Y. Zhang, H. Xu, S. Takagi, L. Chow, *J. Mater. Sci.: Mater. Med.* 17, 437 (2006).
[124] M. Bohner, *Eur. Spine J.* 10 114 (2001).
[125] Y. Fukase, E. D. Eanes, S. Takagi, L. C. Chow, W. E. Brown, *J. Dent. Res.* 69, 1852 (1990).
[126] K. S. TenHuisen, P. W. Brown, *J. Biomed. Mater. Res.* 36, 306 (1997).
[127] M. H. Alkhralsat, F. T. Marino, C. R. Rodriguez, L. B. Jerez, E. L. Cabarcos, *Acta Biomater.* 4, 664 (2008).
[128] T. Kokubo, *Bone-bounding Biomaterials*. Healthcare Communications, Leiderdrop, The Netherlands (1992).
[129] A. C. Tas, *Biomater.* 21, 1429 (2000).
[130] L. L. Hench, *J. Amer. Ceram. Soc.* 81, 1705 (1998).
[131] A. Ueno, Y. Miwa, K. Miyoshi, T. Horiguchi, H. Inoue, I. Ruspita, K. Abe, K. Yamashita, E. Hayashi, T. Noma, *J. Cell. Phys.* 209, 322 (2006).
[132] T. Kokubo and H. Takadama, *Biomater.* 27, 2907 (2006).
[133] C. Ohtsuki, Y. Aoki, T. Kokubo, Y. Fujita, S. Kotani, T. Yamamuro, *Transactions of the 11th annual meeting of Jap Soc Biomater* (1989).
[134] C. Ohtsuki, T. Kokubo, M. Neo, S. Kotani, T. Yamamuro, T. Nakamura, *Phos. Res. Bull.* 1, 191 (1991).
[135] Y. Fujita, T. Yamamuro, T. Nakamura, S. Kotani, *J. Biomed. Mater. Res.* 25, 991 (1991).
[136] S. Kotani, Y. Fujita, T. Kitsugi, T. Nakamura, T. Yamamuro, *J. Biomed. Mater. Res.* 25, 1303 (1991).

[137] M. Neo, T. Nakamura, C. Ohtsuki, T. Kokubo, T. Yamamuro, *J. Biomed. Mater. Res.* 27, 999 (1993).

[138] O. Bermudez, M. G. Boltong, F. C. M. Driessens, J. A. Planell, *J. Mater. Sci. – Mater. Med.* 5, 160 (1994).

[139] D. Apelt, F. Theiss, A.O. El-Warrak, K. Zlinszky, R. Bettschart-Wolfisberger, M. Bohner, S. Matter, J.A. Auer, B. von Rechenberg, *Biomater.* 25, 1439 (2004).

[140] E. M. Ooms, J. G. Wolke, M. T. van de Heuvel, B. Jeschke, J. A. Jansen, *Biomater.* 24, 989 (2003).

[141] R. Z. LeGeros and J. P. LeGeros, *Bioceram.* 240-2, 3 (2003).

[142] H. Schliephake, R. Gruber, M. Dard, R. Wenz, S. Scholz, *J. Biomed. Mater. Res. A* 69, 382 (2004).

[143] H. P. Yuan, Y. B. Li, J. D. de Bruijn, K. de Groot, X. D. Zhang, *Biomater.* 21, 1283 (2000).

[144] A. J. Ambard, L. Mueninghoff, *J. Prosthodont.* 15, 321 (2006).

[145] E. Arthur, in *Primer on the Metabolic Bone Diseases and Disorders of Mineral Metabolism,* J. Murray, Ed., Lippincott-Raven, New York (1993).

[146] J. Eisman, in *Rheumatology,* J. Kippel, P. Dieppe, Eds., Mosby, London (1998).

[147] S. Wallach, *Magnes. Trace Elem.* 10, 281 (1991-1992).

[148] P. Marie, P. Ammann, G. Boivin, C. Rey, *Calcif. Tissue Int.* 69, 121 (2001).

[149] M. Otsuka, Y. Ohshita, S. Marunaka, Y. Matsuda, A. Ito, N. Ichinose, K. Otsuka, W. Higuchi, *J. Biomed. Mater. Res. A* 69, 552 (2004).

[150] B. Vallee, K. Falchuk, *Physiol. Rev.* 72, 79 (1993).

[151] P. Murphy, I. Wadiwala, D. Sharland, G. Rai, *J. Amer. Geriatr. Soc.* 33, 847 (1985).

[152] B. O'Dell, *Nutr. Rev.* 50, 48 (1992).

[153] K. K. Johal, R. G. Hill, I. M. Brook, *J. Mater. Sci. – Mater. Med.* 13, 543 (2002).

[154] G. Lewis, *J. Biomed. Mater. Res. B-Appl. Biomater.* 76, 456 (2006).

[155] X. Wang, J. D. Ye, Y. Wang, *Acta Biomater.* 3, 757 (2007).

[156] M. Watanabe, M. Tanaka, M. Sakurai, M. Maeda, *J. Eur. Ceram. Soc.* 26, 549 (2006).

[157] L. E. Carey, H. H. K. Xu, C. G. Simon, S. Takagi, L. C. Chow, *Biomater.* 26, 5002 (2005).

[158] H. H. Xu, S. Takagi, J. B. Quinn, I. C. Chow, *J. Biomed. Mater. Res. A* 68, 725 (2004).

[159] M. Bohner, H. P. Merkle, J. Lemaitre, *J. Mater. Sci. – Mater. Med.* 11, 155 (2000).

[160] H. H. Xu, J. B. Quinn, S. Takagi, L. C. Chow, *J. Dent. Res.* 81, 219 (2002).

[161] W. Mors, E. Kaminski, *Arch. Oral. Biol.* 20, 365 (1975).

[162] A. Di Martino, M. Sittinger, M. V. Risbud, *Biomater.* 26, 5983 (2005).

[163] R. A. Muzzarelli *et al.*, *Biomater.* 14, 39 (1993).

[164] D. X. Li, H.S. Fan, X.D. Zhu, Y.F. Tan, W.Q. Xiao, J. Lu, Y. Xiao, J. Chen, X. Zhang, *J. Mater. Sci.- Mater. Med.* 18, 2225 (2007).

[165] X. P. Wang, J. D. Ye, H. Wang, *J. Biomed. Mater. Res. B-Appl. Biomater.* 78, 259 (2006).

[166] K.-D. Kuhn, *Bone cements: Up-to-date comparison of physical and chemical properties of commercial materials*, Springer-Verlag, Berlin (2000).

[167] M. Otsuka, Y. Matsuda, D. Yu, J. Wong, J. Fox, W. Higuchi, *Chem. Phar. Bull.* 38, 3500 (1990).

[168] M. Takechi, Y. Miyamoto, K. Ishikawa, M. Nagayama, M. Kon, K. Asaoka, K. Suzuki, *J. Biomed. Mater. Res.* 39, 308 (1998).

[169] L. Chow, S. Takagi, P. Costantino, C. Friedman, *Matl. Res. Symp. Proc.* 179, 3 (1991).
[170] J. Lemaître, *Inn. Tech. Biol. Med.* 16, 109 (1995).
[171] L. Lidgren, H. Drar, J. Möller, *Acta Orthop. Scand.* 55, 536 (1984).
[172] L. Lidgren, B. Bodelind, J. Möller, *Acta Orthop. Scand.* 58, 27 (1987).
[173] C. D. Friedman, P. D. Costantino, S. Takagi, L. C. Chow, *J. Biomed. Mater. Res.* 43, 428 (1998).
[174] T. F. Higgins, S. D. Dodds, S. W. Wolfe, *J.Bone Joint Surg.* 84A, 1579 (2002).
[175] D. B. Thordarson, Y. Miyamoto, K. Ishikawa, M. Nagayama, M. Kon, K. Asaoka, K. Suzuki, *J. Biomech.* 1, 3 (1998).
[176] T. M. Turner, *Spine J.* 8, 482 (2008).
[177] C. Stankewich, M. F. Swiontkowski, A. F. Tencer , D. N. Yetkinler, R. D. Poser, *J. Orthop. Res.* 14, 786 (1996).
[178] S. B. Goodman, T. Bauer, D. Carter, P. Casteleyn, S. Goldstein, R. Kyle, S. Larsson, C. Stankewich, *Clin. Orthop.* 348, 42 (1998).
[179] P. F. Heini, U. Berlemann, M. Kaufmann, K. Lippuner, C. Fankhauser, P. van Landuy, *Eur. Spine J.* 10, 164 (2001).
[180] D. B. Kamerer, C. D. Friedman, P. D. Costantino, C. H. Snyderman, B. F. Hirsch, *Amer. J. Otol.* 15, 47 (1994).
[181] M. Sliindo, P. Costantino, C. Friedman, L. Chow, *Arch. Otolaryngol. Head. Neck. Surg.* 119, 185 (1993).
[182] J. F. Kveton, C. D. Friedman, P. D. Costantino, *Amer. J. Otol.* 16, 465 (1995).
[183] S. A. Gehrke, G. M. Famà, *Implants* 11, 4 (2010).
[184] S. A. Gehrke, *Implants* 10, 5 (2009).
[185] J. Noetzel, K. Özer, B. Reisshauer, A. Anil, R. Rössler, K. Neumann, A. Kielbassa, *Clin. Oral Invest.* 10, 77 (2006).
[186] W. Zhang, X. F. Walboomers, J. A. Jansen, *J. Biomed. Mater. Res. A* 85, 439 (2008).
[187] L. Comuzzi, E. Ooms, J. A. Jansen, *Clin. Oral Implants Res.* 13, 304 (2002).
[188] H. M. Chaung, C. Hong, C. Chiang, S. Lin, Y. Kuo, W. Lan, C. Hsieh, *J. Formos. Med. Assoc.* 95, 545 (1996).
[189] A. El-Ghannam, K. Ahmed, M. Omran, *J. Biomed. Mater. Res. B: Appl. Biomater.* 73, 277 (2005).
[190] S. J. Meraw, C. M. Reeve, C. M. Lohse, T. M. Sioussat, *J. Periodont.* 71, 8 (2000).

In: Calcium Phosphate
Editor: Robert B. Heimann

ISBN 978-1-62257-299-1

Chapter 12

TETRACALCIUM PHOSPHATE-BASED DENTAL CEMENTS

Claus Moseke and Uwe Gbureck*
Department for Functional Materials in Medicine and Dentistry,
University of Würzburg, Würzburg, Germany

ABSTRACT

Monoclinic tetracalcium phosphate (TTCP, $Ca_4(PO_4)_2O$), also known by the mineral name hilgenstockite, is formed in the ($CaO–P_2O_5$) system at temperatures above 1300°C. TTCP is the only calcium phosphate with a Ca/P ratio larger than hydroxyapatite (HA). In aqueous solutions a slow hydrolysis of TTCP occurs due to the lower solubility of HA. Because of the higher Ca/P ratio of TTCP compared to HA, complete hydration is expected to lead to the formation of calcium hydroxide and a strong alkaline pH-value > 12. The low reactivity of TTCP in water is a result of the formation of a thin HA-layer around the TTCP particles, which acts as a diffusion barrier and inhibits further hydrolysis.

The reactivity of TTCP can be strongly increased, when the material is mechanically activated by extensive ball-milling. This process leads to a partial amorphisation of TTCP which is linked to an increase of both solubility and formation enthalpy of the hydrolysis product.

TTCP is widely used in either single- or multicomponent, mainly hydroxyapatite forming cements or in conjunction with functional polymers to obtain organic-inorganic composites. The latter are formed in an acid-base reaction from alkaline TTCP with carboxylic acid or organic phosphate moieties of polymers, such as polyacrylic acid, poly(methyl vinyl ether-maleic acid), poly[bis(carboxylatophenoxy) phosphazene] or poly(vinyl phosphonate). The setting reaction of these cements occurs by deprotonation of the organic acid followed by the formation of Ca^{2+}-acid chelates and the precipitation of nanocrystalline hydroxyapatite.

The high alkaline pH value of the setting reaction of TTCP cements provides a strong antimicrobial potency which is of interest for dental applications, e.g. as pulp capping agents or root canal fillers. Recent studies have demonstrated this higher potency

* Email: uwe.gbureck@fmz.uni-wuerzburg.de

compared to commercial $Ca(OH)_2$/salicylate cements against various oral microbial strains, e.g. *streptococcus salvarius*, *staphylococcus epidermidis* or a clinically isolated plaque mixture.

1. INTRODUCTION

Tetracalcium phosphate ($Ca_4(PO_4)_2O$, TTCP, mineralogical name: hilgenstockite) was first described in 1883 by *Hilgenstock* [1] as tabular crystals found as a component of basic and phosphate-rich slags obtained during steel production by the Thomas process. The slag was extensively used as fertilizer in the 19th and 20th centuries due to the higher solubility of their compounds TTCP and $Ca_3(PO_4)_2 \cdot Ca_2SiO_4$ compared with natural phosphate sources such as apatite minerals [2,3].

TTCP is the only calcium phosphate phase with a Ca/P ratio greater than stoichiometric hydroxyapatite and it is formed in the (CaO-P_2O_5) system at temperatures above 1300°C. The compound is metastable, hence the synthesis of phase-pure TTCP requires either rapid quenching or the absence of moisture to prevent decomposition into hydroxyapatite and lime in the temperature range of 1000-1200°C [4,5] according to

$$3\ Ca_4(PO_4)O + H_2O \rightarrow 2\ Ca_5(PO_4)_3OH + 2\ CaO \qquad (1)$$

TTCP crystallizes in the monoclinic crystal system with space group $P2_1$ with the lattice parameters a = 7.023(1) Å, b = 11.986(4) Å, c = 9.473(2) Å and ß = 90.90(1)° at 25°C [6]. One unit cell contains four formula units, and the calculated density of 3.051 Mg/m^3 is well correlated with the observed density of 3.06 Mg/m^3 [7]. As shown by measurements of lattice parameters and x-ray intensities, the structure of TTCP is pseudo-orthorhombic [5] and has close similarities to the unit cell of hydroxyapatite, which are repeated every 2/3 of TTCP d_{010} along the b-axis.

The same factor connects the unit cell volumes, namely 530 $Å^3$ for HA and 795 $Å^3$ for TTCP. This structural similarity may explain why solid state decomposition of the TTCP phase to HA and CaO occurs practically without enthalpy changes [8]. It also suggests an epitactic relationship between the two calcium phosphates, which enables direct transformation of TTCP to HA crystals [6].

In contrast to other calcium orthophosphates such as hydroxyapatite (HA, $Ca_5(PO_4)_3OH$) or monetite (DCPA, $CaHPO_4$), which are widely used as food ingredients, in tooth pastes, pharmaceutical applications or chromatography [9], phase-pure TTCP has found only an application as ceramic biomaterial. It often appears as an (unwanted) by-product in plasma-sprayed hydroxyapatite coatings, formed by thermal decomposition of HA to the high-temperature phases ß-tricalcium phosphate (ß-TCP, $Ca_3(PO_4)_2$), TTCP, and CaO [10,11].

In contrast, TTCP in particulate form is an important ingredient in self-setting bone cements, forming hydroxyapatite in a continuous dissolution-precipitation process. This chapter reviews the preparation routes of phase-pure TTCP, the application of TTCP in dental cements, and the biological *in vitro* and *in vivo* compatibility of TTCP-based biocements.

2. SYNTHESIS OF TTCP

TTCP cannot be synthesized in aqueous environment, because of the oxygen atom in its formula [12]. Precipitates formed in basic aqueous solutions with a Ca/P ratio of 2 will incorporate hydroxyl ions and therefore always consist of carbonate- and/or hydroxyl-ion-containing apatitic phases. Hence, common methods for the synthesis of TTCP are limited to solid-state reactions at high temperatures [13-15]. These methods are usually based on mixtures of calcium carbonate ($CaCO_3$) and dicalcium phosphate anhydrate ($CaHPO_4$) to obtain the required Ca/P ratio of 2, which are then heated to 1450 –1500°C for 6-12 h. The reaction is given by the following equation:

$$2CaHPO_4 + 2CaCO_3 \rightarrow Ca_4(PO_4)_2O + 2CO_2 + H_2O \qquad (2)$$

After heating the mixtures must be rapidly quenched to room temperature to avoid formation of undesired secondary phases such as hydroxyapatite (HA), calcium oxide (CaO), $CaCO_3$, and ß-tricalcium phosphate (ß-$Ca_3(PO_4)_2$, ß-TCP). Sargin et al. reported several attempts to lower the required process temperature for TTCP synthesis by using different combinations of various starting products [16]. By varying the process temperatures they showed that in all reactions HA or TCP appeared as intermediate products at lower temperatures, while TTCP was the main product at higher temperature. A very pure and crystalline TTCP product was obtained by the following reaction at 1350°C:

$$2\ NH_4H_2PO_4 + 4\ CaCO_3 \rightarrow Ca_4(PO_4)_2O + 4\ CO_2 + 2\ NH_3 + 3\ H_2O \qquad (3)$$

Furthermore, this reaction enabled the formation of TTCP even at 1200°C, a much lower temperature than was reported previously. However, its purity was lower than in the product obtained at 1350°C.

Jalota et al. [11] investigated a method to further reduce the temperature required for the preparation of phase-pure TTCP. A series of grinding and heating procedures transformed a stoichiometric mixture of $NH_4H_2PO_4$ and $Ca(CH_3COO)_2{\cdot}H_2O$ powders to TTCP, requiring a maximum heating temperature of 1230°C. However, the method was only suitable to synthesize HA-seeded TTCP powders, which are very sensitive and will transform into HA when mechanochemically activated [8]. On the other hand, mechanochemical activation can be beneficial when applied to the starting substances for TTCP synthesis. Romeo and Fanovich [17] studied the solid-state reaction of $CaCO_3$ and $(NH_4)_2HPO_4$ powders mixed to obtain a Ca/P ratio of 2, and ball-milled prior to thermal treatment. The smaller crystallite size of the milled powders increased the reactivity and significantly improved the purity of the end product as confirmed by x-ray diffraction (XRD) and Fourier transform infrared spectroscopy (FTIR).

It has been proposed by Ciesla *et al.* [18-20] that an intermediate phase of apatite with a Ca/P ratio exceeding 1.67 is formed following decomposition of TTCP at temperatures below 660°C. At low temperature this apatitic phase may be phosphate-deficient as a result of incomplete decomposition.

Thermal analysis using differential scanning calorimetry (DSC) showed an endothermic peak at 750°C in the pattern of mechanically activated TTCP, which may be attributed to the combined transformations of this non-stoichiometric apatite phase to stoichiometric HA and to crystalline CaO [21].

In order to further confirm that mechanical activation was responsible for the decomposition of TTCP at temperatures as low as 200°C, DSC measurements were performed on mechanically activated TTCP that had been hydrolysed to CaO and HA at room temperature in water. The endothermic peaks thought to be associated with decomposition of TTCP and non-stoichiometric apatite were absent, yet similar thermal behaviour was observed at ~1300°C.

The reduction in the decomposition temperatures of mechanically activated TTCP compared with crystalline TTCP from ~600°C to ~200°C is in agreement with previous investigations on the thermal behaviour of mechanically activated ilmenite and pyrite [22], where it was also demonstrated that mechanical activation led to a decrease in decomposition temperature. The authors proposed that an increase in lattice distortions caused by mechanical activation enabled decomposition to occur at significantly lower temperatures.

Lin *et al.* proposed, on the basis of XRD measurements following thermal decomposition of HA in air, that decomposition proceeded at 1500°C according to equation 4, forming ß-tricalcium phosphate (ß-TCP) and TTCP [23]. The same authors also discovered that TTCP reconverts solely to HA yielding calcium hydroxide at a temperature of 500°C in the presence of moisture. The decomposition of TTCP to an apatitic phase and CaO, respectively appears to be a thermodynamically favoured process.

$$Ca_{10}(PO_4)_6(OH)_2 \rightarrow Ca_4(PO_4)_2{}^{\circ} + 2\ Ca_3(PO_4)_2 + H_2O \qquad (4)$$

3. Self-Setting TTCP Dental Cements

TTCP is widely used as a dental cement reactant in either single- or multicomponent, mainly hydroxyapatite-forming cements or in conjunction with polymeric reactants which react with TTCP in an acid-base reaction to form organic-inorganic composites. Among all calcium phosphates, TTCP has the highest solubility [5] at a pH value < 4. In aqueous solutions slow hydrolysis of TTCP can occur according to equation 4 and due to the lower solubility of HA. Because of the higher Ca/P ratio of TTCP compared to HA, complete hydration is expected to form calcium hydroxide and a strong alkaline pH-value > 12:

$$3Ca_4(PO_4)_2O + 3H_2O \longrightarrow Ca_{10}(PO_4)_6(OH)_2 + 2Ca(OH)_2 \qquad (5)$$

Nevertheless, TTCP remains unreacted in water at room temperature up to four weeks, but at 80°C complete hydrolysis to HA has been reported to occur within 80 h [24]. Another study by Martin and Brown [25] showed that the hydration of phase-pure TTCP in aqueous slurries at temperatures between 25°C and 70°C resulted in a strong initial heat evolution. X-ray diffraction patterns of these water-stored materials were similar to non-treated TTCP, indicating that there was practically no hydrolysis to HA. This effect was probably caused by the formation of a thin HA-layer around the TTCP particles, which acted as a diffusion barrier and inhibited further hydrolysis [21, 26, 27].

3.1. Inorganic TTCP Cements

As mentioned above, hydrolysis of (crystalline) TTCP in water to form hydroxyapatite and calcium hydroxide according to equation 5 is slowed down at ambient temperatures due to the formation of a thin hydroxyapatite layer around the particles. However, single component TTCP cements can be obtained after prolonged ball-milling which leads to partial amorphisation of the material [8, 28, 29]. Mechanical activation, i.e. the formation of amorphous material by introducing mechanical energy, is a well established and investigated method for metals and ceramics [30] to increase reaction rates and lower reaction temperatures of materials. These parameter changes increase strongly the kinetic solubility (rate of dissolution) as well as the thermodynamic solubility product, such that setting is enabled even at room or body temperature. Mechanically activated single component TTCP (MA-TTCP) cements set to form an insoluble matrix of HA and slowly soluble $Ca(OH)_2$ (solubility approx. 1.7 g/l [31]) as a by-product. Hardening times of MA-TTCP cements decrease with milling time from about 5-6 h (1 h milled; 5.3 μm medium particle size) to about 2 h (24 h milled; 1.1 μm medium particle size). Phase composition of MA-TTCP cements after setting for 24 h with water was determined by Rietveld refinement of the obtained diffraction patterns. The only detectable phases were TTCP and HA, while the presence of $Ca(OH)_2$ could not be confirmed by X-ray diffraction analyses. The non-ground TTCP raw material converted to 0.29 % HA with water, while the quantity of precipitated HA dramatically increased with milling time and reached a maximum for the 24 h milled sample with nearly 58 wt% [8]. Using a 2.5 % Na_2HPO_4 solution as cement binder liquid further increased the conversion rate to a maximum of 88 % hydroxyapatite after 24 h setting time [28]. Cement made with MA-TTCP ground for one hour in ethanol (powder to liquid ratio P:L = 2.67), had a mean compressive strength (CS) of ~5 MPa (24 h after setting), but after 24 hours grinding (P:L = 2), strength was ~10 MPa, despite a higher porosity as the result of the lower P:L mixing ratio. An accelerated *in vitro* test comparing the pH of water after prolonged boiling of nanocrystalline TTCP cements and a calcium hydroxide/salicylate composite material showed that hydroxyl ion release from the cement was comparable to that from a calcium salicylate material [28]. Due to their high alkaline pH value during setting and leaching, MA-TTCP cements were proposed to be suitable for dental applications, e.g. as pulp capping agents or root canal fillers. The effect of heat treatment of single-component TTCP on the setting properties when mixed with an $(NH_4)_2HPO_4$ solution at a powder to liquid ratio of 3.3 g/ml was studied by Lin et al. [32] who showed that temperatures in the range of 140-400°C (maintained for 30-120 min in air) prolonged the setting time of the cement from 9.5 min (non-heat treated) to 23 min (300°C), while the mechanical performance was maintained for temperatures up to 250°C. A single component TTCP based on the reaction of TTCP (particle size 5-20 μm) and a solution saturated with respect to dicalcium phosphate dihydrate ($[Ca^{2+}]$=0.72 mol/l, $[HPO_4^{2-}]$=1.88 mol/l, pH = 2.06) was described by Sugawara et al. [33]. This cement showed comparatively low setting times (10.3 ± 0.6 min) which was likely a result of a rapid formation of DCPD when both cement solid and liquid were mixed, followed by subsequent HA formation according to equation 5.

In another study by Wei et al. [34] fluoride-substituted dental cement for direct filling and repair of carious cavities was formed by the reaction of TTCP with sodium fluoride and phosphoric acid. The results demonstrated that the setting reaction of this cement leads to the formation of fluorapatite with a Ca/P ratio of 1.66 and a fluoride amount of 3 wt% after setting for 48 hours. The fluoride substitution decreased the cement solubility in Tris-HCl solution and the cement was strongly bound to the enamel surface due to a chemical reaction between the cement reactants and the apatite of the enamel. The extracts of the cement caused no cytotoxicity on L929 cells thus demonstrating good cytocompatibility of the material and high potential for the reconstitution of carious lesion of dental enamel. A higher fluoride content with a Ca/F ratio of 5 was obtained by the same group by altering the cement composition using a mixture of TTCP, DCPA and ammonium fluoride as cement ingredients [35].

3.2. Polymeric Cement Composites

Polymer-modified TTCP-based cements can be classified into three major groups. The first type is based on the modification with (water-soluble) non-reactive polymers such as hydroxypropylmethylcellulose (HPMC) [36] or chitosan [37]. Alternatively, TTCP particles could be embedded in a photocurable polymeric matrix based on typical dental monomers such as ethoxylated bisphenol A dimethacrylate and pyromellitic glycerol dimethacrylate [38]. These resin-based calcium phosphate cements (RCPC) have been reported as remineralizing pulp-capping or lining cement with only minor cytotoxic response of odontoblast and pulp cells,. The mild cytotoxicity was related to residual unpolymerized dimethacrylate monomer after curing. Other studies aimed at increasing the local calcium and phosphate ion concentration around the cement filling at a cariogenic pH. This was done by incorporating TTCP particles in a photocurable composite which was additionally modified with either glass particles [39] or whiskers [40] for mechanical reinforcement. Ion release and mechanical properties were measured after immersion in solutions with pH of 7, 5.5, and 4. The TTCP/glass composite had flexural strengths in the range of 99 MPa (glass particle reinforced) to 116 MPa (whisker reinforced) which was 2–3 times higher than those of a resin-modified glass ionomer control. The released Ca ion concentration after 28 days was 4.61 ± 0.18 mmol/L at a pH of 4, much higher than 1.14 ± 0.07 at a pH of 5.5, and 0.27 ± 0.01 at a pH of 7 ($p < 0.05$). Similarly, phosphate ion release was also dramatically increased at cariogenic, namely acidic pH.

The "smart" behaviour of the composites whereas calcium and phosphate ion release is increased when the pH drops from neutral to cariogenic pH is beneficial to inhibit tooth caries. In combination with the good mechanical properties, this material may have a great potential as a new filling composite for dental applications.

In a third type of cements, the high reactivity and strongly basic character of TTCP compared with other calcium phosphates is used to react it with carboxylic acid or organic phosphate moieties of polymer such as polyacrylic acid [41-43], poly(methyl vinyl ether-maleic acid) [15,44], poly[bis(carboxylatophenoxy)phosphazene] [45], and poly(vinyl phosphonate) [46-48] (Figure 1). The setting reactions of these cements occur by deprotonating the organic acid followed by formation of Ca^{2+}-acid chelates and precipitation of nanocrystalline hydroxyapatite. Either the polymers can be used as an aqueous solution (at room temperature) or the dry TTCP/polymer mixture can be processed at elevated temperature/pressure via a solid-state reaction [46-48]. Calcium chelates could be formed either by intra- or inter-chained bonding [46]; the latter is thought to be the reason for an enhancement of the mechanical properties compared to non-polymer-modified cements.

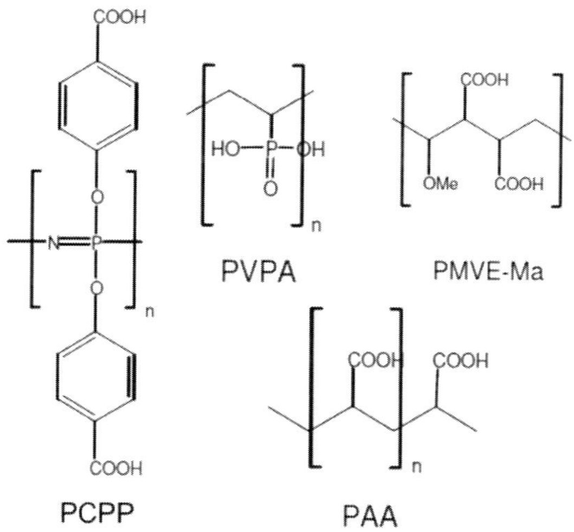

Figure 1. Chemical structures of polymers used for the formation of modified TTCP cements/ceramics; PAA: polyacrylic acid; PVPA: polyvinyl phosphonate; PMVE-Ma: poly(methyl vinyl ether-maleic acid); PCPP: poly[bis(carboxylatophenoxy)phosphazene].

The high reactivity of TTCP with organic acids was also used for manufacturing of calcium phosphate ceramics by 3D-powder printing. Khalyfa et al. [49] produced implant structures by printing TTCP/ß-TCP powder with 25-30 wt% citric acid solution as liquid binder. As a result of partial hydrolysis of TTCP in the aqueous solution and thermal decomposition during thermal post-treatment they obtained ceramics composed of a mixture of hydroxyapatite, ß-TCP, low-crystalline calcium hydroxide and residual amounts of TTCP after sintering at 1200°C. This process was further modified by Vorndran et al. [50] to manufacture phase-pure ß-TCP ceramics using a mixture of TTCP, ß-TCP and DCPA (Ca/P ratio = 1.5) for printing. When TTCP is used in combination with diluted phosphoric acid for printing, the instantly formed brushite phase can be hydrothermally converted with residual TTCP to form a matrix of nanocrystalline hydroxyapatite [51].

4. *In Vitro* and *In Vivo* Compatibility of TTCP-Based Ceramics and Cements

The biocompatibility of tetracalcium phosphate-based ceramics and cements was investigated in various *in vitro* [52-54] and *in vivo* studies [55-57]. Proposed application sites were either bone replacement [52-57], the use in dentistry as pulp-capping agent or root canal filler [58-61] or the applications of TTCP as remineralization filler in chewing gums [62].

Doi et al. [54] in a study on the resorption of various (phase-pure) calcium phosphate ceramics by osteoclasts obtained from long bones of neonatal rabbits found only marginal cell activity on TTCP surfaces compared with HA-based ceramics. This was attributed to changes of calcium and phosphate concentrations in the cell culture medium which did not occur with the immersion of HA. However, similar to ß-TCP, TTCP appears to promote osteogenesis by increasing collagen synthesis and calcification of the extracellular matrix when added in small doses of 1 or 10 μM to a cell culture of MC3T3-E1 cells [53]. Another *in vitro* study by

Yoshimine et al. [52] investigated the interaction of calvarial osteogenic cells with cement composed of TTCP, 40 wt% polyacrylic acid/itaconic acid copolymer, and 10 wt% citric acid. The authors found nodular structures covered with active osteoblastic cells in the periphery of the test samples containing mineralized tissue with features similar to bone. They concluded that the TTCP cement is able to bind directly to newly-formed hard tissue. The same cement was tested *in vivo* in a cranial defect of rats over a period of 30 d [57]. The results demonstrated no foreign-body giant cell reaction; instead, after 30 d the cement was surrounded by newly formed bone. These findings are in accordance with the above mentioned *in vitro* results showing osteoconductive properties of TTCP cement. The histological evaluation of a cement with similar composition exposed to the pulp of rat maxillary incisors over a period of 10 d [58] demonstrated great benefits compared with traditionally used calcium hydroxide cement. While the latter provoked tissue necrosis beneath the cement before new hard tissue was formed, TTCP cement formed a dentin bridge with no evidence of inflammation or necrosis.

A single-component TTCP cement (with $(NH_4)_2HPO_4$ as binder liquid) was tested *in vivo* by Tsai et al. [56] in the cortical and cancellous site of the femur in rabbits over a period of 24 weeks. Histological examination of the retrieved implants showed an excellent cement-bone host bonding with the observation of cement resorption, new blood vessels, osteocytes, osteons and osteoblast-like cells after 4 weeks of implantation. After 24 weeks, only small amounts of remaining cement could be identified and bone remodelling was almost finished at the cortical site. The antimicrobial potency of single component TTCP cements with a comparable composition (Na_2HPO_4 as binder liquid) was proved by the agar diffusion test against various oral microbial strains, e.g. *streptococcus salvarius*, *staphylococcus epidermidis* or a clinically isolated plaque mixture. The results demonstrated a significantly higher antimicrobial potency of the TTCP cements as revealed by inhibition zones of approximately 3-5 mm width, compared with a commercial $Ca(OH)_2$/salicylate cement, which only produced small inhibition zones around the cement specimens of 1.5 mm or less [29].

Conclusion

Tetracalcium phosphate is an important component of low-temperature setting dental biocements. Although it is a metastable compound, various (high temperature) reaction routes are described to obtain it as a phase-pure material in monolithic or particulate form. According to the requirements for certain biomedical implant materials this may also be an interesting field of investigation in the future, particularly regarding the promising results of the studies reviewed herein on the biocompatibility and antimicrobial properties of TTCP cements and ceramics.

References

[1] G. Hilgenstock, *Stahl und Eisen* 498 (1883).
[2] T. Dieckmann, E. Houdremont, *Z Anorg Allg Chem* 120, 129 (1922).

[3] C. Scheibler, *Ber Dtsch Chem Ges* 19, 1883 (1886).
[4] H. Monma, M. Goto, H. Nakajima, H. Hashimoto, *Gypsum and Lime* 202, 17 (1986).
[5] W.E. Brown, E.F. Epstein, *J Res National Bureau Standards – A. Physics and Chemistry* 69A, 547 (1965).
[6] B. Dickens, W.E. Brown, G.J. Kruger, J.M. Stewart, *Acta Cryst* B29, 2046 (1973).
[7] H. Bucking, G. Linck, *Stahl und Eisen* 7, 245 (1887).
[8] U. Gbureck, J.E. Barralet, M. Hoffmann, R. Thull, *J Am Ceram Soc* 87, 311 (2004).
[9] M. Bohner, *Injury* 31, D37 (2000).
[10] J. Weng, X. Liu, X. Zhang, X. Ji, *J Mater Sci Lett* 13, 159 (1994).
[11] D.E. MacDonald, F. Betts, M. Stranick, S. Doty, A.L. Boskey, *J Biomed Mater Res* 54, 480 (2001).
[12] S. Jalota, A.C. Tas, S.B. Bhaduri, *J Am Ceram Soc* 88, 3353 (2005).
[13] W.F. Brown, L.C. Chow, *United states. patent* No. 4,518,430 (1985).
[14] L.C. Chow, S. Takagi, *United states patent* No. 5,525,148 (1996).
[15] Y. Matsuya, S. Matsuya, J.M. Antonacci, S. Takagi, L.C. Chow, A. Akamine, *Biomaterials* 20, 691 (1999).
[16] Y. Sargin, M. Kiziyalli, C. Telli, H. Güler, *J Europ Ceram Soc* 17, 963 (1997).
[17] H.E. Romeo, M.A. Fanovich, *J Mater Sci: Mater Med* 19, 2751 (2008).
[18] K. Ciesla, R. Rudnicki, *Pol J Chem* 61, 719 (1987).
[19] K, Ciesla, R, Rudnicki, *Pol J Chem* 62, 31 (1988).
[20] K. Ciesla, R. Rudnicki, *Pol J Chem* 63, 83 (1989).
[21] U. Gbureck, M. Hofmann, J.E. Barralet, *J. Am. Ceram. Soc.* 88, 1327 (2005).
[22] H. Hu, Q. Chen, Z. Yin, P. Zhang, *Thermochemica Acta* 398, 233 (2002).
[23] F.H. Lin, C.J. Liao, K.S. Chen, J.S. Sun, *Mater. Sci. Engin. C- Biomimetic Supramolec. Systems* 13, 97 (2000).
[24] J.C. Elliott, *Structure and chemistry of the apatites and other calcium orthophosphates*, Elsevier, Amsterdam (1994).
[25] I. Martin, P.W. Brown, *Adv. Cem. Res.* 5, 115 (1993).
[26] W.E. Brown, L.C. Chow, *Cements Research Progress*. Westerville, OH: The Am. Ceram. Soc. 352-379 (1986).
[27] L.C. Chow, M. Markovic, S. Takagi. in: L.J. Struble (editor). *Cements Research Progress*. Westerville, OH: The Am. Ceram. Soc. 215-238 (1987).
[28] U. Gbureck, J.E. Barralet, M. Hofmann, R. Thull, *J. Dent. Res.* 83, 425 (2004).
[29] U. Gbureck, O. Knappe, N. Hoffmann, J.E. Barralet, *J. Biomed. Mater. Res.* 83, 132 (2007).
[30] F.S. Pan, K. Li, A.T. Tang, Y. Wang, J. Zhang, Z.X. Guo, *Adv. Mater. Process. II, Mater. Sci. Forum* 437, 105 (2003).
[31] http://de.wikipedia.org/wiki/Calciumhydroxid
[32] K.L. Lin, C.P. Ju, J.C. Lin, *J. Mater. Sci.* 40, 2631 (2005).
[33] A. Sugawara, S. Hirayama, K. Fujikawa, S. Takagi, L.C. Chow, *J. Dent. Res.* 86(Spec Iss A), Abstract 0045 (2007).
[34] J. Wei, J. Wang, X. Liu, J. Ma, C. Liu, J. Fang, S. Wei. *Appl. Surf. Sci.* 257, 7887 (2011).
[35] J. Wei, J. Wang, W. Shan, X. Liu, J. Ma, C. Liu, J. Fang, S. Wei, *J Mater Sci: Mater Med* 22, 1607 (2011).
[36] A. Cherng, S. Takagi, L.C. Chow, *J. Biomed. Mater. Res.* 35, 273 (1997).

[37] S. Takagi, L.C. Chow, S. Hirayama, F.C. Eichmiller, *Dent. Mater.* 19, 797 (2003).
[38] J. E. Bolanda, M. MacDougallb, D.L. Carnesa, S.H. Dickens, *Dent. Mater.* 22, 338 (2006).
[39] H.H.K. Xu, J.L. Moreau, *J. Biomed. Mater. Res.* 92B, 332 (2010).
[40] H.H.K. Xu, M.D. Weir, L. Sun, *Dent. Mater* 25, 535 (2009).
[41] A.O. Majekodunmi, S. Deb, J.W. Nicholson, *J. Mater. Sci.: Mater. Med.* 14, 747 (2003).
[42] K.E. Watson, K.S. TenHuisen, P.W. Brown, *J. Mater. Sci.: Mater. Med.* 10, 205 (1999).
[43] K.S. TenHuisen, P.W. Brown, *J. Dent. Res.* 73, 598 (1994).
[44] Y. Matsuya, J.M. Antonucci, S. Matsuya, S. Takagi, L.C. Chow, *Dent. Mater.* 12, 2 (1996).
[45] Y.E. Greish, J.D. Bender, S. Lakshmi, P.W. Brown, H.R. Allcock, C.T. Laurencin, *J. Biomed. Mater. Res.* 77, 416 (2006).
[46] Y.E. Greish, P.W. Brown, *Biomaterials* 22, 807 (2001).
[47] Y.E. Greish, P.W. Brown, *J. Mater. Sci.: Mater. Med.* 12, 407 (2001).
[48] Y.E. Greish, P.W. Brown, *J. Am. Ceram. Soc.* 85, 1738 (2002).
[49] A. Khalyfa, S. Vogt, J. Weisser, G. Grimm, A. Rechtenbach, W. Meyer, M. Schnabelrauch, *J. Mater. Sci.: Mater. Med.* 18, 909 (2007).
[50] E. Vorndran, M. Klarner, U. Klammert, L.M. Grover, S. Patel, J.E. Barralet, U. Gbureck, *Adv. Engin. Mater.* 10, B67 (2008).
[51] U. Gbureck, T. Hölzel, R. Thull, F.A. Müller, J.E. Barralet, *Cytotherapy* 8, 14 (2006).
[52] Y. Yoshimine, M. Sumi, R. Isobe, H. Anan, K. Maeda, *Biomaterials* 17, 2241 (1996).
[53] A. Ehara, K. Ogata, S. Imazato, S. Ebisu, T. Nakano, Y. Umakoshi, *Biomaterials* 24, 831 (2003).
[54] Y. Doi, H. Iwanaga, T. Shibutani, Y. Moriwaki, Y. Iwayama, *J. Biomed. Mater. Res.* 47, 424 (1999).
[55] T. Kitsugi, T. Yamamuro, T. Nakamura, M. Oka, *Biomaterials* 16, 1101 (1995).
[56] C.H. Tsai, R.M. Lin, C.P. Ju, J.H. Chern, *Biomaterials* 29, 984 (2008).
[57] Y. Yoshimine, A. Akamine, M. Mukai, K. Maeda, M. Matsukura, Y. Kimura, T. Makishima, *Biomaterials* 14, 403 (1993).
[58] Y. Yoshimine, K. Maeda, *Oral Surg. Oral Med. Oral Pathol.* 79, 351 (1995).
[59] M. Yoshikawa, S. Hayami, I. Tsuji, T. Toda, *J. Endodontics* 23, 162 (1997).
[60] J.D. Regan, J.L. Gutmann, A.M. Iacopino, T. Diekwisch, *Internat. Endodont. J.* 32, 171 (1999).
[61] C.D. Laghios, B.W. Benson, J.L. Gutmann, C.W. Cutler, *Internat. Endodont. J.* 33, 311 (2000).
[62] L.C. Chow, S. Takagi, R.J. Stern, T.H. Chow, K.K. Takagi, B.A. Sieck, *J. Dent. Res.* 73, 26 (1994).

In: Calcium Phosphate
Editor: Robert B. Heimann
ISBN 978-1-62257-299-1

Chapter 13

CALCIUM PHOSPHATE URINARY STONES: PREVALENCE, COMPOSITION AND MANAGEMENT

Maria Luigia Giannossi [*] ***and Vito Summa***
Laboratory of Environmental and Medical Geology,
Institute of Methodologies for Environmental Analysis (IMAA) –
CNR, C.da Santa Loja, Tito Scalo (Pz), Italy

ABSTRACT

This chapter provides a concise summary of the epidemiology, clinical significance, etiology, and compositional analysis of phosphate kidney stone disease. Specific emphasis is placed on the prevalence of phosphate kidney stones and a brief outlook into potential future directions to prevent kidney stone disease is provided. A practical guide to identification of mineralogy and texture of stones is also included.

Comprehensive analysis of calcium phosphate stones, combining morphologic examination and compositional analysis, with identification of stone components, quantification of their respective proportions, and determination of the internal structure, contributes to improved etiologic diagnosis. Finally, an example of a regional study is given followed by brief concluding remarks.

1. INTRODUCTION

The pathogenesis of kidney stone disease has been a topic of discussion among researchers for many years. Numerous general reviews have been published [1-6]. Owing to the vastness of the topic and the complexities of the individual pathological conditions for formation of stones, the literature abounds with more esoteric type reviews that have focused on individual conditions rather than on the entire spectrum of disorders that give rise to different types of stone.

---

[*] Email address: giannossi@imaa.cnr.it

In order to develop methods revealing incipient states and making prophylaxis of the stone diseases, it is necessary to study in details the crystalline substance of biomineral aggregates and mechanisms of their formation. However, detailed investigations with new technology are almost absent. In addition, the mechanisms of the formation of renal stones are unclear at present.

Essential progress in studying of formation of pathogenic biominerals in human organism may be achieved as a result of simulation experiments on crystallization of their analogues under laboratory conditions.

The present work provides a concise summary of the epidemiology, clinical significance, etiology, and compositional analysis of phosphate kidney stone disease. Phosphates minerals (CaP) are among the most frequently occurring components of these pathogenic stones.

In clinical practice, the distinction between the crystalline phases of calcium salts, and especially CaP, is usually not considered relevant for diagnosis and medical management. Thus, the full benefits that might be gained from comprehensive morphologic and compositional examination of stones remain largely ignored or underappreciated.

1.1. Urinary stones

Generation of solid stones inside the kidney (nephrolithiasis) and the ureter (urolithiasis) is an increasing danger to human health. These concretions (calculi) usually arise due to the breakdown of a delicate metabolic balance in the body. Renal stones are common in western industrialized countries. Worldwide, about 10% of the population is affected by this pathology at some stage of their life with the maximum incidence occurring between the ages of 20 and 40.

The most common urinary stone types are calcium oxalate, calcium phosphate, uric acid, struvite (magnesium ammonium phosphate), and cystine [7].

Several studies show a high incidence of calcium oxalate stones, especially calcium oxalate monohydrate (CaOx) but a low percentage of CaP stones (Table 1).

Table 1. Incidence of urinary stone types

Reference	# of samples	% stone containing CaOx	% stone containing CaP	Country
Daudon et al., 1995 [8]	10 617	65.98%	19.46%	France
Mandel and Mandel, 1989 [9]	10 163	61.1%	6.8%	USA
Silva et al., 2009 [10]	630	59.3%	6.8%	Brazil
Takasaki, 1971 [11]	500	68.2%	3.2%	Japan
Jing et al., 2010 [12]	625	95.1%	15%	China
Gault and Chafe, 2000 [13]	14557	67%	13%	Canada
Ansari et al., 2005 [14]	1050	93.04%	1.8%	India

1.1.1. Calcium Stones

About 12% of all calcium stone-formers have an underlying metabolic cause of their stones; the remainder are so-called idiopathic.

Calcium oxalate (CaOx) is the most insoluble of all the stone-forming salts and acids. It is the constituent of stones normal subjects are most at risk of precipitating in their urine. However, stone-formers have higher urinary supersaturation levels with respect to calcium salts than normal subjects, and thus form crystals more frequently and of greater size. Calcium oxalate crystallizes in three hydrate stages, i.e. as calcium oxalate monohydrate (whewellite, COM), calcium oxalate dihydrate (also known as weddellite, COD) [15-16], and calcium oxalate trihydrate ($CaC_2O_4{\cdot}3H_2O$; COT) [17], a less common form in pathological stone formation.

The main urinary risk factors for the formation of calcium-containing stones are, in decreasing order of importance, a low urine volume, mild hyperoxaluria, an increase in urinary pH, hypercalciuria, hypocitraturia, hypomagnesiuria, and hyperuricosuria [18]. In most patients there is no single obvious urinary abnormality cause formation of stone. Usually stone formation is due to a combination of several small changes in urine composition which together cause the patient to be at an increased overall biochemical risk of forming stones.

According to the schematical kidney morphology proposed by Söhnel and Grases [19], calcium oxalate monohydrate calculi can be classified in three groups: COM papillary calculi, no-papillary COM calculi, and COD calculi [20].

COM papillary calculi appear attached to papillae. This group presents a conical morphology and is composed of a core or stone nucleus, situated in the urolith interior, and an outer shell growing out from the core.

The second group of non-papillary COM calculi presents typically ellipsoidal morphology which is clearly different from the papillary calculi. Those stones grow in closed renal cavities. In this type of kidney stone, the material is distributed irregularly in the inner part and may occasionally contain small spheres of hydroxyapatite or can contain a core that is mainly composed of organic matter that functions as a seed for the development of the stone body. Most of the bodies of this stone type are constituted of columnar COM crystals emerging from the core and, because of the absence of an attachment to the epithelium, are found in renal cavities.

1.1.2. Phosphates Stones

Renal calculi containing phosphate(s) can be broadly divided into two categories: (i) phosphate stones composed either of pure hydroxyapatite (HAP) or in a mixture with calcium oxalate and/or dicalcium phosphate (brushite, BR), and (ii) the so-called infectious stones consisting of magnesium ammonium phosphate (struvite, ST).

Very alkaline urine with an unphysiologically high concentration of ammonium ions provides ideal conditions for precipitation of CaP and struvite. Since urine of normal subjects is generally well undersaturated with respect to struvite those individuals are not at risk of struvite crystalluria.

Infection stones are more common in women than in men. It should be noted that a proportion of stone-formers develop an urinary tract infection secondary to their original sterile stone of a totally different composition. Hence sometimes struvite and CaP can form on top of other calculi. In these cases, there is usually an underlying set of problems that led to the original sterile stone, and these have to be investigated. Treatment of infection stones requires antibiotic therapy to control the underlying infection [21]. However, it is often extremely difficult to eradicate the infection completely, particularly if the stones and/or their fragments, which may continue to harbour living micro-organisms, are not completely removed.

1.1.3. Uric acid stones

The majority of these stones are attributable to some metabolic disorder which leads to high excretion of uric acid in urine and/or a persistently acidic urine with a pH < 5.5 [22]. As with all types of urolithiasis, a low urine volume increases the risk of crystalluria and stone formation. All three factors, i.e. hyperuricosuria, persistently acid urine, and highly concentrated urine are important in the genesis of uric acid stones. Treatment is designed to lower the supersaturation of urine with respect to uric acid by correcting the urinary risk factors for uric acid stones.

Uric acid (UA) is not present in renal calculi in a single chemical form. It mainly appears as anhydrous uric acid (thermodynamically stable phase), forming occasionally well-developed crystals, thus demonstrating that they have been formed through slow growth (low supersaturation) or by recrystallization. Uric acid dihydrate that is formed at higher supersaturation has also been detected. Uric acid dihydrate is, however, a very unstable phase that will dehydrate relatively fast to the anhydrous form. It consequently develops fragile structures with cracks and fissures [23-24]. Frequently crystals of calcium oxalate monohydrate are being detected among uric acid crystals, confirming the capacity of the latter to act as a heterogeneous nucleation agent for crystallization of calcium oxalate. Uric acid calculi largely form pebble-like objects with a smooth, but not polished surface, though an uneven and globular surface is also occasionally encountered. A cross-section of uric acid stone usually shows concentric layers of microcrystalline material forming typical laminations with radial striation converging to the core area. Less frequently, some calculi exhibit a coarse granular interior without a sign of lamination, texture or preferred orientation of the constituting crystals. The calculus core is commonly composed of randomly oriented plate-like crystals of uric acid.

1.1.4. Cystine stone

Cystine (CY) is an amino acid that is very insoluble in water at low pH values. For this reason when an abnormally high urinary elimination of cystine takes place and the urinary pH is low, this substance reaches high supersaturation values, precipitate and can eventually generate a renal calculus [25-26]. Normal subjects excrete very low concentrations of cystine in their urine, well below the precipitation threshold of this amino acid. Patients with cystinuria, an autosomal recessive disorder caused by a defect in the renal tubular transport mechanism for reabsorbing cystine, ornithine, lysine and arginine, excrete much higher concentrations of the amino acid [27]. This type of renal calculus is not papillary and its formation is basically caused by a combination of sedimentation processes and crystalline growth.

1.1.5. Infrequent stone

Some types of rare stones are caused by ingestion of inappropriately large doses of medications, and treatment requires merely cessation of the medication to prevent further stone episodes. Other stones are genetic disorders of metabolism, which require a more complex medical decision-making process.

2. Calcium Phosphate Nephrolithiasis

2.1. Calcium phosphate stone prevalence

Calcium phosphate stones account for about 15% of all urinary stones, with a marked female preponderance, and reflect a wide diversity of etiology [13, 28-33]. Variation of the relative prevalence of CaP urolithiasis over time is disputed, and relevance of CaP stone analysis for etiologic diagnosis is underestimated or even negated.

All recent studies mention a marked parallel increase in the proportion of CaP stones made of infrequent forms such as brushite as well as whitlockite and octacalcium phosphate [29, 34-35]. In these studies the CaP stones percentage is sometimes referred to the percentage of stones containing any proportion of CaP, and not only to the percentage of pure CaP stones.

In addition, there is a problem to include in the CaP stones such calculi containing any amount, even minor, of struvite without any differentiation among CaP and magnesium phosphates stones.

Mandel et al. [34] found no significant variation in the proportion of stones containing hydroxyapatite (without associated struvite) among their studies [9] whereas they observed a significant increase in the prevalence of brushite-containing stones [28].

In Canada, Gault and Chafe [13] observed a decline in the percentage of stones containing 50% of CaP without struvite.

In contrast, Parks and co-workers [29] reported an increase in the average content of CaP in calcium stones, in parallel with a marked increase in the prevalence of brushite stones, particularly in men.

Brushite stones, although much less frequent than carbapatite ones, pose an intriguing problem. Because brushite is thermodynamically unstable, it is rapidly converted into the more thermodynamically stable carbapatite [36]. The reasons why brushite persists as the major component of CaP stones in some patients, mostly males, and shows an increasing prevalence in recent years, are presently not understood and thus require further studies. One explanation could be a shortened delay between stone formation and stone removal by urological procedures. Indeed, brushite stones are especially recurrent, resistant to extracorporeal shock wave lithotripsy fragmentation, and poorly responsive to medical metaphylaxis [37].

As recently reported by Evan et al. [38], brushite stones are often associated with some signs of nephropathy with cortical fibrosis, tubule atrophy and plugging of collecting ducts, suggesting that such stones correspond to a more severe disease than other lithogenic conditions. Relationships between brushite and etiopathogenic conditions remain to be determined.

From a study by Mandel et al. [34] the difference emerges in the percent occurrence of the various crystal chemistries for unique and repeat stone formers. The most important of these differences were the increase in CaOx among the unique population compared to the recurrent population, and the increase of CaP among the repeat stone formers compared to the unique population. This can be considered a conversion of CaOx stone formers to CaP stone formers. The full significance of the observations that recurrent CaOx/CaP stone formers appear to shift the composition of stones from CaOx to CaP is not yet known.

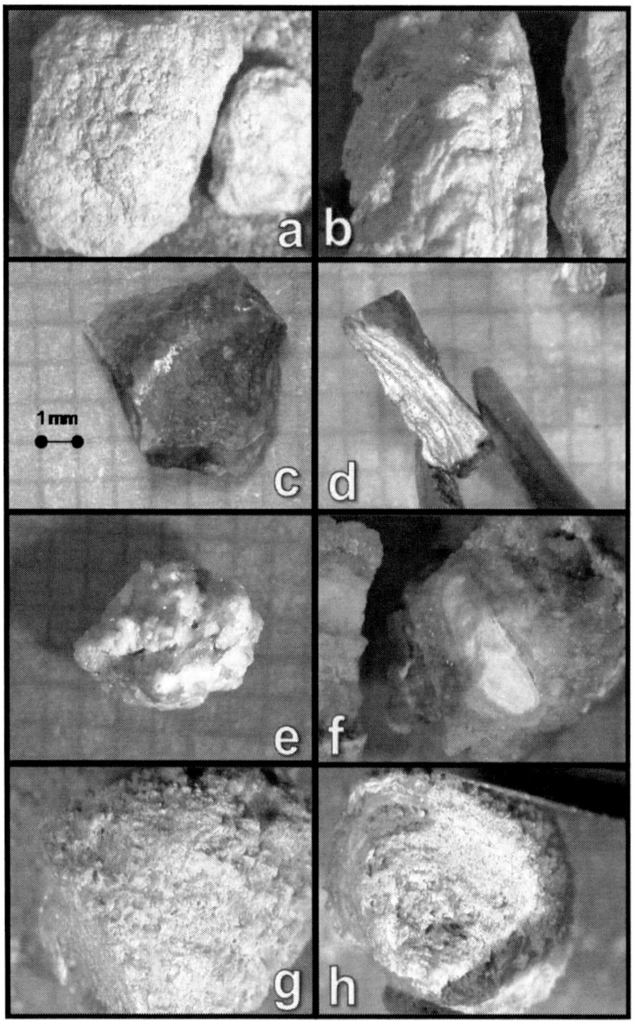

Figure 1. Typical morphological aspects of CaP stones.

2.2. Calcium phosphates minerals

2.2.1. Hydroxyapatite $Ca_5(PO_4)_3(OH)$

Hydroxyapatite, or hydroxylapatite, also known as apatite, is a common mineral in nature. Chemically it is a complex calcium phosphate with varied, attached molecules of hydroxyl (OH) and fluorine (F) ions and sometimes other elements. Apatite is the fundamental mineral component in bones and teeth, and when apatite has fluorine in its crystal structure, it is mechanically stronger. This is why fluorine is added to water and toothpaste in some countries.

In kidney stones, the hydroxyl group is predominant, and carbonate ions (CO_3^{2-}) substituting for some of the phosphate ions form a mineral that is relatively poorly crystallized. Apatite often forms the nucleus on which other urinary minerals are deposited to form the stone (Figure 1f).

Hydroxyapatite appears as layer of amorphous material, called 'aspidinic' hydroxyapatite according to Cifuentes [39]. Aspidinic layers consist of small spheres of 0.1 μm diameter, cemented together. Sphere of hydroxyapatite of 5-10 μm in size are also common.

2.2.2. *Brushite* $CaHPO_4 \cdot 2H_2O$

Calcium hydrogen phosphate dihydrate, also known as brushite, was named in 1864 by G.E. Moore, in honour of George Jarvis Brush (1831–1912). Brushite is a calcium phosphate compound that is structurally very similar to the common mineral gypsum (calcium sulphate). Brushite is found as a common cave mineral in guano deposits and in phosphorites formed at low pH (acidic) by reaction of phosphate-rich solutions with calcite and clay. It is a soft, silky mineral, usually honeybrown and showing a fine radial fibrous structure.

2.2.3. *Monetite* $CaHPO_4$

This triclinic, anhydrous analogue of brushite was identified only three times in this group of calculi. Prien and Frondel [79] reported it in urinary sediments of carnivorous animals.

Beck, Mulvaney and Rhamy (1974) studied the first reported human occurrences attributing its formation to very low urinary pH (4.13-5.08), naturally buffered urines, and high osmolality. Monetite in the calculi appears sometime as laminated and gray-to-brown material, as well as fine granular material associated with oxalates and apatite.

2.2.4. *Whitlockite* $Ca_3(PO_4)_2$

Tricalcium phosphate, or whitlockite (WHI), was named in 1940 for Herbert Percy Whitlock (1868–1948). Whitlockite is very rarely found in the urinary system but usually present in prostate stones. It is a calcium phosphate with small amounts of magnesium, $Ca_9(Mg,Fe)H(PO_4)_7$, and its occurrence may be stabilized by trace amounts of zinc, which perhaps accounts for its predilection for the prostate gland, which has a relatively high zinc content. The mineral is a resinous, brown material.

2.2.5. *Typical morphological aspects of CaP stones*

The fine inner structure of phosphate stones was widely studied [39-42]. Phosphate calculi are composed of structureless amorphous HAP layers or blocks, abundant organic matter and small spherulites disseminated in the calculus volume or in large assemblies in which spherulites are often intergrown (Figures 1e and f).

The highly non-homogeneous inner fine structure of stones signifies that the composition of urine during their development was greatly variable.

Phosphates stones show poorly organized structure, rough and homogenous surface, whitish colour (Figures 1a, b). This type of stone is observed mainly in cases of idiopathic hypercalciuria with or without some amount of struvite, and relatively high urine pH, and in stones induced by carbonic anhydrate inhibitors.

Brushite stones have surface composed of aggregated rod-like crystals of brushite with a radiating organization, cream to pale yellow-brown colour.

Stones of mixed minerals present compact concentric layers with a radiating organization of the crystals (Figures 1c, d, g, h). The surface is made of bipyramidal crystals of weddellite mixed with a rough structure of carbapatite, and pale yellow-brown colour. Such a mixed stone type is commonly found in resorptive hypercalciuria.

2.3. Analytical techniques

Over the first three decades of last century the chemical composition of urinary stones has been investigated with only qualitative methods. These kind of analyses employed qualitative wet chemical reactions to identify the different anions and cations present in the urinary stones.

On the contrary, in the 90s of the last century there have been progressive increases in the use of physical techniques, i.e. infrared spectroscopy and X-ray diffraction. However, so far no method has been found suitable for providing all necessary information on the structure and composition of urinary stones. Only the combination of a refined morphological and structural examination of stones with optical and scanning electron microscopy, supported by compositional analysis using X-ray powder diffraction, can provide a precise and reliable avenue to identify the stone type.

Binocular stereoscopic microscopy (BSM) is an easily applicable, cost-efficient tool, used to obtain accurate and reliable information regarding the stone components. While the method permits to distinguish between calcium oxalate and calcium phosphate stones it is not suited to distinguish among the different phosphate types. BSM was not successful in the analysis of struvite stones.

Scanning electron microscopy (SEM) is a method for ultramicroscopic investigation of the fine structure of urinary stones, including determination of single crystal surface structure, sections of urinary calculi, and the possible presence of unknown components within the calculus [43-44]. However, it also needs specialized equipment. It is well adapted to identify struvite minerals by their micro-structure. With SEM observation is also possible to investigate the different morphology of HAP: aspidinic and spherulitic types.

There is no doubt that X-ray diffraction analysis (XRD) is the most appropriate method to determine the composition of urinary stones. XRD can distinguish among all different crystal types in a particular mixture, and is therefore accepted as the gold standard for stone analysis [45-46]. XRD has advantages of reliability in qualitative analysis and accuracy in quantitative analysis. It operates simply and has a high sensitivity. The XRD data of the most common components of human kidney stones are presented in Tables 2 and 3.

The diffraction data are presented in crystallographic language as interplanar d-spacings in Ångstroms (d(Å)) associated with the distances between atoms in the structure and as diffraction intensities either relative to a weak versus strong scale or on a maximum of 100 scale [47-49].

Several reports have been published on the comparison of IR techniques to wet chemical methods for renal stone and other biological analysis, though these can be somewhat outdated [50-52].

Table 2. XRD data for the most common components of human kidney stones

COM		COD		ST		UA		CY	
d (Å)	I	d (Å)	I	d (Å)	I	d (Å)	I	d (Å)	I
5.93	100	8.70	12	5.90	38	6.56	38	4.70	100
5.79	25	6.31	100	5.60	45	5.63	20	4.68	100
4.64	7	6.15	45	5.38	22	4.91	48	4.63	100
4.52	6	4.40	45	4.60	6	4.76	6	4.45	5
3.78	13	3.89	14	4.25	100	3.86	55	4.20	7
3.65	100	3.09	18	4.14	34	3.70	5	3.32	8
3.00	10	3.07	18	3.29	24	3.28	14	3.18	19
2.97	46	2.81	20	3.02	10	3.18	70	3.13	30
2.91	12	2.77	20	2.96	16	3.09	100	3.06	16

Table 3. XRD data for the phosphate minerals (AP apatite, BR brushite, MO monetite, WHI whitlockite)

AP		BR		MO		WHI	
d (Å)	I	d (Å)	I	d (Å)	I	d (Å)	I
8.15	29	7,56	100	6,75	18	8,05	12
5.26	6	4,21	93	3,48	13	6,43	17
4.08	8	3,03	73	3,37	100	5,16	32
3.89	10	2,90	46	3,35	87	4,02	15
3.44	45	2,84	9	3,33	33	3,42	31
3.17	10	2,60	42	3,12	28	3,17	59
2.81	100	2,59	28	2,95	63	2,98	10
2.78	62	2,42	14	2,95	77	2,85	100
2.72	58	2,40	13	2,93	46	2,73	20

Like X-ray diffraction, infrared spectroscopy provides results on the actual salts, including the different degree of hydration, with an additional advantage of being able to identify non-crystalline compounds. All phosphate-containing calculi show an intense absorption band around 1000 cm^{-1}. This band permits its easy identification even in mixtures with calcium oxalate monohydrate or dihydrate. Pure brushite calculi exhibit characteristic IR spectra that allow to clearly distinguishing them from hydroxyapatite or ammonium magnesium phosphate calculi.

Many modern methods of analysis, also powder XRD, destroy the structure of the calculi when the samples are prepared for introduction into the instrument. Maintaining the structural integrity of the calculi is important to elucidate the chemistry of formation and the etiology of the calculi in the urinary system.

Micro-diffractometer XRD is preferably used when a very limited amount of sample is available but also on the bulk sample, without any type of treatment.

Microcomputed tomography (micro-CT) is a potential method to analyse urinary stone composition and morphology in a non-destructive manner at very high resolution [53-54].

Micro-CT allows non-destructive mapping of the internal and surface structure of urinary stones and permits identification of mineral composition based on X-ray attenuation values. However, Micro-CT cannot differentiate among mineral types when the stone is highly complex and micro-heterogeneous with significant mixing of different mineral types at a scale below the spatial resolution of the instrument.

2.4. Stone formation and etiological factors

Nephrolithiasis can start from sub-micron particles of organic material complexed with amorphous hydroxyapatite that after erosion to the papillary surface provide a platform to which crystals formed in the urine can attach [55].

Much work was done to elucidate this initial phase of nephrolithiasis [56-57]. However, for the transformation of kidney calcifications to a clinically symptomatic stone, further crystal apposition is necessary. It probably occurs during crystalluria by crystal aggregation [58-59].

Randall [60-62] theorized that mineral depositions were initially located on the renal papillae before the development of calcium oxalate calculi, also known as Randall's plaque that is considered an ideal site for an overgrowth of calcium oxalate and calcium phosphate to develop into calculi. Several investigators have previously demonstrated that Randall's plaque is composed of apatite, a common crystalline phase of calcium phosphate. These deposits are thought to serve as initiating nidus for urinary crystal attachment and development into calculi [55, 63].

Several studies have shown that pre-existent attached particles of phosphates, uric acid, mucoprotein aggregates etc., act as powerful inductors of calcium oxalate crystal formation [64-67]. This indicates the high capacity of calcium phosphate salts to promote the formation of calcium oxalate deposits on those crystals.

Many authors have suggested that phosphate stones are formed in kidney cavities with low urodynamic efficacy, combined with urinary pH above 6 and hypomagnesiuria [68]. Under these conditions hydroxyapatite microcrystals are generated in the urine that will settle around the walls of the cavity. Organic and inorganic matters present in a phosphate calculus are in approximately equimolar quantities.

Formation of renal calculi of brushite requires the existence of renal cavities with reduced urodynamics. In this case, however, magnesium urinary concentration is usually normal, the urinary pH oscillates between 6.0 and 7.0, and it is common to detect an inhibitory deficit [68]. Under these conditions, brushite crystals and hydroxyapatite are generated in urine. Due to the supersaturation values reached, growth of brushite crystals is slow, thus reaching considerable sizes, in such a way that the morphology of these crystals allows them to grow in parallel, forming columnar structures. Therefore, in this type of calculi, sedimentation and crystalline growth phenomena are combined.

3. Case Study

The overall probability that an individual will form stones varies in different parts of the world. The incidence of urolithiasis in a given population is dependent on the geographic area, racial distribution, and socio-economic status of the community. Changes in socio-economic conditions over time, and the subsequent changes in dietary habits, have affected not only the incidence but also the site and chemical composition of calculi.

The risk of developing urolithiasis in adults appears to be higher in the western hemisphere (5–9% in Europe, 12% in Canada, 13–15% in the USA) than in the eastern hemisphere (1–5%), although the highest risks have been reported in some Asian countries such as Saudi Arabia (20.1%) [69-70].

In order to better understand the spread and the various existing forms of stones, it is necessary to perform detailed studies in various geographic areas. A number of researchers have conducted surveys of specific zones for a specific period, trying to understand the specificity of the urinary stone for the respective zone [9, 14, 71-72].

In the remainder of this Chapter an example will be presented of a regional study performed on the prevalence of urinary stones and their morpho-compositional analysis to identify the main types. The case study has been conducted for a region of southern Italy, Basilicata [73]. The present authors have undertaken a study on the analysis of constituents of urinary stones from various patients of this region with the expectation that the data from a large number of patients may provide some idea about the factors causing high urinary stone disease of this region.

3.1. Materials and method

One hundred patients hospitalized for nephrolithiasis at the San Carlo Hospital, the main medical facility in Basilicata, were selected as donors and involved in this study. The 12-to-79 years old patients were of both genders. All kidney stones collected were dried in a dissector at room temperature for 2 weeks; then the dry samples were photographed.

Considering that the literature usually refers to phosphate stones in general without indicating what type of phosphate there is, in this study we applied a protocol of analysis which provides the application of XRD technique that allows the correct discrimination of various types of phosphates.

Optical observations were carried out by a stereomicroscope to determine colour, shape, overall appearance, surface features, and any possible occurrence of crystalline layer and/or organic matter at the surface. Special attention was paid to identify the site of stone attachment to the kidney wall.

During stereomicroscopic observations a plane passing as close as possible to the geometric centre of a kidney stone was recognized in order to cleave the stone into two fragments. Then one or more fragments which better displayed the structural features were observed under a high-magnifying Leica 440 Stereoscan Scanning Electron Microscope. Half of the sample was used for XRD analysis. Kidney stones were crushed in an agate mortar to a fine mesh powder for X-ray diffraction to obtain the crystalline phase composition of the kidney stones.

The XRD measurements were performed with a microdiffractometer D/max RAPID Rigaku, using a CuKα tube working at 40 kV and 30 mA. The qualitative interpretation of the diffractometric profiles was achieved by comparing the peaks of reflection with those in the PDF (Powder Diffraction File) [74].

3.2. Results and discussion

The results of the case study show that 66% of the kidney stones studied were composed of calcium oxalates (38% consisting of calcium oxalate monohydrate, whewellite, and 28% presenting in major quantity the dihydrate form of calcium oxalate, weddellite). The calcium oxalates are not pure since trace quantities of other components (organic and inorganic) could be observed.

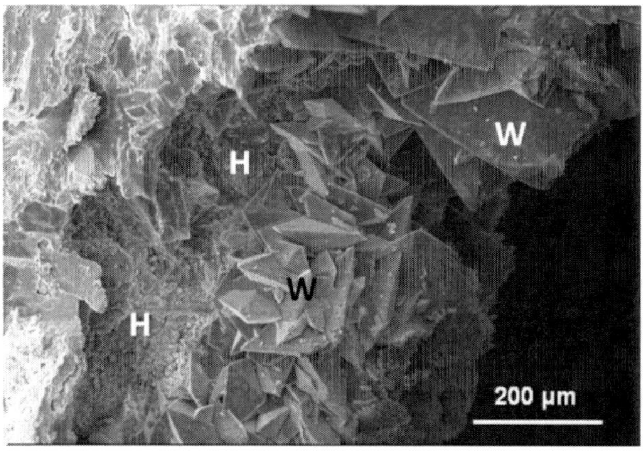

Figure 2. SEM image of hydroxyapatite (H) phase within weddellite (W) crystals.

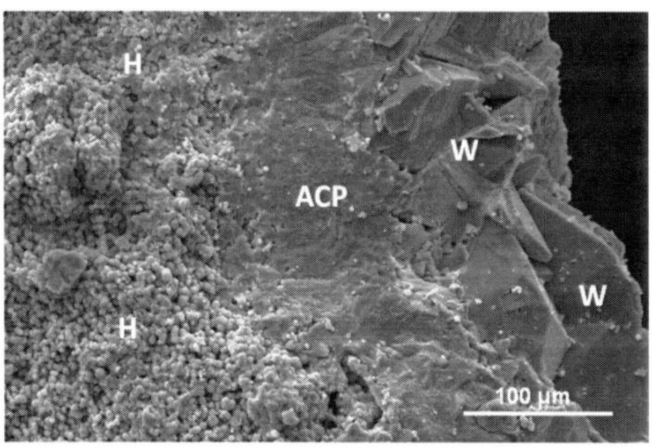

Figure 3. SEM image of the internal part of a phosphate stone. Hydroxyapatite (H) spherulites and amorphous calcium phosphate (ACP) layers can be observed near large weddellite (W) crystals.

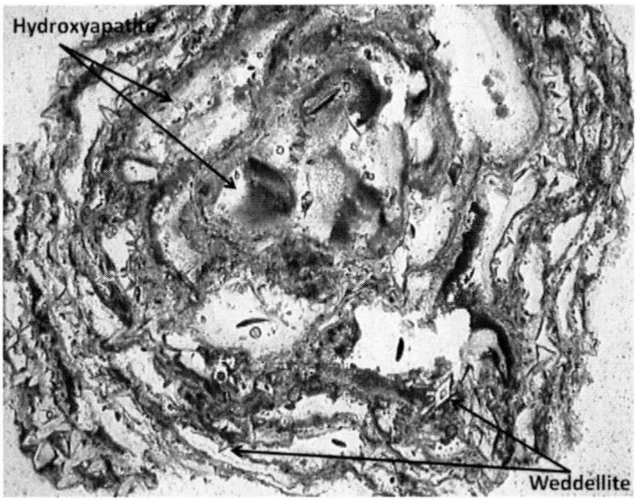

Figure 4. Petrographic thin section of a mixed stone composed of weddellite (well-developed dipyramidal crystals) and hydroxyapatite (dark layers).

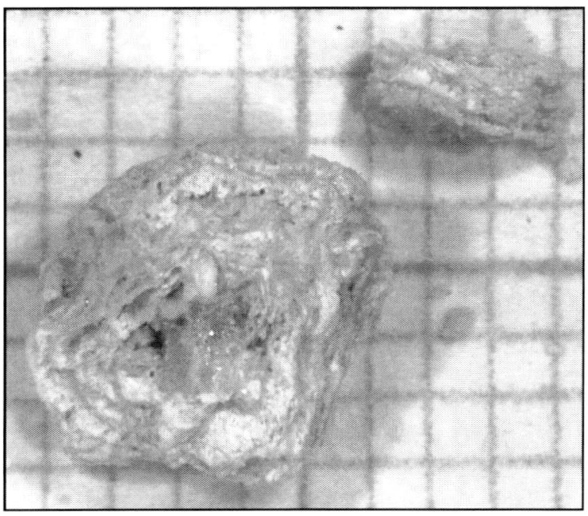

Figure 5. Mixed urinary stone composed of weddellite and hydroxyapatite.

The remaining 34% of the sampling showed other phases such as uric acid (15%), ammonium magnesium phosphates (struvite, 6%), hydroxyapatite (13%) and cystine (1%). 32% of the stones were found to be multi-composed, showing weddellite + hydroxyapatite, weddellite + struvite, and whewellite + uric acid.

During the mineralogical analysis we recognized in 6% of the total stones analyzed the presence of a new mineralogical assemblage consisting of weddellite + struvite [46], a combination not included in the classification of Grases et al. [20, 68].

Only a few stones were found to be composed entirely of struvite. Hydroxyapatite is the phase that occurs much more frequently as secondary phase or as surface deposits (Figure 2). Figure 3 shows a weddellite-type stone with a high degree of crystallinity whereas the calcium phosphate phase appears to be highly amorphous.

In mixed stones consisting of layers of weddellite and hydroxyapatite we observed that the CaP occupies the empty spaces between the large weddellite crystals (Figure 4).

Calcium phosphates occur as fine spherules that were deposited in layers filling the spaces between the larger oxalates crystals. This type of phosphate calculi was composed of fine concentric laminations, alternating between calcium oxalate and phosphates (Figure 5). Calcium phosphate laminations preserved the shape of earlier deposited calcium oxalate surface since phosphate minerals were sedimented onto surface irregularities. As in Figure 4, little porosity was observed in calculi of this type.

It is not clear whether the apatite phases have formed independently or whether the calcium oxalate phase has been transformed into these phases with time. Mandel and Mandel [34] have performed analyses of a large number of stones and suggested a strong trend towards conversion of stone occurrence from calcium oxalate- to calcium phosphate-containing chemistries. This finding indicates the possibility of time-dependent phase transformation from calcium oxalate to other phases. However, this is a complex phenomenon and the cause of phase transformation is not clearly known yet.

Kidney stones composed of weddellite crystals are arranged in a rosette-like form. The surface crystals generally display well-developed crystal faces. In these stones, within coarse surface crystals, a moderately developed porosity was observed, often filled with fine hydroxyapatite spherules or struvite.

Kidney stones composed of whewellite form small spheroids with minor surface irregularities. The interior appears homogenous and consists of closely packed fine whewellite crystals creating concentric laminations to form a rhythmic structure. This type of structure gives the whewellite stones a hardness that makes them more difficult to be fragmented. This is the reason why there is a high percentage of recurrence among people who developed this type of stones. The serious consequence of fragmentation not performed correctly is the inability to spontaneously expel fragments (as too large) which remain inside the cavity and may function as substrate for a new crystallization event [75-76].

Structural analyses allow determining the porosity of different types of kidney stones. There is the misconception that porous structures correspond to softer kidney stones, blurring the hardness of each mineral with the hardness of its aggregates. Despite a scratch hardness of 5 on the Mohs scale, phosphate kidney stones are among the easiest to be destroyed by lithotripsy techniques. Macroporosity was not usually observed in stones, but small pores were present in some mixed stones where different mineral phases were in contact. The type of the mineral affects the porosity. Stones that contain weddellite are porous as well as those consisting of struvite.

Struvite $(NH_4)Mg(PO_4)\cdot 6H_2O$, another mineral entirely constituting some of the stones analyzed, belongs to the group of phosphate so it is difficult to recognize individual struvite crystals with a stereomicroscope since they appear white and look similar to other minerals of the same group. Struvite stones are recognizable thanks to their large size and coral form. Despite their porosity, these stones are among the heaviest samples (average weight 155 mg).

Cystine stones ($C_6H_{12}N_2O_4S_2$) are easily recognizable for their appearance: round form and yellow wax-like colour. The crystals show hexagonal shapes in the SEM and in thin sections, with compact internal structure without porosity.

The internal structure and the relation between major and minor components provided information to classify the 80 kidney stones investigated in eight distinctive types and 13 subtypes. The prevalence percentage of Basilicata stones has been calculated (Table 4) according to the most recent classification proposed by Grases et al. [20, 68]. However, we found it difficult to compare our results with previous studies since the latest classifications [20, 77] use morphoconstitutional information to distinguish seven types of stone instead of only three categories: oxalate, phosphate and urate stones [78-81].

Table 4. Comparison of percentage of prevalence of kidney stones in Italy (Basilicata), Spain (Balearic Islands) and France

Type of kidney stone	Basilicata region (Italy)	Balearic Islands [68]	France [32]
Calcium oxalates	59%	63%	65%
-COM papillary	11%	13%	44%
-COM on cavity	19%	16%	-
-COD	29%	34%	21%
COD + HAP	6%	11%	-
Pure calcium phosphates	0%	8%	14%
-Hydroxyapatite	0%	7%	13%
-Brushite	0%	1%	1%
Struvite	4%	4%	2%
Uric acid	18%	8%	9%
COM + uric acid	6%	3%	-
Cystine	1%	1%	1%
Other types	6%	2%	9%

Therefore, for a comparative assessment of the percentage of prevalence of these stones based on the same level of detail applied, our results were compared with those of studies performed in other areas such as the Balearic Islands [68] and France [32].

This comparison showed that in Basilicata calcium oxalate stones seem to be more frequent and that there is a high prevalence of uric acid kidney stones (18% compared to 8% and 9%, respectively). This abundance is not unusual for Italy [82-83]. A lower prevalence of pure phosphate kidney stones was also observed (0% compared to 8% and 14%, respectively) (Table 4). An abundance of uric acid stones could be observed only in male patients (86% of cases). The most determinant risk factor for the crystallization of uric acid is the presence of a strongly acidic environment ($pH < 5.5$) in which uric acid crystals cannot dissolve [84].

The higher prevalence of uric acid stones in the Basilicata region may be due to diets high in purine and consumption of water with low permanent hardness. In fact, municipalities with a prevalence of uric acid stones are supplied with soft drinking water. All patients with uric acid stones also use bottled water with low bicarbonate content (150-200 mg/L) and low hardness (2-7°f). These characteristics prevent the dissolution of uric acid because they do not favour an alkaline environment. In Basilicata, in areas with a higher prevalence of uric acid stones, there is a deficiency of stones consisting of calcium phosphate. Hydroxyapatite is present as secondary phase or as surface deposits but never as major stone ingredient.

The stone composition is different for the genders, which may relate to the different diet, anatomical structure, and metabolic disorder. Females suffered urinary tract infections more frequently compared with males, which causes urine pH increase, and easy precipitation of magnesium ammonium phosphate.

CONCLUDING REMARKS

Considerable progress has been made in the investigation of the formation of urinary stones, attempting to identify promoting and inhibiting factors for their growth. However, extensive interdisciplinary studies are required in order to correlate the calculi composition with the pathophysiology of illness. A better understanding of the physiochemical principles underlying the formation, promotion and inhibition of calculi needs more accurate information on their chemical content. Imprecise chemical analysis can lead to wrong planning of their treatment and of ways to stop their reoccurrence.

Our experience provides evidence that morpho-constitutional stone analysis is much more informative than simple FTIR or XRD analysis, and is of clinical relevance.

In conclusion, precise identification of the crystalline phases of urinary stones preferably by means of FTIR, combined with careful morphologic examination, provides wider and more precise etiologic information than simple compositional analysis. Clinical relevance of morpho-constitutional analysis, as proposed here, is of particular interest with respect to CaP stones, which constitute a very heterogeneous group with multiple disparate etiologies, including some infrequent diseases. As stone composition and morphology reflect the physico-chemical causes of lithogenesis, comprehensive CaP stone analysis will decisively improve etiologic evaluation of CaP stone-formers.

REFERENCES

[1] F.L. Coe, J.H. Parks, J.R. Asplin, *New Engl. J. Med.* 327, 1141 (1992).
[2] F.L. Coe, A. Evan, *J. Clin. Invest.* 115, 2598 (2005).
[3] F.L. Coe, J.H. Parks, *J. Bone Miner Res.* 12, 522 (1997).
[4] B. Baggio, M. Plebani, G. Gambaro, *Crit. Rev. Clin. Lab. Sci.* 35, 153 (1998).
[5] O. Moe, *Lancet.* 367, 333 (2006).
[6] K. Sakhaee, *Kidney Int.* 75, 585 (2009).
[7] W.G. Robertson (2010). I.S. Shergill et al., "Medical Therapy in Urology", DOI 10.1007/978-1-84882-704-2_11, © Springer-Verlag London Limited
[8] M. Daudon, R. Dosimoni, C. Hennequin, et al., *Urol. Res.* 23, 319 (1995).
[9] N.S. Mandel, C.S. Mandel, *J. Urol.* 142, 1516 (1989).
[10] S.F.R. Silva, S.L. Silva, E.F. Daher, G.B. Silva Junior, R.M.S. Mota, C.A.B. Silva, *Clin. Chem. Lab. Med.* 47, 561 (2009).
[11] B.H. Takasaki, *Tissue Res.* 7, 232 (1971).
[12] Z. Jing, W. GuoZeng, J. Ning, Y. JiaWei, G. Yan, Y. Fang, *Urol. Res.* 38, 111 (2010).
[13] M.H. Gault, L. Chafe, *J. Urol.* 164, 302 (2000).
[14] M.S. Ansari, N.P. Gupta, A.K. Hemal, P.N. Dogra, A. Seth, M. Aron, TP. Singh, *International Journal of Urology.* 12, 12 (2005).
[15] C. Sterling, *Acta Crystallogr.* 18, 917 (1965).
[16] V. Tazzoli, C. Domeneghetti, *American Mineralogist.* 65, 27 (1980).
[17] S. Deganello, R.A. Kampf, B.P. Moore, *American Mineralogist.* 66, 859 (1981).
[18] R.W. Norman, S.S. Bath, W.G. Robertson, M. Peacock, *J. Urol.* 132, 1137 (1984).
[19] O. Söhnel and F. Grases, *Adv Colloid Interface Science.* 59, 1(1995).

[20] F. Grases, L. Garcıa-Ferragut, A. Costa-Bauzà, *Recent Res. Dev. Pure Appl Anal Chem.* 1, 187 (1998).

[21] R.H. Chinn, R. Maskell, J.E. Mead, A. Polak, *B Med. J.* 2, 1411 (1976).

[22] A. Metcalfe-Gibson, F.M. MacCallum, R.B.I. Morrison, O. Wrong, *Clin. Sci.* 28, 325 (1965).

[23] F. Grases, A.I. Villacampa, A. Costa-Bauzá, O. Sohnel, *Clin. Chem. Acta.* 302, 89 (2000).

[24] A. Hesse, W. Berg, C. Bothor, *Int. Urol Nephrol.* 11, 11 (1979).

[25] S.D. Rutchik and M.I. Resnick, *Urologic Clinics of North America.* 24, 163 (1997).

[26] J.A. Orts Costa, A. Zúñiga Cabrera, J. Martínez de la Cára y Salmerón, *An Med. Interna.* 20, 317 (2003).

[27] C.E. Dent, B. Senior, J.M.Walshe, *J. Clin. Invest.* 33, 1216 (1954).

[28] M. Daudon, H. Bouzidi, D. Bazin, *Urol Res.* 38, 459 (2010).

[29] J.H. Parks, E.M. Worcester, F.L. Coe, A.P. Evan, J.E. Lingeman, *Kidney Int.* 66, 777 (2004).

[30] M.H. Gault, L.L. Chafe, J.M. Morgan, et al., *Comparison of patients with idiopathic calcium phosphate and calcium oxalate stones.* Medicine (Baltimore) 70, 345 (1991).

[31] L. Maurice-Estepa, P.Levillain, B. Lacour, M. Daudon, *Scand J. Urol Nephrol.* 33, 299 (1999).

[32] M. Daudon, J.C. Doré, P. Jungers, B. Lacour, *Urol Res.* 32, 241 (2004).

[33] C.Y. Pak, J.R. Poindexter, R.D. Peterson, H.J.Heller, *J. Urol.* 171, 1046 (2004).

[34] N. Mandel, I. Mandel, V K. Fryjo, T. Rejniak, G. Mandel, *J. Urol.* 169, 2026 (2003).

[35] G. Schubert, *Urol Res.* 34, 146 (2006).

[36] R. Tang, G.H. Nancollas, J.L. Giocondi, J.R. Hoyer, C.A. Orme, *Kidney Int.* 70, 71 (2006).

[37] L.W. Klee, C.G. Brito, J.E. Lingeman, *J. Urol.* 145, 715 (1991).

[38] A.P. Evan, J.E.Lingeman, F.L. Coe et al., *Kidney Int.* 67, 5762 (2005).

[39] L. Cifuentes, *Composición y estructura de los cálculos renales.* 1st ed. Barcelona, Salvat, (1984).

[40] A.S. Meyer, B. Finlayson, L. DuBois, *Brit. J. Urol.* 43, 154 (1971).

[41] F. Grases, O. Söhnel, A.I. Vilacampa, J.G.March, *Clin. Chim. Acta.* 244, 45 (1996).

[42] F. Grases, A. Costa-Bauza, R.M. Prieto, I. Gomila, E. Pieras, O. Söhnel, *Scand. J. Clin. Lab. Invest.* 71, 407 (2011).

[43] A. Hesse, W. Hicking, D. Bach, W. Vahlensieck, *Urol Int.* 36, 281 (1981).

[44] P. Hyacinth, K. Rajamohanan, F.Y. Marickar, P. Koshy, S. Krishnamurthy, *Urol Res.* 12, 227 (1984).

[45] S. Ghosh, S. Basu, S. Chakraborty, A.K. Mukherjee, *J. Appl. Cryst.* 42, 629 (2009).

[46] M.L. Giannossi, G. Mongelli, F. Tateo, V. Summa, *Journal of X-ray science and technology* 20, 1 (2012)..

[47] D.J. Sutor, S. Scheidt, *Br. J. Urol.* 40, 22 (1968).

[48] N.S. Mandel, G.S. Mandel, *Structures of crystals that provoke inflammation.* In: Advances in Inflammation Research, (Weissmann G, ed.). Raven Press, New York, NY, (1982).

[49] JCPDS, Powder Diffraction Files. International Centre for Diffraction Data. Swarthmore, PA, (1985)

[50] J.C. Anderson, J.C. Jr Williams, A.P. Evan, K.W. Condon, A.J. Sommer, *Urol Res.* 35,41 (2007).
[51] M.H. Gault, M. Ahmed, J. Kalra, I. Senciall, W. Cohen, D. Churchill, *Clin. Chim. Acta.* 104, 349 (1980).
[52] P. Carmona, J. Bellanato, E. Escolar, *Biospectroscopy.* 3, 331 (1997).
[53] E. Krambeck, N. F. Khan, M. E. Jackson, J. E. Lingeman, J. A. McAteer, J. C. Williams, The *Journal Of Urology.* 184, 1543 (2010).
[54] C.A. Zarse, J.A. McAteer, A.J. Sommer, S.C. Kim, E.K. Hatt, J.E. Lingeman, A.P. Evan, J.C. Jr. Williams, *BMC Urology.* 4, 15 (2004).
[55] B.R. Matlaga, F.L. Coe, A.P. Evan, J.E. Lingeman, *J. Urol.* 177, 31 (2007).
[56] S.R. Khan, *Urol Res.* 23, 71 (1995).
[57] S.R. Khan, S. Thamilselvan, *Mol. Urol.* 4, 305 (2000).
[58] J.M. Baumann, B. Affolter, U. Caprez, U. Henze, *Eur. Urol.* 43, 421 (2003).
[59] N.K. Saw, P.N. Rao, J.P. Kavanagh, *Urol Res.* 36, 11 (2008).
[60] A. Randall, *N. Engl. J. Med.* 214, 234 (1936).
[61] A. Randall, *Ann. Surg.* 105, 1009 (1937).
[62] A. Randall, *J. Urol.* 44, 580 (1940).
[63] A.P. Evan, J. Lingeman, F.L. Coe and E. Worcester, *Kidney Int.* 69, 1313 (2006).
[64] F. Grases, A. Costa-Bauzá, J.G. March, O. Sohnel, *Nephron.* 65, 77 (1993).
[65] F. Grases, A. Costa-Bauzá, J.G. March, *J. Urol.* 74, 298 (1994).
[66] F. Grases, M. Kroupa, A. Costa-Bauzá, *Urol Res.* 22, 39 (1994).
[67] O. Sohnel, F. Grases, J.G. March, *Urol Res.* 21, 95 (1993).
[68] F. Grases, A. Costa-Bauzá, M. Ramis, V. Montesinos and A. Conte, *Clin. Chim. Acta.* 322, 29 (2002).
[69] A. Ramello, C. Vitale, M. Marangella, *J. Nephrol.* 13, S65 (2000).
[70] W.G. Robertson, H. Hughes, *Epidemiology of urinary stone disease in Saudi Arabia.* In: Ryall R, Bais R, Marshall VR, Rofe AM, Smith LH, Walker VR (eds) Urolithiasis 2. Plenum Press, New York London, (1994).
[71] P.J. Carson and D.R. Brewster, *J. Paediatr. Child Health.* 39, 325 (2003).
[72] A. Terai, Y. Okada, T. Ohkawa, O. Ogawa, O. Yoshida, *International Journal of Urology.* 7, 452 (2000).
[73] M.L. Giannossi, G. Mongelli, F. Tateo, V. Summa, *Journal of X-Ray Science and Technology* (under review) (2011).
[74] ICDD, *Powder Diffraction File. International Centre for diffraction Data*, edited by WF McClune, Newtown Square, PA: 19073-3272 (2005).
[75] A. Costa-Bauzà, J. Perello, B. Isern, F. Grases, *Urol. Res.* 33, 51 (2005).
[76] M.L. Giannossi, V. Summa, *Nephrology Dialysis Trasplantation Plus.* 3, 405 (2010).
[77] M. Daudon, C.A. Bader, P. Jungers, *Scanning Microsc.* 7, 1081 (1993).
[78] R.I. Gibson, *Am. Mineral.* 59, 1177 (1974).
[79] E.L. Prien, C. Frondel, *J. Urol.* 57, 949 (1947).
[80] W.G. Robertson and M. Peacock, *Handbook of urology.* Springer Verlag, Berlin, (1985).
[81] D.B. Leusmann, *Scandinavian Journal of Urology and Nephrology.* 25, 141 (1991).
[82] L. Borghi, P.P. Ferretti, G.F. Elia, F. Amato, E. Melloni, M.R. Trapasi, A. Novarini,. *Br. J. Urol.* 65, 231 (1990).

[83] M. Pavone-Macaluso and L. Miano, *Epidemiology of urolithiasis in Italy* – XVII Congres de la Societé Internationale d'Urologie. Pris, Joly, (1979).
[84] P. Ferrari and O. Bonny, *Ther. Umsch.* 61, 571 (2004).

In: Calcium Phosphate
Editor: Robert B. Heimann

ISBN 978-1-62257-299-1

Chapter 14

Transition Metal-Substituted Calcium Orthophosphates with NaSiCON Structure: A Novel Type of Bioceramics

Robert B. Heimann *

Professor emeritus of Applied Mineralogy and Materials Science, Görlitz, Germany

Abstract

Calcium orthophosphates substituted with transition metal ions such as titanium and/or zirconium display remarkable properties including solid state ionic conductivity, low solubility at least one order of magnitude lower than that of hydroxyapatite, and very low thermal expansion. In addition, suitable osteoconductivity render these ceramic materials possible novel biomaterials, for example as long-term stable coatings for endoprosthetic implants or osteosynthetic metallic supports, applied by plasma spraying technology. Owing to their ionic conductivities these ceramics could also be utilized as charge carriers for novel implant devices to speed up bone healing and bone in-growth, respectively in the presence of a stationary electromagnetic field.

1. Introduction

Plasma-sprayed hydroxyapatite (HAp) coatings are still the 'golden standard' of osteoconductive coatings applied to the stems of cementless hip endoprostheses and the roots of dental implants to support osteointegration. Since the composition of such coatings is close to that of the inorganic component of natural bone a strong and lasting osteoconductive bond is supposed to develop between living tissue and biomaterial. Indeed, a host of clinical studies conducted in several countries strongly suggest that HAp-coated cementless hip implants show exceptionally high survival rates around 95% at 10-14 years follow-up [1]. Similar positive results have been reported for the average survival of HAp-coated dental implants

* Email: robert.heimann@ocean-gate.de

[2]. However, these high survival rates apply only to patient older than 70 years whereas for younger patients (< 55 years) they are considerably lower, around 70% at 10 years post implantation [3]. Osteoconductive coatings form spontaneously a bond with bone tissue through deposition of a thin layer of bone-like Ca-deficient hydroxyapatite, the formation of which is controlled by adsorption of non-collageneous proteins such as osteocalcin, osteonectin, glycoproteins, silylated proteins etc. These proteins are known to have high affinity to hydroxyapatite and therefore play an important role as cell signal transmitters during bone formation [4]. In addition, functional groups at the surface of the biomaterials such as Ti-OH, Zr-OH, and Si-OH appear to support bone bonding [5].

Central to the performance of the coating is its lasting adhesion to the surface of the metallic implant. Gaps developing between implant and coating will inevitably be invaded by connective tissue resulting in aseptic loosening, i.e. destabilization due to lacking osteogenetic connection with the surrounding cortical bone. On the other hand the coatings should be able to release Ca^{2+} and phosphate ions in a controlled manner to build up a layer of secondary bone-like hydroxyapatite at the biomaterial-bone interface. Hence the rate of retreat of the coating must match the rate of advance of the bone to avoid gap formation. This requirement is closely related to the still unsolved question how thick a functional coating should be. Thicker coating are more stable against resorption in the body environment but pose the risk of adding up residual quenching and thermal stresses that may lead to cohesive (spallation) and adhesive (coating delamination) failure. Nevertheless, in some situations thick coatings will be required to ensure a more permanent bond thus guaranteeing implant stability by a lasting biological effect [6,7], for example when during an endoprosthetic replacement operation involving an exchange of the implant the cortical bone matter has been previously damaged, often in conjunction with an undesirable geometric configuration of the implant-supporting bone. In this case a thin, rapidly resorbed calcium phosphate coating will not be sufficient to sustain the necessary large-scale bone regeneration. Hence either thicker HAp coatings will be required to stimulate bone reconstruction over longer times, or thin coatings of biomaterials with higher resorption resistance and better adhesion to the implant. Such a favorable material could be transition metal-substituted calcium phosphate with NaSiCON structure as described below. Also, small particles released from the HAp coating may enter the lymphatic system and be distributed throughout the body, possibly causing periprothetic osteolysis, adverse inflammatory response of periprothetic connective tissues [8] as well as, in severe cases, hepatic degeneration ("particle disease"). Hence improvement is required of the notoriously weak adhesion strength of HAp to the surface of titanium implants. While in the past many attempts have been made to design bioceramic coating systems with either suitable bond coats [9] or graded HAp-TiO_2 composite coatings (for example [10]) a different route could be taken by utilizing strongly adhering transition metal-substituted calcium phosphate coatings.

2. STRUCTURE AND PROPERTIES OF NASICONS

The $Ca(Ti,Zr)_4(PO_4)_6$ hexaorthophosphates are members of the NaSiCON (Na superionic conductor) structural family, the prototype of which is $Na_2Zr_4(PO_4)_6$, [NZP] with the rhombohedral space group $R\bar{3}c$ [11].

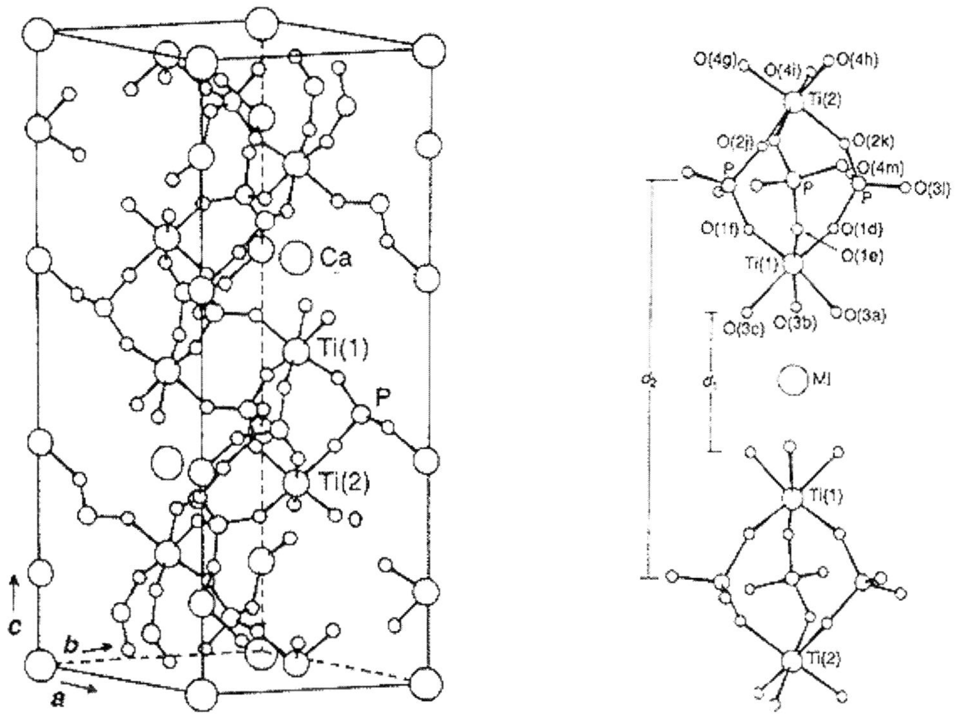

Figure 1. Unit cell of $CaTi_4(PO_4)_6$ (left) and arrangement of $Ti_2(PO_4)_3$ groups along the crystallographic *c*-axis (right) [40].

Its compo-sition can be described by the general formula $MA_2(X_3O_{12})$, where A = small ions such as Ti, Zr, Hf, Nb, Ta or other transition metals, M = large or medium-sized ions such as alkali metals, alkaline earth metals, lanthanides, or lattice vacancies, and X = phosphorus or silicon. The stability and flexibility of the 3D-[NZP] skeleton and the presence of a large number of interconnected structural voids allow the existence of hundreds of compositions with a plethora of ion exchange [12] and redox reactions, and intercalation processes. Based on these unique properties that include exceptionally low leaching rates under a wide range of environmental conditions [13], ceramics were designed for immobilization of radioactive nuclides [14,15] as well as carriers for immobilization and delivery of enzymes [16,17].

The name-giving [NZP] compound [18] is electrically moderately conductive even at room temperature [11, 19, 20]. All members of the NaSiCON family are characterized by a low coefficient of linear thermal expansion [21, 22] and high thermal shock resistance. Figure 1 shows the unit cell of $CaTi_4(PO_4)_6$. The $[Ti_2(PO_4)_3]^-$ groups form a three-dimensional network of two TiO_6 octahedra that are connected through their vertices to three PO_4 tetrahedra (Figure 1, left). These basic units appear as $-O_3TiO_3-O_3TiO_3-$ bands along the *c*-axis of the hexagonal unit cell. Along the *ab* plane perpendicular to the *c*-axis these bands are connected by PO_4 tetrahedra [20]. This atomic arrangement results in a hexagonal skeleton of space group $R\overline{3}c$ for M = monovalent ions such as Na^+, and $R\overline{3}$ for M' = divalent ions such as Ca^{2+}. In the unit cell large interconnected structural vacancy sites exist of octahedral symmetry (M1), and trigonal-prismatic symmetry sites (M2) occupied by vacancies as in $\square Nb_2(PO_4)_3$, by two-valent ions as in † $_{0.5}Ca_{0.5}Ti_2(PO_4)_3$ or three-valent ions as in $_{0.67}La_{0.33}$

$Ti_2(PO_4)_3$ [23]. These ƚ vacancies account for the structural variability of the NaSiCON family as well as their ionic conductivity.

The lattice constants of $CaTi_4(PO_4)_6$ (CT_4P_3[1]) are a = 0.8375(4) nm, c = 2.2061(7) nm, V = 1.374 nm^3, S.G. $R\bar{3}$, Z = 3 [20]. The complex structure may be related to ß-tricalcium phosphate (whitlockite, ß-TCP) with a_0 = 1.033 nm, c_0 = 3.710 nm, S.G. R3c, Z = 6 [24] and/or the meteoritic mineral chladniite, $(CaNa_2)Mg_7(PO_4)_6$ with a_0 = 1.497 nm, c_0 = 4.259 nm, V = 8.264 nm^3, S.G. $R\bar{3}$, Z = 18 [25]. In the chladniite structure slabs parallel (00.1) of $M^{[6]}$ octahedra occupied by Mg^{2+} are surrounded by $Ca^{[8]}$-$Na^{[8]}$ polyhedra and connected by PO_4 tetrahedra. These slabs alternate with more complex configurations of $M^{[5]}$, $M^{[6]}$ and $M^{[7]}$ polyhedra linked by PO_4 tetrahedra. The structure is related to that of glaserite, $K_3Na(SO_4)_2$ [26]. The similarity of the ionic radii of $Ca^{[8]}$ and $Na^{[8]}$ as well as $Ti^{[6]}$, $Zr^{[6]}$ and $Mg^{[6]}$ in $CaTi_4(PO_4)_6$ ($Ca^{[8]}$ 112 pm, $Ti^{[6]}$ 61 pm, $Zr^{[6]}$ 72 pm) and $(CaNa_2)Mg_7(PO_4)_6$ ($Na^{[8]}$ 116 pm, $Mg^{[6]}$ 72 pm) suggests isomorphous replacement and in particular filling of the structural vacancies at Ca positions in CT_4P_3 by equally sized Na ions in chladniite. Hence the unit cell volume of chladniite (V = 8.264 nm^3 [25]) is six times that of CT_4P_3 (V = 1.374 nm^3 [20]), corresponding to their unit cell contents, Z.

Quantitative data of the electric conductivity of NaSiCONs reported in the literature appear to differ considerably depending on the chemical composition and preparation conditions. Silva et al. [27] measured the d.c. conductivity of $CaTi_4(PO_4)_6$ produced by high-energy dry ball milling of a mixture of $Ca(H_2PO_4)_2$ and TiO_2 to be around $\sigma = 4.0 \cdot 10^{-12}$ $S \cdot m^{-1}$, about 30 times the conductivity of hydroxyapatite. The dielectric permittivity was reported as ε = 15.4, somewhat higher than that of hydroxyapatite (ε = 10-12). However, the reaction product still contained large amounts of the starting materials attesting to a very incomplete reaction. Hence the low conductivity value given is presumably related to the presence of residual non-electronically conducting dielectric TiO_2 and calcium dihydrogen phosphate. On the other hand, submicrometer-sized grains of Y-doped $Na_3(Zr,Y)_2(SiO_4)_2(PO_4)$ synthesized by sintering of mechanically activated precursor powders showed an ionic conductivity as high as $\sigma = 2.5 \cdot 10^{-5}$ $S \cdot m^{-1}$ [28], seven orders of magnitude higher than that found by Silva et al. [27]! This high value of electric conductivity may be explained with the large concentration of the charge-carrying Na ions with high electron mobility compared to the less mobile Ca ions.

3. Solubility of $Ca(Ti,Zr)_4(PO_4)_6$ Powder Compacts

Figure 2, left shows solubility data of several selected Ca-Ti-Zr orthophosphates in protein-free simulated body fluid (pf-SBF according to Ref. [29]) [30,31].

The highest solubility was found for tricalcium phosphate (α-TCP), followed by hydroxyapatite. The $CaTiZr_3(PO_4)_6$ (F, CTZ_3P_3) exhibits the lowest solubility. Hence this composition has been considered a biomaterial of the second generation with unusually high biostability in a body environment [31-33]. Similar results were reported for calcium titanium phosphate glass-ceramics [35].

[1] Cement chemical notation: C = CaO, T = TiO_2, Z = ZrO_2, P = P_2O_5.

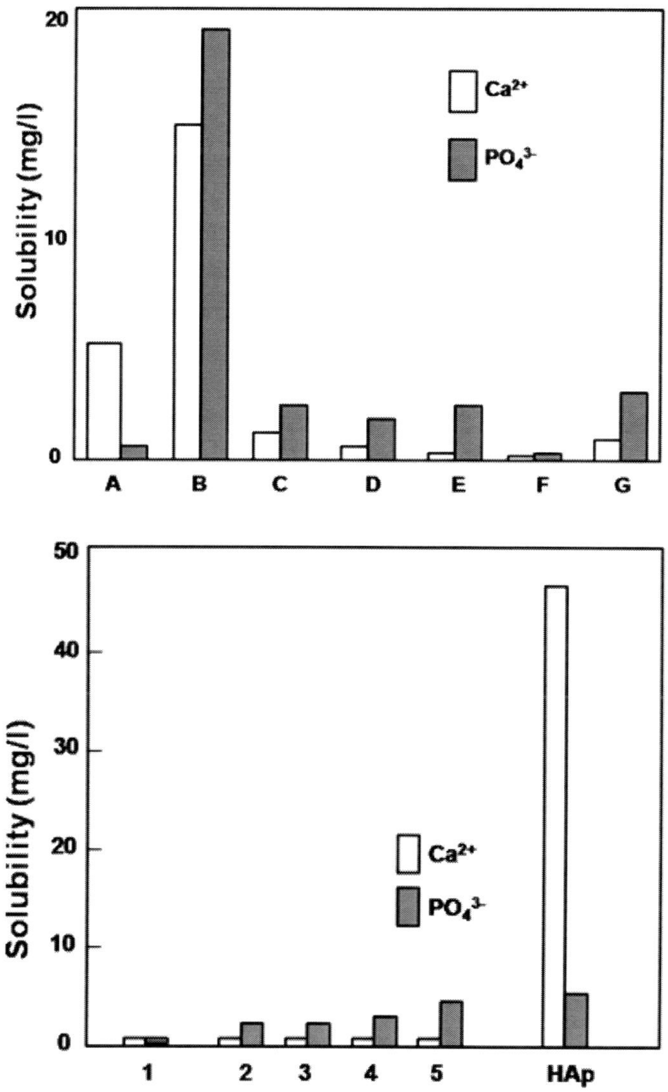

Figure 2. Top: Solubility of selected Ca-Ti-Zr orthophosphates (C-G) as well as hydroxyapatite (A) and α-tricalcium phosphate (B) in protein-free simulated body fluid (pf-SBF) at 37°C and pH = 7.4 [30,31,40]. C: CT_4P_3; D: CT_3ZP_3; E: $CT_2Z_2P_3$; F: CTZ_3P_3; G: CZ_4P_3 (C = CaO, T = TiO_2, Z = ZrO_2, P = P_2O_5). Bottom: Solubility of plasma-sprayed CTZ_3P_3 coatings deposited with different sets of spray parameters (2-5) compared to the starting powder (1) and plasma-sprayed hydroxyapatite coatings (HAp). Solvent: 0.2 molar TRIS-HCl buffer, pH 7.4, 37°C [31, 40].

4. Experimental $CaTiZr_3(PO_4)_6$ Coatings Deposited by APS

Agglomerated $CaTiZr_3(PO_4)_6$ (in the following: CTZ_3P_3, cement chemical notation) powder synthesized by sol-gel technique [31] was deposited by atmospheric plasma spraying (APS) [34] onto Ti6Al4V substrates using statistical design of experiments methodology (12 run-Plackett-Burman design [36]). The seven spray parameters varied and their levels are shown in Table 1.

Table 1. Plasma spray parameter levels. *The first number of the powder feeder settings refers to rotation of the powder feeder disk, the second number to rotation of the stirrer

Parameter	High level (+)	Low level (-)
Plasma power [kW]	35	25
Argon flow [slpm]	50	40
Hydrogen flow [slpm]	12	6
Carrier gas flow [slpm]	6	3
Powder feeder settings [%]*	40/80	20/60
Spray distance [mm]	120	80
Grain size [μm]	+45-70	+25-45

4.1. Structure and Composition of CTZ_3P_3 Coatings

Figure 3 shows cross-sections of a typical coating with an average porosity of 17% that is thought to promote easy in-growth of bone cells. The low crack concentration within the coatings is related to the high thermal shock resistance of the material [21]. Moreover, in conjunction with the low coefficient of thermal expansion ($\alpha_a = 5 \cdot 10^{-6}\ K^{-1}$; $\alpha_c = 0.9 \cdot 10^{-6}\ K^{-1}$; [20]) there is a tendency to generate compressive residual stress at the substrate-coating interface. The magnitude of this compressive stress can be estimated using the Dietzel equation [37]

$$\sigma_c = \{E_c\,(\alpha_c - \alpha_s)\Delta T/(1 - \nu_c)\} + \{[\tfrac{1-\nu_s}{E_s}]d_c/d_s\}, \qquad (1)$$

where α is the coefficient of linear thermal expansion, ν is the Poisson ratio, E is the modulus of elasticity, and d is the thickness. The subscripts 'c' and 's' refer to coating and substrate, respectively. The sign of the stresses depends on the sign of (α_c - α_s). Here $\alpha_c << \alpha_s$ ($\alpha_s \sim 8 \cdot 10^{-6}\ K^{-1}$ for Ti6Al4V), hence compressive stresses develop in the coating that will be compensated by tensile ones in the substrate [38,39]. These compressive stresses support strong adhesion of the coating to the substrate, an important requirement of functional bioceramic coatings for endoprosthetic implants.

The structural integrity of plasma-sprayed CTZ_3P_3 coatings contrasts very favorably with that of hydroxyapatite coatings that almost always show cracks and fissures owing to tensile residual stresses imposed by the positive value of of (α_c- α_s); here $\alpha_c > \alpha_s$ ($\alpha_c \sim 14 \cdot 10^{-6}\ K^{-1}$ for HAp; $\alpha_s \sim 8 \cdot 10^{-6}\ K^{-1}$ for Ti6Al4V).

In addition to zirconium di(pyro)phosphate (ZrP_2O_7) and small amounts of rutile (TiO_2) and baddeleyite (ß-ZrO_2), amorphous decomposition products appear during plasma spraying that may be related either to incongruent melting of the material or to an inherent structural instability caused by a high lattice vacancy concentration of the NaSiCON structure. Formally this thermal decomposition can be described by the equation

$$CaTiZr_3(PO_4)_6 \rightarrow 2\ ZrP_2O_7 + TiO_2 + ZrO_2 + CaP_2O_6, \quad (2)$$

where the calcium monophosphate anhydrate (MPAH) will be incorporated into the amorphous phase.

Figure 3, right shows a highly inhomogeneous coating with at least four chemically different phases, denoted 1 to 4. The dark gray phase ("1") constitutes about 50% of the coating volume and consists of rounded particles of the target compound CTZ_3P_3 embedded in a lighter colored matrix phase ("2") of the approximate composition $Ca(Ti,Zr)_{4.5}O_{4.6}(PO_4)_{3.6}$ and a strongly phosphate-depleted phase ("3") the composition of which was measured by EDX to be $Ca(Ti,Zr)_{4.2}O_{6.85}(PO_4)_{1.7}$. Occasionally ß-$ZrO_2$ (baddeleyite) appears ("4"). Comparable results were obtained previously on plasma-sprayed $CaTi_4(PO_4)_6$ and $CaTi_2Zr_2(PO_4)_6$ coatings [43].

The high temperatures occurring during plasma spraying cause a very substantial loss of the highly volatile P_2O_5 that render the coating composition stoichiometrically unbalanced. Hence ways must be found to suppress this thermal decomposition, for example by using low-temperature deposition techniques such as suspension plasma spraying (SPS) [44]. It is expected that proper spray parameter engineering will lead to coatings optimized in terms of porosity and phase content, and with minimum solubility under physiological conditions in a human body environment.

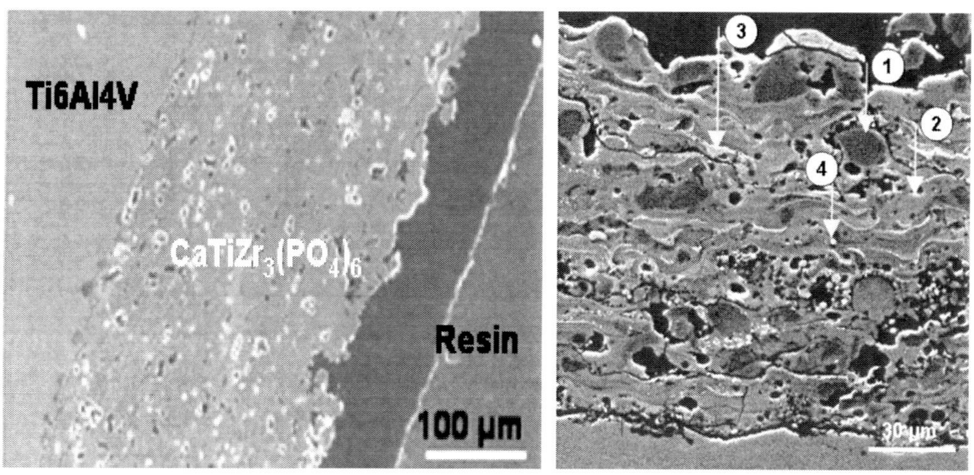

Figure 3. Typical cross-sections of CTZ_3P_3 coatings deposited by plasma spraying on a Ti6Al4V substrate. The spraying parameters were as follows: power 35 kW, Ar flow rate 50 slpm, hydrogen flow rate 12 slpm, spray distance 120 mm, powder feed gas flow rate (Ar) 3 slpm, powder grain size +25-45 µm. Note the absence of microcracks and larger interconnected pores (left), and also the inhomogeneous distribution of thermal decomposition products (right) [40]. For an explanation see text.

4.2. Mechanical Properties and *In Vitro* Solubility of CTZ_3P_3 Coatings

Plasma-sprayed CTZ_3P_3 coatings show mechanical properties that render them suitable for application as well-adhering, rather strong and reasonably porous coatings for titanium-based endoprosthetic implants.

Table 2. Properties of plasma-sprayed CTZ_3P_3 coatings [40]

Coating thickness (μm)	45-300
Average surface roughness R_a (μm)	50
Average porosity (vol%)	17
Pore diameter (μm)	10-40
Adhesive strength (MPa)	
for coating thickness > 200 μm	≤ 10
< 100 μm	≤ 17
Shear strength (MPa)	< 13
Thermal expansion ($10^6 \cdot K^{-1}$)	~ 3
Solubility in pf-SBF (mg/l)	< 5
Solubility in TRIS-HCl (mg/l)	< 1

Table 2 lists some pertinent properties and their ranges. The solubility was obtained as the sum of the concentrations of Ca^{2+} and $(PO_4)^{3-}$ ions ($\Sigma[cCa^{2+} + cPO_4^{3-}]$).

The tensile adhesion strength and the shear strength are both negative functions of the coating thickness, i.e. thinner coating show in general higher adhesion and shear strengths.

Detailed statistical evaluation of a series of coating experiments further revealed that the tensile adhesion strength increases with increasing grain size of the spray powder but decreases with increasing plasma power whereas the shear strength increases with the plasma gas (argon) flow rate but decreases with increasing amount of powder fed into the plasma jet [40].

The *in vitro* solubility of the phosphorus-depleted phases found in plasma-sprayed coatings in 0.2 molar buffer solution consisting of tris(hydroxymethyl) aminomethane (TRIS)-HCl (Figure 2, right) is generally higher compared to stoichiometric CTZ_3P_3, and varies with the spraying conditions. The plasma-sprayed samples 2 to 5 shown in Figure 2, right display solubility slightly higher than those of the starting powder (sample 1) but much lower than the solubility of plasma-sprayed hydroxyapatite coatings (sample 6). In particular, loss of Ca^{2+} ions is much higher in hydroxyapatite than in any CTZ_3P_3 coating observed.

4.3. Cell Proliferation and Cell Vitality Tests

To confirm biocompatibility, osteoconductive potential and, in particular, absence of cytotoxicity, the coatings were tested to determine their cell proliferation and cell vitality capabilities. Compressed powder disks (10 mm diameter) were sintered at 1300°C for 72 hours. Primary cultures of rat bone marrow cells were seeded onto the surface of the sintered disks with a density of $3 \cdot 10^4$cells/cm^2. Before seeding the cells were incubated for 24 hours in 500 μl cell culture medium (α-MEM; minimum essential medium + 15 % fetal bovine serum

+ 10^{-8} mole dexamethason + 50 μg/ml ascorbic acid + 10 nmol ß-glycerophosphate). The culture medium was changed daily; after 2 weeks the cells were fixed and stained with Giemsa stain. Thermanox™ cell cover slips were used as control. The cells growth was investigated by light optical microscopy using a graded evaluation standard that categorizes the cell growth into four classes as follows (Figure 4):

Class I:Areas with no cells, isolated cells or cell detritus.
Class II: Areas with loose reticulary cell associations
Class III: Areas covered with a thick layer of cells
Class IV: Nodulary centers with multilayered cell formation

Figure 4 shows that the CTZ_3P_3 supported cell growth somewhat better (classes III + IV = 75%) than the Thermanox™ control (66%). The percentage area of class I was much lower in case of the former (1%) than observed for the control (7%).

Human bone-derived cells were grown on sintered calcium titanium/zirconium phosphate samples with various Ti/Zr ratios, and tested for expression of various indicators for cell proliferation and vitality such as osteocalcin, osteonectin, osteopotin, alkaline phosphatase, and bone sialoprotein [41]. Compositions conforming to CTZ_3P_3 displayed maximum osteoblastic differentiation including sufficient expression of an array of osteogenic markers thus suggesting a high degree of osteoconductive potential. Similar results were reported for the growth of bone marrow stromal cells cultured on calcium titanium phosphate microspheres to be used as potential scaffolds for bone reconstruction [42].

On the other hand *in vivo* experiments in a rabbit model did not quite support this optimistic view: besides macrophages in soft tissue at the interface between bone and CTZ_3P_3 biomaterial primary trabecular bone was formed with a low degree of mineralization [67]. The reason for this inhibition of mineralization was thought to be related to the presence of Zr ions that was claimed to be responsible for impaired development of matrix vesicles and their function [68,70].

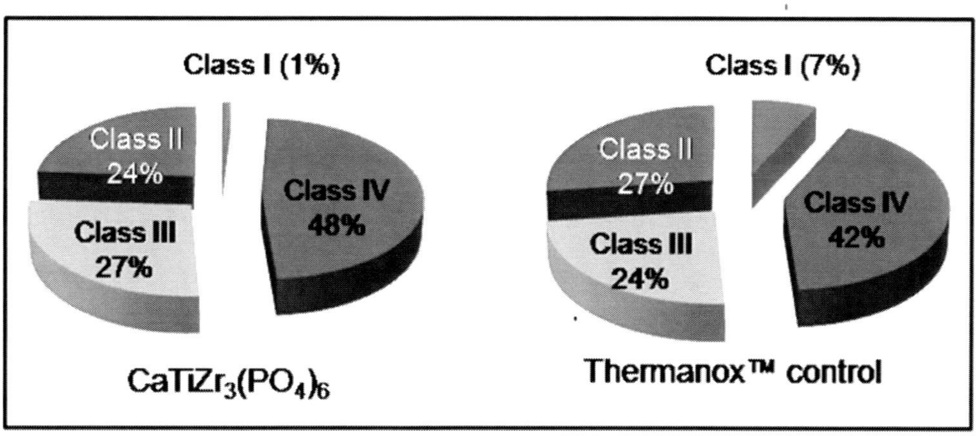

Figure 4. Rating of cell vitality and morphology.

4.4. Osteoconductive Properties of CTZ_3P_3 Coatings

As reported by Szmukler-Moncler et al. [32], *in vivo* tests with $CaZr_4(PO_4)_6$ disks implanted into the distal epiphysis of *femura* and *tibiae* of dogs showed favorable cell response including expression of a direct contact between implant and bone without noticeable resorption of the biomaterial.

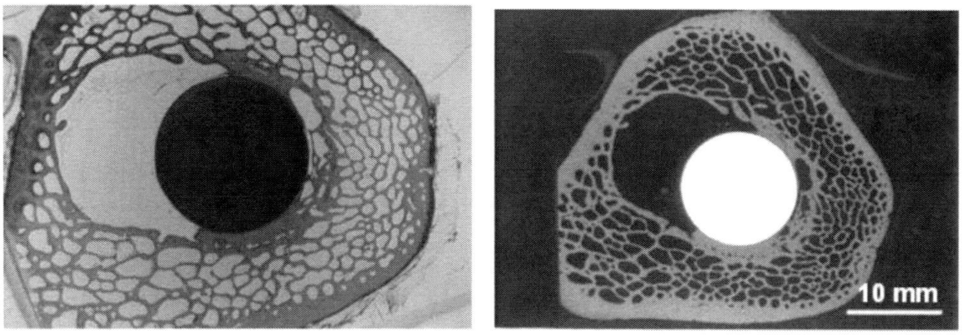

Figure 5. Distal section through the femoral medulla of sheep with an implanted Ti6Al4V rod coated with an approximately 150μm thick CTZ_3P_3 coating. Left: Toluidin blue-stained histological section. Right: Microradiograph [40].

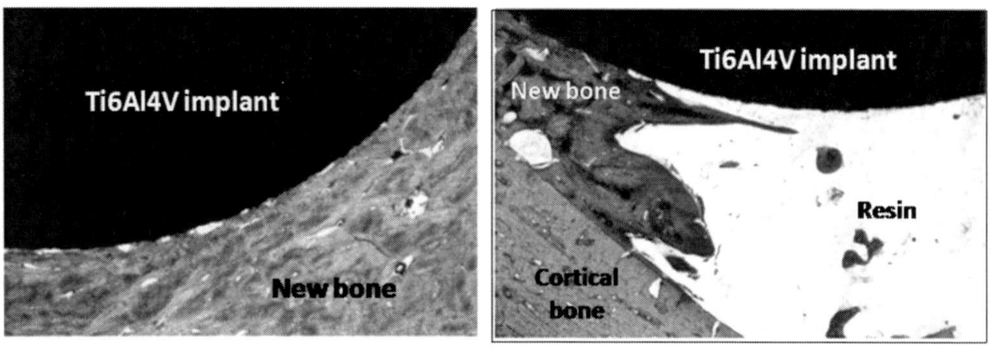

Figure 6. Proximal section through the femoral medulla of sheep with an implanted Ti6Al4V rod coated with an approximately 150μm thick CTZ_3P_3 coating. Left: Higher magnification showing close integration of the coated implant. Right: The gap bridging potential of the newly formed bone is reduced when the distance between coated implant and surrounding cortical bone increases [40].

Compared to hydroxyapatite coatings there was an earlier bone apposition reaction, associated with a large bone contact area. Extensive remodeling of osteons occurred after 9 months in direct contact with the coatings.

A series of survey animal experiments were conducted by the present author to assess the biological performance of plasma-sprayed CTZ_3P_3 coatings [40] and compared to standard hydroxyapatite coatings [5]. Figure 5 shows a distal section of the coated implant (black circular disk) with the surrounding trabecular bone tissue as a histological section (left) and a corresponding microradiograph (right). Larger scale bony integration is shown in the dense toluidin-stained (blue) bridging connections between implant and spongy bone.

However, the distal part of the implant oriented towards the marrow-filled medulla does not show continuous bone apposition but only spotty lamellar bony in-growth. While Figure

6, left reveals at higher magnification close bony integration of the proximal part of the coated implant, the gap bridging potential is being reduced at larger distances between the implanted rod and the surrounding (cortical) bone (Figure 6, right).

5. BONE GROWTH IN AN ELECTRIC FIELD

It has been known for some time that the growth of bone is being affected by the presence of materials with different dielectric behavior as suggested by the so-called 'bioelectric phenomenon' in bone [45-47]. Electrical currents have been already used to heal bone fractures since the 19th century by Julius Wolff (1835-1902) and James Prescott Joule (1818-1889). Since then the effect of electrical stimulation on bone has long been studied and is well documented [48-50]. In principle, mechanically triggered electrical charge separation acts as the prime mechanism during remodeling of a broken bone. In this process the structure and function of growing bone cells and extracellular structures are being influenced by piezoelectric and stream potentials. Clinical studies show that electrical stimulation of bone growth with fields aligned parallel to the axis of a long bone acts to reduce the time required for endosteal callus remodeling and hence speeds up the healing process [47, 51-53]. However, the type of the electric field applied appears to play an important role. It was observed by Watson et al. [54] that only a pulsed, transverse electric field tends to accelerate growth of embryonic chick tibiae while a static field did not show a significant accelerating effect.

In these experiments it was also found that improved bone healing observed at the negative electrode is accompanied by resorption of bone (osteolysis) or osteoporosis at the positive electrode. Consequently, the concept of electrical stimulation of bone healing must be approached judiciously. Unfortunately, at present neither the origin nor the functions of bioelectric potentials are known with certainty. Presumably diffusion gradients, i.e. ionic currents are established that are able to concentrate polarizable molecules, collageneous proteins, non-collageneous polyanionic proteins, and electrolytes at the wound site. For example, there is some indication that the concentration of Ca^{2+} ions at the negative electrode is higher in an electrically stimulated than in a non-stimulated fracture. Ca^{2+} ions originate from depolymerization of mucopolysaccharides, P ions from glycogenolysis of phosphate esters by action of alkaline phosphatase. The ions initially stored in the mitochondria are released by a change of the potential of the mitochondrial membrane via the cytoplasma and the cell membrane into the extracellular matrix [55]. Nucleation at wound site of calcium phosphate occurs that act as crystallization centers of apatite nanocrystals. The important role the membrane potentials play in releasing Ca and P ions suggests that an external electric field may accelerate this process by channeling essential metabolic ions towards the wound site.

A substantially higher maximum shear force was required to push out of its bony bed an implant subjected to electrical stimulation compared to its control [52]. Commercially pure (cp)-titanium cylindrical rods acting as a cathode were implanted into the mid-femoral diaphysis of a dog, and segments of kinked Ti wires, acting as anode, were placed in coils around the bone above and below the implant. Thus the electric field lines would be directed along the long axis of the femur. The connection was made by PTFE-coated wires running subcutaneously from the implant to a battery harnessed to the dog's back. While the interfacial shear strength of the stimulated implants was consistently greater than that of controls no difference could be observed in the quality of bone in-growth. Nevertheless, the data suggest that both the rate and quantity of bone in-growth were significantly enhanced by electrical stimulation. Similar experiments conducted on the hind legs of rabbits showed conclusively after osteotomy of the tibiae that the healing process proceeded much faster in the presence of a direct current of 10 μA as verified by ^{87m}Sr ($t_{1/2}$ = 2.8 h) scintigraphy [47].

The notion of acceleration of bone growth by a poled electrical field is strongly supported by more recent research findings that show that the growth of hydroxyapatite crystals from a simulated body fluid is being dramatically quickened on negatively polarized dielectric and ferroelectric substrates such as calcium and barium titanate [56]. These experiments suggest that a uniform electrical field rather than localized charges is the stimulating factor of bone remodeling during the healing process [57]. There is evidence that growth of bone-like hydroxyapatite in an electric field is accelerated by reorientation of the dipole moments between O^{2-} and H^+ of lattice OH^- ions in response to the electric polarization conditions. In this way the ordered alignment of OH^- columns similar to that present in monoclinic hydroxyapatite is being attained [9]. Positively charged Ca^{2+} cations will be preferentially adsorbed on the negatively charged polarized surface attracting a cloud of the counter-ions such as HPO_4^{2-}, HCO_3^-, and OH^- causing strong nucleation and eventually precipitation of a thick layer of bone-like apatite that bonds directly to the negatively charged hydroxyapatite surface [58]. The polarized state will cause an alignment of the OH^- groups in the hydroxyapatite structure. As observed by Yamashita et al. [56] a polarized HAp ceramic body immersed in 1.5 SBF (1.5 times the ion concentration in human blood plasma) supported the growth of a HAp layer much more strongly (6.0 μm/day) than a non-polarized surface (1.0 μm/day) by inducing apatite mineralization on synthetic surfaces that promote attachment of bone to orthopaedic implants.

In contrast to this the non-polarized state of a hydroxyapatite surface shows randomly oriented OH^- groups and consequently much reduced adsorption of positively charged counter-cations, resulting in decelerated growth of bone-like apatite. This has important implications: although non-periodic fields encountered by Yamashita et al. [56] are weak compared to electrostatic fields, even small changes of the local interfacial energy could be sufficient to markedly influence the mechanism of transformation of incipient nuclei into macroscopic mineral phases.

The results of these findings were contrasted by studies of the biological performance of ferroelectric composite hydroxyapatite-barium titanate (HABT) ceramics (d_{33} = 58 pC/N). They showed that attachment, proliferation, viability, morphology and metabolic activity of cells cultured on non-polarized HABT surfaces were comparable to those observed on commercially available hydroxyapatite at all times. However, there were no striking differences between the properties of poled and non-poled ceramics, at least not during short incubation times [59].

While polarization orientation of the OH^- groups of hydroxyapatite requires an electric field it is also conceivable that CTZ_3P_3 behaves differently in that spontaneous polarization may occur owing to charge imbalances within the TiO_6 octahedral groups. This would impose ferroelectric behavior which, however, has not been reported hitherto for the NaSiCON structure family. On the contrary, ferroelectric ceramics such as PZT (lead zirconate titanate) implanted in the diaphysis of the metatarsus of roosters in either polarized or depolarized state were found to have a negligible effect on rate and quality of bone formation [60, 61]. This is presumably related to the combined effects of (i) insufficient pressure load that produces low signal strength, (ii) high internal electric resistance, and (iii) incompatible elastic moduli of ceramics and bone.

Calcium titanate ($CaTiO_3$) has been implicated in enhanced osteoblastesis *in vitro* on titanium-coated hydroxyapatite surfaces. Hence orthopedic coatings that form $CaTiO_3$ could increase the osseointegration with juxtaposed bone that is needed for increased implant efficacy [62]. It is appealing to speculate that the weak ferroelectric nature of doped calcium titanate may be a reason for its osseoconductive behavior. If this is the case than the notion that a titanium oxide surface is an important prerequisite for osteoblastesis must be modified. The mechanism of control of apatite nucleation by calcium titanate surfaces [5, 62, 63] has been explained by Wei et al. [64] by an epitaxial structural relation between the (022) lattice plane of calcium titanate and the (00.1) lattice plane of hydroxyapatite. The (022) plane of the perovskite (P) structure is defined by the position of the oxygen atoms whereas the (00.1) plane of hydroxyapatite (H) is defined by the position of hydroxyl ions. A 2-dimensional lattice match exists characterized by a close fit $2X_C \approx X_H$ (mismatch: 0.8%) and $3Y_C \approx Y_H$ (mismatch: 0.09%). Hence biomimetic formation of hydroxyapatite during incubation in simulated body fluid and during osseoconductive integration in the bone bed *in vivo* relies on interfacial molecular recognition determined by the degree of epitaxial crystal lattice matching.

Recently a novel hypothetical regulation mechanism for bone modeling and remodeling under the influence of an electromagnetic field has been proposed [65]. Damage induced in cortical bone by load changes during daily activities generate numerous microcracks that will be removed by osteolysis via osteoclasts. In turn growth factors present in latent forms in osteocytes are being activated and released into the extracellular matrix where they stimulate osteoblasts to refill the cavities produced during osteolysis. An electromagnetic field should be able to stimulate the multiplication of growth factors and by this mechanism accelerate indirectly the bone remodeling process. Similar conclusions were drawn in a contribution by Wang et al. [66] suggesting that enhanced bone in-growth by electrical polarization into hydroxyapatite with high interpore connectivity is controlled by cooperative interaction between osteoconductivity of the hydroxyapatite and enhanced osteogenic cell activity induced by large charges stored at pore surfaces.

6. Concept of a Novel Bio'capacitor' Device for Accelerated Bone Healing

Based on the favorable mechanical, electrical and osteoconductive properties the novel bioceramic CTZ_3P_3 coatings could be utilized to design devices for accelerating bone growth and possibly also providing a remedy for osteoporotic damage of long bones.

Such a device may consist conceptually of a succession of (i) a bioinert electrically conducting metal substrate such as Ti6Al4V, (ii) a bioinert dielectric bond coat of about 20 μm thickness such as TiO_2, (iii) a thin (5-10 μm) bioactive, i.e. osteoconductive intermediate coat with CTZ_3P_3 composition, and (iv) possibly a top coat of hydroxyapatite composition as thick as 100 μm. Electrical gold lead wires would connect the Ti6Al4V implant (positive pole) and the intermediate coat (negative pole) to a battery or transducer implanted subcutaneously near the bone growth site, similarly to existing shielded power sources for heart pacemakers.

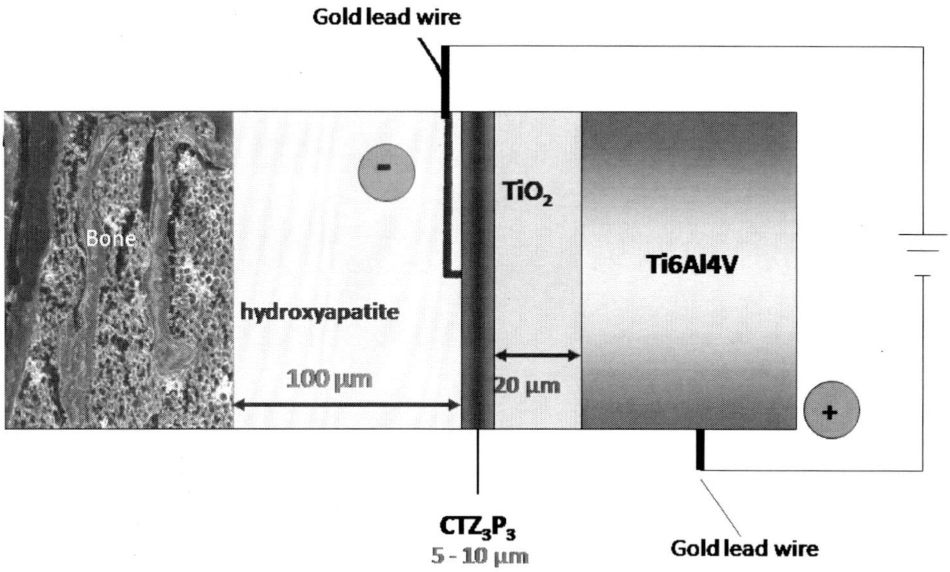

Figure 7. Schematic cross-section of a bio"capacitor" circuit designed to accelerate healing of broken and damaged bone by storing negative electric charges close to the bone.

Figure 7 shows schematically the arrangement of the different coating layers. In effect the device constitutes the equivalent circuit of a capacitor that would store negative electric charges in the immediate vicinity of the growing bone.

However, the current transmitted to the bone growth site will be attenuated by the HAp layer owing to its dielectric nature [69]. This means that the thickness of this layer must be well optimized. The device could be applied to accelerate healing of operational traumata, speed up bone in-growth, and may also be used as potential treatment of osteoporotic bone. All this will result in reduced health care cost as the time of hospitalization of patients could be drastically shortened.

CONCLUSION

Calcium (titanium, zirconium) orthophosphate ceramics of the general composition $CaTi_nZr_{4-n}(PO_4)_6$ ($0 < n < 4$) show solubility in simulated body fluid at least one order of magnitude lower than that of other calcium orthophosphates including hydroxyapatite and, in particular tricalcium phosphate and tetracalcium phosphate. Plasma-sprayed coatings of transition metal-substituted calcium phosphates display sufficiently strong adhesion to Ti6Al4V implant substrates, with reasonable shear strength. However, considerable thermal decomposition due to incongruent melting has been observed as indicated by formation of zirconium di(pyro)phosphate (ZrP_2O_7), rutile (α-TiO_2), and baddeleyite (β-ZrO_2). There is some evidence, however, that these bioinert decomposition products may lead to particle-mediated reinforcement of the microstructure thus improving the cohesive strength of the coatings.

In vitro biocompatibility tests with primary rat bone marrow cells and also human bone-derived cells showed enhanced expression of non-collageneous proteins and thus substantial cell proliferation in the presence of fetal bovine serum. In addition, *in vivo* animal tests proved that thicker coatings based on CTZ_3P_3 applied to Ti6Al4V rods and implanted in the femura of sheep and dogs promoted strong induction of dense bone at a stable interface implant-bioceramic coating without deleterious coating delamination frequently observed with hydroxyapatite.

Future research should also concentrate on development of a composite $TiO_2//CTZ_3P_3//(HAp)$ coating system to design implants with a „bio"capacitor equivalent circuit that by appropriate poling could store negative electrical charges close to the interface with the growing bone thus enhancing bone apposition rate and bone density. For example, utilization of a bone growth stimulator based on CTZ_3P_3 would be required when during an endoprosthetic replacement operation involving an exchange of the implant the cortical bone matter has been previously damaged. This frequently happens in conjunction with an undesirable geometric configuration of the implant-supporting cortical bone. In this situation a thin, rapidly resorbed calcium phosphate-based coating may not be sufficient to sustain the necessary large-scale bone regeneration. Hence either thicker HAp coatings must be used to stimulate bone reconstruction over longer times, or thin coatings of biomaterials with higher resorption resistance and better adhesion to the implant. Such a favorable material could be transition metal-substituted calcium phosphate with NaSiCON structure described in the present contribution.

REFERENCES

[1] J.A. Epinette, in: *Bioceramics in Joint Arthroplasty*, A. Toni, G. Willmann (eds.), Georg Thieme, Stuttgart, New York, pp. 84-91 (2001).

[2] N. Creugers, *Evidence-based Dentistry*, 3(3), 77 (2002).

[3] T.J.S. Puolakka, K.J.J. Pajamäki, P.J. Halonen, P.O. Pulkkinen, P. Paavalainen, J.K. Nevalainen, *Acta Othop. Scand.*, 72, 433 (2001).

[4] Q.Q. Hoang, F. Sikeri, A.J. Howard, D.S.C. Yang, Nature 425, 977 (2003).

[5] T. Kokubo, H.M. Kim, M. Kawashita, *Biomaterials*, 24, 2161 (2003).

[6] E. Dörre, in: *Künstlicher Knochenersatz in der Orthopädie und Traumatologie*, A. Kirgis, W. Noack (eds.), Pontenagel Press, Bochum, Germany, p. 17 (1992).

[7] R.B. Heimann, N. Schürmann, R.T. Müller, *J. Mater. Sci.: Mater. Med.*, 15, 1045 (2004).

[8] J. Gallo, P. Kamínek, V. Tichá, P. Reháková, R. Ditmar, *Biomed. Papers*, 146(2), 21 (2002).

[9] R.B. Heimann, in: *Trends in Biomaterials Research*, Patrick J. Pannone (ed.), Nova Science Publishers Inc., Hauppauge, USA, pp. 1-80 (2007).

[10] X. Nie, A. Leyland, A. Matthews, *Surf. Coat. Technol.*, 125(1-3), 407 (2000).

[11] J. Alamo, *Solid State Ionics*, 63/65, 547 (1993).

[12] N. Hirose, *J. Mater. Chem.*, 4, 9 (1994).

[13] A.C. Lasaga, *J. Geophys. Res.*, 89, 4009 (1984).
[14] R. Roy, E.R. Vance, J. Alamo, *Mater. Res. Bull.*, 17, 585 (1982).
[15] R. Roy, L.J. Yang, J. Alamo, E.R. Vance, in: *Nuclear Waste Management*, D.G. Brookings (ed.), Plenum, New York, 1983.
[16] T. Suzuki, M. Toriyama, H. Hosono, Y. Abe, *J. Ferment. Bioeng.*, 72(5), 384 (1991).
[17] C.C. Ribeiro, C.C. Barrias, M.A. Barbosa, *Biomaterials*, 25(18), 4363 (2004).
[18] H.Y. Hong, *Mater. Res. Bull.*, 11, 173 (1976).
[19] H. Hosono, Y. Sakai, Y. Abe, *J. Non-Cryst. Solids*, 139, 90 (1992).
[20] S. Senbhagaraman, T.N. Guru Row, A.M. Umarji, *J. Mater. Chem.*, 3(3), 309 (1993).
[21] R. Roy, D.K. Agrawal, J. Alamo, R.A. Roy, *Mater. Res. Bull.*, 19(4), 471 (1984).
[22] J. Alamo, J.L. Rodrigo, *Solid State Ionics*, 63/65, 678 (1993).
[23] S. Senbhagaraman, A.M. Umarji, *J. Solid State Chem.*, 85, 169 (1990).
[24] C. Calvo, R. Gopal, *Am. Mineral.*, 60, 120 (1975).
[25] T.J. McCoy, I.M. Steele, K. Keil, B.F. Leonard, M. Endress, *Am. Mineral.*, 79, 375 (1994).
[26] H. Strunz, E.H. Nickel, *Strunz Mineralogical Tables. Chemical Structural Mineral Classification System*, 9th ed., E. Schweizer-bart'sche Verlagsbuchhandlung, Stuttgart, 2001, p. 433.
[27] C.C. Silva, M.P.F. Graça, M.A. Valente, A.S.B. Sombra, *J. Non-Cryst. Solids*, 352, 1490 (2006).
[28] R.O. Fuentes, F.M. Figueiredo, M.R. Soares, F.H.B. Marques, *J. Eur. Ceram. Soc.*, 25(4), 455 (2005).
[29] T. Kokubo, *Thermochim. Acta*, 280/81, 479 (1996)
[30] R.B. Heimann, K. Schneider, Final report, DFG project He 923/9-1 (2000).
[31] K. Schneider, Unpublished Master thesis, TU Bergakademie Freiberg (2002).
[32] S. Szmukler-Moncler, G. Daculsi, J. Delécrin, N. Passuti, C. Deudon. *Adv. Biomater.*, 10, 377 (1992).
[33] K. Schneider, R.B. Heimann, G. Berger, *Mat.-wiss. u. Werkstofftech.* 32, 166 (2001).
[34] R.B. Heimann, *Plasma Spray Coating. Principles and Applications*, 2nd ed., Wiley-VCH, Weinheim, 2008. ISBN 978-3-527-32050-9.
[35] T. Kasuga, T. Kimata, A, Obata, *J. Am. Ceram. Soc.*, 92(8), 1709 (2009).
[36] G.E.P. Box, W.G. Hunter, J.S. Hunter, *Statistics for Experimenters. An Introduction to Design, Data Analysis, and Model Building*. John Wiley and Sons, New York, 1978.
[37] D.R. Levine, R.A. Miller, M.A. Gedwill, in: Proc. 2nd Conf. on Adv. Mater. for Alternative Fuel-Capable Heat Engines, Monterey, CA, 1981.
[38] C. Godoy, E.A. Souza, M.M. Lima, J.C.A. Batista, *Thin Solid Films*, 420/21, 438 (2002).
[39] R.B. Heimann, in: *Plasma Spray Coating, Principles and Applications*, 2nd ed., Wiley-VCH, Weinheim, 2008, p.144.
[40] R.B. Heimann, *BIOmaterialien*, 7(1), 29 (2006).
[41] C. Knabe, G. Berger, R. Gildenhaar, F. Klar, H. Zreiqat, *Biomaterials*, 25(20), 4911 (2004).
[42] C.C. Barrias, C.C. Ribeiro, M. Lamghari, C.S. Miranda, M.A. Barbosa, *J. Biomed. Mater. Res. A*, 72(1), 57 (2005).
[43] G. Reisel, Unpublished Master thesis, Rheinisch-Westfälische Technische Hochschule Aachen (1996).

[44] E. Bouyer, F. Gitzhofer, M.I. Boulos, *J. Mater. Sci. Lett.*, 49(2), 58 (1997).
[45] E. Fukada, *J. Phys. Soc. Jpn.*, 12, 1159 (1957).
[46] M.H. Shamos, L.S. Lavine, M.I. Shamos, *Nature,* 197, 81 (1963).
[47] M. Weigert, *Anregung der Knochenbildung durch elektrischen Strom*, Hefte zur Unfallheilkunde, 115, Springer, Berlin, Heidelberg, New York, 103 pp.
[48] C.A.L. Bassett, R.O. Becker, *Science*, 137, 1065 (1962).
[49] C.A.L. Bassett, *Calcif. Tissue Res.*, 1(4), 252 (1968).
[50] C.A.L. Bassett, R.S. Pawluck, A. Pilla, *Ann. N.Y. Acad. Sci.*, 238, 242 (1974).
[51] A.M. Weinstein, *J. Biomed. Mater. Res.*, 10, 231 (1976).
[52] S.M. Colella, A.G. Miller, R.G. Stang, T.G. Stoebe, D.M. Spengler, *J. Biomed. Mater. Res.*, 15, 37 (1981).
[53] J.L. Berry, J.M. Geiger, J.M. Moran, J.S. Skraba, A.S. Greenwald, *J. Biomed. Mater. Res.*, 20, 65 (1986).
[54] J. Watson, W.G. de Haas, S.S. Hauser, *Nature*, 254, 331 (1975).
[55] B. Rubin, A. Saffir, *Nature*, 225, 78 (1970).
[56] K. Yamashita, N. Oikawa, T. Umegaki, *Chem. Mater.*, 8, 2697 (1996).
[57] P. Calvert, S. Mann, *Nature*, 386, 127 (1997).
[58] S. Nakamura, T. Kobayashi, M. Nakamura, K. Yamashita, *J. Mater, Sci.: Mater. Med.*, 20(1), 99 (2009).
[59] F.R. Baxter, I.G. Turner, C.R. Bowen, J.P. Gittings, J.B. Chaudhuri, *J. Mater. Sci.: Mater. Med.*, 20(8), 1697 (2009).
[60] J.B. Park, B.J. Kelly, G.H. Kenner, A.F. von Recum, M.F. Grether, W.W. Coffeen, *J. Biomed. Mater. Res.*, 15, 103 (1981).
[61] D. Schumacher, V. Strunz, U. Gross, in: *Osteogenese und Knochenwachstum*, M.H. Hackenbroch, H.J. Refior, M. Jäger (eds.), Georg Thieme, Stuttgart, New York, p. 231 (1982).
[62] T.J. Webster, C. Ergun, R.H. Doremus, W.A. Lanford, *J. Biomed. Mater. Res.*, 67A(3), 976 (2003).
[63] N. Ohtsu, K. Saito, K. Asami, T. Hanawa, *Surf. Coat. Technol.*, 200, 5455 (2006).
[64] D. Wei, Y. Zhou, D. Jia, Y. Wang, Surf. *Coat. Technol.*, 201, 8715 (2007).
[65] C. Qu, Q.H. Qin, Y. Kang, *Biomaterials*, 27(21), 4050 (2006).
[66] W. Wang, S. Itoh, Y. Tanaka, A. Nagai, K. Yamashita, *Acta Biomater.*, 5(8), 3132 (2009).
[67] U. Gross, C. Müller-Mai, G. Berger, U. Ploska, *Key Eng. Mater.*, 240/242, 629 (2003).
[68] U. Gross, C. Müller-Mai, G. Berger, U. Ploska, *Key Eng. Mater.*, 254/256, 635 (2004).
[69] S. Hontsu, T. Matsumoto, J. Iishi, M. Nakamori, H. Tabata, T. Kawai, *Thin Solid Films*, 295, 214 (1997).
[70] A. Bernstein, D. Nobel, H.O. Mayr, F. Göbel, G. Berger, U. Ploska, R. Gildenhaar, J. Brandt, *J. Biomed. Mater. Res., part B Appl. Biomater.*, 86(2), 422 (2008).

In: Calcium Phosphate
Editor: Robert B. Heimann

ISBN 978-1-62257-299-1

Chapter 15

CALCIUM PHOSPHATES IN DAIRY PRODUCTS

Frédéric Gaucheron*
[1]INRA, UMR1253, Rennes, France
[2]AGROCAMPUS OUEST, UMR1253, Rennes, France

ABSTRACT

Milk and dairy products contain significant quantities of calcium and phosphate. The quantity and quality of calcium and phosphate present in these products depend on the type of dairy products and especially on their physico-chemical conditions. In milk, the total concentrations of calcium and inorganic phosphate are 30 and 20 mM, respectively. Both compounds are located in the aqueous and micellar phases. In the aqueous phase, calcium is mainly associated with citrate and inorganic phosphate. This phase, containing less than 1 mM of calcium phosphate, is considered supersaturated. In the micellar phase, the calcium concentration is about 20 mM, associated with organic phosphate (phosphoseryl residues of casein molecules) and inorganic phosphate to form nanoclusters of calcium phosphate. Their physical and chemical structures are not completely resolved. It is thought that associations of calcium with inorganic and organic phosphates contribute to the structure and stability of casein micelles. These different forms of calcium phosphate present in milk are strongly modified by physico-chemical conditions. For example, a solubilisation of calcium phosphate from casein micelles is observed during milk acidification or after addition of calcium chelatants. On the other hand, calcium and phosphate co-precipitate when milk or whey liquids are heated and/or made alkaline. In cheese, the contents of calcium and phosphate are functions of the technological steps used during manufacture. In fact, the calcium and inorganic phosphate contents are strongly related to the acidification step. Moreover, since acidification is associated with the draining step during cheese making, the calcium concentration in the curd and final cheese will be reduced. Thus, the calcium and inorganic phosphate contents variy on the order: milks/fermented milks/fresh cheeses < soft cheeses < semi-hard cheeses < hard cheeses. The nature of calcium phosphates occurring in dairy products has received less attention than that present in milk. It has been reported that in some ripened cheeses such as camembert, insoluble calcium phosphate forms particularly in its rind. During dairy technologies such as heat treatment,

* E-mail address: frederic.gaucheron@rennes.inra.fr.

evaporation under vacuum, and membrane technologies, calcium phosphate can precipitate in the processing equipment causing problems of fouling. Several dairy industries recover calcium phosphate contained in different dairy liquids such as cheese whey by heating and alkalinization. The chemical and physical properties of these milk-based calcium phosphates depend on the whey used and the parameters of manufacture. These mineral ingredients are later used for calcium enrichment of different food products.

1. INTRODUCTION

The objective of this chapter is to describe what is known about calcium phosphate in milk and dairy products. In the first section, the concentrations of calcium and phosphate and their distribution among the soluble and micellar phases of milk will be defined. Special discussion will focus on the possible chemical and physical forms of micellar calcium phosphate and its importance for the structure and stability of casein micelles which are the main milk proteins. The modifications of the distributions of calcium and phosphate, called 'milk salt equilibrium', as a function of physico-chemical conditions practiced in dairy industry will be discussed in the second part with particular attention paid to the phenomena of precipitation and solubilization of calcium phosphate. The nature of the different possible calcium phosphate salts will also be addressed under these different conditions. In the third section, the quantities and the nature of calcium phosphate in different cheeses (fresh, soft, semi-hard, hard and processed cheeses) will be described. Evidently the amounts of calcium phosphate can be very different as a function of cheese manufacture, and calcium phosphate can exist in different physical and chemical forms. In the fourth section, the fouling of equipment commonly used in dairy industry such as heat-exchangers, evaporators and filtration membranes will be discussed. In these cases, calcium phosphate can precipitate and thus affect the performance and productivity of the equipment. In the last section, possibilities to recover calcium phosphate from different dairy liquids by playing with the physico-chemical conditions, in particular alkalinization and temperature will be elaborated on. The characteristics and applications of these dairy calcium phosphate ingredients will be described and discussed.

2. CALCIUM PHOSPHATE IN MILK

Milk is a complex liquid rich in proteins, lipids, lactose, vitamins and minerals. The main minerals are phosphate, chloride and citrate of calcium, magnesium, sodium and potassium (Table 1) [1-8].

In particular, calcium and phosphate occur in two main locations, the soluble and micellar phases (Table 2 and Fig. 1) [1-8].

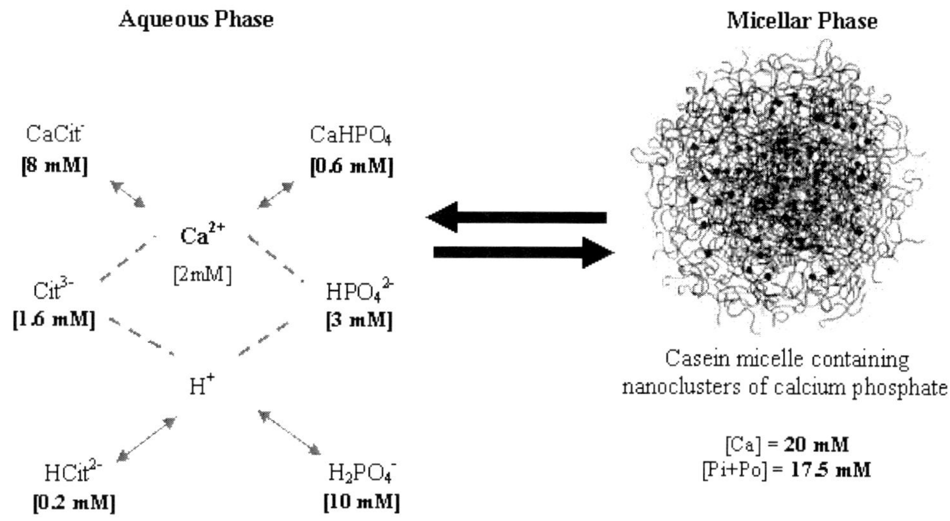

Figure 1. Calcium and phosphate distributions in milk at pH 6.7. In the aqueous phase, calcium is free and associated with inorganic phosphate and citrate. In the micellar phase, calcium is associated with organic phosphate Po (phosphoseryl residues in casein molecules) and inorganic phosphate (Pi) to form nanoclusters (black points in the interior of casein micelles). The concentrations of each association are indicated within brackets in the figure.

The soluble (aqueous or diffusible) phase corresponds to the water of milk (left part of Fig. 1). It contains different ions and molecules (whey proteins, lactose, water soluble vitamins and some minerals). In this phase, calcium exists in three different forms:

- free ion or ionic calcium Ca^{2+} (about 2 mM);
- in association with trivalent citrate Cit^{3-} (about 8 mM);
- in association with inorganic phosphate (as a mixture of $H_2PO_4^-$ and HPO_4^{2-}) (about 0.6 mM).

The low concentration of calcium phosphate in this aqueous phase is due to its low solubility. At the pH of milk (about 6.6-6.7), the soluble phase is considered supersaturated in calcium phosphate.

Table 1. Mineral composition of cow milk [1-8]

Ion	Concentration (mg/kg)	Concentration (mmol/kg)
Calcium	1,043-1,283	26-32
Magnesium	97-146	4-6
Inorganic phosphate	1,805-2,185	19-23
Citrate	1,323-2,079	7-11
Sodium	391-644	17-28
Potassium	1,212-1,681	31-43
Chloride	772-1,207	22-34

Table 2. Distribution of the main ions between soluble and micellar phases of skim milk at pH 6.7. The concentrations in the micellar phase are deduced from the difference between total and soluble concentrations [1-8]

Ion	Concentration (mmol/kg)
Total calcium	30
Soluble calcium	10
Total magnesium	5
Soluble magnesium	3.5
Total inorganic phosphate	21
Soluble inorganic phosphate	11
Total citrate	10
Soluble citrate	9
Total sodium	25.5
Total potassium	37
Total chloride	30

The micellar phase consists of casein micelles, complex supramolecular structures containing different casein molecules (α_{s1}-, α_{s2}-, β- and κ-caseins) (right part of Fig. 1). Calcium and phosphate are also present in this phase. This micellar calcium phosphate can be schematically described as a mixture of calcium caseinate (containing organic phosphate present in the phosphoseryl residues of caseins) and calcium orthophosphate. This type of calcium phosphate is called micellar or colloidal and consists in nanoclusters of calcium phosphate. The chemical and physical natures of this micellar calcium phosphate are not well understood. Studies by calculation of the apparent solubility product [9,10] or performed by different methods such as x-ray absorption and infrared spectroscopies [10-17] show that its composition is close to brushite. Holt [3,4] proposed a composition akin to a modified dicalcium phosphate $Ca(HPO_4)_{0.7}(PO_4)_{0.2}$, and van Dijk [18,19] proposed a complex containing four moieties of inorganic phosphate, two phosphoseryl residues and eight cations (six calcium ions). However, the size of this complex is not compatible with the size of 2.5 to 3 nm determined by cryo-transmission electron microscopy and x-ray scattering [20]. It was found to be amorphous using diffraction methods and high-resolution transmission electron microscopy [12,13]. Choi *et al.* [21] considered two possible shapes of the nanoclusters which would need to be capped by phosphoserine clusters. The nanoclusters were found to be tetrahedra with four faces or (trigonal) bipyramids with six faces. From these considerations, the theoretical molecular weight of the nanocluster were calculated to be 4,898 g/mol for a tetrahedral morphology (36 molecular units of $CaHPO_4$) or as large as 9,757 g/mol for a bipyramidal morphology (72 molecular units of $CaHPO_4$), the next larger form. It is, at the same time, a cross-linking and a neutralizing agent of phosphoseryl residues and for these reasons the micellar calcium phosphate presumably contributes to the structure and stability of casein micelles [20-31].

The mineral fraction, in particular calcium and phosphate has been subjected to extensive investigations for over 60 years and today it is generally assumed that equilibrium of calcium and inorganic phosphate concentrations exist between the aqueous and micellar phases . The distribution of these ions is dynamic with permanent ion exchanges between both phases [32-34].

3. Quantitative and Qualitative Changes of Milk Calcium Phosphate as a Function of Different Physico-chemical Conditions

As a function of physico-chemical conditions, the salt equilibrium in milk previously described is modified [5,6,35]. Indeed, variations in pH (acidification or alkalinization), thermal treatment (increase or decrease in temperatures), and addition of different salts and chelatants change the partition of calcium and inorganic phosphate. The modifications of these equilibria are controlled by the solubility of the calcium phosphate salt in the aqueous phase. Thanks to this important knowledge on the behavior of the mineral fraction, today it is possible to predict by theoretical calculation the salt equilibrium as a function of different physico-chemical conditions [10,36,37]. On the other hand, the consequences of these changes of the salt equilibrium include modifications of the structure and stability of casein micelles.

3.1. Solubilization of Micellar Calcium Phosphate Induced by Acidification

Acidification is a key step in the manufacture of several dairy products including fermented milks, different cheeses, and casein. During acidification, solubilization of micellar calcium phosphate is observed [38-46]. From a mechanistical point of view, the H^+ ions added or biologically produced by fermentation of lactose in lactic acid bind the forms of citrate calcium and calcium phosphate. These bindings of H^+ decrease the interactions between phosphate and calcium in the aqueous phase. Consequently, transfers of inorganic phosphate and calcium from the micellar to the aqueous phase are observed. These solubilizations of calcium and inorganic phosphate of milk can be described in three steps [42] (Fig. 2):

- between pH 6.7 and 6.0: 22 and 34 % of calcium and inorganic phosphate, respectively are solubilized;
- between pH 6.0 and 5.2: 50 and 66 % of calcium and inorganic phosphate, respectively are solubilized. It is noteworthy that (i) in this pH range, the solubilization of calcium is linear; (ii) at pH 5.2, the micellar phosphate is totally solubilized;
- between pH 5.2 and 3.5 : 28 % of calcium are solubilized. At pH 3.5, the totality of calcium is soluble.
- At pH < 2, calcium ions are in ionic form and inorganic phosphate occurs as phosphoric acid (H_3PO_4).

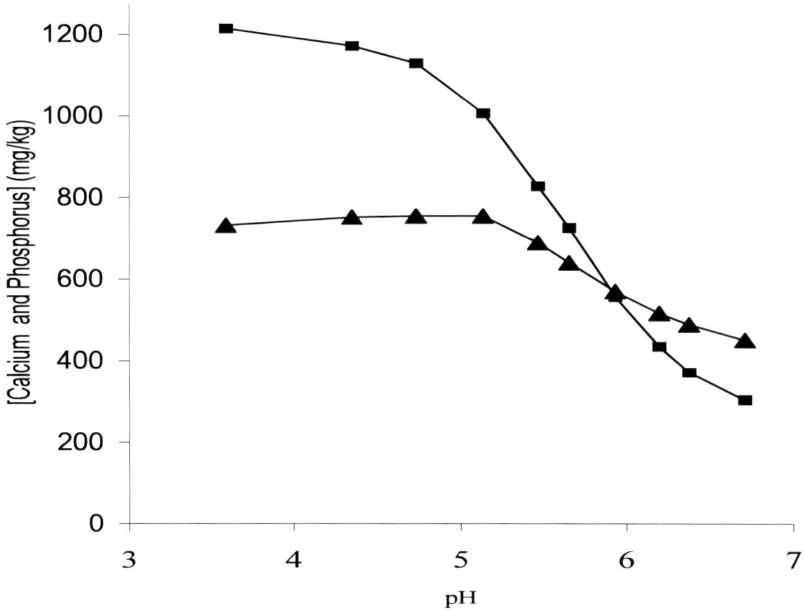

Figure 2. Concentrations of calcium (■) and phosphorus (▲) in the aqueous phase of milk between pH 6.7 and pH 3.5 [42].

3.2. Solubilization of Micellar Calcium Phosphate induced by Calcium Chelatant Additions

Addition of calcium chelatants to dairy liquids is sometimes performed to prevent calcium phosphate precipitation which is deleterious for (i) the performance of dairy equipments like heat-exchangers and filtration membranes and (ii) the stability of some dairy products such as dairy formula. Addition of calcium chelatants such as citrate, EDTA, and oxalate to dairy liquids modifies the salt equilibrium between the aqueous and micellar phases. As these chelatants have high affinities for calcium, they bind to it and consequently the micellar calcium phosphate is progressively solubilized with destruction of the structure of casein micelles in turn [47-49]. In this case, the stability of the modified casein micelles is improved. The effects depend on the type and concentration of chelatants.

3.3. Precipitation of Calcium Phosphate by Alkalinization

Increase in pH from 6.7 to 10.0 of dairy liquids (milk, whey liquids, permeates) induces effects opposite to those described for acidification. The mineral equilibria are strongly modified and from mechanistical point of view, there are two successive steps: first ionization (deprotonation) of phosphate ions ($H_2PO_4^-$ ⇨ HPO_4^{2-} ⇨ PO_4^{3-}) and secondly associations of these ionized anions with calcium ions to form calcium phosphate salts. As the aqueous phase is supersaturated in calcium phosphate, precipitation is observed.

Alkalinization of milk is not a common practice used in dairy industry. During these physico-chemical changes, casein micelles is probably demineralized and strong changes in the micellar structure occur since milk becomes transparent, indicating a decrease in the micellar size [50-53]. Alkalinization of whey liquids and permeates is visually different because casein micelles are not present in these liquids. In this case, a visual precipitation of calcium phosphate is observed. This whey alkalinization is known to eliminate a significant part of calcium phosphate present in the fluid. Indeed, the minerals are sometimes not desired because they influence negatively the process of spray drying. In this context, Fauquant *et al.* [54] alkalinized acid whey to precipitate calcium phosphate and demineralize the whey. After alkalinization and precipitate removal, the contents in calcium and phosphorus decrease by 80 and 75 %, respectively. On the other hand, it is interesting to recover (or eliminate) the calcium phosphate precipitate which can then be used as ingredient for calcium enrichment of different food products (see section 6).

3.4. Precipitation of Calcium Phosphate induced by Thermal Treatment

Thermal treatments are classical operations performed in dairy industry. They are realized to improve the microbiological quality of the dairy liquids for safe consumption. The heat treatments can be classified into four types: thermisation (50 - 60°C for few seconds), pasteurization (70-90°C for a few seconds to few minutes), UHT treatment (140°C for 3-5 seconds), and sterilization (higher than 100°C for several minutes).

During heat treatment of milk, proteins, lactose, vitamins and minerals are more or less modified depending on the intensity of the heat treatments. Concerning calcium and phosphate, heat treatment induces ionization of phosphate as follows: $H_2PO_4^- \Rightarrow HPO_4^{2-} \Rightarrow PO_4^{3-}$. The ionized forms of phosphate react with calcium to form calcium phosphate salts. As the aqueous phase is supersaturated with calcium phosphate salts, precipitation is observed. The consequence is a decrease of calcium and inorganic phosphate in the aqueous phase [55-60]. These decreases depend on the couple time-temperature and it is admitted that thermisation or pasteurization do not irreversibly affect the mineral equilibrium. Indeed, the quantification of this precipitation is often difficult to evaluate for these heat treatments because these reactions are reversible. During more intensive heat treatment like sterilization, the mineral equilibria are strongly and irreversibly modified with precipitation of calcium phosphate. By direct recovery of the aqueous phase by ultrafiltration at high temperatures, Pouliot *et al.* [57-60] showed that a heat treatment of milk at 93°C for 11 min induces precipitation of calcium phosphate with decreases of 50 and 18 % of calcium and phosphorus in the aqueous phase, respectively. The exact nature of the precipitate and its possible association with milk proteins are not well established. Depending on the authors, it could be poorly crystallized tricalcium calcium phosphate [7, 61,62], hydroxyapatite or amorphous calcium phosphate [38] or dicalcium phosphate [57].

As described for thermal treatments of milk, one consequence of heat treatment of whey liquids or milk permeate (these liquids do not contain casein molecules) is also a decrease of calcium phosphate solubility [63,64]. Brulé *et al.* [63] studied the influence of different heat treatments on the stability of mineral fraction of milk permeate. Thus, heat treatment between 60 and 90°C induces a decrease in soluble calcium and inorganic phosphate suggesting calcium phosphate precipitation. These decreases of concentrations in the soluble phase increase with the intensity of the heat treatment. Analysis of precipitates by x-rays indicates the presence of di- and tricalcium phosphate.

3.5. Precipitation of Calcium Phosphate by Addition of Supplementary Calcium

Calcium addition to milk and whey liquids is a common practice in dairy industry. Calcium can be added for technological or nutritional reasons [65]. Calcium enrichment of milk modifies the salt equilibrium and the effects depend on the concentration and type of added calcium.

Addition of calcium chloride to milk induces an association of calcium ions added to inorganic phosphate and citrate ions present in the aqueous phase. As this phase is supersaturated in calcium phosphate, precipitation is indicated [66-69]. The chemical and physical nature of the newly formed calcium phosphate precipitate is not precisely known. On the other hand, modification of the phospho-calcic equilibrium induces strong changes of the structure of casein micelles including decreases in their hydration and zeta-potentials, and increases in the lightness, turbidity and protein hydrophobicity [66-69]. As consequences, the thermal stability of casein micelles decreases and the rennet coagulation performed in cheese technology is improved. Similar effects are observed after addition of lactate or gluconate of calcium.

Addition of calcium to whey liquids and permeate of filtration modifies also the mineral interactions [69-73]. Globally, the calcium added reacts with inorganic phosphate to form insoluble salts of calcium phosphate. The formation of this salt is increased if the pH is higher than 7. Addition of solid calcium salts can also be performed to initiate precipitation of calcium phosphate [72]. These different methods are used in dairy industry to recover calcium phosphate which can be later used as fortifying ingredient rich in calcium [65, 73].

4. Calcium Phosphate in Cheeses

The quantities of calcium and phosphate in cheeses are as function of the technological steps used during their manufacture [5, 7, 8, 74, 75]. In fact, the calcium content is strongly related to the acidification step. Moreover, if acidification is associated with the draining step (elimination of whey liquid), the calcium content in the curd and final cheese will be reduced. Thus, the calcium content varies as follows: milks/fermented milks/fresh cheeses < soft cheeses < semi-hard cheeses <hard cheeses (Table 3) [75].

The chemical forms and associations of calcium and phosphate are less well described in cheeses compared to milk. However, it was reported that in some ripened cheeses the formation of insoluble calcium phosphate occurs especially in the cheese rind [76-83]. The interpretation of this precipitation of calcium phosphate at the cheese surface is related to water evaporation and pH increase at the cheese surface due to metabolic changes such as degradation of proteins with ammoniac production and consumption of lactate ions. As a consequence of these precipitations at the surface, a concentration reduction of these same ions is observed in the aqueous phase after four days of ripening and beyond [81]. Similar results were obtained with model cheeses stored in ammoniacal atmosphere without development of surface flora [82]. Brooker [80] observed the presence of crystalline aggregates at the surface of soft cheese by polarized light microscopy or after coloration by the von Kossa technique (specific staining of phosphate compound). More recently, Hannon *et al.* [83] indicated the presence of a mineral layer at the surface of UF-cheeses ripened containing *Penicillium camemberti* as the surface flora (Fig. 3). The analysis of the rind reveals the presence of calcium phosphate which could induce defects of the surface.

Table 3. Calcium and phosphorus concentrations in different cheeses [75]. The concentrations are expressed in mg / 100 g of product

Minerals	Calcium	Phosphorus
Blue (USA)	528	387
Brick	674	451
Brie	184	188
Camembert	387	346
Cheddar	721	512
Cheshire	643	464
Cottage	60	132
Cream	80	104
Edam	731	535
Feta	492	337
Gouda	700	546
Gruyere	1,011	605
Limburger	498	393
Mozarella (whole milk)	517	371
Munster	717	468
Neufchatel	75	136
Parmesan (grated)	1,376	807
Pasteurized Process Cheese	723	526
Port Salut	650	360
Provolone	756	496
Ricotta (whole milk)	207	158
Roquefort	662	392
Tilsit	700	500

Other authors indicated the presence of microcrystals of calcium phosphate and calcium pyrophosphate dihydrate in processed cheeses [84,85]. Frau *et al.* [86] determined also the presence of microscopic crystalline inclusions in 32 samples of Mahón cheese (non-cooked pressed cheese manufactured from cow's milk) at different ripening times. These inclusions are circular, oval or kidney-shaped structures approximately 20 μm in diameter (range 4-65 μm). The quantity of crystalline inclusions per unit surface area observed in Mahón cheese is heterogeneous, with an average of 30 crystals/m^2. Calcium and phosphorus are found to be the most common elements present in these inclusions.

5. CALCIUM PHOSPHATE IN DAIRY EQUIPMENT

5.1. Fouling of Heat-exchangers

During heat treatment of dairy liquids, a deposit can be observed in the exchanger that induces fouling [87-95]. This fouling is related to denaturation of whey proteins that starts at

50-60°C and to precipitation of calcium phosphate owing to decreasing solubility with increasing temperature. In fact, it is described as precipitation of calcium phosphate at stainless steel surfaces of equipment, an association of calcium phosphate precipitates with β-lactoglobulin aggregates and /or with casein micelles. The fouling is more important for whey liquid (absence of casein micelles) compared to milk. Acid whey (strongly mineralized) is more problematic than sweet whey.

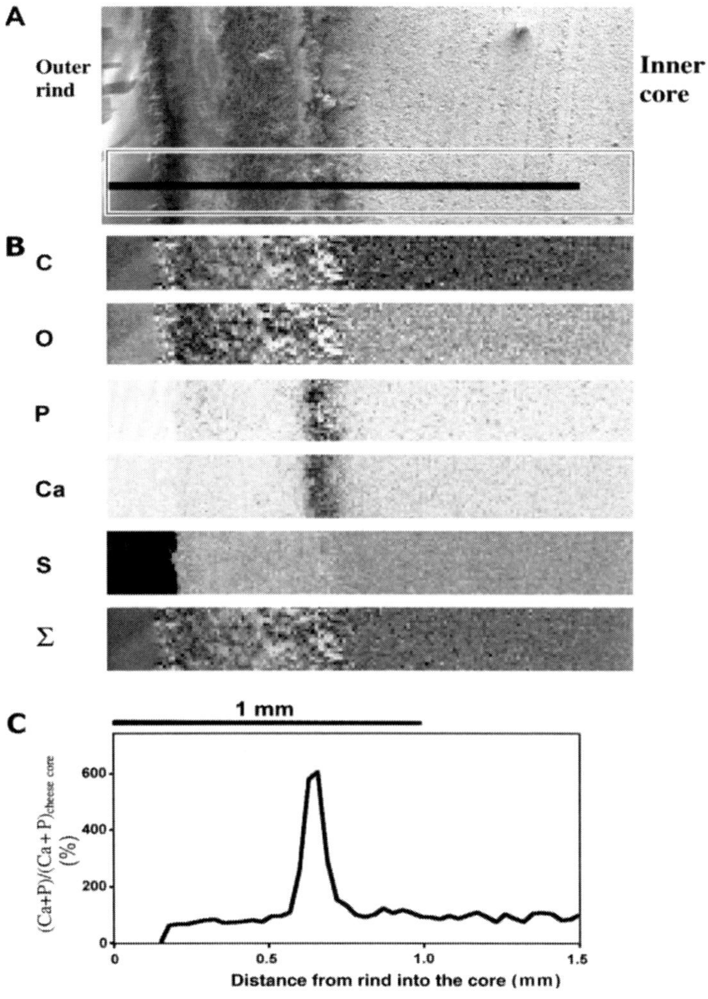

Figure 3. X-ray analysis of UF-cheese at day 50 of ripening. (A) High resolution image of secondary electrons showing the area of analysis. (B) Maps of carbon (C), oxygen (O), phosphorus (P), calcium (Ca), sulphur (S) and of total X photons (Σ). (C) Changes of (Ca + P) along the black line drawn in (A) as a percentage of (Ca + P) in the core zone [83].

The mineral analysis of deposits indicates that they are composed of mixtures of calcium phosphate dehydrate, $CaHPO_4{\cdot}2H_2O$ (brushite) and octacalcium phosphate, $Ca_8H_2(PO_4)_6{\cdot}5\ H_2O$ which can transformed to hydroxyapatite, $Ca_5OH(PO_4)_3$ [93]. Citrate and phosphate additions to milk before heat treatment decrease precipitation of fouling agents. Consequently, the problem of fouling is reduced and the performances of this type of treatment improved.

5.2. Fouling of Evaporators

Evaporation of dairy liquids under vacuum is used to eliminate water. This operation is mainly used on milk and whey liquids before spray drying. The concentration by this process induces an increase in the ionic strength, a decrease in pH, and an increase in the activity of ionic calcium [96]. The main consequence of these changes is precipitation of calcium phosphate. The evaporation of acid or sweet wheys to yield dry matter close to 30 % solid content induces calcium phosphate precipitation. Precipitation is more important for acid compared to sweet wheys. The precipitates are mainly composed of calcium phosphate and calcium citrate. Analyses of the precipitate using x-ray diffraction and infrared spectroscopy showed the presence of brushite [97].

5.3. Fouling of Filtration Membranes

In dairy industry, four different types of filtration are used: ultrafiltration, reverse osmosis, microfiltration, and nanofiltration [98-106]. The two dairy liquids most treated by these methods are milk and wheys. The aims of these operations are to concentrate milk protein or to separate casein micelles from whey proteins. In general, during treatment of dairy fluids by membrane technologies, a progressive decrease of productivity is observed as a function of time. These decreases are related to fouling due to absorption of proteins to the surface membrane and, as previously described, to precipitation of calcium phosphate. This precipitation is more intensive at higher values of temperature, pH and calcium concentration, and also when casein micelles (cases of whey liquids) and proteins (ultrafiltrate, nanofiltrate compared to milk and whey liquids) have been removed. Thus, the contribution of calcium phosphate to fouling is more important with whey liquids compared to milk. To avoid or to limit the problem of fouling, different physical (heat, separation by centrifugation or microfiltration) and chemical (adjustment of pH and ionic strength, addition of calcium chelatants) pretreatments can be performed before the use of membrane technologies [94].

6. Dairy Calcium Phosphate as Ingredient

Calcium ions participate in the structure of bones and teeth by providing rigidity, strength and hardness. At the same time, they contribute to blood clotting, neurotransmission, neuromuscular excitability, cardiac conduction, stimulation of the sympathic system, intestinal mobility, cell signaling with a role as second messenger, fertility, cell proliferation, muscle contraction, Secretion or cell death, hormone secretion, and structure and activation of certain proteins. For all these reasons, calcium-fortified foods exist on the market. Different possible sources of calcium to enrich food are indicated in Table 4.

Their contents in calcium depend on the type of salts, and dairy calcium phosphate constitutes a good and natural source of calcium. For this reason, it is of interest to isolate calcium phosphate from dairy liquids. The most interesting dairy liquids used to recover calcium are milk ultrafiltrates and cheese whey liquids [73]. Different methods are described in the literature. One principle of recovery is to reduce solubility of calcium phosphate by increasing the temperature and pH. The yield of recovery depends on temperature, pH, time of precipitation, and composition of the dairy liquids. One of possible conditions is a temperature higher than 45°C, and a pH above 7. The precipitate obtained can be separated by centrifugation or filtration, and then dried. For some applications, the powder should be micronized (reduction of the particle size) to limit the sedimentation in enriched products. One other described method to produce a milk-derived magnesium/calcium material is chromatography. In the protocol [107], calcium and magnesium contained in milk or milk substances are separated first on a weakly acidic cation exchange resin and then on a strongly acidic cation-exchange resin. Subsequently, magnesium and calcium adsorbed onto the strongly acidic cation-exchange resin are eluted and recovered as magnesium and calcium salts.

Table 4. Calcium content of different ingredients used for calcium food enrichment

Compound	Calcium %
Calcium phosphate	40
Dibasic calcium phosphate	29
Tribasic calcium phosphate	38
Calcium chloride	36
Calcium citrate	21
Calcium lactate	13
Calcium gluconate	9
Milk calcium phosphate	25

Table 5. Composition in calcium and phosphorus of different commercial preparations milk calcium

Suppliers	Product name	Calcium %	Phosphorus %
Fonterra	Alamin	28	15
DMV	Lactoval	Not indicated	Not indicated
Armor Proteins	Vitalarmor	24	11
Glanbia	Trucal	24	12.5

Depending on the origin of the initial materials and conditions of industrial extraction, the powders obtained are mainly composed of minerals with different contents in calcium and phosphorus (Table 5).

The products obtained are mixtures of different forms of calcium phosphate (mono-, di- and tricalcium phosphate) which are insoluble at neutral pH. Due to their insolubilities [108], they are also considered chemically inert. Their tastes are neutral. They are commonly used as ingredients in calcium-enriched food products such as processed cheeses, fresh cheeses, yoghurts, ice creams, fortified acid beverages, sport beverages, snacks, bakery products, meat and fish products, tablets and dietary supplements.

CONCLUSION

Milk is a liquid rich in calcium and phosphate. To localize these compounds is complex because they can be free, associated with other ions to form salts, and associated with proteins. This distribution is in thermodynamic equilibrium and all changes in the physico-chemical conditions of milk affect the physico-chemistry of this salt in terms of ionization, stoichiometry, solubilities and crystallization. In general, increase in pH, temperature and addition of supplementary calcium ions induces precipitation of calcium phosphate salts. On the other hand, acidification, cooling, or adding calcium chelatants to milk induce solubilization of micellar calcium phosphate. Consequently and depending on these conditions, multiple forms of calcium phosphate salts exist in milk, dairy products, and dairy equipment. They can be

- soluble in the aqueous phase of milk, cheese whey, or aqueous phase of cheese; in these cases, calcium and phosphate co-exist but saturation is not reached to yield precipitation of calcium phosphate;
- associated with organic and inorganic phosphate in casein micelles or in the cheese network;
- precipitated at the surface of heat-exchanger, filtration membranes or evaporator walls;
- precipitated in the rind of some cheeses.

In spite of this important knowledge concerning the calcium phosphate in dairy products, it is of further interest to precisely determine

- the structural and stabilizing roles calcium phosphate plays in casein micelles;
- the implication of this calcium phosphate for the functional properties of casein micelles such as their stability and capacity to form gels, emulsions and foams; and
- the effect of other compounds present in food products on the absorption of calcium and phosphate.

All these questions are being studied in research projects that are in progress or planned in the future.

REFERENCES

[1] H. Tessier and D. Rose, *J Dairy Sci.* 41, 351 (1958).
[2] G. Brulé, *Rev. Lait. Fr.* 400, 61 (1981).
[3] C. Holt, *The milk salts: their secretion, concentration and physical chemistry*, In : Fox P.F. (ed.), Developments in Dairy Chemistry, Vol. 3: Lactose and minor constituents, Elsevier Science, London, pp 143-181(1985).
[4] C. Holt, *The milk salts and their interaction with casein*, In : Fox P.F. (Ed.), Advanced Dairy Chemistry, Vol. 3: Lactose, Water, Salts and Vitamins, Chapman and Hall, London, pp 233-254 (1997).

[5] F. Gaucheron, *Minéraux et Produits Laitiers*, Tec & Doc, Paris (2004).
[6] F. Gaucheron, *Reprod. Nutr. Dev.* 45, 473 (2005).
[7] P. Walstra and R. Jenness, *Dairy chemistry and physics*, Wiley Intersciences, New York, (1984).
[8] P. Walstra, J. T. M. Wouters and T. J. Geurts, *Dairy Science and Technology* (2nd ed.), CRC Taylor and Francis, Boca Raton, USA (2006).
[9] L. C. Chaplin, *J. Dairy Res.* 51, 251 (1984).
[10] C. Holt, D. G. Dalgleish and R. Jenness, *Anal. Biochem.* 113, 154 (1981).
[11] C. Holt, M. J. J. M. van Kemenade, L. S. Nelson, L. Sawyer, J. E. Harries, R. T. Bailey and D. W. Hukins, *J. Dairy Res.* 56, 411 (1989).
[12] R. L. J. Lyster, S. Mann, S. B. Parker and R. J. P. Williams, *Biochim. Biophys. Acta* 801, 315 (1984).
[13] T. C. A. McGann, W. Buchheim, R. D. Kearney and T. Richardson, *Biochim. Biophys. Acta* 760, 415 (1983).
[14] T. C. A. McGann, R. D. Kearney, W. Buchheim, A. S. Posner, F. Betts and N. C. Blumenthal, *Calcif. Tissue Int.* 35, 821 (1983).
[15] C. Holt, S .S. Hasnain and D. W. L. Hukins, *Biochim Biophys. Acta* 719, 299 (1982).
[16] C. Holt and D. W. L. Hukins, *Int. Dairy J.* 1, 151 (1991).
[17] C. Holt, *Eur. Biophys.* J. 33, 421 (2004).
[18] H. J. M. Van Dijk, *Neth. Milk Dairy J.* 44, 65 (1990).
[19] H. J. M. Van Dijk, *Neth. Milk Dairy J.* 44,111 (1990).
[20] S. Marchin, J. L. Putaux, F. Pignon and J. Léonil, *J. Chem. Phys.* 126, 045101 (2007).
[21] J. Choi, D. S. Horne and J. A. Lucey, *J. Dairy Sci.* 94, 3250 (2011).
[22] C. Holt, *Adv. Protein Chem.* 43, 63 (1992).
[23] C. Holt and D. S. Horne, *Neth. Milk Dairy J.* 50, 85 (1996).
[24] D. S. Horne, *Int. Dairy J.* 8, 171 (1998).
[25] R. Tuinier and C. G. de Kruif, *J. Chem. Phys.* 117, 1290 (2002).
[26] D. S. Horne, *Curr. Opin. Colloid Interface Sci.* 11, 148 (2006).
[27] D. G. Dalgleish, *Lait* 87, 385 (2007).
[28] P. X. Qi, *Lait* 87, 363 (2007).
[29] D. J. McMahon and B. S. Oommen, *J. Dairy Sci.* 91, 1709 (2008).
[30] K. Nagy, A. M. Pilbat, G. Groma, B. Szalontai and F. J. G. Cuisinier, *J. Biol Chem.* 285, 38811(2010).
[31] D. G. Dalgleish, *Soft Matter* 7, 2265 (2011).
[32] A. Pierre, G. Brulé, J. Fauquant, *Lait* 63, 473 (1983).
[33] M. Bak, L .K. Rasmussen, T. E. Petersen and N. C. Nielsen, *J. Dairy Sci.* 84, 1310 (2001).
[34] Z. I. Kolara , T. G. Verburg and H. J. M. van Dijk, *J. Inorg. Biochem.* 90, 61 (2002).
[35] M. A. De la Fuente, *Trends Food Sci. Techn.* 9, 417 (1998).
[36] O. Mekmene, Y. Le Graët and F. Gaucheron, *Food Chem.* 116, 233 (2009).
[37] O. Mekmene, Y. Le Graët and F. Gaucheron. *J. Agric. Food Chem.* 58, 4440 (2010).
[38] G. Brulé and J. Fauquant, *J. Dairy Res.* 48, 91 (1981).
[39] A. C. M. Van Hooydonk, H. G. Hagedoorn and I. J. Boerrigter, *Neth. Milk Dairy J.* 40, 281 (1986).
[40]

J. Visser, A. Minihan, P. Smits, S. B. Tjan and I. Heertje, *Neth. Milk Dairy J.* 40, 351 (1986).
[41] D. G. Dalgleish and A. J. R. Law, *J. Dairy Res.* 56, 727 (1989).
[42] Y. Le Graët and G. Brulé, *Lait* 73, 51 (1993).
[43] M. H. Famelart, F. Lepesant, F. Gaucheron, Y. Le Graët and P. Schuck, *Lait* 76, 445 (1996).
[44] F. Gaucheron, Y. Le Graët, M. Piot and E. Boyaval, *Lait* 76, 433 (1996).
[45] E. Gastaldi, A. Lagaude and B. Tarodo De la Fuente, *J. Food Sci.* 62, 671 (1996).
[46] Y. Le Graët and F. Gaucheron, *J. Dairy Res.* 66, 215 (1999).
[47] S. Odagiri and T. A. Nickerson, *J. Dairy Sci.* 48, 19 (1965).
[48] C. V. Morr, *J. Dairy Sci.* 50, 1038 (1967).
[49] C. Le Ray, J. L. Maubois, F. Gaucheron, G. Brulé, P. Pronnier and F. Garnier, *Lait* 78, 375 (1998).
[50] S. Odagiri and T. A. Nickerson, *J. Dairy Sci.* 48, 1157 (1965).
[51] H. J. M. Van Dijk, *Neth. Milk Dairy J.* 46 101 (1992).
[52] B. Vaia, M. A. Smiddy, A. L. Kelly and T. Huppertz, *J. Agric. Food Chem.* 54, 8288 (2006).
[53] S. Ahmad, M. Piot, F. Rousseau, J. F. Grongnet and F. Gaucheron, *Dairy Sci. Technol.* 89, 387 (2009).
[54] J. Fauquant, A. Pierre and G. Brulé *Techn. Laitière* 1003, 37 (1985).
[55] D. G. Dalgleish, *Int. Dairy Fed. Bull.* 238, 31 (1988).
[56] C. Holt, *Effect of heating and cooling on the milk salts and their interaction with casein*, In : Fox P.F. (ed.), Heat Induced Changes in Milk, International Dairy Federation, No. 9501, Brussels, pp 105-133 (1995).
[57] Y. Pouliot, M. Boulet and P. Paquin, *J. Dairy Res.* 56, 185 (1989).
[58] Y. Pouliot, M. Boulet and P. Paquin, *J. Dairy Res.* 56, 193 (1989).
[59] Y. Pouliot, M. Boulet and P. Paquin, *J Dairy Res.* 56, 513 (1989).
[60] Y. Pouliot, M. Boulet and P. Paquin, *J Dairy Sci.* 72, 36 (1989).
[61] R. L. J. Lyster, *J Dairy Res.* 32, 208 (1965).
[62] L. S. Nelson, C. Holt and D. W. L. Hukins, *Physica B* 158, 103 (1989).
[63] G. Brulé, E. Real del Sol, J. Fauquant and C. Fiaud, *J Dairy Sci* 61,1225 (1978).
[64] Y. Pouliot and J. Landry, *J. Dairy Sci.* 73, 91 (1990).
[65] M. A. Augustin and R. P. W. William, *Food Australia* 54, 132 (2002).
[66] A. C. M. Van Hooydonk, H. G. Hagedoorn and I. J. Boerrigter, *Neth. Milk Dairy J.* 40,369 (1986).
[67] M. Philippe., F. Gaucheron, Y. Le Graët, F. Michel and A. Garem, *Lait* 83, 45 (2003).
[68] M. Philippe, F. Gaucheron and Y. Le Graët. *Milchwissenchaft* 59, 498 (2004).
[69] M. Philippe, Y. Le Graët and F. Gaucheron, *Food Chem.* 90, 673 (2005).
[70] H. W. Modler and D. B. Emmons, *J. Dairy Sci.* 61, 294 (1978).
[71] C. F. Brothersen, N. F. Olson and T. Richardson, *J. Dairy Sci.* 65, 23 (1982).
[72] Y. Pouliot, J. Landry and J. Giasson, *Lait* 71, 313 (1991).
[73] M. Harju, *Int. J. Dairy Techn.* 54, 61 (2001).
[74] J. A. Lucey and P. F. Fox, *J Dairy Sci* 76, 1714 (1993).
[75] F. V. Kosikowski and V. V. Mistry, in : Cheese and Fermented Milk Foods, Volume 1, origins and principles, 3nd edn, Westport, Connecticut, USA, pp 684-697, (1997).
[76] M. Metche and J. Fanni, *Lait* 58, 337 (1978).

[77] Y. Le Graët, A. Lepienne, G. Brulé and P. Ducruet, *Lait* 63, 317 (1983).
[78] Y. Le Graët and G. Brulé, *Lait* 68, 219 (1988).
[79] C. Karahadian and R. C. Lindsay, *J. Dairy Sci.* 70, 909 (1987).
[80] B. E. Brooker, *Food Microstruct.* 6, 25 (1987).
[81] R. Boutrou, F. Gaucheron, M. Piot, F. Michel, J. L. Maubois and J. Léonil, *Lait* 79, 503 (1999).
[82] F. Gaucheron, Y. Le Graët, F. Michel, V. Briard and M. Piot. *Lait* 79, 553 (1999).
[83] J. A. Hannon, S. Lortal, J. P. Tissier and M. H. Famelart, *Dairy Sci. Technol.* 89, 555 (2009).
[84] J. F. Pommert, A. Klaebe, J. Perie, A. Lebugle and J. Puech, *J. Food Sci.* 53, 1367 (1988).
[85] L. G. Scharpf Jr. and T. P. Kichline, *J. Dairy Sci.* 51, 853 (1968).
[86] M. Frau, A. Mulet, S. S. Simal, J. Massanet and C. Rossello, *Food Sci. Technol. Int.* 3, 43 (1997).
[87] M. Burdett, *J. Dairy Res.* 41, 123 (1974).
[88] M. Lalande and G. Corrieu, *Fouling of a plate heat exchanger by milk,* in : B. Hallstrom, D. Lund, C. Tragardh (Eds), Fundamentals and Applications of Surface Phenomena Associated with Fouling and Cleaning in Food Processing, Food chemical Center, University of Lund (Sweden) pp 279-288 (1981).
[89] R. M. Joshi and A. A Patel, *J. Dairy Res.* 53, 323 (1986).
[90] P. J. Fryer, M. T. Belmar-Belny and P. J. R. Schreier, *Fouling and cleaning in milk processing*, In : Fox P.F. (ed.), Heat Induced Changes in Milk, International Dairy Federation, No. 9501, Brussels, pp 364-395 (1995).
[91] T. J. M. Jeurnink and C. G. de Kruif, *Neth. Milk Dairy J.* 49,151 (1995).
[92] T. J. M. Jeurnink, P. Walstra and C. G. de Kruif, *Neth. Milk Dairy J.* 50, 407 (1996).
[93] H. Visser, T.J.M. Jeurnink, J.E. Schraml, P. Fryer and F. Delplace, *Bull. IDF* 328, 7 (1997).
[94] G. Daufin and J. P. Labbé, *Equipment fouling in the dairy application: problem and pretreatment,* In : Zahid Amjad (ed.), Calcium Phosphate in Biological and Industrial Systems, Kluwer Academic Publishers, Boston, pp 437-463 (1998).
[95] B. Bansal and X. D. Chen, *Comprehensive Rev. Food Sci. & Food Safety* 5, 27 (2006).
[96] Y. Le Graët and G. Brulé, *Lait* 62, 113 (1982).
[97] F. Saulnier, F. Ferrero, J. Y. Bottero and G. Linden, *Lait* 75, 93 (1995).
[98] N. Andritsos, S. G. Yiantsios and A. J. Karabelas, *Trans IChemE* 80, 223 (2002).
[99] K. R. Morison and S. H. Tie, *Tran IChemE* 80, 326 (2002).
[100] R. Rosmaninho and L. F. Melo, *J. Food Engin* 73, 379 (2006).
[101] N. Spanos, A. Patis, D. Kanellopoulou, N. Andritsos and P. G. Koutsoukos, *Crystal Growth Design* 7, 25 (2007).
[102] G. Rice, S. Kentish, A. O'Connor, G. Stevens, N. Lawrence and A. Barber, *Desalination* 199, 239 (2006).
[103] G. Rice, A. Barber, A. O'Connor, G. Stevens and S. Kentish, *J. Membrane Sci.* 330, 117(2009).
[104] R. Rosmaninho and L. F. Melo, *Int. Dairy J.* 16, 81 (2006).
[105] R. Rosmaninho and L. F. Melo, *Int. Dairy J.* 18, 72 (2008).

[106] G. Rice, A. Barber, A. O'Connor, G. Stevens and S. Kentish, *Int. Dairy J.* 20, 694 (2010).
[107] Snow brand Milk Product, European Patent Specification No. 99124052.4, EP 1008903B1, April 20, 2005.
[108] S. G. Anema, *Dairy Sci. Technol.* 89, 269 (2009).

In: Calcium Phosphate
Editor: Robert B. Heimann

ISBN 978-1-62257-299-1

Chapter 16

ANALYSIS OF A NEW ACTIVE INGREDIENT FOR SUNSCREENS BASED ON TRICALCIUM PHOSPHATE NANOPARTICLES

T. S. de Araujo[1,*], W. Miyakawa[2], J. B. dos Santos-Filho[3], C. Otani[4], E. M. B. de Sousa[5] and S. O. de Souza[6]

[1]Instituto Federal de Ciências e Tecnologia de Sergipe,
Aracaju, SE, Brasil
[2]Divisão de Fotônica, Instituto de Estudos Avançados,
São José dos Campos, SP, Brasil
[3]Instituto Federal de Ciências e Tecnologia de Sergipe,
São Cristóvão, SE, Brasil
[4]Instituto Tecnológico de Aeronáutica,
São José dos Campos, SP, Brasil
[5]Centro de Desenvolvimento de Tecnologia Nuclear,
Belo Horizonte, MG, Brasil
[6]Universidade Federal de Sergipe, São Cristovão, SE, Brasil

ABSTRACT

Inorganic sunscreens constitute a safe way to protect the skin from noxious effects of solar radiation. Tricalcium phosphate (β-TCP) has potential as an active ingredient in sunscreens since it is biocompatible and non-toxic. Moreover when doped with appropriate metals it is able to absorb UVB (290-320 nm) and UVA (320-400 nm) radiations. The objective of this work was to chemically and physically analyze Fe^{3+}-doped tricalcium phosphate synthesized via a chemical precipitation route. X-ray diffractometry and energy dispersive x-ray fluorescence spectroscopy results showed that tricalcium phosphate was formed without any toxic species. Optical absorption spectra indicated that the incorporation of Fe^{3+} ions in the tricalcium phosphate matrix generated absorption in the ultraviolet region. Atomic force microscopy revealed nanoparticles with

* E-mail address: tatiana.araujo@ifs.edu.br.

dimensions on the order of 42 nm. It is concluded that Fe^{3+}-doped tricalcium phosphate can be used as active medium for sunscreens since it combines excellent biocompatibility and the capability to absorb UV radiation.

1. INTRODUCTION

The number of new cases of skin cancer is continuously growing worldwide, particularly in Brazil. Consequently, there is an urgent need to develop safer sunscreen formulations using biocompatible materials. Sunscreens have to be capable of absorbing or reflecting the incident UV radiation, thus protecting the individual [1]. Inorganic sunscreens are best and more effective due to the low level of skin irritability, being also recommended for children and people with sensitive skins [2].

Titanium dioxide and zinc oxide are popular inorganic sunscreen agents, which, however, can make the skin look unnaturally white when incorporated into the products. Additionally, their high photocatalytic activity promotes the generation of reactive oxygen species, which can oxidize and degrade other ingredients in the formulation, raising safety concerns [3].

Tricalcium phosphate ($Ca_3(PO_4)_6$, TCP) is an inorganic compound that occurs in three polymorphic crystal forms, β-TCP, stable below 1120°C; α-TCP, stable between 1120°C and 1470°C, and α'-TCP, stable above 1470°C [4]. The physical properties and the biocompatibility of ß-TCP are well known [5]. In β-TCP, calcium ions can be partially replaced by a variety of metal ions such as manganese (Mn^{2+}), zinc (Zn^{2+}), and chromium (Cr^{3+}) [6,7], though these substitutions can alter the crystalline structure, the stability and other features. Nevertheless, β-TCP has a potential for use as a broad-spectrum inorganic sunscreen in personal-care products since it is biocompatible and its optical absorption in the ultraviolet region can be enhanced by introducing appropriate doping elements such as Fe^{3+} that can generate optical absorption bands in the UV range [8,9].

It is worth mentioning that the optical behavior of nanomaterials depends strongly on particle size and shape [9,10]. Nanoparticles of the same material may exhibit different optical properties and lend themselves to various applications, provided they have different shapes and sizes. The particle size is extremely important in sunscreens. For example, formulations incorporating micrometer particles, should be avoided as they reduce the effectiveness of sunscreen and leave an unnatural white appearance of the product [11], as it is the case for titanium dioxide and zinc oxide. In contrast, β-TCP produced in nanometer size is transparent on the skin.

Many studies have reported the synthesis of nano-sized particles of tricalcium phosphate for different purposes [7, 12 -14]. However, to the best of our knowledge, no study was accomplished to use it as a sunscreen material. Hence, the objective of this work was to synthesize ß-FeTCP via a chemical precipitation route, to determine the optical and structural changes caused by the presence of Fe^{3+} in the material matrix, and to investigate the effect of heat treatment at various temperatures on the parameters of interest for use in sunscreen.

2. Experimental Setup

Powders were prepared by the aqueous precipitation method, adding 0.4 mol/L diammonium phosphate solution $(NH_4)_2HPO_4$ to a mixture of 0.6 mol/L calcium nitrate solution $Ca(NO_3)_2.4H_2O$ and 0.01 mol/L iron nitrate $Fe(NO_3)_3.9H_2O$, at a rate of 1.5 ml/min. The pH of the solution was maintained at 10.8 by continuous addition of ammonium hydroxide. The temperature was fixed at 30°C and the solution was allowed to mature for 50 h. [12]. After maturation, the white precipitate was washed in distilled water and filtered. The material obtained was divided into four parts for calcination in an oven for 2 hours at temperatures of 700°C, 800°C, 900°C and 1000°C. A finer powder was obtained by grinding in an agate mortar.

The reaction of the ß-TCP formation can be expressed as follows:

$$3Ca(NO_3)_2.4H_2O + 2(NH_4)_2HPO_4 + 2NH_4OH \rightarrow Ca_3(PO_4)_2 + 6NH_4NO_3 + 6H_2O \quad (1)$$

The precipitated ß-FeTCP powder was chemical and physically characterized by powder x-ray diffractometry (XRD), UV-VIS spectroscopy, and atomic force microscopy (AFM). For characterization, a powder grain size range between -150+75 μm was selected.

X-ray powder diffraction and energy dispersive x-ray fluorescence techniques were used to study the effect of dopants on phase composition. The samples were placed in the specimen holder of a Rigaku DMAX2000 diffractometer, using Cu Kα radiation at 40 kV/40 mA with a monochromator detector. and scanned in the $^{o}2\theta$ range of 10°–80°. Energy dispersive x-ray fluorescence spectra were obtained with an EDX-720 Shimadzu instrument.

Optical absorption was recorded in the wavelength range from 200 to 800 nm using a Shimadzu UV 2401, UV–VIS recording spectrophotometer, calibrated with barium sulphate reference material.

Atomic force microscopy images were obtained with a Shimadzu SPM9500J3 microscope. The powder was pressed into pellets of approximately 6 mm diameter and 1 mm thick. Scanning areas of 500 nm x 500 nm were imaged at positions randomly chosen on the samples surfaces. Based on these images, particle analyses were performed using the AFM software facility (Particle Analysis Software, Shimadzu SPM-9500 series, Shimadzu Corp. Inc., Japan). The following parameters were measured: maximum diameter (largest distance between two opposite points of the contour of the particle); pattern width (greatest distance between two points orthogonally to the contour line that defines the maximum diameter), average radius (average distance from the centre of gravity of the contour particle), distance to nearest neighbour and degree of circularity (which is equal to 1 when the particle contour is circular, and the higher this value, less circular is the contour).

3. Results and Discussion

The XRD patterns of the ß-FeTCP samples calcined at different temperatures are shown in Figure 1. Samples calcined at 700°C were found to contain low crystalline HAP ($Ca_{10}(PO_4)_6(OH)_2$, hydroxyapatite). After calcination at 800°C, crystallization of ß-FeTCP occurred. From 900°C to 1000°C, small amounts of HAP were identified, though ß-FeTCP was the predominant phase. At these temperatures, no significant differences were observed in the degree of crystallinity. The peak centered at 31.5°, indicative of the presence of HAP, was less intense in the material calcined at 1000°C, suggesting that an increase in temperature discourages the formation of this phase. It is worth mentioning that HAP, the basic mineral component of hard tissues (bone, teeth), is biocompatible and can also be used as active ingredient in sunscreens [10, 13 -17].

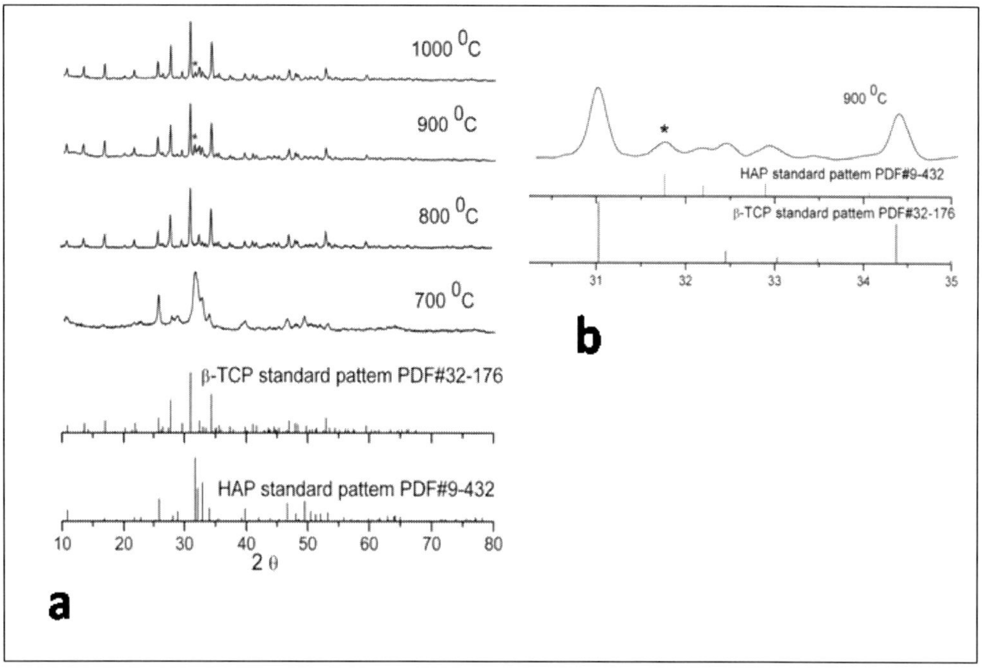

Figure 1. (a) X-ray diffraction patterns of ß-FeTCP samples after heat treatment at various temperatures. The symbol (*) indicates the presence of hydroxyapatite. (b) High-resolution pattern of the angular range between 31 and 35 °2Θ.

Formation of minority phases, which could not be identified by XRD, was investigated by EDXRF. This study was extremely important, as the formation of toxic species in these materials jeopardizes its application as sunscreen. The results obtained for ß-FeTCP after heat treatment at various temperatures are shown in Table 1. In all samples, calcium oxide (CaO) was present. As this new phase is also biocompatible [17], its presence does not preclude the application of these materials in sunscreens. However, despite the low concentration of CaO during production of the sunscreens care should be taken to avoid a higher pH by formation of $Ca(OH)_2$, which can occur in the presence of water.

Based on these results, it was assumed that an extra amount of Ca^{2+} remained in the mixture and reacted with the oxygen (O_2) from the environment, due to absorption of water vapor in the air. The formation of FeHAP in ß-FeTCP calcined at 900°C and 1000°C was confirmed by XRD analysis [18-20]. Only the sample calcined at 1000°C revealed the presence of Fe_2O_3.

Table 1. Phase composition in mass % as a function of calcination temperature

Temperature (°C)	Phases	Mass %
700	FeHAP	96.9 ±0.2
	CaO	3.1 ±0.1
800	β-FeTCP	96.8 ±0.1
	CaO	3.186 ±0.002
900	β-FeTCP	91.6 ±0.1
	CaO	2.88 ±0.04
	FeHAP	5.52 ±0.09
1000	β-FeTCP	92.6 ±0.1
	CaO	2.19 ±0.05
	FeHAP	3.15 ±0.006
	Fe_2O_3	2.1 ±0.2

This study focuses on exploring the absorption capacity in the UV region of the ß-FeTCP. Reports in the literature on optical absorption studies of ß-TCP and/or ß-FeTCP are rarely found. Figure 2 shows the optical absorption spectrum of pure ß-TCP calcined at 800°C for 2 hours and 0.01mol/L ß-FeTCP calcined at different temperatures for 2 hours. The optical absorption of ß-TCP has undergone significant changes due to the incorporation of Fe^{3+} to form ß-FeTCP. In contrast to this, phase-pure ß-TCP, revealed no optical absorption in any region of the spectrum, confirming its inability to act as an active medium in sunscreens.

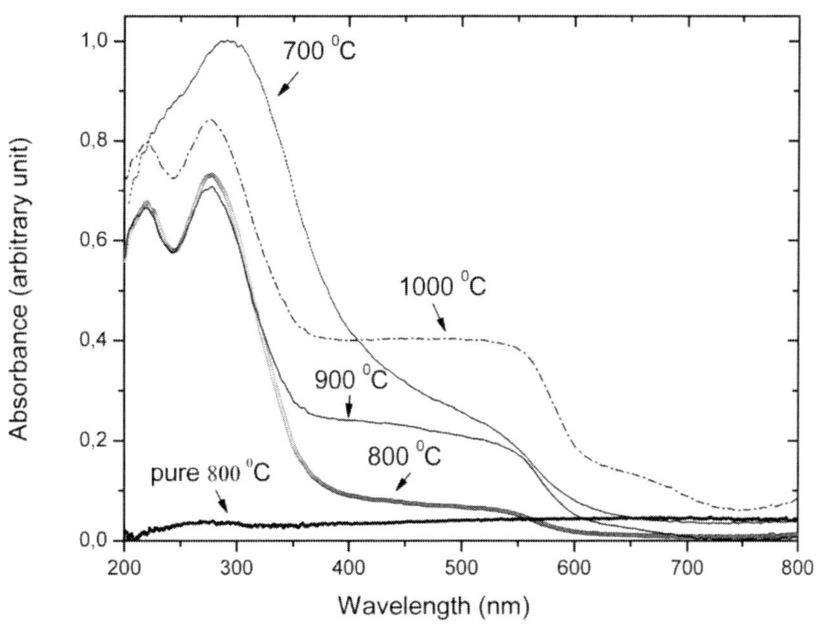

Figure 2. UV-Vis absorption spectra of pure ß-TCP calcined at 800°C and ß-FeTCP after heat treatment at various temperatures. All samples were calcined for 2 hours.

The Fe-doped sample calcined at 700°C displayed the most intense absorption from the UV to the visible (200-624 nm) range with a maximum at 293 nm. In the material calcined at 800°C, two very close absorption bands in the UV region (200-240, 243-375 nm) were observed. The calcination at 900°C produced absorption bands similar to those at 800°C, however with a very intense band in the visible (up to about 598 nm) range. The sample calcined at 1000°C presented a more intense absorption in the UV and visible ranges in a similar spectrum. Differences in the spectra of these samples after treatment at various temperatures were expected, due to the formation of HAP, as identified by XRD and EDXRF analysis. The comparable optical behavior of samples calcinated between 900°C and 1000°C can be explained by the similarity of the crystalline structures, except for the presence of Fe_2O_3 at 1000°C. The results suggested that HAP and Fe_2O_3 were responsible for absorption in the visible region.

The presence of other phases is very common in samples produced by the precipitation method, as many variables can influence the synthesis process, such as the reaction time, the temperature gradient inside the oven, and others. However, in this work, formation of toxic species was not observed.

Optical absorption bands in the visible range can yield undesirable visual color effect in sunscreens, which, however, might be interesting in other cosmetic products containing sunscreen, such as makeup. In the latter case, materials could be exploited according to their optical behavior: FeHAP (700°C), for high absorption of UV radiation, and ß-FeTCP (800°C) for absorbing UV but no visible radiation. Optical absorption intensity for the sample calcined at 800°C could probably be promoted by simply increasing the Fe^{3+} concentration, however how much a solubility limit of Fe^{3+} in the ß-TCP has not been studied.

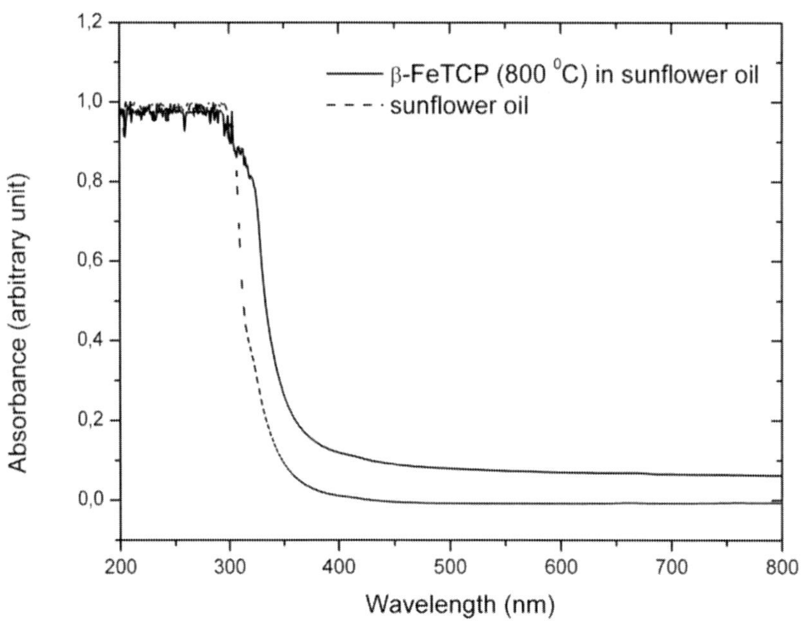

Figure 3. UV-Vis absorption spectrum of sunflower oil and ß-FeTCP (800°C) diluted in sunflower oil.

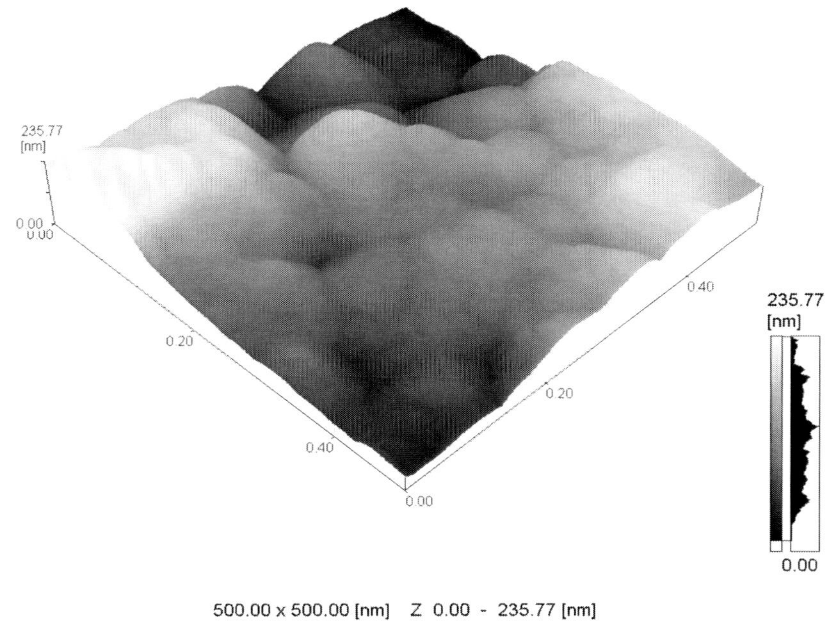

(a)

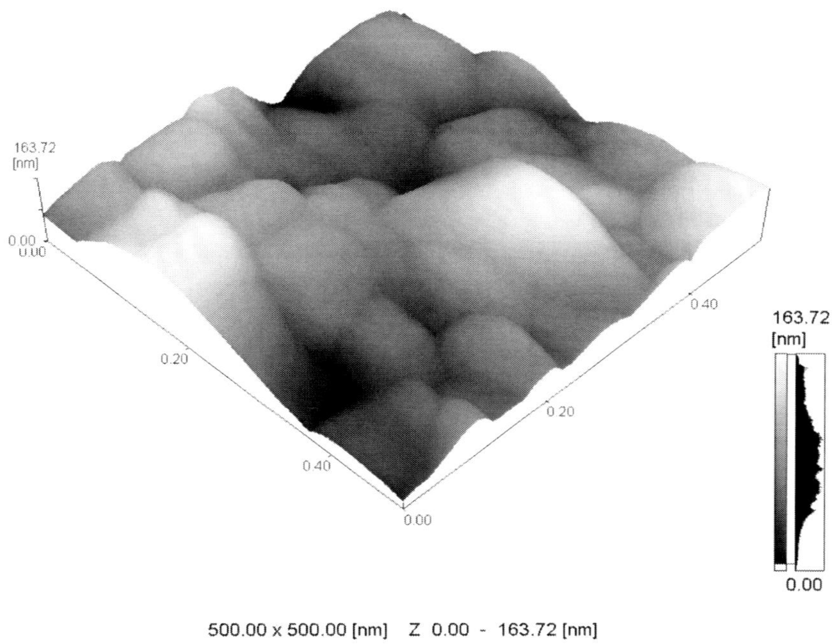

(b)

Figure 4. AFM micrographs of 500 nm x 500 nm scanning area of (a) pure TCP and (b) ß-FeTCP (800°C).

The UV-VIS absorption spectrum of ß-FeTCP (800°C) diluted in sunflower oil (a common ingredient in sunscreens) is shown in Figure 3. Compared to the pure sunflower oil spectrum, the edge of the UV absorption band had shifted about 20 nm to longer wavelengths, i.e., there is more absorption in the region of interest. A small increase (less than 10%) of absorbance in the visible and near infrared range was also observed. In addition, the mixture was homogeneous and transparent when applied to the skin, which would ensure an effective protection against the harmful effects of the sun and provide a good market acceptance.

These results demonstrate, in comparison to ZnO and TiO_2, that ß-FeTCP absorbs appreciably more UVB and UVA light. ZnO and TiO_2 sunscreen ingredients approved for use in the United States are capable of absorbing only a fraction of the UV spectrum: the ZnO protects against UVA radiation, the TiO_2 can absorb mostly UVB, and depending on the particle size, it reflects the UVA [17]. An efficient sunscreen should strongly absorb both the UVB radiation, which is responsible for the formation of erythema and long-term skin cancer, and the UVA radiation, responsible for photoaging and skin cancer [11, 17].

Atomic force microscopy (AFM) is an excellent and unique approach for observing crystal growth dynamics, imaging and measuring surface features as well as studying defect formations on the nanometer scale [21]. For example, Monkawa et al. [22] used the AFM technique to characterize the particle sizes of hydroxyapatite. Typical AFM micrographs in three-dimensional visualization can be viewed in Figure 4 for (a) pure ß-TCP and (b) ß-FeTCP calcined at 800°C. The images suggest that the precipitates have various sizes and shapes. According to the data of Table 2, particles are nanometric and ellipsoidal, as the circularity degree is >1 for both samples. In ß-FeTCP, the mean radii are greater than in pure ß-TCP, suggesting that Fe^{3+} ion incorporation augmented the particle size. In Figure 4 (a), 85 particles with 26 nm of mean radius were identified. In Figure 4 (b), only 20 particles with approximately 42 nm of mean radius were identified. The nearest-neighbor distance in ß-FeTCP was greater than in pure ß-TCP, which partially reflects the larger size of ß-FeTCP particles, but it also reveals that particles were more densely packed in pure ß-TCP.

The distinctive properties of nanoparticles are of particular interest to the skin care industry. In cosmetics products, nano-sized particles improve texture, produce more vibrant colors, and increase skin penetration. The ß-FeTCP particle sizes are satisfactory, since inorganic sunscreens should be made preferably with particles smaller than 400 nm. Some types of sunscreens even involve particle sizes in the range 70-200 nm [5].

Table 2. Mean radius (r) and respective standard deviation, distance to nearest neighbor (DNN), circularity degree (c) and total number of particles (N) measured from images displayed in Figure 4

	Mean radius (nm)	DNN (nm)	c	N
Pure β-TCP	26±8	34.04	2.451	85
β-FeTCP	42±10	69.67	1.978	20

CONCLUSION

The synthesis method and the optical properties of nano-sized ß-FeTCP powders were investigated. From the x-ray diffractometry and the energy dispersive x-ray fluorescence results, it was concluded that no toxic species were formed. The presence of FeHAP is not relevant to the performance of the product, as it is biocompatible, and Fe_2O_3 was observed as a minority phase only at a calcination temperature of 1000°C. Further studies are still necessary to evaluate the effect of a mass percentage so tiny (2% of the total mass) in sunscreens.

ß-TCP does not naturally absorb radiation in the ultraviolet and in the visible ranges of the electromagnetic spectrum. As expected, optical spectroscopic studies allowed to conclude that incorporation of Fe^{3+} ions in the β-TCP matrix generated appropriate optical absorbance.

AFM results showed that the β-FeTCP (800°C) grains have an ellipsoidal shape, the measured dimensions of which (mean radius of approximately 42 nm) are suitable for its use in sunscreens.

It was also demonstrated that β-FeTCP has a broader attenuation spectrum than both titanium dioxide and zinc oxide. Furthermore, the nano-sized particles of β-FeTCP are significantly more acceptable for cosmetic applications than titanium dioxide and zinc oxide since at the same particle size the latter involve the risk of free radicals formation. Hence, we conclude that nano-sized particles of β-FeTCP can be suitable as a cosmetic ingredient for a variety of products aiming at protection of skin against ultraviolet radiation.

REFERENCES

[1] S. Oladepo, G. R. Loppnow, *Anal. Chim. Acta* 628, 57 (2008).

[2] N. Serpone, D. Dondi, A. Albini, *Inorg. Chim. Acta* 360, 794 (2007).

[3] T. Masui, H. Hirai, N. Imanaka, G. Adachi, *J. All. Comp.* 408, 1141 (2006).

[4] M. Descamps, J. C. Hornez, A. Leriche, *J. Eur. Ceram. Soc.* 27, 2401 (2007).

[5] J. Flor, M. R. Davolos, *Quím. Nova* 30, 153 (2007).

[6] K. Ishikawa, Y. Miyamoto, T. Toh, *Biomaterials* 23, 423 (2002).

[7] A.Ito, M. Otsuka, H. Kawamura, *Cur. Appl. Phys.* 5, 402 (2005).

[8] R. P. S. Chakradhar, G. Sivaramaiah, J.L. Rao, N.O. Gopal, *Spectrochimica Acta Part A* 62, 51 (2005).

[9] S. Basu, D. Chakravorty, *J. Non-Cryst. Sol.* 352, 380 (2006).

[10] T. S. Araujo, Z. S. Macedo, P. A. S. C. Oliveira, M. E. G. Valerio, *J. Mater. Sci.* 42, 2236 (2007).

[11] M. V. R. V. De Paola, *Cosmetics and toiletries* 13, 74 (2001).

[12] K. Lin, J. Chang, J. Lu, W. Wu, Y. Zeng, *Ceram. Inter.* 33 (2007) 979-985.

[13] W. Acchar, E.G. Ramalho, Mater. *Sci. and Eng. C.* 28 (2008) 248-252.

[14] X. Li, A. Ito, Y. Sogo, X. Wang, R Z. LeGeros, *Acta Biomaterialia* 5 (2009) 508–517.

[15] T. S Araujo, S. O Souza, *Sci. Plena* 4 1 (2008).

[16] T. S Araujo, S. O Souza, W. Miyakawa, E. M. B. Sousa, *Mater. Chem. Phys* 124, 1071 (2010).

[17] E. S Ahn, N. J Gleason, A. Nakahira, J. Y Ying, *Nano Letters* 1, 149 (2001).

[18] R. Morrissey, L. M. Rodriguez-Lorenzo, K. A. Gross, *J. Mater. Sci.: Mater. Med.* 16(5) 387 (2005).

[19] W. Pon-on, S. Meejoo, I. M. Tang, *Intern. J. Nanoscience*, 6(1) 9 (2007).

[20] T. K. Anee, M. Palanichamy, M. Ashok, N.M. Sundaram, S.N. Kalkura, Mater. Lett. 58 478 (2004).

[21] E. Meyer, H. J Hug, R. Bennewitz, *Scanning probe microscopy: the lab on a tip*, Springer-Verlag, Berlin (2004).

[22] A. Monkawa, T. Ikoma, S. Yunoki, K. Ohta, J. Tanaka, *Materials Letters* 60, 3647 (2006).

In: Calcium Phosphate
Editor: Robert B. Heimann

ISBN 978-1-62257-299-1

Chapter 17

THE USE OF APATITE-TYPE MINERALS FOR IMMOBILIZATION PURPOSES

Herbert Pöllmann*
Institut für Geowissenschaften, Martin-Luther-Universität Halle,
Halle, Germany

ABSTRACT

Apatite with the general formula $(A)_5[(XO_4)_3/(Y)]$ are minerals widely occurring in nature. The crystallo-chemical variation of apatite structure-type minerals leads to a wide application in immobilization of hazardous waste materials. Due to this structural variability the following four possibilities for crystallo-chemical exchange exist: (i) variation of metal cations A, (ii) variation of anion type I, XO_4, (iii) variation of anion type II, Y, and (iv) complex combined exchange of cations and anions.

Apatite minerals can occur under many different formation conditions in our geological surrounding, ranging from the melting state to metamorphic crystallization, weathering conditions, and sedimentary formation (see Refs. [43, 70, 73, 82, 97, 217, 235, 255, 300, 336, 338]). Heavy metal-containing minerals with apatite structure such as pyromorphite, mimetite or vanadinite crystallize in the oxidizing zone of ore mineral deposits and lead to immobilization of heavy metal cations but also some anions.

For toxic element fixation, the Refs. [13], [14], [68], [69], [71], [72], [220], [222], [223], [242], [243], [249], [270], [275], [276], [277], [278], [295], [296], [342], [347] and [348] describe the optimum conditions and possibilities to apply apatite minerals for immobilization purposes.

In this chapter, different processes such as XLTITE (Crystal Tight), MRP (Mineral reservoir processes), PIMS (Phosphate-induced metal stabilization), and SUPER CALCINE (can contain apatite) will be described that use apatite mineral structures for immobilization. The crystallo-chemical background, the formation conditions, the variability, and some applications will be described. The crystallization of apatite and the apatite structure-type mineral ellestadite will be explained that occur in many different industrial residues subjected to combined thermal and hydraulic processes. In addition, apatite minerals in nature will be compared to synthetic apatites.

* Email address: herbert.poellmann@geo.uni-halle.de.

1. INTRODUCTION

Apatite is the most abundant phosphate mineral in rocks and can even form large phosphate deposits of different geological origin [339]. Most of the industrial use of apatite is concerned with agricultural fertilizers, production of phosphoric acid, and as a precursor of different phosphorous-containing compounds. Due to the high stability under environmental conditions apatite can also be used for many different immobilization applications [219].

Natural examples for high immobilization capacity for different cations and anions are minerals of the apatite group that show fixation of many heavy metals, and different cations and anions. Famous and well-known apatite minerals are among those abundant lead minerals pyromorphite, mimetite and vanadinite crystallizing in the oxidation zones of lead-containing ore deposits that prevent lead, and arsenate and vanadate ions from being leached out. Leaching data for apatite were obtained by Guidry et al. [141], Valsami-Jones et al. [329], Stefánsson [305], and Stefánsson et al. [306].

In addition to ore deposits, apatite minerals do occur in other environments also including

1. Igneous rocks
2. Metamorphic rocks
3. Sedimentary rocks
4. Hydrothermal deposits
5. Low temperature formations

Belousova et al. [22] used apatite as an indicator mineral for mineral exploration by studying their trace-element compositions and relationship to the type of host rocks. Naray-Szabo [239] determined its crystal structure. In particular, the structure of OH-apatite was determined by Sudarsanan [311]. The different formation conditions and occurrences mentioned above influence the chemical composition of apatite. Thus to understand the various diadochic substitutions, the crystal structure of apatite is shown in Figure 1. Petrogenetic implications were explored by Gromet [138].

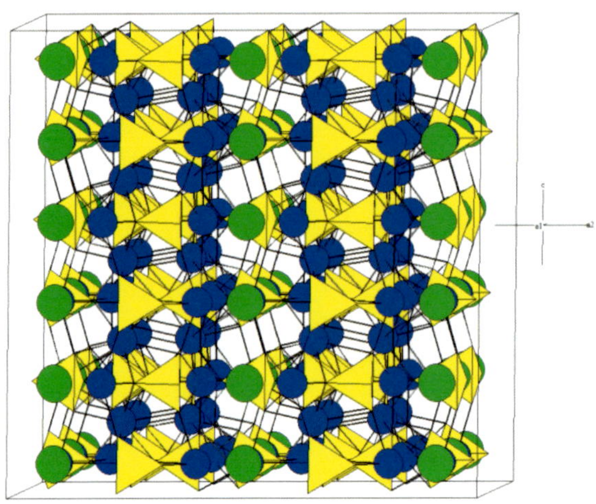

Figure 1. Crystal structure of fluoroapatite (yellow: phosphate tetrahedra, blue: calcium. green: fluorine).

The variability in chemical composition of apatites and their stability under geochemical oxidation conditions lead to their wide application in immobilization technology. Their high capacity for fixation of heavy metals is already known from natural sources in oxidation zones of mineral deposits, especially the lead minerals pyromorphite, $Pb_5[(OH)/(PO_4)_3]$, mimetite, $Pb_5[(OH)/(AsO_4)_3]$ and vanadinite, $Pb_5[(OH)/(VO_4)_3]$ (Table 1). The famous britholites were already discussed by Bambauer [11, 12] and McCarthy [218] for their capability to immobilize radionuclides. In britholites the replacement of $[PO_4]^{3-}$ by $[SiO_4]^{4-}$ is balanced by a replacement of Ca^{2+} with another metal cation, Me^{3+} [250]. Calcium also can be replaced by magnesium, cadmium, strontium, barium, and lead [304]. For silicate-rich industrial residues the minerals of the ellestadite group are very important in which phosphate ions are replaced by sulfate and silicate ions [220]. Many different apatites were described by Schwarz [291-296], Liu et al. [194] and McConnel [221].

Figure 2. Natural apatites. (a) Hydrothermal apatite of basaltic origin, (b) Blue apatite in a pegmatitic feldspar matrix, (c) Apatite single crystal (pegmatite, Brazil), (d) Secondary green apatite on feldspar, (e) Apatite crystal from Ipira, Brazil, (f) Apatite from igneous rock, Kola Peninsula, Russia.

The different occurrences in nature of apatite lead to these varying compositions. As sedimentary apatites (phosphorites) can contain slightly elevated cadmium levels, their use for agricultural fertilizer applications is now discouraged.

Some natural apatites from different geological origin - igneous, metamorphic and sedimentary - are shown in Figure 2.

Other minerals with apatite structure mainly occurring in the oxidation zone of heavy metal mineral deposits are displayed in Figure 3. Owing to their hexagonal prismatic form and their intensive typical green, amber and brown colours they can be identified rather easily.

Figure 3. Lead-, arsenate- and vanadate-containing minerals of the apatite group. (a) Vanadinite, Marocco, (b) Pyromorphite, Nassau, (c) Pyromorphite (oxidation zone), VR China, (d) Mimetite (oxidation zone), Tsumeb, Namibia.

Besides the typical incorporation of lead [175], and phosphate, arsenate and vanadate ions many other substitutions can take place in the lattice of apatite-type structures [76,77]. Anderson and Kostiner [2] described Co-substituted apatite, and Arden et al. [3], Arey et al. [4], Audubert et al. [5], Badraoui et al. [9], Badrour et al. [10], and Baud et al. [16, 17] other apatite structures. Bigi et al. [28, 29] synthesized Ba- and Cd-substituted calcium apatites, Borisov et al. [39] described the structure of REE-Sr apatites. Sr-apatite was synthesized by Chakhmouradian et al. [53]. Buvaneswari et al. [49], Calvo et al. [50], and Dihn et al. [85] described new apatite-related phosphates. The thermochemistry of some apatites was described by Khattech et al. [178]. Besides simple diadochic replacements many coupled substitutions including differently charged and sized cations and anions are possible

[101,102]. The introduction of silicate in the apatite structure even opens up possibilities to immobilize many different industrial residues and many more silicatic materials. The existence of apatite-type minerals containing $[Si_2O_7]^{6-}$ - groups is one of these possibilities. Even protons can be localized [64].

Table 1. Crystallo-chemical variabilities in different apatite structure-type minerals

Chemical formula	Constant	Variation	Examples for exchange
$(Ca)_5[(PO_4)_3/(Y)]$ **Apatites**	Metal = Ca	Anion = Y	Y = OH^-, F^-, Cl^-, CO_3^{2-}, I^-
$(Ca)_5[(PO_4)_3/(Y)]$	Metal = Ca	Anion = Y main anion = variable	Y = OH^-, F^-, Cl^-, CO_3^{2-}, I^- Replacement of $(PO_4)_3$ by $(AsO_4)_3$ or $(VO_4)_3$
$(Ca)_5[(PO_4)_3/(Y)]$ **Pyromorphite**	Phosphate	Replacement of Ca; Y = Anion	Y = OH^-, F^-, Cl^-, CO_3^{2-}, I^- Replacement of Ca by Pb^{2+},Sr^{2+}, Cd^{2+}, Ba^{2+}
$(Ca)_5[(PO_4)_3/(Y)]$ **Mimetite, vanadinite**	_____	Replacement of Ca; Y = Anion; main anion = variable	Y = OH^-, F^-, Cl^-, CO_3^{2-}, I^- Replacement of $(PO_4)_3$ by $(AsO_4)_3$ or $(VO_4)_3$ Replacement of Ca by Pb^{2+},Sr^{2+}, Cd^{2+}, Ba^{2+}
$(Ca)_{10}[(SiO_4)_3/(SO_4)_3(Y)_2]$ **Ellestadite**	Metal = Ca	Anion = Y	Y = OH^-, F^-, Cl^-, CO_3^{2-}, I^-
$(Ca)_{10}[(SiO_4)_3/(SO_4)_3(Y)_2]$ **Ellestadite**	_____	Replacement of Ca Y = Anion Main anion = variable	Y = OH^-, F^-, Cl^-, CO_3^{2-}, I^- Replacement of $(PO_4)_3$ by $(AsO_4)_3$, $(VO_4)_3$, $(SiO_4)_3$, $(SO_4)_3$, $(CrO_4)_3$ Replacement of Ca by Pb^{2+},Sr^{2+}, Cd^{2+}, Ba^{2+}
$Pb_9Ca_5Mn^{2+}[SiO_4/Si_2O_7]_3$. **Ganomalite**	_____	Replacement of Ca Main anion = variable	Replacement of $(PO_4)_3$ by $[Si_2O_7)_2(SiO_4)_2]$ Replacement of Ca by Pb^{2+},Sr^{2+}, Cd^{2+}, Ba^{2+}
$Pb_6Ca_4[(Si_2O_7)_3/Cl_2]$ **Nasonite**	_____	Replacement of Ca Main anion = variable	Replacement of $(PO_4)_3$ by $[Si_2O_7)_3Cl_2]$ Replacement of Ca by Pb^{2+},Sr^{2+}, Cd^{2+}, Ba^{2+}

CRYSTALLO-CHEMICAL VARIABILITY OF APATITE GROUP MINERALS

Many different replacements are described in the literature, based on the basic structure type of the calcium hydroxylphosphate as given in Figure 1. In Table 1 numerous crystallo-chemical variabilities in different apatite structure- type minerals are summarized. Also some of the natural mineral analogues are mentioned. In apatites two different types of anions exist. Besides the PO_4^{3-} anion other anions such as Cl^-, Br^-, I^- and OH^- can be replaced as described by Baumer et al. [18,20], Beevers et al. [21], Berndt et al. [23], Böhlke et al. [34], Brenan [46], Cho et al. [64,65], Christy et al. [66], Doi et al. [86], Dong et al. [87], Elliott et al. [104, 105], Gross et al. [140], Houndslow et al. [154], Hughes et al. [156;157], Ikoma et al. [165], Kreidler et al. [185], Mackie et al. [200], Prener [273], Ruszala et al. [285], Vignoles et al. [331], and Young et al. [353]. Zhang et al. [353] described iodine-containing lead-vanadate apatites. Uptake of CN^- was mentioned by Dowker et al. [88], and NO_2^- - incorporation by Dugas et al. [90]. Cadmium apatites with different halogenides were synthesised by Sudarsanan et al. [312-315]. Some other compositions were mentioned by Brophy et al. [47], Machatschki [199], and Mackie et al. [201].

Many different substitutional mechanisms have been proposed in the literature. The main mechanisms for calcium substitution in the apatite mineral group are included in Table 2.

Many other special apatites incorporating REE, Mg, alkali ions, Mn, Si, Sb, B, Ba, Cr, Sr, Fe, Cd, Rb, Be, K, Ge, and Re were described in Refs. [6], [8], [15], [25,26], [27], [39], [54], [55-57], [59,60], [61], [63], [74], [78], [79], [80], [81], [91], [96], [98], [99], [106], [107-110], [111], [112], [114], [115-121], [122], [124], [125], [126], [129], [131], [132], [135], [137], [146-148], [149], [153], [155], [156-162], [171], [172], [173], [181-183], [184], [177], [179], [187], [190], [207,208], [209], [211-215], [216], [227], [229], [230], [236], [232], [240], [244], [245], [246], [248], [251], [257], [258,259], [260], [261], [263], [274], [281], [283], [286], [289], [297], [298], [299], [302], [303], [307], [312-315], 319], [320], [321], [325,326], [327], [328], [330], [332], [333-335], [337], [340], and [351].

Anion substitution of carbonate for phosphate in tetrahedral position together with OH^-, F^- and other halogenides were proposed as shown in Table 3.

Table 2. Mechanisms of calcium substitution in apatite

Substitutions on Ca sites
$Me^{2+} \leftrightarrow Ca^{2+}$
$Me^{+} + Me^{3+} \leftrightarrow 2Ca^{2+}$
$Me^{4+} + \square \leftrightarrow Ca^{2+}$
$2Me^{3+} + \square \leftrightarrow 3Ca^{2+}$
$Me^{3+} + An^{2-} \leftrightarrow Ca^{2+} + An^{1-}$
$2Me^{+} \leftrightarrow Ca^{2+} + \square$
$Me^{3+} + SiO_4^{4-} \leftrightarrow Ca^{2+} + PO_4^{3-}$

Table 3. Mechanisms of carbonate incorporation in apatite-structure

$CO_3OH^{3-} \leftrightarrow PO_4^{3-}$
$CO_3F^{3-} \leftrightarrow PO_4^{3-}$
$CO_3^{2-} + SiO_4^{4-} \leftrightarrow 2PO_4^{3-}$

Ames [1] described the genesis of carbonate apatites. Many other anion-related substitutes replacing CO_3^{2-} were described in Refs. [19], [24], [33], [35, 38], [40,41], [89], [103], [139], [144], [169], [170], [188,189], [192], and [317]. Hitme et al. [151] described carbonate-containing strontium arsenate-apatites, Nadal et al. [238] strontium-containing carbonate-apatites, Okazaki [252], Regnier et al. [279] and Tõnsuaadu et al. [324] fluorine-

and carbonate- containing apatites. Bothe et al. [42] described arsenic calcium compounds. Johnbaumite was described by Dunn et al. [93].

A selection of different exchange possibilities is given in Table 4.

Likewise the incorporation of sulfate in the apatite structure was examined in many publications, frequently according to the exchange reaction:

$$SO_4^{2-} + Na^+ \leftrightarrow PO_4^{3-} + Ca^{2+}$$

Schneider [287,288], Suitch et al. [318], Perez-Mendez et al. [262], and Cavaretta et al. [52] studied sulfate-containing apatites. Brauns [45] mentioned a natural sulfate apatite from the Laacher See caldera in the Eifel mountains, Germany.

Table 4. Crystal chemical variability of apatite structures

$$A_4\, A_6[(B)_6/X_2]$$

	Ca^{2+}	Na^+				SO_4^{2-}
	Cd^{2+}	Li^+				CrO_4^{2-}
	Zn^{2+}	K^+				SeO_4^{2-}
	Mg^{2+}	Rb^+		OH^-		PO_4^{3-}
	Sr^{2+}	Ag^+		Cl^-		$B(OH)_4^-$
A =	Fe^{2+}	Tl^+	X =	$½\ CO_3^{2-}$	B =	SiO_4^{4-}
	Mn^{2+}	Sn^{2+}		F^-		WO_4^{2-}
	Ba^{2+}	Pb^{2+}		Br^-		MnO_4^-
	Co^{2+}	Th^{4+}		O^{2-}		VO_4^{3-}
	Ni^{2+}	REE^{3+}				AsO_4^{3-}
	Al^{3+}	Tl^{3+}				
		Y^{3+}				

The replacement of phosphate anions by silicate anions was investigated extensively by many authors, especially due to the possibility to use these apatites for high temperature immobilization ([7], [30-32], [48], [127], [128], [130], [133], [142], [143], [145], [167,168], [176], [193], [204], [228], [242], [256], [271], [280], [282], [290], [291-295], [308], [346], [350,352]). Figure 4 shows the condensation of silicate tetrahedra [269, 308].

Technical apatite powders and natural fine grained phosphorites are shown in Figure 5. These fine grained apatites can be used for exchange reaction of heavy metals. In natural systems these apatites contain very often cadmium at the crystallographic Ca position which precludes the use of these apatites in agriculture. Biomaterials and synthetic apatites were described by Bonel [36,37]. Brasseur et al. [44] described synthesis of apatites. Dental apatites were described by Elliott [100]. Characterisation of apatites for use as microfiltration membranes was performed by Masmoudi et.al. [205]. Many of these apatite materials are incorporating different amounts of halogenides, but also carbonate and hydroxide.

Crystalline lead silicates were investigated by Grimmer et al. [136] and Kato [174]. Other apatite-like minerals were mentioned by Fayos et al. [113] and Fletcher [123]. The system $PbO\text{-}SiO_2$ was described by Smart et al. [301]. Dunn [79] mentioned lead silicates from Franklin, N.J. Other systems including PbO were described by McMurdie [224], Merker et al. [225,226], and Moore et al. [233,234].

The main natural minerals of the apatite group with their approved mineral names, formulae and symmetry groups are summarized in Table 5 (Hendricks et al. [150], Hogarth et al. [152]).

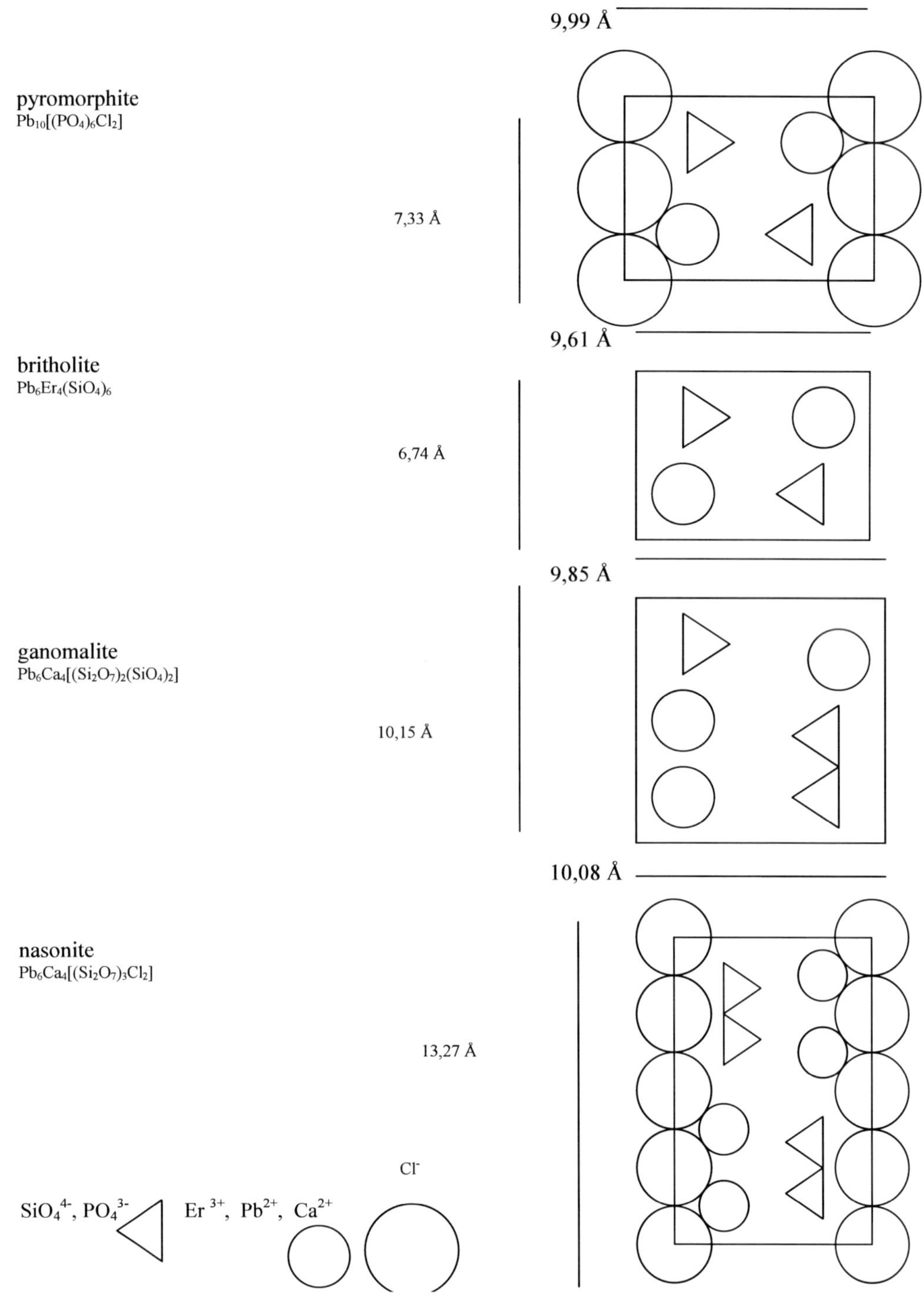

Figure 4. Condensation of silicate tetraedra in apatite structures [308].

Figure 5. Synthetic hydroxylapatite granules (left) and natural phosphate (phosphorite concentrate) sand. (right)

Table 5. Natural members of the apatite group

Apatite	$Ca_5(PO_4)_3(OH,F,Cl)$	$P\,6_3/m$
Apatite-(CaCl)	$Ca_5(PO_4)_3Cl$	$P\,6_3/m$
Apatite-(CaOH)	$Ca_5(PO_4)_3(OH)$	$P\,6_3/m$
Apatite-(CaOH)-M	$(Ca,Na)_5[(P,S)O_4]_3(OH,Cl)$	$P\,2_1/b$
Apatite-(CaF)	$Ca_5(PO_4)_3F$	$P\,6_3/m$
Carbonate-fluorapatite?	$Ca_5(PO_4,CO_3)_3F$	$P\,6_3/m$
Carbonate-hydroxylapatite?	$Ca_5(PO_4,CO_3)_3(OH)$	$P\,6_3/m$
Belovite-(Ce)	$(Sr,Ce,Na,Ca)_5(PO_4)_3(OH)$	P 3
Kuannersuite-(Ce)	$Ba_6Na_2REE_2(PO_4)_6FCl$	P 3
Fluorcaphite	$(Ca,Sr,Ce,Na)_5(PO_4)_3F$	$P\,6_3$
Alforsite	$Ba_5(PO_4)_3Cl$	$P\,6_3/m$
Belovite-(Ce)	$(Sr,Ce,Na,Ca)_5(PO_4)_3(OH)$	P 3
Belovite-(La)	$(Sr,La,Ce,Ca)_5(PO_4)_3(F,OH)$	P 3
Fermorite	$(Ca,Sr)_5(AsO_4,PO_4)_3(OH)$	$P\,6_3/m$
Deloneite-(Ce)	$NaCa_2SrCe(PO_4)_3F$	P 3
Johnbaumite	$Ca_5(AsO_4)_3(OH)$	$P6_3/m, P6_3$
Clinomimetite	$Pb_5(AsO_4)_3Cl$	$P\,2_1/b$
Hedyphane	$Ca_2Pb_3(AsO_4)_3Cl$	$P\,6_3/m$
(Britholite) Abukumalite	$((Y,Ca)_5[(OH/F)/(SiO_4/PO_4)]_3$	
Mimetite	$Pb_5(AsO_4)_3Cl$	$P\,6_3/m$
Apatite-(SrOH)	$(Sr,Ca)_5(PO_4)_3(F,OH)$	$P\,6_3/m$
Morelandite	$(Ba,Ca,Pb)_5(AsO_4,PO_4)_3Cl$	$P\,6_3/m, P\,6_3$
Pyromorphite	$Pb_5(PO_4)_3Cl$	$P\,6_3/m$
Vanadinite	$Pb_5(VO_4)_3Cl$	$P\,6_3/m$
IMA2008-068	$Ca_2Pb_3(PO_4)_3F$	$P\,6_3/m$
Svabite	$Ca_5(AsO_4)_3F$	$P\,6_3/m$
Turneaureite	$Ca_5[(As,P)O_4]_3Cl$	$P\,6_3/m$
Phosphohedyphane	$Ca_2Pb_3(PO_4)_3Cl$	$P\,6_3/m$
IMA2008-009	$Sr_5(PO_4)_3F$	$P\,6_3/m$
Hydroxylpyromorphite	$Pb_5(PO_4)_3OH$	$P\,6_3/m$

Very important minerals with apatite structures belong to the silico-sulfate ellestadite group of apatites [62] mainly used in high temperature immobilization of industrial residues ([270], [343-345]). The typical replacement scheme in ellestadites can be expressed by the following equation:

$$SO_4^{2-} + SiO_4^{4-} \leftrightarrow 2PO_4^{3-}$$

This replacement is very important for the immobilization process of siliceous systems since despite generally low amount of phosphate in industrial residues, complex solid solution can occur.

The main natural minerals of the ellestadite group with apatite structure indicating their approved mineral names and formulae are collected in Table 6. The uptake of silicium was described by Chen et al. [54]. Ito et al. [166] investigated the incorporation of boron in silicate apatites. Khorari et al. [180] worked on the replacement of phosphate by silicate and sulfate. The relations between apatite and silico-sulfates were investigated by Pliego-Cuervo et al. [265]. Many sorosilicates with apatite structure were investigated by Stemmermann et al. [209,210].

Table 6. Minerals of the ellestadite group

Ellestadite	$Ca_5(SiO_4,PO_4,SO_4)_3(F,OH,Cl)$	P 63/m
Britholite-(Ce)	$(Ce,Ca,Th,La,Nd)_5(SiO_4,PO_4)_3(OH,F)$	P 63/m
Britholite-(Y)	$(Y,Ca)_5(SiO_4,PO_4)_3(OH,F)$	P 63/m
Ellestadite-(F)	$Ca_5(SiO_4,PO_4,SO_4)_3(F,OH,Cl)$	P63/m
Ellestadite-(OH)	$Ca_5(SiO_4,SO_4)_3(OH,Cl,F)$	P 63/m
Ellestadite-(Cl)	$Ca_5(SiO_4,PO_4,SO_4)_3(Cl,OH,F)$	P 63/m
Mattheddleite	$Pb_{20}(SiO_4)_7(SO_4)_4Cl_4$	P 63/m
Karnasurtite-(Ce)	$(Ce,La,Th)(Ti,Nb)(Al,Fe)(Si,P)_2O_7(OH)_4 \cdot 3(H_2O)$(?)	
Fluorbritholite-(Ce)	$(Ca,Ce,La,Na)_5(SiO_4,PO_4)_3(OH,F)$	P 63/m
Fluorcalciobritholite	$(Ca,REE)_5[(Si,P)O_4]_3F$	P 63/m
Britholite-(Ce)	$(Ce,Ca,Th,La,Nd)_5(SiO_4,PO_4)_3(OH,F)$	P 63/m
Britholite-(Y)	$(Y,Ca)_5(SiO_4,PO_4)_3(OH,F)$	P 63/m
Ellestadite-(Cl)	$Ca_5(SiO_4,PO_4,SO_4)_3(Cl,OH,F)$	P 63/m
Fluorbritholite-(Ce)	$(Ca,Ce,La,Na)_5(SiO_4,PO_4)_3(OH,F)$	P 63/m
Ellestadite-(F)	$Ca_5(SiO_4,PO_4,SO_4)_3(F,OH,Cl)$	P 63/m
Ellestadite-(OH)	$Ca_5(SiO_4,SO_4)_3(OH,Cl,F)$	P 63/m
Fluorcalciobritholite	$(Ca,REE)_5[(Si,P)O_4]_3F$	P 63/m

Ganomalite is a rare mineral containing isolated SiO_4 tetrahedra and Si_2O_7 groups. During condensation of the silicate anions first Si_2O_7 groups will be formed and subsequently other apatite structure-type minerals with mixed SiO_4 and Si_2O_7 groups. The chemical composition can be expressed by $Pb_9Ca_5Mn^{2+}[SiO_4/Si_2O_7]_3$.

Some other synthetic products are summarized in Table 7.

Another silicate apatite with higher condensation of the SiO_4 tetrahedra is a mineral with apatite structure called nasonite (Figure 7) [128]. This higher condensation of the silicate tetrahedra leads to typical sorosilicate apatite minerals containing only Si_2O_7 groups ([108], [128], [134]). The chemical composition of nasonite can be expressed by $Pb_6Ca_4[Cl_2/(Si_2O_7)_3]$.

Table 7. Ganomalite type compounds [108]

$Pb_3BiNaSi_3O_{11}$
$Pb_3Ca_2Si_3O_{11}$
$Pb_3Cd_2Si_3O_{11}$
$Pb_3Ca_2Ge_3O_{11}$
$Pb_3BiNaGe_3O_{11}$
$Pb_5Ge_3O_{11}$

The crystal structure of ganomalite as shown in Figure 6 was determined by Dunn et al. [94,95] and Carlson et al. [51].

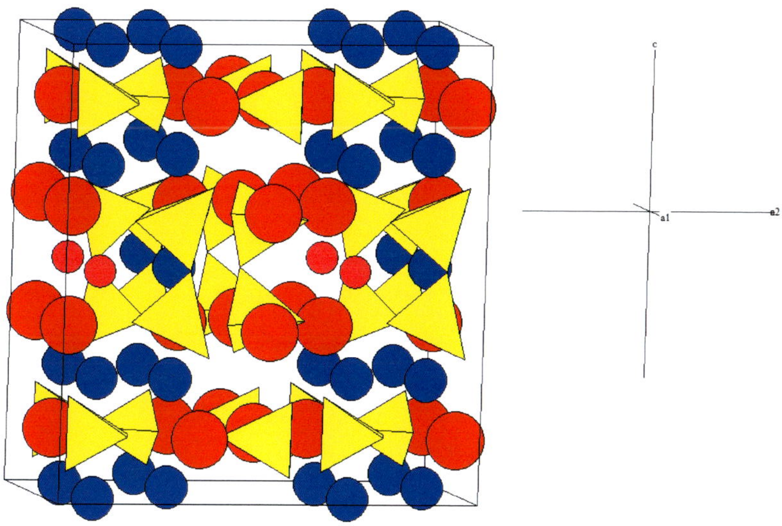

Figure 6. Crystal structure of ganomalite containing Si_2O_7 groups and isolated SiO_4 tetrahedra (Ca = blue, lead = red, manganese = pink, silicate tetrahedra = yellow).

3. Immobilization Technologies

All minerals with apatite structure-type can be used and indeed are being used to immobilize hazardous cations and anions. While these minerals are capable of many different crystallo-chemical replacements they can be used in low temperature applications as well as in hydrothermal or sintering processes whereby silicate-containing apatite minerals are crystallizing. Some of these technologies are described below in detail by the relevant reaction process but also by the applied technology.

The bonding of Pb^{2+} in OH-apatite was reported by Miyake et al. [231]. Pb^{2+}, Cd^{2+}, Sr^{2+} and Hg^{2+} can be incorporated by ionic exchange at room temperature [276].

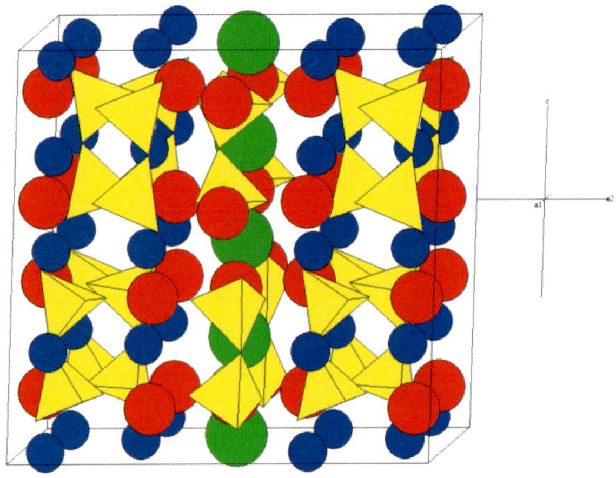

Figure 7. Crystal structure of nasonite showing Si_2O_7 groups = yellow tetrahedra, chlorine = green, lead = red, calcium = blue.

According to Posner et al. [272] crystalline structures of OH-apatites with the following crystal chemistry are formed. Incorporation of relevant metals in hydroxylapatite can be expressed by the general equation:

$Me_{(10-x)}M_{2x}\,(PO_4)_6\,(OH)_2$
$Me = Pb^{2+}, Ca^{2+}, Mg^{2+}$ ....
$M = H^+, Na^+$

The upper limit for this exchange reaction is $x \approx 2$.

Transformation of amorphous phosphates to apatite is influenced by kinetic and thermodynamic factors. The reaction with phosphorous compounds and Ca^{2+}, Al^{3+}, Fe^{2+}, Fe^{3+} was experimentally performed using the following reagents [323]:

1) $Al_2(SO_4)_3$	pH-decrease, alkalinity decrease sulfate flocculation, pH = 5.5 - 6.5
2) $NaAl(OH)_4$	no alkali use, no sulfate
3) $FeCl_3$	low pH (± 1) in comparison to Al anaerobic Fe, P dissolution
4) $FeCl_2$/HCl	low pH needs neutralisation, produces excess solution
5) $Ca(OH)_2$	pH = 9.5 - 11.5, neutralisation by CO_2

The sorption of lead was investigated by Lower et al. [195,196] and Cd sorption by Marchat et al. [203].

Some of these laboratory experiments were later utilized to develop industrial-scale immobilization technologies. Some selected applications are as follows.

3.1. PIMS (Phosphate-Induced Metal Stabilization)

The reaction of apatite with different metals is rather instantaneous ([58], [186], [197], [198], [202], [206]). Typical reactions occur as follows:

Apatite II

$$Ca_{10-x}Na_x(PO_4)_{6-x}(CO_3)_x(OH)_2 + 14\ H^+ \rightarrow (10-x)\ Ca^{2+} + x\ Na^+ + $$
$$+ (6-x)\ [H_2(PO_4)]^- + x\ H_2CO_3 + 2\ H_2O$$

Pyromorphite

$$10\ Pb^{2+} + 6\ H_2(PO_4)^- + 2\ OH^- \rightarrow Pb_{10}(PO_4)_6(OH)_2 + 12\ H^+$$

Sodium-containing apatite with carbonate substitution, often almost amorphous, is produced from fish bones as a waste product [71]. This carbonated sodium-calcium apatite can be used readily in remediation of heavy metals.

3.2. MRS (Mineral Reservoir Technology) - Apatite Formed by Solid State Reaction

Apatite can be formed easily by a high temperature reaction of starting materials like oxides, silicates, sulfates, halogenides, phosphates, and others. The relevant apatites formed after thermal treatment are based on a wide variety of chemical compositions and solid solutions. Many attempts to transform and immobilize different wastes were given by Stemmermann [308] and Neubauer [241]. In these reactions the normal phosphatic apatites were replaced by ellestadites.

To initiate solid state reactions to form apatites temperatures of 1000°C are preferentially applied. Representative reactions to form phosphate-containing apatite or silicosulfate-apatite are given in Table 8.

SEM micrographs of synthesized ellestadite are shown in Figure 8.

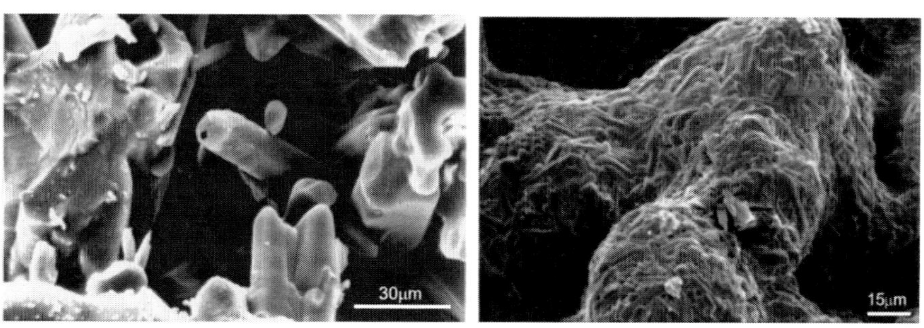

Figure 8. Ellestadites synthesized by thermal (left) and hydrothermal (right) treatment.

To transform and thus immobilize different types of hazardous wastes some test mixtures must be produced to form the essential new mineral type ellestadite. This reaction is based on a thermal treatment (reaction A) to form basically the apatite-structure type, but also some cement minerals such as mayenite which react in a second step to form cement hydration products such as ettringite (reaction B). The relevant reactions of A and B are shown in Figure 9.

Table 8. Examples of starting materials (left) and compositions of apatites formed (right)

CaO, SrO, PbO, SiO_2,$CaSO_4$,$SrSO_4$,$CaCl_2$, BaO, Me_2O_3, Me[SiO_4]	$(Ca,Pb,Sr)_{10}[Cl_2/(SiO_4)(SO_4)]_3$
CaO, SrO, PbO, SiO_2, $CaSO_4$, $SrSO_4$, $CaCl_2$, BaO,Me_2O_3, Me[PO_4], Me[SiO_4]	($(Ca,Pb,Sr)_{10}[Cl_2/(SiO_4)_{(3-x)}(SO_4)_{(3-y)}/(PO_4)_{(x+y)}]$, $x+y \leq 3$

A. Thermal treatment

1 part fly ash + 3 parts municipal solid waste + 1 part foundry sand + 4 parts lime stone + 3 parts FGD-gypsum

⇒ ellestadite + mayenite + anhydrite + quartz

B. Hydraulic reaction

ellestadite + mayenite + anhydrite + quartz + water

⇒ ellestadite (primary mineral reservoir) +
ettringite (secondary mineral reservoir) +
quartz (inert)

Figure 9. Reaction schemata for thermal and hydraulic reaction in treatment of wastes according the mineral reservoir concept.

3.3. The SUPERCALCINE Concept

The typical mineral assemblages of the "Supercalcine" process mainly used for radioactive waste immobilization contain apatite, pollucite, scheelite, fluorite, spinel and corundum [218]. Among the mineral products formed, the apatites preferentially incorporate the lanthanide elements and strontium. Immobilization of low level radwaste according to this was described by Olson et al. [253].

3.4. Direct Precipitation

The precipitation of strontium and barium apatites from solutions was investigated by Denk and Boppel [83]. The direct precipitation of apatite from solution is easier to be accomplished for strontium apatites than for barium apatites since barium more easily forms tertiary barium phosphate. Some relevant reaction equations are given for chloro- and hydroxylapatites as follows:

$$5\ SrCl_2 + 3\ KH_2PO_4 + 7\ KOH \rightarrow Sr_5(PO_4)_3OH + 10\ KCl + 6H_2O$$

$$5\ SrCl_2 + 3\ KH_2PO_4 + 6\ KOH \rightarrow Sr_5(PO_4)_3Cl + 9\ KCl + 6H_2O$$

Larsen et al. [191] described the uptake of fluorine from solutions using OH-apatite. The exchange reaction can be formulated as follows:

$$Ca_{10}[(PO_4)_6(OH)_{(2-x)}] \rightarrow Ca_{(10)}[(PO_4)_6(F)_2]$$

Mavropoulos et al. [210] used Brazilian phosphate rock to immobilize cadmium, lead and copper. In the case of lead immobilization, formation of $(Ca,Pb)_{10}\ (PO_4,CO_3)_6\ (OH,F,Cl)_{2.56}\cdot 1.5\ H_2O$ from aqueous solutions was confirmed, in addition to other accompanying minerals that were participating in the immobilization reactions. Precipitation of phosphate minerals in waste water was investigated by Omson et al. [254].

3.5. The XTALTITE Procedure

To immobilize residues from the smelting and refining industry a concept was developed to transform these industrial residues to insoluble compounds that can be subsequently dumped in tailings [343,344]. These stabilized and immobilized wastes were based on high As_2O_3, Sb_2O_3 and PbO contents. The arsenic and lead contents of these mixtures were transformed into apatite-like minerals, for example svabite (Table 5). Antimony was immobilized in the mineral $Ca_2Sb_2O_7$. The reactions taking place are based on a two-step process. To avoid volatilization of waste elements, the fine powders were mixed primarily with water to form a slurry, and arsenic oxide was subsequently transformed to calcium arsenate hydrate. Important processing steps include a slight drying and spraying of sodium silicate solution which acts as a binder in the following step. The newly formed stable minerals are then treated at high temperature to be transformed to apatite-like minerals such as svabite and hedyphane. The reaction schemes at calcination temperatures of 1100°C can be expressed as follows (Figure 10).

$$3\ As_2O_3 + 9\ CaO + CaF_2 + 3O_2 \rightarrow Ca_5(AsO_4)_3F$$
Arsenolite Lime Fluorite Svabite

$$9\ PbO + 3\ As_2O_3 + CaF_2 + 3O_2 \rightarrow 2(Pb_{4.5}Ca_{0.5})(AsO_4)_3F$$
Litharge Arsenolite Fluorite Hedyphane

$$Sb_2O_3 + 2\ CaO + O_2 \rightarrow Ca_2Sb_2O_7$$
Valentinite Lime Romeite

Figure 10. Reaction schemes of immobilization of industrial residues to apatite-like structures at 1100 - 1200°C.

Several related immobilization technologies were developed through the SMITE (Synthetic Mineral Immobilization Technology) concept [343-345]. Wright et al. [341] described the *in situ* immobilization of heavy metals in apatite. Ruby et al. [284] reported *in situ* formation of Pb in soils. Chen et al. [59, 60] used sorption of heavy metals on apatite for immobilization. Immobilization of uranium and thorium in apatite were studied by Hughson et al. [163,164], Clarke et al. [67], Murrel et al. [237], and Wronkiewicz et al. [349]. In addition to apatite-like minerals other related structures were found to crystallize in cementitious systems, mainly in the CaO-SiO_2-sulfate-chloride system ([75], [247], [264], [266-268], [316], [322]).

REFERENCES

[1] Ames, L.L. Jr.: The genesis of carbonate apatites. *Econ Geol* 54:829-841 (1959).

[2] Anderson, J.B. & Kostiner, E.: The crystal structure of cobalt-substituted calcium chlorapatite. *J Solid State Chem.* 66:343-349 (1987).

[3] Arden, K.M. & Halden, N.M.: Crystallization and alteration history of britholite in rare-earth-element-enriched pegmatitic segregations associated with the Eden Lake Complex, Manitoba, *Canada Can Mineral* 37:1239-1253 (1999).

[4] Arey, J.S., Seaman, J.C. & Bertsch, P.M.: Immobilization of uranium in contaminated sediments by hydroxyapatite addition. *Environ Sci. Tech* 33:337-342 (1999).

[5] Audubert, F., Savariault, J.M. & Lacout, J.L.: Penta (vanadate) iodide, a defect vanadinite-type compound. *Acta Crystallogr* C55~271-273 (1999).

[6] Ayers, J.C. & Watson, E.B.: Apatite/fluid partitioning of rare earth elements and strontium: Experimental results at 1.0 G Pa and 1000°C and application to models of fluid-rockinteraction. *Chem. Geol* ll0:299-314 (1993).

[7] Azimov, S.V., Ismatov, A.A. & Fedorov, N.F.: Synthetic silicophosphates, silicovanadates and silicoarsenates with the apatite structure. *Inorg Mater* 17:1384-1387 (1981).

[8] Bacquet, G., Vo, Q.T., Vignoles, M. & Bonel, G.: ESR of the F^+ centre in B-type carbonated hydroxyapatite. *Phys Status Solidi* 68:71-74 (1981).

[9] Badraoui, B., Aissa, A. & Debbabi, M.**:** Structure refinements of cadmium substituted lead fluoroapatites by powder X-ray pattern fitting. Journal of Physics and Chemistry of solids, Vol.68, Pages211-216 (2007).

[10] Badrour, L., Sadel, A., Zahir, M., Kimakh, L. & El Hajbi, A.: Synthesis and physical and chemical characterization of $Ca_{10-x}Ag_x(PO_4)_6(OH)_{2-xx}$ apatites. Annales *de Chimie Science des Matériaux,* Vol. 23, Issues 1-2, January-February, Pages 61-64, (1998).

[11] Bambauer, H.U.: The application of mineralogy to environmental management - An overview - Int. *Con. Appl. Min.,* Paper 3, (1991).

[12] Bambauer, H.U.: *Das innere Barrieresystem als Strategie zur Immobilisierung von Schadstoffen in Deponien aus Braunkohlenaschen und REA-Produkten* - Springer Verlag, (1991).

[13] Banks, E., Greenblatt, M. & Mcgarvey, B.R.: Electron spin resonance of CrO_4^{3-} in chlorapatite $Ca_5(PO_4)_3Cl$. *J Solid State Chem.* 3:308-313 (1971).

[14] Banks, E. & Jaunarajs, K.L.: Chromium analogs of apatite and spodiosite. *Inorg Chem* 4:78-83 (1965).

[15] Baran, E.J., Baud, G. & Besse, J-P.: Vibrational spectra of some rhenium-apatites containing ReO_5-groups. *Spectrochim Acta* 39A:383-386 (1983).

[16] Baud, G., Besse, J-P., Capestan, M., Sueur, G. & Chevalier, R.: Etude comparative d'apatites contenant I'ion $(ReO_5)^{3-}$. Structure des fluoro et carbonatoapatites. *Ann Chim. Sci. Mater* 5:575-583 (1980).

[17] Baud, G., Besse, J-P., Sueur, G. & Chevalier, R.: Structure de nouvelles apatites au rhenium contenant des anions volumineux: $Ba_{10}(ReO_5)_6X_2$ (X = Br, I). *Mater Res Bull* 14:675-682 (1979).

[18] Baumer, A., Caruba, R., Bizzouard, H. & Peckett, A.: Chlorapatite de synthesis: substitution et inclusions de Mn, Ce, U et Th. traces. *Can Mineral* 21:567-573 (1983).

[19] Baumer, A., Caruba, R. & Ganteaume, M.: Carbonate-fluorapatite: Mise en évidencede la substitution $2PO_4^{3-}$ $SiO_4^{4-} + SO_4^{2-}$ par spectrometrie infrarouge. *Eur. J Mineral* 2, 297-304 (1990).

[20] Baumer, A., Ganteaume, M. & Klee, W.: Determination of OH ions in hydroxyl-fluorapatites by infrared spectroscopy. *Bull. Mineral.,* 108, 145-152 (1985).

[21] Beevers, C. A. & McIntyre, D. B.: The atomic structure of fluor-apatite and its relation to that of tooth and bone material. *Min.Mag.,*27,254-257 (1946).

[22] Belousova, E.A., Griffin, W.L., O'Reilly, S.Y. & Fisher, N.I.: Apatite as an indicator mineral for mineral exploration: trace-element compositions and their relationship to host **rock type**. *Journal of Geochemical Exploration,* Vol. 76, Pages 45-69 (2002).

[23] Berndt, M.E. & Seyfried, W.E. Jr.: Calibration of Br/Cl fractionation during phase separation of seawater: Possible halite at 9 to 10°N East Pacific Rise. *Geochim. Cosmochim. Acta* 61:2849-2854 (1997).

[24] Beshah, K., Rey, C., Glimcher, M.J., Schimizu, M. & Griffin, R.G.: Solid state carbon-13 and proton NMR studies of carbonate-containing calcium phosphates and enamel. *J Solid State Chem* 84:71-81 (1990).

[25] Besse, J-P., Baud, G., Chevallier, R. & Zarembowitch, J.: Mise en revidence de l'ion O_2^- dans l'apatite au rhenium Bas $(ReO_5)_3O_2$. *Mater Res Bull* 15:1255-1261 (1980).

[26] Besse, J-P., Baud, G., Levasseur, G. & Chevallier, R.: Structure crystalline de Bas $(ReO_5)_3Cl$: une nouvelle apatite contenant I'ion $(ReO_5)^{3-}$. *Acta Crystallogr* B35:1756-1759 (1979).

[27] Bigi, A., Falini, G., Gazzano, M., Roveri, N. & Tedesco, E.: Structural refinements of strontium substituted hydroxyapatites. *Mater Sci Forum* 278:814-819 (1998).

[28] Bigi, A., Foresti, E., Marchetti, F., Ripamonti, A. & Roveri, N.: Barium calciumhydroxyapatite solid solutions. *J Chem. Soc Dalton Trans,* p 1091-1093 (1984).

[29] Bigi, A., Gazzano, M., Ripamonti, A., Foresti, E. & Roveri, N.: Thermal stability of cadmium-calcium hydroxyapatite solid solutions. *J Chem. Soc Dalton Trans,* p 241-244 (1986).

[30] Billhard, H.W.: Untersuchungen im System $PbO-PbSiO_3-PbSO_4$.- *Dissertation,* Karlsruhe (1968).

[31] Billhardt, H.W.: Phasenuntersuchungen im System $PbO-PbSiO_3$.-*Glastech. Ber.* 42, 12, 498-505 (1969).

[32] Billhardt, H.W.: Synthesis of lead pyrosilicate and other barysilite-like compounds. - *Am. Min.,* 54, 510-521 (1969).

[33] Binder, G. & Troll, G.: Coupled anion substitution in natural carbon-bearing apatites. *Contrib. Mineral Petrol* 101:394-401 (1989).

[34] Böhlke, J.K. & Irwin, J.J.: Laser microprobe analysis of CI, Br, I and K in fluid inclusions: Implications for sources of salinity in some hydrothermal fluids. *Geochim. Cosmochim. Acta 56*: 203-225 (1992).

[35] Bonel, G.: Contribution a l'etude de la carbonatation des apatites: I. *Ann Chim* 7:65-87 (1972).

[36] Bonel, G., Heughebaert, J-C., Hughebaert, M., Lacout, J.L. & Lebugle, A.: Apatitic calciumortho-phosphates and related compounds for biomaterials preparation. *Ann New York Acad. Sci.* 523: 115-130 (1988).

[37] Bonel, G., Labarthe, J. C. & Vignoles, C.: Contribution a l'étude structurale des apatites carbonatées de type B. *Colloq Intern CNRS* 230:117-125 (1973).

[38] Bonel, G. & Montel, G.: Sur une nouvelle apatitecarbonatée synthétique. *Compt Rend Acad Sci* 258:923-926 (1964).

[39] Borisov, S. V. & Klevtsova, R.F.: the crystal structure of REE-Sr apatite. *Zh Struct Khim* 4:629-631 (1963).

[40] Borneman-Starynkevich, I.D.: On some isomorphic substitutions in the apatite group. *Dokl. Akad. Nauk. SSSR* 19:253-255 (1938).

[41] Borneman-Starynkevich, I.D. & Belov, N.V.: Carbonate-apatites. Doklady Akad. *Nauk SSSR* 90:89-92 (1953).

[42] Bothe, N.Jr. & Brown, P.W.: Arsenic immobilization by calcium arsenate formation. *Environ Sci. Tech* 33:3806-3811 (1999).

[43] Boudreau, A.E.: Fluid evolution in layered intrusions: Evidence from the chemistry of halogen-hearing minerals. In Magmas, Fluids, and Ore Deposits. Thompson JFH (ed) *Mineral Assoc Can Short Course Series* 23:25-45 (1995).

[44] Brasseur, H. & Dallmagne, M.J.: Synthese de l'apatites. *Bull Soc. Chim. France* 135-137 (1949).

[45] Brauns, R.: Über den Apatit aus dem Laacher Seegebiet. *Sulfatapatit und Carbonatapatit.*- N.J.f.M.,41, 60-90 (1916).

[46] Brenan, J.M.: Partitioning of fluorine and chlorine between apatite and aqueous fluids at high pressure and temperature: Implications for the F and Cl contents of high P-T fluids. *Earth Planet Sci Lett* 117:251-263 (1993).

[47] Brophy, G.P. & Nash, J.T.: Compositional, infrared, and X-ray analysis of fossil bone. *Am. Min.* 53: 445-454 (1968).

[48] Bush, A.A. & Venevtsev, Y.N.: New data from an investigation of ferroelectric crystals of solid solutions $Pb_5(Ge_{1-x}Si_x)_3O_{11}$.-*Kristallografiya,* 26, 2, 349-355 (1981).

[49] Buvaneswari, G. & Varadaraju, U.V.: Synthesis and characterization of new apatite-relatedphosphates. *J Solid State Chem.* 149:133-136 (2000).

[50] Calvo, C., Faggiani, R. & Krishnamachari, N.: Crystal structure of $Sr_{9.402}Na_{0.209}(PO_4)_6B_{0.996}O_2$- a deviant apatite. *Acta Crystallogr* B31: 188-192 (1975).

[51] Carlson, S., Norrestam, R., Holstam, D. & Spengler, R.: *The crystal structure of ganomalite,* $Pb_9Ca_{5.44}Mn_{0.56}Si_9O_{33}$, *Z.Krist.* (1977).

[52] Cavaretta, G., Mottana, A. & Tecce, E.: Cesanite, $Ca_2Na_3[(OH)(SO_4)_3]$, a sulfate isotypic to apatite from the Cesano geothermal field (Latium Italy). *Mineral Mag.* 44:269-273 (1981).

[53] Chakhmouradian, A., Reguir, E. & Mitchell, R: Strontium-apatite: New occurrences and the extent of Ca-Sr substitution in apatite-group minerals. *Can Mineral* 40:121-136 (2002).

[54] Chen, M. & Fang, Y.: The Chemical Composition and Crystal Parameters of Calcium Chlorosulfato-silicate. *Cem.Concr.Res.,* 19, 184-188 (1989).

[55] Chen, N., Pan, Y. & Weil, J.A.: Electron paramagnetic resonance spectroscopic study of synthetic fluorapatite: Part I. Legal structural environment and substitution mechanism of Gd at the Ca2 site. *Am. Min.* 87: 37-46 (2002).

[56] Chen, N., Pan, Y., Weil, J.A. & Nilges, M.J.: Electron paramagnetic resonance spectroscopic study of synthetic fluorapatite: Part 11. Gd^{3+} at the Cal site, with a neighboring Ca2 vacancy. *Am Mineral* 87:47-55 (2002).

[57] Chen, N., Pan, Y., Weil, J.A. & Nigles, M.J.: EPR study of ^{157}Gd-doped fluorapatite: Hyperfine and nuclear quadrupole splitting anisotropy. *Geol Assoc Can Mineral Assoc Can Abstr.* 27: 19-20 (2002).

[58] Chen, X.B., Wright, J., Conca, I. & Peurrung, L.M.: Evaluation of heavy metal remediation using mineral apatite. *Water Air Soil Pollut* 98:57-78 (1997).

[59] Chen, X., Wright, J., Conca, J.L. & Peurrung, L.M.: "Effects of pH on heavy metal sorption on mineral apatite." *Environ. Sci. Technol.* 31: 624-631 (1997).

[60] Chen, X., Wright, J., Conca, J.L. & Peurrung, L.M.: "Evaluation of heavy metal remediation using mineral apatite." *Water, Air and Soil Pollution.* 98: 57-78. (1997).

[61] Cherniak, D.J.: Rare earth element diffusion in apatite. *Geochim Cosmochim Acta* 64:3871-3885 (2000).

[62] Chesnokov, B.V., Bazhenova, L.F. & Bushmakin, A.F.: Fluorellestadite $Ca_{10}[(SO_4),(SiO_4)]_6F_2$-a new mineral. *Zap Vser Mineral Obsh.* 1167:743-746 (1987).

[63] Chiranjeevirao, S.V., Hemmerle, J., Voegel, J.C. & Frank, R.M.: A method of preparation and characterization of magnesium-apatites. *Inorg. Chim. Acta* 67:183-187 (1982).

[64] Cho, G. & Yesinowski, J.P.: Multiple-quantum NMR dynamics in the quasi-one-dimensional distribution of protons in hydroxyapatite. *Chem Phys Lett* 205:1-5 (1993).

[65] Cho, G., Yesinowski, J.P.: ^{1}H and ^{19}F multiple-quantum NMR dynamics in quasi-one-dimensional spin clusters in apatites. *J Chem. Phys* 100:15716-15725 (1996).

[66] Christy, A.G., Alberius-Henning, P. & Lidin, S.A.: Computer modelling and description of nonstoichiometric apatites $Cd_{5-/2}(VO_4)_3I_{1-}$ and $Cd_{5-/2}(PO_4)_3Br_{1-}$ as modified chimney-ladder structures with ladder-ladder and chimney-ladder coupling. *Journal of Solid State Chem.* 156:88-100 (2001).

[67] Clarke, R.S. & Altschuler, Z.S.: Determination of the oxidation state of uranium in apatite and phosphorite deposits. *Geochim. Cosmochim. Acta.* 13:127-142 (1958).

[68] Cockbain, A.G. & Smith, G.V.: Alkaline-earth - rare-earth silicate and germanate apatites. *Mineral Mag* 36:411-42I (1967).

[69] Cockbain, A.G.: The crystal chemistry of apatites.- *Min.Mag.* 36, 654-660 (1968).

[70] Comodi, P., Liu, Y., Stoppa, F. & Woolley, A.R.: A multi-method analysis of Si-, S- and REE-rich apatite from a new kind of kalsilite-bearing leucitite (Abruzzi, Italy). *Mineral Mag* 63:661-672 (1999).

[71] Conca, J.L., Lu, N., Parker, G., Moore, B., Adams, A., Wright, J. & Heller, P.: "PIMS–Remediation of Metal Contaminated Waters and Soils." (2000). In G.B. Wickramanayake, A. R. Gavaskar, J.T. Gibbs and J.L. Means (Eds.), Remediation of

Chlorinated and Recalcitrant Compounds, vol. 7. p. 319-326. Battelle Memorial Inst., Columbus, OH. *Florida Institute of Phosphate Research.* (2003). http://fipr.state.fl.us/publications

[72] Conca, J.L. & Wright, J.: An Apatite II permeable reactive barrier to remediate groundwater containing Zn, Pb and Cd. *Applied Geochemistry*, Vol. 21, Pages 2188-2200 (2006).

[73] Creaser, R.A. & Gray, C.M.: Preserved initial $^{87}Sr/^{86}Sr$ in apatites from altered felsic igneous rocks: A case study from the Middle Proterozoic of South Australia. *Geochim. Cosmochim. Acta.* 56:2789-2795 (1992).

[74] Cruft, E.F., Ingamells, C.O. & Muysson, J.: Chemical analysis and stoichiometrie of apatite. *Geochim. Cosmochim. Acta,* 29,581-597 (1965).

[75] Czaya, R. & Bissert, G.: Die Kristallstruktur von Tricalciummonosilikatdichlorid ($Ca_2SiO_4{\cdot}CaCl_2$). - *Acta Cryst.* B27, 747-752 (1970).

[76] Dai, Y. & Hughes, I.M.: Crystal structure refinements of vanadinite and pyromorphite. *Can Mineral* 27: 189-192 (1989).

[77] Dai, Y., Hughes, I.M. & Moore, P.B.: The crystal structure of mimetite and clinomimetite, $Pb_5(AsO_4)_3Cl$. *Can Mineral* 29:369-376 (1991).

[78] Dardenne, K., Vivien, D. & Huguenin, D.: Color of Mn(V)-Substituted Apatites A10((B, Mn)O4)6F2, A=Ba, Sr, Ca; B=P, V. *Journal of Solid State Chemistry,* Vol. 146, Issue 2, Pages 464-472 (1999).

[79] Davis, T.S., Kreidler, E.R., Parodi, J.A. & Soules, T.F.: The luminescent properties of antimony in calcium halophosphates, *J Lumines* 4:48-62 (1971).

[80] Deboer, B.G., Sakthivel, A., Cagle, I.R. & Young, R.A.: Determination of the antimony substitution site in calcium fluorapatite from powder X-ray diffraction data. *Acta Crystallogr* 847:683-692 (1991).

[81] Deganello, S.: The crystal structure of ceanite at 21 and 236°C. *N Jahrb Mineral Monatsh* 305-313 (1983).

[82] Della-Ventura, G., Williams, C.T., Cabella, R., Oberti, R., Caprilli, E. & Bellatreccia, F.: Britholite-hellandite intergrowths and associated REE-minerals from the alkali syenitic ejecta of the Vico volcanic complex (Latium, ltaly): petrological implications bearing on REE mobility in volcanic systems. *Eur J Mineral* 11:843-854 (1999).

[83] Denk, G. & Boppel, B.: Über die Fällung von Phosphaten des Strontiums und Bariums, *Z.An.Allg.Ch.* 310, 185 – 194 (1961).

[84] Dickens, B. & Brown, W.E.: The crystal structure of $Ca_5(PO_4)2SiO_4$ (Silicocarnotite).- *Tscher.Min. Petr.,* 16, 1-27 (1971).

[85] Dihn, P. & Klement, R.: Isomorphe Apatitarten. Zeit.f.elektr.Chem. u. *angew.phys. Chem.,*48,331-333 (1942).

[86] Doi, Y., Moriwaki, Y., Aoba, T., Takahashi, J. & Joshin, K.: ESR and IR studies of carbonate-containing hydroxyapatites. *Calcif Tissue Intl* 34:178-181 (1982).

[87] Dong, P. & Pan, Y.: F-CI-Br partitioning between apatites and halide-rich melts: Experimental studies and applications. *Geol Ass Can Mineral Ass Can Abstr.* 27:29 (2002).

[88] Dowker, S.E.P. & Elliott, J.C.: Infrared study of the formation, loss and location of cyanate and cyanamide in thermally treated apatites. *J Solid State Chem.* 49: 334-340 (1983).

[89] Driessens, F.C.M., Verbeeck, R.M.H. & Kiekens, P.: Mechanism of Substitution in Carbonated Apatites. - *Z. anorg. allg. Chemie* 504, 195-200 (1983).

[90] Dugas, J., Bejjaji, B., Sayah, D. & Trombe, J.C.: Etude par RPE de l'ion NO_2^{2-} dans une apatite nitrée. *J Solid State Chem.* 24:143-151 (1978).

[91] Dugas, J. & Rey, C.: Electron spin resonance characterization of superoxide ions in some oxygenated apatites. *J Phys Chem.* 81:1417-1419 (1977).

[92] Dunn, P.J.: The lead silicates from Franklin, New Jersey: occurrence and composition. - *Min. Mag.,* 49, 721-727 (1985).

[93] Dunn, P.J., Peacor, D.R. & Newberry, N.: Johnbaumite, a new member of the apatite group from Franklin, New Jersey. *Am Mineral* 65:1143-1145 (1980).

[94] Dunn, P.J., Peacor, D.R., Valley, J.W. & Randall, C.A.: Ganomalite from Franklin, New Jersey, and Jakobsberg, Sweden; new chemical and crystallographic data. *Mineral Mag.* 49:579-582 (1985).

[95] Dunn, P.J., Petersen, E.U. & Peacor, D.R.: Turneaureite, a new member of the apatite group from Franklin, New Jersey, Balmant, New York and Laangban, Sweden. *Can Mineral* 23:251-254 (1985).

[96] Dunn, P.J. & Rouse, R.C.: Morelandite, a new barium arsenate chloride member of the apatite group. *Can Mineral* 16:601-604 (1978).

[97] Dymek, R.F. & Owens, B.E.: Petrogenesis of apatite-rich rocks (nelsonites and oxide-apatite gabbronorites) associated with massif anorthosites. *Econ Geol96*:797-815 (2001).

[98] Edgar, A.D.: Barium- and strontium-enriched apatites in lamproites from West Kimberley, Western Australia. *Am Mineral* 74:889-895 (1989).

[99] Efimov, A.S., Kravchenko, S.M. & Vasil'eva, Z.V.: Strontium-apatite a new mineral. *Dokl. Akad. Nauk.* SSSR 142:439-442 (1962).

[100] Elliott, J.C.: The crystallographic structure of dental enamel and related apatites. *PhD dissertation,* University of London, London, UK (1964).

[101] Elliott, J.C.: Recent progress in the chemistry, crystal chemistry and structure of the apatites. *Calcif Tissue Res* 3:293-307 (1969).

[102] Elliott, J.C.: *Structure and Chemistry of the Apatites and Other Calcium Orthophosphates.* Elsevier, Amsterdam (1994).

[103] Elliott, J.C., Bonel, G. & Trombe, J.C.: Space group and lattice constants of Ca_{10} $(PO_4)_6CO_3$, *J Appl. Crystallogr.* 13:618-621 (1980).

[104] Elliott, J.C., Dykes, E. & Mackie, P.E.: Structure of bromapatite, $Ca_5(PO_4)_3Br$, and the radius of the bromide ion. *Acta Crystallogr.* B37:435-438 (1981).

[105] Elliott, J.C., Mackie, P.E. & Young, R.A.: Monoclinic hydroxyapatite. *Science* 180:1055-1057 (1973).

[106] Ellison, A.J.G. & Hess, P.C.: Solution properties of rare earth elements: Inference from immiscible liquids. *Geochim Cosmochim Acta* 53:1965-1974 (1989).

[107] Engel, G.: Hydrothermalsynthese von Bleihyoxylapatiten $Pb_5(XO_4)_3OH$ mit X = P, As, V. *Naturwissenschaften* 57:355 (1970).

[108] Engel, G.: Ganomalite, an intermediate between the nasonite and apatite types. - *Naturwissenschaften* 59, 121-122 (1972).

[109] Engel, G.: Infrarotspektroskopische und röntgenographische Untersuchungen von Bleihydroxylapatit, Bleioxyapatit und Bleialkaliapatit *J Solid State Chem.* 6:286-292 (1973).

[110] Engel, G.: Fluoroberyllate mit Apatitstruktur und ihre Beziehungen zu Sulfaten und Silicaten. *Mater Res Bull* 13:43-48 (1978).

[111] Engel, G., Krieg, F. & Reif, G.: Mischkristallbildung und Kationenordnung im System Bleihydroxylapatit-Calciumhydroxylapatit. *J Solid State Chem.* 15:117-126 (1975).

[112] Ercit, T.S., Cerny, P. & Groat, L.A.: The crystal structure of Mn-rich fluorapatite and the role of Mn in the apatite structure. *Geol Assoc Can Mineral Ass Can Abstr.* 19:A34 (1994).

[113] Fayos, J., Watkin, D.J. & Pérez-Méndez, M.: Crystal structure of the apatite-like compound $K_3Ca_2(SO_4)_3F$. *Am Mineral* 72:209-212 (1987).

[114] Felsche, F.: Rare earth silicates with apatite structure. *J Solid State Chem.* 5:266-275 (1972).

[115] Fleet, M.E. & Pan, Y.: Site preference of Nd in fluorapatite [$Ca_{10}(PO_4)_6F_2$] *J Solid State Chem.* 111: 78-81 (1994).

[116] Fleet, M.E. & Pan. Y.: Site preference of rare earth elements in fluorapatite. *Am. Min.* 80:329-335 (1995).

[117] Fleet, M.E. & Pan. Y.: Crystal chemistry of rare earth elements in fluorapatite and calc-silicateminerals. *Eur J Mineral* 7:591-605 (1995).

[118] Fleet, M.E. & Pan, Y.: Site preference of rare earth elements in fluorapatite: Binary (LREE+HREE) substituted crystals. *Am. Min.* 82:870-877 (1997).

[119] Fleet, M.E. & Pan, Y.: Rare earth elements in apatite: Uptake from H_2O-bearing phosphate-fluoride melts and the role of volatile components. *Geochim. Cosmochim. Acta* 61:4745-4760 (1997).

[120] Fleet, M.E., Liu, X. & Pan, Y.: Rare earth elements in chlorapatite [$Ca_{10}(PO_4)_6Cl_2$]: Uptake, site preference and degradation of monoclinic structure. *Am. Min.* 85:1437-1446 (2000).

[121] Fleet, M.E., Liu, X. & Pan, Y.: Site preference of rare earth elements in hydroxyapatite [$Ca_{10}(PO_4)_6(OH)_2$]. *J Solid State Chem.* 149:391-398 (2000).

[122] Fleischer, M. & Altschuler, Z.S.: The lanthanides and yttrium in minerals of the apatite group - an analysis of the available data. *N. Jahrb. Mineral. Monatsh.* 467-480 (1986).

[123] Fletcher, L.: On Crystals of Percylite Caracolite and an Oxychloride of Lead (Daviesite), from Mina Beatriz Sierra Gorda, Atacama, South America. *-Min.Mag.* 8, 172 (1889).

[124] Förtsch, E. & Freiburg, I.B.: Untersuchungen von Mineralien der Pyromorphit Gruppe. *N. Jahrb. Mineral. Abh.* 113:219-250 (1970).

[125] Fowler, B.O.: Infrared studies of apatites. II. Preparation of normal and isotopically substituted calcium, strontium, and barium hydroxyapatite and spectra structure-composition-correlations. *Inorg. Chem.* 13: 207-214 (1974).

[126] Fransolet, A.-M. & Schreyer, W.: Unusual, iron-hearing apatite from a garnetiferous pegmatoid, Northampton Block, Western Australia. *N Jahrh. Mineral. Monatsh.* 317-327 (1981).

[127] Fröhlich, R.: $Pb_8[O_2(SO_4)(Si_4O_{13})]$, a new tetrasilicate. - *Acta Crystallogr.* A40, Supplement, C224 (1984).

[128] Frondel, C. & Bauer, L.H.: Nasonite and ist relation to Pyromorphite. - *Am. Min.,* 36, 534 (1951).

[129] Frondel, C. & Ito, J.: Geochemistry of germanium in the oxidized zone of the Tsumeb Mine, *South-West Africa. Am. Min.* 42:743-753 (1957).

[130] Furukawa, T., Brawer, S. A. & White, W.B.: Raman and Infrared Spectroscopic Studies of the Crystalline Phases in the system Pb_2SiO_4-$PbSiO_3$ - *J. Am. Ceram. Soc.,* 62, 351-356 (1979).

[131] Gaetani, G.A. & Grove, T.L.: Partitioning of rare earth elements between clinopyroxene and silicate melt: CrystaJ-chemical controls. Geochim. Cosmochim. *Acta* 59: 1951-1962 (1995).

[132] Gaft, M., Reisfeld, D.R., Ranczer, G., Shoval, S., Champagnon, B. & Boulon, G.: Eu^{3+} luminescence in high symmetry sites of natural apatite. *J Luminesc.* 72-74:572-574 (1997).

[133] Gaudé, J.L., Haridon, P., Hamond, C., Marchand, R. & Laurent, Y.: Composes á structure apatite. I. Structure de I'oxynitrure $Sm_{10}Si_6N_2O_{24}$. *Bull Soc fr Mineral Cristal,* 98:214-217 (1975).

[134] Giuseppetti, G., Rossi, G. & Tadini, C.: The crystal structure of Nasonite. - *Am. Min..,* 56, 1174-1179 (1971).

[135] Grandjean-Lécuyer, P., Feist, R. & Albarede, F.: Rare earth elements in old biogenic apatites. *Geochim Cosmochim Acta* 57:2507-2514 (1993).

[136] Grimmer, A.R., Hoebbel, D., Goetz, M. & Mägl, M.: Si-MAS-NMR-Untersuchungen an kristallinen Bleisilicaten. - *Z. anorg. allg. Chem.,* 547, (4), 45-56 (1987).

[137] Grisafe, D.A. & Hummel, F.A.: Crystal chemistry and color in apatites containing cobalt, nickel and rare-earth ions. *Am. Min.* 55:1 131-1145 (1970).

[138] Gromet, L.P. & Silver, L.T.: Rare earth element distributions among minerals in a granodiorite and their petrogenic implications. *Geochim Cosmochim Acta* 47:925-939 (1983).

[139] Groner, J.W. & McConnell, D.: The problem of the carbonate-apatites. *Z. Krist.* 97A:208-215 (1937).

[140] Gross,K.A. & Rodríguez-Lorenzo, L.M.: Sintered hydroxyfluorapatites. Part I: Sintering ability of precipitated solid solution powders. *Biomaterials,* Vol. 25, Issues 7-8, Pages1375-1384 (2004).

[141] Guidry, M.W. & Mackenzie, F.T.: Experimental study of igneous and sedimentary apatite dissolution: Control of pH, distance from equilibrium, and temperature on dissolution rates. *Geochim. Cosmochim.* Acta,Vol. 67, Issue 16, Pages 2949-2963 (2003).

[142] Guyader, J., Grekov, F.F., Marchand, R. & Lang, J.: Nouvelles séties de silicoapatites enrichies en azote. *Rev Chim. Mineral* 15:431-438 (1978).

[143] Harada, K., Nagashima, K., Nakao, K. & Kato, A.: Hydroxylellestadite, a new apatite from Chichibu mine, Saitama prefecture, Japan. *Am. Min.* 56, 1507-1518 (1971).

[144] Harries, J.E., Hasnain, S.S. & Shah, J.S.: EXAFS study of structural disorder in carbonate-containing hydroxyapatite. *Calcif Tissue* Intl41:346-350 (1987).

[145] Hasegawa, H., Shimada, M., Kanamaru, F. & Koizumi, M.: The Crystallization of $Pb_5Si_3O_{11}$ from the Glass in the PbO-SiO_2-System. *Bull. Chem. Soc.* Japan, 50, 2, 529-530 (1977).

[146] Hata, M., Mammo, F., Iwai, S. & Aoki, H.: Structure of barium chlorapatite. *Acta Crystallogr* B35:2382-2384 (1979).

[147] Hata, M., Marumo, F., Iwai, S. & Aoki, H.: Structure of a lead apatite $Pb_9(PO_4)_6$ *Acta Crystallogr* B36:2128-2130 (1980).

[148] Hata, M., Okada, K., Iwai, S., Akao, M. & Aoki, H.: Cadmium hydroxyapatite. *Acta Crystallogr* B34:3062-3064 (1978).

[149] Heijligers, H.I.M., Verbeeck, R.M.H. & Driessens, F.C.M.: Cation distribution in calcium- strontium-hydroxyapatites. *J Inorg Nuclear Chem.* 41:763-764 (1979).

[150] Hendricks, S.B., Jefferson, M.E. & Mosley, V.M.: The crystal structure of some natural and synthetic apatite-like substances. *Z. Krist.* 81, 352-369 (1932).

[151] Hitmi, N., Lacabanne, C., Bonel, G., Roux, P. & Young, R.A.: Dipole cooperative motions in an A-type carbonated apatite, $Sr_{10}(AsO_4)_6CO_3$ *J Phys Chem. Solids* 47:507-515 (1986).

[152] Hogarth, D.D., Staecy, H.R., Semenov, E.I., Proshchenko, E.G., Kazakova, M.E. & Kataeva, Z.T.: New occurrences and data of spencite. *Can Mineral* 12:66-71 (1973).

[153] Holmden, C., Creaser, R.A., Muehlenbachs, K., Leslie, S.A. & Bergstrom , S.M.: Isotopic and elemental systematics of Sr and Nd in 454 Ma biogenic apatites: Implications for paleo seawater studies. *Earth Planet Sci Lett.* 142:425-437 (1996).

[154] Houndslow, I. & Chao, F.: Chlorapatite - Specimen from Bob's Lake Ontario, Canada. *Can.Mineral.,*10,252 (1970).

[155] Huang, J. & Sleight, A.W.: The apatite structure without an inversion center in a new bismuth calcium vanadium oxide $BiCa_4V_3O_{13}$. *J Solid State Chem.* 104:52-58 (1993).

[156] Hughes, J.M., Cameron, M. & Crowley, K.D.: Structural variations in natural F, OH, and Cl apatites. *Am Mineral* 74:870-876 (1989).

[157] Hughes, J.M., Cameron, M. & Crowley, K.D.: Crystal structures of natural ternary apatites: solid solutions in the $Ca_5(PO_4)_3X$ (X=F, OH, Cl) system. *Am. Min. 75*:295-304 (1990).

[158] Hughes, J.M., Cameron, M. & Crowley, K.D.: Ordering of divalent cations in apatite structure: crystal structure refinements of natural Mn- and Sr-bearing apatites. *Am. Min.* 76:1857-1862 (1991).

[159] Hughes, J.M., Cameron, M. & Mariano, A.N.: Rare-earth element ordering and structural variations in natural rare-earth-hearing apatite. *Am. Min.* 76: 1165-1173 (1991).

[160] Hughes, J.M. & Drexler, J.W.: Cation substitution in the apatite tetrahedral site: crystal structures of type hydroxylel1stadite and type fermorite. *N Jahrb Mineral Monatsh* 327-336 (1991).

[161] Hughes, J.M., Fransolet, A.-M. & Schreyer, W.: The atomic arrangement of iron-hearing apatite. *N. Jahrb Mineral Monatsh.* 504-510 (1993).

[162] Hughes, J.M., Mariano, A.N. & Drexler, J.W.: Crystal structures of synthetic Na-REE-Si oxyapatites, synthetic monoclinic britholite. *N Jahrb Mineral Monatsh.* 311-319 (1992).

[163] Hughson, M.R. & Gupta, G.S.: A thorian intermediate member of the britholite-apatite series. *Am.Min.,49*: 937-951 (1964).

[164] Hughson, M.R. & Sen Gupta, J.G.: A thorian intermediate member of the britholite-apatite series. *Am. Min.* 49:937-951 (1964).

[165] Ikoma, T., Yamazaki, A., Nakamura, S. & Akao, M.: Preparation and structure refinement of monoclinic hydroxyapatite. *J Solid State Chem.* 144:272-276 (1999).

[166] Ito, A., Aoki, H., Akao, M., Miura, N., Otsuk, A.R. & Tsutsumi, S.: Structure of borate groups in borate-containing apatite. *J Ceram Soc Japan* 96:695-697 (1988).

[167] Ito, J.: Silicate apatites and oxyapatites. Am. Min., 53:890-907 (1968).

[168] Ito, J.: Synthesis of some lead calcium zinc silicates. *Am. Min.,* 53, 231-240, (1968).

[169] Ivanova, T.I., Frank-Kamenetskaya, O.V., Koltsov, V. & Ugolkov, L.: Crystal structure of calcium-deficient carbonated hydroxyapatite. Thermal decomposition. *J Solid State Chem.* 160:340-349 (2001).

[170] Jahnke, R.A.: The synthesis and solubility of carbonate fluorapatite. *Am J Sci* 284:58-78 (1984).

[171] Johnson, W.: Two synthetic compounds containing chromium in different valency states.- *Min.Mag*.,32,408-411 (1960).

[172] Jolliff, B.L., Haskin, L.A., Colson, R.O. & Wadhwa, M.: Partitioning in REE-saturating minerals: Theory, experiment and modelling of whitlockite, apatite, and evolution of lunar residual magmas. *Geochim Cosmochim Acta* 57:4069-4094 (1993).

[173] Joris, S.J. & Amberg, C.H.: The nature of deficiency in nonstoichiometric hydroxyapatite. II. Spectroscopic studies of calcium and strontium hydroxyapatite. *J Phys Chem.* 75:3172-3178 (1971).

[174] Kato, K.: Die Kristallstruktur des Bleisilicats $Pb_{11}Si_3O_{17}$. *Acta Cryst,* B38, 57-62 (1982).

[175] Kautz, K. & Gubser, R.: Untersuchungen mit der Elektronenmikrosonde an zonargebauten Mineralen der Pyromorphitgruppe. *Contrib Mineral Petrol* 20:298-305 (1969).

[176] Kazak, V.F., Lyutin, V.I., Ilyukhin, V.V. & Belov, N.V.: *Crystal structure of monoclinic acid sodium calcium orthosilicate. - Kristallografija,* 2, 19, 954-957, Moskau (1974).

[177] Kazin, P.E., Gazizova, O.R., Karpov, A.S., Jansen, M. & Tretyakov, Y.D.: Incorporation of 3d-metal ions in the hexagonal channels of the $Sr_5(PO_4)_3OH$ apatite. Solid State Sciences, online (2007).

[178] Khattech, I. & Jernal, M.: Thermochemistry of phosphate products. II. Standard enthalpies of formation and mixing of calcium and strontium fluorapatites. *Thermochim Acta* 298:23-30 (1997).

[179] Khomyakov, A.P., Lisitsyn, D.V., Kulikova, I.M. & Rastsvetayeva, R.K.: Deloneite-(Ce), $NaCa_2SrCe(PO_4)_3F$, a new mineral with a belovite-like structure. *Zap Vser Mineral Obsh* 125:83-94 (1996).

[180] Khorari, S., Cahay, R., Rulmont, A. & Tarte, P.: The coupled isomorphic substitution 2 $(PO_4)^{3-}$ = $(SO_4)^{2-}$ + $(SiO_4)^{4-}$ in synthetic apatite $Ca_{10}(PO_4)_6F_2$: a study by X-ray diffraction and vibrational spectroscopy. *Eur J Solid State Inorg Chem.* 31:921-934 (1994).

[181] Khudolozhkin, V.O., Urusov, V.S. & Tobelko, K.I.: Ordering of Ca and Sr in cation positions in the hydroxylapatite-belovite isomorphous series. Geochem. *Intl* 9:827-833 (1972).

[182] Khudolozhkin, V.O., Urusov, V.S. & Tobelko, K.I.: Distribution of cations between sites in the structure of Ca, Sr, Ba - apatites. *Geochem. Intl* 10:266-269 (1973).

[183] Khudolozhkin, V.O., Urusov, V.S., Tobelko, K.I. & Vemadskiy, V.I.: Dependence of structural ordering of rare earth atoms in the isomorphous series apatite-britholite (abukumalite) on composition and temperature. *Geochem Indl* 10: 1171-1177 (1973).

[184] Kingsley, J.D., Prener, J.S. & Segall, B.: Spectroscopy of $(MnO_4)^{3-}$ in calcium halophosphates. *Phys Rev* A137:189-202 (1965).

[185] Kreidler, E.R. & Hummel, F.A.: The crystal chemistry of apatite: structure fields of fluor- and chlorapatite. *Am. Min.* 55: 170-184 (1970).
[186] Koeppenkastrop, D. & De Carlo, E.J.: Sorption of rare earth elements from seawater onto synthetic mineral phases. *Chem. Geol.* 95: 251-263 (1990).
[187] von Kutoglu, A.: *Structure refinement of the apatite* $Ca_5(VO_4)_3(OH)$. N Jahrb Mineral Monatsh. 210-218 (1974).
[188] Labarthe, J.-C., Bonel, G. & Montel, G.: Sur la localisation des carbonate dans le réseau des apatites calciques. *Compt Rend Acad. Sci* 273:349-351 (1971).
[189] Labarthe, J.-C., Therasse, M., Bonel, G. & Montel, G.: Sur la structure des apatites phosphocalciques carbonatées de type B. *Compt. Rend Acad. Sci* 276:1175-1178 (1973).
[190] Larsen, L.M.: Distribution of REE and other trace elements between phenocrysts and peralkaline under saturated magmas, exemplified by rocks from the Gardar igneous province, south Greenland. *Lithos.* 12:303-315 (1979).
[191] Larsen, M.J. & Pearce, E.I.: Partial defluoridation of drinking water using fluorapatite precipitation, *Caries Res.* 26 (1):22-8, (1992).
[192] Le Geros, R.Z.: Effect of carbonate on the lattice parameters of apatite. *Nature* 206:403-404 (1965).
[193] Liebau, F.K.F.: *Structural Chemistry of Silicates.* 347 S., Springer Verlag, Berlin-Heidelberg-New York-Tokio (1985).
[194] Liu Y. & Comodi, P.: Some aspects of the crystal chemistry of apatites. *Mineral Mag.* 57:709-719 (1993).
[195] Lower, S.K., Maurice, P.A., Traina, S.J. & Carlson, E.H.: Aqueous lead sorption by hydroxylapatite: Applications of atomic force microscopy to dissolution, nucleation and growth studies. *Am. Min.,* 83: 147-158 (1998).
[196] Lower, S. K., Maurice, P.A. & Traina, S.J.: Simultaneous dissolution of hydroxylapatite and precipitation of hydroxypyromorphite: Direct evidence of homogeneous nucleation. Geochim Cosmochim. *Acta.* 62: 1773-1780 (1998).
[197] Lu, N., Xu, H., Wright, J. & Conca, J.L.: PIMS - Remediation of metal-contaminated groundwater and soil using a special reactive form of the mineral apatite. In J. Rammlmair; Th. O. Mederer; R.B. Heimann and H. Pentinghaus (Eds.), Applied Miner. In Research, Economy, Techy., *Ecology and Culture,* vol. 2, p. 603-606. A. A. Balkema Publ., Rotterdam, Germany (2001).
[198] Ma, Q.Y., Logan, T.J. & Traina, S.J.: Lead immobilization from aqueous solutions and contaminated soils using phosphate rocks. *Environ. Sci. Technol.* 29: 1118-1126 (1995).
[199] Machatschki, F.: Sind Abukumalit und Britholith Glieder der Apatitreihe? Zentralblatt f. Mineralogie, *Abhandlungen,* 161-167 (1939).
[200] Mackie, P.E., Elliott, J.C. & Young, R.A.: Monoclinic structure of synthetic $Ca_5(PO_4)_3Cl$ chlorapatite. *Acta Crystallogr* A28:1840-1848 (1972).
[201] Mackie, P.E. & Young, R.A.: Location of Nd dopant in fluorapatite, $Ca_5(PO_4)_3F$: Nd. *J Appl. Crystallogr.* 6: 26-31 (1973).
[202] Manecki, M., Maurice, P.A. & Traina, S.J.: Kinetics of aqueous Pb reaction with apatites. *Soil Science.* 165(12): 920-933 (2000).

[203] Marchat, D., Bernache-Assollant, D. & Champion, E.: Cadmium fixation by synthetic hydroxyapatite in aqueous solution - Thermal behaviour. *Journal of Hazardous Materials,* Vol.139, Issue3, Pages 453-460 (2007).

[204] Masaoka, M., Kyono, A., Hatta, T. & Kimata, M.: Single crystal growth of $Pb_5(P_xAs_{1-x}O_4)_3Cl$ solid solution with apatite type structure. *Journal of Crystal Growth,*Vol. 292, Issue1, Pages 129-135 (2006).

[205] Masmoudi, S., Larbot, A., El Feki, H. & Ben Amar, R.: Elaboration and characterisation of apatite based mineral supports for microfiltration and ultrafiltration membranes, *Ceramics International,* 33, Issue 3, 337-344 (2007).

[206] Matheson, L.J., Goldberg, W.C., Huddleston, G.J. & Bostick, W.D.: Remediation of metal-contaminated soil and groundwater using Apatite II. In A.R. Gavaskar and A. Chen (Eds.), *Remediation of Chlorinated and Recalcitrant Compounds, p. 2H-13. Battelle Memorial Inst.,* Columbus, OH (2002).

[207] Mathew, M., Brown, W.E., Austin, M. & Negas, T.: Lead alkali apatites without hexad anion: the crystal structure of $Pb_8K_2(PO_4)_6$ *J Solid State Chem.* 35:69-76 (1980).

[208] Mathew, M., Mayer, I., Dickens, B. & Schroeder, L.W.: Substitution in barium-fluoride apatite: the crystal structures of $Ba_{10}(PO_4)_6F_2$, $Ba_6La_2Na_2(PO_4)_6F_2$ and $Ba_4La_3Na_3(PO_4)_6F_2$. *J Solid State Chem.* 28:79-95 (1979).

[209] Maunaye, M., Hamon, C., L'haridon, P. & Laurent, Y.: Composés á structure apatite. IV. Etude structurale de l'oxynitrure $Sm_8Cr_2Si_6N_2O_{24}$. *Bull Soc Fr Minéral Cristal* 99:203-205 (1976).

[210] Mavropoulos, E., Nilce, C., Da Rocha, C., Moreira, J.C., Bertolino, L.C. & Rossi, A.M.: Pb $^{2+}$, Cu $^{2+}$and Cd $^{2+}$ ions uptake by Brazilian phosphate rocks, *J. Braz. Chem. Soc.* vol.16 no.1 São Paulo Jan./Feb. (2005).

[211] Mayer, I. & Cohen, S.: The crystal structure of $Ca_6Eu_2Na_2(PO_4)_6F_2$. *J Solid State Chem.* 48: 17-20 (1983).

[212] Mayer, I., Fischbein, E. & Cohen, S.: Apatites of divalent europium. *J Solid State Chem.*14:307-312 (1975).

[213] Mayer, I., Roth, R.S. & Brown, W.E.: Rare earth substituted fluoride-phosphate apatites. *J Solid State Chem.* II, 33-37 (1974).

[214] Mayer, I. & Semadja, A.: Bismuth-substituted calcium, strontium, and lead apatites. *J Solid State Chem.* 46:363-366 (1983).

[215] Mayer, I. & Swissa, S.: Lead and strontium phosphate apatites substituted by rare earth and silver ions. *J Less Common Metals* 110:411-414 (1985).

[216] Mazza, D., Tribaudino, M., Delmstro, A. & Lebech, B.: Synthesis and neutron structure of $La_5Si_2BO_{13}$, an analogue of the apatite mineral. *J Solid State Chem.* 155:389-393 (2000).

[217] McArthur, M.J.: Francolite geochemistry-compositional controls during formation, diagenesis, metamorphism, and weathering. *Geochim Cosmochim Acta* 49:23-35 (1985).

[218] McCarthy, G.J.: High level waste ceramics: materials Considerations, process simulation and product characterization - *Nucl. Techn.* 32, 91-105 (1977).

[219] McCarthy, G.J., White, W.B., Roy, R., Scheetz, B.E., Komarneni, S., Smith, D.K. & Roy, D.M.: Interactions between nuclear waste and surrounding rock. *Nature,* 273; 216-217, London (1978).

[220] McConnel, D.: The Substitution of SiO_4- and SO_4-Groups for PO_4-Groups in the Apatite Structure; Ellestadite, the endmember. *Am.Min.* 22, 977-986 (1937).

[221] McConnel, D.: A Structural investigation of the isomorphism of the apatite group. *Am.Min.* 23, 1-19 (1938).

[222] McConnell, D.: The problem of the carbonate apatites. IV. Structural substitutions involving CO_3 and OH. *Bull Soc Fr Mineral Cristall.* 75:428-445 (1952).

[223] McConnell, D.: Apatite. *Its Crystal Chemistry, Mineralogy, Utilization, and Geologic and Biologic Occurrences.* Springer, New York (1973).

[224] McMurdie, H.F.: X-ray studies of compounds in the systems PbO-B_2O_3 and K_2O-PbO-SiO_2. - *J. Res. Nat. Bur. Stand.,* 26, 489-494 (1941).

[225] Merker, L.: Zum Verhalten von Halogeniden in Bleigläsern. - *Angew. Chem., 75*, 798 (1963).

[226] Merker, L., Engel, G., Wondratschek, H. & Ito, J.: Lead ions and empty halide sites in apatites. *Am. Min.,* 55:1435-1436 (1970).

[227] Mishra, K.C., Patton, R.J., Dale, E.A. & Das, T.P.: Location of antimony in a halophosphate phosphor. *Phys Rev* B35:1512-1520 (1987).

[228] Mingyuan C. & Fang, Y.: The chemical composition and crystal parameters of calcium chlorosulfato-silicate. *Cem.Concr.Res.* 19, 184-188 (1988).

[229] Mitchell, L., Faust, G.T., Hendricks, S.B. & Reynolds, D.S.: The mineralogy and genesis of hydroxylapatite. *Am. Min.* 28:356-371 (1943).

[230] Miyake, M., Ishigaki, K. & Suzuki, T.: Structure refinements of Pb^{2+} ion-exchanged apatites by X-ray powder pattern-fitting. *J Solid State Chem.* 61 230-235 (1986).

[231] Miyake, M., Komarneni, S. & Roy, R.: Immobilization of Pb^{2+}, Cd^{2+}, Sr^{2+} and Ba^{2+-} ions using calcite and aragonite - *Cem.Concr.Res.* 18, 485-490, (1988).

[232] Mohseni-Koutchesfehani, S.: Contribution á l' étude des apatites barytiques. *Ann Chim.* 463-479 (1961).

[233] Moore, E. & Eitel, W.: *A Borosilicate of the Apatite Group.-Naturwissenschaften,* 44, 8, 259 (1957).

[234] Moore, P.B. & Araki, T.: Samuelsonite: its crystal structure and relation to apatite and octacalcium phosphate. *Am. Min.* 62:229-245 (1977).

[235] Morse, J.W. & Casey, W.H.: Ostwald processes and mineral paragenesis in sediments. *American Journal Science*. 288: 537-560 (1988).

[236] Morss, L.S.: Thermochemical properties of yttrium, lanthanum, and lanthanide elements and ions. *Chem. Rev* 76:827-841 (1976).

[237] Murrell, M.T., Brandriss, M., Woolum, D.S. & Burnett, D.S.: Pu-REE-Y partitioning between apatite and whitlockite. *Lunar Planet Sci XV:* 579-580 (1984).

[238] Nadal, M., Le Geros, R.Z., Bonel, G. & Montel, G.: Mise en évidence d'un phénoméne d'order-désordre dans le réseau des carbonate-apatites strontique. *Compt Rend Acad. Sci.* 272:45-48 (1971).

[239] Naray-Szabo, ST.: The structure of apatite (CaF)$Ca_4(PO_4)_3$.- *Zeit.Krist.,*75, 387-398 (1930).

[240] Nassau, K., Shiever, J.W., Joy, D.C. & Glass, A.M.: The crystallization of vitreous and metastable $Pb_5Ge_3O_{11}$. - *J. Cryst. growth,* 42, 574-578 (1977).

[241] Neubauer, J.: Ph.D. *dissertation,* Universität Erlangen (1992).

[242] Neubauer, J. & Pöllmann, H.: Einbau von Schwermetallen in Silikatapatite. *Eur. J. Min. Bh.,* 191 (1990).

[243] Neubauer, J. & Pöllmann, H.: Synthesis and characterization of silicosulfate-apatites, - *N. Jb. Min., Mh.*7, 295 – 310 (1992).

[244] Neuman, W.F. & Mulryan, B.J.: Synthetic hydroxyapatite crystals IV. Magnesium incorporation. *Calcif Tissue Intl* 7: 133-138 (1971).

[245] Newberry, N.G., Essene, E.J. & Peacor, D.R.: Alforsite, a new member of the apatite group: The barium analogue of chlorapatite. *Am. Min.* 66: 1050-1053 (1981)

[246] Newnham, R.E., Wolfe, R.W. & Darligton, C.N.W.: Prototype Structure of $Pb_5Ge_3O_{11}$. - *J. Solid State Chem.,* 6, 378-383 (1973).

[247] Nobes, R.H., Akhmatskaya, E.V., Milman, V., White, J.A., Winkler, B. & Pickard, C.J.: An ab initio study of hydrogarnets. *Am. Min.* 85:1706-1715 (2000).

[248] Nriagu, J.O.: Formation and stability of base metal phosphates in soils and sediments. In *Phosphate Minerals.* Nriagu, J.O., Moore, P.B. (eds) Springer-Verlag, New York, p. 318-329 (1984).

[249] Nriagu, J.O. & Moore, P.B.: *Phosphate Minerals.* Springer-Verlag, New York (1984).

[250] Oberti, R., Ottolini, L., Della Ventura, G. & Pardon, G.C.: On the symmetry and crystal chemistry of britholites: New structural and microanalytical data. *Am. Min.* 86:1066-1075 (2001).

[251] Ohkubo, Y.: EPR spectra of manganese (II) ions in synthetic calcium chloride fluoride phosphates. *J Appl. Phys* 39:5344-5345 (1968).

[252] Okazaki, M.: FCO_3^{2-} interactions in IR spectra of fluoridated CO_3-apatites. *Calcif Tissue Intl* 35:78-81 (1983).

[253] Olson, R.A., Tennis, P.D., Bonen, D., Jennings, H.M., Mason, T.O., Christensen, B.J., Brough, A.R., Sun, G.K. & Young, J.F.: Early containment of high-alkaline solution simulating low-level radioactive waste in blended cement. *Journal of Hazardous Materials,* Vol. 52, Issues 2-3, Pages 223-236 (1997).

[254] Omson, M.B. & Vignona, I.: *Precipitatation of phosphate minerals in waste water treatment systems* – in Nriagu & Moore (1984).

[255] O'reilly, S.Y. & Griffin, W.L.: Apatite in the mantle: implications for metasomatic processes and high heat production in Phanerozoic mantle. *Lithos.* 53 217-232 (2000).

[256] Ott, W.R. & McLaren, M.G.: Subsolidus Studies in the System $PbO-SiO_2$. - *J. Amer. Ceram. Soc.,* 53, 7, 374-375 (1970).

[257] Otto, H.H.: Zur Kristallchemie metastabiler Bleigermanate im Konzentrationsbereich PbO/GeO_2 = 5/3 bis 3/2. - *Z. Kristallogr.,* 160, 93-107 (1982).

[258] Pan, Y. & Fleet, M.E.: Intrinsic and external controls on rare earth elements in calc-silicate minerals. *Can Mineral* 34: 147-159 (1996).

[259] Pan, Y., Fleet, M.A., Chen, N. & Weil, J.A.: Site preference of Gd in synthetic fluorapatite by single-crystal W-band EPR and X-ray refinement of structure: A comparative study. *Can Mineral* 40 (in press) (2002).

[260] Patel, P.N.: Magnesium calcium hydroxylapatite solid solutions. *J. Inorg. Nuclear Chem.* 42: 1129-1132 (1980).

[261] Pekov, I.V., Kulikova, I.M., Kabalov, Y.U.K., Yeletskaya, O.V., Chukanov, N.V., Men'shikov, Y.U.P. & Khomyakov, A.P.: Belovite-(La), $Sr_3Na(La,Ce)(PO_4)_3(F,OH)$, a new rare earth mineral in the apatite group. *Zap Vser Mineral Obsh.* 125: 101-109 (1996).

[262] Perez-Mendez, E. & Fayos, J.: A very modified apatite-type structure $K_3CA_2(SO_4)3$; F Perret R, Bouillet AM The sulfate apatites $Na_3Cd_2(SO_4)_3Cl$ and $Na_3Pb_2(SO_4)_3Cl$. *Bull Soc fr. Minéral. Cristallogr.* 98:254-255 (1975).

[263] Piriou, B., Fahmi, D., Dexpert-Ghys, J., Taitai, A. & Lacout, J.L.: Unusual fluorescent properties of Eu^{3+} in oxyapatites. *J Lumines.* 39:97-103 (1987).

[264] Pliego-Cuervo, Y.B. & Glasser, F.P.: Role of sulfates in cement clinkering reactions: Phase formation and melting in the system CaO-Ca_2SiO_4-$CaSO_4$-K_2SO_4.- *Cem.Concr.Res.,*7,477-482 (1977).

[265] Pliego-Cuervo, Y.B. & Glasser, F.P.: Phase relations and crystal chemistry of apatite and silicocarnotite solid solutions. *Cem.Con.Res.,*8,519-523 (1978).

[266] Pliego-Cuervo, Y.B. & Glasser, F.P.: Role of sulfates in cement clinkering: The Calcium Silicosulfate phase. *Cem.Con.Res.,*8, 455-460 (1978).

[267] Pliego-Cuervo, Y.B. & Glasser, F.P.: Role of sulfates in cement clinkering: Subsolidus phase relations in the system CaO-Al_2O_3-SiO_2-SO_3.- *Cem.Con.Res.,* 9,51-56 (1979).

[268] Pliego-Cuervo, Y.B. & Glasser, F.P.: Role of sulfates in cement clinkering: Phase formation in the system CaO-Al_2O_3-Fe_2O_3-SiO_2-$CaSO_4$-K_2SO_4 - *Cem.Con. Res.,* 9, 573-581 (1979).

[269] Pöllmann, H.: *Immobilization of Pollutants in Waste Disposals by Forming Storing up Mineral Reservoirs.*- ICAM, Pretoria,2,Paper 43 (1991).

[270] Pöllmann, H.: *Mineralization of Industrial waste*, Shaker Verlag. (2010).

[271] Pöllmann, H., Auer, S., Neubauer, J. & Stemmermann, P.: Bildung eines Innere-Barriere Systems in Abfalldeponien. *Immobile Bindung von Chlorid und Schwermetallen in Speichermineralen.- Buch der Umweltanalytik,* Band 3,82-83,GIT-Verlag (1991).

[272] Posner, A.S., Perloff, A. & Diorio, A.F.: Refinement of the hydroxyapatite structure, *Acta Cryst.* 11, 4, 308-309 (1958).

[273] Prener, J.S.: The growth and crystallographic properties of calcium fluor- and chlorapatite crystals. *J Electrochem. Soc* 114:77-83 (1967).

[274] Pushcharovskii, D.Y.U., Nadezhina, T.N. & Khomyakov, A.P.: Crystal structure of strontium apatite from Khibiny. *Soviet Phys. Crystallogr.* 32:524-526 (1987).

[275] Rakovan, J. & Hughes, J.M.: Strontium in the apatite structure: strontian fluorapatite and belovite-(Ce). *Can Mineral* 38:839-845 (2000).

[276] Rakovan, J. & Reeder, R.J.: Differential incorporation of trace elements and dissymmetrization in apatite: The role of surface structure during growth. *Am. Min.* 79:892-903 (1994).

[277] Rakovan, J. & Reeder, R.J.: Intra crystalline rare earth element distributions in apatite: Surface structural influences on incorporation during growth. *Geochim Cosmochim Acta* 60:4435-4445 (1996).

[278] Rakovan, J., Reeder, R.J., Elzinga, E.J., Cherniak, D.J., Tait, C.D. & Morris, D.E.: Structural characterization of U(VI) in apatite by X-ray absorption spectroscopy. *Environ Sci Technol* 36:3114-3117 (2002).

[279] Regnier, P., Lasaga, A.C., Berner, R.A., Han, O.H. & Zilm, K.W.: Mechanism of CO_3^{2-} substitution in carbonate fluorapatite: evidence from FTIR spectroscopy, ^{13}C NMR and quantum mechanical calculations. *Am. Min.* 79:809-818 (1994).

[280] Riebe, H.-J. & Keller, H.L.: Darstellung und Kristallstruktur der ersten Vertreter des neuen Verbindungstyps $Pb_4(SiO_4)X_4$ mit X = Cl, *Br. - Z. anorg. allg. Chem.,* 574: 182-190 (1989).

[281] Roeder, P.L., Macarthur, D., Ma X.P. & Palmer, G.R.: Cathodoluminescence and microprobe study of rare-earth elements in apatites. *Am. Min.* 72:801-81 1 (1987).

[282] Rouse, R.C. & Dunn, P.J.: A contribution to the crystal chemistry of ellestadite and the silicate sulfate apatites. *Am. Min.* 67:90-96 (1982).

[283] Roy, D.M., Drafall, L.E. & Roy, R.: Crystal chemistry, crystal growth, and phase equilibria of apatites. In *Phase Diagrams, Material Sciences and Technology* 6-V. Alper AM (ed) Academic Press, New York, p 186-239 (1978).

[284] Ruby, M.V., Davis, A. & Nicholson, A.: "In situ formation of lead phosphates in soils as a method to immobilize lead." *Environ. Sci. Technol.* 28: 646-654 (1994).

[285] Ruszala, F. & Kostiner, E.: Preparation and characterization of single crystals in the apatite system $Ca_{10}(PO_4)_6(Cl,OH)_2$. *J Crystal Growth* 30:93-95 (1975).

[286] Ryan, F.M., Hopkins, R.H. & Warren, R.W.: The optical properties of divalent manganese in strontium fluorophosphate: a comparison with calcium fluorophosphate. *J Lumines* 5:313-333 (1972).

[287] Schneider, W.: Caracolit, das $Na_3Pb_2(SO_4)_3Cl$ mit Apatitstruktur.- N. *J. f. Min., Monatshefte,* 84-289 (1967).

[288] Schneider, W.: Bestimmung einer Überstruktur am Caracolit.- *N. J. f. Min.*, Monatshefte, 58-64 (1968).

[289] Schriewer, M.S. & Jeitschko, W.: Preparation and crystal structure of the isotypic orthorhombic strontium perrhenate halides $Sr_5(ReO_5)_3X$ (X = chloride, bromide, iodide) and structure refinement of the related hexagonal apatite-like compound barium perrhenate chloride ($Ba_5(ReO_5)_3Cl$). *J Solid State Chem.* 107:1-17 (1993).

[290] Schroeder, L.W. & Mathew, M.: Cation ordering in $Ca_2La_8(SiO_4)_6O_2$. *J Solid State Chem.* 26:383-387 (1978).

[291] Schwarz, H.: Apatite des Typs $Pb_6K_4(X^VO_4)_2(X^V$ = P,As; X^{VI} = S, Se). *Z Anorg Allgem. Chem.* 356:29-35 (1967).

[292] Schwarz, H.: Apatite des Typs $M^{11}{}_{10}(X^{VI}O_4)_3(X^{IV}O_4)_3(M^{11}$ = Sr,Pb; X^{VI} = S, Cr; X^{IV} = Si ,Ge). *Z Anorg. Allgem. Chem.* 356:36-45 (1967).

[293] Schwarz, H.: Verbindungen mit Apatitstruktur. 1.Ungewöhnliche Silikatapatite. *Inorg. Nucl. Chem. Letters* 3, 231-236 (1967).

[294] Schwarz, H.: Verbindungen mit Apatitstruktur. 2.Apatite des Typs $Pb_6K_4(X^VO_4)_4(X^{VI}O_4)_2$ mit X^V=P,As und X^{VI}=S,Se.- *Zeit.f.anorg.u. allgem.Chem.,*356,29-35 (1967).

[295] Schwarz, H.: Verbindungen mit Apatitstruktur. 3.Apatite des Typs $M^{II}10(X^{IV}O_4)_3(X^{VI}O_4)_3F_2$ mit X^{IV}=Si,Ge und X^{VI}=S, *Cr.- Zeit.f. anorg.u. allgem.Chem.*,356, 36-45 (1967).

[296] Schwarz, H.: Strontiumapatite des Typs $Sr_{10}(PO_4)_4(X^{IV}O_4)_2$ (X^{IV} = Si ,Ge). *Z. Anorg. Allgem. Chem.* 357: 43-53 (1968).

[297] Serret, A., Cabarias, M.V. & Vallet-Regi, M.: Stabilization of calcium oxyapatites with lanthanum(III)-created anionic vacancies. *Chem. Mater* 12:3836-3841 (2000).

[298] Sery, A., Manceau, A. & Greaves, G.N.: Chemical state of Cd in apatite phosphate ores as determined by EXAFS spectroscopy. *Am. Min.* 81:864-873 (1996).

[299] Simpson, D.R.: Substitutions in apatite: I. Potassium-bearing apatite. *Am. Min.* 53:432-444 (1968).

[300] Slansky, M.: Géologie des phosphates sedimentaires.- *Mém,*114 (1980).

[301] Smart, R.M. & Glasser, F.P.: Compound formation and phase equilibria in the system PbO-SiO_2. - *J. Am. Cer. Soc.,* 57, 9, 378-382 (1974).

[302] Smith, G.F.H. & Prior, G.T.: On fermorite, a new arsenate and phosphate of lime and strontia, and tilasite, from the manganese ore deposit of India. *Mineral Mag.* 16:84-96 (1911).

[303] Soules, T.F., Davis, T.S. & Kreidler, E.R.: Molecular orbital model for antimony luminescent centers in fluorophosphates. *J. Chem. Phys.* 55: 1056-1064 (1971).

[304] Speer, J.A. & Ribbe, P.H.: Miscellaneous Orthosilicates. - In: Ribbe, P. H. (Editor): *Reviews in Mineralogy,* 5, 383-427 (1982).

[305] Stefánsson, A.: Dissolution of primary minerals of basalt in natural waters: I. Calculation of mineral solubilities from 0°C to 350°C. *Chemical Geology,* Vol. 172, Issues 3-4,225-250 (2001).

[306] Stefánsson, A., Gíslason, S.R. & Arnórsson, S.: Dissolution of primary minerals in natural waters: II. Mineral saturation state. *Chemical Geology,* Vol. 172, Iss. 3-4, 251-276 (2001).

[307] Steinbruegge, K.B., Hermingsen, T., Hopkins, R.H., Mazelsky, R., Melamed, N.T., Riedel, E.P. & Roland, D.W.: Laser properties of Nd^{+3} and Ho^{+3} doped crystals with the apatite structure. *Appl. Optics* 11:999-1012 (1972).

[308] Stemmermann, P.: Silikatapatite: Struktur, Chemismus und Anwendung als Speichermineral zur Konditionierung von Rauchgasreinigungsrückständen. *Ph.D.dissertation,* Universität Erlangen (1992).

[309] Stemmermann, P. & Pöllmann, H.: The system CaO-SiO_2-$CaCl_2$-phase equilibria and polymorphs below 1000°C - An interpretation on garbage combustion ashes.- *N.Jb.Miner.Mh.,*9,409-431 (1992).

[310] Stemmermann, P., Pöllmann, H. & Kuzel, H.J.: Apatitähnliche Sorosilikate mit 2-wertigen Kationen. - Eur. J. Min., 3,1,259 (1991).

[311] Sudarsanan, K.: Structure of hydroxylellestadite. *Acta Crystallogr.*B36: 1636-1639 (1980).

[312] Sudarsanan, K., Mackie, P.E. & Young, R.A.: Comparison of synthetic and mineral fluorapatite, $Ca_5(PO_4)_3F$, in crystallographic detail. *Mater Res Bull* 7:1331-1338 (1972).

[313] Sudarsanan, K. & Young, R.A.: Structure refinement and random error analysis for strontium "chlorapatite", $Sr_5(PO_4)_3Cl$. *Acta Crystallogr.* B30:1381-1386 (1974).

[314] Sudarsanan, K. & Young, R.A.: Structure of partially substituted chlorapatite $(Ca,Sr)_5(PO_4)_3Cl$. *Acta Crystallogr.* B36: 1525-1530 (1980).

[315] Sudarsanan, K., Young, R.A. & Wilson, A.J.C.: The structures of some cadmium "apatites" $Cd_5(MO_4)_3X$: I. Determination of the structures of $Cd_5(VO_4)_3I$, $Cd_5(PO_4)_3Br$, $Cd_5(AsO_4)_3Br$ and $Cd_5(VO_4)_3Br$. *Acta Crystallogr.* B33:3136-3142 (1977).

[316] Sudoh, G., Ohta, T. & Harada, H.: High Strength Cement in the CaO-Al_2O_3-SiO_2-SO_3 System and its Application.- *7th International Congress on the Chemistery of Cement*,Paris,III,V152-V157 (1980).

[317] Suetsugu, Y., Takahashi, Y., Okamura, F.P. & Tanaka, J.: Structure analysis of A-type carbonate apatite by a single-crystal X-ray diffraction method. *J Solid State Chem.* 155:292-297 (2000).
[318] Suitch, P.R., Lacout, J.L., Hewat, A. & Young, R.A.: The structural location and role of Mn^{2+} partially substituted for Ca^{2+} in fluorapatite. *Acta Crystallogr.* B41: 173-179 (1985).
[319] Suitch P.R., Taitai, A., Lacout, J.L. & Young, R.A.: Structural consequences of the coupled substitution of Eu, S, in calcium sulfoapatite. *J Solid State Chem.* 63:267-277 (1986).
[320] Taitai, A. & Lacout, I.L.: On the coupled introduction of Eu^{3+} and S^{2+} ions into strontium apatites. *J. Phys Chem. Solids* 50:851-855 (1989).
[321] Takahashi, M., Uematsu, K., Ye, Z.G. & Sato, M.: Single-crystal growth and structure determination of a new oxide apatite, $NaLa_9(GeO_4)_6O_2$. *J Solid State Chem.* 139:304-309 (1998).
[322] Teoreanu, I. & Andronescu, E.: Relations of compatibility (of 13 compounds) in BaO-CaO-SiO_2. - *Silic. Ind.,* 51(5-6), 87-90 (1986).
[323] Tomson, M.B. & Vignona, L.: Precipitation of Phosphate Minerals in Waste Water Treatment Systems, J.O. Nriagu, P.B. Moore, Editors, *Phosphate Minerals*, SYSTAT, Inc, pp. 386–399. (1984).
[324] Tõnsuaadu, K., Gruselle, M., Villain, F., Thouvenot, R., Peld, M., Mikli, V., Traksmaa, R., Gredin, P., Carrier, X. & Salles, L.: A new glance at ruthenium sorption mechanism on hydroxy, carbonate, and fluor apatites: Analytical and structural studies. *Journal of Colloid and Interface Science,* Vol. 304, Issue 2, Pages 283-291 (2006).
[325] Trombe, J.-C. & Montel, G.: Some features of the incorporation of oxygen in different oxidation states in the apatite lattice-II. On the synthesis and properties of calcium and strontium peroxiapatites. *J. Inorg. Nuclear Chem.* 40:23-26 (1978).
[326] Trombe, J.-C. & Montel, G.: On the existence of bivalent ions in the apatite channels. A new example phosphocalcium cyanamido-apatite. *J. Solid State Chem.* 40: 152-160 (1981).
[327] Trueman, N.A.: Substitutions for phosphate ions in apatite. Nature 210:937-938 (1966).
[328] Urusov, V.S. & Khudolozhkin, V.O.: An energy analysis of cation ordering in apatite. *Geochem. Intl* 11: 1048-1053 (1974).
[329] Valsami-Jones, E., Ragnarsdottir, K.V., Putnis, A., Bosbach, D., Kemp, A.J. & Cressey, G.: The dissolution of apatite in the presence of aqueous metal cations at pH 2–7. *Chemical Geology,* Vol. 151, Issues 14, Pages 215-233 (1998).
[330] Verbeeck, R.M.H., Lassuyt, C.J., Heijligers, H.J.M., Driessens, F.C.M. & Vrolijk, J.W.G.A.: Lattice parameters and cation distribution of solid solutions of calcium and lead hydroxyapatites. *Calcif Tissue Intl* 33:243-247 (1981).
[331] Vignoles, M. & Bonel, G.: Sur la localization des ions fluorure dans les carbonate-apatites de type B. *Compt Rend Acad Sci* 287:321-324 (1978).
[332] Wallaeys, R.: Contribution á l'étude des apatites phosphocalciques. *Ann Chim* 7:808-848 (1952).
[333] Warren, R.W.: EPR of Mn^{+2} in calcium fluorophosphate: I. The Ca(II) site. *Phys Rev* B2:4383-4388 (1970).

[334] Warren, R.W.: Defect centres in calcium fluorophosphate. *Phys Rev* B6:4679-4689 (1972).
[335] Warren, R.W. & Mazelsky, R.: EPR of Mn^{+2} in calcium fluorophosphate. II. Modified Ca(II) site. *Phys Rev* B10:19-25 (1974).
[336] Watson, E.B.: Two-liquid partition coefficients: Experimental data and geochemical implications. *Contribut Mineral Petrol* 56: 119-134 (1976).
[337] Watson, E.B., Harrison, T.M. & Ryerson, F.J.: Diffusion of Sm, Sr, and Pb in fluorapatite. *Geochim Cosmochim Acta* 49:1813-1823 (1985).
[338] Watson, E.B. & Oreen, T.H.: Apatite/liquid partition coefficients for the rare earth elements and strontium. *Earth Planet Sci Lett* 56:405-421 (1981).
[339] Welin, E.: X-ray powder data for minerals from Längban and the related mineral deposits of Central Sweden. *Arkiv Mineral Geol.* 4:499-541 (1968).
[340] Wilson, A.J.C., Sudarsanan, K. & Young, R.A.: The structures of some cadmium "apatites" $Cd_5(MO_4)_3X$: II. The distribution of the halogen atoms in $Cd_5(VO_4)_3X$, $Cd_5(PO_4)_3Br$, $Cd_5(AsO_4)_3Br$ and $Cd_5(VO_4)_3Br$. *Acta Crystallogr.* B33:3142-3154 (1977).
[341] Wright, J.V., Peurrung, L.M., Moody, T.E., Conca, J.L., Chen, X., Didzerekis, P.P. & Wyse, E: In Situ Immobilization of Heavy metals in Apatite Mineral Formulations, *Technical Report to the Strategic Environmental Research and Development Program,* Department of Defense, Pacific Northwest Laboratory, Richland, WA, 154 p. (1995).
[342] White, T.J.: *Design, testing an economics of XTALTITE toxic metal ceramics* - 100th Northwest Mining Ass., Spokane, Washington (1994).
[343] White, T.J., Poniatowski, G., Kawashima, N., Lincoln, F., Rodier, B., Easton, G.F. & Kyle, J.H.: Xtaltite ceramics for toxic metal immobilization - *Proc. of Sudbury 95 Conference on mining and the environment* (1995).
[344] White, T.J. & Toor, I.A.: Synthetic mineral immobilization technology for the stabilization of incineration ashes - *Proc. Int. Incin. Conference,* Seattle, WA (1995).
[345] White, T.J. & Toor, I.A.: *Stabilizing toxic metal concentrates by using SMITE* - JOM, 54- 58 (1996).
[346] Winkler, A., Hoebbel, D., Grimmer, A.-R., Wieker, W., Ujszaszy, K. & Mägi, M.: $Ba_7[Si_7O_{21}]$. - $10BaCl_2$- a compound with a new type of cyclic silicate anions. - *Revue de Chimie minerale,* 20, 801-806 (1983).
[347] Wondratschek, H.: Untersuchungen zur Kristallchemie der Blei-Apatite (Pyromorphite). - *N. Jb. Miner. Abh.* 99, 2, 113-160 (1963).
[348] Wondratschek, H.: Other compounds with the apatite structure. – *Manuskript* (1984).
[349] Wronkiewicz, D.J., Wolf, S.F. & Disanto, T.S.: Apatite- and monazite-hearing glass-crystal compositions for the immobilization of low-level nuc1ear and hazardous wastes. *Mater Res Soc* 412:345-352 (1996).
[350] Yamaguchi, O., Oka, H. & Shimizu, K.: Formation and Transformation of Pb_5SiO_4. - *Polyhedron* 4, 591 (1985).
[351] Yamaguchi, O., Sugiura, K. & Muto, M.: Metastable form of $Pb_5Ge_3O_{11}$. - *Mater. Lett.* 2, 4A, 286-8 (1984).
[352] Yamaguchi, O., Sugiura, K. & Tunekawa, T.: Chemical Formula of the phase X in the system $PbO-SiO_2$. - *Mater. Lett.,* 3(11), 429-31 (1985).

[353] Young, E.J., Myers, A.T., Munson, E.L. & Conklin, N.M.: Mineralogy and geochemistry of fluorapatite from Cerro de Mercado, Durango, Mexico. *U S Geol Surv Paper* 650-D: 84-93 (1969).

[354] Zhang, M., Maddrell, E.R., Abraitis, P.K. & Salje, E.K.H.: Impact of leach on lead vanado-iodoapatite [Pb5(VO4)3/ I]: An infrared and Raman spectroscopic study. *Materials Science and Engineering:* B, Volume 137, Issues 1-3, 149-155 (2007).

In: Calcium Phosphate
Editor: Robert B. Heimann

ISBN 978-1-62257-299-1

Chapter 18

CALCIUM PHOSPHATE MATERIALS FOR RADIOACTIVE WASTE IMMOBILISATION

E. R. Vance[*] and D. J. Gregg
Institute for Materials Engineering, ANSTO, Kirrawee DC NSW, Australia

ABSTRACT

Certain calcium phosphate based materials have long been proposed as candidates for immobilising high- and intermediate-level radioactive waste. The main structure types of interest in this regard are the apatite-bearing ceramics and the sodium zirconium phosphate (NZP) structure type ceramics. Apatite, $Ca_{10}(PO_4)_6X_2$ (X = OH, Cl, F), is a complex structure with moderate radiation tolerance and several examples are found in nature with radioactive elements trapped within the crystal structure. However because of its greater flexibility in terms of solid solution substitutions, the sodium zirconium phosphate (NZP) structure has more recently been advocated for this purpose. The special attraction of the NZP structure is that it can accommodate almost all radionuclides present in high-level waste streams within its crystal structure. The substitution of alkaline earth ions for alkalis in NZP increases the leach resistance considerably, thus $CaZr_4(PO_4)_6$ and $CaTi_4(PO_4)_6$ are preferred options. The general background to the development of these materials as ceramic nuclear waste forms is given, and the crystal chemistry, aqueous durability and radiation tolerances are indicated. The potential use of crandallite ores as waste forms and as scavenging agents of specific ions in the nuclear waste stream is briefly reviewed. Inert matrix nuclear fuels for the burning of surplus actinides are also of recent interest for advanced nuclear reactors and NZP-structured materials for this application are briefly discussed.

1. INTRODUCTION

The focus of this chapter is to explain the role of calcium phosphate candidate materials in the immobilisation of high- and intermediate-level radioactive waste arising from the

[*] Lou.Vance@ansto.gov.au

nuclear fuel cycle. Figure 1 shows the distribution of fission products produced from ^{235}U. Many of these fission products are highly radioactive. While storage of nuclear fuel can allow the short-lived activity to decay away, significant proportions of fission products and waste actinides which are also produced in reactors have very long half-lives (see Table 1).

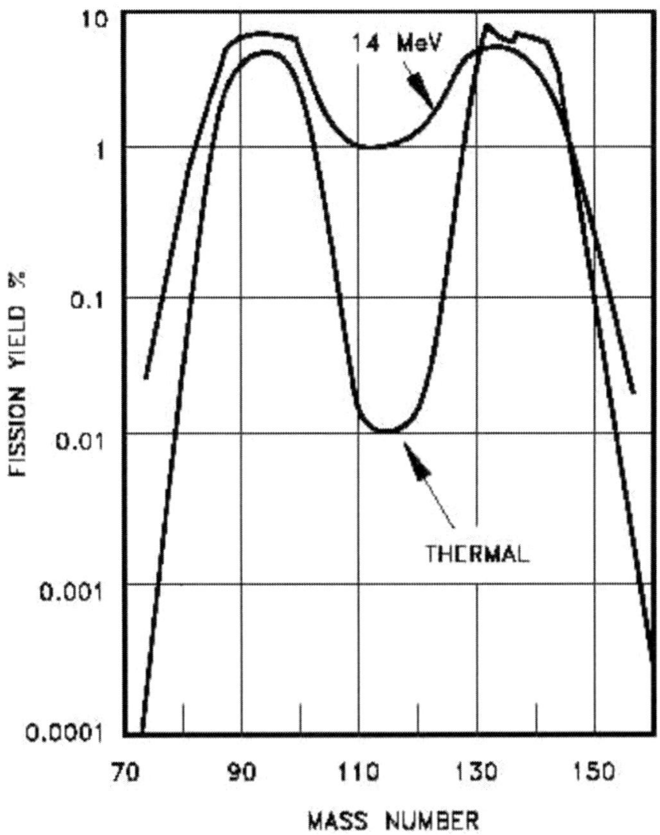

Figure 1. Histogram showing relative abundances of fission products from thermal and fast neutron fission of uranium.

There are different classes of nuclear waste. High-level nuclear waste (HLW) has activities of roughly thousands of curies/litre (1 Curie = 37 GBq, where the Becquerel, Bq, corresponds to 1 decay/s), arising from nuclear power plants in the form of used nuclear fuel or the fission product-rich material derived from reprocessing of used fuel. Worldwide inventories of HLW from nuclear power and weapons production, other than used fuel, constitute hundreds of thousands and perhaps millions of tonnes [1] and are still stored on site, although some progress is being made on repository construction and qualification. Figure 2 shows the time dependence of the radioactivity of HLW from nuclear fuel reprocessing. Dilution of HLW does not change its classification. The HLW will remain active for millions of years and the immobilisation target period ranges between 10^4 and 10^6 yr.

Disposal of HLW other than spent fuel itself basically involves adding specific material to the waste, and processing it to transform it to a dense refractory solid - known as the waste form. In a large majority of cases, the solid would be an oxide, but metallic components may also be present, depending on the waste and the processing conditions.

The solid waste form would be a glass, ceramic, glass-ceramic or cementitious material, depending on the nature of the waste to be dealt with. It would then be transferred in a metal disposal canister to a geological repository, in which the radioactive and toxic components would be immobilised for long periods, up to millions of years if necessary, depending on the half-lives of the dangerous species. The long-term performance criteria for solids designed for HLW incorporation include (a) minimal reactivity in aqueous media, and good understanding of the aqueous dissolution processes, including the effects of self-irradiation from the incorporated HLW; (b) fabrication by relatively simple, extensively demonstrated technology; and (c) high waste concentrations, consistent with radiogenic heat loading restrictions on the standard canisters for HLW containment. How such materials meet regulatory and economic criteria, and simplicity or otherwise of processing methods, will be outlined in sections 3-5.

In this Chapter, the focus of the immobilisation is on the waste form for HLW, but of course the repository would also be expected to play a significant role in preventing the transport of radionuclides and toxic ions to the biosphere. It seems unlikely that used nuclear fuel in the form of irradiated UO_2, a ceramic, will be chemically processed, but rather simply allowed to cool as the short-lived fission products decay, and then containerised for eventual repository disposal.

Intermediate-level waste (ILW) is roughly 1000 times less active than high-level waste and can derive from radioisotope production or general waste arising from nuclear power plant operations. Low-level waste from nuclear operations or spent radioactive sources has activities below about 1 millicurie/litre. Extant volumes of LLW and ILW far exceed those of HLW, although HLW contains the major radioactivity. Disposal of ILW and LLW is normally seen as a near-surface operation.

Table 1. Approximate compositions (wt%) and halflives$^{\&}$ of main fission product and actinide oxides in Purex fuel reprocessing HLW that has been stored for > 10 years

FP Oxide (wt%)$^{\$}$	Halflife (yr.)
Cs_2O (6)	30
SrO (3)	30
BaO (4)	-
RE_2O_3 (15)	100$^{\&}$
ZrO_2 (15)	-
MoO_3 (15)	-
TcO_2 (6)	210000
An*O_2 (6)	>10000
RuO_2 (10)	-
PdO (6)	-
Rh_2O_3 (2)	-

Water excluded; *An = actinide. RE= rare earth. $^{\$}$Contains additional stainless steel corrosion products. Group halflives$^{\&}$ are very approximate as they range from short to long times for different components. Absence of halflife value = stable elements.

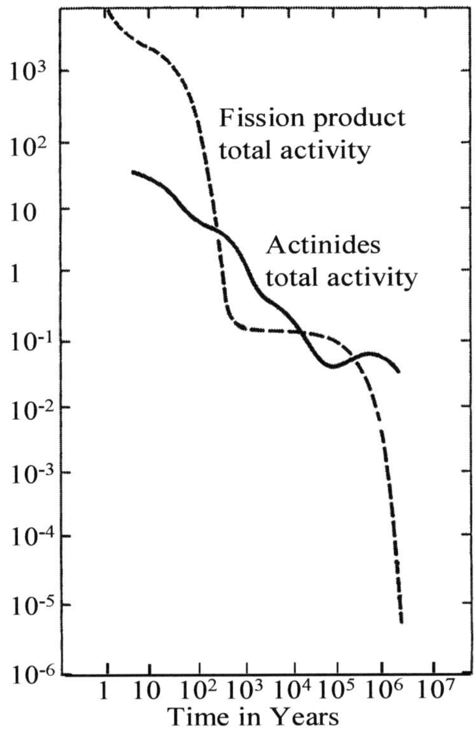

Figure 2. Relative time dependence of HLW liquid activity [1].

2. Historical Development and the Role of Ceramics for HLW

Not long after the first demonstration of nuclear reactors in 1942, attention was paid to the disposition of radioactive waste from nuclear power plants. The first recorded works aimed to capture the radionuclides in clay or nepheline syenite minerals [2, 3]. However in the 1960s, the US AEC had demonstrated the incorporation of HLW in borosilicate glass at full scale with actual radioactive waste. The glass was seen as a solvent for nearly all fission products as either network formers or network modifiers and could be processed by vitrification at temperatures of ~1100°C. But in the 1970s, it was shown that incorporation of radionuclides in synthetic minerals which could be shown to have survived for millions of years in hot, wet conditions was a potentially superior way to deal with HLW/ILW [4, 5].

To amplify this point, some minerals contain radioactive elements and from measurements of the abundance of U and Th and their radioactive daughters in the minerals it can be deduced that some of these minerals have geological ages of millions, even billions of years in some cases. If such minerals are found in warm, wet environments, as is not uncommon, noting that these are approximately the conditions eventually expected in a geological nuclear waste repository, this can be interpreted as showing long survival times in repositories. Thus ever since the late 1970s there has been competition between ceramics and borosilicate glasses for immobilisation of HLW (e.g. the "Atlanta shoot-out" in which borosilicate glass was seen as the then best available demonstrated technology for Savannah

River (SRL) HLW in 1981 [6,7], and the US Department of Energy decision on impure US/Russian surplus Pu in 1998 in which a ceramic synroc derivative option was preferred [8,9]). However there are now so many types of HLW/ILW having a large variety of chemical compositions remaining to be immobilised around the world, it is becoming clear that glass, glass-ceramics, ceramics and on occasion cementitous materials all have their place for different HLWs and ILWs.

The final destination of HLW is generally agreed to be a deep (0.5-3 km) geological repository so the principal object of fabricating a waste form for HLW is, using the simplest and cheapest possible processing methods, to incorporate the waste in a solid which is minimally porous, has high leaching resistance when exposed to water, adequate strength, and a high fraction (> 20 wt%) of waste per unit volume. A limiting feature for HLW is the radiogenic heat output [10] as it is desirable to keep the repository temperature to 100°C or less to minimise radionuclide transport from the waste form by reaction with groundwater.

Many of the ceramic phases of interest in the context of nuclear waste immobilisation are refractory and are chemically and structurally similar to the long-lived minerals. A partial list of ceramic host phases for various radionuclides is given in Table 2. These synthetic mineral-modelled ceramics are able to be processed in the convenient temperature range of 1000-1200°C to produce them with low porosity to prevent bulk intrusion of water.

Table 2. Partial 'library' of ceramic host phases

Phase	Radionuclide
Pollucite, $CsAlSi_2O_6$	Cs
Hollandite, $(Cs,Sr,Ba,Rb)_{1.14}(Al,Ti^{3+},Fe)_{2.28}TiO_{16}$	Cs, Rb, Sr, Ba
Apatite, $Ca_{10}([P,Si]O_4)_6(OH,F,Cl)_2$	RE, An, Sr, Ba
NZP $(Na,Ca_{0.5})(Zr,Ti)_2(PO_4)_3$	Many
Monazite, $REPO_4$	RE, An
Zircon, $ZrSiO_4$	RE, An
Zirconolite, $CaZrTi_2O_7$	RE, An
Perovskite, $CaTiO_3$	Sr, RE, Tc, An
$(RE,An)O_2$	RE, An
Pyrochlore, $RE_2Ti_2O_7$	RE, Zr, An
Titanite, $CaTiSiO_5$	RE, An, Sr
Rutile, TiO_2	Tc
Sodalite, $Na_4Al_3Si_3O_{12}I$	I

RE = Rare Earth; An = Actinide.

More generally ceramic assemblages need to be multi-phase so that the waste/additives ratio can be varied somewhat without changing the phase assemblage. This brings phase compatibility into question which needs experimental verification for a given theoretical assemblage. The tolerance to variations in waste chemistry and additive/waste ratios is an important factor since individual ceramic phases generally have rigid stoichiometries and such variations should not give rise to the formation of water-soluble phases.

Important criteria in waste form suitability include their resistance to dissolution in ground water and self-irradiation behaviour as well as ease of processing so we will review these factors before we discuss the use of calcium phosphate-based ceramics for the immobilisation of ILW/HLW.

3. Aqueous Dissolution

Reactivity with water of particular solids depends on the state of aggregation of the solid and the dissolution rate of a solid body will be a lot less than that of a fine powder. The dissolution rate itself can be quantified by measuring the concentration of dissolved species in the water in relation to the original inventory in the solid before the onset of leaching. Dissolution rates can then be expressed as elemental loss by mass per unit surface area (expressed as geometrical or Branauer-Emmett-Teller {BET} [11] specific surface area values) per unit time. These rates however can depend critically on surface area/liquid volume, pH, temperature, presence of salts in the water etc. The measurement of dissolution rates can be further complicated by the presence of colloids and adherence of primarily dissolved species to vessel walls. However separation of the solids from the liquid, followed by acidification of the liquid, can dissolve species attached to the leach container walls as well as colloids. Colloids themselves can be detected by light scattering measurements. Ultimately, the dissolution rate of a solid in water is controlled by thermodynamics and various software programs are available to describe the dissolution process, although they are usually limited by lack of basic data for some of the ions in the solid.

Many laboratories now use apparatus in which the liquid of interest flows over the solid at a rate such that the rate is not too fast to prevent a measurable concentration of dissolved species to accumulate and not too slow as to allow the buildup of high concentrations of dissolved materials and consequent complications by solution saturation. As indicated previously, in most repositories, the limiting temperature would be designed as ~ $100^{o}C$ so measurements on candidate waste forms are experimentally relatively simple, except of course for the use of realistic exposure times.

As mentioned in the Introduction, the required time of immobilisation for real HLW is variously targeted as 10^4-10^6 years. It is very difficult to perform accelerated tests as the use of higher temperatures or more aggressive solutions is easily compromised by the thermodynamics of the solid-liquid interaction. Key regulatory leaching tests are the MCC-1 (1984) and PCT-B (1997) protocols [12,13] which employ polished flat samples and sized powders respectively that are exposed to hot water. From these tests, short-term (a few days) leach rates of ~ 1 $g/m^2/day$, normalised to take account of the fractional elemental extraction, rather than the absolute quantities, are considered as satisfactory and these rates correspond to ~ 0.1 mm/yr., further noting that the leach rates of solids almost always decrease with increasing leaching time even at high degrees of dilution. This decrease is generally attributed to initial preferential attack at "active surface sites" on the cut or polished prepared surface of a candidate solid and/or the development of adherent, relatively impervious surface alteration layers on the solid which inhibit leaching at extended leaching times.

4. Radiation Effects

In a ceramic waste form the principal sources of radiation from the incorporated HLW are β-decay of the fission products (e.g., ^{137}Cs and ^{90}Sr) and α-decay of the actinides (e.g., U, Np, Pu, Am, and Cm) [14]. In general, the α-decay events dominate after approximately 500 years of storage. Self-irradiation of solids containing radionuclides has the potential to cause lattice defects and a general compromise of the material's crystallinity. These features can lead to increased aqueous dissolution rates. Whereas the lattice defects and crystallisation breakdown are generally easy to measure by X-ray or electron diffraction techniques and volume expansion, effects on the aqueous dissolution rate are more difficult to quantify. The effects of alpha radiation from substitutional actinides in waste form materials are particularly marked, insofar as ~2000 atoms are displaced by an alpha-recoil ion and ~100 atoms by the alpha particle. The ranges of the recoil atom and the alpha-particle in dense ceramics are ~20 nm and ~20 μm respectively. As an example, complete amorphisation by U/Th substitutional impurities in the mineral zircon gives rise to a volume expansion of ~16% [15]. Damage effects by beta/gamma radiation are normally much less serious than those for alpha particles. In multi-phase ceramics subjected to radiation damage from alpha-emitters located in certain phases, but not others, the expansions of the alpha-bearing phases attendant upon radiation damage have a tendency to induce microcracking, but this can be minimised by maintaining a very small grain size by appropriate processing methods.

Reviews of the effects of radiation of different types and experimental ways to simulate them have been given several times-see e.g. [14,16,17]. It is worthy of note that some titanate ceramics can be amorphised by alpha radiation without much effect on their dissolution behaviour [18].

5. Ceramic Processing

The principal options for the production of refractory ceramic or glass-ceramic materials are sintering, uniaxial hot-pressing, hot isostatic pressing and melting. As mentioned above, it is clearly advantageous if these materials are fully dense or at least devoid of open porosity to prevent ingress of water into the interior of the material. Because of radioactivity hazards there is a high premium on making a waste form by a very simple process.

All HLWs slated for ceramic processing would be calcined in air at ~700°C to remove organics, nitrates, water, carbonates, hydroxides etc. If relatively small amounts of material are to be dealt with (say a few tens of tons only), it might be advantageous to mix the waste with the additives before calcination.

Though sintering is the baseline method in industry for making (inactive) ceramics, making dense ceramics via sintering is not easy. Metal oxide phases, preferably with multiple cation sites to allow substitution of a variety of fission products, actinides and process chemicals, form the main constituents of ceramic waste forms and it is necessary to achieve good mixing at an early stage of processing. Best mixing would utilise water-soluble liquid precursors such as metal nitrates to achieve atomic-scale mixing for high-quality homogeneous ceramics but this is a relatively complex process. So it is not surprising that the synroc derivative ceramics (see section 2) for surplus US/Russian impure Pu utilised dry feeds that were attrition milled to achieve mixing and reactivity with the dry precursors. These dry operations are most useful if radioactive volatile losses upon sintering are very small, as in the case just mentioned, but if volatile losses are potentially significant, hot isostatic pressing of waste form ceramics has advantages (see below).

5.1. Hot Uniaxial Pressing

Hot Uniaxial Pressing (HUP) of a powder contained in a die is a batch process. An inductively heated graphite die can be used and this imposes reducing conditions on the sample, even if the sample is contained in a collapsible metal container or can within the bore of the graphite die. Only ~20 MPa of pressure can be exerted with graphite, but very high temperatures (> 2000°C) are possible. However for waste forms, it is usual to keep the temperature to a maximum of around 1400°C to minimise volatile losses. The use of alumina dies allows the use of more pressure and oxidising conditions, but it is more difficult to extricate the sample, especially when it is radioactive. Of course in principle, the hot-pressed sample may be left in the graphite or alumina die, which then would constitute a fairly expensive transport container insert. However hot isostatic pressing(HIP) has basically superseded HUP as a processing option for oxide-based ceramics.

5.2. Hot Isostatic Pressing

In hot isostatic pressing (HIP) of ceramics or glass-ceramics the reactive calcined waste form (waste + additives) material is first packed by vibratory means inside a relatively thin-walled metal can. This is then evacuated and heated to 300-600°C for several hours to remove adsorbed gases, sealed, and then consolidated to full density by compressing it with several tens or even hundreds of MPa of argon gas during the heating cycle. The use of a suitable metal container, which may be stainless or mild steel, nickel or other metals, can help to achieve the desired redox conditions, minimise any potentially deleterious reaction between the waste form and the container, and of course prevents offgas escape. So the entire process produces offgas only in the calcination stage where temperatures are much lower than those in the final consolidation (roughly the same as those used for vitrification, i.e. 1000-1400°C). Figure 3 shows stainless steel HIP cans before consolidation and as a diametrically cut section of the consolidated can.

The HIP process was invented by the Battelle company in the US in the 1950s and has been used since the 1960s in preparing nuclear fuel for submarines. Moreover the method is widely used in industry for preparing inactive ceramics.

A large advantage is also the relatively small footprint of HIP equipment, compared to that for vitrification, mainly because of the absence of off-gas in the hot-consolidation step. Moreover the main part of the HIP equipment can be located outside the hot cell so that the HIP does not experience significant radioactive contamination and therefore require disposal at the end of the waste treatment campaign. Also, the HIP process can be used for encapsulation in metal for some wastes.

For radioactive ceramic waste forms a prime advantage of hot-pressing in general is to achieve theoretical density with minimum temperature and therefore minimum grain size, thereby adding to the overall strength and reducing the potential of microcracking via radiation damage (see section 4) when the waste form contains a substantial amount of alpha-emitting waste actinides.

Figure 3. Stainless Steel HIP can before consolidation and diametrically cut section of a HIPed sample. Figure provided by ANSTOsynroc (www.synrocansto.com) with permission.

5.3. Melting

Joule melters for large-scale melting employ large refractory ceramic baths (several m^2 in area and ~ 1 m deep) containing ceramic electrodes to directly heat mixtures of glass frit and nuclear waste to a molten state. The melt is then poured into steel transport canisters in which it cools slowly to yield a borosilicate glass. Such melters are only viable in the longer term if temperatures do not exceed ~ 1150°C. However for waste form ceramics or glass-ceramics we have argued elsewhere [19] that HIPing is preferred in many cases, notably for wastes which are problematic for vitrification.

6. Calcium Phosphate Based Ceramics for Radioactive Waste Immobilisation

6.1. Apatites

Apatites have been heavily studied around the world as candidate ceramics for immobilisation of nuclear waste since the original work of McCarthy [20] on supercalcines, which are sintered mixtures of crystalline phases targeted to incorporate fission product-rich waste arising from used nuclear fuel reprocessing.

The structure of apatite, $Ca_{10}(PO_4)_6(X)_2$ where X = OH, O, Cl, F, CO_3, SO_4 etc., is hexagonal with phosphate tetrahedra surrounding two Ca sites. The four 4f Ca sites are 9-fold coordinated by oxygen ions. The six 6h Ca sites are bonded to six oxygens plus an anion, and the anions are in *c* axis directed channels. Na, K, Sr, Ba and rare earths are readily incorporated in apatite phases [21,22]. Trivalent actinides are readily incorporated and while tetravalent actinides have restricted entry into phosphate apatites [23], charge compensation with silicon allows the formation of apatite-structured britholites, $Ca_4An_6(SiO_4)_6(X)_2$, where An = tetravalent actinide. Hexavalent ions can substitute as XO_4^{2-} in the anion sites. However a key uncertainty is the disposition of Cs in apatite-rich wasteforms, noting that the ionic size of Cs is much larger than that of Ca. This is important because Cs^+ is an abundant fission product and being an alkali metal ion is very soluble in water. While it has been asserted [24] that Cs can be incorporated in chlorapatite on the basis of XRD and SEM respectively, Campayo et al. [25] were unable to verify this, and recent work also suggests that Cs cannot be significantly incorporated [26], although very recent work does claim limited incorporation of Cs in hydroxyapatite [27].

In the supercalcine ceramics, any Cs was taken up by pollucite, $CsAlSi_2O_6$, and the leach resistance of these sintered ceramics, containing substantial amounts of apatite phases, was adequate [20]. Since then many investigators have studied the crystal chemistry of apatite and since actinides can readily be incorporated, the incorporation and effects of alpha-decay radiation damage in synthetic and natural apatites have been widely investigated [21,28-30].

6.2. CaZr/Ti phosphates, $Ca_{0.5}(Zr/Ti)_2(PO_4)_3$

6.2.1. Crystal Structure

Calcium zirconium phosphate ($Ca_{0.5}Zr_2(PO_4)_3$, CZP) and calcium titanium phosphate ($Ca_{0.5}Ti_2(PO_4)_3$, CTP) belong to the orthophosphate group of $NaZr_2(PO_4)_3$ (sodium zirconium phosphate or NZP) family which have long been proposed as candidates for the immobilisation of radioactive waste [31,32]. The properties of this family of compounds include low thermal expansion, high chemical, radiation and temperature stability, and superionic conductivity [33]. The NZP type structure was first reported as a potential host for nuclear waste by Roy et al. [31]. It has the structural formula $[M1][M2][A^{VI}{}_2][B^{IV}{}_3]O_{12}$, where M1 and M2 are interstitial sites (cavities) either partially or fully occupied by cations, the A-site is normally occupied by Zr or Ti, and the B-site by P, although a variety of coupled substitutions can occur in these positions. NZP crystallises in the rhombohedral space group R-3c with a structure consisting of vertex linked ZrO_6 octahedra and PO_4 tetrahedra. The basic unit of the network consists of two octahedra and three tetrahedra ($Zr_2P_3O_{12}$) which form chains aligned along the *c* axis. These chains are joined perpendicular to the *c* axis by PO_4 tetrahedra to produce a three-dimensional network. As a result there are two kinds of cavities, M1 and M2, and the Na cation occupies the M1 cavities in NZP, which also align along the *c* axis (see Figure 4). The symmetry is lowered to R-3 when the monovalent Na cation is substituted by a divalent cation such as Ca as an ordering of cations and vacancies in the M1 cavities occurs [34]. The M2 cavities are located between the chains and are only populated when additional Na or K is required for charge compensation. A unique feature of this structure is its flexibility as it can incorporate an enormous array of cations within framework or inter-framework positions, with oxidation states from 1+ to 5+, Shannon ionic radii [35] from 0.53 to 1.65 Å and Pauling electronegativity from 0.9 to 1.7 [36]. In fact it is a near universal (solid state) solvent in the context of radioactive waste immobilisation with the potential to accommodate not only the fission products within its crystal structure but also the actinides.

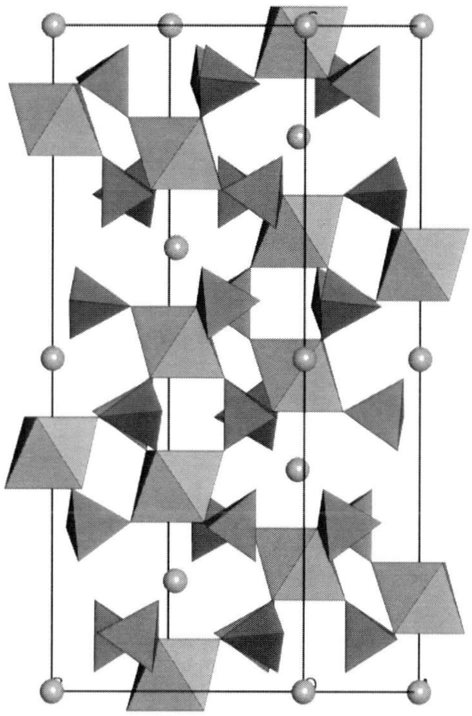

Figure 4. The structure of $NaZr_2(PO_4)_3$ aligned perpendicular to the c-axis showing the ZrO_6 octahedra and the PO_4 tetrahedra. The Na ions (spheres) occupy the M1 cavity.

6.2.2. Synthesis of NZP Type Ceramics

In general, three synthetic methods have been applied in the early studies for the preparation of NZP structure type compounds. These methods are: (a) Sol–gel [37-42]; (b) Solid state [43-47] and (c) Hydrothermal [48,49]. Typically, the sol-gel and solid state methods require temperatures in the range of 800-1400°C, long sintering times, and cold isostatic pressing of the calcined powder is often employed to achieve a maximum particle contact. Moreover, conventional solid state sintering procedures usually result in NZP type compounds with < 85% of theoretical density even after firing at 1200–1400°C for 24–72 h. Hydrothermal treatment is carried out in teflon or glass-lined hydrothermal vessels at 200-300°C for several days. More recently, the combustion reaction method has been developed [50] as well as the microwave-assisted procedure [51,52] for the simple, fast and low temperature synthesis of these ceramics. The latter approach leads to samples with > 97% densification. Hot isostatic pressing has also been successfully employed to obtain NZP structure type ceramic samples with > 95% densification [53].

Table 3. Waste element Ionic Substitutions in NZP structure type ceramics: $[M1][M2][A^{VI}_2][B^{IV}_3]O_{12}$.

Site	Ionic substitutions
MI	Cs, Sr, Ba, Pd*, RE, An
MII	Na, K
A^{VI}	Zr, Ti, Mo, Ru, Rh*, Tc*, RE, An
B^{IV}	P, Mo

* Speculated substitution.

6.2.3.Systematic Substitutions

In order to describe the specific ionic substitutions that can occur in CZP and CTP we have summarised relevant studies which investigate the versatility of the NZP type structure matrix. Table 3 summarises the substitutions within the NZP structure framework for the main waste elements following Purex fuel reprocessing of HLW which has been stored for > 10 years.

Cs and Sr incorporation into the NZP structure has been proven [54] and the structures elucidated. It has be shown that up to ~9.0 wt% of Cs can be loaded into NZP formulations without significant changes to the three-dimensional framework structure and at higher loadings, Cs forms a separate $CsZr_2(PO_4)_3$ phase [55]. The radionuclide ^{135}Cs exhibits a half-life of ~three million years [56] and is believed to be one of the first radionuclides likely to reach the biosphere from a deep repository due to its high mobility, therefore efficient immobilisation is essential [57]. Of course long halflives imply low specific activities. CZP and CTP can be fully substituted with Sr to form $Sr_{0.5}Zr_2(PO_4)_3$ and $Sr_{0.5}Ti_2(PO_4)_3$ with R-3 symmetry [58-60]. Similarly, $Ba_{0.5}Zr_2(PO_4)_3$ and $Ba_{0.5}Ti_2(PO_4)_3$ have been formed with R-3 and R-3c symmetry, respectively [34,61]. Apparently the titanium analogue has half occupied but disordered M1 cavities.

The possibility for two different arrangements of lanthanide cations in the NZP structure type has been shown experimentally. The rare earth elements can be incorporated into the interframework positions of the M1 cavity to give a structure of general formula $M_{0.33}Zr_2(PO_4)_3$, where M = Y, La, Ce, Pr, Nd, Sm, Eu, Gd, Dy, Ho, Er, Tm, Yb, Lu [62-64]. These systems still belong to the NZP structure type but there is a further reduction in symmetry from R-3c to either P-3 or P-3c1, depending on the RE [65]. Interestingly, $La_{0.33}Ti_2(PO_4)_3$ belongs to the R-3 space group [66]. Compounds where M = Sc [64] also crystallize in the NZP structural type. The substitution of lanthanide cations for Zr into the framework to provide the limited solid solutions of the form of $Na_{1+x}Zr_{2-x}Ln_x(PO_4)_3$, where Ln = In, Yb, Er, Y, Dy, Tb, and Gd [67], has been established. The maximum achievable concentration of lanthanide within the solid solutions was found to depend on the ionic radius of the lanthanide and to decrease with increasing radius.

A limited number of NZP structure type compounds of the form $(M1)An_2(PO_4)_3$ (An = tetravalent actinide element) have been reported, for example M1-An = K-U, Rb-U, Rb-Np, K-Pu, Rb-Pu, Na-Np, Na-Pu) [68-72]. In addition, the existence of several compounds with interframework tetravalent actinides, $An_{0.25}Zr_2(PO_4)_3$ (An = Th, U, Pu) and trivalent actinides $An_{0.33}Zr_2(PO_4)_3$ (An = Am) have been described [73,74]. The $Am_{0.33}Zr_2(PO_4)_3$ and $Pu_{0.25}Zr_2(PO_4)_3$ systems show a reduction in lattice symmetry from rhombohedral (space group R-3c) to simple trigonal and the thorium analogue $Th_{0.25}Zr_2(PO_4)_3$ has been structurally characterised recently and refined with the hexagonal space group P-3c [74].

Depending on its oxidation state, molybdenum can occupy either the A^{VI} or B^{IV} positions of the NZP structure type framework [75-79]. In particular, Mo^{6+} can be substituted at the tetrahedral position, to form air and chemically stable compounds. It should be noted however that the isotopes of molybdenum are essentially stable. Results from X-ray diffraction experiments [80] suggest Ru^{4+} is incorporated, as expected, onto the A^{VI} site and Tc^{4+}, Rh^{4+} and 3d metal ions would very probably be incorporated into this site also, assuming suitable charge compensation. An ionic substitution model has been proposed [81] which assigns Pd^{2+} to the M1 cavity and Tc^{7+} to the tetrahedral B^{IV} position.

6.2.4. Immobilisation of Simulated Nuclear Waste and Durability

Several studies have been carried out to demonstrate the feasibility of an NZP type structure waste form by using a simulated waste mixture. It has been shown that simulated HLW from Purex reprocessing can be incorporated into an NZP structure type ceramic in the range of several wt% [82]. In this example monazite coexists as a minor secondary phase and Ru and Pd segregate from the main NZP matrix. In another study, waste loadings of up to 20 wt% have been incorporated into the structure using sol-gel and sintering techniques while maintaining a single phase material [83,84] with monazite formed as a second phase for loadings > 20%. It has been reported that substitution of Ti for Zr to form the NTP type structure allows up to 60% waste loadings under suitable conditions [85]. In this study, leach rates based on Cs, Na or P were found to be of the order of 10^{-1} g/m^{2}d (at 100°C), which are comparable to those from borosilicate glass compositions [85,86]. More recently, fast breeder reactor simulated waste has been incorporated and phase pure NZP formulations could be prepared with simulated waste loadings up to 20 wt% [80]. Again, beyond this limit a minor secondary phase of monazite was also observed. Zyryanov and Vance [53] assessed the leachability of dense NZP based ceramics containing simulated Purex waste using 7-day MCC-1 tests at 90°C and found it to be within only an order of magnitude higher than that for Synroc-C. It has been further demonstrated that processing conditions are of vital importance, with a 10- to 20-fold decrease in leachability achieved by higher processing temperatures [85]. The leach resistance of $CsZr_2(PO_4)_3$ appears to be comparable with that of other phases under consideration for radio-Cs immobilization [54]. There is also evidence that substitution of alkaline earth ions for alkalis in NZP increases the leach resistance considerably. For example strontium appears totally immobilised in $Sr_{0.5}Zr_2(PO_4)_3$ as it could not be detected in solutions even after hydrothermal treatment at 300°C [87]. It is expected therefore, that $CaZr_4(PO_4)_6$ and $CaTi_4(PO_4)_6$ would provide preferred waste form options [32].

6.2.5. Radiation Tolerance

Actinide incorporation is utilised to examine the long-term effects of α-decay events. These are simulated by incorporating in the crystal structure short-lived actinides, such as ^{238}Pu which has a half-life of 87.7 years (compared to ^{239}Pu which has a half-life of 24,100 years) [88], and provides up to ~ 10^{19} α-decay events/g in a matter of a several years. The most significant study of this kind on NZP structure type materials was carried out by Orlova et al. [72]. They have investigated $NaPu_2(PO_4)_3$ with variations in the isotopic ratio of the Pu, providing different dose rates and integrated doses. Trigonal $NaPu_2(PO_4)_3$ was formed after heat treatment at 1100°C and was found using X-ray diffraction to become amorphous after 9.3×10^{18} α-decay events/g. No leaching experiments were reported in this study.

There are limited studies on the radiation effects in NZP structure type materials using ion irradiation. The Cs-analogue, $CsZr_2(PO_4)_3$, was irradiated by 3 MeV Ar ions and found to become X-ray amorphous at ~ 0.1 displacements per atom [89]. Differential thermal analysis on the damaged samples suggested recrystallization at ~ 900°C. Further, the radiation damage was shown to increase the dissolution rates of the material in both deionised water and brine at 100°C, with enhancement factors of 2 to 25. These values are in broad agreement with those obtained for nuclear fuel waste glasses after heavy ion irradiation [90-92]. The influence of 50 MeV Li^{3+} ion irradiation was investigated on the series $Ca_{0.5-x}Sr_xZr_2(PO_4)_3$ for $0 \leq x \leq 0.5$ [93]. X-ray diffraction showed the irradiation had produced some amount of amorphous material in amongst the crystalline material at a fluence of 1×10^{14} ions/cm^2. The degree of amorphisation depended on the Ca/Sr substitution.

It has been shown by X-ray diffraction and fourier transform infrared spectroscopy that the action of external γ-radiation from a ^{60}Co source (dose up to 5×10^9 Gy) on NZP, does not result in phase degradation or changes in the composition and structure [94]. Similarly, no structural changes were observed for the Cs analogue after irradiation by a ^{60}Co γ-source up to a dose of 3×10^8 Gy [72]. In another study, the phosphates $M1A_2(PO_4)_3$ (M1 = Li, Cs, Ba, Sr, Ca; A = Ti, Zr, Hf) were demonstrated to show excellent stability under γ-radiation up to a dose of 5×10^8 Gy [95]. The water resistance of the $CsZr_2(PO_4)_3$ structure as well as the Cs leaching were shown not to be affected by the γ-radiation.

6.2.6. Candidacy as an Inert Matrix Fuel

There has been considerable recent interest in the use of inert matrix fuels (IMF) for the burning of plutonium in a reactor or as targets for the transmutation of actinides in reactors or accelerator systems [96-98]. Inert matrices are non-fertile materials and as such do not produce further actinides under irradiation, unlike the conventional UO_2 fuel pellets. This concept could potentially be used to reduce both the inventory of weapons grade plutonium and the quantity of the radioactive waste that must be safely disposed from commercial reactors. A further advantage is that it utilises the energetic value of plutonium. Although there is extensive literature on the use of IMFs, there is limited information on the fate of the fission products generated and on the disposition of the spent IMF [99]. Similar to the rock-like oxide IMF developed in Japan [100], CZP has been suggested as a potential IMF material. Not only does CZP have a low thermal expansion, a low neutron absorption cross-section and a high chemical, radiation and temperature stability [32,33,72,89,101] but as mentioned above it is a "near-universal solvent" with the potential to not only encapsulate the actinides for burning but also the resulting fission products. If the composition of the IMF was designed so that the fission products could be incorporated into a water-insoluble NZP type phase within the fuel, direct geological disposal would then be facilitated. One of the critical requirements for potential IMF materials is a high thermal conductivity as this allows the heat due to fission to escape from the fuel, and prevents excessive centerline temperatures inside the material. In this regard the thermal stability, heat capacity and thermal conductivity of CZP have been investigated recently [102].

CZP was found to be stable in air up to 1600°C but it partially decomposed around 1400°C in an inert atmosphere of helium through reduction of the phosphate and attendant loss of phosphorus and associated oxygen. The thermal conductivity was calculated from experimental measurement of thermal diffusivity, heat capacity, and density, according to the relationship:

$$K = D\rho C_p; \qquad (1)$$

where K is the thermal conductivity (W m^{-1} K^{-1}), D is the thermal diffusivity (m^2 s^{-1}), ρ is the density (kg m^{-3}), and C_p is the heat capacity (J kg^{-1} K^{-1}). The thermal conductivity of CZP is plotted in Figure 5.

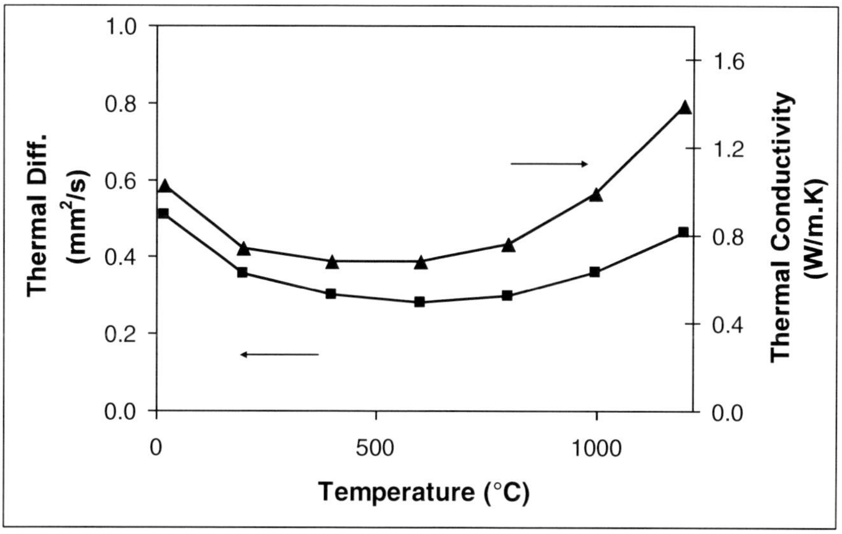

Figure 5. Temperature dependence of thermal diffusivity (■) and thermal conductivity (▲) of CZP.

The thermal conductivity coefficient was found to exhibit a minimum value of ~ 0.75 W m^{-1} K^{-1} in the 400 to 600°C region, and reach a value of ~ 1.45 W m^{-1} K^{-1} at 1200°C. The thermal conductivity is 1.0 W m^{-1} K^{-1} at 1000°C which is slightly lower than for UO_2 at this temperature [103]. A composite concept such as a cercer (ceramic-ceramic) or a cermet (ceramic-metallic) would improve the candidacy of this material as a host matrix for minor actinide transmutation.

6.2.7. Separation and Encapsulation

Amorphous zirconium phosphate or its crystalline form ($Zr(HPO_4)_2.nH_2O$, ZP) has cation exchange capacity and a high selectivity for Cs ions [104-106]. Potentially, a ZP medium (e.g. filter cake or ion-exchange column) could be used to selectively remove radionuclides from a waste stream and the loaded ZP converted into a NZP structure type durable waste form by a composition adjusting process. This approach was first proposed by Komarneni and Roy [105] and demonstrated for Cs ions, and has been reported more recently for molten salt reactor waste [107].

6.3. Chlorspodiosite, $Ca_2(PO_4)Cl$

Chlorspodiosite has a higher Cl loading than chlorapatite so has enjoyed some favour for immobilising Cl-bearing waste [108,109]. It can be prepared synthetically by reacting $Ca_3(PO_4)_2$ with excess $CaCl_2$ and then removing the excess $CaCl_2$ by washing in water [110], but cannot be made phase-pure from stoichiometric mixtures of dry $CaCl_2$ and $Ca_3(PO_4)_2$, chlorapatite being formed as a second phase with concomitant Cl_2 losses [27]. The leaching behaviour has been found to be poor in MCC-1 tests [27].

6.4. Crandallites

Crandallites are natural minerals containing Ca and phosphate that belong to the rhombohedral structure type alunite [111,112]. They have been considered by some of the radioactive waste community as sorbents for certain radioactive waste ions, particularly ^{90}Sr. Crandallite has the formula $CaAl_3(PO_4)_2(OH)_5.2H_2O$ and has been found near the Oklo natural nuclear reactors [113] in the Gabon area in Africa to have immobilised radiogenic ^{90}Sr, presumably as evidenced by the current presence of the ^{90}Zr decay product. On this ground, it has been argued by some that crandallite could be a good immobilising agent for Sr and perhaps other radionuclides such as K, Th, U and REE [114]. However crandallite is not a refractory ceramic insofar as it contains substantial amounts of hydrous material.

Early studies have demonstrated that certain members of the crandallite group exhibit low solubilities [115-117]. In experiments in our laboratory however, we have found that finely ground natural material (sample D56949, supplied by the Australian Museum) with a particle size of ~10 microns, exhibited substantial Ca and Al leaching in 7-day PCT tests utilising 20 mg of material in 10 mL of DIW for 7 days at only 40°C. It was found that ~ 90% of the tested material had dissolved at this temperature making it unsuitable as a potential waste form. Attempts to synthesise crandallite by the method of [118] yielded only near-amorphous material that did not form crandallite on moderate heating.

Conclusion

Calcium phosphate minerals, notably apatite and NZP-structured $Ca_{0.5}(Ti/Zr)_2(PO_4)_3$, have shown significant candidacy in playing a role in the immobilisation of high level nuclear waste. However, after 50-60 years of international research on nuclear waste immobilisation, it is considered that the nuclear industry is ready to immobilise a large fraction of extant high- and intermediate-level nuclear waste. It is likely that the processed waste will be stored for 100-200 years before actually disposing it in engineered geological repositories. The Ca zirconium/titanium phosphates also have candidacy as inert matrices for minor actinide "burning" in reactors.

ACKNOWLEDGMENTS

We thank R. Pogson of the Australian Museum, Sydney for supply of natural crandallite samples and K. Olufson of the Institute for Materials Engineering, ANSTO for leaching experiments on crandallite.

REFERENCES

[1] R. C. Ewing, *MRS Bull.* 33, 338 (2008).
[2] L. P. Hatch, *Am. Sci.* 41, 410 (1953).
[3] G. J. McCarthy, W. B. White and D. E. Pfoertsch, *Mater. Res. Bull.* 12, 1239 (1978).
[4] A. E. Ringwood, S. E. Kesson, N. G. Ware, W. Hibberson, and A. Major, *Nature*, 278, 219 (1979).
[5] A. E. Ringwood, S. E. Kesson, K. D. Reeve, D. M. Levins and E. J. Ramm, *Radioactive Waste Forms for the Future*, North-Holland, Amsterdam (1988) p 233.
[6] *The evaluation and selection of candidate high-level waste forms*, T. A. Bernadzikowski (ed), US Dept of Energy Report DOE/TIC 11611, National Technical Information Service, Springfield, VA , (1982).
[7] L. L. Hench, D. E. Clark and J. Campbell, *Nuc. Chem. Waste Man.* 5, 149 (1984).
[8] E. R. Vance, A. Jostsons, S. Moricca, M. W. A. Stewart, R. A. Day, B. D. Begg, M. J. Hambley, K. P. Hart and B. B. Ebbinghaus, *Ceramics Transactions (Environmental Issues and Waste Management Technologies IV)*, American Ceramic Society, Westerville, OH, (1999) p 323.
[9] S. G. Cochran, W. H. Dunlop, T. A. Edmunds, L. M. MacLean and T. H. Gould, *Record of Decision for the Surplus Plutonium Disposition. Final Environmental Impact Statement,* UCRL-ID-128705 (1997), US Department of Energy, Jan 4, 2000.
[10] G. D. Sizgek, *Nucl. Eng. Des.* 235, 101 (2005).
[11] S. Brunauer, P. H. Emmett and E. Teller, *J. Am. Chem. Soc.* 60, 309 (1938).
[12] J. E. Mendel (Compiler), *Nuclear Waste Materials Handbook, MCC-1 Static Test*, US Dept of Energy Report DOE/TIC-11400, Material Characterization Centre, Richland, WA (1984).
[13] *Standard Test Methods for Determining Chemical Durability of Nuclear, Hazardous, and Mixed Waste Glasses: The Product Consistency Test (PCT)*; American Society for Testing and Materials (ASTM), ASTM Designation C1285-97, West Conshohoken, PA (1997).
[14] W. J. Weber, R. C. Ewing, C. R. A. Catlow, T. Diaz de la Rubia, L. W. Hobbs, C. Kinoshita, Hj. Matzke, A. T. Motta, M. Nastasi, E. K. H. Salje, E. R. Vance and S. J. Zinkle, *J. Mater. Res* 13, 1434 (1998).
[15] H. D. Holland and D. Gottfried, *Acta Crystallogr.* 8, 291 (1955).
[16] R. C. Ewing, W. J. Weber and W. Lutze, *Disposal of Weapons Plutonium: approach and prospects*, Kluwer Academic Publishers, Netherlands (1996) p. 65.
[17] E. R. Vance, *Ceramic Interfaces: Properties and Applications V*, Trans. Tech Publications Ltd., Switzerland (2003) p 267.

[18] D. M. Strachan, R. D. Scheele, E. C. Buck, J. P. Icenhower, A. E. Kozelisky, R. L. Sell, R. J. Elovich and W. C. Buchmiller, *J. Nucl. Mater.* 345, 109 (2005).
[19] E. R. Vance, S. Moricca, B. D. Begg, M. W. A. Stewart, Y. Zhang and M. L. Carter, *5th Forum on New Materials Part B*, Trans Tech Publications Ltd., Switzerland (2010) p 130.
[20] G. J. McCarthy, *Nucl. Tech.* 32, 92 (1977).
[21] R. C. Ewing, *Canad. Mineral.* 39, 697 (2001).
[22] J. Carpena, L. Boyer, M. Fialin, J-R. Kienast and J-L. Lacout, *C. R. Acad. Sci. Paris/Earth and Planetary Sciences*, 333, 373 (2001).
[23] J. Chaumont, S. Soulet, J. C. Krupa and J. Carpena, *J. Nucl. Mater.301*, 122 (2002).
[24] N. Senamaud, D. Bernache-Assollant, J. Carpena and M. Fialin, *Phosphorus Res. Bull.* 10, 353 (1999).
[25] N. Senamaud, D. Bernache-Assollant, J. Carpena and C. Pin, *Scientific Basis for Nuclear Waste Management XXII, Mater. Res. Soc. Symp. Proc.* 556, 93 (1999).
[26] L. Campayo, F. Audubert, J-E. Lartigue, S. Botuha and D. Bernache-Assollant, *J. Nucl. Mater.* 374, 101 (2008).
[27] E. R. Vance, J. Davis, K. Olufson, I. Chironi, I. Karatchetvseva and I. Farnan, *J. Nucl. Mater.* 420, 396 (2012).
[28] H. Jena, R. Asuvathraman and K. V. G. Kutty, *J. Mater. Eng. Perform.* 20, 108 (2011).
[29] W. J. Weber, *J. Am. Ceram. Soc.* 65, 544 (1982).
[30] T. S. Yudintseva, *Scientific Basis for Nuclear Waste Management XXIX, Mater. Res. Soc. Symp. Proc.* 932, 1049 (2006).
[31] R. Roy, L. J. Yang, J. Alamo and E. R. Vance, *Scientific Basis for Nuclear Waste Management VI, Mater. Res. Soc. Symp. Proc.* 15, 15 (1983).
[32] B. E. Scheetz, D. K. Agrawal, E. Breval and R. Roy, *Waste Manage.* 14, 489 (1994).
[33] N. Anantharamulu, K. K. Rao, G. Rambabu, B. V. Kumar, V. Radha and M. Vithal, *J. Mater. Sci.* 46, 2821 (2011).
[34] D. A. Woodcock, P. Lightfoot and R. I. Smith, *J. Mater. Chem.* 9, 2631 (1999).
[35] R. D. Shannon, *Acta Cryst.* A32, 751 (1976).
[36] A. I. Orlova, *Radiokhimiya* 44, 385 (2002).
[37] J. Alamo and R. Roy, *J. Am. Cer. Soc.* 67, C-78 (1984).
[38] J. Alamo and R. Roy, *J. Mater. Sci.* 21, 444 (1986).
[39] A. V. Kuriakose, T. A. Wheat, A. Ahmad and J. Pirocco, *J. Am. Cer. Soc.* 67, 179 (1984).
[40] D. K. Agrawal and J. H. Adair, *J. Am. Cer. Soc.* 73, 2153 (1990).
[41] B. E. Yoldas and I. K. Lloyd, *Mater. Res. Bull.* 18, 1171 (1983).
[42] H. Perthuis and P. H. Colomban, *Mater. Res. Bull.* 19, 621 (1984).
[43] A. Clearfield, P. Jirustithipong, R. N. Cotman and S. P. Pack, *Mater. Res. Bull.* 15, 1603 (1980).
[44] A. Clearfield, B. D. Roberts and M. A. Subramanian, *Mater. Res. Bull.* 19, 219 (1984).
[45] A. Clearfield, M. A. Subramanian, W. Wang and P. Jerus, *Solid State Ionics* 9-10, 895 (1983).
[46] M. Kinoshita and M. Inoue, *Chem. Soc. Japan* 8, 1219 (1980).
[47] D. H. Quon, T. A. Wheat and W. Nesbitt, *Mater. Res. Bull.* 15, 1533 (1980).
[48] S. Komarneni, *Int. J. High Tech. Ceram.* 4, 31 (1988).
[49] S. Yamanaka, K. Yoshioka and M. Hattori, *Solid State Ionics* 40-41, 43 (1990).

[50] E. Breval, H. A. MacKinstry and D. K. Agrawal, *J. Mater. Sci.* 35, 3359 (2000).
[51] B. Vaidhyanathan, D. K. Agrawal and R. Roy, *J. Am. Ceram. Soc.* 87, 834 (2004).
[52] A. H. Naik, S. B. Deb, A. B. Chalke, M. K. Saxena, K. L. Ramakumar, V. Venugopal and S. R. Dharwadkar, *J. Chem. Sci.* 122, 71 (2010).
[53] V. N. Zyryanov and E. R. Vance, *Scientific Basis for Nuclear Waste Management XX, Mater. Res. Soc. Symp. Proc.* 465, 409 (1997).
[54] R. Roy, E. R. Vance and J. Alamo, *Mater. Res. Bull.* 17, 585 (1982).
[55] R. Chourasia, A. Bohre, R. D. Ambastha, O. P. Shrivastava and P. K. Wattal, *J. Mater. Sci.* 45, 533 (2010).
[56] D. R. Lide, *Handbook of chemistry and physics* 73rd ed, CRC Press, Boca Raton, FL, (1992–1993).
[57] P. Renaud, K. Beaugelin, H. Maubert and P. Ledenvic, *Consequences radiologiques et dosimetriques de l'accident de Tchernobyl en France*, Rapport IPSN 97-03, IPSN, France (1997).
[58] S. Y. Limaye, D. K. Agrawal, R. Roy and Y. Mehrottra, *J. Mater. Sci.* 26, 93 (1991).
[59] S. Y. Limaye, D. K. Agrawal, H. A. McKinstry, *J. Am. Ceram. Soc.* 70, C232 (1987).
[60] D. A. Woodcock, P. Lightfoot and C. Ritter, *Chem. Commun.* 1, 107 (1998).
[61] E. R. Gobechiya, Yu. K. Kabalov, V. I. Pet'kov and M. V. Sukhanov, *Crystallogr. Rep.* 49, 741 (2004).
[62] M. Alami Talbi, R. Brochu, C. Parent, L. Rabardel and G. Le Flem, *J. Solid State Chem.* 110, 350 (1994).
[63] J. M. Heintz, L. Rabardel, M. Al Qaraoui, M. Alami Talbi, R. Brochu and G. Le Flem, *J. Alloys Comp.* 250, 515 (1997).
[64] S. Tamura, N. Imanaka and G. Adachi, *Solid State Ionics* 154-155, 767 (2002).
[65] V. S. Kurazhkovskaya, D. M. Bykov, E. Yu. Borovikova, N. Yu. Boldyrev, L. Mikhalitsyn and A. I. Orlova, *Vib. Spectrosc.* 52, 137 (2010).
[66] P. Lightfoot, D. A. Woodcock, J. D. Jorgensen and S. Short, *Int. J. Inorg. Mater.* 1, 53 (1999).
[67] Y. Miyajima, T. Miyshi, J. Tamaki, M. Matsuoka, Y. Yamamoto, C. Masquelier, M. Tabuchi, Y. Saito and H. Kageyama, *Solid State Ionics* 124, 201 (1999).
[68] A. A. Burnaeva, Yu. F. Volkov, A. I. Kryukova, I. A. Korshunov and O. V. Skiba, *Radiokhimiya* 34, 12 (1992).
[69] Yu. F. Volkov, R. F. Melkaya, V. I. Spiryakov and G. A. Timofeev, *Radiokhimiya* 36, 205 (1994).
[70] H. T. Hawkins, D. R. Spearing, D. K. Veirs, J. A. Danis, D. M. Smith, C. D. Tait, W. H. Runde, M. N. Spilde and B. E. Scheetz, *Chem. Mater.* 11, 2851 (1999).
[71] Yu. F. Volkov, S. V. Tomilin, A. I. Orlova, A. A. Lizin, V. I. Spiryakov and A. N. Lukinykh, *Radiochemistry* 45, 319 (2003).
[72] A. I. Orlova, Yu. F. Volkov, R. F. Melkaya, L. Y. Masterova, I. A. Kulikov, V. A. Alferov, *Radiochemistry* 36, 322 (1995).
[73] D. M. Bykov, A. I. Orlova, S. V. Tomilin, A. A. Lizin and A. N. Lukinykh, *Radiochemistry* 48, 234 (2006).
[74] A. I. Orlova, V. Yu. Volgutov, G. R. Castro, S. García-Granda, S. A. Khainakov and J. R. García, *Inorg. Chem.* 48, 9046 (2009).
[75] K. H. Lii, J. J. Chen and S. L. Wang, *J. Solid State Chem.* 78, 93 (1989).

[76] L. Bennouna, S. Arsalane, R. Brochu, M. R. Lee, J. Chassaing and M. Quarton, *J. Solid State Chem.* 114, 224 (1995).

[77] A. Laclaire, M. M. Borel, A. Grandin and B. Raveau, *Eur. J. Solid State Inorg. Chem.* 26, 45 (1989).

[78] N. M. Kozhevnikova and I. Yu. Kotova, *Zh. Neorg. Khim.* 45, 102 (2000).

[79] B. I. Lazarev and V. A. Efremov, *Zh. Neorg. Khim.* 32, 652 (1987).

[80] R. Chourasia, O. P. Shrivastava, R. D. Ambashta and P. K. Wattal, *Ann. Nucl. Energy* 37, 103 (2010).

[81] Y. Hirose, T. Fukasawa, D. K. Agrawal, B. E. Scheetz, R. Nageswaran, J. A. Curtis and S. Y. Limaye, *WM'99 Proceedings*; Tucson, AZ, (1999) p. 2873.

[82] Y. Seida, M. Yuki, K. Suzuki and T. Sawa, *Scientific Basis for Nuclear Waste Management XXVI, Mater. Res. Soc. Symp. Proc.* 757, 329 (2003).

[83] I. W. Donald, B. L. Metcalfe and R. N. J. Taylor, *J. Mater. Sci.* 32, 5851 (1997).

[84] H. T. Hawkins, B. E. Scheetz and G. D. Guthrie Jr., *Scientific Basis for Nuclear Waste Management XX, Mater. Res. Soc. Symp. Proc.* 465, 387 (1997).

[85] L. J. Yang, S. Komarneni and R. Roy, *Advances in ceramics* 8, American Ceramic Society, Columbus, OH, (1984) p. 255.

[86] R. C. Ewing and LuMin Wang, *Rev. Mineral. Geochem.* 48, 673 (2002).

[87] B. E. Scheetz, S. Komarneni, W. Fajun, L. J. Yang, M. Ollinen and R. Roy, *Scientific Basis for Nuclear Waste Management VIII, Mater. Res. Soc. Symp. Proc.* 44, 903 (1985).

[88] E. Browne, J. M. Dairiki and R. E. Doebler, *Table of Isotopes,* 7th ed, John Wiley and Sons, New York, NY (1978).

[89] E. R. Vance, L. Cartz and F. G. Karioris, *J. Mater.Sci.* 19, 2943 (1984).

[90] J. C. Dran, M. Maurette and J. C. Petit, *Science* 209, 1518 (1980).

[91] J. C. Dran, M. Maurette, J. C. Petit and B. Vassent, *Scientific Basis for Nuclear Waste Management, Mater. Res. Soc. Symp. Proc.* 3, 449 (1981).

[92] W. Primak, *Nucl. Sci. Eng.* 80, 689 (1982).

[93] B. Angadi, V. M. Jali, M. T. Lagare, N. S. Kini, A. M. Umarji, R. Kumar, S. K. Arora and D. Kanjilal, *Nucl. Instrum. Meth. B* 187, 87 (2002).

[94] C-R. Aída, F. V. Suilma M., O-R. Eduardo and C. A. Héctor, *Proceedings of the 17th Pacific Basin Nuclear Conference*, 1 (2010).

[95] A. I. Kryukova, I. A. Kulikov, G. Yu. Artem'eva, O. V. Pechenevskaya and V. A. Alferov, *Radiokhimiya* 34, 82 (1992).

[96] H. Kleykamp, *J. Nucl. Mater.* 275, 1 (1999).

[97] C. Degueldre and J. M. Paratte, *J. Nucl. Mater.* 274, 1 (1999).

[98] Y-W. Lee, *Advanced Reactors with Innovative Fuels*, NEA/OECD, Paris (2002) p. 103.

[99] M. W. A. Stewart and E. R. Vance, *J. Aust. Ceram. Soc.* 42, 50 (2006).

[100] T. Yamashita, K-I. Kuramoto, H. Akie, Y. Nakano, N. Nitani, T. Nakamura, K. Kusagaya and T. Ohmichi, *J. Nucl. Sci. Technol.* 39, 865 (2002).

[101] A. B. Yaroslavtsev and I. A. Stenina, *Russ. J. Inorg. Chem.* 51, Suppl. 1, S97 (2006).

[102] D. J. Gregg, I. Karatchevtseva, G. Triani, G. R. Lumpkin and E. R. Vance, *Scientific Basis for Nuclear Waste Management XXXV, Mater. Res. Soc. Symp. Proc.* in press (2012).

[103] D. L. Hargman, *A Handbook of Materials Properties for Use in the Analysis of Light Water Reactor Fuel Rod Behavior*, MATPRO-Version11, Idaho National Eng. Lab (1981).
[104] S. Komarneni and R. Roy, *Nature* 299, 707 (1982).
[105] S. Komarneni and R. Roy, *Scientific Basis for Nuclear Waste Management VI, Mater. Res. Soc. Symp. Proc.* 15, 77 (1983).
[106] A. Clearfield, W. L. Duax, A. S. Medina, G. D. Smith and J. R. Thomas, *J. Phys. Chem.* 75, 3424 (1969).
[107] H-S. Park, I-T. Kim, H-Y. Kim, S-K. Ryu and J-H. Kim, *J. Radioanal. Nucl. Ch.* 268, 617 (2006).
[108] I. W. Donald, B. L. Metcalfe and R. S. Greedharee, *Scientific Basis for Nuclear Waste Management XXV, Mater. Res. Soc. Symp. Proc.* 713, 287 (2002).
[109] S. K. Fong, I. W. Donald and B. L. Metcalfe, *J. Alloy Compd.* 444-445, 424 (2007).
[110] M. Greenblatt, E. Banks and B. Post, *Acta Crystallogr.* 23, 166 (1967).
[111] R. Wang, W. F. Bradley and H. Steinfink, *Acta Crystallogr.* 18, 249 (1965).
[112] S. Menchetti and C. Sabelli, *Neues JB. Miner. Monat.* 406 (1976).
[113] H. Hidaka, *J. Nucl. Radiochem. Sci.* 8, 99 (2007).
[114] U. Kolitsch and A. Pring, *J. Miner. Petrol. Sci.* 96, 67 (2001).
[115] K. Norris, *Trans.* 9th *International Congress of Soil Science*, Vol. II, Adelaide, Australia (1968) p. 713.
[116] R. G. Schwab, C. Götz, H. Herold and N. P. de Oliveira, *Neues JB. Miner. Monat.* 551 (1993).
[117] A. Roca, J. Vinals, M. Arranz and J. Calero, *Can. Metall. Quart.* 38, 93 (1999).
[118] J. M. Monteagudo, A. Durán, M. S. Carmona, R. G. Schwab and P. Higueras, *J. Chem. Technol. Biotechnol.* 78, 399 (2003).

In: Calcium Phosphate
Editor: Robert B. Heimann
ISBN 978-1-62257-299-1

Chapter 19

CALCIUM PHOSPHATE IN SOFT-PASTE PORCELAIN: BONE CHINA

Robert B. Heimann*
Professor emeritus of Applied Mineralogy and Materials Science, Görlitz, Germany

ABSTRACT

Bone china was produced in Britain from the late 18th century CE onwards. It is characterized by a high proportion of calcined animal bone that provides ß-TCP to the ceramic paste able to react during firing beyond 1,000°C with silica and alumina released from decomposing clay minerals, forming anorthite. Hence the ceramic body of bone china consists of about 40% ß-TCP, 20% calcic plagioclase (anorthite), and 40% vitreous phase. In contrast to European hard-paste porcelain of Meissen type, the rather low proportion of vitreous phase confers to this hybrid soft-paste porcelain ware several advantageous mechanical and thermal properties, making it a popular tableware throughout the world.

1. INTRODUCTION

The invention of bone china was a unique English answer to the quest to reinvent the cherished, rare, and expensive Chinese import porcelain during the 18th and 19th centuries CE in Europe. The solution English potters came up with is a hybrid soft-paste porcelain containing bone ash to replace some of the feldspar used in Meissen hard-paste porcelain, and coating the biscuit-fired ceramic body, frequently decorated by underglaze transfer printing, by a lead-based glaze. Whilst hard-paste porcelain chips fairly easily and, unless specially treated, is usually tinged with blue or gray color hues, bone china to the contrary is easier to manufacture, is strong, does not chip easily, withstands reasonable thermal shock, and has a pleasing ivory-white appearance. It is thus no surprise that this ware still enjoys widespread use in Britain and beyond as a popular ceramic material for exclusive tableware.

* E-mail address: robert.heimann@ocean-gate.de

1.1. Historical Development

Who actually introduced this type of pottery is rather contentious. Whereas the initial development of densely fired bone china is attributed to Josiah Spode I (1733-1797) who founded his pottery business in 1770 in Stoke-on-Trent, Staffordshire, UK, a patent for using bone ash in ceramics had already been taken out in 1748 by Thomas Frye of Bow manufactory who produced a soft-paste bone ash-containing 'porcelain' that, however, was porous, and lacked the luster, translucency and appeal of true porcelain.

Initially Spode's company produced earthenware pottery from local clay deposits, called *creamware* and *pearlware*. Cream-colored earthenware was developed in the 1740s by the Staffordshire potter Enoch Booth. Later known as *Queensware*, it was manufactured from white clay and flint, similar to the mixture first developed for white salt-glazed stoneware. The lead oxide used to glaze the body was contaminated with iron that imparted a creamy tint to the glaze thus giving it its descriptive name. Creamware was refined and improved, until by the 1760s it was suitable for manufacturing fashionable, elegant tableware. In turn this pottery prized all over Europe and beyond propelled Britain from the backwater of ceramic production into the limelight of attention that was later reinforced by the invention and development of ironstone china and bone china.

Pearlware or china glaze ware was introduced around 1775 as a successful attempt to improve the whiteness of creamware and thus to make it appear more like true porcelain. Cobalt oxide was added to the glaze mix to optically mask the yellow iron tone thus yielding earthenware with a distinct gray-blue finish. Since this glaze was ideal for blue transfer printed earthenware, potters continued to use it long after a clear, colorless glaze was achieved in the early years of the 19th century CE.

2. Phosphatic Porcelains

2.1. Early Developments

Phosphatic porcelains were invented in the mid-18th century CE in Britain and conquered the public taste quickly. Several manufacturers produced this pottery, among them Bow manufactory to which the original patent was issued in 1748 to use calcined bone in their ceramic ware (Figure 1). The main advantage over previous soft-paste porcelain wares was the increased durability, fracture toughness, hardness and thermoshock resistance of the novel material. Typical ceramic masses consisted of 10-25% clay of different origin, 30-45% bone ash, 30-60% quartz, and a few percent of alkali carbonate derived from plant or kelp ashes as a fluxing agent. Figure 1 shows an SEM image of soft-paste bone ash porcelain, so-called Bow porcelain. Whereas most of the pores in the bone ash (white phase, B) are filled with melt (medium gray, m), a few large black pores (P) remain that may act as stress concentration sites thus reducing toughness. Silica polymorphs (slightly darker gray than the melt phase, Q) are still angular, i.e. largely unresorbed.

Conspicuously absent in the microstructure are anorthite ($CaAl_2Si_2O_8$) crystals that should have formed by reaction of CaO released from decomposing hydroxyapatite,

$Ca_{10}[(OH)_2/(PO_4)_6]$ and/or tricalcium phosphate, $Ca_3(PO_4)_2$ with metakaolin, $Al_2O_3{\cdot}2SiO_2$ according to

$$Ca_{10}[(OH)_2/(PO_4)_6] \rightarrow 3\ Ca_3(PO_4)_2 + CaO + H_2O \quad (1)$$

$$Al_2[(OH)_4/Si_2O_5] \rightarrow Al_2O_3{\cdot}2SiO_2 + 2\ H_2O \quad (2)$$

$$Al_2O_3{\cdot}2SiO_2 + CaO \rightarrow CaAl_2Si_2O_8 \quad (3)$$

The absence of anorthite can be related to early resorption of calcium phosphate grains by the small amounts of melt phase present in insufficiently vitrified samples, thus lacking excess lime required to form anorthite [1] (see section 3). It should also be noted that the Ca/P-ratio of the bone ash (phase B in Figure 1) is closer to that of hydroxyapatite (Ca/P = 1.67) than to whitlockite (Ca/P = 1.5). It is assumed that during calcining of the bone the temperatures applied were insufficient to completely dehydroxylate the hydroxyapatite of the bone matrix but produced only oxyapatite ($Ca_{10}[O/(PO_4)_6]$ (for example [2], see also Chapters 4 and 7 of this volume) that in the presence of moisture readily reverted back to fully hydroxylated apatite.

The rather imperfect and immature early bone ash-containing Bow and Vauxhall wares actually predated true bone china that today is considered a hybrid ware somewhat intermediate between soft-paste and hard-paste porcelains[1].

2.2. The Invention of Bone China

After having manufactured cream- and pearlware as well as a range of stoneware products including black basalt, caneware and jasperware invented and popularized earlier by Josiah Wedgewood, Josiah Spode I proceeded to develop a formulation for bone china between 1789 and 1793 that was initially kept a close industrial secret. This formulation called originally for six parts of bone ash (45%), four parts of Cornish or china stone[2] (30%), and three and a half parts of china clay or kaolinite (25%) that constituted the standard English ceramic body during the 19th century CE and is still being used at present.

Figure 2 represents bone ash porcelain of somewhat different nature as large amounts of glass (G1) were added to ball clay, bone ash (B) and some quartz (Q). Firing generated a melt phase (G2), and the resulting microstructure still contains large pores (P), shown in black.

---

1 European porcelain falls into two categories: true or hard-paste (*pâte dure*) porcelain (also called 'triaxial' porcelain) and artificial or soft-paste (*pâte tendre*) porcelain, the distinguishing criterion being the higher amount of feldspathic flux of the latter. Chinese soft-paste porcelain is produced from kaolinite (china clay), *petuntse* (from 白墩子, in pinyin: bai dun zi) or china stone, the historic term for a wide range of micaceous or feldspathic rocks, and fine quartz sand. The classic hard-paste 'triaxial' composition is kao/fsp/qz = 50/25/25 whereas the content of fluxing, i.e. alkali-containing feldspar is higher in soft-paste porcelain to yield ratios of approximately 30/50/20. However, in the ceramic praxis there are many deviations from these standard values.

2 Cornish stone is a feldspar-rich but iron-poor, partially weathered granite found particularly in Cornwall. In its composition it is approximately akin to the Chinese *petuntse*. The feldspar content of both Cornish (china) stone and *petuntse* has been transformed to kaolinite by weathering.

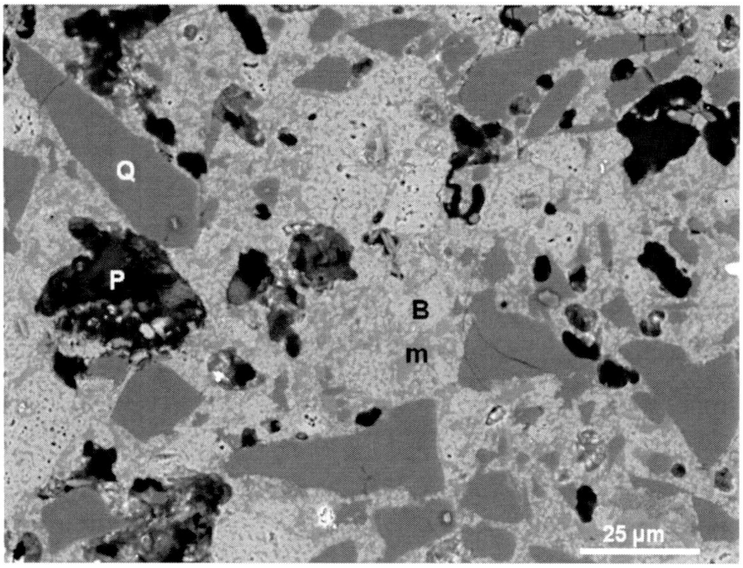

Figure 1. SEM image of porous historical bone ash porcelain presumably manufactured around 1760 by the Bow manufactory of London (Courtesy: Prof. J. Victor Owen, Saint Mary's University, Halifax, Nova Scotia, Canada). B bone ash, m melt, P pore, Q quartz

As evident from Figures 1 and 2, the microstructure of early bone ash porcelain is rather porous since the low degree of vitrification, caused by the low concentration of fluxes in the ceramic mass and a low firing temperature, precluded the formation of copious amounts of melt phase and thus densification during firing. This drawback was overcome when the brilliant idea by Josiah Spode I caught on to include fluxing Cornish stone[2] into the raw materials mix, resulting in true bone china as a hybrid between bone ash porcelain and Chinese-type porcelain.

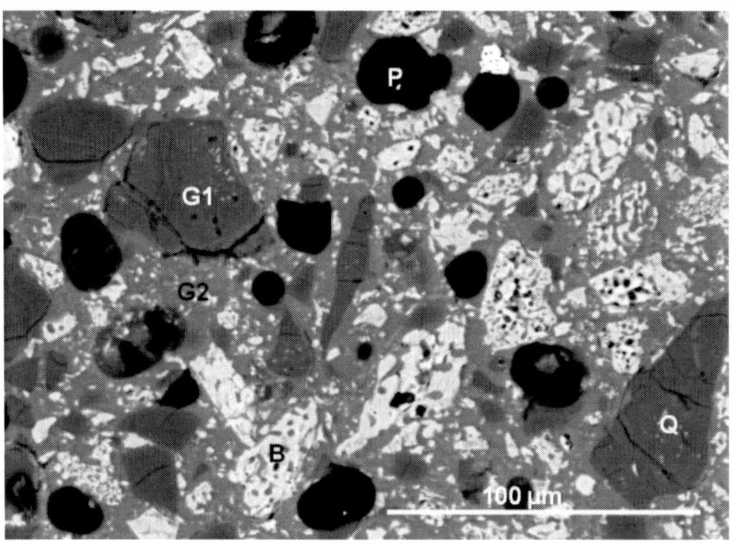

Figure 2. Microstructure of historical bone ash porcelain made around 1750 by the Vauxhall manufactury, London and produced from ball clay, glass and bone ash (Courtesy: Prof. Ian Freestone, University College London). B bone ash, G1 glass, G2 melt phase, P pore, Q quartz.

However, very soon the secret of Spode's successful bone china recipe leaked out and was copied by the Minton, Coalport, Davenport, Derby, Worcester, and Herculaneum factories at Liverpool. Still later on it was used by New Hall (1810), Wedgwood (1812), and Rockingham (1820). The quality, style and decoration of the bone china ware varied from factory to factory; some tended, after about 1820, toward brilliant color, lavish gilding, and overcrowded gaudy designs while others such as Chelsea manufactory produced tasteful, simply ornamented tableware. It should be mentioned in passing that Josiah Spode II (1754-1827) introduced in his new bone china pottery in 1797 a somewhat different raw material formulation, consisting of 25% china clay, 25% Cornish stone, and 50% bone ash. It very quickly became a popular ceramic body for several reasons. The declining import of original Chinese porcelain owing to very heavy import duties levied on porcelain (108% in 1799), and the apparent shortness of merchant ships owing to the need of the British Empire to sustain naval and military forces overseas were strong driving forces in this development. In addition, the patronage of the Prince of Wales, unchallenged leader of public taste at the beginning of the 19th century CE was influential for increased sales of bone china tableware to the professional and merchant classes that provided a healthy market for the new product. It was now easy for existing factories to convert to making bone china because the sequence of processes, the available kilns as well as the techniques of glost and enamel firing, and most of all, the required temperatures were close to identical to those required to produce classic Staffordshire earthenware.

3. Microstructure and Properties of Bone China

The strength of bone china is provided by the fusion of body ingredients during firing. Like hard-paste porcelain[1] this ceramic is fired twice: biscuit firing is done at 1200-1300°C, followed after glazing by a second or glost firing at 1050-1100°C. When overglaze color pigments are applied, a third firing step below 1000°C is needed. These temperatures are considerably below those required for hard-paste porcelain of Meissen type. Thus the production of bone china is considered simpler and more economical.

The crucial ingredient of bone china is calcined bone ash that is making up about 50% by weight of the final body recipe. It is produced from animal bones that are first processed to remove any adhering meat that is generally sold as pet food. The bone is then treated to remove collagen that makes up about 30% of the living bone mass. Collagen is processed and upgraded for use in normal applications where glue is used, and also for the sizing of expensive paper. The raw bone left after the meat and collagen have been extracted is then heated to about 1000°C to burn off any residual organic material. During heating the structure of the bone is changed to a state suitable for the manufacture of bone china. Prior to use the bone is finely ground with water before including it in the bone china body. It is this calcined bone that gives traditional English bone china its unique translucency and appealing whiteness.

The traditional preparation of pure tricalcium orthophosphate was described in a contemporary encyclopaedia as follows: "*Phosphate of lime, proper. As this salt constitutes the basis of bones, it is not necessary to prepare it artificially. It may be obtained in a state of purity by the following process: Calcine the bones to whiteness, reduce them to powder, and wash them repeatedly with water, to separate several soluble salts which are present. Dissolve the whole in muriatic acid, and precipitate by means of ammonia. The precipitate, when well washed and dried, is pure phosphate of lime*" [3].

Several studies have revealed that the microstructure of bone china is far from thermodynamic equilibrium. Modern bone china consists of distinct regions of lath-like (0.4–10 μm) anorthite crystals and spheroidal (1–3 μm) β-tricalcium phosphate (β-TCP) nodules embedded in a heterogeneous composition glass along with small amounts of isolated irregular (≤30 μm) α-quartz crystals. The composition and morphology of the phases formed on firing suggest that (i) anorthite crystallised in clay relicts by the reaction of metakaolinite with CaO (see Figure 4), (ii) β-TCP is the result of decomposition of hydroxyapatite of the bone ash, and that (iii) the liquid formed on vitrification has variable composition depending on the composition of adjacent phases. Interestingly, P_2O_5 was never detected in the glass suggesting that any P_2O_5 that dissolves in the liquid on firing is transient [4]. This is in contrast to more recent findings that the resorption of phosphate by the matrix melt virtually precludes recognition of anorthite formed by the 'non-phosphate glass equation' [1]. An explanation for this dichotomy is presented below. Cracks sometimes observed in and around β-TCP clusters are thought to arise from thermal expansion mismatch between β-TCP and the surrounding glass and anorthite phases [4, 5].

Figure 3 shows the typical microstructure of a polished biscuit-fired bone china body [5]. The product was obtained by experimental firing to between 1220 and 1250°C of commercially available spray-dried bone china body granules (BCP 4 grade, Jesse Shirley & Sons Ltd, Etruria, Stoke-on-Trent, England). The ß-TCP crystals show a high degree of microporosity presumably related to the expulsion of residual structural water during thermal conversion of hydroxyapatite. Reaction-related anorthite and other calcic plagioclases usually occur in the form of tiny microlites crystallized from the vitreous phase (G) also visible throughout the microstructure.

Minor amounts of quartz (Q) occur surrounded by thin reaction rims of a calcium-rich silicate glass. The microstructure of this true bone china may be compared to that of the early bone ash porcelain shown in Fig. 1 with its much higher quartz content and also higher macroporosity related to the lower degree of vitrification.

The development of the microstructure of bone china starts beyond 1,000°C by progressive thermal decomposition of the Ca-deficient hydroxyapatite of the bone to form ß-TCP (whitlockite) and free CaO (eq. 1). The latter reacts quickly with active alumina and silica released from decomposing kaolinite and metakaolinite (eq. 2), respectively to crystalline anorthite (eq. 3). The conversion of dehydrated clay particles such as kaolinite into calcic plagioclase, e.g. anorthite according to eq. 3 is shown in the enlarged part of Fig. 4. Stacks of clay lamellae transform topotactically into calcic plagioclase showing a rim of newly formed plagioclase (A) surrounding a core of still unreacted clay (C). This is an excellent example of the incompleteness of the reaction sequence leading to a porcelain microstructure, owing to the rather low firing temperature applied by the ancient potters.

The composition of the average bone china body thus consists of about 40% tricalcium phosphate (ß-TCP), 20% anorthite and 40% calcium aluminium silicate glass phase (Fig. 3). This ought to be compared to the average composition of the hard-paste porcelain of Meissen type that consists of about 30% mullite and 70% potassium aluminium silicate glass. The much lower proportion of the glass phase in bone china renders this ceramic ware mechanically stronger than the classical Meissen ware.

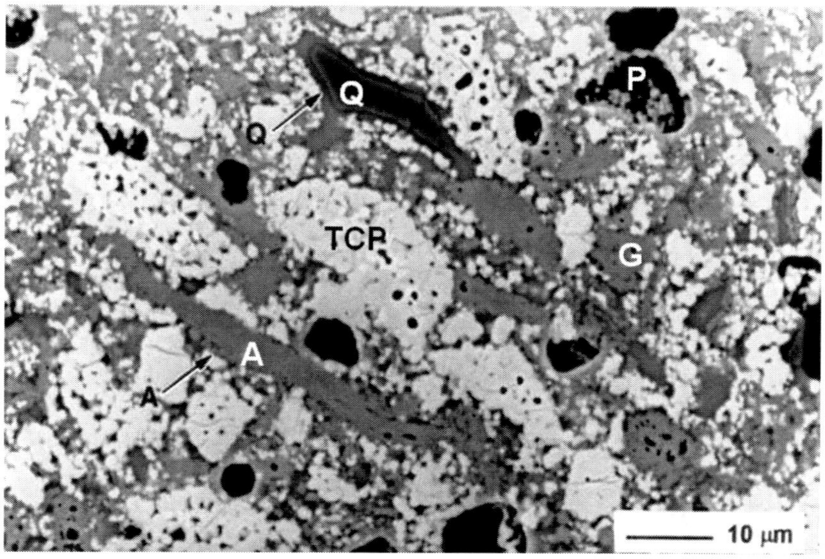

Figure 3. SEM micrograph of a polished modern bone china body. A anorthite, G glassy phase, P pores, Q quartz, TCP ß-tricalcium phosphate (Courtesy: Dr. Alpagut Kara, Anadolu University, Eskişehir, Turkey).

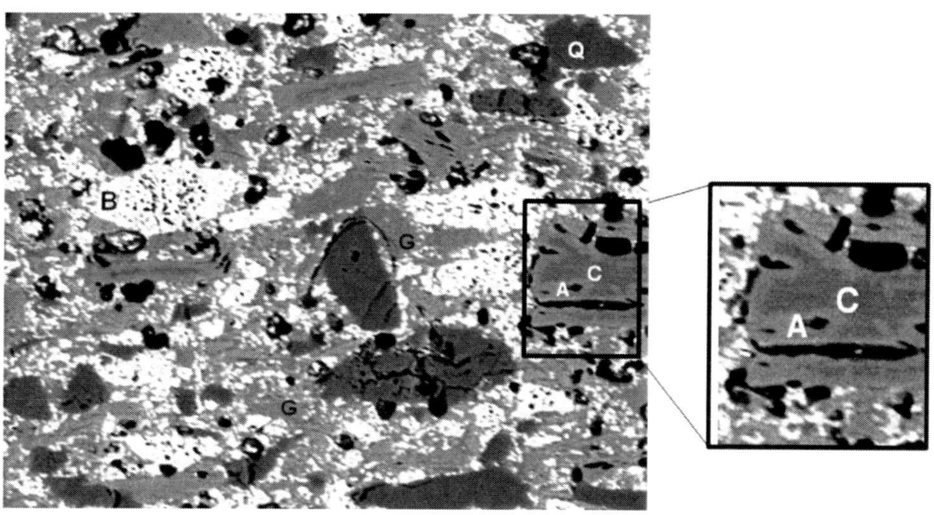

Figure 4. Microstructure of early bone china pottery from Coalport, Shropshire (c. 1825). The enlarged section shows the conversion of stacks of clay lamellae (C) into calcic plagioclase (A). B bone ash, G glass, Q quartz (Courtesy: Prof. Ian Freestone, University College London).

The fact that several studies of ancient and modern bone china show deviating chemical compositions of the vitreous phase, either being virtually free of phosphorus or containing phosphate contents up to 15% has been explained by Owen et al. [1] as follows. Anorthite formation during firing can occur by two mechanisms, the so-called 'non-phosphate glass' route (eq. 4) or the 'phosphate glass' route (eq. 5):

$$\text{hydroxyapatite} + \text{kaolinite} \rightarrow 3\ \text{TCP (whitlockite)} + \text{anorthite} + 3\ \text{water} \quad (4)$$

$$3\ \text{hydroxyapatite} + 6\ \text{kaolinite} \rightarrow 8\ \text{TCP (whitlockite)} + 6\ \text{anorthite} + P_2O_5 + 15\ \text{water} \quad (5)$$

Equation 4 involves implicitly the reaction of matrix kaolinite with excess lime produced by thermal decomposition of hydroxyapatite according to the sequence shown in eqs. 1 to 3. This reaction is limited by the availability of lime and thus a function of the degree of vitrification and as such the firing temperature: low fired bone china would be expected to develop microstructures consistent with eq. 4. On the other hand, progress of reaction according to eq. 5 is limited by the availability of alumina, producing both TCP and anorthite in a 4:3 ratio and liberating P_2O_5 that becomes dissolved in the vitreous phase and can be detected as such. This type of reaction would be preferred in highly vitrified bone china in which the composition of the melt phase develops towards the thermal minimum in the ternary system anorthite-TCP-silica. However, Owen et al. [1] caution to use the criterion of P-content in the glassy phase to distinguish between the two reaction paths since simultaneous resorption of TCP in the melt phase tends to obscure the recognition of either reaction paths and thus the anorthite formed by the 'phosphate glass' equation.

CONCLUSION

Tricalcium phosphate ('bone ash') formed by thermal treatment at around 1,000°C of biological hydroxyapatite derived from the inorganic matrix of animal bone is one of the main constituents of bone china. Bone china is considered a hybrid between soft-paste and hard-paste porcelains. It was invented in Britain in the mid-18th century CE as an alternative to the European hard-paste porcelain of Meissen type. The 'bone ash' acts to replace some of the feldspar used in Meissen hard-paste porcelain.

The original formulation of the ceramic paste consisted of 45-50 % bone ash (ß-TCP and whitlockite, respectively), 25-30% Cornish or china stone, as well as 25% kaolinite. This composition constitutes the standardized English ceramic body during the 19th century CE and is still being used at present. The microstructure of modern bone china consists of about 40% porous ß-TCP, 20% newly formed calcic plagioclase (anorthite), 40% calcium aluminum silicate glass, and a few percent of residual quartz.

Owing to the fairly low proportion of vitreous phase compared to hard-paste porcelain that chips rather easily, bone china to the contrary is stronger, easier to manufacture, and withstands reasonable mechanical and thermal shocks. These properties make bone china popular tableware that today enjoys widespread use worldwide.

REFERENCES

[1] J.V. Owen, R. Hunter, R. Jellicoe and M. Zierden, *Geoarchaeology*, 26, 292 (2011).

[2] G. Montel, G. Bonel, J.C. Trombe, J.C. Heughebaert and C. Rey, *Pure Appl. Chem.*, 52, 973 (1980).

[3] J.M. Good, G. Olinthus and B. Newton. *Pantologia: A new cyclopædia, comprehending a complete series of Essays, Treatises, and Systems, alphabetically arranged; with a general dictionary of arts, sciences, and words: the whole presenting a distinct survey of Human Genius, Learning, and Industry*. Vol. IX. P – PYX. London. 1813.

[4] Y. Iqbal, P.F. Messer and W.E. Lee, *Brit. Ceram. Trans.*, 99(3), 110 (2000).

[5] A. Kara and R. Stevens, *J. Eur. Ceram. Soc.*, 22(5), 731 (2002).

INDEX

A

B

C

D

E

F

G

H

I

J

K

L

M

N

O

P

Q

R

S

T

X

Y

Z